Volvo 240 Series Owners Workshop Manual

Rob Maddox
and Steve Churchill

Models covered
All Volvo 240 Series models,
1986 cc, 2127 cc & 2316 cc petrol engines

Does not cover Diesel or ohv engine models

ABCDE
FGHIJ
KLMNO
PQR
2

Haynes Publishing
Sparkford Nr Yeovil
Somerset BA22 7JJ England

Haynes North America, Inc
861 Lawrence Drive
Newbury Park
California 91320 USA

Acknowledgements

Thanks are due to Champion Spark Plug, who supplied the illustrations of various spark plug conditions, and Volvo North America Corporation, for assistance with technical information and certain illustrations, and to all those people at Sparkford and Newbury Park who helped in the production of this manual. Technical writers who contributed to this project include Larry Warren and Mike Stubblefield.

A book in the **Haynes Owners Workshop Manual Series**

Printed by J. H. Haynes & Co. Ltd, Sparkford, Nr Yeovil, Somerset BA22 7JJ, England

ISBN 1 85960 074 3

British Library Cataloguing in Publication Data
A catalogue record for this book is available from the British Library.

Restoring and Preserving our Motoring Heritage

Few people can have had the luck to realise their dreams to quite the same extent and in such a remarkable fashion as John Haynes, Founder and Chairman of the Haynes Publishing Group.

Since 1965 his unique approach to workshop manual publishing has proved so successful that millions of Haynes Manuals are now sold every year throughout the world, covering literally thousands of different makes and models of cars, vans and motorcycles.

A continuing passion for cars and motoring led to the founding in 1985 of a Charitable Trust dedicated to the restoration and preservation of our motoring heritage. To inaugurate the new Museum, John Haynes donated virtually his entire private collection of 52 cars.

Now with an unrivalled international collection of over 210 veteran, vintage and classic cars and motorcycles, the Haynes Motor Museum in Somerset is well on the way to becoming one of the most interesting Motor Museums in the world.

A 70 seat video cinema, a cafe and an extensive motoring bookshop, together with a specially constructed one kilometre motor circuit, make a visit to the Haynes Motor Museum a truly unforgettable experience.

Every vehicle in the museum is preserved in as near as possible mint condition and each car is run every six months on the motor circuit.

Enjoy the picnic area set amongst the rolling Somerset hills. Peer through the William Morris workshop windows at cars being restored, and browse through the extensive displays of fascinating motoring memorabilia.

From the 1903 Oldsmobile through such classics as an MG Midget to the mighty 'E' type Jaguar, Lamborghini, Ferrari Berlinetta Boxer, and Graham Hill's Lola Cosworth, there is something for everyone, young and old alike, at this Somerset Museum.

Haynes Motor Museum

Situated mid-way between London and Penzance, the Haynes Motor Museum is located just off the A303 at Sparkford, Somerset (home of the Haynes Manual) and is open to the public 7 days a week all year round, except Christmas Day and Boxing Day.

Telephone 01963 440804.

Contents

1981 model Volvo 245 GL Estate

1988 model Volvo 240 GL Saloon

About this manual

Its aim

The aim of this manual is to help you get the best value from your vehicle. It can do so in several ways. It can help you decide what work must be done (even should you choose to get it done by a garage), provide information on routine maintenance and servicing, and give a logical course of action and diagnosis when random faults occur. However, it is hoped that you will use the manual by tackling the work yourself. On simpler jobs it may even be quicker than booking the car into a garage and going there twice, to leave and collect it. Perhaps most important, a lot of money can be saved by avoiding the costs a garage must charge to cover its labour and overheads.

The manual has drawings and descriptions to show the function of the various components so that their layout can be understood. Then the tasks are described and photographed in a clear step-by-step sequence.

Its arrangement

Note: UK readers should also refer to the "Notes for UK readers" Section.

The manual is divided into Chapters, each covering a logical sub-division of the vehicle. The Chapters are each divided into Sections, numbered with single figures, eg 5; and the Sections are divided into numbered paragraphs.

It is freely illustrated, especially in those parts where there is a detailed sequence of operations to be carried out. The reference numbers used in illustration captions pinpoint the pertinent Section and the paragraph within that Section. That is, illustration 3.2 means that the illustration refers to Section 3, and paragraph 2 within that Section.

Procedures, once described in the appropriate place in the text, are not normally repeated. When it is necessary to refer to another Chapter, the reference will be given as Chapter and Section number. Cross-references given without the word "Chapter" apply to other Sections in the same Chapter. For example, a cross-reference such as "see Section 8" means Section 8 in that Chapter.

There is an alphabetical index at the back of the manual as well as a contents list at the front. Each Chapter is also preceded by its own individual contents list.

References to the "left" or "right" of the vehicle are in the sense of a person in the driver's seat facing forward.

Unless otherwise stated, nuts and bolts are removed by turning anti-clockwise, and tightened by turning clockwise.

Vehicle manufacturers continually make changes to specifications and recommendations, and these, when notified, are incorporated into our manuals at the earliest opportunity.

We take great pride in the accuracy of information given in this manual, but vehicle manufacturers make alterations and design changes during the production run of a particular vehicle of which they do not inform us. No liability can be accepted by the authors or publishers for loss, damage or injury caused by any errors in, or omissions from, the information given.

Notes for UK readers

General

Because this manual was written in the US, it will be noticed that references to components are in the US style. To avoid confusion, the preliminary Sections (ie up to Chapter 1) have been re-written specifically for the UK market but keeping to the same US style so that the reader can identify with the components described in the main Chapters of the manual. The UK equivalent of US components and various other US words is given in the Section headed "Use of English".

The following information should be considered before refering to the main Chapters of this manual.

Model differences

The B200 (1986cc) engine was not fitted to the 240 Series destined for the US market. The specifications for this, and other UK engines, can be found in the UK Supplement at the back of this manual.

Specifications

Where capacities or volumes appear in the Specifications Sections of each Chapter, note that references to quarts means US quarts, and the correct conversion factor should be used accordingly (see "Conversion factors"). All specifications in the main Chapters of the manual appear in imperial form; the equivalent metric values can be calculated using the Conversion factors page.

Maintenance intervals

The main maintenance interval for UK models is 12 000 miles or 12 months (whichever comes first), and there are several differences which make it difficult to relate the UK schedule to the US schedule given in Chapter 1. For this reason, the UK schedule is shown separately in the UK Supplement at the back of this manual.

Air conditioning system

The latter Sections of Chapter 3 describe work on the air conditioning system, and the Specifications Section includes the refrigerant capacity, etc. Although air conditioning is becoming more common in the UK, you are strongly advised to have a specialist carry out repairs to any part of it. The refrigerant used in the air conditioning system is potentially hazardous to health, so the evacuating and recharging of the air conditioning system must be performed by a specialist in any case. It therefore makes sense to have any other repairs carried out by that specialist at the same time.

Emission control information

US emission control laws call for an information label to be attached in a prominent position in the engine compartment. At the present time, there is no requirement for this in the UK. US laws also call for the renewal of emission control components (such as the oxygen sensor) at specific intervals. Although renewal of these components could very well be beneficial in terms of engine efficiency, there are no equivalent laws in the UK at the time of writing.

Where reference is made to California, Federal and Canadian systems, the UK system is in most cases equivalent to the Canadian type.

Fuel injection system and fault codes

Because of the different emission control laws in the US, the instrument panel on some US models includes a malfunction warning light, to warn the driver of a fault in the engine management system. This is not normally fitted to UK models, and therefore some of the information given in Chapter 6 will not apply. In particular, the method for extracting fault codes from the ECU using an analog voltmeter is not applicable to UK models.

Notes, cautions and warnings

A note provides information necessary to properly complete a procedure, or information which will make the procedure easier to understand.

A caution provides a special procedure or special steps which must be taken while completing the procedure where the caution is found. Not heeding a caution can result in damage to the assembly being worked on.

A warning provides a special procedure or special steps which must be taken while completing the procedure where the warning is found. Not heeding a warning can result in personal injury.

Introduction to the Volvo 240 Series

The Volvo 240 Series of Saloons and Estate cars represents all that has become synonymous with the word Volvo - safety, reliability and rugged longevity. Volvo do not change things for the sake of change, but for improvement to an already high standard of build. Despite being well equipped with the more modern vehicle innovations, the Volvo 240 retains a certain simplicity and ease of maintenance which the home mechanic will find to his or her liking.

All engine options are derived from the basic B21 overhead camshaft (OHC) unit, with both carburettor and fuel-injection versions available. Transmission options include a four-speed (with overdrive) and five-speed manual or a three-speed automatic and three-speed with lock-up fourth (overdrive) automatic. The transmission is mounted to the back of the engine, and power is transmitted to the solid rear axle by a two-piece driveshaft.

The front suspension is MacPherson strut design, with the coil spring/shock absorber unit making up the upper suspension link. The rear suspension is made up of coil springs and conventional shock absorbers with trailing arms and a track bar locating the axle. Braking is by disc brakes all-round with an anti-lock braking system (ABS) available on some models.

Retaining the in-line engine, rear wheel drive configuration, with rigid safety cage and proven suspension and brakes, the Volvo 240 Series is one of the safest, most reliable vehicles around today.

General dimensions and weights

Note: *All figures are approximate, and may vary according to model. Refer to manufacturer's data for exact figures.*

Dimensions

Overall length:	
1975 to 1980	193 in (490 cm)
1981-on	189 in (480 cm)
Overall width	67 in (170 cm)
Overall height (unladen):	
Saloon models	56 in (142 cm)
Estate models	57 in (145 cm)
Wheelbase	104 in (264 cm)
Front track	56.3 in (143 cm)
Rear track	53.5 in (136 cm)

Weights

Kerb weight	2830 to 2977 lb (1285 to 1350 kg)
Maximum gross vehicle weight	3921 to 4295 lb (1780 to 1950 kg)
Maximum roof rack load	220 lb (100 kg)
Maximum trailer weight	3307 lb (1500 kg)

Jacking, towing and wheel changing

Jacking and wheel changing

The jack supplied with the vehicle should be used only for raising the vehicle when changing a tire or placing jackstands under the frame. **Warning**: *Never crawl under the vehicle or start the engine when this jack is being used as the only means of support.*

The vehicle should be on level ground with the wheels blocked and the transmission in Park (automatic) or Reverse (manual). Pry off the hub cap (if equipped) using the tapered end of the lug wrench. Loosen the lug bolts one-half turn and leave them in place until the wheel is raised off the ground.

Position the arm of the jack under the side of the vehicle, making sure to engage the hook with the attachment pin made for this purpose (just behind the front wheel or forward of the rear wheel). Turn the jack handle clockwise until the wheel is raised off the ground. Remove the lug bolts, pull off the wheel and replace it with the spare.

Install the lug bolts and tighten them until snug. Lower the vehicle by turning the jackscrew counterclockwise. Remove the jack and tighten the bolts in a diagonal pattern to the torque listed in the Chapter 1 Specifications. If a torque wrench is not available, have the torque checked by a service station as soon as possible. Replace the hubcap.

When raising the vehicle with the jack, make sure to securely engage the hooked portion of the arm of the jack with the horizontal pin

Towing

Manual transmission-equipped vehicles can be towed with all four wheels on the ground. Automatic transmission-equipped vehicles can be towed with all four wheels on the ground if speeds do not exceed 10 mph and the distance is not over 20 miles, otherwise transmission damage can result.

Towing equipment specifically designed for this purpose should be used and should be attached to the main structural members of the vehicle, not the bumper or brackets. Sling-type towing equipment must **not** be used on these vehicles.

Safety is a major consideration while towing and all applicable state and local laws must be obeyed. A safety chain system must be used for all towing.

While towing, the parking brake should be released and the transmission should be in Neutral. The steering must be unlocked (ignition switch in the Off position). Remember that power steering and power brakes will not work with the engine off.

Buying spare parts and vehicle identification numbers

Buying spare parts

Spare parts are available from many sources, including maker's appointed garages, accessory shops, and motor factors. To be sure of obtaining the correct parts, it will sometimes be necessary to quote the vehicle identification number. If possible, it can also be useful to take the old parts along for positive identification. Items such as starter motors and alternators may be available under a service exchange scheme - any parts returned should always be clean.

Our advice regarding spare part sources is as follows.

Officially-appointed garages

This is the best source of parts which are peculiar to your car, and are not otherwise generally available (eg badges, interior trim, certain body panels, etc). It is also the only place at which you should buy parts if the vehicle is still under warranty.

Accessory shops

These are very good places to buy materials and components needed for the maintenance of your car (oil, air and fuel filters, spark plugs, light bulbs, drivebelts, oils and greases, brake pads, touch-up paint, etc). Components of this nature sold by a reputable shop are of the same standard as those used by the car manufacturer.

Besides components, these shops also sell tools and general accessories, usually have convenient opening hours, charge lower prices, and can often be found not far from home. Some accessory shops have parts counters where the components needed for almost any repair job can be purchased or ordered.

Motor factors

Good factors will stock all the more important components which wear out comparatively quickly and can sometimes supply individual components needed for the overhaul of a larger assembly (eg brake seals and hydraulic parts, bearing shells, pistons, valves, alternator brushes). They may also handle work such as cylinder block reboring, crankshaft regrinding and balancing, etc.

Tyre and exhaust specialists

These outlets may be independent or members of a local or national chain. They frequently offer competitive prices when compared with a main dealer or local garage, but it will pay to obtain several quotes before making a decision. When researching prices, also ask what 'extras' may be added - for instance, fitting a new valve and balancing the wheel are both commonly charged on top of the price of a new tyre.

Other sources

Beware of parts or materials obtained from market stalls, car boot sales or similar outlets. Such items are not invariably sub-standard, but there is little chance of compensation if they do prove unsatisfactory. In the case of safety-critical components such as brake pads there is the risk not only of financial loss but also of an accident causing injury or death.

Second-hand components or assemblies obtained from a car breaker can be a good buy in some circumstances, but this sort of purchase is best made by the experienced DIY mechanic.

Vehicle identification numbers

Modifications are a continuing and unpublicised process in vehicle manufacture, quite apart from major model changes. Spare parts manuals and lists are compiled upon a numerical basis, the individual vehicle identification numbers being essential to correct identification of the component concerned.

When ordering spare parts, always give as much information as possible. Quote the car model, year of manufacture, body and engine numbers as appropriate.

The *engine number* is stamped on the left-hand front end of the engine cylinder block **(see illustration)**.

The *vehicle identification number (VIN)* and *chassis number* is stamped on the VIN plate located under the bonnet by the right-hand front suspension strut turret and also stamped on the right-hand front door pillar **(see illustration)**.

The *paint code number* is stamped on the VIN plate located under the bonnet by the right-hand front suspension strut turret.

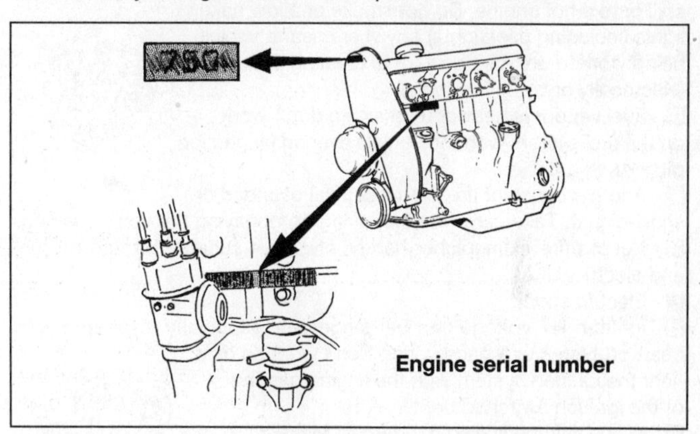

Engine serial number

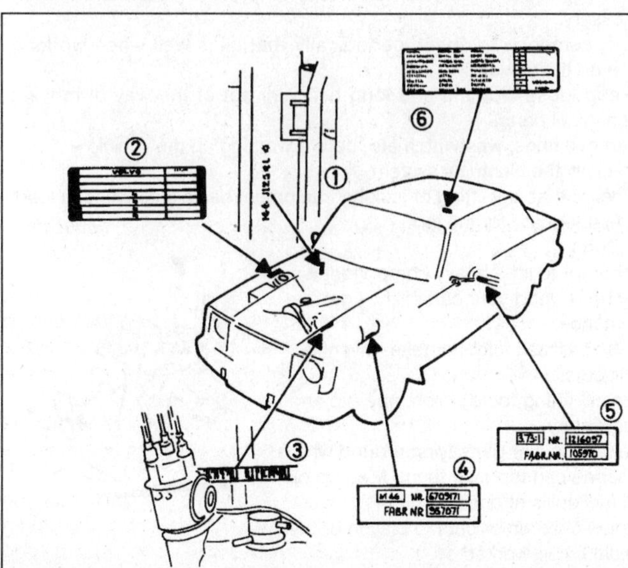

Typical vehicle identification plates

1. Type and model year designation and chassis number
2. Type designation, maximum permitted load and colour code
3. Type designation and part number of engine
4. Type designation, serial number and part number of transmission unit
5. Final drive reduction ratio, part number and serial number of rear axle
6. Service plate - below the rear window on right-hand (Saloon) or in large storage compartment (Estate

Safety First!

Working on your car can be dangerous. This page shows just some of the potential risks and hazards, with the aim of creating a safety-conscious attitude.

General hazards

■ Scalding

☐ Don't remove the radiator or expansion tank cap while the engine is hot.
Engine oil, automatic transmission fluid or power steering fluid may also be dangerously hot if the engine has recently been running.

■ Burning

☐ Beware of burns from the exhaust system and from any part of the engine. Brake discs and drums can also be extremely hot immediately after use.

■ Crushing

☐ When working under or near a raised vehicle, always supplement the jack with axle stands, or use drive-on ramps. ***Never venture under a car which is only supported by a jack.***

☐ Take care if loosening or tightening high-torque nuts when the vehicle is on stands.
Initial loosening and final tightening should be done with the wheels on the ground.

■ Fire

☐ Fuel is highly flammable; fuel vapour is explosive. Don't let fuel spill onto a hot engine. Do not smoke or allow naked lights (including pilot lights) anywhere near a vehicle being worked on. Also beware of creating sparks (electrically or by use of tools).

☐ Fuel vapour is heavier than air, so don't work on the fuel system with the vehicle over an inspection pit.

☐ Another cause of fire is an electrical overload or short-circuit. Take care when repairing or modifying the vehicle wiring.

☐ Keep a fire extinguisher handy, of a type suitable for use on fuel and electrical fires.

■ Electric shock

☐ Ignition HT voltage can be dangerous, especially to people with heart problems or a pacemaker. Don't work on or near the ignition system with the engine running or the ignition switched on.

☐ Mains voltage is also dangerous. Make sure that any mains-operated equipment is correctly earthed. Mains power points should be protected by a residual current device (RCD) circuit breaker.

■ Fume or gas intoxication

☐ Exhaust fumes are poisonous; they often contain carbon monoxide, which is rapidly fatal if inhaled. Never run the engine in a confined space such as a garage with the doors shut.

☐ Fuel vapour is also poisonous, as are the vapours from some cleaning solvents and paint thinners.

■ Poisonous or irritant substances

☐ Avoid skin contact with battery acid and with any fuel, fluid or lubricant, especially antifreeze, brake hydraulic fluid and Diesel fuel. Don't syphon them by mouth. If such a substance is swallowed or gets into the eyes, seek medical advice.

☐ Prolonged contact with used engine oil can cause skin cancer. Wear gloves or use a barrier cream if necessary. Change out of oil-soaked clothes and do not keep oily rags in your pocket.

☐ Air conditioning refrigerant forms a poisonous gas if exposed to a naked flame (including a cigarette). It can also cause skin burns on contact.

■ Asbestos

☐ Asbestos dust can cause cancer if inhaled or swallowed. Asbestos may be found in gaskets and in brake and clutch linings. When dealing with such components it is safest to assume that they do contain asbestos.

Special hazards

■ Hydrofluoric acid

☐ This extremely corrosive acid is formed when certain types of synthetic rubber, found in some O-rings, oil seals, fuel hoses etc, are exposed to temperatures above 400°C. The rubber changes into a charred or sticky substance containing the acid. *Once formed, the acid remains dangerous for years. If it gets onto the skin, it may be necessary to amputate the limb concerned.*

☐ When dealing with a vehicle which has suffered a fire, or with components salvaged from such a vehicle, wear protective gloves and discard them after use.

■ The battery

☐ Batteries contain sulphuric acid, which attacks clothing, eyes and skin. Take care when topping-up or carrying the battery.

☐ The hydrogen gas given off by the battery is highly explosive. Never cause a spark or allow a naked light nearby. Be careful when connecting and disconnecting battery chargers or jump leads.

■ Air bags

☐ Air bags can cause injury if they go off accidentally. Take care when removing the steering wheel and/or facia. Special storage instructions may apply.

■ Diesel injection equipment

☐ Diesel injection pumps supply fuel at very high pressure. Take care when working on the fuel injectors and fuel pipes.
Never expose the hands, face or any other part of the body to injector spray; the fuel can penetrate the skin with potentially fatal results.

A few tips

■ Do...

... use eye protection when using power tools, and when working under the vehicle.

... wear gloves or use barrier cream to protect your hands when necessary.

... get someone to check periodically that all is well when working alone on the vehicle.

... keep loose clothing and long hair well out of the way of moving mechanical parts.

... remove rings, wristwatch etc, before working on the vehicle – especially the electrical system.

... ensure that any lifting or jacking equipment has a safe working load rating adequate for the job.

■ Don't...

... attempt to lift a heavy component which may be beyond your capability – get assistance.

... rush to finish a job, or take unverified short cuts.

... use ill-fitting tools which may slip and cause injury.

... leave tools or parts lying around where someone can trip over them. Mop up oil and fuel spills at once.

... allow children or pets to play in or near a vehicle being worked on.

General repair procedures

Whenever servicing, repair or overhaul work is carried out on the car or its components, it is necessary to observe the following procedures and instructions. This will assist in carrying out the operation efficiently and to a professional standard of workmanship.

Joint mating faces and gaskets

When separating components at their mating faces, never insert screwdrivers or similar implements into the joint between the faces in order to prise them apart. This can cause severe damage which results in oil leaks, coolant leaks, etc. upon reassembly. Separation is usually achieved by tapping along the joint with a soft-faced hammer in order to break the seal. However, note that this method may not be suitable where dowels are used for component location.

Where a gasket is used between the mating faces of two components, ensure that it is renewed on reassembly and fit it dry unless otherwise stated in the repair procedure. Make sure that the mating faces are clean and dry with all traces of old gasket removed. When cleaning a joint face, use a tool which is not likely to score or damage the face, and remove any burrs or nicks with an oilstone or fine file.

Make sure that tapped holes are cleaned with a pipe cleaner and keep them free of jointing compound, if this is being used, unless specifically instructed otherwise.

Ensure that all orifices, channels or pipes are clear and blow through them, preferably using compressed air.

Oil seals

Oil seals can be removed by levering them out with a wide flat-bladed screwdriver or similar implement. Alternatively, a number of self-tapping screws may be screwed into the seal and these used as a purchase for pliers or some similar device in order to pull the seal free.

Whenever an oil seal is removed from its working location, either individually or as part of an assembly, it should be renewed.

The very fine sealing lip of the seal is easily damaged and will not seal if the surface it contacts is not completely clean and free from scratches, nicks or grooves. If the original sealing surface of the component cannot be restored, and the manufacturer has not made provision for slight relocation of the seal relative to the sealing surface, the component should be renewed.

Protect the lips of the seal from any surface which may damage them in the course of fitting. Use tape or a conical sleeve where possible. Lubricate the seal lips with oil before fitting and, on dual-lipped seals, fill the space between the lips with grease.

Unless otherwise stated, oil seals must be fitted with their sealing lips toward the lubricant to be sealed.

Use a tubular drift or block of wood of the appropriate size to install the seal and, if the seal housing is shouldered, drive the seal down to the shoulder. If the seal housing is unshouldered, the seal should be fitted with its face flush with the housing top face (unless otherwise instructed).

Screw threads and fastenings

Seized nuts, bolts and screws are quite a common occurrence where corrosion has set in, and the use of penetrating oil or releasing fluid will often overcome this problem if the offending item is soaked for a while before attempting to release it. The use of an impact driver may also provide a means of releasing such stubborn fastening devices when used in conjunction with the appropriate screwdriver bit or socket. If none of these methods works, it may be necessary to resort to the careful application of heat, or the use of a hacksaw or nut splitter device.

Studs are usually removed by locking two nuts together on the threaded part and then using a spanner on the lower nut to unscrew the stud. Studs or bolts which have broken off below the surface of the component in which they are mounted can sometimes be removed using a proprietary stud extractor. Always ensure that a blind tapped hole is completely free from oil, grease, water or other fluid before installing the bolt or stud. Failure to do this could cause the housing to crack due to the hydraulic action of the bolt or stud as it is screwed in.

When tightening a castellated nut to accept a split pin, tighten the nut to the specified torque, where applicable, and then tighten further to the next split pin hole. Never slacken the nut to align the split pin hole unless stated in the repair procedure.

When checking or retightening a nut or bolt to a specified torque setting, slacken the nut or bolt by a quarter of a turn, and then retighten to the specified setting. However, this should not be attempted where angular tightening has been used.

For some screw fastenings, notably cylinder head bolts or nuts, torque wrench settings are no longer specified for the latter stages of tightening, "angle-tightening" being called up instead. Typically, a fairly low torque wrench setting will be applied to the bolts/nuts in the correct sequence, followed by one or more stages of tightening through specified angles.

Locknuts, locktabs and washers

Any fastening which will rotate against a component or housing in the course of tightening should always have a washer between it and the relevant component or housing.

Spring or split washers should always be renewed when they are used to lock a critical component such as a big-end bearing retaining bolt or nut. Locktabs which are folded over to retain a nut or bolt should always be renewed.

Self-locking nuts can be reused in non-critical areas, providing resistance can be felt when the locking portion passes over the bolt or stud thread. However, it should be noted that self-locking stiffnuts tend to lose their effectiveness after long periods of use, and in such cases should be renewed as a matter of course.

Split pins must always be replaced with new ones of the correct size for the hole.

When thread-locking compound is found on the threads of a fastener which is to be re-used, it should be cleaned off with a wire brush and solvent, and fresh compound applied on reassembly.

Special tools

Some repair procedures in this manual entail the use of special tools such as a press, two or three-legged pullers, spring compressors etc. Wherever possible, suitable readily available alternatives to the manufacturer's special tools are described, and are shown in use. In some instances, where no alternative is possible, it has been necessary to resort to the use of a manufacturer's tool and this has been done for reasons of safety as well as the efficient completion of the repair operation. Unless you are highly skilled and have a thorough understanding of the procedures described, never attempt to bypass the use of any special tool when the procedure described specifies its use. Not only is there a very great risk of personal injury, but expensive damage could be caused to the components involved.

Environmental considerations

When disposing of used engine oil, brake fluid, antifreeze etc, give due consideration to any detrimental environmental effects. Do not, for instance, pour any of the above liquids down drains into the general sewage system or onto the ground to soak away. Many local council refuse tips provide a facility for waste oil disposal as do some garages. If none of these facilities are available, consult your local Environmental Health Department for further advice.

With the universal tightening-up of legislation regarding the emission of environmentally harmful substances from motor vehicles, most current vehicles have tamperproof devices fitted to the main adjustment points of the fuel system. These devices are primarily designed to prevent unqualified persons from adjusting the fuel/air mixture with the chance of a consequent increase in toxic emissions. If such devices are encountered during servicing or overhaul, they should, wherever possible, be renewed or refitted in accordance with the vehicle manufacturer's requirements or current legislation.

Tools and working facilities

Introduction

A selection of good tools is a fundamental requirement for anyone contemplating the maintenance and repair of a motor vehicle. For the owner who does not possess any, their purchase will prove a considerable expense, offsetting some of the savings made by doing-it-yourself. However, provided that the tools purchased meet the relevant national safety standards and are of good quality, they will last for many years and prove an extremely worthwhile investment.

To help the average owner to decide which tools are needed to carry out the various tasks detailed in this manual, we have compiled three lists of tools under the following headings: Maintenance and minor repair, Repair and overhaul, and Special. Newcomers to practical mechanics should start off with the Maintenance and minor repair tool kit and confine themselves to the simpler jobs around the vehicle. Then, as confidence and experience grow, more difficult tasks can be undertaken, with extra tools being purchased as, and when, they are needed. In this way, a Maintenance and minor repair tool kit can be built up into a Repair and overhaul tool kit over a considerable period of time without any major cash outlays. The experienced do-it-yourselfer will have a tool kit good enough for most repair and overhaul procedures and will add tools from the Special category when it is felt that the expense is justified by the amount of use to which these tools will be put.

Maintenance and minor repair tool kit

The tools given in this list should be considered as a minimum requirement if routine maintenance, servicing and minor repair operations are to be undertaken. We recommend the purchase of combination spanners (ring one end, open-ended the other); although more expensive than open-ended ones, they do give the advantages of both types of spanner.

Combination spanners:
Metric - 8, 9, 10, 11, 12, 13, 14, 15, 16, 17, 19, 21, 22 & 26 mm
Adjustable spanner - 35 mm jaw (approx)
Engine sump/gearbox drain plug key
Set of feeler gauges
Brake bleed nipple spanner
Screwdrivers:
- *Flat blade - approx 100 mm long x 6 mm dia*
- *Cross blade - approx 100 mm long x 6 mm dia*

Combination pliers
Hacksaw (junior)
Tyre pump
Tyre pressure gauge
Oil can
Oil filter removal tool
Fine emery cloth
Wire brush (small)
Funnel (medium size)

Repair and overhaul tool kit

These tools are virtually essential for anyone undertaking any major repairs to a motor vehicle, and are additional to those given in the Maintenance and minor repair list. Included in this list is a comprehensive set of sockets. Although these are expensive, they will be found invaluable as they are so versatile - particularly if various drives are included in the set. We recommend the half-inch square-drive type, as this can be used with most proprietary torque wrenches. If you cannot afford a socket set, even bought piecemeal, then inexpensive tubular box spanners are a useful alternative.

The tools in this list will occasionally need to be supplemented by tools from the Special list.

Sockets (or box spanners) to cover range in previous list
Reversible ratchet drive (for use with sockets) **(see illustration)**
Extension piece, 250 mm (for use with sockets)
Universal joint (for use with sockets)
Torque wrench (for use with sockets)
Self-locking grips
Ball pein hammer
Soft-faced mallet (plastic/aluminium or rubber)
Screwdrivers:
- *Flat blade - long & sturdy, short (chubby), and narrow (electricians) types*
- *Cross blade - Long & sturdy, and short (chubby) types*

Pliers:
- *Long-nosed*
- *Side cutters (electricians)*
- *Circlip (internal and external)*

Cold chisel - 25 mm
Scriber
Scraper
Centre punch
Pin punch
Hacksaw
Brake hose clamp
Brake bleeding kit
Selection of twist drills
Steel rule/straight-edge
Allen keys (inc. splined/Torx type) **(see illustrations)**
Selection of files
Wire brush
Axle-stands
Jack (strong trolley or hydraulic type)
Light with extension lead

Special tools

The tools in this list are those which are not used regularly, are expensive to buy, or which need to be used in accordance with their manufacturers' instructions. Unless relatively difficult mechanical jobs are undertaken frequently, it will not be economic to buy many of these tools. Where this is the case, you could consider clubbing together with friends (or joining a motorists' club) to make a joint purchase, or borrowing the tools against a deposit from a local garage or tool hire specialist. It is worth noting that many of the larger DIY superstores now carry a large range of special tools for hire at modest rates.

The following list contains only those tools and instruments freely available to the public, and not those special tools produced by the vehicle manufacturer specifically for its dealer network. You will find occasional references to these manufacturers' special tools in the text of this manual. Generally, an alternative method of doing the job without the vehicle manufacturers' special tool is given. However, sometimes there is no alternative to using them. Where this is the case and the relevant tool cannot be bought or borrowed, you will have to entrust the work to a franchised garage.

Valve spring compressor **(see illustration)**
Valve grinding tool
Piston ring compressor **(see illustration)**
Piston ring removal/installation tool **(see illustration)**
Cylinder bore hone **(see illustration)**
Balljoint separator
Coil spring compressors (where applicable)
Two/three-legged hub and bearing puller **(see illustration)**
Impact screwdriver
Micrometer and/or vernier calipers **(see illustrations)**
Dial test indicator **(see illustration)**
Tachometer
Universal electrical multi-meter
Stroboscopic timing light **(see illustration)**
Compression testing gauge **(see illustration)**
Vacuum pump and gauge **(see illustration)**
Clutch plate alignment set **(see illustration)**
Brake shoe steady spring cup removal tool **(see illustration)**

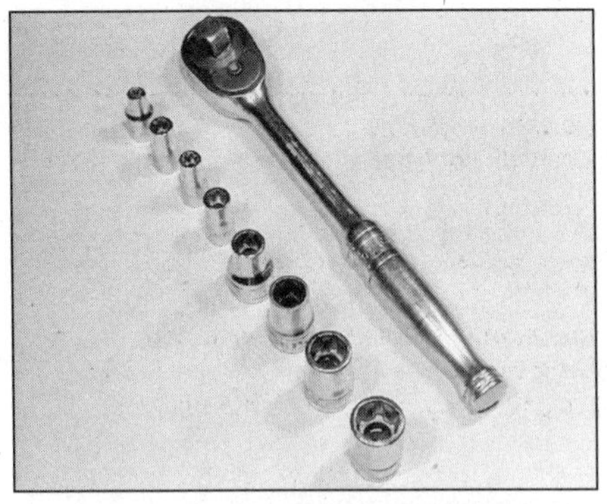
Sockets and reversible ratchet drive

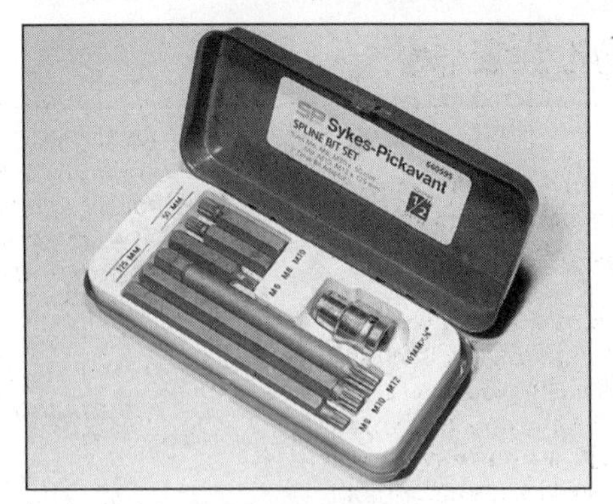

Spline bit set

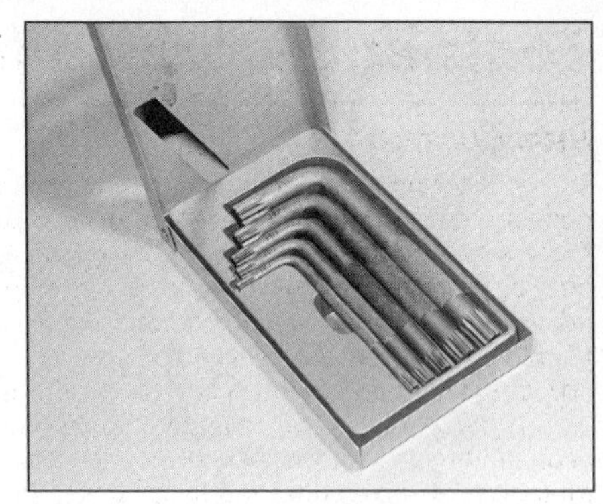
Spline key set

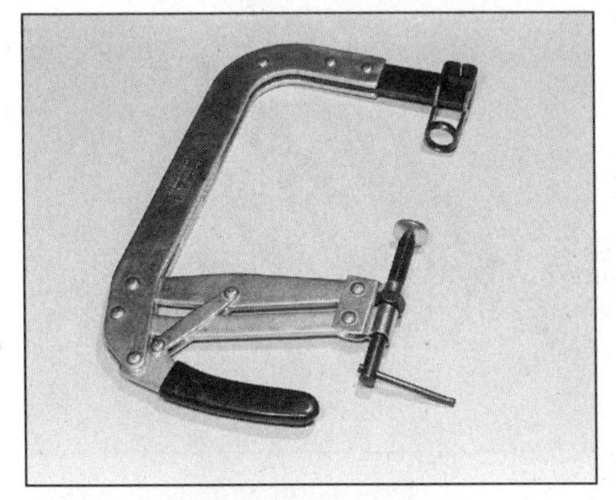
Valve spring compressor

Piston ring compressor

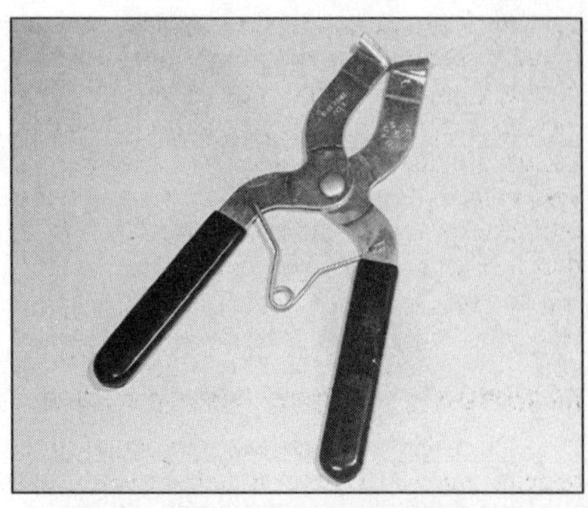
Piston ring removal/installation tool

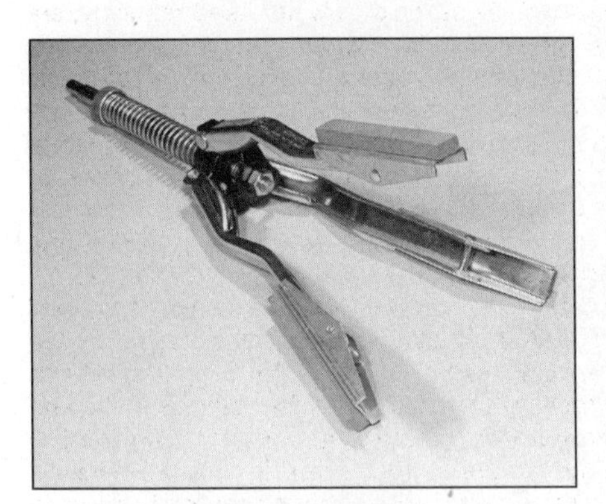
Cylinder bore hone

Two/three-legged hub and bearing puller

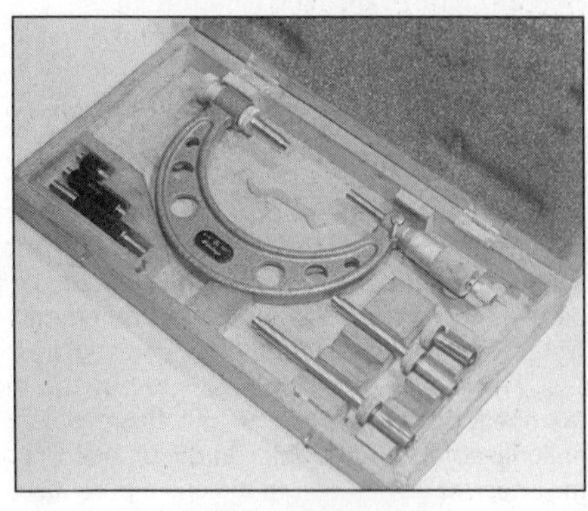
Micrometer set

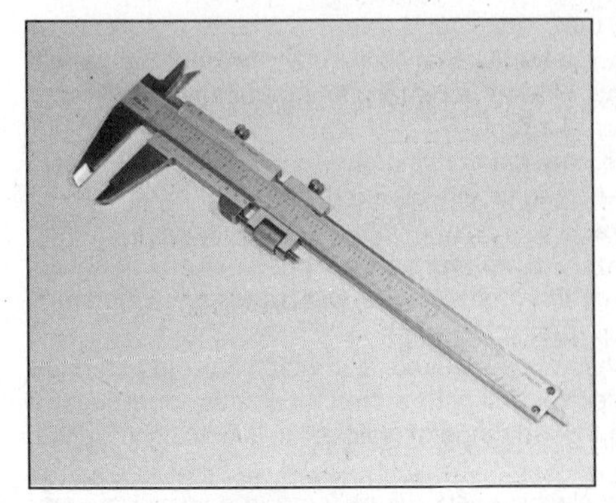
Vernier calipers

Dial test indicator

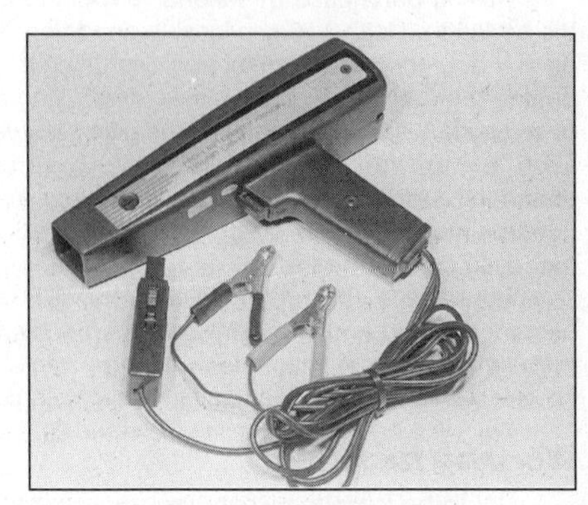
Stroboscopic timing light

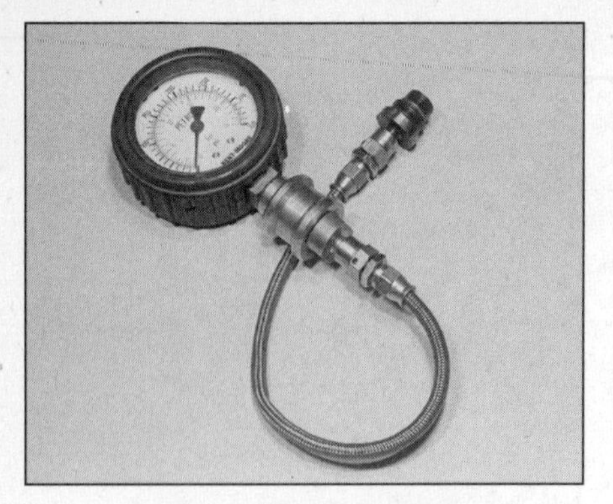
Compression testing gauge

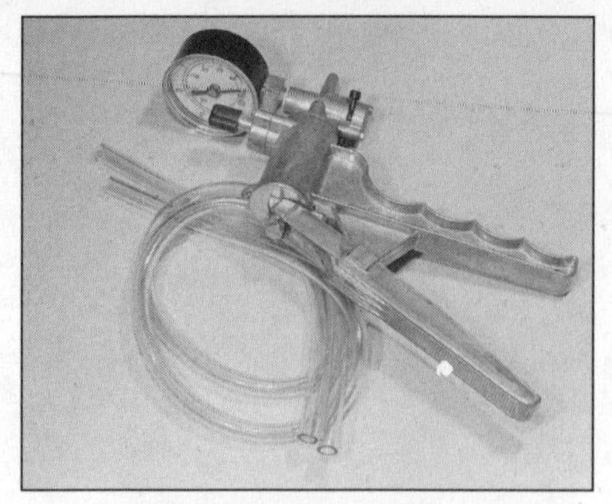
Vacuum pump and gauge

Clutch plate alignment set

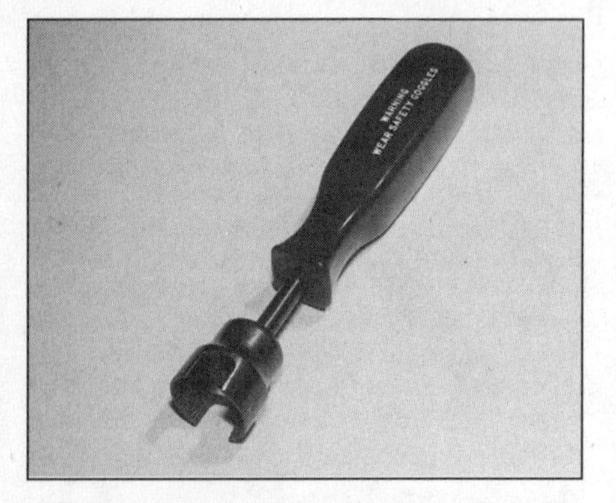

Brake shoe steady spring cup removal tool

Bush and bearing removal/installation set

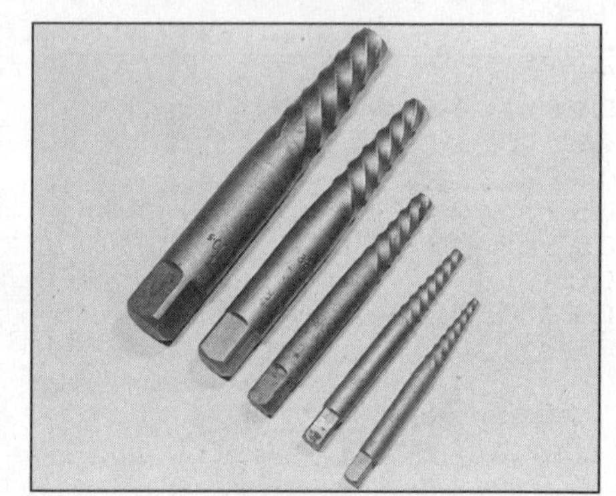
Stud extractors

Bush and bearing removal/installation set **(see illustration)**
Stud extractors **(see illustration)**
Tap and die set **(see illustration)**
Lifting tackle
Trolley jack

Buying tools

Reputable accessory shops offer excellent quality tools at reasonable prices. Beware of very cheap tools, especially those on sale at market stalls, car boot sales or similar outlets. Badly-made or badly-finished tools are at best frustrating and disappointing to use; at worst they can cause damage or personal injury.

Remember, you don't have to buy the most expensive items on the shelf, but always aim to purchase tools which meet the relevant national safety standards. If in doubt, ask the proprietor or manager of the shop for advice before making a purchase.

Care and maintenance of tools

Having purchased a reasonable tool kit, it is necessary to keep the tools in a clean and serviceable condition. After use, always wipe off any dirt, grease and metal particles using a clean, dry cloth, before putting the tools away. Never leave them lying around after they have been used. A simple tool rack on the garage or workshop wall for items such as screwdrivers and pliers is a good idea. Store all normal spanners and sockets in a metal box. Any measuring instruments, gauges, meters, etc, must be carefully stored where they cannot be damaged or become rusty.

Take a little care when tools are used. Hammer heads inevitably become marked and screwdrivers lose the keen edge on their blades from time to time. A little timely attention with emery cloth or a file will soon restore items like this to a good serviceable finish.

Working facilities

Not to be forgotten when discussing tools is the workshop itself. If anything more than routine maintenance is to be carried out, some form of suitable working area becomes essential.

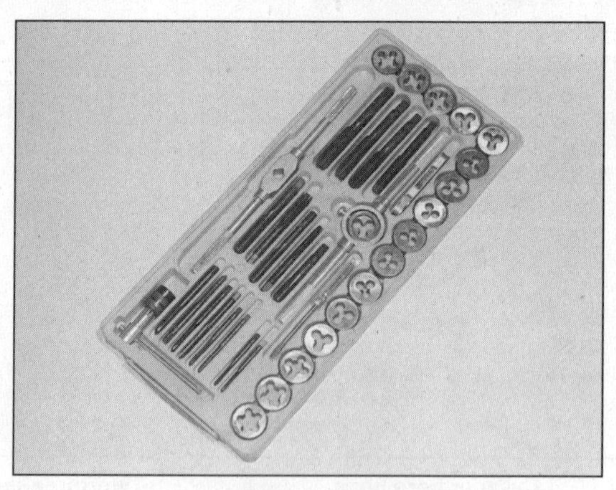
Tap and die set

It is appreciated that many an owner mechanic is forced by circumstances to remove an engine or similar item without the benefit of a garage or workshop. Having done this, any repairs should always be done under the cover of a roof.

Wherever possible, any dismantling should be done on a clean, flat workbench or table at a suitable working height.

Any workbench needs a vice; one with a jaw opening of 100 mm is suitable for most jobs. As mentioned previously, some clean dry storage space is also required for tools, as well as for any lubricants, cleaning fluids, touch-up paints and so on, which become necessary.

Another item which may be required, and which has a much more general usage, is an electric drill with a chuck capacity of at least 8 mm. This, together with a good range of twist drills, is virtually essential for fitting accessories.

Last, but not least, always keep a supply of old newspapers and clean, lint-free rags available, and try to keep any working area as clean as possible.

Spanner jaw gap and bolt size comparison table

Jaw gap - in (mm)	Spanner size	Bolt size
0.197 (5.00)	5 mm	M 2.5
0.216 (5.50)	5.5 mm	M 3
0.218 (5.53)	7/32 in AF	
0.236 (6.00)	6 mm	M 3.5
0.250 (6.35)	1/4 in AF	
0.275 (7.00)	7 mm	M 4
0.281 (7.14)	9/32 in AF	
0.312 (7.92)	5/16 in AF	
0.315 (8.00)	8 mm	M 5
0.343 (8.71)	11/32 in AF	
0.375 (9.52)	3/8 in AF	
0.394 (10.00)	10 mm	M 6
0.406 (10.32)	13/32 in AF	
0.433 (11.00)	11 mm	M 7
0.437 (11.09)	7/16 in AF	1/4 in SAE
0.468 (11.88)	15/32 in AF	
0.500 (12.70)	1/2 in AF	5/16 in SAE
0.512 (13.00)	13 mm	M8
0.562 (14.27)	9/16 in AF	3/8 in SAE
0.593 (15.06)	19/32 in AF	
0.625 (15.87)	5/8 in AF	7/16 in SAE
0.669 (17.00)	17 mm	M 10
0.687 (17.44)	11/16 in AF	
0.709 (19.00)	19 mm	M 12
0.750 (19.05)	3/4 in AF	1/2 in SAE
0.781 (19.83)	25/32 in AF	
0.812 (20.62)	13/16 in AF	
0.866 (22.00)	22 mm	M 14
0.875 (22.25)	7/8 in AF	9/16 in SAE
0.937 (23.79)	15/16 in AF	5/8 in SAE
0.945 (24.00)	24 mm	M 16
0.968 (24.58)	31/32 in AF	
1.000 (25.40)	1 in AF	11/16 in SAE
1.062 (26.97)	1 1/16 in AF	3/4 in SAE
1.063 (27.00)	27 mm	M 18
1.125 (28.57)	1 1/8 in AF	
1.182 (30.00)	30 mm	M 20
1.187 (30.14)	1 3/16 in AF	
1.250 (31.75)	1 1/4 in AF	7/8 in SAE
1.260 (32.00)	32 mm	M 22
1.312 (33.32)	1 5/16 in AF	
1.375 (34.92)	1 3/8 in AF	
1.418 (36.00)	36 mm	M 24
1.437 (36.49)	1 7/16 in AF	1 in SAE
1.500 (38.10)	1 1/2 in AF	
1.615 (41.00)	41 mm	M 27

Booster battery (jump) starting

When jump starting a car using a booster battery, observe the following precautions.

a) *Before connecting the booster battery, make sure that the ignition is switched off.*

b) *Ensure that all electrical equipment (lights, heater, wipers etc) is switched off.*

c) *Make sure that the booster battery is the same voltage as the discharged one in the vehicle.*

d) *If the battery is being jump started from the battery in another vehicle, the two vehicles MUST NOT TOUCH each other.*

e) *Make sure that the gearbox is in Neutral.*

Connect one jump lead between the positive (+) terminals of the two batteries. Connect the other jump lead first to the negative (-) terminal of the booster battery, and then to a good earthing point on the vehicle to be started, such as a bolt or bracket on the engine block, at least 18 in (45 cm) from the battery if possible (see illustration). Make sure that the jump leads will not come into contact with the fan, drivebelts or other moving parts of the engine.

Start the engine using the booster battery, then with the engine running at idle speed, disconnect the jump leads in the reverse order of connection.

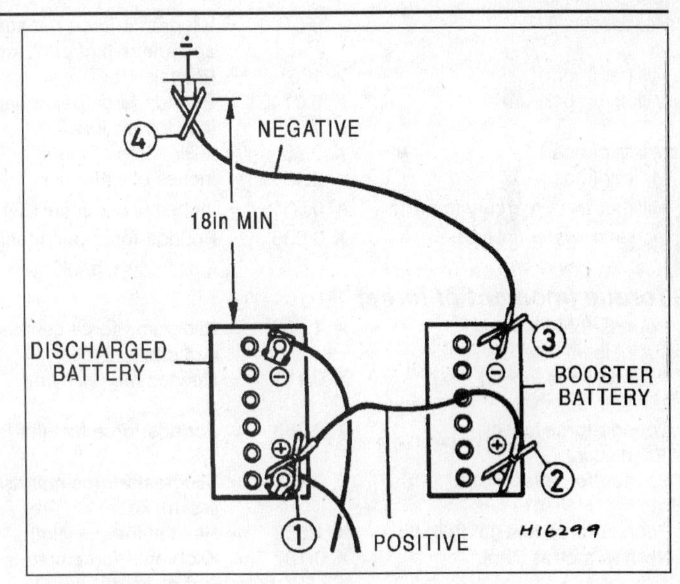

Jump start lead connections for negative earth vehicles - connect leads in order shown

Conversion factors

Length (distance)

Inches (in)	X 25.4	= Millimetres (mm)	X 0.0394	= Inches (in)
Feet (ft)	X 0.305	= Metres (m)	X 3.281	= Feet (ft)
Miles	X 1.609	= Kilometres (km)	X 0.621	= Miles

Volume (capacity)

Cubic inches (cu in; in^3)	X 16.387	= Cubic centimetres (cc; cm^3)	X 0.061	= Cubic inches (cu in; in^3)
Imperial pints (Imp pt)	X 0.568	= Litres (l)	X 1.76	= Imperial pints (Imp pt)
Imperial quarts (Imp qt)	X 1.137	= Litres (l)	X 0.88	= Imperial quarts (Imp qt)
Imperial quarts (Imp qt)	X 1.201	= US quarts (US qt)	X 0.833	= Imperial quarts (Imp qt)
US quarts (US qt)	X 0.946	= Litres (l)	X 1.057	= US quarts (US qt)
Imperial gallons (Imp gal)	X 4.546	= Litres (l)	X 0.22	= Imperial gallons (Imp gal)
Imperial gallons (Imp gal)	X 1.201	= US gallons (US gal)	X 0.833	= Imperial gallons (Imp gal)
US gallons (US gal)	X 3.785	= Litres (l)	X 0.264	= US gallons (US gal)

Mass (weight)

Ounces (oz)	X 28.35	= Grams (g)	X 0.035	= Ounces (oz)
Pounds (lb)	X 0.454	= Kilograms (kg)	X 2.205	= Pounds (lb)

Force

Ounces-force (ozf; oz)	X 0.278	= Newtons (N)	X 3.6	= Ounces-force (ozf; oz)
Pounds-force (lbf; lb)	X 4.448	= Newtons (N)	X 0.225	= Pounds-force (lbf; lb)
Newtons (N)	X 0.1	= Kilograms-force (kgf; kg)	X 9.81	= Newtons (N)

Pressure

Pounds-force per square inch (psi; lbf/in^2; lb/in^2)	X 0.070	= Kilograms-force per square centimetre (kgf/cm^2; kg/cm^2)	X 14.223	= Pounds-force per square inch (psi; lbf/in^2; lb/in^2)
Pounds-force per square inch (psi; lbf/in^2; lb/in^2)	X 0.068	= Atmospheres (atm)	X 14.696	= Pounds-force per square inch (psi; lbf/in^2; lb/in^2)
Pounds-force per square inch (psi; lbf/in^2; lb/in^2)	X 0.069	= Bars	X 14.5	= Pounds-force per square inch (psi; lbf/in^2; lb/in^2)
Pounds-force per square inch (psi; lbf/in^2; lb/in^2)	X 6.895	= Kilopascals (kPa)	X 0.145	= Pounds-force per square inch (psi; lbf/in^2; lb/in^2)
Kilopascals (kPa)	X 0.01	= Kilograms-force per square centimetre (kgf/cm^2; kg/cm^2)	X 98.1	= Kilopascals (kPa)
Millibar (mbar)	X 100	= Pascals (Pa)	X 0.01	= Millibar (mbar)
Millibar (mbar)	X 0.0145	= Pounds-force per square inch (psi; lbf/in^2; lb/in^2)	X 68.947	= Millibar (mbar)
Millibar (mbar)	X 0.75	= Millimetres of mercury (mmHg)	X 1.333	= Millibar (mbar)
Millibar (mbar)	X 0.401	= Inches of water (inH$_2$O)	X 2.491	= Millibar (mbar)
Millimetres of mercury (mmHg)	X 0.535	= Inches of water (inH$_2$O)	X 1.868	= Millimetres of mercury (mmHg)
Inches of water (inH$_2$O)	X 0.036	= Pounds-force per square inch (psi; lbf/in^2; lb/in^2)	X 27.68	= Inches of water (inH$_2$O)

Torque (moment of force)

Pounds-force inches (lbf in; lb in)	X 1.152	= Kilograms-force centimetre (kgf cm; kg cm)	X 0.868	= Pounds-force inches (lbf in; lb in)
Pounds-force inches (lbf in; lb in)	X 0.113	= Newton metres (Nm)	X 8.85	= Pounds-force inches (lbf in; lb in)
Pounds-force inches (lbf in; lb in)	X 0.083	= Pounds-force feet (lbf ft; lb ft)	X 12	= Pounds-force inches (lbf in; lb in)
Pounds-force feet (lbf ft; lb ft)	X 0.138	= Kilograms-force metres (kgf m; kg m)	X 7.233	= Pounds-force feet (lbf ft; lb ft)
Pounds-force feet (lbf ft; lb ft)	X 1.356	= Newton metres (Nm)	X 0.738	= Pounds-force feet (lbf ft; lb ft)
Newton metres (Nm)	X 0.102	= Kilograms-force metres (kgf m; kg m)	X 9.804	= Newton metres (Nm)

Power

Horsepower (hp)	X 745.7	= Watts (W)	X 0.0013	= Horsepower (hp)

Velocity (speed)

Miles per hour (miles/hr; mph)	X 1.609	= Kilometres per hour (km/hr; kph)	X 0.621	= Miles per hour (miles/hr; mph)

Fuel consumption*

Miles per gallon, Imperial (mpg)	X 0.354	= Kilometres per litre (km/l)	X 2.825	= Miles per gallon, Imperial (mpg)
Miles per gallon, US (mpg)	X 0.425	= Kilometres per litre (km/l)	X 2.352	= Miles per gallon, US (mpg)

Temperature

Degrees Fahrenheit = (°C x 1.8) + 32

Degrees Celsius (Degrees Centigrade; °C) = (°F - 32) x 0.56

** It is common practice to convert from miles per gallon (mpg) to litres/100 kilometres (l/100km), where mpg (Imperial) x l/100 km = 282 and mpg (US) x l/100 km = 235*

Use of English

As this book was written in the US, it uses the appropriate US component names, phrases, and spelling. Some of these differ from those used in the UK. Normally, these cause no difficulty, but to make sure, a glossary is printed below. In ordering spare parts remember the parts list may use some of these words:

AMERICAN	ENGLISH
Aluminum	Aluminium
Antenna	Aerial
Authorized	Authorised
Auto parts stores	Motor factors
Axleshaft	Halfshaft
Back-up	Reverse
Barrel	Choke/venturi
Block	Chock
Bushing	Bush
Carburetor	Carburettor
Center	Centre
Coast	Freewheel
Color	Colour
Drop head coupe	Convertible
Cotter pin	Split pin
Counterclockwise	Anti-clockwise
Countershaft (of gearbox)	Layshaft
Dashboard	Facia
Denatured alcohol	Methylated spirit
Dome lamp	Interior light
Driveaxle	Driveshaft
Driveshaft	Propeller shaft
Fender	Wing/mudguard
Firewall	Bulkhead
Flashlight	Torch
Float bowl	Float chamber
Floor jack	Trolley jack
Freeway, turnpike etc	Motorway
Freeze plug	Core plug
Frozen	Seized
Gas tank	Petrol tank
Gasoline (gas)	Petrol
Gearshift	Gearchange
Generator (DC)	Dynamo
Ground (electrical)	Earth
Header	Exhaust manifold
Heat riser	Hot spot
High	Top gear
Hood (engine cover)	Bonnet
Installation	Refitting
Intake	Inlet
Jackstands	Axle stands
Jumper cable	Jump lead
Keeper	Collet
Kerosene	Paraffin
Knock pin	Roll pin
Lash	Clearance, Free-play
Latch	Catch
Latches	Locks
License plate	Number plate
Light	Lamp
Lopes	Hunts
Lug nut/bolt	Wheel nut/bolt
Metal chips or debris	Swarf
Misses	Misfires
Muffler	Silencer
Odor	Odour
Oil pan	Sump
Open flame	Naked flame
Owner's manual	Owner's handbook
Panel wagon/van	Van
Parking brake	Handbrake
Parking light	Sidelight
Pinging	Pinking
Piston pin or wrist pin	Gudgeon pin
Pitman arm	Drop arm
Power brake booster	Servo unit
Primary shoe (of brake)	Leading shoe (of brake)
Prussian blue	Engineer's blue
Pry	Prise (force apart)
Prybar	Lever
Prying	Levering
Quarter window	Quarterlight
Recap	Retread
Release cylinder	Slave cylinder
Repair shop	Garage
Replacement	Renewal
Ring gear (of differential)	Crownwheel
Rocker panel (beneath doors)	Sill panel (beneath doors)
Rod bearing	Big-end bearing
Rotor/disk	Disc (brake)
Secondary shoe (of brake)	Trailing shoe (of brake)
Sedan	Saloon
Setscrew	Grub screw
Shock absorber, shock	Damper
Snap-ring	Circlip
Soft top	Hood
Spacer	Distance piece
Spare tire	Spare wheel
Spark plug wires	HT leads
Spindle arm	Steering arm
Stabilizer or sway bar	Anti-roll bar
Station wagon	Estate car
Stumbles	Hesitates
Tang or lock washer	Tab washer
Throw-out bearing	Thrust bearing
Tie-rod	Trackrod
Tire	Tyre
Transmission	Gearbox
Troubleshooting	Fault finding/diagnosis
Trunk	Boot (luggage compartment)
Turn signal	Indicator
TV (throttle valve) cable	Kickdown cable
Valve cover	Rocker cover
Valve lifter	Tappet
Valve lifter or tappet	Cam follower or tappet
Vapor	Vapour
Vise	Vice
Vise grips	Self-locking pliers
Wheel cover	Roadwheel trim
Whole drive line	Transmission
Windshield	Windscreen
Wrench	Spanner

Fault diagnosis

Contents

Introduction

The vehicle owner who does his or her own maintenance according to the recommended service schedules should not have to use this section of the manual very often. Modern component reliability is such that, provided those items subject to wear or deterioration are inspected or renewed at the specified intervals, sudden failure is comparatively rare. Faults do not usually just happen as a result of sudden failure, but develop over a period of time. Major mechanical failures in particular are usually preceded by characteristic symptoms over hundreds or even thousands of miles. Those components which do occasionally fail without warning are often small and easily carried in the vehicle.

With any fault finding, the first step is to decide where to begin investigations. Sometimes this is obvious, but on other occasions a little detective work will be necessary. The owner who makes half a dozen haphazard adjustments or replacements may be successful in curing a fault (or its symptoms), but will be none the wiser if the fault recurs and ultimately may have spent more time and money than was necessary. A calm and logical approach will be found to be more satisfactory in the long run. Always take into account any warning signs or abnormalities that may have been noticed in the period preceding the fault - power loss, high or low gauge readings, unusual smells, etc - and remember that failure of components such as fuses or relays may only be pointers to some underlying fault.

The pages which follow provide an easy reference guide to the more common problems which may occur during the operation of the vehicle. These problems and their possible causes are grouped under headings denoting various components or systems, such as Engine, Cooling system, etc. The Chapter and/or Section which deals with the problem is also shown in brackets. Whatever the fault, certain basic principles apply. These are as follows:

Verify the fault. This is simply a matter of being sure that you know what the symptoms are before starting work. This is particularly important if you are investigating a fault for someone else who may not have described it very accurately.

Don't overlook the obvious. For example, if the vehicle won't start, is there fuel in the tank? (Don't take anyone else's word on this particular point, and don't trust the fuel gauge either!) If an electrical fault is indicated, look for loose or broken wires before digging out the test gear.

Cure the disease, not the symptom. Substituting a flat battery with a fully charged one will get you off the hard shoulder, but if the underlying cause is not attended to, the new battery will go the same way.

Don't take anything for granted. Particularly, don't forget that a "new" component may itself be defective (especially if it's been rattling around in the boot for months), and don't leave components out of a fault diagnosis sequence just because they are new or recently fitted. When you do finally diagnose a difficult fault, you'll probably realise that all the evidence was there from the start.

Engine

1 Engine will not rotate when attempting to start

1 Battery terminal connections loose or corroded (Chapter 1).
2 Battery discharged or faulty (Chapter 1).
3 Automatic transmission not completely engaged in Park (Chapter 7B) or clutch not completely depressed (Chapter 8).
4 Broken, loose or disconnected wiring in the starting circuit (Chapters 5 and 12).
5 Starter motor pinion jammed in flywheel ring gear (Chapter 5).
6 Starter solenoid faulty (Chapter 5).
7 Starter motor faulty (Chapter 5).
8 Ignition switch faulty (Chapter 12).
9 Starter pinion or flywheel teeth worn or broken (Chapter 5).
10 Internal engine problem (Chapter 2B).

2 Engine rotates but will not start

1 Fuel tank empty.
2 Battery discharged (engine rotates slowly) (Chapter 5).
3 Battery terminal connections loose or corroded (Chapter 1).
4 Leaking fuel injector(s), faulty fuel pump, pressure regulator, etc. (Chapter 4).
5 Fuel not reaching carburetor or fuel injection system (Chapter 4).
6 Ignition components damp or damaged (Chapter 5).
7 Worn, faulty or incorrectly gapped spark plugs (Chapter 1).
8 Broken, loose or disconnected wiring in the starting circuit (Chapter 5).
9 Loose distributor is changing ignition timing (Chapter 1).
10 Broken, loose or disconnected wires at the ignition coil or faulty coil (Chapter 5).

3 Engine hard to start when cold

1 Battery discharged or low (Chapter 1).
2 Fuel system malfunctioning (Chapter 4).
3 Injector(s) leaking (Chapter 4).
4 Distributor rotor carbon tracked (Chapter 5).

4 Engine hard to start when hot

1 Air filter clogged (Chapter 1).
2 Fuel not reaching the carburetor or fuel injection system (Chapter 4).
3 Injector(s) leaking (Chapter 4).
4 Corroded battery connections, especially ground (Chapter 1).

5 Starter motor noisy or excessively rough in engagement

1 Pinion or flywheel gear teeth worn or broken (Chapter 5).
2 Starter motor mounting bolts loose or missing (Chapter 5).

6 Engine starts but stops immediately

1 Loose or faulty electrical connections at distributor, coil or alternator (Chapter 5).
2 Insufficient fuel reaching the carburetor or fuel injector(s) (Chapters 1 and 4).
3 Faulty ballast resistor (Chapter 5).
4 Malfunctioning fuel injection system (Chapter 4).
5 Faulty fuel injection relays (Chapter 4).

7 Oil puddle under engine

1 Oil pan gasket and/or oil pan drain bolt seal leaking (Chapter 2).
2 Oil pressure sending unit leaking (Chapter 2).
3 Valve cover gasket leaking (Chapter 2).
4 Engine oil seals leaking (Chapter 2).

8 Engine lopes while idling or idles erratically

1 Vacuum leakage (Chapter 4).
2 Air filter clogged (Chapter 1).
3 Fuel pump not delivering sufficient fuel to the fuel injection system (Chapter 4).
4 Leaking head gasket (Chapter 2).
5 Timing belt or sprockets worn (Chapter 2).
6 Camshaft lobes worn (Chapter 2).

9 Engine misses at idle speed

1 Spark plugs worn or not gapped properly (Chapter 1).
2 Faulty spark plug wires (Chapter 1).
3 Vacuum leaks (Chapter 1).
4 Incorrect ignition timing (Chapter 5).
5 Uneven or low compression (Chapter 2).

10 Engine misses throughout driving speed range

1 Fuel filter clogged and/or impurities in the fuel system (Chapter 1).
2 Low fuel output at the fuel injection system (Chapter 4).
3 Faulty or incorrectly gapped spark plugs (Chapter 1).
4 Incorrect ignition timing (Chapter 5).
5 Cracked distributor cap, disconnected distributor wires or damaged distributor components (Chapter 1).
6 Leaking spark plug wires (Chapter 1).
7 Faulty emission system components (Chapter 6).
8 Low or uneven cylinder compression pressures (Chapter 2).
9 Weak or faulty ignition system (Chapter 5).
10 Vacuum leak in carburetor or fuel injection system, intake manifold or vacuum hoses (Chapter 4).

11 Engine stumbles on acceleration

1 Spark plugs fouled (Chapter 1).
2 Carburetor or fuel injection system malfunctioning (Chapter 4).
3 Fuel filter clogged (Chapters 1 and 4).
4 Incorrect ignition timing (Chapter 5).
5 Intake manifold air leak (Chapter 4).

12 Engine surges while holding accelerator steady

1 Intake air leak (Chapter 4).
2 Fuel pump or carburetor faulty (Chapter 4).
3 Loose fuel injector harness connections (Chapters 4 and 6).
4 Defective ECU (Chapter 6).

13 Engine stalls

1 Idle speed incorrect (Chapter 1).
2 Fuel filter clogged and/or water and impurities in the fuel system (Chapter 1).
3 Distributor components damp or damaged (Chapter 5).
4 Faulty emissions system components (Chapter 6).
5 Faulty or incorrectly gapped spark plugs (Chapter 1).
6 Faulty spark plug wires (Chapter 1).
7 Vacuum leak in carburetor or fuel injection system, intake manifold or vacuum hoses (Chapter 4).

14 Engine lacks power

1 Incorrect ignition timing (Chapter 5).
2 Excessive play in distributor shaft (Chapter 5).
3 Worn rotor, distributor cap or wires (Chapters 1 and 5).
4 Faulty or incorrectly gapped spark plugs (Chapter 1).
5 Carburetor or fuel injection system malfunctioning (Chapter 4).
6 Faulty coil (Chapter 5).
7 Brakes binding (Chapter 1).
8 Automatic transmission fluid level incorrect (Chapter 1).
9 Clutch slipping (Chapter 8).
10 Fuel filter clogged and/or impurities in the fuel system (Chapter 1).
11 Emission control system not functioning properly (Chapter 6).
12 Low or uneven cylinder compression pressures (Chapter 2).

15 Engine backfires

1 Emissions system not functioning properly (Chapter 6).
2 Ignition timing incorrect (Chapter 1).
3 Faulty secondary ignition system (cracked spark plug insulator, faulty plug wires, distributor cap and/or rotor) (Chapters 1 and 5).
4 Carburetor or fuel injection system malfunctioning (Chapter 4).
5 Vacuum leak at fuel injector(s), intake manifold or vacuum hoses (Chapter 4).

16 Pinging or knocking engine sounds during acceleration or uphill

1 Incorrect grade of fuel.
2 Ignition timing incorrect (Chapter 5).
3 Carburetor in need of adjustment or fuel injection system malfunctioning (Chapter 4).
4 Improper or damaged spark plugs or wires (Chapter 1).
5 Worn or damaged distributor components (Chapter 5).
6 Faulty emission system (Chapter 6).
7 Vacuum leak (Chapter 4).

17 Engine runs with oil pressure light on

1 Low oil level (Chapter 1).
2 Idle rpm too low (Chapter 1).
3 Short in wiring circuit (Chapter 12).
4 Faulty oil pressure sending unit (Chapter 2).
5 Worn engine bearings and/or oil pump (Chapter 2).

18 Engine diesels (continues to run) after switching off

1 Idle speed too high (Chapter 1).
2 Excessive engine operating temperature (Chapter 3).
3 Incorrect fuel octane grade.

Engine electrical system

19 Battery will not hold a charge

1 Alternator drivebelt defective or not adjusted properly (Chapter 1).
2 Electrolyte level low (Chapter 1).
3 Battery terminals loose or corroded (Chapter 1).
4 Alternator not charging properly (Chapter 5).
5 Loose, broken or faulty wiring in the charging circuit (Chapter 5).
6 Short in vehicle wiring (Chapters 5 and 12).
7 Internally defective battery (Chapters 1 and 5).

20 Discharge warning light fails to go out

1 Faulty alternator or charging circuit (Chapter 5).
2 Alternator drivebelt defective or out of adjustment (Chapter 1).
3 Alternator voltage regulator inoperative (Chapter 5).

21 Discharge warning light fails to come on when key is turned on

1 Warning light bulb defective (Chapter 12).
2 Fault in the printed circuit, dash wiring or bulb holder (Chapter 12).

Fuel system

22 Excessive fuel consumption

1 Dirty or clogged air filter element (Chapter 1).
2 Incorrectly set ignition timing (Chapter 5).
3 Emissions system not functioning properly (Chapter 6).
4 Carburetor or fuel injection system malfunctioning (Chapter 4).
5 Low tire pressure or incorrect tire size (Chapter 1).

23 Fuel leakage and/or fuel odor

1 Leak in a fuel feed or vent line (Chapter 4).
2 Tank overfilled.
3 Carburetor or fuel injection system malfunctioning (Chapter 4).

Cooling system

24 Overheating

1 Insufficient coolant in system (Chapter 1).
2 Water pump drivebelt defective or out of adjustment (Chapter 1).
3 Radiator core blocked or grille restricted (Chapter 3).
4 Thermostat faulty (Chapter 3).
5 Radiator cap not maintaining proper pressure (Chapter 3).
6 Ignition timing incorrect (Chapter 5).

25 Overcooling

1 Faulty thermostat (Chapter 3).

26 External coolant leakage

1 Deteriorated/damaged hoses; loose clamps (Chapters 1 and 3).
2 Water pump seal defective (Chapters 1 and 3).
3 Leakage from radiator core or header tank (Chapter 3).
4 Engine drain or water jacket core plugs leaking (Chapter 2).

27 Internal coolant leakage

1 Leaking cylinder head gasket (Chapter 2).
2 Cracked cylinder bore or cylinder head (Chapter 2).

28 Coolant loss

1 Too much coolant in system (Chapter 1).
2 Coolant boiling away because of overheating (Chapter 3).
3 Internal or external leakage (Chapter 3).
4 Faulty radiator cap (Chapter 3).

29 Poor coolant circulation

1 Inoperative water pump (Chapter 3).
2 Restriction in cooling system (Chapters 1 and 3).
3 Water pump drivebelt defective/out of adjustment (Chapter 1).
4 Thermostat sticking (Chapter 3).

Clutch

30 Pedal travels to floor - no pressure or very little resistance

1 Master or slave cylinder faulty (Chapter 8).
2 Hose/pipe burst or leaking (Chapter 8).
3 Connections leaking (Chapter 8).
4 No fluid in reservoir (Chapter 1).
5 If fluid is present in master cylinder dust cover, rear master cylinder seal has failed (Chapter 8).
6 Broken release bearing or fork (Chapter 8).
7 Broken release cable (Chapter 8).

31 Fluid in area of master cylinder dust cover and on pedal

Rear seal failure in master cylinder (Chapter 8).

32 Fluid on slave cylinder

Slave cylinder plunger seal faulty (Chapter 8).

33 Pedal feels "spongy" when depressed

Air in system (Chapter 8).

34 Unable to select gears

1 Faulty transmission (Chapter 7).
2 Faulty clutch plate (Chapter 8).
3 Fork and bearing not assembled properly (Chapter 8).
4 Faulty pressure plate (Chapter 8).
5 Pressure plate-to-flywheel bolts loose (Chapter 8).

35 Clutch slips (engine speed increases with no increase in vehicle speed)

1 Clutch plate worn (Chapter 8).
2 Clutch plate is oil soaked by leaking rear main seal (Chapter 8).
3 Clutch plate not seated. It may take 30 or 40 normal starts for a new one to seat.
4 Warped pressure plate or flywheel (Chapter 8).
5 Weak diaphragm spring (Chapter 8).
6 Clutch plate overheated. Allow to cool.

36 Grabbing (chattering) as clutch is engaged

1 Oil on clutch plate lining, burned or glazed facings (Chapter 8).
2 Worn or loose engine or transmission mounts (Chapters 2 and 7).
3 Worn splines on clutch plate hub (Chapter 8).
4 Warped pressure plate or flywheel (Chapter 8).

37 Noise in clutch area

1 Fork improperly installed (Chapter 8).
2 Faulty release bearing (Chapter 8).

38 Clutch pedal stays on floor

1 Fork binding in housing (Chapter 8).
2 Broken release bearing or fork (Chapter 8).

39 High pedal effort

1 Fork binding in housing (Chapter 8).
2 Pressure plate faulty (Chapter 8).
3 Incorrect size master or slave cylinder installed (Chapter 8).
4 Clutch release cable worn (Chapter 8).

Manual transmission

40 Vibration

1 Rough wheel bearing (Chapters 1 and 10).
2 Damaged axleshaft (Chapter 8).
3 Out-of-round tires (Chapter 1).
4 Tire out-of-balance (Chapters 1 and 10).
5 Worn U-joint (Chapter 8).

41 Noisy in Neutral with engine running

Damaged clutch release bearing (Chapter 8).
Damaged transmission input shaft bearing (Chapter 7A).

42 Noisy in one particular gear

1 Damaged or worn constant mesh gears.
2 Damaged or worn synchronizers.

43 Noisy in all gears

1 Insufficient lubricant (Chapter 1).
2 Damaged or worn bearings.
3 Worn or damaged input gear shaft and/or output gear shaft.

44 Slips out of gear

1 Worn or improperly adjusted linkage (Chapter 7).
2 Transmission loose on engine (Chapter 7).
3 Shift linkage does not work freely, binds (Chapter 7).
4 Input shaft bearing retainer broken or loose (Chapter 7).
5 Worn shift fork (Chapter 7).

45 Leaks lubricant

1 Excessive amount of lubricant in transmission (Chapters 1 and 7).
2 Loose or broken input shaft bearing retainer (Chapter 7).
3 Input shaft bearing retainer O-ring and/or lip seal damaged (Chapter 7).

Automatic transmission

Note: *Due to the complexity of the automatic transmission, it is difficult for the home mechanic to properly diagnose and service this component. For problems other than the following, the vehicle should be taken to a dealer or transmission shop.*

46 Fluid leakage

1 Automatic transmission fluid is a deep red color. Fluid leaks should not be confused with engine oil, which can easily be blown by airflow to the transmission.

2 To pinpoint a leak, first remove all built-up dirt and grime from the transmission housing with degreasing agents and/or steam cleaning. Then drive the vehicle at low speeds so airflow will not blow the leak far from its source. Raise the vehicle and determine where the leak is coming from. Common areas of leakage are:

a) Fluid pan (Chapters 1 and 7)
b) Filler pipe (Chapter 7)
c) Transmission fluid cooler lines (Chapter 7)
d) Speedometer sensor (Chapter 7)

47 Transmission fluid brown or has a burned smell

Transmission fluid burned (Chapter 1).

48 General shift mechanism problems

1 Chapter 7 Part B deals with checking and adjusting the shift linkage on automatic transmissions. Common problems which may be attributed to poorly adjusted linkage are:

a) Engine starting in gears other than Park or Neutral.
b) Indicator on shifter pointing to a gear other than the one actually being used.
c) Vehicle moves when in Park.

2 Refer to Chapter 7 Part B for the shift linkage adjustment procedure.

49 Transmission will not downshift with accelerator pedal pressed to the floor

Throttle valve (downshift) cable out of adjustment (Chapter 7).

50 Engine will start in gears other than Park or Neutral

Neutral start switch malfunctioning (Chapter 7).

51 Transmission slips, shifts roughly, is noisy or has no drive in forward or reverse gears

There are many probable causes for the above problems, but the home mechanic should be concerned with only one possibility - fluid level. Before taking the vehicle to a repair shop, check the level and condition of the fluid as described in Chapter 1. Correct the fluid level as necessary or change the fluid if needed. If the problem persists, have a professional diagnose the probable cause.

Brakes

Note: *Before assuming that a brake problem exists, make sure that:*

a) The tires are in good condition and properly inflated (Chapter 1).
b) The front end alignment is correct (Chapter 10).
c) The vehicle is not loaded with weight in an unequal manner.

52 Vehicle pulls to one side during braking

1 Incorrect tire pressures (Chapter 1).
2 Front end out of line (have the front end aligned).
3 Unmatched tires on same axle.
4 Restricted brake lines or hoses (Chapter 9).
5 Malfunctioning caliper assembly (Chapter 9).
6 Loose suspension parts (Chapter 10).
7 Loose calipers (Chapter 9).

53 Noise (high-pitched squeal when the brakes are applied)

Front and/or rear disc brake pads worn out. The noise comes from the wear sensor (if equipped) rubbing against the disc. Replace pads with new ones immediately (Chapter 9). Also, check the brake discs for scoring. Resurfacie or replace as necessary (Chapter 9).

54 Brake roughness or chatter (pedal pulsates)

1 Excessive lateral disc runout (Chapter 9).
2 Parallelism not within specifications (Chapter 9).
3 Uneven pad wear caused by caliper not sliding due to improper clearance or dirt (Chapter 9).
4 Defective disc (Chapter 9).

55 Excessive pedal effort required to stop vehicle

1 Malfunctioning power brake booster (Chapter 9).
2 Partial system failure (Chapter 9).
3 Excessively worn pads (Chapter 9).
4 Piston in caliper stuck or sluggish (Chapter 9).
5 Brake pads contaminated with oil or grease (Chapter 9).
6 New pads installed and not yet seated. It will take a while for the new material to seat against the disc.

56 Excessive brake pedal travel

1 Partial brake system failure (Chapter 9).
2 Insufficient fluid in master cylinder (Chapters 1 and 9).
3 Air trapped in system (Chapters 1 and 9).

57 Dragging brakes

1 Master cylinder pistons not returning correctly (Chapter 9).
2 Restricted brakes lines or hoses (Chapters 1 and 9).
3 Incorrect parking brake adjustment (Chapter 9).

58 Grabbing or uneven braking action

1 Malfunction of power brake booster unit (Chapter 9).
2 Binding brake pedal mechanism (Chapter 9).
3 Disc or drum and brake lining contaminated by fluid leakage (Chapter 9).
4 Brake linings worn out (see Chapter 1 for inspection, Chapter 9 for replacement).

59 Brake pedal feels spongy when depressed

1 Air in hydraulic lines (Chapter 9).
2 Master cylinder mounting bolts loose (Chapter 9).
3 Master cylinder defective (Chapter 9).

60 Brake pedal travels to the floor with little resistance

Little or no fluid in the master cylinder reservoir caused by leaking caliper piston(s), loose, damaged or disconnected brake hoses or lines (Chapter 9).

61 Parking brake does not hold

Parking brake linkage improperly adjusted (Chapter 9).

Suspension and steering systems

Note: *Before attempting to diagnose the suspension and steering systems, perform the following preliminary checks:*
a) Tires for wrong pressure and uneven wear.
b) Steering universal joints from the column to the steering gear for loose connectors or wear.
c) Front and rear suspension and the rack-and-pinion assembly for loose or damaged parts.
d) Out-of-round or out-of-balance tires, bent rims and loose and/or rough wheel bearings.

62 Vehicle pulls to one side

1 Mismatched or uneven tires (Chapter 10).
2 Broken or sagging springs (Chapter 10).
3 Front wheel or rear wheel alignment (Chapter 10).
4 Front brake dragging (Chapter 9).

63 Abnormal or excessive tire wear

1 Front wheel or rear wheel alignment (Chapter 10).
2 Sagging or broken springs (Chapter 10).
3 Tire out-of-balance (Chapter 10).
4 Worn shock absorber (Chapter 10).
5 Overloaded vehicle.
6 Tires not rotated regularly.

64 Wheel makes a "thumping" noise

1 Blister or bump on tire (Chapter 10).
2 Improper shock absorber action (Chapter 10).

65 Shimmy, shake or vibration

1 Tire or wheel out-of-balance or out-of-round (Chapter 10).
2 Loose, worn or out-of-adjustment wheel bearings (Chapter 1).
3 Worn tie-rod ends (Chapter 10).
4 Worn balljoints (Chapter 10).
5 Excessive wheel runout (Chapter 10).
6 Blister or bump on tire (Chapter 10).
7 Bent axleshaft (Chapter 8).

66 Hard steering

1 Lack of lubrication at balljoints, tie-rod ends and rack-and-pinion assembly (Chapter 1).
2 Front wheel alignment (Chapter 10).
3 Low tire pressure(s) (Chapter 1).

67 Poor returnability of steering to center

1 Lack of lubrication at balljoints and tie-rod ends (Chapter 1).
2 Binding in balljoints (Chapter 10).
3 Binding in steering column (Chapter 10).
4 Lack of lubricant in rack-and-pinion assembly (Chapter 10).
5 Front wheel alignment (Chapter 10).

68 Abnormal noise at the front end

1 Lack of lubrication at balljoints and tie-rod ends (Chapter 1).
2 Damaged shock absorber mounting (Chapter 10).
3 Worn control arm bushings or tie-rod ends (Chapter 10).
4 Loose stabilizer bar (Chapter 10).
5 Loose wheel nuts (Chapter 1).
6 Loose suspension bolts (Chapter 10).

69 Wander or poor steering stability

1 Mismatched or uneven tires (Chapter 10).
2 Lack of lubrication at balljoints and tie-rod ends (Chapter 1).
3 Worn shock absorbers (Chapter 10).
4 Loose stabilizer bar (Chapter 10).
5 Broken or sagging springs (Chapter 10).
6 Front or rear wheel alignment (Chapter 10).

70 Erratic steering when braking

1 Wheel bearings worn (Chapter 1).
2 Broken or sagging springs (Chapter 10).
3 Leaking wheel cylinder or caliper (Chapter 9).
4 Warped discs (Chapter 9).

71 Excessive pitching and/or rolling around corners or during braking

1 Loose stabilizer bar (Chapter 10).
2 Worn shock absorbers or mounts (Chapter 10).
3 Broken or sagging springs (Chapter 10).
4 Overloaded vehicle.

72 Suspension bottoms

1 Overloaded vehicle.
2 Worn shock absorbers (Chapter 10).
3 Incorrect, broken or sagging springs (Chapter 10).

73 Cupped tires

1 Wheels out of alignment (Chapter 10).
2 Worn shock absorbers (Chapter 10).
3 Wheel bearings worn (Chapter 10).
4 Excessive tire or wheel runout (Chapter 10).
5 Worn balljoints (Chapter 10).

74 Excessive tire wear on outside edge

1 Inflation pressures incorrect (Chapter 1).
2 Excessive speed in turns.
3 Front end alignment incorrect (excessive toe-in). Have professionally aligned.
4 Suspension arm bent or twisted (Chapter 10).

75 Excessive tire wear on inside edge

1 Inflation pressures incorrect (Chapter 1).
2 Front end alignment incorrect (toe-out). Have professionally aligned.
3 Loose or damaged steering components (Chapter 10).

76 Tire tread worn in one place

1 Tires out-of-balance.
2 Damaged or buckled wheel. Inspect and replace if necessary.
3 Defective tire (Chapter 1).

77 Excessive play or looseness in steering system

1 Wheel bearing(s) worn (Chapter 10).
2 Tie-rod end loose or worn (Chapter 10).
3 Steering gear loose (Chapter 10).

78 Rattling or clicking noise in steering gear

1 Insufficient or improper lubricant in rack-and-pinion assembly (Chapter 10).
2 Steering gear loose (Chapter 10).

MOT test checks

Introduction

Motor vehicle testing has been compulsory in Great Britain since 1960 when the Motor Vehicle (Tests) Regulations were first introduced. At that time testing was only applicable to vehicles ten years old or older, and the test itself only covered lighting equipment, braking systems and steering gear. Current vehicle testing is far more extensive and, in the case of private vehicles, is now an annual inspection commencing three years after the date of first registration. Test standards are becoming increasingly stringent; for details of changes consult the latest edition of the MOT Inspection Manual (available from HMSO or bookshops).

This section is intended as a guide to getting your vehicle through the MOT test. It lists all the relevant testable items, how to check them yourself, and what is likely to cause the vehicle to fail. Obviously it will not be possible to examine the vehicle to the same standard as the professional MOT tester who will be highly experienced in this work and will have all the necessary equipment available. However, working through the following checks will provide a good indication as to the condition of the vehicle and will enable you to identify any problem areas before submitting the vehicle for the test. Where a component is found to need repair or renewal, reference should be made to the appropriate Chapter in the manual where further information will be found.

The following checks have been sub-divided into four categories as follows.

a) *Checks carried out from the driver's seat.*
b) *Checks carried out with the vehicle on the ground.*
c) *Checks carried out with the vehicle raised and with the wheels free to rotate.*
d) *Exhaust emission checks.*

In most cases the help of an assistant will be necessary to carry out these checks thoroughly.

Checks carried out from the driver's seat

Handbrake

Test the operation of the handbrake by pulling on the lever until the handbrake is in the normal fully-applied position. Ensure that the travel of the lever (the number of clicks of the ratchet) is not excessive before full resistance of the braking mechanism is felt. If so this would indicate incorrect adjustment of the rear brakes or incorrectly adjusted handbrake cables.

With the handbrake fully applied, tap the lever sideways and make sure that it does not release which would indicate wear in the ratchet and pawl. Release the handbrake and move the lever from side to side to check for excessive wear in the pivot bearing. Check the security of the lever mountings and make sure that there is no corrosion of any part of the body structure within 30 cm of the lever mounting. If the lever mountings cannot be readily seen from inside the vehicle, carry out this check later when working underneath.

Footbrake

Check that the brake pedal is sound without visible defects such as excessive wear of the pivot bushes or broken or damaged pedal pad. Check also for signs of fluid leaks on the pedal, floor or carpets which would indicate failed seals in the brake master cylinder.

Depress the brake pedal slowly at first, then rapidly until sustained pressure can be held. Maintain this pressure and check that the pedal does not creep down to the floor which would again indicate problems with the master cylinder. Release the pedal, wait a few seconds then depress it once until firm resistance is felt. Check that this resistance occurs near the top of the pedal travel. If the pedal travels nearly to the floor before firm resistance is felt, this would indicate incorrect brake adjustment resulting in "insufficient reserve travel" of the footbrake. If firm resistance cannot be felt, ie the pedal feels spongy, this would indicate that air is present in the hydraulic system which will necessitate complete bleeding of the system.

Check that the servo unit is operating correctly by depressing the brake pedal several times to exhaust the vacuum. Keep the pedal depressed and start the engine. As soon as the engine starts, the brake pedal resistance will be felt to alter. If this is not the case, there may be a leak from the brake servo vacuum hose, or the servo unit itself may be faulty.

Steering wheel and column

Examine the steering wheel for fractures or looseness of the hub, spokes or rim. Move the steering wheel from side to side and then up and down, in relation to the steering column. Check that the steering wheel is not loose on the column, indicating wear in the column splines or a loose steering wheel retaining nut. Continue moving the steering wheel as before, but also turn it slightly from left to right. Check that there is no abnormal movement of the steering wheel, indicating excessive wear in the column upper support bearing, universal joint(s) or flexible coupling.

Windscreen and mirrors

The windscreen must be free of cracks or other damage which will seriously interfere with the driver's field of view, or which will prevent the windscreen wipers from operating properly. Small stone chips are acceptable. Any stickers, dangling toys or similar items must also be clear of the field of view.

Rear view mirrors must be secure, intact and capable of being adjusted. The nearside (passenger side) door mirror is not included in the test unless the interior mirror cannot be used - for instance, in the case of a van with blacked-out rear windows.

Seat belts and seats

Note: *The following checks are applicable to all seat belts, front and rear. Front seat belts must be of a type that will restrain the upper part of the body; lap belts are not acceptable. Various combinations of seat belt types are acceptable at the rear.*

Carefully examine the seat belt webbing for cuts or any signs of serious fraying or deterioration. If the seat belt is of the retractable type, pull the belt all the way out and examine the full extent of the webbing.

Fasten and unfasten the belt ensuring that the locking mechanism holds securely and releases properly when intended. If the belt is of the retractable type, check also that the retracting mechanism operates correctly when the belt is released.

Check the security of all seat belt mountings and attachments which are accessible, without removing any trim or other components, from inside the vehicle **(see illustration).** Any serious corrosion, fracture or distortion of the body structure within 30 cm of any mounting point will cause the vehicle to fail. Certain anchorages will not be accessible or even visible from inside the vehicle; in this

Check the security of all seat belt mountings

instance further checks should be carried out later, when working underneath. If any part of the seat belt mechanism is attached to the front seat, then the seat mountings are treated as anchorages and must also comply as above.

The front seats themselves must be securely attached so that they cannot move unexpectedly and the backrests must lock in the upright position.

Doors

Both front doors must be able to be opened and closed from outside and inside, and must latch securely when closed. In the case of a pick-up, the tailgate must be securely attached and capable of being securely fastened.

Electrical equipment

Switch on the ignition and operate the horn. The horn must operate and produce a clear sound audible to other road users. Note that a gong, siren or two-tone horn fitted as an alternative to the manufacturer's original equipment is not acceptable.

Check the operation of the windscreen washers and wipers. The washers must operate with adequate flow and pressure and with the jets adjusted so that the liquid strikes the windscreen near the top of the glass.

Operate the windscreen wipers in conjunction with the washers and check that the blades cover their designed sweep of the windscreen without smearing. The blades must effectively clean the glass so that the driver has an adequate view of the road ahead and to the front nearside and offside of the vehicle. If the screen smears or does not clean adequately, it is advisable to renew the wiper blades before the MOT test.

Depress the footbrake with the ignition switched on and have your assistant check that both rear stop lights operate, and are extinguished when the footbrake is released. If one stop light fails to operate it is likely that a bulb has blown or there is a poor electrical contact at, or near, the bulbholder. If both stop lights fail to operate, check for a blown fuse, faulty stop light switch or possibly two blown bulbs. If the lights stay on when the brake pedal is released, it is possible that the switch is at fault.

Checks carried out with the vehicle on the ground

Vehicle identification

Front and rear number plates must be in good condition, securely fitted and easily read. Letters and numbers must be correctly spaced, with the gap between the group of numbers and the group of letters at least double the gap between adjacent numbers and letters.

The vehicle identification number on the plate under the bonnet must be legible. It will be checked during the test as part of the measures taken to prevent the fraudulent acquisition of certificates.

Electrical equipment

Switch on the side lights and check that both front and rear side lights and the number plate lights are illuminated and that the lenses and reflectors are secure and undamaged. This is particularly important at the rear where a cracked or damaged lens would allow a white light to show to the rear, which is unacceptable. Note in addition that any lens that is excessively dirty, either inside or out, such that the light intensity is reduced, could also constitute a fail.

Switch on the headlamps and check that both dipped beam and main beam units are operating correctly and at the same light intensity. If either headlamp shows signs of dimness, this is usually attributable to a poor earth connection or severely corroded internal reflector. Inspect the headlamp lenses for cracks or stone damage. Any damage to the headlamp lens will normally constitute a fail, but this is very much down to the tester's discretion. Bear in mind that with all light units they must operate correctly when first switched on. It is not acceptable to tap a light unit to make it operate.

The headlamps must not only be aligned so as not to dazzle other road users when switched to dipped beam, but also so as to provide adequate illumination of the road. This can only be accurately checked using optical beam setting equipment so if you have any doubts about the headlamp alignment, it is advisable to have this professionally checked and if necessary reset, before the MOT test.

With the ignition switched on, operate the direction indicators and check that they show amber lights to the front and to the rear, that they flash at the rate of between one and two flashes per second and that the "tell-tale" on the instrument panel also functions. Operation of the side lights and stop lights must not affect the indicators - if it does, the cause is usually a bad earth at the rear light cluster. Similarly check the operation of the hazard warning lights, which must work with the ignition on and off. Examine the lenses for cracks or damage as described previously.

Check the operation of the rear foglight(s). The test only concerns itself with the statutorily required foglight, which is the one on the offside. The light must be secure and emit a steady red light. The warning light on the instrument panel or in the switch must also work.

Footbrake

From within the engine compartment examine the brake pipes for signs of leaks, corrosion, insecurity, chafing or other damage and check the master cylinder and servo unit for leaks, security of their mountings or excessive corrosion in the vicinity of the mountings. The master cylinder reservoir must be secure; if it is of the translucent type, the fluid level must be between the upper and lower level markings.

Turn the steering as necessary so that the right-hand front brake flexible hose can be examined. Inspect the hose carefully for any sign of cracks or deterioration of the rubber. This will be most noticeable if the hose is bent in half and is particularly common where the rubber portion enters the metal end fitting **(see illustration).** Turn the steering onto full left then full right lock and ensure that the hose does not contact the wheel, tyre, or any part of the steering or suspension mechanism. While your assistant depresses the brake pedal firmly, check the hose for any bulges or fluid leaks under pressure. Now repeat these checks on the left-hand front hose. Should any damage or deterioration be noticed, renew the hose.

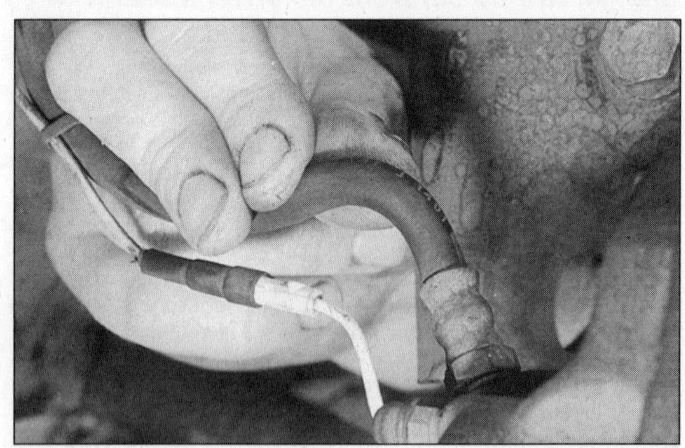

Check the flexible brake hose for cracks or deterioration

Steering mechanism and suspension

Have your assistant turn the steering wheel from side to side slightly, up to the point where the steering gear just begins to transmit this movement to the roadwheels. Check for excessive free play between the steering wheel and the steering gear which would indicate wear in the steering column joints, wear or insecurity of the steering column to steering gear coupling, or insecurity, incorrect adjustment, or wear in the steering gear itself. Generally speaking, free play greater than 1.3 cm for vehicles with rack and pinion type steering or 7.6 cm for vehicles with steering box mechanisms should be considered excessive.

Have your assistant turn the steering wheel more vigorously in each direction up to the point where the roadwheels just begin to turn.

As this is done, carry out a complete examination of all the steering joints, linkages, fittings and attachments. Any component that shows signs of wear, damage, distortion, or insecurity should be renewed or attended to accordingly. On vehicles equipped with power steering also check that the power steering pump is secure, that the pump drivebelt is in satisfactory condition and correctly adjusted, that there are no fluid leaks or damaged hoses, and that the system operates correctly. Additional checks can be carried out later with the vehicle raised when there will be greater working clearance underneath.

Check that the vehicle is standing level and at approximately the correct ride height. Ensure that there is sufficient clearance between the suspension components and the bump stops to allow full suspension travel over bumps.

Shock absorbers

Depress each corner of the vehicle in turn and then release it. If the shock absorbers are in good condition the corner of the vehicle will rise and then settle in its normal position. If there is no noticeable damping effect from the shock absorber, and the vehicle continues to rise and fall, then the shock absorber is defective and the vehicle will fail. A shock absorber which has seized will also cause the vehicle to fail.

Exhaust system

Start the engine and with your assistant holding a rag over the tailpipe, check the entire system for leaks which will appear as a rhythmic fluffing or hissing sound at the source of the leak. Check the effectiveness of the silencer by ensuring that the noise produced is of a level to be expected from a vehicle of similar type. Providing that the system is structurally sound, it is acceptable to cure a leak using a proprietary exhaust system repair kit or similar method.

Checks carried out with the vehicle raised and with the wheels free to rotate

Jack up the front and rear of the vehicle and securely support it on axle stands positioned at suitable load bearing points under the vehicle structure. Position the stands clear of the suspension assemblies and ensure that the wheels are clear of the ground and that the steering can be turned onto full right and left lock.

Steering mechanism

Examine the steering rack rubber gaiters for signs of splits, lubricant leakage or insecurity of the retaining clips **(see illustration).** If power steering is fitted, check for signs of deterioration, damage, chafing or leakage of the fluid hoses, pipes or connections. Also check for excessive stiffness or binding of the steering, a missing split pin or locking device or any severe corrosion of the body structure within 30 cm of any steering component attachment point.

Have your assistant turn the steering onto full left then full right lock. Check that the steering turns smoothly without undue tightness or roughness and that no part of the steering mechanism, including a wheel or tyre, fouls any brake flexible or rigid hose or pipe, or any part of the body structure.

On vehicles with four-wheel steering, similar considerations apply to the rear wheel steering linkages. However, it is permissible for a rear wheel steering system to be inoperative, provided that the rear wheels are secured in the straight-ahead position and that the front wheel steering system is operating effectively.

Front and rear suspension and wheel bearings

Starting at the front right-hand side of the vehicle, grasp the roadwheel at the 3 o'clock and 9 o'clock positions and shake it vigorously. Check for any free play at the wheel bearings, suspension ball joints, or suspension mountings, pivots and attachments. Check also for any serious deterioration of the rubber or metal casing of any mounting bushes, or any distortion, deformation or severe corrosion of any components **(see illustration).** Look for missing split pins, tab washers or other locking devices on any mounting or attachment, or any severe corrosion of the vehicle structure within 30 cm of any suspension component attachment point.

If any excess free play is suspected at a component pivot point, this can be confirmed by using a large screwdriver or similar tool and levering between the mounting and the component attachment. This will confirm whether the wear is in the pivot bush, its retaining bolt or in the mounting itself (the bolt holes can often become elongated).

Now grasp the wheel at the 12 o'clock and 6 o'clock positions, shake it vigorously and repeat the previous inspection **(see illustration).** Rotate the wheel and check for roughness or tightness of the front wheel bearing such that imminent failure of the bearing is indicated.

Carry out all the above checks at the other front wheel and then at both rear wheels.

Roadsprings and shock absorbers

On vehicles with strut type suspension units, examine the strut assembly for signs of serious fluid leakage, corrosion or severe pitting of the piston rod or damage to the casing. Check also for security of the mounting points.

If coil springs are fitted check that the spring ends locate correctly in their spring seats, that there is no severe corrosion of the spring and that it is not cracked, broken or in any way damaged.

If the vehicle is fitted with leaf springs, check that all leaves are intact, that the axle is securely attached to each spring and that there is no wear or deterioration of the spring eye mountings, bushes, and shackles.

The same general checks apply to vehicles fitted with other suspension types, such as torsion bars, hydraulic displacer units etc. In all cases ensure that all mountings and attachments are secure, that there are no signs of excessive wear, corrosion, cracking, deformation or damage to any component or bush, and that there are no fluid leaks or damaged hoses or pipes (hydraulic types).

Inspect the shock absorbers for signs of serious fluid leakage. (Slight seepage of fluid is normal for some types of shock absorber and is not a reason for failing.) Check for excessive wear of the mounting bushes or attachments or damage to the body of the unit **(see illustration).**

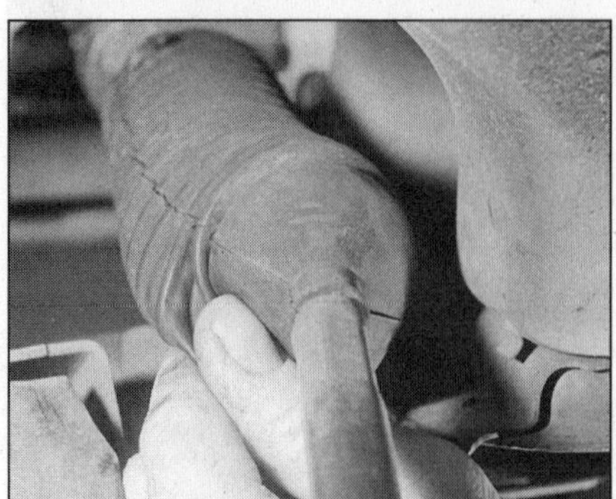

Examine the steering rack rubber gaiters for condition and security

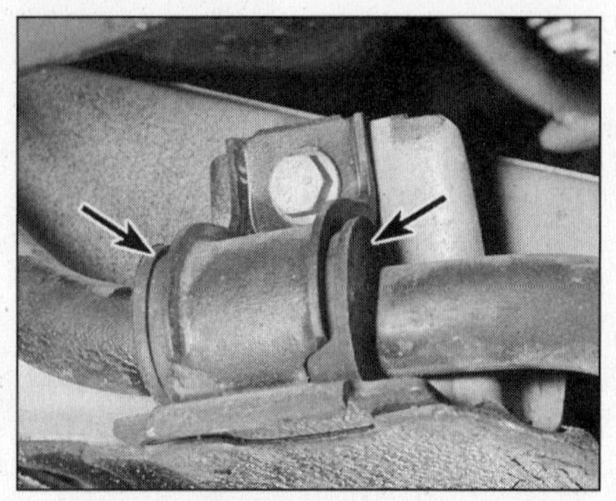

Check all rubber suspension mounting bushes (arrowed) for damage or deterioration

Shake the roadwheel vigorously to check for excess play in the wheel bearings and suspension components

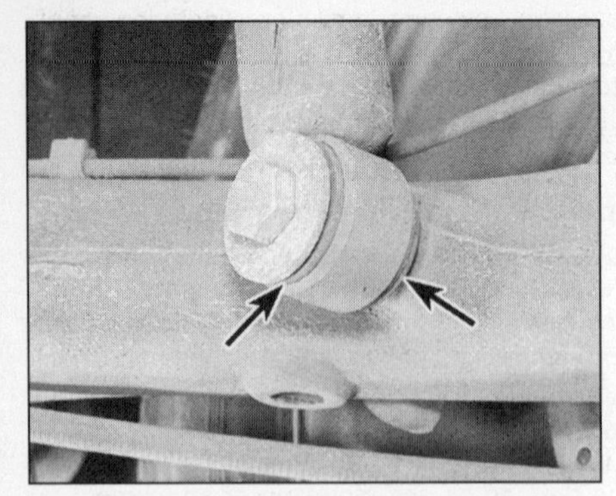

Check the condition of the shock absorber mountings and bushes (arrowed)

Inspect the constant velocity joint gaiters for splits or damage (front-wheel drive shown)

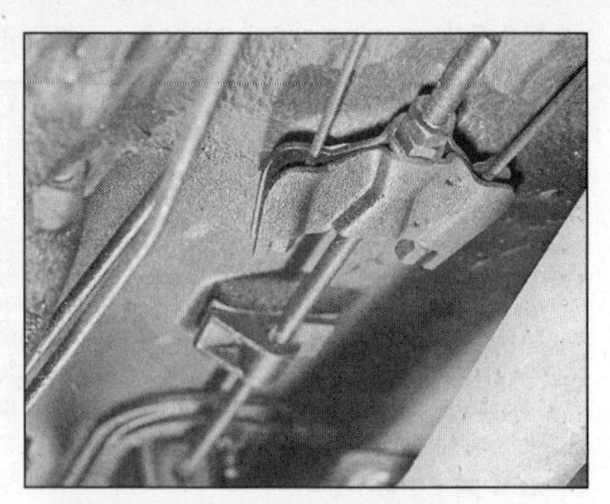

Check the handbrake mechanism for signs of frayed or broken cables or insecurity of the linkage

Driveshafts

With the steering turned onto full lock (front-wheel drive only), rotate each front wheel in turn and inspect the constant velocity joint gaiters for splits or damage **(see illustration).** Also check the gaiter is securely attached to its respective housings by clips or other methods of retention.

Continue turning the wheel and check that each driveshaft is straight with no sign of damage.

Braking system

If possible, without dismantling, check for wear of the brake pads and the condition of the discs. Ensure that the friction lining material has not worn excessively and that the discs are not fractured, pitted, scored or worn excessively.

Carefully examine all the rigid brake pipes underneath the vehicle and the flexible hoses at the rear. Look for signs of excessive corrosion, chafing or insecurity of the pipes and for signs of bulging under pressure, chafing, splits or deterioration of the flexible hoses.

Look for signs of hydraulic fluid leaks at the brake calipers or on the brake backplates indicating failed hydraulic seals in the components concerned.

Slowly spin each wheel while your assistant depresses the footbrake then releases it. Ensure that each brake is operating and that the wheel is free to rotate when the pedal is released. It is not possible to test brake efficiency without special equipment, but (traffic and local conditions permitting) a road test can be carried out to check that the vehicle pulls up in a straight line.

Examine the handbrake mechanism and check for signs of frayed or broken cables, excessive corrosion or wear or insecurity of the linkage **(see illustration).** Have your assistant operate the handbrake while you check that the mechanism works on each relevant wheel and releases fully without binding.

Fuel and exhaust systems

Inspect the fuel tank, fuel pipes, hoses and unions (including the unions at the pump, filter and carburettor). All components must be secure and free from leaks. The fuel filler cap must also be secure and of an appropriate type.

Examine the exhaust system over its entire length checking for any damaged, broken or missing mountings, security of the pipe retaining clamps and condition of the system with regard to rust and corrosion **(see illustration).**

Wheels and tyres

Carefully examine each tyre in turn on both the inner and outer walls and over the whole of the tread area and check for signs of cuts, tears, lumps, bulges, separation of the tread and exposure of the ply or cord due to wear or other damage. Check also that the tyre bead is correctly seated on the wheel rim and that the tyre valve is sound and properly seated. Spin the wheel and check that it is not excessively distorted or damaged particularly at the bead rim.

Check that the tyres are of the correct size for the vehicle and that they are of the same size and type on each axle. (Having a "space saver" spare tyre in use is not acceptable.) The tyres should also be inflated to the specified pressures.

Using a suitable gauge check the tyre tread depth. The current legal requirement states that the tread pattern must be visible over the whole tread area and must be of a minimum depth of 1.6 mm over at least three-quarters of the tread width. It is acceptable for some wear of the inside or outside edges of the tyre to be apparent but this wear must be in one even circumferential band and the tread must be visible. Any excessive wear of this nature may indicate incorrect front wheel alignment which should be checked before the tyre becomes excessively worn. See the appropriate Chapters for further information on tyre wear patterns and front wheel alignment.

Body corrosion

Check the condition of the entire vehicle structure for signs of corrosion in any load bearing areas. For the purpose of the MOT test all chassis box sections, side sills, crossmembers, pillars, suspension, steering, braking system and seat belt mountings and anchorages should all be considered as load bearing areas. As a general guide, any corrosion which has seriously reduced the metal thickness of a load bearing area to weaken it, is likely to cause the vehicle to fail. Should corrosion of this nature be encountered, professional repairs are likely to be needed.

Body damage or corrosion which causes sharp or otherwise dangerous edges to be exposed will also cause the vehicle to fail.

Exhaust emission checks

Have the engine at normal operating temperature and make sure that the preliminary conditions for checking idle speed and mixture (ignition system in good order, air filter clean, etc) have been met.

Before any measurements are carried out, raise the engine speed to around 2500 rpm and hold it at this speed for 20 seconds. Allow the

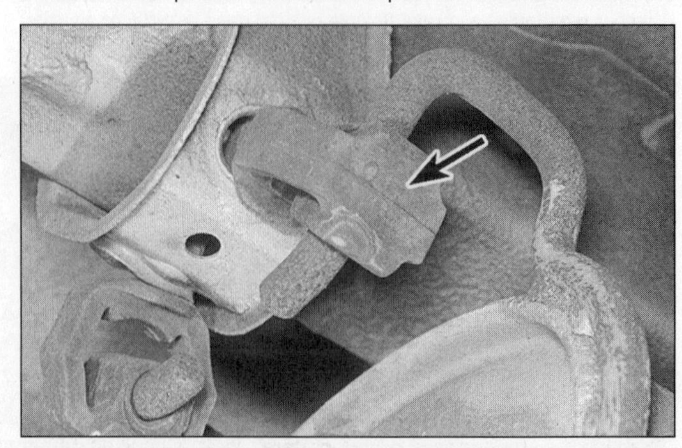

Check the condition of the exhaust system paying particular attention to the mountings (arrowed)

engine speed to return to idle and watch for smoke emissions from the exhaust tailpipe. If the idle speed is obviously much too high, or if dense blue or clearly visible black smoke comes from the tailpipe for more than 5 seconds, the vehicle will fail. As a rule of thumb, blue smoke signifies oil being burnt (worn valve stem oil seals, valve guides, piston rings or bores) while black smoke signifies unburnt fuel (dirty air filter element, mixture extremely rich, or other carburettor or fuel injection system fault).

If idle speed and smoke emission are satisfactory, an exhaust gas analyser capable of measuring carbon monoxide (CO) and hydrocarbons (HC) is now needed. The following paragraphs assume that such an instrument can be hired or borrowed - it is unlikely to be economic for the home mechanic to buy one. Alternatively, a local garage may agree to perform the check for a small fee.

CO emissions (mixture)

Current MOT regulations specify a maximum CO level at idle of 4.5% for vehicles first used after August 1983. The CO level specified by the vehicle maker is well inside this limit.

If the CO level cannot be reduced far enough to pass the test (and assuming that the fuel and ignition systems are otherwise in good condition) it is probable that the carburettor is badly worn, or that there is some problem in the fuel injection system. On carburettors with an automatic choke, it may be that the choke is not releasing as it should.

It is possible for the CO level to be within the specified maximum for MOT purposes, but well above the maximum specified by the manufacturer. The tester is entitled to draw attention to this, but it is not in itself a reason for failing the vehicle.

HC emissions

With the CO emissions within limits, HC emissions must be o more than 1200 ppm (parts per million). If the vehicle fails this test at idle, it can be re-tested at around 2000 rpm; if the HC level is then 1200 ppm or less, this counts as a pass.

Excessive HC emissions can be caused by oil being burnt, but they are more likely to be due to unburnt fuel. Possible reasons include:

a) Spark plugs in poor condition or incorrectly gapped.
b) Ignition timing incorrect.
c) Valve clearances incorrect.
d) Engine compression low.

Chapter 1
Tune-up and routine maintenance

Contents

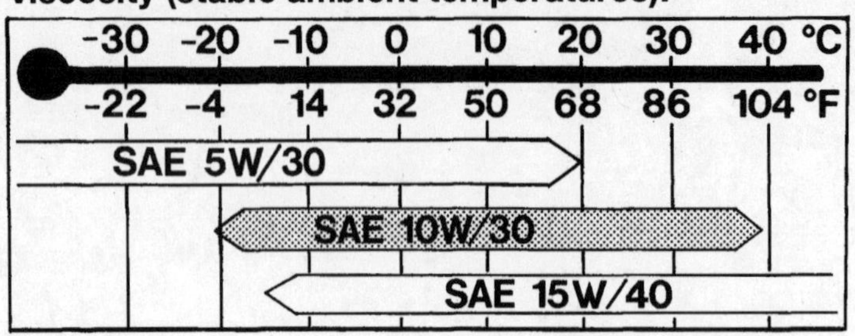

Oil viscosity chart

Specifications

Recommended lubricants and fluids

Engine oil	
Type	10W-30 API grade SG or SF/CD
Viscosity	See accompanying chart
Power steering fluid type	DEXRON II automatic transmission fluid
Brake fluid type	DOT 4 heavy duty brake fluid
Automatic transmission fluid type*	DEXRON II D or E automatic transmission fluid
Manual transmission lubricant type	
M40 and M41	SAE 80W/90 or 80/90 hypoid gear oil
M45, M46 and M47	Type F automatic transmission fluid
Differential lubricant	API GL-5 SAE 90*** hypoid gear lubricant
Coolant type	50/50 mixture of ethylene glycol-based antifreeze and water

**** Below temperatures of 15-degrees F, use SAE 80: If equipped with "limited-slip" use friction modifier additive.*

** 1982 and 1983 Aisin Warner 70/Aisin Warner 71 transmissions were filled at the factory with type F automatic transmission fluid. Don't add Dexron II to these transmissions unless you are certain that it is a new or rebuilt transmission that was filled with Dexron II to begin with.*

Capacities *

Engine oil with filter change	4 quarts
Differential	
Model 1030	1.4 quarts
Models 1031, 1031F and 1041	1.7 quarts
Manual transmission	
M41	1.4 quarts
M40, M45	0.8 quarts
M46	2.4 quarts
M47	1.6 quarts
Automatic transmission (drain and refill)**	
1982 and earlier	4.6 quarts
1983 and later	5.2 quarts
Cooling system	
Manual transmission	9.9 quarts
Automatic transmission	9.7 quarts

** All capacities approximate. Add as necessary to bring to appropriate level.*
*** Capacity is for the transmission drained but not the torque converter.*

Brakes

Disc brake pad thickness (minimum)	1/8-inch
Drum brake shoe lining thickness (minimum)	1/16-inch

Spark plug type and gap

B21F, B21F Turbo and B230 engines

1976 through 1979	
Type	Bosch W7DC or equivalent
Gap	0.028 to 0.032 inch

1980 through 1984	
Type	Bosch W7DC or equivalent
Gap	0.028 to 0.032 inch
1985 and later	
Type	Bosch WR7DC or equivalent
Gap	0.028 to 0.032 inch
B23F engines	
1983 through 1985	
Type	Bosch W7DC or equivalent
Gap	0.028 to 0.032 inch

Cylinder location and distributor rotation

Idle speed (LH-Jetronic 1980 through 1988)

1980	720 rpm
1981 and later	700 rpm

Ignition timing

See Chapter 5

Firing order

1-3-4-2

Dwell/point gap (breaker-point type distributors)

Dwellmeter	60-degrees
Feeler gauge	0.016 inches

Valve clearances (intake and exhaust)

Cold	0.012 to 0.016 inch
Warm	0.014 to 0.018 inch

Torque specifications

	Ft-lbs (unless otherwise indicated)
Automatic transmission pan bolts	
Three-speed	72 to 84 in-lbs
Four-speed	48 to 60 in-lbs
Spark plugs	120 to 168 in-lbs
Oxygen sensor	41 to 44
Overdrive pan bolts	84 in-lbs
Wheel lug bolts	81

1 Introduction

This Chapter is designed to help the home mechanic maintain his or her Volvo 240 with the goals of maximum performance, economy, safety and reliability in mind. Included is a master maintenance schedule, followed by procedures dealing specifically with each item on the schedule. Visual checks, adjustments, component replacement and other helpful items are included. **Refer to the accompanying illustrations** of the engine compartment and the underside of the vehicle for the locations of various components. Servicing the vehicle, in accordance with the mileage/time maintenance schedule and the step-by-step procedures will result in a planned maintenance program that should produce a long and reliable service life. Keep in mind that it is a comprehensive plan, so maintaining some items but not others at specified intervals will not produce the same results.

As you service your Volvo, you will discover that many of the procedures can - and should - be grouped together because of the nature of the particular procedure you're performing or because of the close proximity of two otherwise unrelated components to one another. For example, if the vehicle is raised for chassis lubrication, you should inspect the exhaust, suspension, steering and fuel systems while you're under the vehicle. When you're rotating the tires, it makes good sense to check the brakes since the wheels are already removed. Finally, let's suppose you have to borrow or rent a torque wrench. Even if you only need it to tighten the spark plugs, you might as well check the torque of as many critical fasteners as time allows.

The first step in this maintenance program is to prepare yourself before the actual work begins. Read through all the procedures you're planning to do, then gather up all the parts and tools needed. If it looks like you might run into problems during a particular job, seek advice from a mechanic or an experienced do-it-yourselfer.

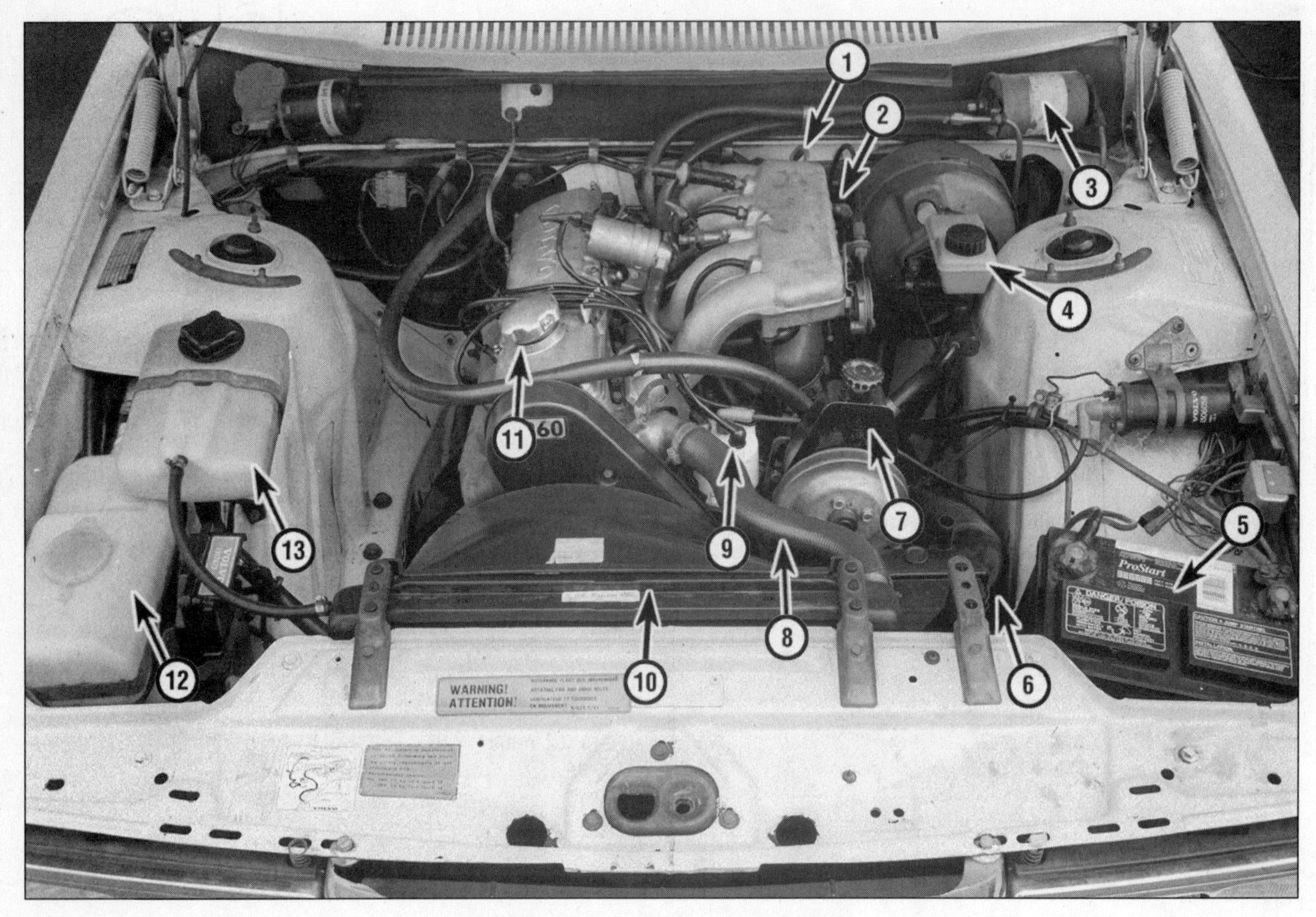

Typical engine compartment components

1. *Engine oil dipstick*
2. *Automatic transmission fluid dipstick location*
3. *Fuel filter (early models)*
4. *Brake fluid reservoir*
5. *Battery*
6. *Air cleaner housing location*
7. *Power steering fluid reservoir*
8. *Radiator hose (upper)*
9. *Distributor*
10. *Radiator*
11. *Oil filler cap*
12. *Windshield washer reservoir*
13. *Coolant expansion tank*

Typical engine compartment underside components

1 *Alternator*
2 *Heated air intake hose*
3 *Drivebelt*
4 *Air conditioning compressor*
5 *Steering gear boot*
6 *Front disc brake*
7 *Engine oil drain plug*
8 *Exhaust system*
9 *Lower balljoint*
10 *Suspension strut*
11 *Spring*

Typical rear underside components

1 *Fuel tank*
2 *Exhaust system*
3 *Rear spring*
4 *Rear disc brake*
5 *Muffler*
6 *Differential check/fill plug*
7 *Driveshaft*
8 *Shock absorber*

2 Volvo 240 Series Maintenance schedule

The following maintenance intervals are based on the assumption that the vehicle owner will be doing the maintenance or service work, as opposed to having a dealer service department do the work. Although the time/mileage intervals are loosely based on factory recommendations, most have been shortened to ensure, for example, that such items as lubricants and fluids are checked/changed at intervals that promote maximum engine/driveline service life. Also, subject to the preference of the individual owner interested in keeping his or her vehicle in peak condition at all times, and with the vehicle's ultimate resale in mind, many of the maintenance procedures may be performed more often than recommended in the following schedule. We encourage such owner initiative.

When the vehicle is new it should be serviced initially by a factory authorized dealer service department to protect the factory warranty. In many cases the initial maintenance check is done at no cost to the owner (check with your dealer service department for more information).

Every 250 miles or weekly, whichever comes first

Check the engine oil level (Section 4)
Check the engine coolant level (Section 4)
Check the brake fluid level (Section 4)
Check the clutch fluid level (Section 4)
Check the washer fluid level (Section 4)
Check the tires and tire pressures (Section 5)

Every 3,000 miles or 3 months, whichever comes first

All items listed above, plus:
Change the engine oil and oil filter (Section 6)
Check the power steering fluid level (Section 7)
Check the automatic transmission fluid level (Section 8)

Every 6,000 miles or 6 months, whichever comes first

All items listed above, plus:
Rotate the tires (Section 9)
Inspect/replace the underhood hoses (Section 10)
Check/adjust the drivebelts (Section 11)

Every 15,000 miles or 12 months, whichever comes first

All items listed above, plus:
Check/service the battery (Section 12)
Check/replace the spark plug wires, distributor cap and rotor (Section 13)
Check/replenish the manual transmission lubricant (Section 14)
Check the differential lubricant level (Section 15)
Check and adjust, if necessary the ignition timing (if applicable) (Section 16)
Idle speed (if applicable) (Section 17)
Check and adjust/replace the ignition points, if equipped (Section 18)
Check and lubricate the throttle linkage (Section 19)
Check the fuel system (Section 20)
Inspect the cooling system (Section 21)
Inspect the exhaust system (Section 22)
Inspect the steering and suspension components (Section 23)
Inspect the brakes (Section 24)
Inspect/replace the windshield wiper blades (Section 25)
Replace the air filter (Section 26)
Check and regap or replace the spark plugs (all 1980 and earlier models and 1984 and earlier Canadian models) (Section 26)

Every 30,000 miles or 24 months, whichever comes first

All items listed above plus:
Check and regap or replace the spark plugs (Section 27)
Check and adjust, if necessary, the valve clearances (Section 28)
Change the automatic transmission fluid and filter (Section 29)
Service the cooling system (drain, flush and refill) (Section 30)
Change the manual transmission lubricant (Section 31)
Change the differential lubricant (Section 32)
Check the evaporative emissions system (Section 33)
Replace the fuel filter (Section 35)
Replace the unheated (single-wire) oxygen sensor (Section 34)

Every 45,000 miles:

Replace the timing belt (see Chapter 2, Part A)

Every 50,000 miles:

Replace the heated (multi-wire) oxygen sensor (Section 34)

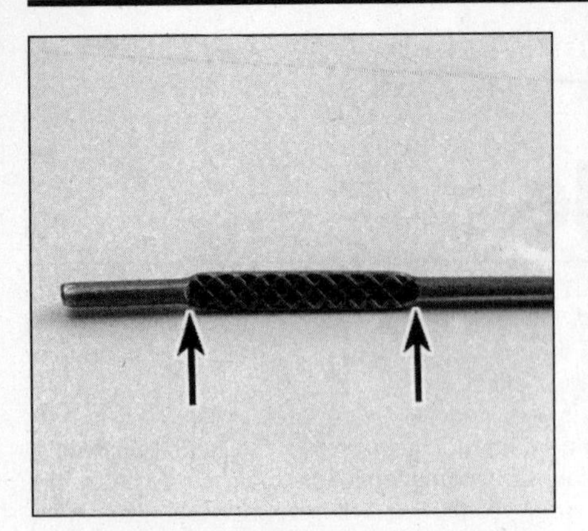

4.4 The oil level should be kept between the two arrows, at or near the upper one, if possible - if it isn't, add enough oil to bring the level to the upper arrow (it takes one full quart to raise the level from the lower to the upper mark)

4.6 The oil filler cap is located in the valve cover - always make sure the area around the opening is clean before removing the cap

4.9 The expansion tank (coolant reservoir) is mounted on the side of the engine compartment - remove the cap to add coolant

3 Tune-up general information

The term tune-up is used in this manual to represent a combination of individual operations rather than one specific procedure.

If, from the time the vehicle is new, the routine maintenance schedule is followed closely and frequent checks are made of fluid levels and high wear items, as suggested throughout this manual, the engine will be kept in relatively good running condition and the need for additional work will be minimized.

More likely than not, however, there will be times when the engine is running poorly due to a lack of regular maintenance. This is even more likely if a used vehicle, which has not received regular and frequent maintenance checks, is purchased. In such cases, an engine tune-up will be needed outside of the regular maintenance intervals.

The first step in any tune-up or diagnostic procedure to help correct a poor running engine is a cylinder compression check. A compression check (see Chapter 2, Part B) will help determine the condition of internal engine components and should be used as a guide for tune-up and repair procedures. If, for instance, a compression check indicates serious internal engine wear, a conventional tune-up will not improve the performance of the engine and would be a waste of time and money. Because of its importance, the compression check should be done by someone with the right equipment and the knowledge to use it properly.

The following procedures are those most often needed to bring as generally poor running engine back into a proper state of tune.

Minor tune-up

Check all engine related fluids (Section 4)
Check all underhood hoses (Section 10)
Check and adjust the drivebelts (Section 11)
Clean, inspect and test the battery (Section 12)
Inspect the spark plug wires, distributor cap and rotor (Section 14)
Check the cooling system (Section 20)
Check the air filter (Section 26)
Replace the spark plugs (Section 27)

Major tune-up

All items listed under minor tune-up, plus . . .
Check the ignition system (see Chapter 5)
Check the charging system (see Chapter 5)
Check the fuel system (see Chapter 4)
Replace the spark plug wires, distributor cap and rotor (Section 13)

4 Fluid level checks (every 250 miles or weekly)

Note: *The following are fluid level checks to be done on a 250 mile or weekly basis. Additional fluid level checks can be found in specific maintenance procedures which follow. Regardless of intervals, be alert to fluid leaks under the vehicle which would indicate a fault to be corrected immediately.*

1 Fluids are an essential part of the lubrication, cooling, brake and windshield washer systems. Because the fluids gradually become depleted and/or contaminated during normal operation of the vehicle, they must be periodically replenished. See *Recommended lubricants and fluids* at the beginning of this Chapter before adding fluid to any of the following components. **Note:** *The vehicle must be on level ground when fluid levels are checked.*

Engine oil

Refer to illustrations 4.4 and 4.6

2 Engine oil is checked with a dipstick, which is located on the side of the engine **(refer to the underhood illustration at the front of this Chapter for dipstick location).** The dipstick extends through a metal tube down into the oil pan.

3 The engine oil should be checked before the vehicle has been driven, or about 15 minutes after the engine has been shut off. If the oil is checked immediately after driving the vehicle, some of the oil will remain in the upper part of the engine, resulting in an inaccurate reading on the dipstick.

4 Pull the dipstick out of the tube and wipe all of the oil away from the end with a clean rag or paper towel. Insert the clean dipstick all the way back into the tube and pull it out again. Note the oil at the end of the dipstick. At its highest point, the oil should be between the two **(see illustration).**

5 It takes one quart of oil to raise the level from the lower mark to the upper mark on the dipstick. Do not allow the level to drop below the lower mark or oil starvation may cause engine damage. Conversely, overfilling the engine (adding oil above the upper mark) may cause oil fouled spark plugs, oil leaks or oil seal failures.

6 To add oil, remove the filler cap located on the valve cover **(see illustration)**. After adding oil, wait a few minutes to allow the level to stabilize, then pull the dipstick out and check the level again. Add more oil if required. Install the filler cap and tighten it by hand only.

7 Checking the oil level is an important preventive maintenance step. A consistently low oil level indicates oil leakage through damaged seals, defective gaskets or past worn rings or valve guides. The condition of the oil should also be noted. If the oil looks milky in color or has water droplets in it, the cylinder head gasket may be blown or the head or block may be cracked. The engine should be repaired

4.16 The brake fluid level should be kept above the MIN mark on the translucent reservoir - remove the cap to add fluid

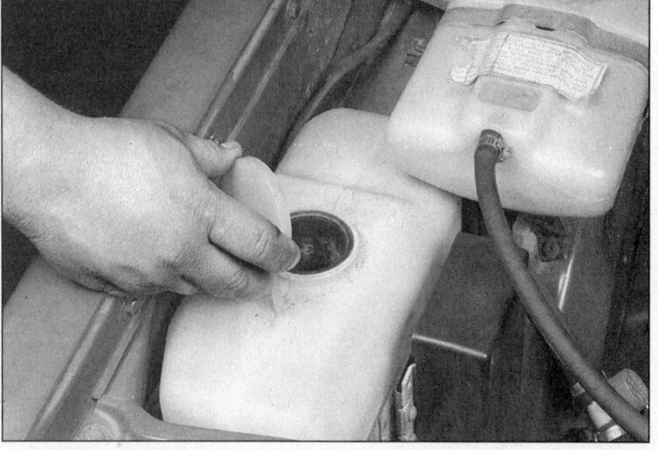
4.22 The windshield washer fluid reservoir is located at the right front corner of the engine compartment on most models

immediately. Whenever you check the oil level, slide your thumb and index finger up the dipstick before wiping off the oil. If you see small dirt or metal particles clinging to the dipstick, the oil should be changed (see Section 6).

Engine coolant

Refer to illustration 4.9

Warning: *Do not allow antifreeze to come in contact with your skin or painted surfaces of the vehicle. Rinse off spills immediately with plenty of water. Antifreeze is highly toxic if ingested. Never leave antifreeze lying around in an open container or in puddles on the floor; children and pets are attracted by it's sweet smell and may drink it. Check with local authorities about disposing of used antifreeze. Many communities have collection centers which will see that antifreeze is disposed of safely.*

8 All vehicles covered by this manual are equipped with a pressurized coolant recovery system. On most models, a white plastic expansion tank (or coolant reservoir) located in the engine compartment is connected by a hose to the radiator. As the engine heats up during operation, the expanding coolant fills the tank. As the engine cools, the coolant is automatically drawn back into the cooling system to maintain the correct level.

9 The coolant level in the expansion tank **(see illustration)** should be checked regularly. **Warning:** *Do not remove the expansion tank cap or radiator cap (if equipped) to check the coolant level when the engine is warm! The level in the expansion tank varies with the temperature of the engine. When the engine is cold, the coolant level should be above the LOW mark on the expansion tank. Once the engine has warmed up, the level should be at or near the FULL mark. If it isn't, allow the engine to cool, then remove the cap from the expansion tank and add a 50/50 mixture of ethylene glycol based antifreeze and water. Don't use rust inhibitors or additives.*

10 Drive the vehicle and recheck the coolant level. If only a small amount of coolant is required to bring the system up to the proper level, water can be used. However, repeated additions of water will dilute the antifreeze and water solution. In order to maintain the proper ratio of antifreeze and water, always top up the coolant level with the correct mixture. An empty plastic milk jug or bleach bottle makes an excellent container for mixing coolant.

11 If the coolant level drops consistently, there may be a leak in the system. Inspect the radiator, hoses, filler cap, drain plugs and water pump (see Section 30). If no leaks are noted, have the expansion tank cap or radiator cap pressure tested by a service station.

12 If you have to remove the cap, wait until the engine has cooled completely, then wrap a thick cloth around the cap and turn it to the first stop. If coolant or steam escapes, let the engine cool down longer, then remove the cap.

13 Check the condition of the coolant as well. It should be relatively clear. If it's brown or rust colored, the system should be drained, flushed and refilled. Even if the coolant appears to be normal, the corrosion inhibitors wear out, so it must be replaced at the specified intervals.

Brake and clutch fluid

Refer to illustration 4.16

Warning: *Brake fluid can harm your eyes and damage painted surfaces, so use extreme caution when handling or pouring it. Do not use brake fluid that has been standing open or is more than one year old. Brake fluid absorbs moisture from the air, which can cause a dangerous loss of brake effectiveness. Use only the specified type of brake fluid. Mixing different types (such as DOT 3 or 4 with DOT 5) can cause brake failure.*

14 The brake master cylinder is mounted at the left (driver's side) rear corner of the engine compartment. The clutch fluid reservoir (used on models with manual transmissions) is mounted adjacent to it.

15 To check the clutch fluid level, observe the level through the translucent reservoir. The level should be at or near the step molded into the reservoir. If the level is low, remove the reservoir cap to add the specified fluid.

16 The brake fluid level is checked by looking through the plastic reservoir mounted on the master cylinder **(see illustration)**. The fluid level should be between the MAX and MIN lines on the reservoir. If the fluid level is low, wipe the top of the reservoir and the cap with a clean rag to prevent contamination of the system as the cap is unscrewed. Top up with the recommended brake fluid, but do not overfill.

17 While the reservoir cap is off, check the master cylinder reservoir for contamination. If rust deposits, dirt particles or water droplets are present, the system should be drained and refilled by a dealer service department or repair shop.

18 After filling the reservoir to the proper level, make sure the cap is seated to prevent fluid leakage and/or contamination.

19 The fluid level in the master cylinder will drop slightly as the disc brake pads wear. A very low level may indicate worn brake pads. Check for wear (see Section 24).

20 If the brake fluid level drops consistently, check the entire system for leaks immediately. Examine all brake lines, hoses and connections, along with the calipers, wheel cylinders and master cylinder (see Section 24).

21 When checking the fluid level, if you discover one or both reservoirs empty or nearly empty, the brake or clutch hydraulic system should be checked for leaks and bled (see Chapters 8 and 9).

Windshield washer fluid

Refer to illustration 4.22

22 Fluid for the windshield washer system is stored in a plastic

5.2 Use a tire tread depth indicator to monitor tire wear - they're available at auto parts stores and service stations and cost very little

reservoir in the engine compartment **(see illustration)**.

23 In milder climates, plain water can be used in the reservoir, but it should be kept no more than two-thirds full to allow for expansion if the water freezes. In colder climates, use windshield washer system antifreeze, available at any auto parts store, to lower the freezing point of the fluid. This comes in concentrated or pre-mixed form. If you purchase concentrated antifreeze, mix the antifreeze with water in accordance with the manufacturer's directions on the container. **Caution:** *Do not use cooling system antifreeze - it will damage the vehicle's paint.*

5 Tire and tire pressure checks (every 250 miles or weekly)

Refer to illustrations 5.2, 5.3, 5.4a, 5.4b and 5.8

1 Periodic inspection of the tires may save you the inconvenience of being stranded with a flat tire. It can also provide you with vital information regarding possible problems in the steering and suspension systems before major damage occurs.

2 Tires are equipped with 1/2-inch wide bands that will appear when tread depth reaches 1/16-inch, at which time the tires can be considered worn out. Tread wear can be monitored with a simple, inexpensive device known as a tread depth indicator **(see illustration)**.

3 Note any abnormal tire wear **(see illustration)**. Tread pattern irregularities such as cupping, flat spots and more wear on one side that the other are indications of front end alignment and/or balance problems. If any of these conditions are noted, take the vehicle to a tire shop or service station to correct the problem.

4 Look closely for cuts, punctures and embedded nails or tacks. Sometimes a tire will hold air pressure for a short time or leak down very slowly after a nail has embedded itself in the tread. If a slow leak persists, check the valve stem core to make sure it is tight **(see illustration)**. Examine the tread for an object that may have embedded itself in the tire or for a "plug" that may have begun to leak (radial tire punctures are repaired with a plug that is installed in the puncture). If a puncture is suspected, it can be easily verified by spraying a solution of soapy water onto the puncture **(see illustration)**. The soapy solution will bubble if there is a leak. Unless the puncture is unusually large, a tire shop or service station can usually repair the tire.

5 Carefully inspect the inner sidewall of each tire for evidence of brake fluid leakage. If you see any, inspect the brakes immediately.

6 Correct air pressure adds miles to the life span of the tires, improves mileage and enhances overall ride quality. Tire pressure cannot be accurately estimated by looking at a tire, especially if it's a

Condition	Probable cause	Corrective action	Condition	Probable cause	Corrective action
Shoulder wear	• Underinflation (both sides wear) • Incorrect wheel camber (one side wear) • Hard cornering • Lack of rotation	• Measure and adjust pressure. • Repair or replace axle and suspension parts. • Reduce speed. • Rotate tires.	Feathered edge Toe wear	• Incorrect toe	• Adjust toe-in.
Center wear	• Overinflation • Lack of rotation	• Measure and adjust pressure. • Rotate tires.	Uneven wear	• Incorrect camber or caster • Malfunctioning suspension • Unbalanced wheel • Out-of-round brake drum • Lack of rotation	• Repair or replace axle and suspension parts. • Repair or replace suspension parts. • Balance or replace. • Turn or replace. • Rotate tires.

5.3 This chart will help you determine the condition of the tires, the probable cause(s) of abnormal wear and the corrective action necessary

5.4a If a tire loses air on a steady basis, check the valve core first to make sure it's snug (special inexpensive wrenches are commonly available at auto parts stores for this purpose)

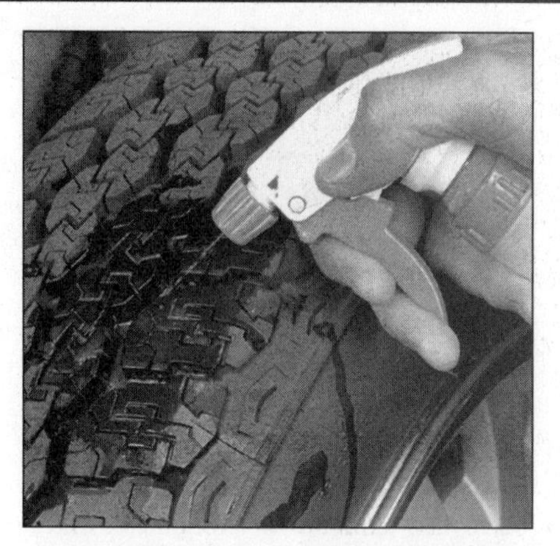

5.4b If the valve core is tight, raise the corner of the vehicle with the low tire and spray a soapy water solution onto the tread as the tire is turned slowly - a leak will cause small bubbles to appear in the area of the puncture

5.8 To extend the life of the tires, check the air pressure at least once a week with an accurate gauge (don't forget to check the spare!)

radial. A tire pressure gauge is essential. Keep an accurate gauge in the glove compartment. The pressure gauges attached to the nozzles of air hoses at gas stations are often inaccurate.

7 Always check tire pressure when the tires are cold. Cold, in this case, means the vehicle has not been driven over a mile in the three hours preceding a tire pressure check. A pressure rise of four to eight pounds is not uncommon once the tires are warm.

8 Unscrew the valve cap protruding from the wheel or hubcap and push the gauge firmly onto the valve stem **(see illustration)**. Note the reading on the gauge and compare the figure to the recommended tire pressure shown in your owner's manual or on the tire placard on the passenger side door or door pillar. Be sure to reinstall the valve cap to keep dirt and moisture out of the valve stem mechanism. Check all four tires and, if necessary, add enough air to bring them to the recommended pressure.

9 Don't forget to keep the spare tire inflated to the specified pressure (refer to your owner's manual or the placard attached to the door pillar).

6 Engine oil and filter change (every 3,000 miles or 3 months)

Refer to illustrations 6.2, 6.7 and 6.16

1 Frequent oil changes are the most important preventive maintenance procedures that can be done by the home mechanic. As engine oil ages, it becomes diluted and contaminated, which leads to premature engine wear.

2 Make sure that you have all the necessary tools before you begin this procedure **(see illustration)**. You should also have plenty of rags or newspapers handy for mopping up oil spills

3 Start the engine and allow it to reach normal operating temperature - oil and sludge will flow more easily when warm. If new oil, a filter or tools are needed, use the vehicle to go get them and warm up the engine oil at the same time. Park on a level surface and shut off the engine when it's warmed up. Remove the oil filler cap from the valve cover.

4 Access to the oil drain plug and filter will be improved if the vehicle can be lifted on a hoist, driven onto ramps or supported by jackstands. **Warning:** *DO NOT work under a vehicle supported only by a bumper, hydraulic or scissors-type jack - always use jackstands!*

5 Raise the vehicle and support it on jackstands. Make sure it is safely supported!

6 If you haven't changed the oil on this vehicle before, get under it and locate the drain plug and the oil filter. The exhaust components

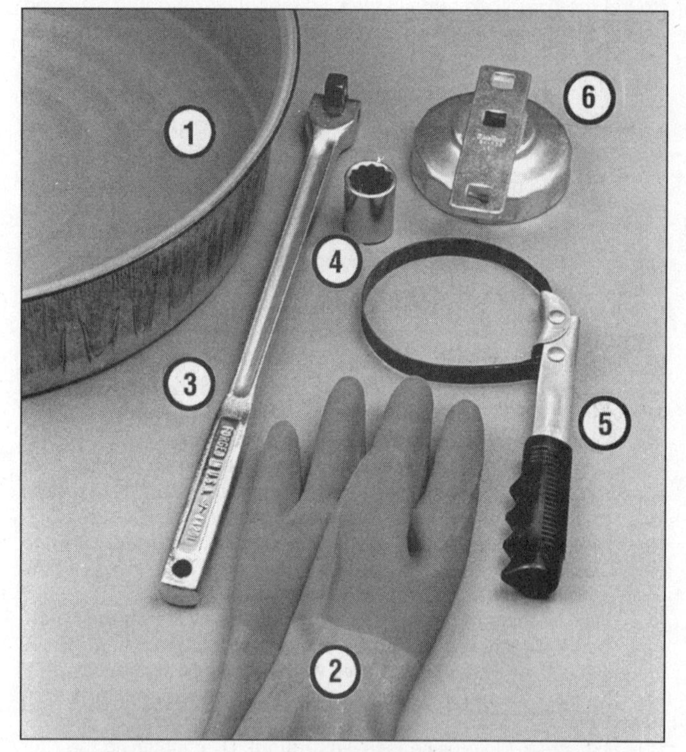

6.2 These tools are required when changing the engine oil and filter

1 **Drain pan** - *It should be fairly shallow in depth, but wide to prevent spills*
2 **Rubber gloves** - *When removing the drain plug and filter, you will get oil on your hands (the gloves will prevent burns)*
3 **Breaker bar** - *Sometimes the oil drain plug is tight, and a long breaker bar is needed to loosen it*
4 **Socket** - *To be used with the breaker bar or a ratchet (must be the correct size to fit the drain plug - six-point preferred)*
5 **Filter wrench** - *This is a metal band-type wrench, which requires clearance around the filter to be effective*
6 **Filter wrench** - *This type fits on the bottom of the filter and can be turned with a ratchet or breaker bar (different-size wrenches are available for different types of filters)*

6.7 Use a proper size box-end wrench or socket to remove the oil drain plug to avoid rounding it off

6.16 Lubricate the oil filter gasket with clean engine oil before installing the filter on the engine

7.2 The power steering fluid reservoir (arrow) is located on the left side of the engine compartment

will be hot as you work, so note how they are routed to avoid touching them.

7 Being careful not to touch the hot exhaust components, position a drain pan under the plug in the bottom of the engine. Clean the area around the plug, then remove the plug **(see illustration)**. It's a good idea to wear a rubber glove while unscrewing the plug the final few turns to avoid being scalded by hot oil. It will also help to hold the drain plug against the threads as you unscrew it, then pull it away from the drain hole suddenly. This will place your arm out of the way of the hot oil, as well as reducing the chances of dropping the drain plug into the drain pan.

8 It may be necessary to move the drain pan slightly as oil flow slows to a trickle. Inspect the old oil for the presence of metal particles.

9 After all the oil has drained, wipe off the drain plug with a clean rag. Any small metal particles clinging to the plug would immediately contaminate the new oil.

10 Reinstall the plug and tighten it securely, but don't strip the threads.

11 Move the drain pan into position under the oil filter.

12 Loosen the spin-off type oil filter by turning it counterclockwise with a filter wrench. Any standard filter wrench will work.

13 Sometimes the spin-off type oil filter is screwed on so tightly that it can't be loosened. If it is, punch a metal bar or long screwdriver directly through it and use it as a T-bar to turn the filter. Be prepared for oil to spurt out of the canister as it's punctured.

14 Once the filter is loose, use your hands to unscrew it from the block. Just as the filter is detached from the block, immediately tilt the open end up to prevent oil inside the filter from spilling out.

15 Using a clean rag, wipe off the mounting surface on the block. Also, make sure that none of the old gasket remains stuck to the mounting surface. It can be removed with a scraper if necessary.

16 Compare the old filter with the new one to make sure they are the same type. Smear some engine oil on the rubber gasket of the new filter and screw it into place **(see illustration)**. Overtightening the filter will damage the gasket, so don't use a filter wrench. Most filter manufacturers recommend tightening the filter by hand only. Normally, they should be tightened 3/4-turn after the gasket contacts the block, but be sure to follow the directions on the filter or container.

17 Remove all tools and materials from under the vehicle, being careful not to spill the oil in the drain pan, then lower the vehicle.

18 Add new oil to the engine through the oil filler cap in the valve over. Use a funnel to prevent oil from spilling onto the top of the engine. Pour four quarts of fresh oil into the engine. Wait a few minutes to allow the oil to drain into the pan, then check the level on the dipstick (see Section 4 if necessary). If the oil level is in the SAFE range, install the filler cap.

19 Start the engine and run it for about a minute. While the engine is running, look under the vehicle and check for leaks at the oil pan drain plug and around the oil filter. If either one is leaking, stop the engine and tighten the plug or filter slightly.

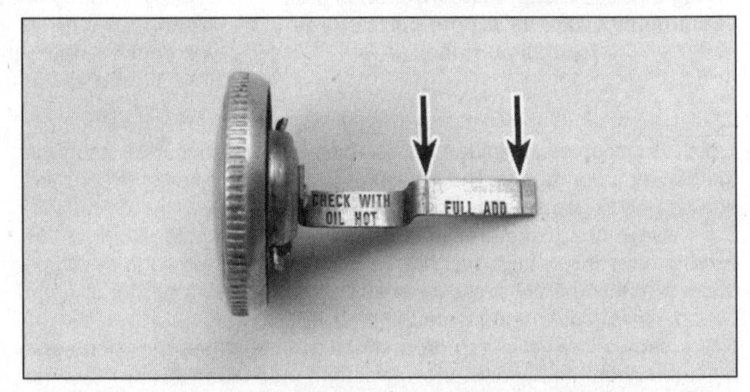

7.5 The power steering fluid level should be kept between the two arrows near the upper step on the dipstick

20 Wait a few minutes, then recheck the level on the dipstick. Add oil as necessary to bring the level into the SAFE range.

21 During the first few trips after an oil change, make it a point to check frequently for leaks and proper oil level.

22 The old oil drained from the engine cannot be reused in its present state and should be discarded. Oil reclamation centers and some auto repair shops and gas stations will accept the oil, which can be recycled. After the oil has cooled, it can be drained into a container (plastic jugs, bottles, milk cartons, etc.) for transport to a disposal site.

7 Power steering fluid level check (every 3,000 miles or 3 months)

Refer to illustrations 7.2 and 7.5

1 Check the power steering fluid level periodically to avoid steering system problems, such as damage to the pump. **Caution:** *DO NOT hold the steering wheel against either stop (extreme left or right turn) for more than five seconds. If you do, the power steering pump could be damaged.*

2 The power steering fluid reservoir is located on the left side of the engine compartment, and is equipped with a twist-off cap with an integral fluid level dipstick **(see illustration)**.

3 Park the vehicle on level ground and apply the parking brake.

4 Run the engine until it has reached normal operating temperature. With the engine at idle, turn the steering wheel back-and-forth several times to get any air out of the steering system. Shut the engine off, remove the cap by turning it counterclockwise, wipe the dipstick clean and reinstall the cap.

5 Remove the cap again and note the fluid level. It must be between the two lines **(see illustration)**.

6 Add small amounts of fluid until the level is correct. **Caution:** *Do*

8.5 The automatic transmission fluid dipstick (arrow) is located near the firewall on the left side of the engine compartment

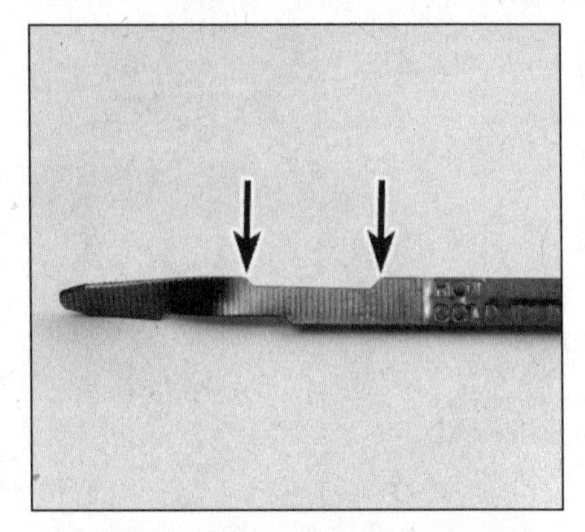

8.6 With the fluid hot, the level should be kept between the two dipstick notches, near the upper one

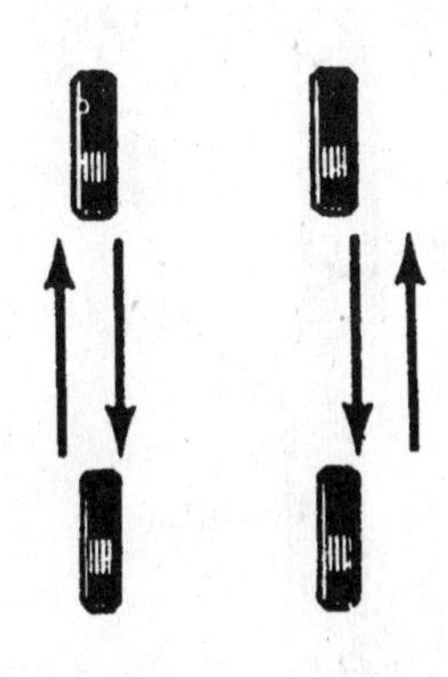

9.2 The tire rotation pattern for models with radial tires

not overfill the reservoir. If too much fluid is added, remove the excess with a clean syringe or suction pump. Insert the dipstick and tighten the cap.

7 Check the power steering hoses and connections for leaks and wear (see Section 10).

8 Check the condition and tension of the drivebelt (see Section 11).

8 Automatic transmission fluid level check (every 3,000 miles or 3 months)

Refer to illustrations 8.5 and 8.6

Caution: *The use of transmission fluid other than the type listed in this Chapter's Specifications could result in transmission malfunctions or failure.*

1 The automatic transmission fluid should be carefully maintained. Low fluid level can lead to slipping or loss of drive, while overfilling can cause foaming and loss of fluid. Either condition can cause transmission damage.

2 Since transmission fluid expands as it heats up, the fluid level should only be checked when the transmission is warm (at normal operating temperature). If the vehicle has just been driven over 20 miles (32 km), the transmission can be considered warm. **Caution:** *If the vehicle has just been driven for a long time at high speed or in city traffic, in hot weather, or if it has been pulling a trailer, an accurate fluid level reading cannot be obtained. Allow the transmission to cool down for about 30 minutes. You can also check the transmission fluid level when the transmission is cold. If the vehicle has not been driven for over five hours and the fluid is about room temperature (70 to 95-degrees F), the transmission is cold. However, the fluid level is normally checked with the transmission warm to ensure accurate results.*

3 Immediately after driving the vehicle, park it on a level surface, set the parking brake and start the engine. While the engine is idling, depress the brake pedal and move the selector lever through all the gear ranges, beginning and ending in Park.

4 Locate the automatic transmission dipstick tube in the left rear corner of the engine compartment.

5 With the engine still idling, pull the dipstick out of the tube **(see illustration)**, wipe it off with a clean rag, push it all the way back into the tube and withdraw it again, then note the fluid level.

6 The level should be between the two marks **(see illustration)**. If the level is low, add the specified automatic transmission fluid through the dipstick tube - use a clean funnel to prevent spills.

7 Add just enough of the recommended fluid to fill the transmission to the proper level. It takes about one pint to raise the level from the low mark to the high mark when the fluid is hot, so add the fluid a little at a time and keep checking the level until it's correct.

8 The condition of the fluid should also be checked along with the level. If the fluid is black or a dark reddish-brown color, or if it smells burned, it should be changed (see Section 29). If you are in doubt about its condition, purchase some new fluid and compare the two for color and smell.

9 Tire rotation (every 6,000 miles or 6 months)

Refer to illustration 9.2

1 The tires should be rotated at the specified intervals and whenever uneven wear is noticed. Since the vehicle will be raised and the tires checked anyway, check the brakes also (see Section 24). **Note:** *Even if you don't rotate the tires, at least check the lug nut tightness.*

2 It is recommended that the tires be rotated in a specific pattern **(see illustration)**.

3 Refer to the information in *Jacking and towing* at the front of this manual for the proper procedure to follow when raising the vehicle and changing a tire. If the brakes must be checked, don't apply the parking brake as stated.

4 The vehicle must be raised on a hoist or supported on jackstands to get all four tires off the ground. Make sure the vehicle is safely supported!

5 After the rotation procedure is finished, check and adjust the tire pressures as necessary and be sure to check the lug bolt tightness.

10 Underhood hose check and replacement (every 6,000 miles or 6 months)

Warning: *Replacement of air conditioning hoses must be left to a dealer service department or air conditioning shop that has the equipment to depressurize the system safely. Never disconnect air conditioning hoses or components until the system has been depressurized.*

General

1 High temperatures under the hood can cause deterioration of the rubber and plastic hoses used for engine, accessory and emission systems operation. Periodic inspection should be made for cracks, loose clamps, material hardening and leaks.

2 Information specific to the cooling system can be found in Section 21.

3 Most (but not all) hoses are secured to fittings with clamps. Where clamps are used, check to be sure they haven't lost their tension, allowing the hose to leak. If clamps aren't used, make sure the hose has not expanded and/or hardened where it slips over the fitting, allowing it to leak.

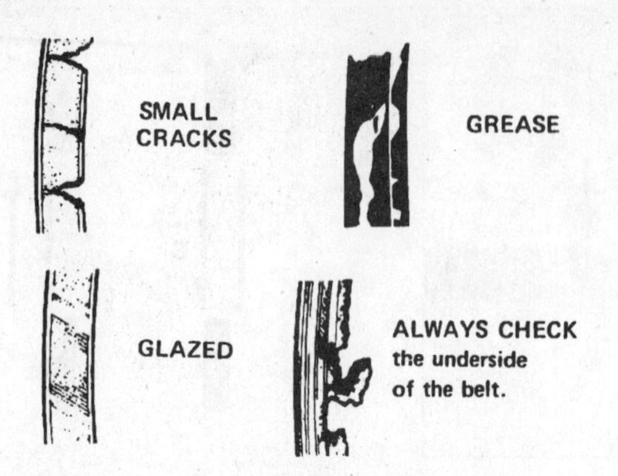

11.3 Here are some of the more common problems associated with drivebelts (check the belts very carefully to prevent an untimely breakdown)

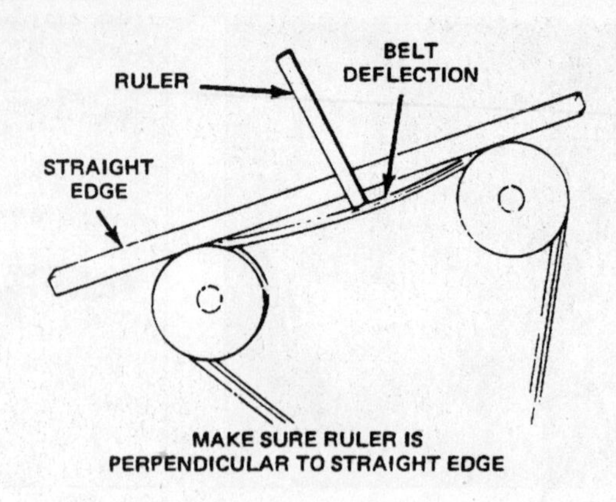

11.4 Measuring drivebelt deflection with a straightedge and ruler

Vacuum hoses

4 It's quite common for vacuum hoses, especially those in the emissions system, to be color coded or identified by colored stripes molded into them. Various systems require hoses with different wall thickness, collapse resistance and temperature resistance. When replacing hoses, be sure the new ones are made of the same material.

5 Often the only effective way to check a hose is to remove it completely from the vehicle. If more than one hose is removed, be sure to label the hoses and fittings to ensure correct installation.

6 When checking vacuum hoses, be sure to include any plastic T-fittings in the check. Inspect the fittings for cracks and the hose where it fits over each fitting for distortion, which could cause leakage.

7 A small piece of vacuum hose can be used as a stethoscope to detect vacuum leaks. Hold one end of the hose to your ear and probe around vacuum hoses and fittings, listening for the "hissing" sound characteristic of a vacuum leak. **Warning:** *When probing with the vacuum hose stethoscope, be careful not to come into contact with moving engine components such as the drivebelt, cooling fan, etc.*

Fuel hoses

Warning: *There are certain precautions which must be taken when servicing or inspecting fuel system components. Work in a well ventilated area and do not allow open flames (cigarettes, appliance pilot lights, etc.) or bare light bulbs near the work area. Mop up any spills immediately and do not store fuel-soaked rags where they could ignite.*

8 The fuel lines are usually under pressure, so if any fuel lines are to be disconnected be prepared to catch spilled fuel. **Warning:** *If your vehicle is equipped with fuel injection, you must relieve the fuel system pressure before servicing the fuel lines. Refer to Chapter 4 for the fuel system pressure relief procedure.*

9 Check all rubber fuel lines for deterioration and chafing. Check especially for cracks in areas where the hose bends and just before fittings, such as where a hose attaches to the fuel pump, fuel filter and fuel injection system.

10 Only high quality fuel line, made specifically for use with high-pressure fuel injection systems, should be used for fuel line replacement. Never, under any circumstances, use unreinforced vacuum line, clear plastic tubing or water hose for fuel lines.

11 Band-type clamps are commonly used on fuel lines. These clamps often lose their tension over a period of time, and can be "sprung" during removal. Replace all band-type clamps with screw clamps whenever a hose is replaced

Metal lines

12 Sections of metal line are often used for fuel line between the fuel pump and fuel injection system. Check carefully to make sure the line isn't bent, crimped or cracked.

13 If a section of metal fuel line must be replaced, use seamless steel tubing only, since copper and aluminum tubing do not have the strength necessary to withstand the vibration caused by the engine.

14 Check the metal brake lines where they enter the master cylinder and brake proportioning unit (if used) for cracks in the lines and loose fittings. Any sign of brake fluid leakage calls for an immediate thorough inspection of the brake system.

Power steering hoses

15 Check the power steering hoses for leaks, loose connections and worn clamps. Tighten loose connections. Worn clamps or leaky hoses should be replaced.

11 Drivebelt check, adjustment and replacement (every 6,000 miles or 6 months)

Refer to illustrations 11.3, 11.4 and 11.6

Check

1 The drivebelts, sometimes called V-belts or simply "fan" belts, are located at the front of the engine and play an important role in the overall operation of the vehicle and its components. Due to their function and material make up, the belts are prone to failure after a period of time and should be inspected and adjusted periodically to prevent major engine damage.

2 The number of belts used on a particular vehicle depends on the accessories installed. Drivebelts are used to turn the alternator, power steering pump, water pump and air conditioning compressor. Depending on the pulley arrangement, a single belt may be used to drive more than one of these components.

3 With the engine off, open the hood and locate the various belts at the front of the engine. Using your fingers (and a flashlight, if necessary), move along the belts checking for cracks and separation of the belt plies. Also check for fraying and glazing, which gives the belt a shiny appearance **(see illustration)**. Both sides of the belts should be inspected, which means you will have to twist each belt to check the underside.

4 The tension of each belt is checked by pushing firmly with your thumb and see how much the belt moves (deflects). Measure the deflection with a ruler **(see illustration)**. A good rule of thumb is that the belt should deflect 1/4-inch if the distance from pulley center-to-pulley center is between seven and eleven inches. The belt should deflect 1/2-inch if the distance from pulley center-to-pulley center is between 12 and 16 inches.

11.6 Loosen the nut on the other end of the adjuster bolt and turn the bolt to increase or decrease tension on the drivebelt

Adjustment

5 If it is necessary to adjust the belt tension, either to make the belt tighter or looser, it is done by moving the belt-driven accessory on the bracket.

6 For each component there will be an adjusting bolt and a pivot bolt. Both bolts must be loosened slightly to enable you to move the component. On some components the drivebelt tension can be adjusted by turning an adjusting bolt after loosening the lock-bolt **(see illustration)**.

7 After the two bolts have been loosened, move the component away from the engine to tighten the belt or toward the engine to loosen the belt. Hold the accessory in position and check the belt tension. If it is correct, tighten the two bolts until just snug, then recheck the tension. If the tension is correct, tighten the bolts.

8 It will often be necessary to use some sort of prybar to move the accessory while the belt is adjusted. If this must be done to gain the proper leverage, be very careful not to damage the component being moved or the part being pried against.

Replacement

9 To replace a belt, follow the instructions above for adjustment, however completely remove the belt from the pulleys.

10 In some cases you will have to remove more then one belt because of their arrangement on the front of the engine. Due to this and the fact that belts will tend to fail at the same time, it is wise to replace all belts together. Mark each belt and its appropriate pulley groove so all replacement belts can be installed in their proper positions.

11 It is a good idea to take the old belts with you when buying new ones in order to make a direct comparison for length, width and design.

12 Battery check, maintenance and charging (every 15,000 miles or 12 months)

Refer to illustrations 12.1, 12.8a, 12.8b, 12.8c and 12.8d

Check and maintenance

Warning: *Certain precautions must be followed when checking and servicing the battery. Hydrogen gas, which is highly flammable, is always present in the battery cells, so keep lighted tobacco and all other flames and sparks away from it. The electrolyte inside the battery is actually dilute sulfuric acid, which will cause injury if splashed on your skin or in your eyes. It will also ruin clothes and painted surfaces. When removing the battery cables, always detach the negative cable first and hook it up last!*

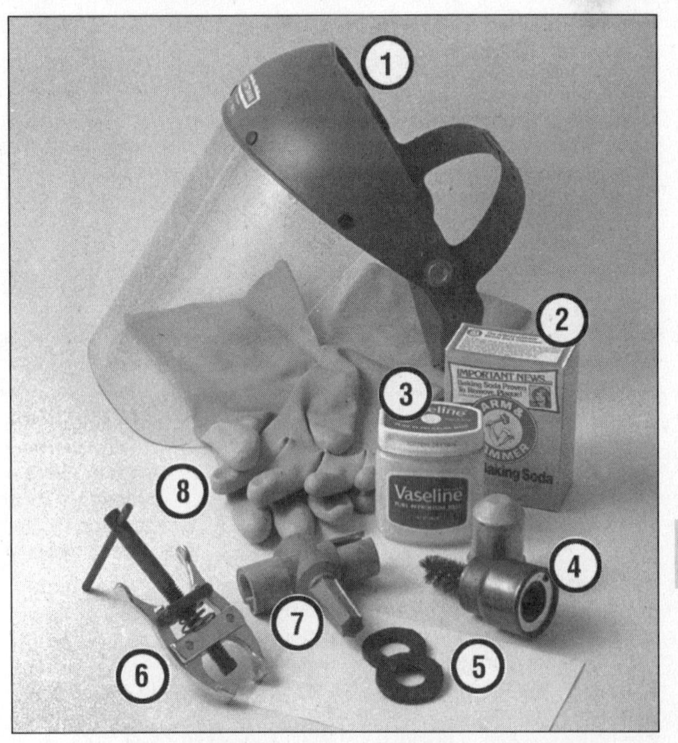

12.1 Tools and materials required for battery maintenance

1 ***Face shield/safety goggles*** - *When removing corrosion with a brush, the acidic particles can easily fly up into your eyes*
2 ***Baking soda*** - *A solution of baking soda and water can be used to neutralize corrosion*
3 ***Petroleum jelly*** - *A layer of this on the battery posts will help prevent corrosion*
4 ***Battery post/cable cleaner*** - *This wire brush cleaning tool will remove all traces of corrosion from the battery posts and cable clamps*
5 ***Treated felt washers*** - *Placing one of these on each post, directly under the cable clamps, will help prevent corrosion*
6 ***Puller*** - *Sometimes the cable clamps are very difficult to pull off the posts, even after the nut/bolt has been completely loosened. This tool pulls the clamp straight up and off the post without damage*
7 ***Battery post/cable cleaner*** - *Here is another cleaning tool which is a slightly different version of Number 4 above, but it does the same thing*
8 ***Rubber gloves*** - *Another safety item to consider when servicing the battery; remember that's acid inside the battery!*

1 Battery maintenance is an important procedure which will help ensure that you are not stranded because of a dead battery. Several tools are required for this procedure **(see illustration)**.

2 Before servicing the battery, always turn the engine and all accessories off and disconnect the cable from the negative terminal of the battery.

3 A low-maintenance battery is standard equipment. The cell caps can be removed and distilled water can be added, if necessary.

4 Remove the caps and check the electrolyte level in each of the battery cells. It must be above the plates. There's usually a split-ring indicator in each cell to indicate the correct level. If the level is low, add distilled water only, then install the cell caps. **Caution:** *Overfilling the cells may cause electrolyte to spill over during periods of heavy charging, causing corrosion and damage to nearby components.*

5 If the positive terminal and cable clamp on your vehicle's battery is equipped with a rubber protector, make sure that it's not torn or

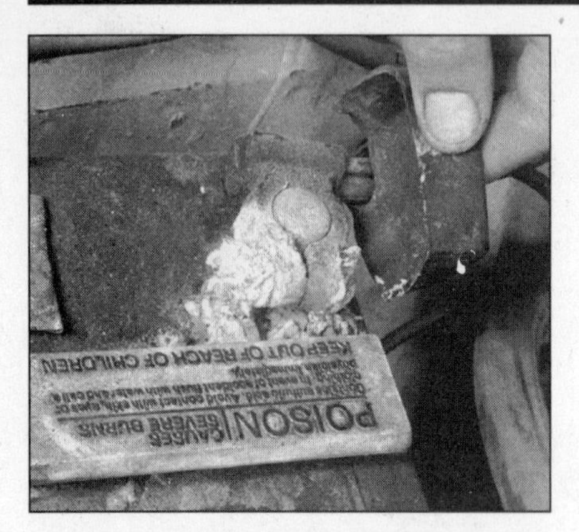

12.8a Battery terminal corrosion usually appears as light, fluffy powder

12.8b Removing a cable from the battery post with a wrench - sometimes special battery pliers are required for this procedure if corrosion has caused deterioration of the hex nut (always remove the ground cable first and hook it up last!)

12.8c Regardless of the type of tool used on the battery posts, a clean, shiny surface should be the result

damaged. It should completely cover the terminal.

6 The external condition of the battery should be checked periodically. Look for damage such as a cracked case.

7 Check the tightness of the battery cable clamps to ensure good electrical connections and inspect the entire length of each cable, looking for cracked or abraded insulation and frayed conductors.

8 If corrosion (visible as white, fluffy deposits) is evident, remove the cables from the terminals, clean them with a battery brush and reinstall them **(see illustrations)**. Corrosion can be kept to a minimum by installing specially treated washers available at auto parts stores or by applying a layer of petroleum jelly or grease to the terminals and cable clamps after they are assembled.

9 Make sure that the battery carrier is in good condition and that the hold-down clamp bolt is tight. If the battery is removed (see Chapter 5 for the removal and installation procedure), make sure that no parts remain in the bottom of the carrier when it's reinstalled. When reinstalling the hold-down clamp, don't overtighten the bolt.

10 Corrosion on the carrier, battery case and surrounding areas can be removed with a solution of water and baking soda. Apply the mixture with a small brush, let it work, then rinse it off with plenty of clean water.

11 Any metal parts of the vehicle damaged by corrosion should be coated with a zinc-based primer, then painted.

12 Additional information on the battery and jump starting can be found in Chapter 5 and the front of this manual.

Charging

13 Remove all of the cell caps (if equipped) and cover the holes with a clean cloth to prevent spattering electrolyte. Disconnect the negative battery cable and hook the battery charger leads to the battery posts (positive to positive, negative to negative), then plug in the charger. Make sure it is set at 12-volts if it has a selector switch.

14 If you're using a charger with a rate higher than two amps, check the battery regularly during charging to make sure it doesn't overheat. If you're using a trickle charger, you can safely let the battery charge overnight after you've checked it regularly for the first couple of hours.

15 If the battery has removable cell caps, measure the specific gravity with a hydrometer every hour during the last few hours of the charging cycle. Hydrometers are available inexpensively from auto parts stores - follow the instructions that come with the hydrometer. Consider the battery charged when there's no change in the specific gravity reading for two hours and the electrolyte in the cells is gassing (bubbling) freely. The specific gravity reading from each cell should be very close to the others. If not, the battery probably has a bad cell(s).

16 Some batteries with sealed tops have built-in hydrometers on the top that indicate the state of charge by the color displayed in the hydrometer window. Normally, a bright-colored hydrometer indicates a full charge and a dark hydrometer indicates the battery still needs charging. Check the battery manufacturer's instructions to be sure you know what the colors mean.

12.8d When cleaning the cable clamps, all corrosion must be removed (the inside of the clamp is tapered to match the taper on the post, so don't remove too much material)

17 If the battery has a sealed top and no built-in hydrometer, you can hook up a voltmeter across the battery terminals to check the charge. A fully charged battery should read 12.6-volts or higher.

18 Further information on the battery and jump starting can be found in Chapter 5 and at the front of this manual.

13 Spark plug wire, distributor cap and rotor check and replacement (every 15,000 miles or 12 months)

Refer to illustrations 13.11a, 13.11b and 13.11c

1 The spark plug wires should be checked at the recommended intervals and whenever new spark plugs are installed in the engine.

2 Begin this procedure by making a visual check of the spark plug wires while the engine is running. In a darkened garage (make sure there is ventilation) start the engine and observe each plug wire. Be careful not to come into contact with any moving engine parts. If there is a break in the wire, you will see arcing or a small spark at the damaged area. If arcing is noticed, make a note to obtain new wires,

13.11a After removing the distributor cap, the rotor can be pulled straight up and off on most models

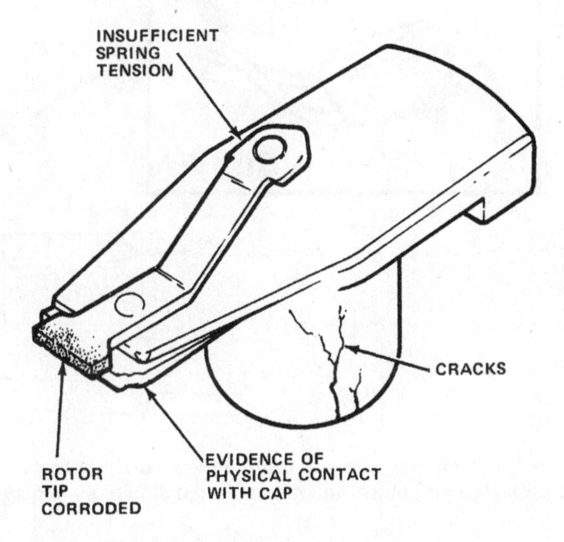

13.11c The ignition rotor should be checked for wear and corrosion as indicated here (if in doubt about its condition, buy a new one)

then allow the engine to cool.

3 Disconnect the negative cable from the battery.

4 The wires should be inspected one at a time to prevent mixing up the order, which is essential for proper engine operation.

5 Disconnect the plug wire from the spark plug. A removal tool can be used for this purpose or you can grab the plastic boot, twist slightly and pull the wire free. Do not pull on the wire itself, only on the boot.

6 Inspect inside the boot for corrosion, which will look like a white crusty powder. Push the wire and boot back onto the end of the spark plug. It should be a tight fit on the plug end. If it is not, remove the wire and use pliers to carefully crimp the metal connector inside the boot until it fits securely on the end of the spark plug.

7 Using a clean rag, wipe the entire length of the wire to remove any built-up dirt and grease. Once the wire is clean, check for burns, cracks and other damage. Do not bend the wire excessively, since the conductor might break.

8 Disconnect the wire from the distributor. Again, pull only on the boot. Check for corrosion and a tight fit in the same manner as the spark plug end. Replace the wire in the distributor.

9 Check the remaining spark plug wires, making sure they are

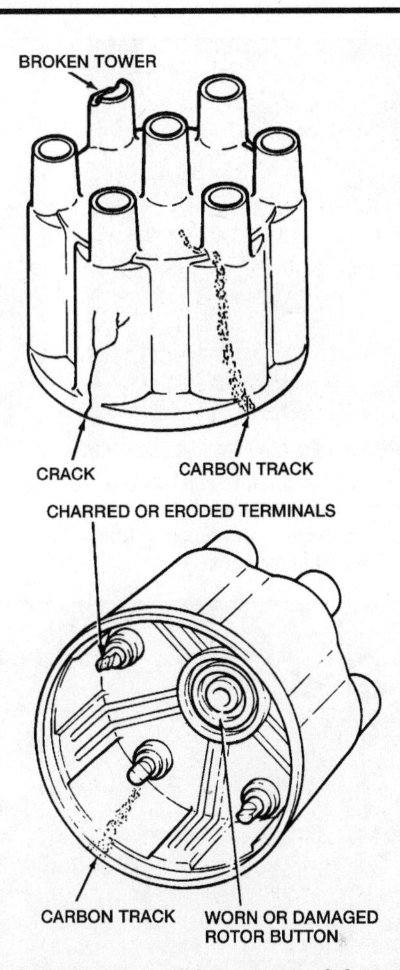

13.11b Shown here are some of the common defects to look for when inspecting the distributor cap (if in doubt about its condition, install a new one)

securely fastened at the distributor and spark plug when the check is complete.

10 If new spark plug wires are required, purchase a set for your specific engine model. Wire sets are available pre-cut, with the boots already installed. Remove and replace the wires one at a time to avoid mix-ups in the firing order.

11 Check the distributor cap and rotor for wear **(see illustration)**. Detach the clips and remove the distributor cap. Remove the rotor off the shaft. Look for cracks, carbon tracks and worn, burned or loose contacts **(see illustrations)**. Replace the cap and rotor with new parts if defects are found. It is common practice to install a new cap and rotor whenever new spark plug wires are installed. When installing a new cap, remove the wires from the old cap one at a time and attach them to the new cap in the exact same location - do not simultaneously remove all the wires or firing order mix-ups may occur.

14 Manual transmission lubricant level check (every 15,000 miles or 12 months)

Refer to illustration 14.2

1 The transmission has a check/fill plug which must be removed to check the lubricant level. If the vehicle is raised to gain access to the plug, be sure to support it safely on jackstands - DO NOT crawl under a vehicle which is supported only by a jack!

2 Remove the plug from the side of the transmission and use your little finger to reach inside plug from the housing and feel the lubricant

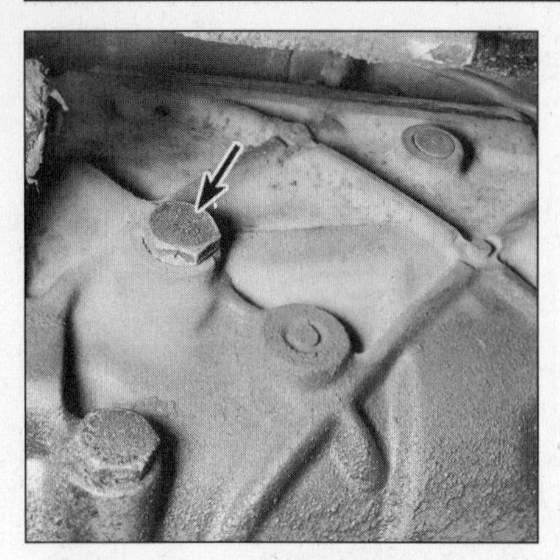

14.2 Use a large wrench to remove the check/fill plug (arrow), (the fluid should be even with the bottom of the hole) - if it's low, add lubricant

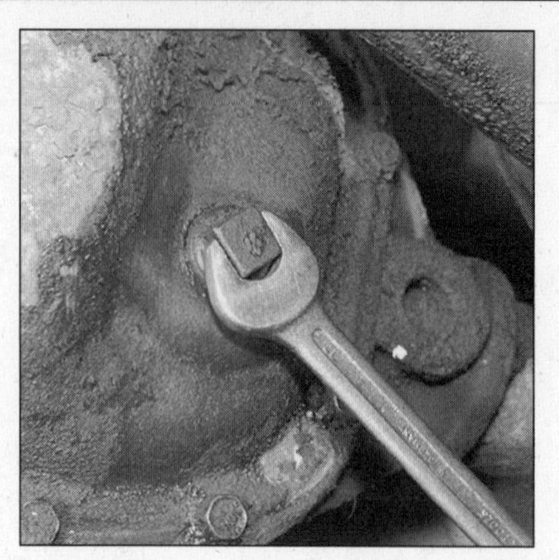

15.2 Remove the differential fill plug with a wrench

15.3a Use your finger as a dipstick to make sure the lubricant level is even with the bottom of the hole

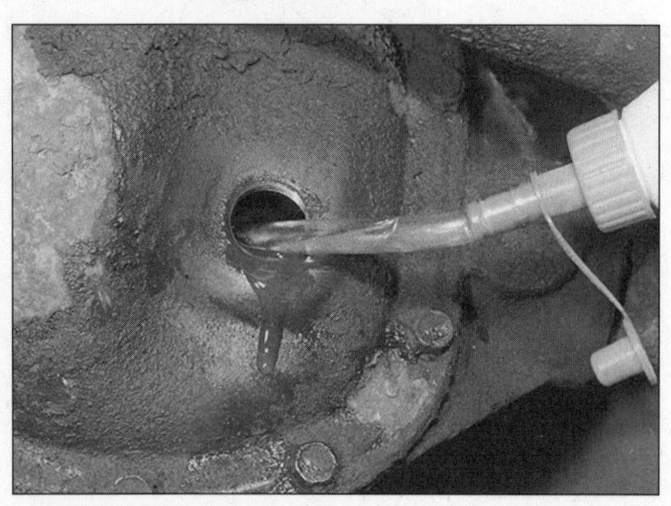

15.3b Add lubricant to the differential until it begins to run out the fill hole

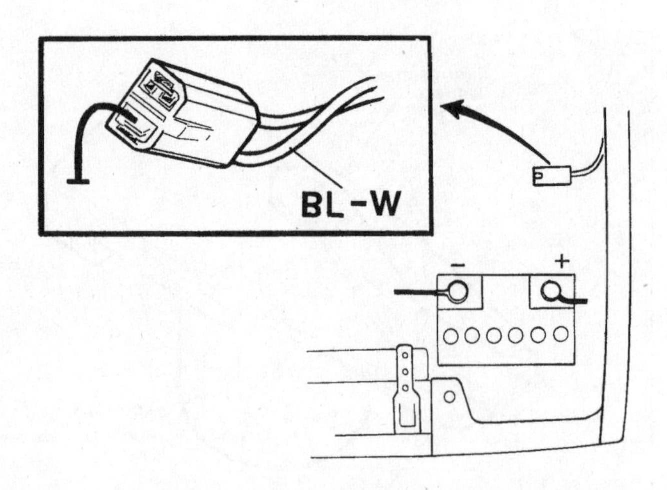

17.4 Use a jumper wire and ground the blue/white test connector to close the air control valve before adjusting base idle speed

level. It should be at or very near the bottom of the plug hole **(see illustration)**.

3 If it isn't, add the recommended lubricant through the plug hole with a syringe or squeeze bottle.

4 Install the plug securely and check for leaks after the first few miles of driving.

15 Differential lubricant level check (every 15,000 miles or 12 months)

Refer to illustrations 15.2, 15.3a and 15.3b

1 The differential has a check/fill plug which must be removed to check the lubricant level. If the vehicle is raised to gain access to the plug, be sure to support it safely on jackstands - DO NOT crawl under the vehicle when it's supported only by the jack!

2 Remove the lubricant check/fill plug from the differential **(see illustration)**.

3 Use your little finger as a dipstick to make sure the lubricant level is even with the bottom of the plug hole. If it's not, use a syringe or squeeze bottle to add the recommended lubricant until it just starts to run out of the opening **(see illustrations)**.

4 Install the plug and tighten it securely.

16 Ignition timing (1976 through 1988 models) - check and adjustment (every 15,000 miles or 12 months)

At the specified intervals, or when the distributor has been removed, the ignition timing must be checked and adjusted if necessary. Refer to Chapter 5 for the correct timing procedures and specifications for your vehicle.

17 Idle speed - check and adjustment (LH-Jetronic models only) (every 15,000 miles or 12 months)

Refer to illustrations 17.4 and 17.5

Note: *Information on CIS equipped models can be found in Chapter 4A. Information on carbureted models can be found in Chapter 4B.*

1 Engine idle speed is the speed at which the engine operates when no accelerator pedal pressure is applied, as when stopped at a traffic light. This speed is critical to the performance of the engine itself, as well as many engine subsystems.

2 A hand-held tachometer must be used when adjusting idle speed to get an accurate reading. The exact hook-up for these meters varies with the manufacturer, so follow the particular directions included.

3 All vehicles covered in this manual should have a tune-up decal or

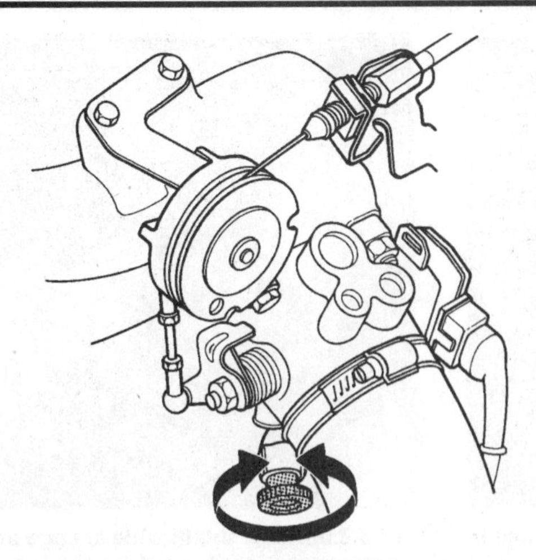

17.5 Adjust the idle speed by turning the knob (arrow) clockwise to decrease the idle speed and counterclockwise to increase idle speed

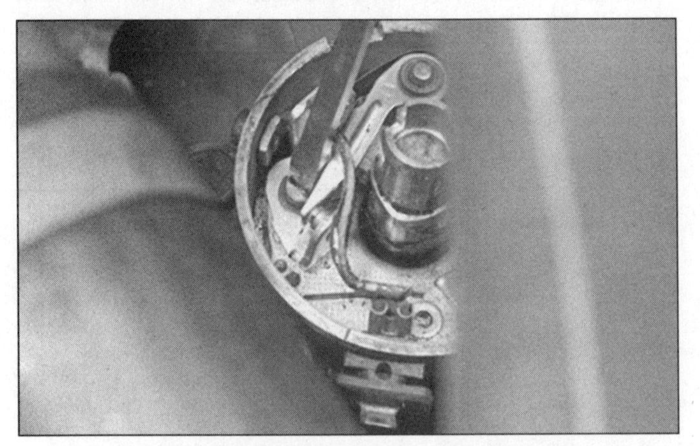

18.4 Remove the ignition points mounting screw

Emissions Control Information label located on the inside of the hood. The printed instructions for setting idle speed can be found on this decal, and should be followed since they are for your particular engine.

4 With the engine at normal operating temperature, the parking brake firmly set and the wheels blocked to prevent the vehicle from rolling, ground the blue/white test lead **(see illustration)** located near the battery. This will close the air control valve thereby not allowing the valve to interfere with the base setting. **Note:** *Do not ground the oxygen sensor test lead (red wire) by mistake or the oxygen sensor will become damaged.*

5 Turn the idle bypass screw to adjust the correct rpm level **(see illustration)**. Check the engine idle speed, following the instructions on the decal. Turn off all lights and accessories.

6 Remove the jumper wire from the test terminal and observe that the rpm increase slightly. If the engine speed does not change, diagnose the problem with the air control valve and/or circuit.

18 Ignition point check and replacement (non-electronic distributor models) (every 15,000 miles or 12 months)

Refer to illustrations 18.1, 18.4, 18.12 and 18.16

1 The ignition points must be replaced at regular intervals on vehicles not equipped with electronic ignition. Occasionally, the rubbing block on the points will wear sufficiently to require readjustment. Several special tools are required to replace the points **(see illustration)**.

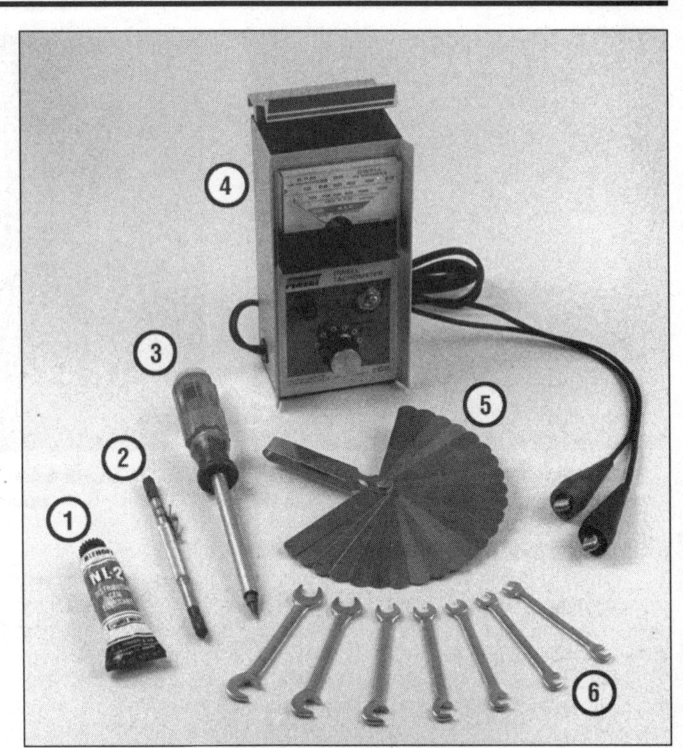

18.1 Tools and materials needed for ignition point replacement and dwell angle adjustment

1 **Distributor cam lube** - *Sometimes this special lubricant comes with the new points; however, its a good idea to buy a tube and have it on hand*
2 **Screw starter** - *This tool has special claws which hold the screw securely as it is started, which helps prevent accidental dropping of the screw*
3 **Magnetic screwdriver** - *Serves the same purpose as no. 2 above. If you do not have one of these special screwdrivers, you risk dropping the points set mounting screws down into the distributor body*
4 **Dwell meter** - *A dwell meter is the only accurate way to determine the point setting (gap). Connect the meter according to the instructions supplied with the meter*
5 **Blade-type feeler gauges** - *These are required to set the initial point gap (space between the points where they are open)*
6 **Ignition wrenches** - *These special wrenches are made to work within the tight confines of the distributor. Specifically, they are needed to loosen the nut/bolt which secures the leads to the points*

2 After removing the distributor cap and rotor, the point and condenser assembly are plainly visible. The points may be examined by gently prying them open to reveal the condition of the contact surfaces. If they are rough, pitted or dirty they should be replaced. **Caution:** *The following procedure requires the removal and installation of small screws which can easily fall into the distributor. To retrieve them, the distributor would have to be removed and disassembled. Use a magnetic or spring loaded screwdriver and exercise caution.*

3 To replace the condenser, which should be replaced along with the points, remove the screw that secures the condenser to the point plate. Loosen the nut or screw retaining the condenser lead and the primary lead to the point assembly. Remove the condenser and mounting bracket.

4 Remove the point assembly mounting screw **(see illustration)**.

5 Attach the new point assembly to the distributor plate with the mounting screw.

18.12 Pry at the point shown with a screwdriver to adjust the points gap

18.16 Use a feeler gauge to check the points gap

19.3 Lubricate the throttle linkage at the points shown (arrows)

6 Install the new condenser and tighten the mounting screw.

7 Attach the primary ignition lead and the condenser lead to the point assembly. Make sure that the forked connectors for the primary ignition and condenser leads do not touch the distributor plate or any other grounded surface.

8 Two adjusting methods are available for setting the points. The first and most effective method involves an instrument called a dwell meter.

9 Connect one dwell meter lead to the primary ignition terminal of the point assembly or to the distributor terminal of the coil. Connect the other lead of the dwell meter to an engine ground. Some dwell meters may have different connecting instructions, so always follow the instrument manufacturer's directions.

10 Have an assistant crank the engine over or use a remote starter and crank the engine with the ignition switch turned to On.

11 Note the dwell reading on the meter and compare it to the Specifications found at the front of this chapter or on the engine tune-up decal. If the dwell reading is incorrect, adjust it by first loosening the point assembly mounting screw a small amount.

12 Move the point assembly plate with a screwdriver inserted into the slot provided next to the points **(see illustration)**. Closing the gap on the points will increase the dwell reading, while opening the gap will decrease the dwell.

13 Tighten the screw after the correct reading is obtained and recheck the setting.

14 If a dwell meter is unavailable, a feeler gauge can be used to set the points.

15 Have a helper crank the engine in short bursts until the rubbing block of the points rests on a high point of the cam assembly. The rubbing block must be exactly on the apex of one of the cam lobes for correct point adjustment. It may be necessary to rotate the front pulley of the engine with a socket and breaker bar to position the cam lobe exactly.

16 Measure the gap between the contact points with the correct size feeler gauge **(see illustration)**. If the gap is incorrect, loosen the mounting screw and move the point assembly until the correct gap is achieved.

17 Retighten the screw and recheck the gap one more time.

18 Apply a small amount of distributor cam lube to the point cam on the distributor shaft.

19 Throttle linkage - check and lubrication (every 15,000 miles or 12 months)

Refer to illustration 19.3

Note: *Information on carbureted models can be found in Chapter 4B.*

1 The throttle linkage should be checked and lubricated periodically to ensure its proper operation.

2 Check the linkage to make sure it isn't binding.

3 Inspect the linkage joints for looseness and the connections for corrosion and damage, replacing parts as necessary **(see illustration)**.

4 Lubricate the connections with spray lubricant or white lithium grease.

20 Fuel system check (every 15,000 miles or 12 months)

Warning: *Gasoline is extremely flammable, so take extra precautions when you work on any part of the fuel system. Don't smoke or allow open flames or bare light bulbs near the work area, and don't work in a garage where a natural gas-type appliance (such as a water heater or a clothes dryer) with a pilot light is present. Since gasoline is carcinogenic, wear latex gloves when there's a possibility of being exposed to fuel, and, if you spill any fuel on your skin, rinse it off immediately with soap and water. Mop up any spills immediately and do not store fuel-soaked rags where they they could ignite. When you perform any type of work on the fuel system, wear safety glasses and have a class B type fire extinguisher on hand.*

1 If you smell gasoline while driving or after the vehicle has been sitting in the sun, inspect the fuel system immediately.

2 Remove the fuel filler cap and inspect it for damage and corrosion. The gasket should have an unbroken sealing imprint. If the gasket is damaged or corroded, install a new cap.

3 Inspect the fuel feed and return lines for cracks. Make sure that the connections between the fuel lines and the fuel injection system and between the fuel lines and the in-line fuel filter are tight. **Warning:** *On fuel-injected models, the fuel system pressure must be relieved before servicing fuel system components. The fuel system pressure relief procedure is outlined in Chapter 4.*

4 Since some components of the fuel system - the fuel tank and some of the fuel feed and return lines, for example - are underneath the vehicle, they can be inspected more easily with the vehicle raised on a hoist. If that's not possible, raise the vehicle and support it on jackstands.

5 With the vehicle raised and safely supported, inspect the gas tank and filler neck for punctures, cracks or other damage. The connection between the filler neck and the tank is particularly critical. Sometimes a rubber filler neck will leak because of loose clamps or deteriorated rubber. Inspect all fuel tank mounting brackets and straps to be sure the tank is securely attached to the vehicle. **Warning:** *Do not, under any circumstances, try to repair a fuel tank (except rubber components). A welding torch or any open flame can easily cause fuel vapors inside the tank to explode.*

6 Carefully check all flexible hoses and metal lines leading away from the fuel tank. Check for loose connections, deteriorated hoses, crimped lines and other damage. Repair or replace damaged sections as necessary (see Chapter 4).

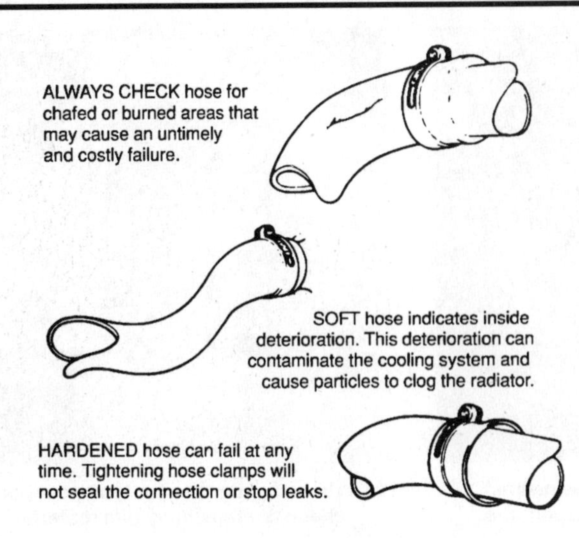

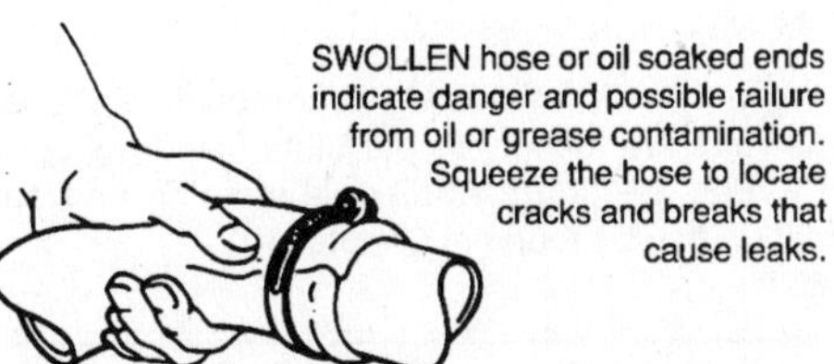

SWOLLEN hose or oil soaked ends indicate danger and possible failure from oil or grease contamination. Squeeze the hose to locate cracks and breaks that cause leaks.

21.4 Hoses, like drivebelts, have a habit of failing at the worst possible time - to prevent the inconvenience of a blown radiator or heater hose, inspect them carefully as shown here

21 Cooling system check (every 15,000 miles or 12 months)

Refer to illustration 21.4

1 Many major engine failures can be attributed to a faulty cooling system. If the vehicle is equipped with an automatic transmission, the cooling system also plays an important role in prolonging transmission life because it cools the fluid.

2 The engine should be cold for the cooling system check, so perform the following procedure before the vehicle is driven for the day or after it has been shut off for at least three hours.

3 Remove the expansion tank cap and clean it thoroughly, inside and out, with clean water. Also clean the filler neck on the tank. The presence of rust or corrosion in the filler neck means the coolant should be changed (see Section 30). The coolant should be relatively clean and transparent. If it's rust colored, drain the system and refill with new coolant.

4 Carefully check the radiator hoses and smaller diameter heater hoses. Inspect each coolant hose along its entire length, replacing any hose which is cracked, swollen or deteriorated **(see illustration)**. Cracks will show up better if the hose is squeezed. Pay close attention to hose clamps that secure the hoses to cooling system components. Hose clamps can pinch and puncture hoses, resulting in coolant leaks.

5 Make sure all hose connections are tight. A leak in the cooling system will usually show up as white or rust colored deposits on the area adjoining the leak. If wire-type clamps are used on the hoses, it may be a good idea to replace them with screw-type clamps.

6 Clean the front of the radiator and air conditioning condenser with compressed air, if available, or a soft brush. Remove all bugs, leaves, etc. embedded in the radiator fins. Be extremely careful not to damage the cooling fins or cut your fingers on them.

7 If the coolant level has been dropping consistently and no leaks are detectable, have the radiator cap and cooling system pressure checked at a service station.

22.2a Check the exhaust system rubber hangers for cracks

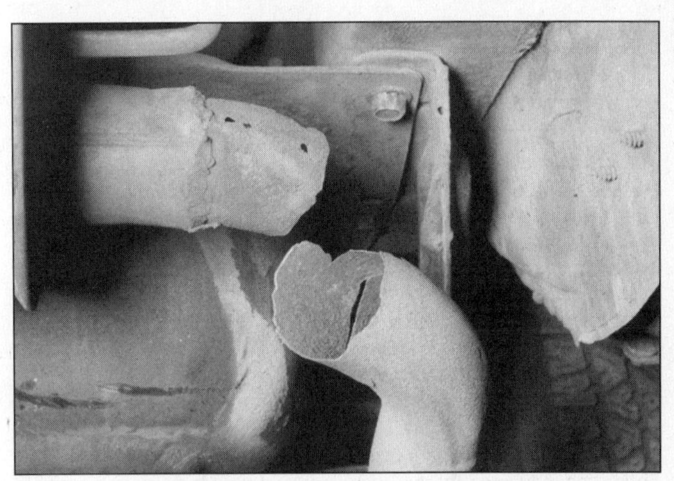

22.2b Replace any rusted-through exhaust components immediately

22 Exhaust system check (every 15,000 miles or 12 months)

Refer to illustrations 22.2a and 22.2b

1 With the engine cold (at least three hours after the vehicle has been driven), check the complete exhaust system from the engine to end of the tailpipe. Ideally, the inspection should be done with the vehicle on a hoist to permit unrestricted access. If a hoist isn't available, raise the vehicle and support it securely on jackstands.

2 Check the exhaust pipes and connections for evidence of leaks, severe corrosion and damage. Make sure that all brackets and hangers are in good condition and are tight **(see illustrations)**.

3 At the same time, inspect the underside of the body for holes, corrosion, open seams, etc. which may allow exhaust gases to enter the passenger compartment. Seal all body openings with silicone or body putty.

4 Rattles and other noises can often be traced to the exhaust system, especially the mounts, hangers and heat shields. Try to move the pipes, muffler and catalytic converter. If the components can come in contact with the body or suspension parts, secure the exhaust system with new mounts.

5 Check the running condition of the engine by inspecting inside the end of the tailpipe. The exhaust deposits here are an indication of engine state-of-tune. If the pipe is black and sooty or coated with white deposits, the engine may need a tune-up, including a thorough fuel system inspection.

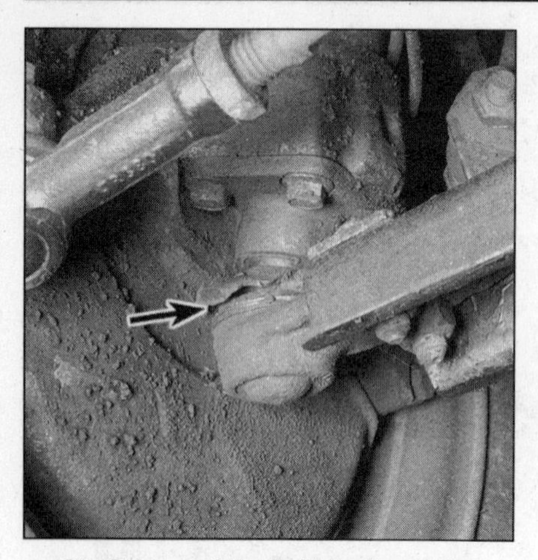
23.10 Inspect the balljoint boots for tears (arrow)

24.11a Look through the caliper inspection window to inspect the brake pads

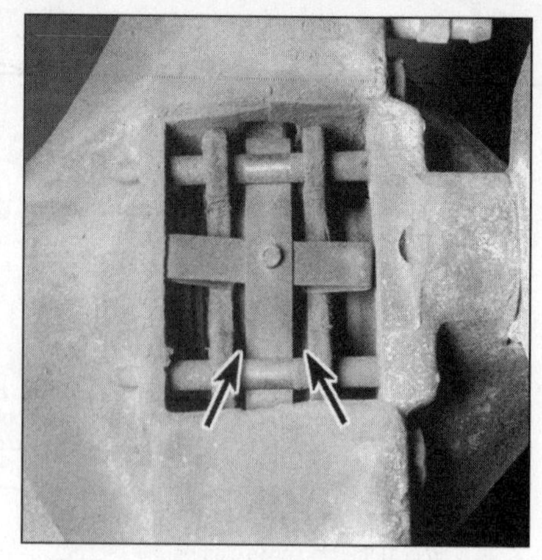
24.11b Look into the rear brake caliper to check the remaining pad material

23 Steering and suspension check (every 15,000 miles or 12 months)

Refer to illustration 23.10

Note: *The steering linkage and suspension components should be checked periodically. Worn or damaged suspension and steering linkage components can result in excessive and abnormal tire wear, poor ride quality and vehicle handling and reduced fuel economy. For detailed illustrations of the steering and suspension components, refer to Chapter 10.*

Strut/shock absorber check

1 Park the vehicle on level ground, turn the engine off and set the parking brake. Check the tire pressures.

2 Push down at one corner of the vehicle, then release it while noting the movement of the body. It should stop moving and come to rest in a level position with one or two bounces.

3 If the vehicle continues to move up-and-down or if it fails to return to its original position, a worn or weak strut or shock absorber is probably the reason.

4 Repeat the above check at each of the three remaining corners of the vehicle.

5 Raise the vehicle and support it on jackstands.

6 Check the struts/shock absorbers for evidence of fluid leakage. A light film of fluid is not a cause for concern. Make sure that any fluid noted is from the struts/shocks and not from any other source. If leakage is noted, replace the struts or shocks as a set.

7 Check the struts/shock absorbers to be sure that they are securely mounted and undamaged. Check the upper mounts for damage and wear. If damage or wear is noted, replace the struts or shock absorbers as a set.

8 If the struts or shock absorbers must be replaced, refer to Chapter 10 for the procedure.

Steering and suspension check

9 Visually inspect the steering system components for damage and distortion. Look for leaks and damaged seals, boots and fittings.

10 Clean the lower end of the steering knuckle. Have an assistant grasp the lower edge of the tire and move the wheel in-and-out while you look for movement at the steering knuckle-to-control arm balljoints. Inspect the balljoint boots for tears **(see illustration)**. If there is any movement, or the boots are torn or leaking, the balljoint(s) must be replaced.

11 Grasp each front tire at the front and rear edges, push in at the front, pull out at the rear and feel for play in the steering linkage. If any freeplay is noted, check the steering gear mounts and the tie-rod balljoints for looseness. If the steering gear mounts are loose, tighten them. If the tie-rods are loose, the balljoints may be worn (check to make sure the nuts are tight). Additional steering and suspension system information can be found in Chapter 10.

24 Brake system check (every 15,000 miles or 12 months)

Refer to illustrations 24.11a and 24.11b

Warning: *Dust produced by lining wear and deposited on brake components may contain asbestos, which is hazardous to your health. DO NOT blow it out with compressed air and DO NOT inhale it! DO NOT use gasoline or solvents to remove the dust. Brake system cleaner should be used to flush the dust into a drain pan. After the brake components are wiped with a damp rag, dispose of the contaminated rag(s) and brake cleaner in a covered and labeled container. Try to use non-asbestos replacement parts whenever possible.*

Note: *In addition to the specified intervals, the brake system should be inspected each time the wheels are removed or a malfunction is indicated. Because of the obvious safety considerations, the following brake system checks are some of the most important maintenance procedures you can perform on your vehicle.*

Symptoms of brake system problems

1 The disc brakes have built-in electrical wear indicators which cause a red warning light on the dash to come on when they're worn to the replacement point. When the light comes on, replace the pads immediately or expensive damage to the brake discs could result.

2 Any of the following symptoms could indicate a potential brake system defect:

- *a) The vehicle pulls to one side when the brake pedal is depressed.*
- *b) The brakes make squealing or dragging noises when applied.*
- *c) Brake pedal travel is excessive.*
- *d) The brake pedal pulsates when applied.*
- *e) Brake fluid leaks are noted (usually on the inner side of the tire or wheel).*

If any of these conditions are noted, inspect the brake system immediately.

Brake lines and hoses

Note: *Steel tubing is used throughout the brake system, with the exception of flexible, reinforced hoses at the front wheels and as connectors at the rear axle. Periodic inspection of these lines is very important.*

3 Park the vehicle on level ground and turn the engine off.

4 Remove the wheel covers. Loosen, but do not remove, the lug bolts on all four wheels.

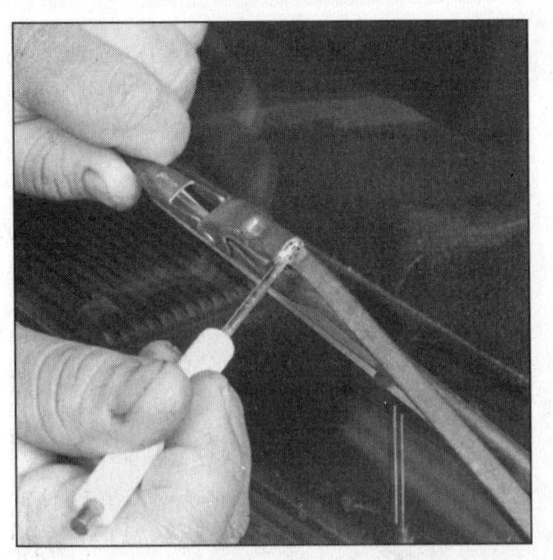

25.5 Lift up on the lever to release the wiper blade from the arm

26.1 Detach the air cleaner housing clips with a screwdriver

26.3 Pull the cover away and lift the air cleaner element out

5 Raise the vehicle and support it securely on jackstands.
6 Remove the wheels (see *Jacking and towing* at the front of this book, or refer to your owner's manual, if necessary).
7 Check all brake lines and hoses for cracks, chafing of the outer cover, leaks, blisters and distortion. Check the brake hoses at front and rear of the vehicle for softening, cracks, bulging, or wear from rubbing on other components. Check all threaded fittings for leaks and make sure the brake hose mounting bolts and clips are secure.
8 If leaks or damage are discovered, they must be fixed immediately. Refer to Chapter 9 for detailed brake system repair procedures.

Disc brakes

9 If it hasn't already been done, raise the vehicle and support it securely on jackstands. Remove the front wheels.
10 The disc brake calipers, which contain the pads, are now visible. Each caliper has an outer and an inner pad - all pads should be checked.
11 Note the pad thickness by looking through the inspection area in the caliper **(see illustrations)**. If the lining material is 1/8-inch thick or less, or if it is tapered from end-to-end, the pads should be replaced (see Chapter 9). Keep in mind that the lining material is riveted or bonded to a metal plate or shoe - the metal portion is not included in this measurement.
12 Check the condition of the brake disc. Look for score marks, deep scratches and overheated areas (they will appear blue or discolored). If damage or wear is noted, the disc can be removed and resurfaced by an automotive machine shop or replaced with a new one. Refer to Chapter 9 for more detailed inspection and repair procedures.
13 Remove the calipers without disconnecting the brake hoses (see Chapter 9).

Parking brake

14 The easiest, and perhaps most obvious, method of checking the parking brake is to park the vehicle on a steep hill with the parking brake set and the transmission in Neutral (stay in the vehicle while performing this check). If the parking brake doesn't prevent the vehicle from rolling, refer to Chapter 9 and adjust it.

25 Wiper blade check and replacement (every 15,000 miles or 12 months)

1 Road film can build up on the wiper blades and affect their efficiency, so they should be washed regularly with a mild detergent solution.

Check

2 The wiper and blade assembly should be inspected periodically. Even if you don't use your wipers, the sun and elements will dry out the rubber portions, causing them to crack and break apart. If inspection reveals hardened or cracked rubber, replace the wiper blades. If inspection reveals nothing unusual, wet the windshield, turn the wipers on, allow them to cycle several times, then shut them off. An uneven wiper pattern across the glass or streaks over clean glass indicate that the blades should be replaced.
3 The operation of the wiper mechanism can loosen the fasteners, so they should be checked and tightened, as necessary, at the same time the wiper blades are checked (see Chapter 12 for further information regarding the wiper mechanism).

Wiper blade replacement

Refer to illustration 25.5

4 Pull the wiper/blade assembly away from the glass.
5 Press the retaining lever and slide the blade assembly down the wiper arm **(see illustration)**.
6 If the element must removed from the frame, detach the end of the element from the wiper frame, then slide the element out of the frame.
7 Compare the new element with the old for length, design, etc.
8 Slide the new element into place and insert the end in the wiper frame to lock it in place.
9 Reinstall the blade assembly on the arm, wet the glass and check for proper operation.

26 Air filter replacement (every 15,000 miles or 12 months)

Refer to illustrations 26.1 and 26.3

Note: *Information on carbureted models can be found in Chapter 4B.*

1 Pry the clamps off the air filter housing using a large screwdriver **(see illustration)**.
2 Separate the cover from the housing by pulling it back several inches.
3 Remove the air filter from the housing, noting the direction it faces **(see illustration)**.
4 Wipe the inside of the air cleaner housing with a clean cloth. If the element is marked TOP, be sure the marked side faces up.
5 Reinstall the cover and secure the clips.
6 Connect the air hose and tighten the clamp screw.

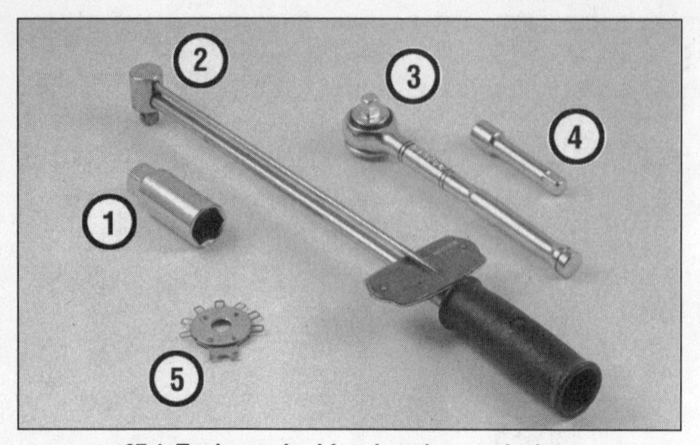

27.1 Tools required for changing spark plugs

1 ***Spark plug socket*** *- This will have special padding inside to protect the spark plug's porcelain insulator*
2 ***Torque wrench*** *- Although not mandatory, using this tool is the best way to ensure the plugs are tightened properly*
3 ***Ratchet*** *- Standard hand tool to fit the spark plug socket*
4 ***Extension*** *- Depending on model and accessories, you may need special extensions and universal joints to reach one or more of the plugs*
5 ***Spark plug gap gauge*** *- This gauge for checking the gap comes in a variety of styles. Make sure the gap for your engine is included*

27 Spark plug check and replacement (every 30,000 miles or 24 months)

Refer to illustrations 27.1, 27.4a, 27.4b, 27.5, 27.7 and 27.9

Note: *The spark plugs on all 1976 through 1980 models and Canadian models through 1984 must be replaced at 15,000 miles. On all other 240 models the spark plugs must be replaced at 30,000 miles.*

1 Before beginning, obtain the necessary tools, which will include a spark plug socket and a gap gauge **(see illustration)**.

2 The best procedure to follow when replacing the spark plugs is to purchase the new spark plugs beforehand, adjust them to the proper gap, and then replace each plug one at a time. When buying the new spark plugs it is important to obtain the correct plugs for your specific engine. This information can be found on the Vehicle Emissions Control Information label located under the hood, in the Specifications section in the front of this Chapter or in the owner's manual. If differences exist between these sources, purchase the spark plug type specified on the Emissions Control label, because the information was printed for your specific engine.

27.4a Spark plug manufacturers recommend using a wire-type gauge when checking the gap - if the wire does not slide between the electrodes with a slight drag, adjustment is required

3 With the new spark plugs at hand, allow the engine to cool completely before attempting plug removal. During this time, each of the new spark plugs can be inspected for defects and the gaps can be checked.

4 The gap is checked by inserting the proper thickness gauge between the electrodes at the tip of the plug **(see illustration)**. The gap between the electrodes should be the same as that given in the Specifications or on the Emissions Control label. The wire should just touch each of the electrodes. If the gap is incorrect, use the notched adjuster to bend the curved side of the electrode slightly until the proper gap is achieved **(see illustration)**. **Note:** *When adjusting the gap of a new plug, bend only the base of the ground electrode, do not touch the tip. If the side electrode is not exactly over the center electrode, use the notched adjuster to align the two. Check for cracks in the porcelain insulator, indicating the spark plug should not be used.*

5 With the engine cool, remove the spark plug wire from one spark plug. Do this by grabbing the boot at the end of the wire, not the wire itself. Sometimes it is necessary to use a twisting motion while the boot and plug wire are pulled free **(see illustration)**.

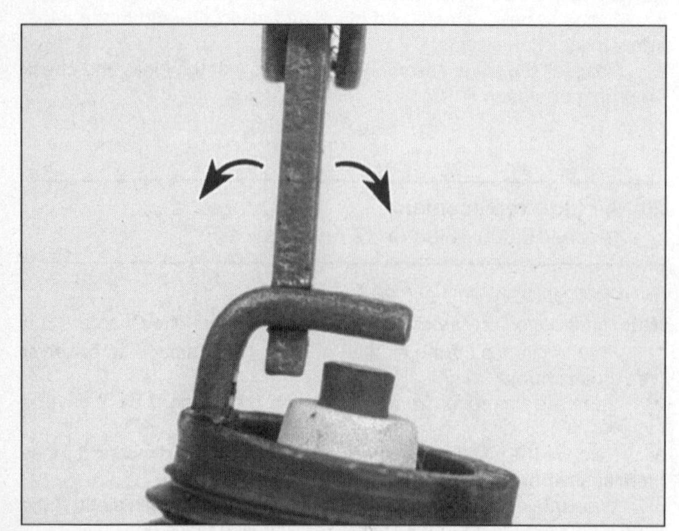

27.4b To change the gap, bend the side electrode only, as indicated by the arrows, and be very careful not to crack or chip the porcelain insulator surrounding the center electrode

27.5 When removing the spark plug wires, pull only on the boot and twist it back-and-forth

For a COLOR version of this spark plug diagnosis page, please see the inside rear cover of this manual

CARBON DEPOSITS

Symptoms: Dry sooty deposits indicate a rich mixture or weak ignition. Causes misfiring, hard starting and hesitation.

Recommendation: Check for a clogged air cleaner, high float level, sticky choke and worn ignition points. Use a spark plug with a longer core nose for greater anti-fouling protection.

OIL DEPOSITS

Symptoms: Oily coating caused by poor oil control. Oil is leaking past worn valve guides or piston rings into the combustion chamber. Causes hard starting, misfiring and hesition.

Recommendation: Correct the mechanical condition with necessary repairs and install new plugs.

TOO HOT

Symptoms: Blistered, white insulator, eroded electrode and absence of deposits. Results in shortened plug life.

Recommendation: Check for the correct plug heat range, over-advanced ignition timing, lean fuel mixture, intake manifold vacuum leaks and sticking valves. Check the coolant level and make sure the radiator is not clogged.

PREIGNITION

Symptoms: Melted electrodes. Insulators are white, but may be dirty due to misfiring or flying debris in the combustion chamber. Can lead to engine damage.

Recommendation: Check for the correct plug heat range, over-advanced ignition timing, lean fuel mixture, clogged cooling system and lack of lubrication.

HIGH SPEED GLAZING

Symptoms: Insulator has yellowish, glazed appearance. Indicates that combustion chamber temperatures have risen suddenly during hard acceleration. Normal deposits melt to form a conductive coating. Causes misfiring at high speeds.

Recommendation: Install new plugs. Consider using a colder plug if driving habits warrant.

GAP BRIDGING

Symptoms: Combustion deposits lodge between the electrodes. Heavy deposits accumulate and bridge the electrode gap. The plug ceases to fire, resulting in a dead cylinder.

Recommendation: Locate the faulty plug and remove the deposits from between the electrodes.

NORMAL

Symptoms: Brown to grayish-tan color and slight electrode wear. Correct heat range for engine and operating conditions.

Recommendation: When new spark plugs are installed, replace with plugs of the same heat range.

ASH DEPOSITS

Symptoms: Light brown deposits encrusted on the side or center electrodes or both. Derived from oil and/or fuel additives. Excessive amounts may mask the spark, causing misfiring and hesitation during acceleration.

Recommendation: If excessive deposits accumulate over a short time or low mileage, install new valve guide seals to prevent seepage of oil into the combustion chambers. Also try changing gasoline brands.

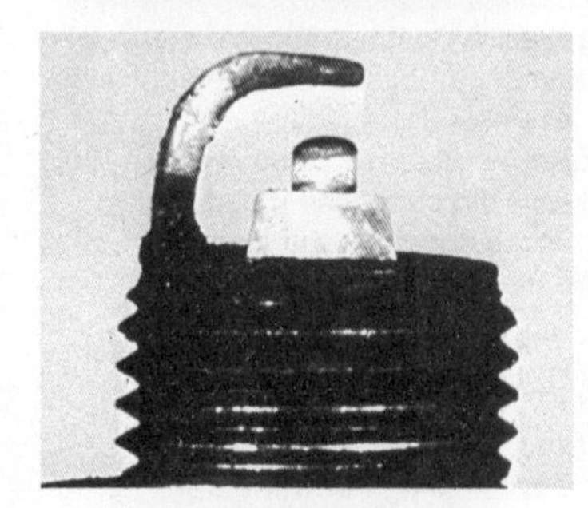

WORN

Symptoms: Rounded electrodes with a small amount of deposits on the firing end. Normal color. Causes hard starting in damp or cold weather and poor fuel economy.

Recommendation: Replace with new plugs of the same heat range.

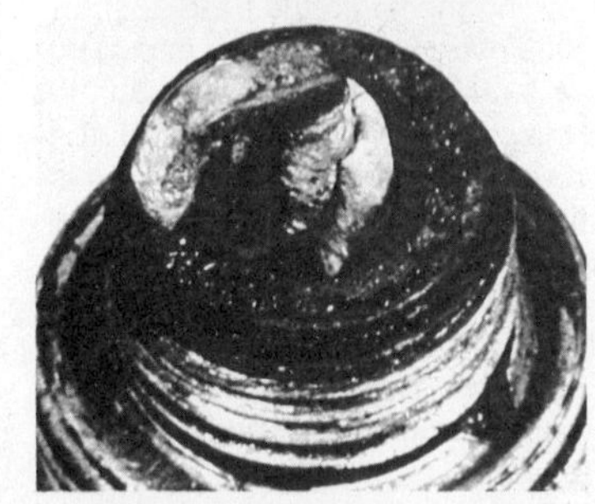

DETONATION

Symptoms: Insulators may be cracked or chipped. Improper gap setting techniques can also result in a fractured insulator tip. Can lead to piston damage.

Recommendation: Make sure the fuel anti-knock values meet engine requirements. Use care when setting the gaps on new plugs. Avoid lugging the engine.

SPLASHED DEPOSITS

Symptoms: After long periods of misfiring, deposits can loosen when normal combustion temperature is restored by an overdue tune-up. At high speeds, deposits flake off the piston and are thrown against the hot insulator, causing misfiring.

Recommendation: Replace the plugs with new ones or clean and reinstall the originals.

MECHANICAL DAMAGE

Symptoms: May be caused by a foreign object in the combustion chamber or the piston striking an incorrect reach (too long) plug. Causes a dead cylinder and could result in piston damage.

Recommendation: Remove the foreign object from the engine and/or install the correct reach plug.

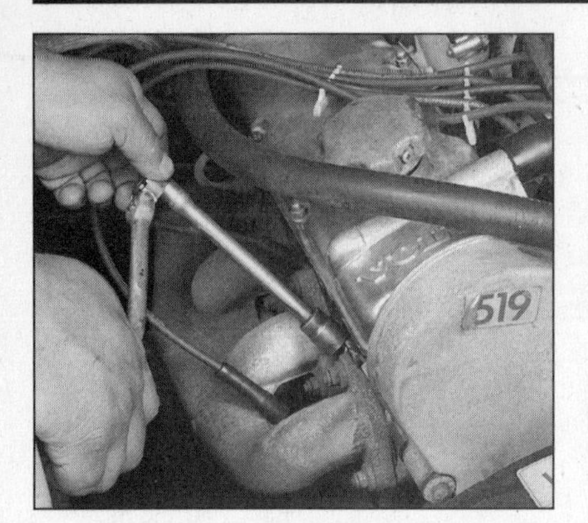

27.7 Use a socket wrench with an extension to unscrew the spark plugs

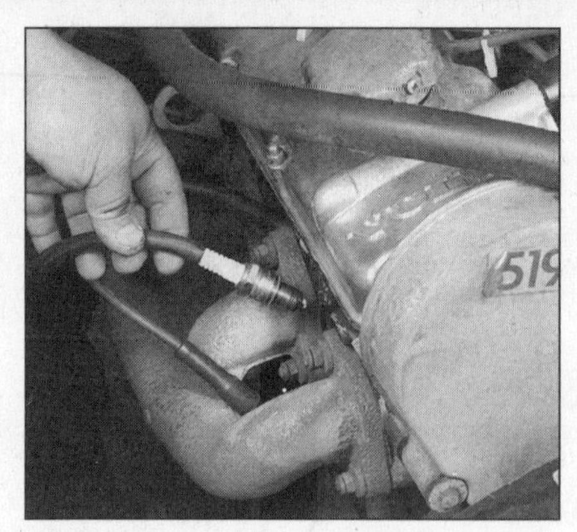

27.9 Use a length of 3/8-inch ID hose to start the spark plug into the plug hole

28.5 Measure the clearance for each valve with a feeler gauge of the specified thickness - if the clearance is correct, you should feel a slight drag on the gauge as you pull it out

6 If compressed air is available, use it to blow any dirt or foreign material away from the spark plug area. A common bicycle pump will also work. The idea here is to eliminate the possibility of debris falling into the cylinder as the spark plug is removed.

7 Place the spark plug socket over the plug and remove it from the engine by turning in a counterclockwise direction **(see illustration)**.

8 Compare the spark plug with those shown in the accompanying photos to get an indication of the overall running condition of the engine.

9 Apply a thin coat of anti-seize compound to the threads of the spark plugs. Install the plug into the head, turning it with your fingers until it no longer turns, then tighten it with the socket. Where there might be difficulty in inserting the spark plugs into the spark plug holes, or the possibility of cross threading them into the head, a short piece of 3/8-inch I.D. hose can be fitted over the end of the spark plug **(see illustration)**. The hose will act as a universal joint to help align the plug with the plug hole, and should the plug begin to cross thread, the hose will slip on the spark plug, preventing thread damage. If one is available, use a torque wrench to tighten the plug to ensure that it is seated correctly. The correct torque figure is included in this Chapter's Specifications.

10 Before pushing the spark plug wire onto the end of the plug, inspect it following the procedures outlined in Section 13.

11 Attach the plug wire to the new spark plug, again using a twisting motion on the boot until it is firmly seated on the spark plug.

12 Follow the above procedure for the remaining spark plugs, replacing them one at a time to prevent mixing up the spark plug wires.

28 Valve clearance check and adjustment (every 30,000 miles or 24 months)

Refer to illustrations 28.5, 28.6a, 28.6b, 28.8a and 28.8b

Note: *The following procedure requires the use of special valve adjusting tools. It is impossible to perform this task without them.*

1 Disconnect the negative cable from the battery. **Caution:** *Make sure the radio is turned off before the battery is disconnected or the microprocessor in the radio will be damaged.*

2 Blow out the recessed area above the valve cover with compressed air, if available, to remove any debris that might fall into the cylinders, then remove the spark plugs (see Section 27).

3 Remove the valve cover (refer to Chapter 2).

4 Refer to Chapter 2 and position the number 1 piston at TDC on the compression stroke.

5 Measure the clearances of the number 1 cylinder valves with a feeler gauge of the specified thickness **(see illustration)**. Record the measurements which are out of specification. They will be used later to determine the required replacement shims.

6 Turn the crankshaft 180-degrees (1/2-turn) and check the valves on number 3 cylinder **(see illustration)**. Rotate the engine again (180-degrees) and check the valves on number 4 cylinder, then rotate it again (180-degrees) and check the valves on number 2 cylinder **(see illustration)**. Write down on a piece of paper all the measurements and the particular valves (number 1 exhaust, number 3 intake etc.) which will need to be adjusted. Remember the firing order is 1-3-4-2 and if

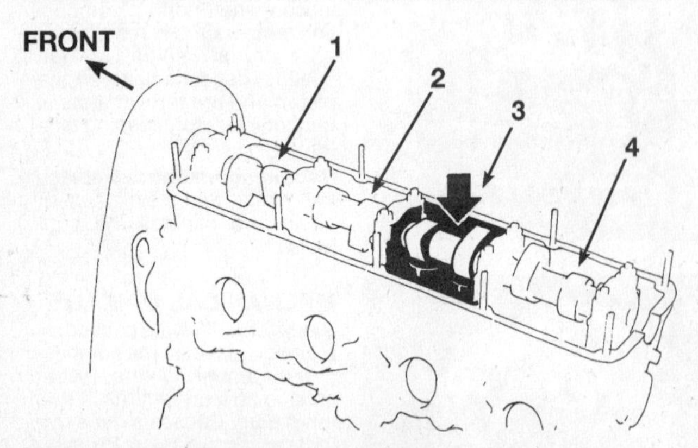

28.6a Rotate the crankshaft 180-degrees and check the valve clearance on cylinder number 3

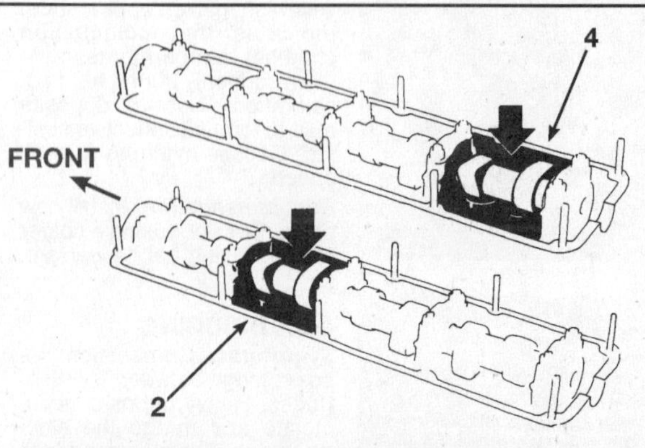

28.6b Rotate the crankshaft 180-degrees and check the valve clearance on cylinder number 4; then rotate the crankshaft another 180 degrees and check the valve clearance on cylinder number 2

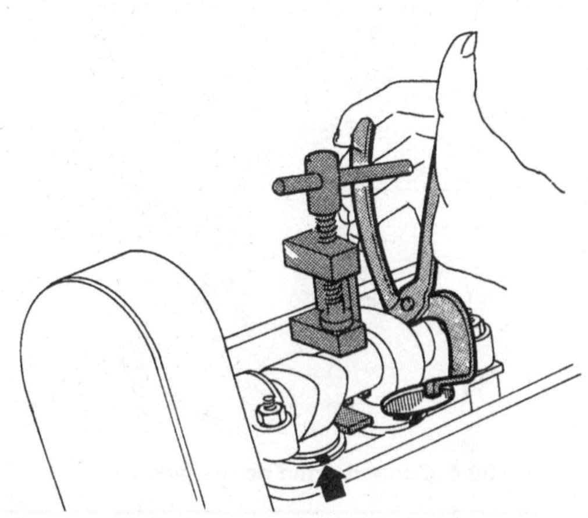

28.8 Use special tools to depress the cam follower into the cylinder head and remove the shim

you are in doubt to exactly which cylinder is selected, remove the distributor cap and follow the alignment of the rotor to the cap and determine the plug wire that is represented.

7 After all the valve clearances have been measured, turn the crankshaft pulley until the camshaft lobe above the first valve which you intend to adjust is pointing upward, away from the shim.

8 Position the notch in the cam follower toward the spark plug. Install the special tool and press down the cam follower **(see illustration)**. Place the special pliers in position as shown, and remove the adjusting shim.

9 Measure the thickness of the shim with a micrometer. To calculate the correct thickness of a replacement shim that will place the valve clearance within the specified value, use the following formula:

Engine hot: N = T + (A – 0.016-inch)
Engine cold: N = T + (A – 0.018-inch)
T = thickness of the old shim
A = valve clearance measured
N = thickness of the new shim

10 Select a shim with a thickness as close as possible to the calculated value. Shims are available in a thickness of 0.130-inch (3.300 mm) to 0.177-inch (4.500 mm) and in increments of 0.0020-inch (0.050 mm). **Note:** *Through careful analysis of the shim sizes needed to bring all the out-of-specification valve clearances within specification, it is often possible to simply move a shim that has to come out anyway to another cam follower requiring a shim of that particular size, thereby reducing the number of new shims that must be purchased.*

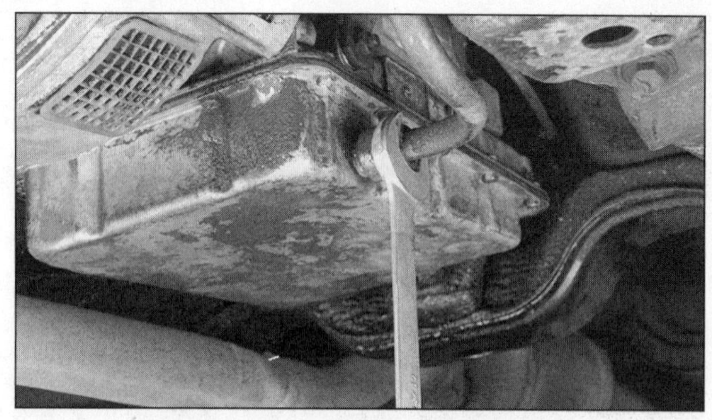

29.5a Unscrew the dipstick tube collar

11 Install the new adjusting shim with the markings downward, remove the cam follower press and measure the clearance with a feeler gauge to make sure that your calculations are correct.

12 Repeat this procedure until all the valves which are out of clearance have been corrected. Lubricate the new shims with clean engine oil, rotate the engine several revolutions and check the valve clearance again.

13 Installation of the spark plugs, valve cover, spark plug wires and boots, accelerator cable bracket, etc. is the reverse of removal.

29 Automatic transmission fluid and filter change (every 30,000 miles or 24 months)

Refer to illustrations 29.5a, 29.5b, 29.6, 29.7 and 29.10

1 At the specified time intervals, the transmission fluid should be drained and replaced. Since the fluid will remain hot long after driving, perform this procedure only after the engine has cooled down completely.

2 Before beginning work, purchase the specified transmission fluid (see *Recommended lubricants and fluids* at the beginning of this Chapter) and a new filter.

3 Other tools necessary for this job include jackstands to support the vehicle in a raised position, a drain pan capable of holding at least eight quarts, newspapers and clean rags.

4 Raise the vehicle and support it securely on jackstands.

5 Loosen the dipstick tube collar and let the fluid drain **(see illustrations)**.

6 Remove the pan mounting bolts and brackets **(see illustration)**.

7 Detach the pan from the transmission and lower it, being careful not to spill the remaining fluid **(see illustration)**.

29.5b Detach the tube and let the fluid drain

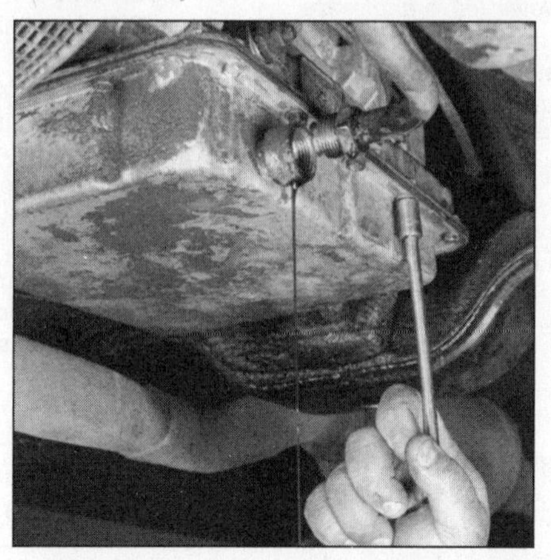

29.6 Use a socket and extension to remove the bolts and brackets

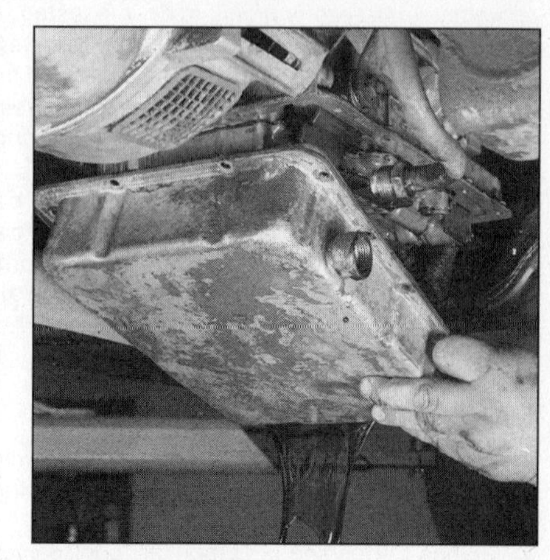

29.7 Lower the pan from the transmission

29.10 Remove the filter bolts and lower the fluid filter from the transmission

8 Carefully clean the gasket contact surface of the transmission.
9 Drain the fluid from the transmission pan, clean it with solvent and dry it with compressed air. Be sure to clean the metal filings from the magnet, if equipped.
10 Remove the filter from the mount inside the transmission **(see illustration)**.
11 Install the O-ring and a new filter, being sure to tighten the bolts securely.
12 Make sure the gasket surface on the transmission pan is clean, then install the gasket. Put the pan in place against the transmission and install the brackets and bolts. working around the pan, tighten each bolt a little at a time until the torque listed in this Chapter's Specifications is reached. Don't overtighten the bolts! Connect the dipstick tube and tighten the collar securely.
13 Lower the vehicle and add the specified amount of fluid through the filler tube (see Section 8).
14 With the transmission in Park and the parking brake set, run the engine at fast idle, but don't race it.
15 Move the gear selector through each range and back to Park. Check the fluid level.
16 Check under the vehicle for leaks after the first few trips.

30 Cooling system servicing (draining, flushing and refilling) (every 30,000 miles or 24 months)

Refer to illustration 30.4

Warning: *Do not allow antifreeze to come in contact with your skin or painted surfaces of the vehicle. Rinse off spills immediately with plenty of water. Antifreeze is highly toxic if ingested. Never leave antifreeze lying around in an open container or in puddles on the floor; children and pets are attracted by it's sweet smell and may drink it. Check with local authorities about disposing of used antifreeze. Many communities have collection centers which will see that antifreeze is disposed of safely.*

1 Periodically, the cooling system should be drained, flushed and refilled to replenish the antifreeze mixture and prevent formation of rust and corrosion, which can impair the performance of the cooling system and cause engine damage. When the cooling system is serviced, all hoses and the radiator cap should be checked and replaced if necessary.

Draining

2 Apply the parking brake and block the wheels. If the vehicle has just been driven, wait several hours to allow the engine to cool down before beginning this procedure.
3 Once the engine is completely cool, remove the expansion tank cap or radiator cap.
4 Move a large container under the radiator drain to catch the

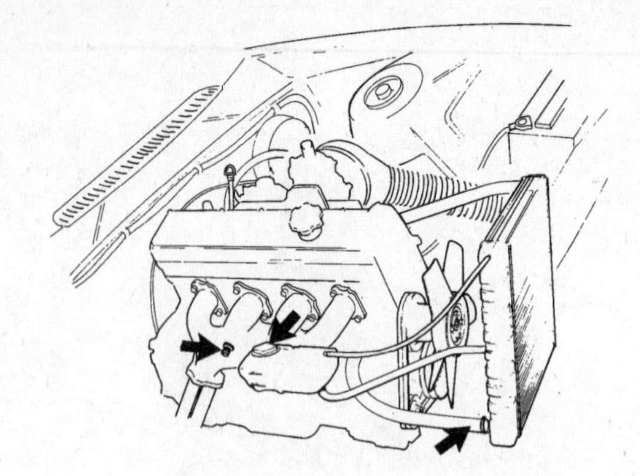

30.4 Coolant drain points (arrows)

coolant, then open the drain fitting if equipped (a pair of pliers or screwdriver may be required to turn it, depending on the model). If no drain fitting is present, loosen the hose clamp on the lower radiator hose and slide a screwdriver between the hose and the fitting **(see illustration)**. Also open the engine block drain plug.
5 While the coolant is draining, check the condition of the radiator hoses, heater hoses and clamps (see Section 21 if necessary).
6 Replace any damaged clamps or hoses (see Chapter 3 for detailed replacement procedures).

Flushing

7 Once the system is completely drained, flush the radiator with fresh water from a garden hose until the water runs clear at the drain. If it doesn't have a drain, tighten the lower hose clamp, fill the system with water, then drain it again. Do this as many times as necessary until the water comes out clean. The flushing action of the water will remove sediments from the radiator but will not remove rust and scale from the engine and cooling tube surfaces.
8 These deposits can be removed by a chemical cleaner. Follow the procedure outlined in the manufacturer's instructions. If the radiator is severely corroded, damaged or leaking, it should be removed (see Chapter 3) and taken to a radiator repair shop.
9 On models so equipped, remove the overflow hose from the coolant expansion tank. Drain the reservoir and flush it with clean water, then reconnect the hose.

Refilling

10 Close and tighten the radiator drain or reconnect the hose. Install and tighten the block drain plug(s). Place the heater temperature control in the maximum heat position.
11 Slowly add new coolant (a 50/50 mixture of water and antifreeze) to the expansion tank until it is full. Add coolant to the reservoir up to the lower mark.
12 Leave the cap off and run the engine in a well-ventilated area until the thermostat opens (coolant will begin flowing through the radiator and the upper radiator hose will become hot).
13 Turn the engine off and let it cool. Add more coolant mixture to bring the coolant level back up to the proper level in the expansion tank.
14 Squeeze the upper radiator hose to expel air, then add more coolant mixture if necessary. Replace the expansion tank cap.
15 Start the engine, allow it to reach normal operating temperature and check for leaks.

31 Manual transmission lubricant change (every 30,000 miles or 24 months)

Refer to illustrations 31.6 and 31.9

1 At the specified time intervals the transmission lubricant should

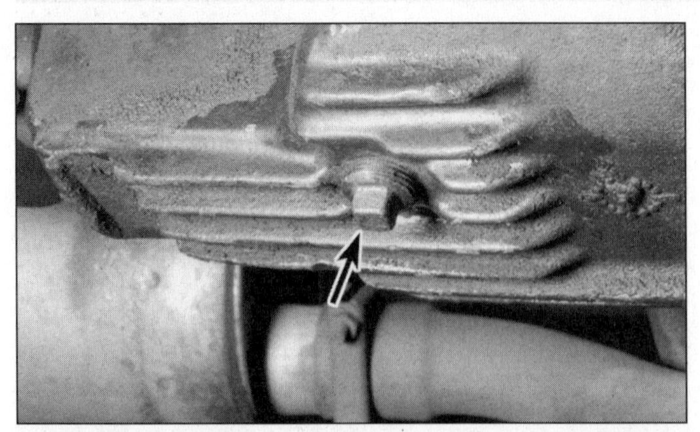

31.6 Use a wrench to remove the drain plug (arrow) from the bottom of the transmission

be changed to ensure trouble free operation. On overdrive-equipped models, the overdrive unit lubricant strainer should be removed and cleaned as part of the this procedure. Before proceeding, purchase the specified type of lubricant.

2 Tools necessary for this job include jackstands to support the vehicle in a raised position, a wrench to remove the drain plugs, a drain pan, newspapers and clean rags.

3 The lubricant should be drained immediately after the vehicle has been driven. This will remove any contaminants better than if the lubricant were cold. Because of this, it would be wise to wear rubber gloves while removing the drain plug.

4 After the vehicle has been driven to warm up the oil, raise it and place it on jackstands. Make sure it is safely supported and as level as possible.

5 Move the necessary equipment under the vehicle, being careful not to touch any of the hot exhaust components.

6 Place the drain pan under the transmission and remove the check/fill plug from the side of the transmission. Loosen the drain plug **(see illustration)**.

7 Carefully unscrew the plug. Be careful not to burn yourself on the lubricant.

8 Allow the lubricant to drain completely.

Overdrive-equipped models

9 Remove the bolts, detach the cover from the overdrive unit and lower it, being careful not to spill any remaining lubricant and remove the filter **(see illustration)**.

10 Carefully clean the contact surface of the overdrive housing.

11 Clean the cover with solvent and dry it with compressed air.

12 Wash the strainer thoroughly in solvent. Be sure to clean the metal filings from the magnet, if equipped.

13 Make sure the gasket surface on the cover is clean, then install the gasket. Put the cover in place against the overdrive unit and install the bolts. Working around the cover, tighten each bolt a little at a time until the torque listed in this Chapter's Specifications is reached. Don't overtighten the bolts!

All models

14 Clean the drain plug then reinstall and tighten it securely.

15 Refer to Section 14 and fill the transmission with new lubricant, then install the check/fill plug, tightening it securely.

32 Differential lubricant change (every 30,000 miles or 24 months)

1 Drive the vehicle for several miles to warm up the differential lubricant, then raise the vehicle and support it securely on jackstands.

2 Move a drain pan, rags, newspapers and an Allen wrench under the vehicle.

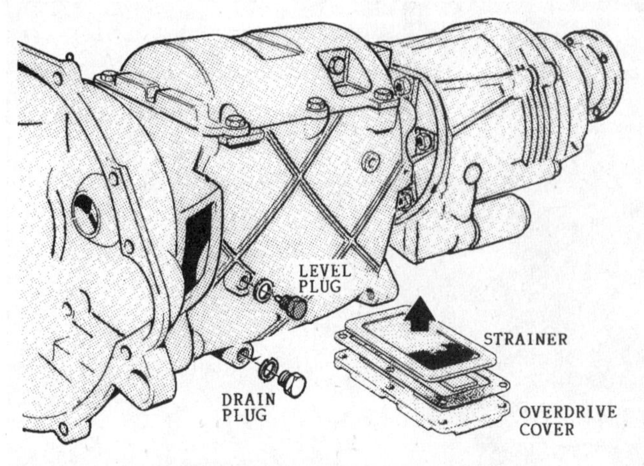

31.9 Overdrive unit strainer details

3 Remove the check/fill plug from the differential. It's the upper of the two plugs.

4 With the drain pan under the differential, use the tool to loosen the drain plug. It's the lower of the two plugs.

5 Once loosened, carefully unscrew it with your fingers until you can remove it from the case. Since the lubricant will be hot, wear a rubber glove to prevent burns.

6 Allow all of the oil to drain into the pan, then replace the drain plug and tighten it securely.

7 Refer to Section 15 and fill the differential with lubricant.

8 Reinstall the fill plug and tighten it securely.

9 Lower the vehicle. Check for leaks at the drain plug after the first few miles of driving.

33 Evaporative Emissions Control (EVAP) system check (every 30,000 miles or 24 months)

Refer to illustration 33.2

1 The function of the Evaporative Emissions Control system is to draw fuel vapors from the tank and fuel system, store them in a charcoal canister and then burn them during normal engine operation.

2 The most common symptom of a fault in the evaporative emissions system is a strong fuel odor in the engine compartment. If a fuel odor is detected, inspect the charcoal canister and system hoses for cracks. The canister is located in the front corner of the engine compartment on most models **(see illustration)**.

33.2 Inspect the hoses (arrows) that connect to the top of the evaporative emissions charcoal canister for damage

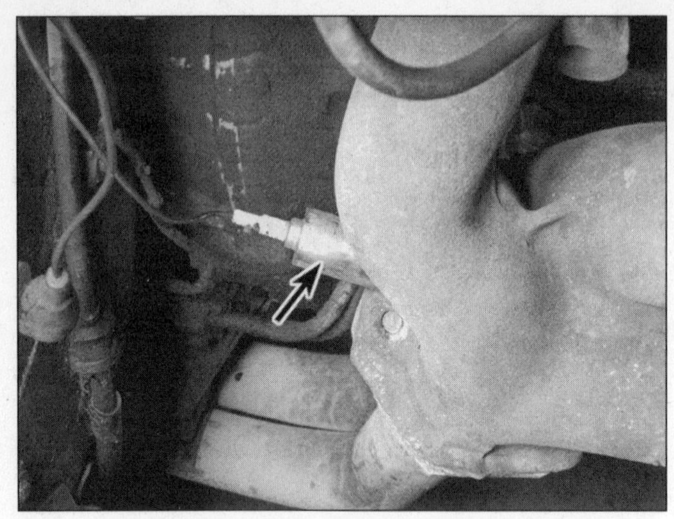

34.2 On most models, the oxygen sensor (arrow) is threaded into the exhaust pipe

3 Refer to Chapter 6 for more information on the evaporative emissions system.

34 Oxygen sensor replacement (every 30,000 or 50,000 miles, depending on model)

Refer to illustration 34.2

Note: *This procedure is best performed when the engine and exhaust system are completely cold.*

1 The oxygen sensor is located in the exhaust pipe. Locate the oxygen sensor and follow the wire back to the connector. Disconnect the oxygen sensor wire at the connector.

2 Use a wrench to remove the oxygen sensor from the exhaust pipe **(see illustration).**

3 Install the new oxygen sensor and tighten it securely.

4 Plug in the electrical connector.

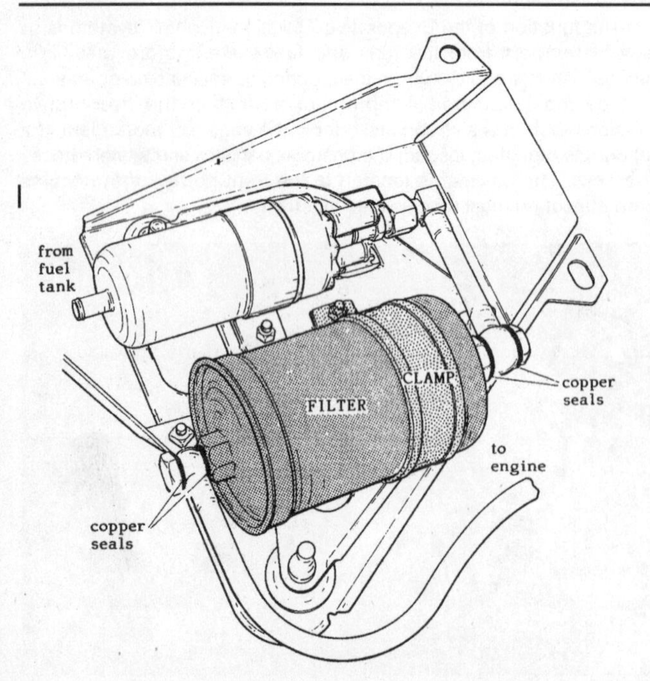

35.4b Under-vehicle mounted fuel filter details

35.4a Use two wrenches to remove the banjo bolts, then disconnect the hoses and detach the filter from the bracket

35 Fuel filter replacement (every 30,000 miles or 24 months)

Refer to illustrations 35.4a and 35.4b

Warning: *Gasoline is extremely flammable, so take extra precautions when you work on any part of the fuel system. Don't smoke or allow open flames or bare light bulbs near the work area, and don't work in a garage where a natural gas-type appliance (such as a water heater or a clothes dryer) with a pilot light is present. Since gasoline is carcinogenic, wear latex gloves when there's a possibility of being exposed to fuel, and, if you spill any fuel on your skin, rinse it off immediately with soap and water. Mop up any spills immediately and do not store fuel-soaked rags where they they could ignite. When you perform any type of work on the fuel system, wear safety glasses and have a class B type fire extinguisher on hand.*

Note: *Information on carbureted model fuel filters can be found in Chapter 4B.*

1 On early fuel-injected models the fuel filter is located on the engine compartment firewall while on later models it is found under the left rear side of the vehicle. On later models it will be necessary to raise the rear of the vehicle and support it securely on jackstands for access to the filter.

2 Depressurize the fuel system (Chapter 4).

3 Disconnect the negative battery cable. **Caution:** *Make sure the radio is turned off before the battery is disconnected or the microprocessor in the radio will be damaged.*

4 Using a backup wrench to steady the filter, remove the threaded banjo bolts at both ends of the fuel filter **(see illustrations).**

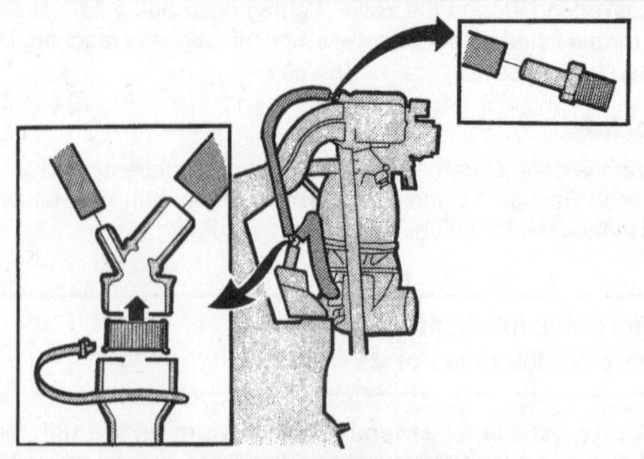

36.1 PCV system details

36.2 Use a straightened paper clip to clean the orifice opening

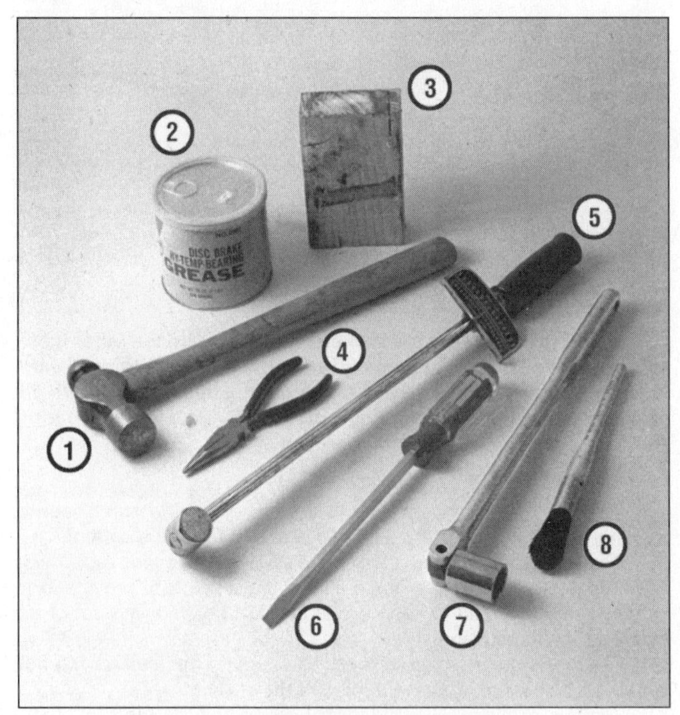

37.1 Tools and materials needed for front wheel bearing maintenance

1 ***Hammer*** *- A common hammer will do just fine*
2 ***Grease*** *- High-temperature grease that is formulated for front wheel bearings should be used*
3 ***Wood block*** *- If you have a scrap piece of 2x4, it can be used to drive the new seal into the hub*
4 ***Needle-nose pliers*** *- Used to straighten and remove the cotter pin in the spindle*
5 ***Torque wrench*** *- This is very important in this procedure; if the bearing is too tight, the wheel won't turn freely - if it's too loose, the wheel will "wobble" on the spindle. Either way, it could mean extensive damage*
6 ***Screwdriver*** *- Used to remove the seal from the hub (a long screwdriver is preferred)*
7 ***Socket/breaker bar*** *- Needed to loosen the nut on the spindle if it's extremely tight*
8 ***Brush*** *- Together with some clean solvent, this will be used to remove old grease from the hub and spindle*

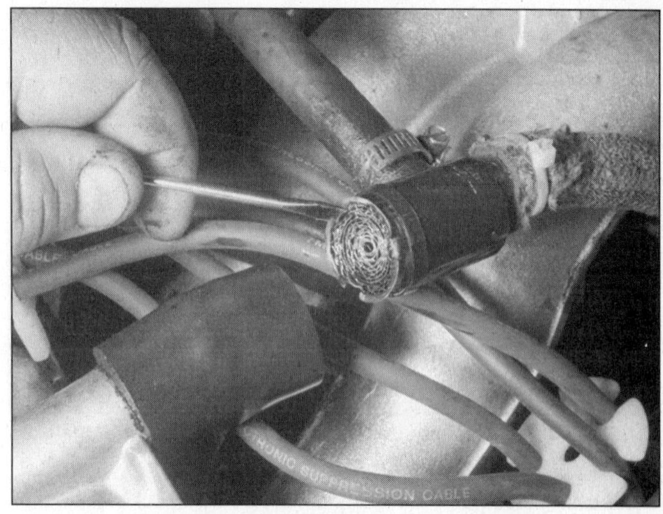

36.3 Use a small screwdriver to pry the flame arrestor out of the end of the hose

5 Remove both bracket bolts/nuts and remove the old filter and the filter support bracket assembly.
6 Make sure the new filter is installed so that it's facing the proper direction **(refer to illustrations 35.4a and 35.4b).**
7 Using the new sealing washers provided by the filter manufacturer, install the inlet and outlet banjo fittings and tighten them securely.
8 The remainder of installation is the reverse of the removal procedure.

36 Positive Crankcase Ventilation (PCV) system check (every 60,000 miles or 48 months)

Refer to illustrations 36.1, 36.2 and 36.3

Note: *The flame trap can easily become plugged and cause crankcase pressure build-up within the engine block. This excessive pressure will cause oil leaks from the seals (camshaft, intermediate shaft, rear main seal) and gaskets. Refer to Chapter 6 for additional information concerning the PCV system and the locations of the flame traps for the different models.*

1 The PCV system hoses, valve cover orifice and flame arrestor should be periodically checked for deterioration and clogging and the flame arrestor should be removed and cleaned (**see illustration**).
2 Disconnect the hose and use a straightened paper clip to clean out the valve cover orifice **(see illustration).**
3 Disconnect the hose from the oil separator and remove the flame arrestor from the hose **(see illustration)**. Wash the flame arrestor in clean solvent, then reinstall it and connect the hose.
4 Replace any plugged or deteriorated hoses.

37 Front wheel bearing check, repack and adjustment (every 60,000 miles or 48 months)

Refer to illustrations 37.1, 37.5, 37.6, 37.15, 37.16, 37.17, 37.19, 37.20 and 37.22

1 In most cases the front wheel bearings will not need servicing until the brake pads are changed. However, the bearings should be checked whenever the front of the vehicle is raised for any reason. Several items, including a torque wrench and special grease, are required for this procedure **(see illustration)**.
2 With the vehicle securely supported on jackstands, spin each wheel and check for noise, rolling resistance and freeplay.
3 Grasp the top of each tire with one hand and the bottom with the other. Move the wheel in-and-out on the spindle. If there's any

37.5 Use a piece of wire to hang the brake caliper out of the way

37.6 Use a screwdriver or large pair of pliers to pry the grease cap off

noticeable movement, the bearings should be checked and then repacked with grease or replaced if necessary.

4 Remove the wheel.

5 Remove the brake caliper (see Chapter 9) and hang it out of the way on a piece of wire **(see illustration)**. Remove the bolts and detach the brake disc.

6 Pry the dust cap off the hub using pliers, a screwdriver or hammer and chisel **(see illustration)**.

7 Straighten the bent ends of the cotter pin, then pull the cotter pin out of the nut. Discard the cotter pin and use a new one during reassembly.

8 Remove the nut and washer from the end of the spindle.

9 Pull the hub/disc assembly out slightly, then push it back into its original position. This should force the outer bearing off the spindle enough so it can be removed.

10 Pull the hub/disc assembly off the spindle.

11 Use a screwdriver to pry the seal out of the rear of the hub. As this is done, note how the seal is installed.

12 Remove the inner wheel bearing from the hub.

13 Use solvent to remove all traces of the old grease from the bearings, hub and spindle. A small brush may prove helpful; however make sure no bristles from the brush embed themselves inside the bearing rollers. Allow the parts to air dry.

14 Carefully inspect the bearings for cracks, heat discoloration, worn rollers, etc. Check the bearing races inside the hub for wear and damage. If the bearing races are defective, the hubs should be taken to a machine shop with the facilities to remove the old races and press new ones in. Note that the bearings and races come as matched sets and old bearings should never be installed on new races.

15 Use high-temperature front wheel bearing grease to pack the bearings. Work the grease completely into the bearings, forcing it between the rollers, cone and cage from the back side **(see illustration)**.

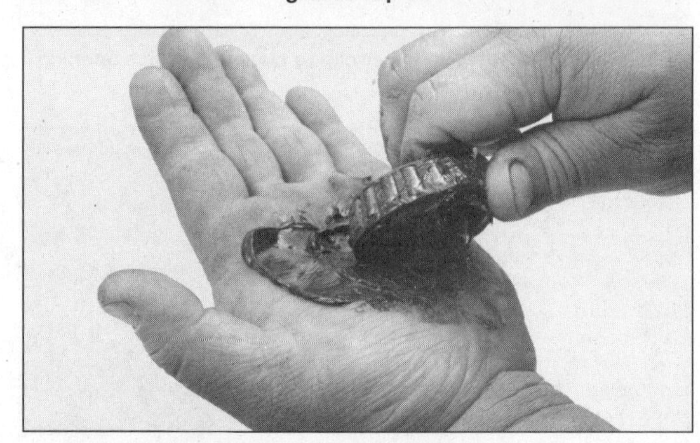

37.15 Put grease in the palm of one hand and use the other hand to force the large-diameter-end edge of each bearing onto the rollers (you should see grease come out the other end) - work all the way around the bearing edge, completely packing the bearing with grease

16 Apply a thin coat of grease to the spindle at the outer bearing seat, inner bearing seat, shoulder and seal seat **(see illustration)**.

17 Put a small quantity of grease inboard of each bearing race inside the hub. Using your finger, form a dam at these points to provide extra grease availability and to keep thinned grease from flowing out of the bearing **(see illustration)**.

18 Place the grease-packed inner bearing into the back of the hub and put a little more grease outboard of the bearing.

19 Place a new seal over the inner bearing and tap the seal evenly into place with a hammer until it's flush with the hub **(see illustration)**.

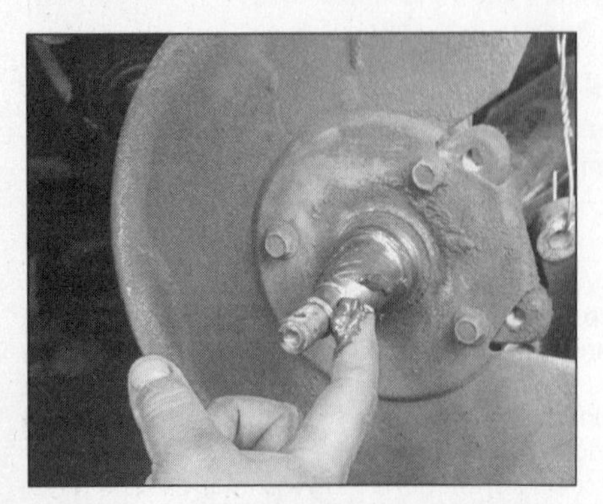

37.16 Apply grease to the spindle contact surface

37.17 Lubricate the inner bearing outer race

37.19 Use a hammer to seat the grease seal - tap all around the circumference, seating it evenly

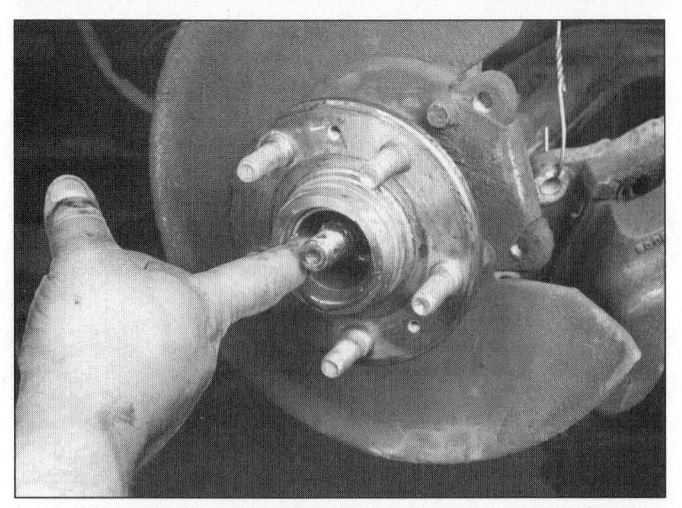

37.20 Push the hub into place on the spindle and grease the outer bearing race

37.22 Tighten the nut while spinning the hub to seat the bearing

20 Carefully place the hub assembly onto the spindle and push the grease-packed outer bearing into position **(see illustration)**.
21 Install the washer and spindle nut. Tighten the nut only slightly (no more than 12 ft-lbs of torque).
22 Spin the hub in a forward direction while tightening the spindle nut to approximately 20 ft-lbs to seat the bearings and remove any grease or burrs which could cause excessive bearing play later **(see illustration)**.
23 Loosen the spindle nut 1/4-turn, then using your hand (not a wrench of any kind), tighten the nut until it's snug. Install a new cotter pin through the hole in the spindle and the slots in the nut. If the slots don't line up, remove the nut and turn it slightly until they do.
24 Bend the ends of the cotter pin until they're flat against the nut. Cut off any extra length which could interfere with the dust cap.
25 Install the dust cap, taping it into place with a hammer.
26 Install the brake disc and install the caliper (see Chapter 9).
27 Install the wheel on the hub and tighten the lug nuts.
28 Grasp the top and bottom of the tire and check the bearings in the manner described earlier in this Section.
29 Lower the vehicle.

38 Service light resetting

Refer to illustrations 38.1a, 38.1b, 38.4 and 38.5

Service interval reminder light

1 Later models are equipped with a service indicator light on the dash which automatically illuminates at 5,000 mile intervals, reminding the driver an oil and filter change is needed. The light will come on for two minutes, each time the engine is started, until the service is performed and the light is reset. The light reset switch is located in the instrument cluster **(see illustrations)**.

Oxygen sensor service light

Note: *Early model CIS engines are equipped with an EGR reminder light that automatically illuminates after 15,000 miles. Follow the same reset procedure for the oxygen sensor reminder light.*

2 Certain models are equipped with an oxygen sensor service light which will illuminate every 30,000 miles as a reminder to replace the

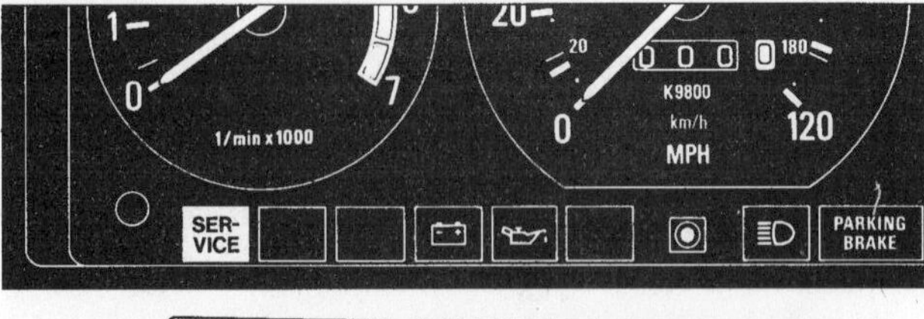

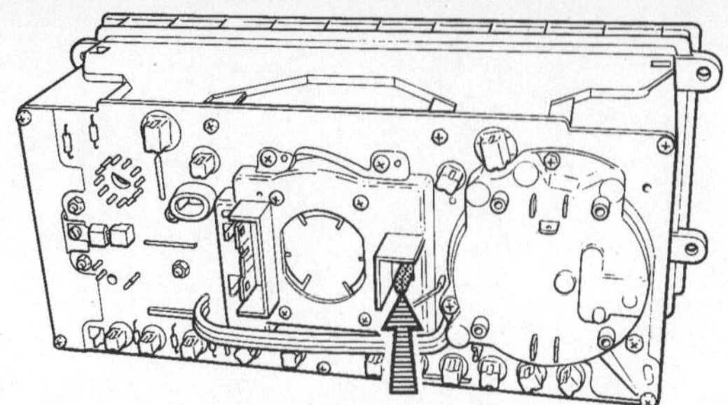

38.1a 1985 through 1992 models are equipped with an engine oil change light that comes on every 5,000 miles. Locate the reset button (lever) behind the instrument cluster and reach up into the dash area behind the cluster and push up on the lever

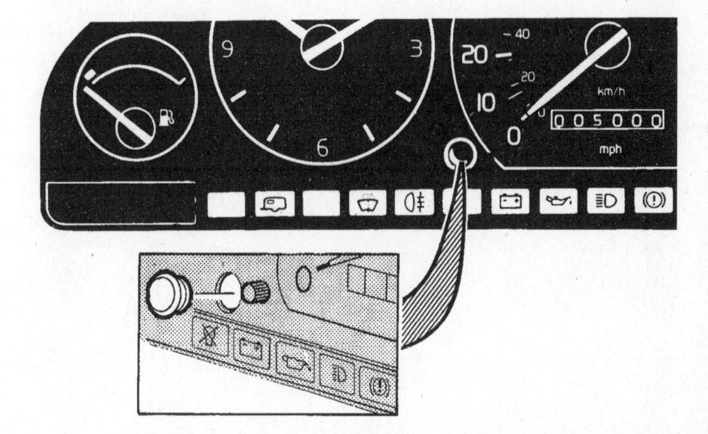

38.1b 1993 models are equipped with a button that is located behind a rubber grommet. Remove the grommet and depress the button using a small screwdriver tip

38.4 Detach the timer from the wiring harness and remove the cover

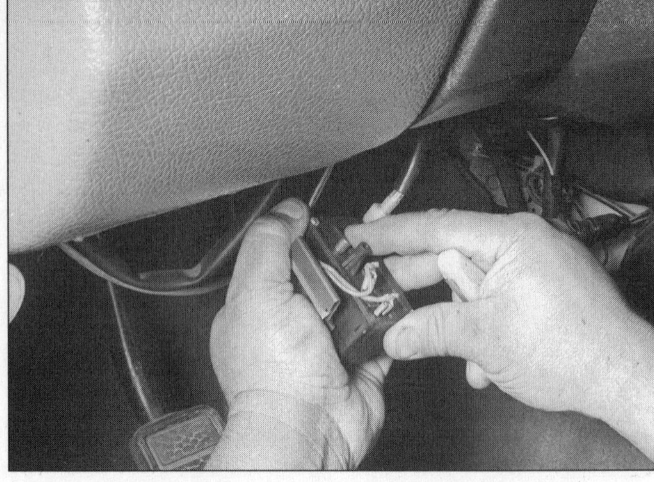

38.5 Press the button to reset the timer

oxygen sensor.

3 The service light timer is located under the dash above the throttle pedal.

4 Reach up under the dash, unclip the timer and remove the cover **(see illustration)**.

5 Press the reset button, install the cover and reinstall the timer under the dash **(see illustration)**.

Chapter 2 Part A Engine

Contents

Specifications

General

Firing order	1-3-4-2
Cylinder numbers (front-to-rear)	1-2-3-4
Displacement	
B21F and B21FT	129.7 cubic inches (2.127 liters)
B23F and B230F	141.2 cubic inches (2.316 liters)
Intake/exhaust manifold warpage limit	0.006 inch

Cylinder head

Maximum warpage	
Lengthwise	0.040 inches
Crosswise	0.020 inches
Cylinder head height	
B21F, B21FT and B23F	
New	5.760 inches
Minimum after machining	5.740 inches
B230F	
New	5.752 inches
Minimum after machining	5.732 inches

Cylinder location and distributor rotation diagram

Camshaft

Number of bearings	5
Endplay	
Standard	0.004 to 0.016 inch
Maximum	Not available

Camshaft (continued)

Runout	0.0006 inch (maximum)
Camshaft journal diameter	
B21F and B21FT	1.1820 to 1.1828 inches
B23F and B230F	1.1791 to 1.1799 inches
Journal oil clearance	
Standard	0.0012 to 0.0028 inch
Maximum	0.006 inch

Oil pump

Oil pump gear clearances	
Backlash	0.006 to 0.014 inches
Axial play	0.0008 to 0.0047 inches
Radial play	0.0008 to 0.0035 inches

Torque specifications

	Ft-lbs (unless otherwise indicated)
Camshaft bearing cap bolts	15
Valve cover bolts	108 in-lbs
Intermediate shaft sprocket bolt	37
Timing belt tensioner nut	37
Camshaft sprocket bolt	37
Crankshaft pulley bolt	
First step	44
Second step	Rotate an additional 60-degrees
Cylinder head bolts	
Early version bolts	
First step	43
Second step	80
Third step	Warm up engine to normal operating temperature, then turn the engine OFF and allow it to cool completely
Fourth step	Turn head bolt number 1 out (counterclockwise) 30-degrees and retorque number 1 bolt to 80 ft. lbs.
Fifth step	Retorque remainder of the cylinder head bolts to 80 ft.lbs.
Late version bolts	
First step	15
Second step	45
Third step	Additional 1/4-turn (90-degrees)
Flywheel-to-crankshaft bolts	52
Driveplate-to-crankshaft bolts	52
Intake manifold bolts	16
Exhaust manifold bolts	23
Oil pan-to-engine bolts	96 in-lbs
Oil pump pick-up tube bolts	96 in-lbs
Oil pump housing bolts	96 in-lbs
Rear main oil seal housing bolts	120 in-lbs
Timing belt cover bolts	
M6	120 in-lbs
M8	14
Water pump bolts	See Chapter 3

7 3 1 5 9
8 4 2 6 10

Cylinder head bolt tightening sequence

Early type

Late type

Cylinder head bolts

1 General information

Refer to illustration 1.4

This Part of Chapter 2 is devoted to in-vehicle repair procedures for the B21F, B23F and B230F engines. All information concerning engine removal and installation and engine block and cylinder head overhaul can be found in Part B of this Chapter.

The following repair procedures are based on the assumption that the engine is installed in the vehicle. If the engine has been removed from the vehicle and mounted on a stand, many of the steps outlined in this Part of Chapter 2 will not apply.

The Specifications included in this Part of Chapter 2 apply only to the procedures contained in this chapter. Chapter 2B contains the Specifications necessary for cylinder head and engine block rebuilding.

The Volvo 240 models covered in this manual are equipped with either a B21F, B23F or B230F engine (fuel-injected). Some Canadian models (1978 through 1984) are equipped with the B21A engine (carbureted) **(see illustration)**. The engine block is similar throughout all the models with slight variations in design. Here is a brief description of differences:

a) *The B230 crankshaft has eight counterweights while the B21 and B23 engines have four counterweights.*
b) *The B230 crankshaft has smaller bearing surfaces to reduce friction.*

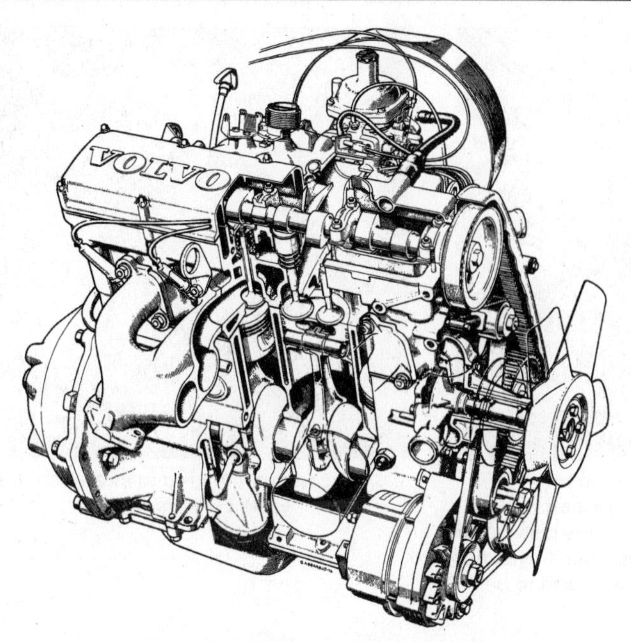

1.4 Cutaway view of the B21A Canadian (carbureted) engine

c) *The B230 pistons are lighter and have thinner rings. Also, the piston pins are located higher up in the piston to accommodate the longer connecting rods.*
d) *The B21 and B23 engines are non-interference designed. These engines are designed to not allow any valve/piston damage in the event the timing belt breaks or if the timing belt is installed improperly. The B230 engine **is** an interference design. Take special caution when performing any type of work on the timing belt.*
e) *The B21F-T and B230F engines are equipped with Stellite coating on the exhaust valves. These valve surfaces are much harder and wear less over a longer period of time. Do not grind these valves on any type of machinery but instead use valve lapping compound. B21F-T engines also use sodium filled exhaust valves that stay cooler at high engine temperatures. These type of valves must not be disposed with other metal because of their explosive properties.*

Note: *The specifications listed for the B21F are the same as the B21A engine for Canada*

These four cylinder engines are equipped with a single overhead camshaft that controls two valves per cylinder (total 8-valves). The engine is also equipped with an aluminum cylinder head. An intermediate shaft turns at half crankshaft speed and controls the distributor and oil pump.

The pistons have two compression rings and one oil control ring. The semi-floating piston pins are press fitted into the small end of the connecting rod. The connecting rod big ends are also equipped with renewable insert-type plain bearings.

The engine is liquid-cooled, utilizing a centrifugal impeller-type pump, driven by a belt, to circulate coolant around the cylinders and combustion chambers and through the intake manifold.

Lubrication is handled by a gear-type oil pump mounted on the bottom of the engine. It is driven by a shaft that is gear driven by the intermediate shaft. The oil is filtered continuously by a cartridge-type filter mounted on the right side of the engine.

2 Repair operations possible with the engine in the vehicle

Clean the engine compartment and the exterior of the engine with some type of degreaser before any work is done. It will make the job easier and help keep dirt out of the internal areas of the engine.

Depending on the components involved, it may be helpful to remove the hood to improve access to the engine as repairs are performed (refer to Chapter 11 if necessary). Cover the fenders to prevent damage to the paint. Special pads are available, but an old bedspread or blanket will also work.

If vacuum, exhaust, oil or coolant leaks develop, indicating a need for gasket or seal replacement, the repairs can generally be made with the engine in the vehicle. The intake and exhaust manifold gaskets, oil pan gasket, crankshaft oil seals and cylinder head gasket are all accessible with the engine in place.

Exterior engine components, such as the intake and exhaust manifolds, the oil pan, the water pump, the starter motor, the alternator, the distributor and the fuel system components can be removed for repair with the engine in place.

Since the cylinder head can be removed without pulling the engine, camshaft and valve component servicing can also be accomplished with the engine in the vehicle. Replacement of the timing chain and sprockets is also possible with the engine in the vehicle.

In extreme cases caused by a lack of necessary equipment, repair or replacement of piston rings, pistons, connecting rods and rod bearings is possible with the engine in the vehicle. However, this practice is not recommended because of the cleaning and preparation work that must be done to the components involved.

3 Top Dead Center (TDC) for number one piston - locating

Refer to illustrations 3.7 and 3.8

Note: *The following procedure is based on the assumption that the spark plug wires and distributor are correctly installed. If you are trying to locate TDC to install the distributor correctly, piston position must be determined by feeling for compression at the number one spark plug hole, then aligning the ignition timing marks as described in Step 8.*

1 Top Dead Center (TDC) is the highest point in the cylinder that each piston reaches as it travels up-and-down when the crankshaft turns. Each piston reaches TDC on the compression stroke and again on the exhaust stroke, but TDC generally refers to piston position on the compression stroke.

2 Positioning the piston(s) at TDC is an essential part of many procedures such as camshaft and timing belt/sprocket removal and distributor removal.

3 Before beginning this procedure, be sure to place the transaxle in Neutral and apply the parking brake or block the rear wheels. Also, disable the ignition system by detaching the primary (low voltage) electrical connectors from the ignition coil. Remove the spark plugs (see Chapter 1).

4 In order to bring any piston to TDC, the crankshaft must be turned using one of the methods outlined below. When looking at the front of the engine, normal crankshaft rotation is clockwise.

a) *The preferred method is to turn the crankshaft with a socket and ratchet attached to the bolt threaded into the front of the crankshaft.*
b) *A remote starter switch, which may save some time, can also be used. Follow the instructions included with the switch. Once the piston is close to TDC, use a socket and ratchet as described in the previous paragraph.*
c) *If an assistant is available to turn the ignition switch to the Start position in short bursts, you can get the piston close to TDC without a remote starter switch. Make sure your assistant is out of the vehicle, away from the ignition switch, then use a socket and ratchet as described in Paragraph a) to complete the procedure.*

5 Note the position of the terminal for the number one spark plug wire on the distributor cap. If the terminal isn't marked, follow the plug wire from the number one cylinder spark plug to the cap.

6 Detach the cap from the distributor and set it aside (see Chapter 1 if necessary).

3.7 Mark the distributor housing directly beneath the number one spark plug wire terminal (double check the distributor cap to verify that the rotor points to the number 1 spark plug wire)

3.8 Align the mark in the flywheel/driveplate with the notch in the pointer then check to see if the distributor rotor is pointing to the number 1 cylinder (if not, the crankshaft will have to be rotated 360-degrees)

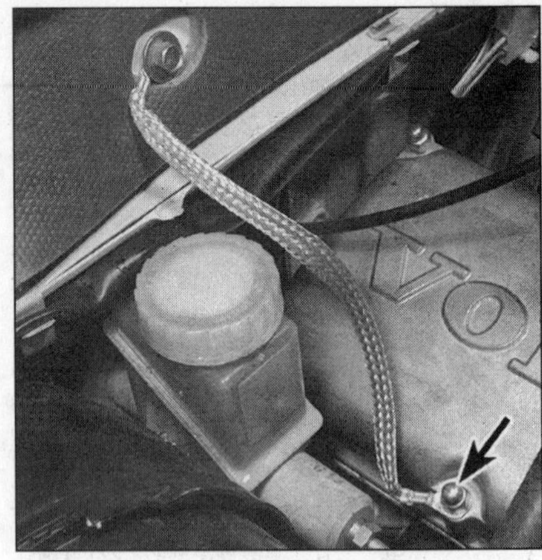

4.3 Disconnect the ground strap from the valve cover (arrow)

7 Mark the distributor body directly under the rotor terminal **(see illustration)** for the number 1 cylinder.

8 Locate the timing marks on the timing cover. You'll see the timing increments cast into the timing cover near the crankshaft pulley. Turn the crankshaft (see Paragraph 3 above) until the TDC mark (zero) on the timing cover is aligned with the groove in the pulley **(see illustration)**.

9 Look at the distributor rotor - it should be pointing directly at the mark you made on the distributor body (cover). If the rotor is pointing at the mark, go to Step 12. If it isn't, go to Step 10.

10 If the rotor is 180-degrees off, the number one piston is at TDC on the exhaust stroke.

11 To get the piston to TDC on the compression stroke, turn the crankshaft one complete turn (360-degrees) clockwise. The rotor should now be pointing at the mark on the distributor. When the rotor is pointing at the number one spark plug wire terminal in the distributor cap and the ignition timing marks are aligned, the number one piston is at TDC on the compression stroke.

12 After the number one piston has been positioned at TDC on the compression stroke, TDC for any of the remaining pistons can be located by turning the crankshaft and following the firing order. Mark the remaining spark plug wire terminal locations on the distributor body just like you did for the number one terminal, then number the marks to correspond with the cylinder numbers. As you turn the crankshaft, the rotor will also turn. When it's pointing directly at one of the marks on the distributor, the piston for that particular cylinder is at TDC on the compression stroke.

4 Valve cover - removal and installation

Removal

Refer to illustration 4.3

1 Detach the cable from the negative battery terminal. **Caution:** *It is necessary to make sure the radio is turned OFF before disconnecting the battery cables to avoid damaging the microprocessor in the radio.*

2 Remove the distributor cap and wires from their cylinder head and valve cover connections (see Chapter 1). Be sure to mark each wire for correct installation.

3 Mark and detach any hoses or ground wires from the throttle body or valve cover area that will interfere with the removal of the valve cover **(see illustration)**.

4 Disconnect the vacuum hose from the PCV valve.

5 Wipe off the valve cover thoroughly to prevent debris from falling onto the exposed cylinder head or camshaft/valve train assembly.

6 Remove the valve cover nuts.

7 Carefully lift off the valve cover and gasket. Do not pry between the cover and cylinder head or you'll damage the gasket mating surfaces. If the gasket is stuck to the cylinder head, scrape it with a gasket tool. Be careful not to remove any metal from the aluminum cylinder head while in the process of removing the gasket material.

Installation

Refer to illustrations 4.10a and 4.10b

8 Use a gasket scraper to remove any traces of old silicon or old gasket material from the gasket mating surfaces of the cylinder head and the valve cover. Clean the surfaces with a rag soaked in lacquer thinner or acetone.

9 Apply beads of RTV sealant to the corners where the cylinder head mates with the rocker arm assembly. Wait five minutes or so and let the RTV "set-up" (slightly harden).

10 Install a new gasket onto the cylinder head **(see illustration)**. Install the molded rubber grommet (half-moon shape) by pushing it into the "half-moon" cutout on the rear of the cylinder head **(see illustration).** Install the valve cover and nuts and tighten them to the torque listed in this Chapter's Specifications. **Note:** *Make sure the RTV sealant has slightly hardened before installing the valve cover. If the weather is damp and cold, the sealant will take some extra time to harden.*

11 The remainder of installation is the reverse of removal.

5 Intake manifold - removal and installation

Warning: *Gasoline is extremely flammable, so take extra precautions when you work on any part of the fuel system. Don't smoke or allow open flames or bare light bulbs near the work area, and don't work in a garage where a natural gas-type appliance (such as a water heater or a clothes dryer) with a pilot light is present. Since gasoline is carcinogenic, wear latex gloves when there's a possibility of being exposed to fuel, and, if you spill any fuel on your skin, rinse it off immediately with soap and water. Mop up any spills immediately and do not store fuel-soaked rags where they could ignite. The fuel system is under constant pressure, so, if any fuel lines are to be disconnected, the fuel pressure in the system must be relieved first* (see Chapter 4 for more information). *When you perform any kind of work on the fuel system, wear safety glasses and have a Class B type fire extinguisher on hand.*

4.10a Install the valve cover gasket on the head and dab a small amount of RTV sealant into the corners where the cam bearing cap meets the cylinder head

4.10b Install a new rubber grommet into the cutout section at the rear of the cylinder head

6.3a Remove the exhaust manifold bolts (be sure to note the location of the engine lifting eye) . . .

Removal

1 Detach the cable from the negative battery terminal. **Caution:** *It is necessary to make sure the radio is turned OFF before disconnecting the battery cables to avoid damaging the microprocessor in the radio.*
2 Drain the cooling system (see Chapter 1).
3 Remove the intake air ducts from the air cleaner assembly (see Chapter 4).
4 Clearly label and detach any vacuum lines and electrical connectors which will interfere with the removal of the manifold.
5 Detach the accelerator cable from the throttle lever (see Chapter 4).
6 Remove the coolant hoses from the intake manifold.
7 Disconnect the fuel feed and (on fuel-injected models) return lines at the fuel rail or carburetor (see Chapter 4).
8 Working under the engine compartment, remove the brace that supports the intake manifold.
9 Remove the intake manifold bolts and remove the manifold from the engine.

Installation

10 Clean the manifold nuts with solvent and dry them with compressed air, if available. **Warning:** *Wear eye protection!*
11 Check the mating surfaces of the manifold for flatness with a precision straightedge and feeler gauges. Refer to this Chapter's Specifications for the warpage limit.
12 Inspect the manifold for cracks and distortion. If the manifold is cracked or warped, replace it or see if it can be resurfaced at an automotive machine shop.
13 Check carefully for any stripped or broken intake manifold bolts/studs. Replace any defective bolts with new parts.
14 Using a scraper, remove all traces of old gasket material from the cylinder head and manifold mating surfaces. Clean the surfaces with lacquer thinner or acetone.
15 Install the intake manifold with a new gasket and tighten the bolts finger-tight. Starting at the center and working out in both directions, tighten the bolts in a criss-cross pattern until the torque listed in this Chapter's Specifications is reached.
16 The remainder of the installation procedure is the reverse of removal. Refer to Chapter 1 and refill the cooling system.

6 Exhaust manifold - removal and installation

Refer to illustrations 6.3a and 6.3b

Removal

1 Disconnect the negative battery cable from the battery. **Caution:** *It is necessary to make sure the radio is turned OFF before disconnecting the battery cables to avoid damaging the microprocessor in the radio.*
2 Raise the front of the vehicle and support it securely on jackstands. Detach the exhaust pipe from the exhaust manifold (see Chapter 4). Apply penetrating oil to the fastener threads if they are difficult to remove.
3 Remove the exhaust manifold nuts **(see illustrations)** and detach the exhaust manifold from the cylinder head. **Note:** *Be sure to remove the bolts from the lower brace located near the flange of the exhaust manifold.*

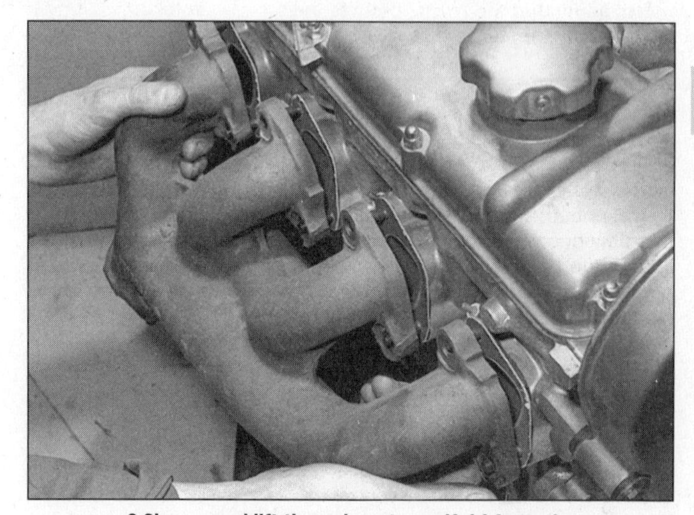

6.3b . . . and lift the exhaust manifold from the engine compartment

Installation

4 Discard the old gaskets and, if necessary, use a scraper to clean the gasket mating surfaces on the manifold and head, then clean the surfaces with a rag soaked in lacquer thinner or acetone. **Caution:** *Be careful not to gouge the soft aluminum.*
5 Install new gaskets over the exhaust manifold studs. Place the exhaust manifold in position on the cylinder head and install the nuts. Starting at the center, tighten the nuts in a criss-cross pattern until the torque listed in this Chapter's Specifications is reached.
6 The remainder of installation is the reverse of removal.
7 Start the engine and check for exhaust leaks between the manifold and the cylinder head and between the manifold and the exhaust pipe.

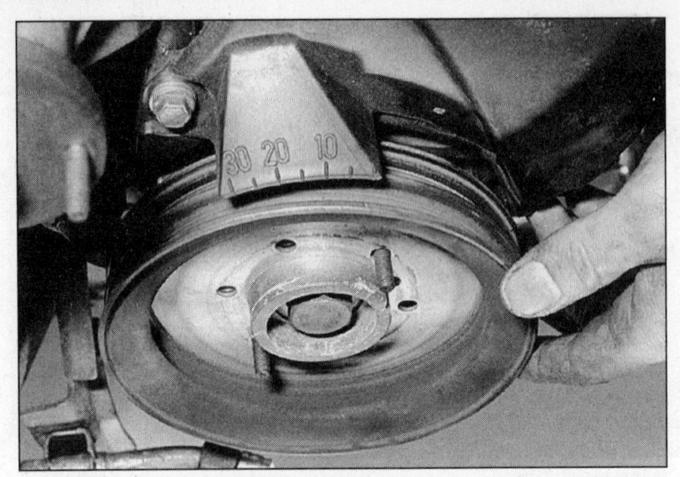

7.3 Remove the large bolt and separate the crankshaft pulley from the crankshaft

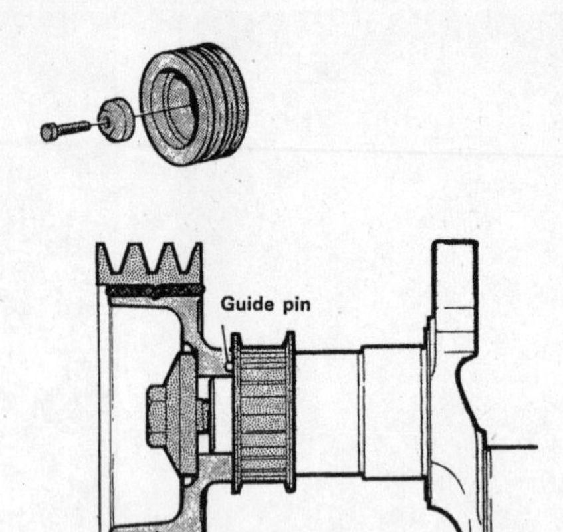

7.5 Make sure the guide pin is in place and not bent or damaged

7 Crankshaft pulley - removal and installation

Refer to illustrations 7.3 and 7.5

1 Remove the drivebelts (see Chapter 1).

2 Remove the fan and the fan shroud from the engine compartment (see Chapter 3).

3 Remove the crankshaft pulley bolt **(see illustration)**. It will probably be necessary to prevent the crankshaft from turning as you loosen the bolt. To do this, remove the torque converter/flywheel access cover and wedge a large screwdriver between the teeth of the starter ring gear and the transmission case.

4 Remove the crankshaft pulley from the crankshaft. If necessary, use a puller to detach the pulley. **Caution:** *Only use a puller that threads into the pulley hub - a puller that bears on the outer circumference of the pulley can cause damage.*

5 Installation is the reverse of removal. Be sure the guide pin is in place **(see illustration)**.

8 Timing belt cover - removal and installation

1 Remove the drivebelts (see Chapter 1).

2 Remove the fan and the fan cover from the engine compartment (see Chapter 3).

3 Remove the water pump pulley.

4 Remove the timing belt cover bolts and remove the cover. **Note:** *Some models are equipped with an upper and a lower timing belt cover.*

5 Installation is the reverse of removal.

9 Timing belt and sprockets - removal, inspection and installation

Caution: *The B21 and B23 engines are designed so the valves do not touch the piston crowns (non-interference cylinder head) in the event the timing belt breaks or is installed improperly. The B230 engine is designed with closer valve/piston clearance and WILL become damaged (interference cylinder head) in the event of timing belt problems.*

Removal

Refer to illustrations 9.4a, 9.4b, 9.4c, 9.5a, 9.5b, 9.6a, 9.6b, 9.6c, 9.6d and 9.7

1 Remove the timing belt cover(s) for access to the timing belt (see Section 8).

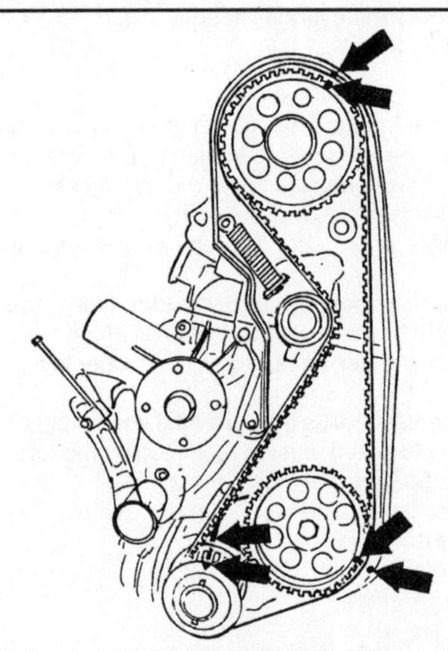

9.4a Timing marks for the camshaft, intermediate shaft and crankshaft sprockets

2 Position the number one cylinder at TDC (see Section 3). Disconnect the cable from the negative terminal of the battery. **Caution:** *It is necessary to make sure the radio is turned OFF before disconnecting the battery cables to avoid damaging the microprocessor in the radio.*

3 Remove the crankshaft pulley (see Section 7).

4 Before removing the timing belt, check the timing gear alignment notches on the camshaft sprocket, the intermediate shaft sprocket and the crankshaft sprocket in relation to the corresponding marks on the rear cover **(see illustration)**. **Note:** *The alignment indicators are notches* **(see illustrations)**, *a circle or a casting mark in the cover.*

5 Loosen the timing belt tensioner nut **(see illustration)** and release the tension on the timing belt. Install a pin into the tensioner hole to prevent the tensioner from snapping back into tensioning position once the belt has been removed **(see illustration)**. Once the tension has been relieved, the timing belt can be slipped off the sprockets. **Note:** *If you intend to reinstall the same timing belt, mark the direction of rotation on the belt so it can be installed correctly.* **Caution:** *On B230 engines, do not turn the crankshaft with the timing belt removed, because the valves may become damaged when the piston crown contacts the them.*

6 The camshaft sprocket and the intermediate shaft sprocket can be removed by placing a screwdriver or large punch between the cylinder head and the sprocket casting hole and carefully removing the bolt with a breaker bar and socket **(see illustrations)**.

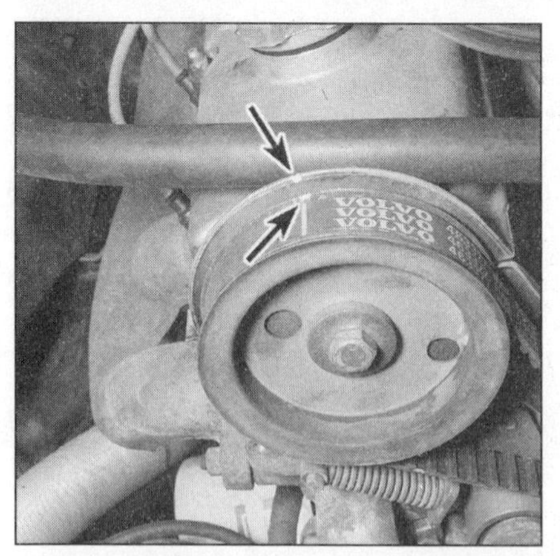

9.4b Align the mark on the camshaft sprocket with the notch in the valve cover. Note: *On factory replacement timing belts, the TDC mark is clearly painted across the timing belt*

9.4c Intermediate shaft and crankshaft timing marks (arrows)

9.5a Loosen the nut on the tensioner

9.5b Install a pin in the tensioner to prevent the tensioner from snapping back to its original position

9.6a Install a screwdriver through the sprocket or use a special locking tool to prevent the camshaft sprocket from turning while loosening the sprocket bolt

9.6b If equipped, remove the sprocket cover . . .

9.6c . . . and separate the sprocket from the camshaft

9.6d Details of the intermediate shaft sprocket

9.7 Removing the crankshaft sprocket and guide plate

7 If necessary, remove the crankshaft sprocket from the crankshaft **(see illustration)**.

Inspection

8 Inspect the sprocket teeth for wear and damage. Check the timing belt for any cracks or excessive oil coating. Also check the camshaft for excessive endplay (see Chapter 2B). Check the timing belt tensioner for smooth operation. Replace any worn parts with new ones.

10.4 Bearing cap numbers on the camshaft

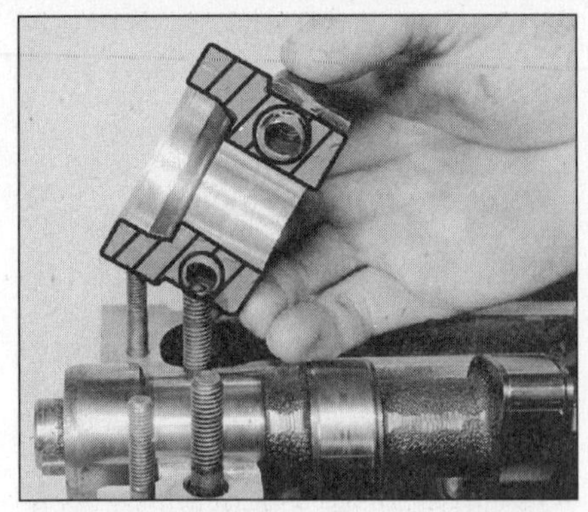
10.5 Lift the bearing caps from the cylinder head

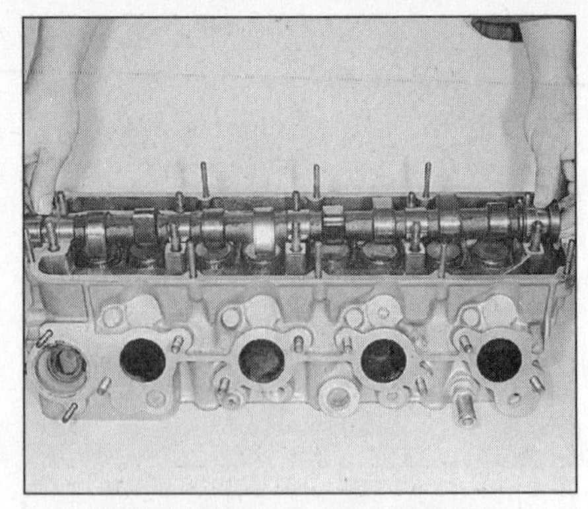
10.6 Lift the camshaft from the cylinder head

10.7a Remove the adjusting shim from the cam follower

10.7b Lift the cam follower (bucket) from the cylinder head

10.8 Use a feeler gauge to check the camshaft endplay

Installation

9 If the camshaft sprocket was removed, install it and tighten the bolt to the torque listed in this Chapter's Specifications.

10 Check to make sure the number one piston is still at Top Dead Center (TDC) (see Section 3).

11 Align all the timing marks on the engine block with the marks on the crankshaft and camshaft sprockets **(see illustration 9.4a)**. Install the timing belt onto the sprockets.

12 Remove the pin from the tensioner and allow the timing belt to become tensioned. Tighten the nut on the tensioner to the torque listed in this Chapter's Specifications.

13 After the timing belt has been tensioned, it is a good idea to rotate the crankshaft 90-degrees to both sides of the TDC mark and then back to TDC just to make sure the timing marks all stay in alignment (use the crankshaft pulley bolt for this test).

14 The remainder of installation is the reverse of removal.

10 Camshaft and cam followers - removal, inspection and installation

Removal

Refer to illustrations 10.4, 10.5, 10.6, 10.7a and 10.7b

1 Remove the valve cover (see Section 4).

2 Set the engine at TDC for cylinder number one (see Section 3). Remove the timing belt (see Section 9).

3 If it is necessary to separate the sprocket from the camshaft, remove the camshaft sprocket bolt (see Section 9).

4 If the camshaft bearing caps don't have numbers on them, mark them before removal **(see illustration)**. Be sure to put the marks on the same ends of all the caps to prevent incorrect orientation of the caps during installation.

5 Remove the camshaft bearing cap nuts, leaving the two middle caps in place but slightly loose **(see illustration)**. Next, loosen the four cap bolts evenly to prevent any damage to the camshaft. **Caution:** *Failure to follow this procedure exactly as described could tilt the camshaft in the bearings or bend the camshaft.*

6 Lift out the camshaft **(see illustration)**, wipe it off with a clean shop towel and set it aside.

7 Wipe off each valve adjustment shim with a clean shop towel and number it with a clean felt tip marker. Remove each cam follower and valve adjustment shim **(see illustrations)** and set them aside in clean labeled plastic bags or an egg carton to keep them from getting mixed up. **Note:** *The cam followers must be returned to their original bores.*

Inspection

Refer to illustrations 10.8, 10.10 and 10.11

8 To check camshaft endplay:

a) *Install the camshaft and secure it with the caps.*
b) *Mount a dial indicator on the head. If a dial indicator is not available, use a feeler gauge of the correct specification positioned between the end bearing cap and the camshaft casting* **(see illustration)**.
c) *Using a large screwdriver as a lever at the opposite end, move the camshaft forward-and-backward and note the dial indicator reading.*
d) *Compare the reading with the endplay listed in this Chapter's Specifications.*
e) *If the indicated reading is higher, either the camshaft or the head is worn. Replace parts as necessary.*

9 To check camshaft runout:

a) *Support the camshaft with a pair of V-blocks and set up a dial indicator with the plunger resting against the center bearing journal on the camshaft.*
b) *Rotate the camshaft and note the indicated runout.*
c) *Compare the results to the camshaft runout listed in this Chapter's Specifications.*

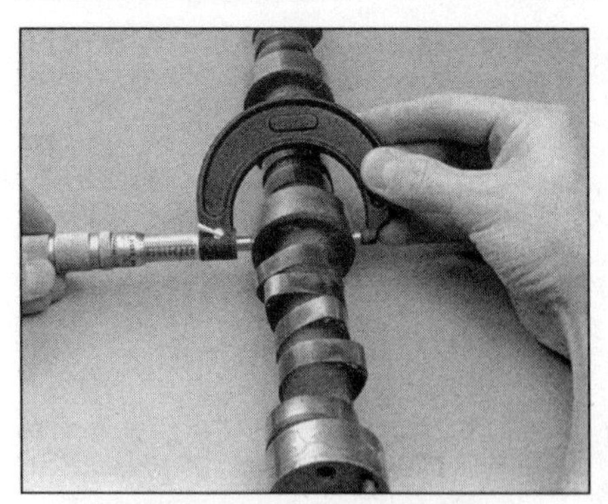

10.10 Check the diameter of each camshaft bearing journal to pinpoint excessive wear and out-of-round conditions

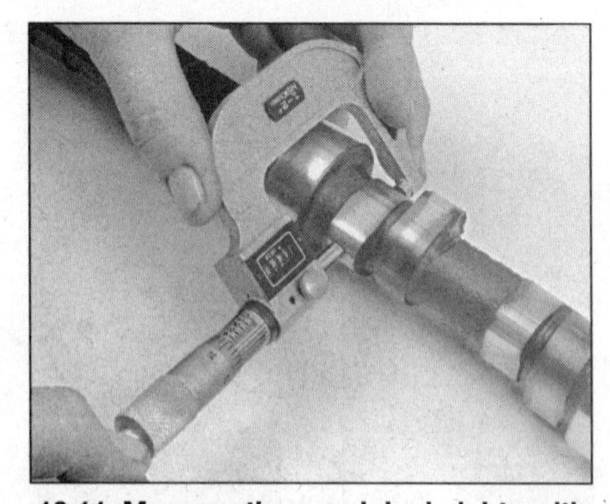

10.11 Measure the cam lobe heights with a micrometer

10.17 Be sure to apply camshaft installation lube or engine assembly lube to the cam lobes and bearing journals before installing the camshaft

d) *If the indicated runout exceeds the specified runout, replace the camshaft.*

10 Check the camshaft bearing journals and caps for scoring and signs of wear. If they are worn, replace the cylinder head with a new or rebuilt unit. Measure the journals on the camshaft with a micrometer **(see illustration)**, comparing your readings with this Chapter's Specifications. If the diameter of any of the journals is out of specification, replace the camshaft.

11 Check the cam lobes for wear:

a) *Check the toe and ramp areas of each cam lobe for score marks and uneven wear. Also check for flaking and pitting.*
b) *If there's wear on the toe or the ramp, replace the camshaft, but first try to find the cause of the wear. Look for abrasive substances in the oil and inspect the oil pump and oil passages for blockage. Lobe wear is usually caused by inadequate lubrication or dirty oil.*
c) *Using a micrometer, measure the cam lobe height* **(see illustration)**. *If the lobe heights vary more than 0.010-inch between any two intake valves or any two exhaust valves, replace the camshaft.*

12 Check the valve adjustment shims. They are usually worn if the camshaft lobes are worn. If the shims are worn, replace them.

13 Inspect the cam followers for galling and signs of seizure. If aluminum from the cylinder head is adhering to the cam followers, replace them. If any of the cam follower bores are rough, scored or worn, replace the cylinder head.

14 If any of the conditions described above are noted, the cylinder head is probably getting insufficient lubrication or dirty oil, so make sure you track down the cause of this problem (low oil level, low oil pump capacity, clogged oil passage, etc.) before installing a new head, camshaft or cam followers.

Installation

Refer to illustration 10.17

15 Thoroughly clean the camshaft, the bearing surfaces in the head and caps, the cam followers and shims (if used). Remove all sludge and dirt. Wipe off all components with a clean, lint-free cloth.

16 Lightly lubricate the cam follower bores with assembly lube or moly base grease. Refer to the numbers marked on the shims and install the cam followers in the and valve adjustment shims in the head. Lubricate the upper side of the shims with assembly lube or moly-base grease.

17 Lubricate the camshaft bearing surfaces in the head and the bearing journals and lobes on the camshaft with camshaft assembly lube or moly-base grease **(see illustration)**. Slide the new oil seal onto the front of the camshaft, then carefully lower the camshaft into position with the lobes for the number one cylinder pointing away from the cam followers. **Caution:** *Failure to adequately lubricate the camshaft and related components can cause serious damage to bearing and friction surfaces during the first few seconds after engine start-up, when the oil pressure is low or nonexistent.*

18 Apply a thin film of camshaft installation lube or engine assembly lube to the bearing surfaces of the camshaft bearing caps and install the camshaft bearing caps into their original locations.

19 Start from the center camshaft bearing cap and work toward the outside caps using a circular pattern while torquing the nuts. **Note:** *If the valve timing was disturbed, align the sprockets and install the belt as described in Section 9.*

20 Remove the spark plugs and rotate the crankshaft by hand to make sure the valve timing is correct. After two revolutions, the timing marks on the sprockets should still be aligned. If they're not, remove the timing belt and set all the timing marks again (see Section 9). **Note:** *If you feel resistance while rotating the crankshaft, stop immediately and check the valve timing by referring to Section 9.*

21 Adjust the valve clearances (see Chapter 1).

22 The remainder of installation is the reverse of removal.

11 Camshaft oil seal - replacement

Refer to illustration 11.5

1 Remove the timing belt and camshaft sprocket from the camshaft (see Section 9).

2 Use a special seal removal tool or a blunt screwdriver to pry the seal from the cylinder head. Don't nick or scratch the camshaft journal or the new seal will leak.

3 Thoroughly clean and inspect the seal bore and the seal journal on the camshaft. Both must be clean and smooth. Use emery cloth or 400-grit sandpaper to remove small burrs.

4 If a groove has been worn into the camshaft (from contact with the seal lip), installing a new seal probably will not stop the oil leak. Such wear normally indicates that the camshaft or the bearing surfaces in the caps are worn. It is time to overhaul the cylinder head (see Chapter 2, Part B) or replace the head or camshaft.

5 Coat the lip of the new seal with clean engine oil or multi-purpose grease **(see illustration)** and carefully tap the seal into place with a

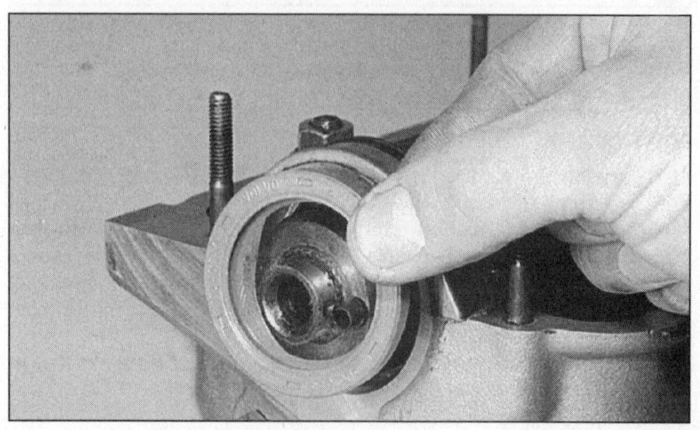

11.5 Install the camshaft seal into the cylinder head

13.15a Install a new head gasket (look for the word TOP on the gasket)

13.15b Install a new O-ring into the groove on the water pump

large socket and piece of pipe and hammer. If you don't have a large socket the size of the seal, carefully tap the outer edges of the seal with a large punch.

6 Install the camshaft sprocket and the timing belt (see Section 9).

7 Start the engine and check for oil leaks.

12 Intermediate shaft oil seal - replacement

1 Remove the timing belt and the intermediate shaft sprocket (see Section 9).

2 Use a special seal removal tool or a blunt screwdriver to pry the seal from the engine block. Don't nick or scratch the inner journal or the new seal will leak.

3 Thoroughly clean and inspect the seal bore. It must be clean and smooth. Use emery cloth or 400-grit sandpaper to remove small burrs.

4 If a groove has been worn into the intermediate shaft (from contact with the seal lip), installing a new seal probably will not stop the oil leak. Such wear normally indicates that the intermediate shaft or the rear cover seal surfaces are worn. It is time to replace the rear timing belt cover (see Chapter 2, Part B) and/or the intermediate shaft.

5 Coat the lip of the new seal with clean engine oil or multi-purpose grease and carefully tap the seal into place with a large socket and piece of pipe and hammer. If you don't have a large socket the size of the seal, carefully tap the outer edges of the seal with a large punch.

6 Install the intermediate shaft sprocket and the timing belt (see Section 9).

7 Start the engine and check for oil leaks.

13 Cylinder head - removal and installation

Refer to illustrations 13.15a, 13.15b, 13.16a and 13.16b

Caution: *Allow the engine to cool completely before beginning this procedure.*

Removal

1 Position the number one piston at Top Dead Center (see Section 3).

2 Disconnect the negative cable from the battery. **Caution:** *It is necessary to make sure the radio is turned OFF before disconnecting the battery cables to avoid damaging the microprocessor in the radio.*

3 Drain the cooling system and remove the spark plugs (see Chapter 1).

4 Remove the intake manifold brace and exhaust manifold flange bolts (see Sections 5 and 6).

5 Remove the valve cover (see Section 4).

6 Remove the distributor (see Chapter 5), including the cap and wires.

7 Remove the timing belt (see Section 9) and the camshaft (see Section 10).

8 Loosen the head bolts in 1/4-turn increments until they can be removed by hand. Work in a pattern that's the *reverse* of the tightening sequence **(see illustration 13.16b)** to avoid warping the head. Note where each bolt goes so it can be returned to the same location on installation.

9 Lift the head off the engine. If resistance is felt, don't pry between the head and block gasket mating surfaces - damage to the mating surfaces will result. Instead, pry between the power steering pump bracket and the engine block. Set the head on blocks of wood to prevent damage to the gasket sealing surfaces.

10 Cylinder head disassembly and inspection procedures are covered in detail in Chapter 2, Part B. It's a good idea to have the head checked for warpage, even if you're just replacing the gasket.

Installation

Note: *Replace the head bolts if they have been torqued down five or more times or if the center section shows signs of stretch (extension).*

11 The mating surfaces of the cylinder head and block must be perfectly clean when the head is installed.

12 Use a gasket scraper to remove all traces of carbon and old gasket material, then clean the mating surfaces with lacquer thinner or acetone. If there's oil on the mating surfaces when the head is installed, the gasket may not seal correctly and leaks may develop. When working on the block, stuff the cylinders with clean shop rags to keep out debris. Use a vacuum cleaner to remove material that falls into the cylinders. Since the head and block are made of aluminum, aggressive scraping can cause damage. Be extra careful not to nick or gouge the mating surfaces with the scraper.

13 Check the block and head mating surfaces for nicks, deep scratches and other damage. If damage is slight, it can be removed with a file; if it's excessive, machining may be the only alternative.

14 Use a tap of the correct size to chase the threads in the head bolt holes. Mount each head bolt in a vise and run a die down the threads to remove corrosion and restore the threads. Dirt, corrosion, sealant and damaged threads will affect torque readings.

15 Place a new gasket on the block and a new O-ring into the groove on top of the water pump **(see illustrations)**. Check to see if there are any markings (such as TOP) on the gasket that say how it is to be installed. Those identification marks must face UP. Also, apply sealant to the edges of the timing chain cover where it mates with the engine block. Set the cylinder head in position.

16 Lubricate the threads and the seats of the cylinder head bolts, then install them. They must be tightened in a specific sequence **(see illustrations)**, in three stages and to the torque listed in this Chapter's Specifications.

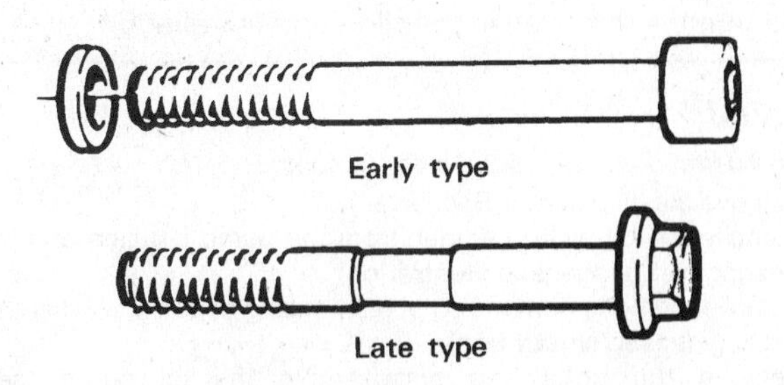

13.16a In the event the cylinder head bolts must be replaced, be sure to install the correct type

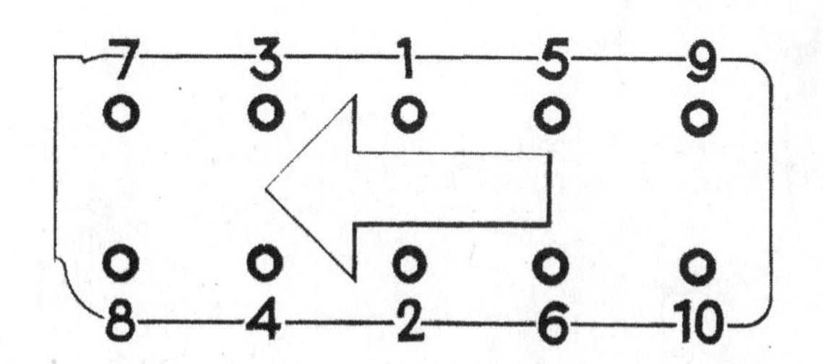

13.16b Cylinder head TIGHTENING sequence (arrow designates the front of the engine)

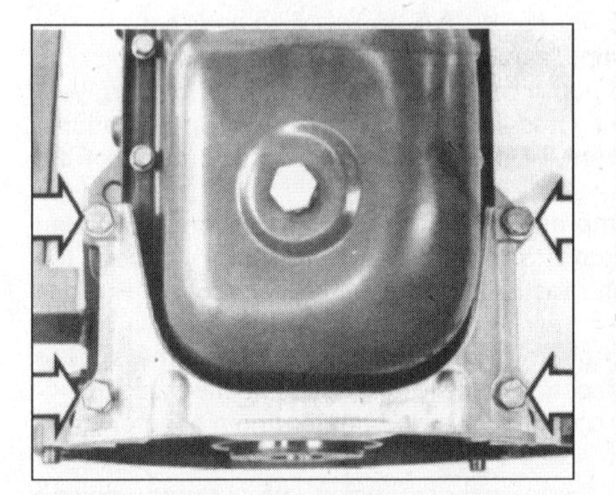

14.4a Remove the bellhousing/engine brace bolts . . .

14.4b . . . then lower the brace from the engine

14.6 Remove the oil pan from the engine

2A

17 Install the camshaft (see Section 10), camshaft sprocket and timing belt (see Section 9).
18 Reinstall the remaining parts in the reverse order of removal.
19 Be sure to refill the cooling system and check all fluid levels.
20 Rotate the crankshaft clockwise slowly by hand through two complete revolutions. **Caution:** *If you feel any resistance while turning the engine over, stop and re-check the camshaft timing. The valves may be hitting the pistons.*
21 Start the engine and check the ignition timing (see Chapter 1).
22 Run the engine until normal operating temperature is reached. Check for leaks and proper operation.

14 Oil pan - removal and installation

Refer to illustrations 14.4a, 14.4b, 14.6 and 14.7

1 Warm up the engine, then drain the oil and replace the oil filter (see Chapter 1).
2 Detach the cable from the negative battery terminal. **Caution:** *It is necessary to make sure the radio is turned OFF before disconnecting the battery cables to avoid damaging the microprocessor in the radio.*
3 Raise the vehicle and support it securely on jackstands.
4 Remove the bellhousing/engine brace **(see illustrations)**.
5 Remove the bolts securing the oil pan to the engine block.
6 Tap on the pan with a soft-face hammer to break the gasket seal, then detach the oil pan from the engine **(see illustration)**. Don't pry between the block and oil pan mating surfaces.
7 Remove the oil pick-up screen from the bottom of the engine **(see illustration)**. Clean the screen and reinstall it.
8 Using a gasket scraper, remove all traces of old gasket and/or sealant from the engine block and oil pan. Remove the seals from each end of the engine block or oil pan. Clean the mating surfaces with lacquer thinner or acetone. Make sure the threaded bolt holes in the block are clean.

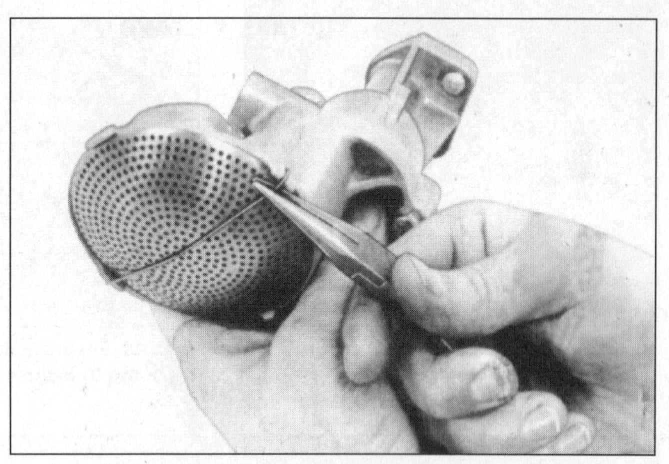

14.7 Remove the screen from the oil pump (oil pump removed for clarity)

9 Clean the oil pan with solvent and dry it thoroughly. Check the gasket flanges for distortion, particularly around the bolt holes. If necessary, place the pan on a block of wood and use a hammer to flatten and restore the gasket surfaces.
10 Apply a 1/8-inch wide bead of RTV sealant to the oil pan gasket surfaces. Make sure the sealant is applied to the inside edge of the bolt holes.
11 Carefully place the oil pan in position.
12 Install the bolts and tighten them in small increments to the torque listed in this Chapter's Specifications. Start with the bolts closest to the center of the pan and work out in a spiral pattern. Don't overtighten them or leakage may occur.
13 Fill the crankcase with the proper type and amount of engine oil (see Chapter 1), then run the engine and check for oil leaks.

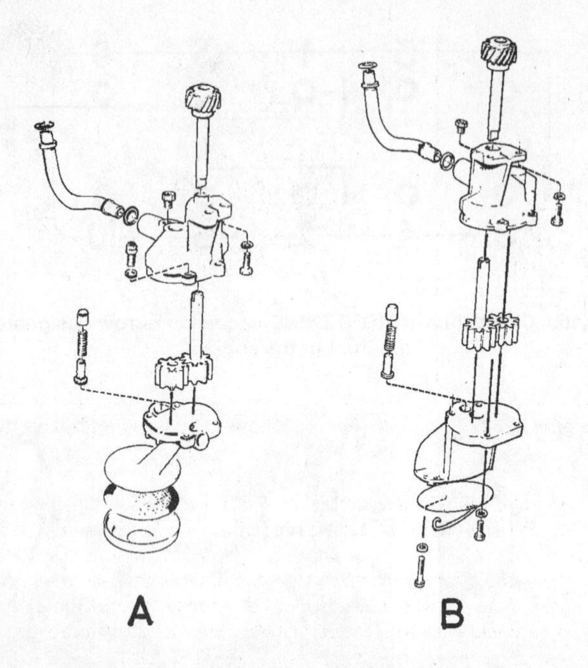

15.2 Exploded view of the oil pump assembly

A Later type *B Early type*

15 Oil pump - removal, inspection and installation

Removal

Refer to illustrations 15.2, 15.3, 15.4a, 15.4b and 15.4c

1 Remove the oil pan (see Section 14).

2 Remove the oil pickup screen from the pump housing and the main bearing cap bridge **(see illustration)**.

3 Remove the bolts from the oil pump housing and lift the assembly from the engine **(see illustration)**.

4 Remove the bolts and disassemble the oil pump **(see illustrations).** You may need to use an impact screwdriver to loosen the pump cover screws without stripping the heads out.

Inspection

Refer to illustrations 15.5a, 15.5b, 15.6, 15.7a, 15.7b and 15.8

5 Check the oil pump clearances on both inner and outer rotors to each other and to the pump body **(see illustrations)**. Compare your measurements to the figures listed in this Chapter's Specifications. Replace the pump if any of the measurements are outside of the specified limits.

6 Unscrew the pump and extract the spring and pressure relief valve plunger **(see illustration)** from the pump housing. Check the spring for distortion and the relief valve plunger for scoring. Replace parts as necessary.

7 Install the pump gears **(see illustrations)**. Pack the spaces between the gears with petroleum jelly (this will prime the pump).

8 Install the pump cover screws and tighten them securely **(see**

15.3 Remove the oil pump from the engine

15.4a Separate the sections of the oil pump to expose the oil pump gears

15.4b Remove the transfer tube from the oil pump

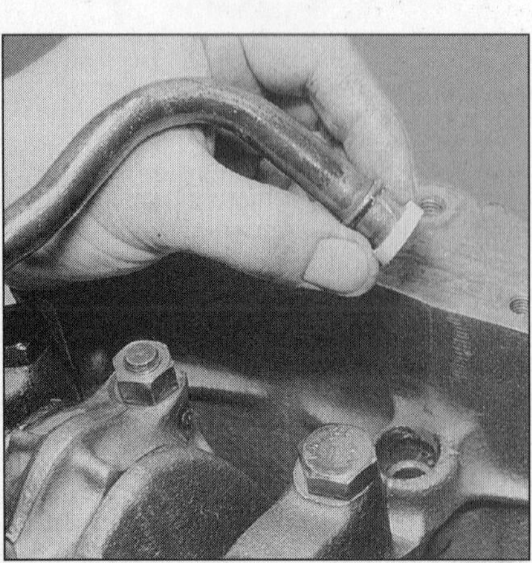

15.4c Install a new seal on the end of the transfer tube

15.5a Checking the gear-to-oil pump body clearance

15.5b Checking the gear(s)-to-oil pump cover clearance

15.6 Remove the relief valve plunger

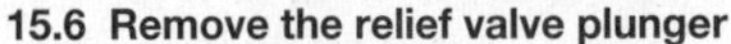

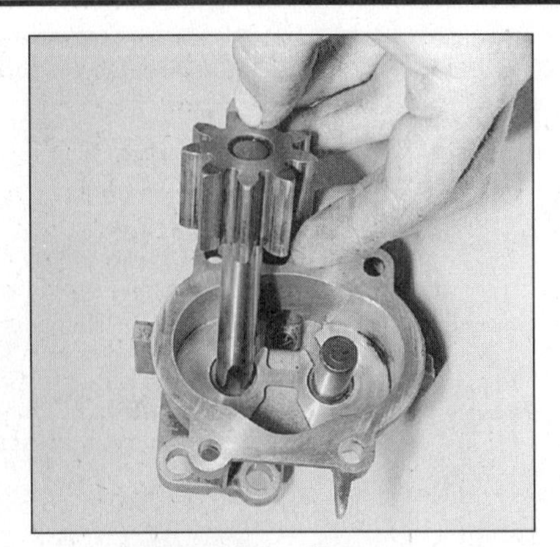
15.7a Install the drive gear . . .

15.7b . . . followed by the companion gear

illustration). Install the oil pressure relief valve and spring assembly. Use a new sealing washer on the plug and tighten the plug securely.

Installation

9 Using new O-rings, install the pump. Tighten the bolts to the torque listed in this Chapter's Specifications.

11 Install the oil pick-up screen.

12 Install the oil pan (see Section 14).

13 Remainder of installation is the reverse of removal. Change the oil filter, add the specified type and quantity of oil (see Chapter 1), run the engine and check for leaks.

16 Flywheel/driveplate - removal and installation

Removal

1 Raise the vehicle and support it securely on jackstands, then refer to Chapter 7 and remove the transmission.

2 Remove the pressure plate and clutch disc (see Chapter 8) (manual transmission equipped vehicles). Now is a good time to check/replace the clutch components and pilot bearing.

3 Mark the relationship of the flywheel/driveplate to the crankshaft to ensure correct installation. Remove the bolts that secure the flywheel/driveplate to the crankshaft. If the crankshaft turns wedge a screwdriver in the ring gear teeth (manual transaxle models), or insert a long punch through one of the holes in the driveplate and allow it to rest against a projection on the engine block (automatic transaxle models).

15.8 Assemble the oil pump sections together

4 Remove the flywheel/driveplate from the crankshaft. Since the flywheel is fairly heavy, be sure to support it while removing the last bolt.

5 Clean the flywheel to remove grease and oil. Inspect the surface for cracks, rivet grooves, burned areas and score marks. Light scoring can be removed with emery cloth. Check for cracked and broken ring gear teeth. Lay the flywheel on a flat surface and use a straightedge to check for warpage. **Note:** *If the surface of the flywheel is scored or shows signs of overheating (blue spots), have it resurfaced by an automotive machine shop. Also, if the ring gear is damaged, it can be replaced by an automotive machine shop.*

6 Clean and inspect the mating surfaces of the flywheel/driveplate and the crankshaft. If the rear main oil seal is leaking, replace it before reinstalling the flywheel/driveplate (see Section 17).

Installation

Refer to illustration 16.7

7 Position the flywheel/driveplate against the crankshaft. Note that some engines have an alignment dowel or staggered bolt holes to ensure correct installation **(see illustration)**. Before installing the bolts, apply thread locking compound to the threads.

8 Prevent the flywheel/driveplate from turning by using one of the methods described in Step 3. Using a crossing pattern, tighten the

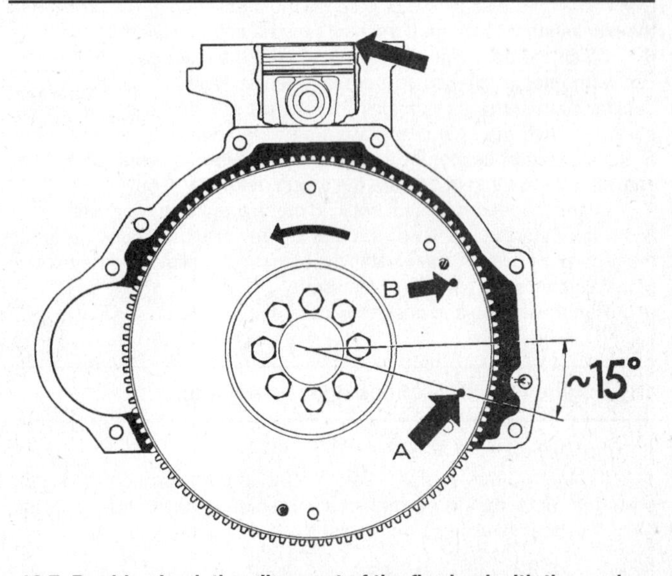

16.7 Double check the alignment of the flywheel with the engine before tightening the flywheel bolts

17.5a Carefully pry the seal out of the seal retainer

17.5b If you can't pry the seal out, it is possible to remove the complete rear main seal housing to replace the seal and gasket

bolts to the torque listed in this Chapter's Specifications.

9 The remainder of installation is the reverse of the removal procedure. It's a good idea to install a new pilot bearing at this time (see Chapter 8).

17 Crankshaft oil seals - replacement

Refer to illustration 17.5a, 17.5b and 17.5c

1 Remove the drivebelts (see Chapter 1).

2 Remove the crankshaft pulley.

3 Remove the transmission and clutch/flywheel assembly (manual transmission) or torque converter/flywheel assembly (automatic transmission) to replace the rear main oil seal (see Chapter 7B).

4 To remove the front oil seal, remove the crankshaft sprocket from the crankshaft (see Section 9). Pry the seal out with a seal removal tool or a screwdriver, being careful not to nick the crankshaft in the process.

5 To remove the rear main oil seal, carefully pry the seal out of the housing with a seal removal tool or a screwdriver. If necessary, the entire seal housing can be unbolted and removed **(see illustrations)**. Don't scratch the seal bore or damage the crankshaft in the process (if the crankshaft is damaged, the new seal will end up leaking).

6 Clean the bore in the housing and coat the outer edge of the new seal with engine oil or multi-purpose grease. Using a socket with an outside diameter slightly smaller than the outside diameter of the seal, carefully drive the seal into place with a hammer. If a socket is not available, a short section of a large diameter pipe will work. Check the seal after installation to be sure the spring did not pop out.

7 Install the rear main seal housing or the crankshaft sprocket.

8 Lubricate the sleeve of the crankshaft pulley with engine oil or multi-purpose grease, then install the crankshaft pulley. The remainder of installation is the reverse of removal.

9 Run the engine and check for leaks.

18 Engine mounts - check and replacement

Refer to illustration 18.7

1 Engine mounts seldom require attention, but broken or deteriorated mounts should be replaced immediately or the added strain placed on the driveline components may cause damage or wear.

Check

2 During the check, the engine must be raised slightly to remove the weight from the mounts.

3 Raise the vehicle and support it securely on jackstands, then position a jack under the engine oil pan. Place a large block of wood between the jack head and the oil pan, then carefully raise the engine just enough to take the weight off the mounts. **Warning:** *DO NOT place any part of your body under the engine when it's supported only by a jack!*

4 Check the mount insulators to see if the rubber is cracked, hardened or separated from the metal plates. Sometimes the rubber will split right down the center.

5 Check for relative movement between the mount plates and the engine or frame (use a large screwdriver or prybar to attempt to move the mounts). If movement is noted, lower the engine and tighten the mount fasteners.

Replacement

6 Disconnect the negative battery cable from the battery, then raise the vehicle and support it securely on jackstands (if not already done). Support the engine as described in Step 3. **Caution:** *It is necessary to make sure the radio is turned OFF before disconnecting the battery cables to avoid damaging the microprocessor in the radio.*

7 Remove the fasteners **(see illustration)**, raise the engine with the jack and detach the mount from the frame bracket and engine.

8 Install the new mount, making sure it is correctly positioned in its bracket. Install the fasteners and tighten them securely.

9 Rubber preservative should be applied to the insulators to slow deterioration.

18.7 Remove the front engine mount bolts (arrows)

Chapter 2 Part B
General engine overhaul procedures

Contents

Specifications

General

Bore (stamped on top of engine block)

B21F engines	
C	3.6248 to 3.6252 inches
D	3.6252 to 3.6256 inches
E	3.6256 to 3.6260 inches
G	3.6264 to 3.6268 inches
OS 1	3.6445 inches
OS 2	3.6642 inches
B23F engines	
C	3.7824 to 3.7828 inches
D	3.7828 to 3.7832 inches
E	3.7832 to 3.7836 inches
G	3.7840 to 3.7844 inches
OS 1	3.7942 inches
OS 2	3.8060 inches

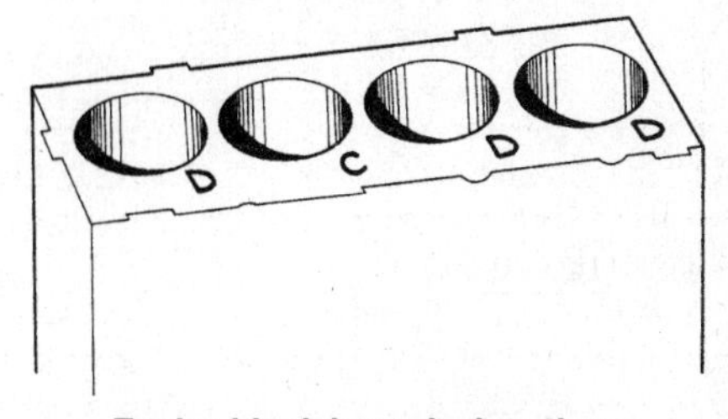

Engine block bore designations

General

Bore (stamped on top of engine block)
  B230F engines
    C .... 3.7795 to 3.7799 inches
    D .... 3.7799 to 3.7803 inches
    E .... 3.7803 to 3.7807 inches
    G .... 3.7811 to 3.7815 inches
    OS 1 .... 3.7913 inches
    OS 2 .... 3.8031 inches
Compression ratio
  B21F and B21FT 7.5:1
  B23F
    Manual transmission (1984) 9.5:1
  All others 10.3:1
  B230F 9.8:1
Displacement
  B21F and B21FT .... 129.7 cubic inches (2.127 liters)
  B23F and B230F .... 141.2 cubic inches (2.316 liters)
Cylinder compression pressure
  Standard .... 170 psi
  Minimum .... 128 psi
Oil pressure
  At idle .... 10 psi
  At 2,000 rpm
    Minimum .... 36 psi
    Maximum .... 87 psi

Engine block

Cylinder taper limit .... 0.002 inch
Cylinder out-of-round limit .... 0.002 inch

Pistons and rings

Piston diameter
Note 1: *(measured 0.28 inches from the bottom of the piston skirt)*
Note 2: *This specification is derived from measuring the engine block bore [diameter] and subtracting the piston/block clearance)*
  B21F and B21FT .... Not available
  B23F .... Not available
  B230F .... Not available
Piston ring side clearance
  B21F, B21FT and B23F
    Top compression ring
      Standard .... 0.0016 to 0.0028 inch
      Service limit .... Not available
    Second compression ring
      Standard .... 0.0016 to 0.0028 inch
      Service limit .... Not available
    Oil scraper ring
      Standard .... 0.0012 to 0.0024 inch
      Service limit .... Not available
  B230F
    Top compression ring
      Standard .... 0.0024 to 0.0036 inch
      Service limit .... Not available
    Second compression ring
      Standard .... 0.0016 to 0.0028 inch
      Service limit .... Not available
    Oil scraper ring
      Standard .... 0.0012 to 0.0025 inch
      Service limit .... Not available
Piston ring end gap
  B21F, B21FT and B23F
    Top compression ring
      Standard .... 0.014 to 0.026 inch
      Service limit .... Not available
    Second compression ring
      Standard .... 0.014 to 0.022 inch
      Service limit .... Not available

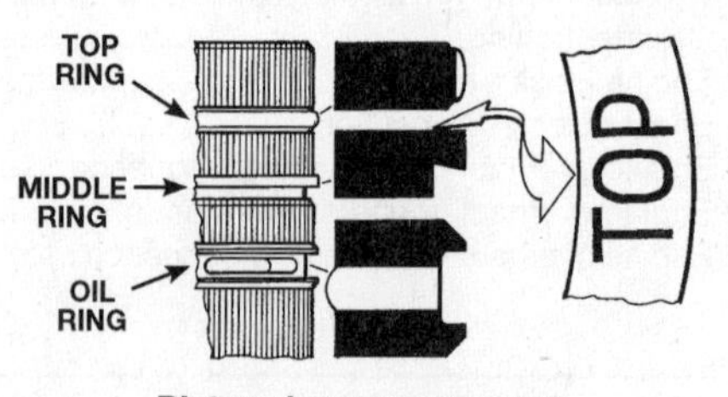

Piston ring arrangement

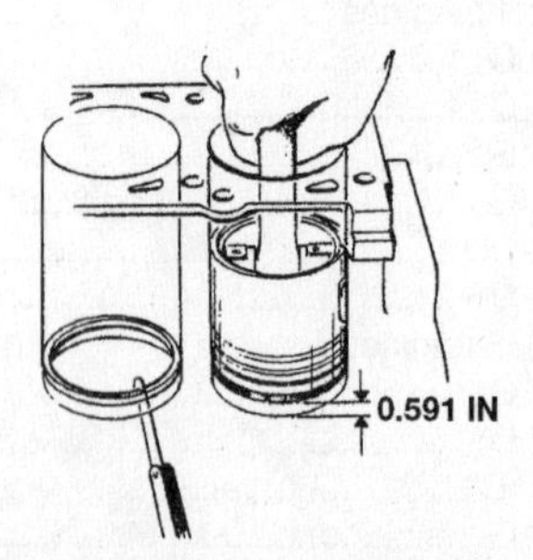

Piston ring end gap check

Oil scraper ring	
Standard	0.010 to 0.024 inch
Service limit	Not available
B230F	
Top compression ring	
Standard	0.012 to 0.022 inch
Service limit	Not available
Second compression ring	
Standard	0.012 to 0.022 inch
Service limit	Not available
Oil scraper ring	
Standard	0.012 to 0.024 inch
Service limit	Not available
Piston-to-cylinder wall clearance	
B21F	
Standard	0.0004 to 0.0016 inch
Service limit	Not available
B21FT	
Standard	0.0008 to 0.0016 inch
Service limit	Not available
B23F	
Standard	0.0004 to 0.0016 inch
Service limit	Not available
B230F	
Standard	0.0004 to 0.0012 inch
Service limit	Not available

2B

Crankshaft and connecting rods

Endplay	
B21F, B21FT and B23	0.0098 inch
B230F	0.0031 to 0.0106 inch
Crankshaft runout	
B21F, B21FT and B23	
Standard	0.0020 inch
Service limit	Not available
B230F	
Standard	0.0010 inch
Service limit	Not available
Main bearing journals	
Diameter (standard)	
B21F, B21FT and B23F	
Standard	2.4981 to 2.4986 inches
First undersize	2.4881 to 2.4886 inches
Second undersize	2.4781 to 2.4786 inches
B230F (1985 through 1988)	
Standard	2.1648 to 2.1653 inches
First undersize	2.1550 to 2.1555 inches
Second undersize	2.1451 to 2.1457 inches
B230F (mid-year 1988 through 1993)	
Standard	2.4798 to 2.4803 inches
First undersize	2.4700 to 2.4705 inches
Second undersize	2.4601 to 2.4606 inches
Taper (maximum)	
B21F, B21FT and B23F	0.0020 inches
B230F	0.00016 inches
Out-of-round	
B21F, B21FT and B23F	0.0028 inches
B230F	0.00016 inches
Main bearing oil clearance (standard)	
B21F, B21FT and B23F	0.0011 to 0.0033 inches
B230F (1985 through 1988)	0.0009 to 0.0028 inches
B230F (1988 through 1993)	0.0009 to 0.0024 inches
Connecting rod journal	
Diameter	
B21F, B21FT and B23F	
Standard	2.1255 to 2.1260 inches
First undersize	2.1155 to 2.1160 inches
Second undersize	2.1055 to 2.1060 inches

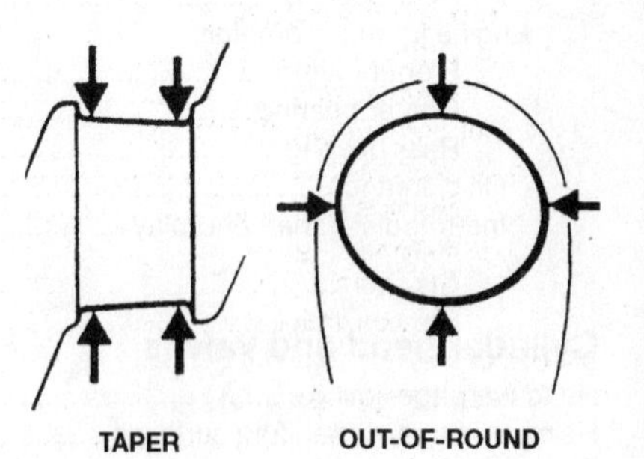

Crankshaft measurements

Crankshaft and connecting rods (continued)

Connecting rod journal (continued)	
Diameter	
B230F (1985 through 1988)	
Standard	1.9285 to 1.9293 inches
First undersize	1.9187 to 1.9195 inches
Second undersize	1.9088 to 1.9095 inches
B230F (mid-year 1988 through 1993)	
Standard	1.9285 to 1.9293 inches
First undersize	1.9187 to 1.9195 inches
Second undersize	1.9088 to 1.9096 inches
Taper	
B21F, B21FT and B23F	0.0020 inches
B230F	0.00016 inches
Out-of-round	
B21F, B21FT and B23F	0.0020 inches
B230F	0.00016 inches
Connecting rod bearing oil clearance	
B21F, B21FT and B23F	0.0009 to 0.0028 inches
B230F	0.0009 to 0.0026 inches
Connecting rod endplay (side clearance)	
Standard	0.006 to 0.012 inch
Service limit	0.016 inch

Intermediate shaft

B21F, B21FT and B23F	
Intermediate shaft diameter	
Front bearing	1.8508 to 1.8518 inches
Center bearing	1.6952 to 1.6962 inches
Rear bearing	1.6912 to 1.6922 inches
Engine journal diameter	
Front bearing	1.8526 to 1.8538 inches
Center bearing	1.6970 to 1.6981 inches
Rear bearing	1.6930 to 1.6942 inches
Oil clearance	0.0008 to 0.0030 inches
Intermediate shaft end play	0.008 to 0.020 inches
B230F	
Intermediate shaft diameter	
Front bearing	1.8494 to 1.8504 inches
Center bearing	1.6939 to 1.6949 inches
Rear bearing	1.6900 to 1.6909 inches
Engine journal diameter	
Front bearing	1.8512 to 1.8524 inches
Center bearing	1.6957 to 1.6968 inches
Rear bearing	1.6917 to 1.6929 inches
Oil clearance	0.0008 to 0.0030 inches
Intermediate shaft end play	0.008 to 0.020 inches

Cylinder head and valves

Head warpage limit	See Chapter 2A
Head warpage at manifold surfaces	0.006 inch
Cylinder head height	
B21F, B21FT and B23F	
New	5.760 inches
Minimum after machining	5.740 inches
B230F	
New	5.752 inches
Minimum after machining	5.732 inches
Valve seat angle	45-degrees
Valve face angle	44.5 degrees
Valve stem-to-guide clearance	
New valve in new guide	
Intake	0.0012 to 0.0021 inch
Exhaust	0.0024 to 0.0035 inch
Service limit (all)	
Intake	0.006 inch
Exhaust	0.006 inch

Valve guide length	
B21F, B21FT and B23F	
Intake	2.0488 inches
Exhaust	2.0488 inches
B230F	
Intake	2.0470 inches
Exhaust	2.0490 inches
Valve stem diameter	
Non-turbocharged engines	
Intake	0.3132 to 0.3138 inch
Exhaust	0.3128 to 0.3134 inch
Turbocharged engines	
Intake	0.3132 to 0.3138 inch
Exhaust*	0.3128 to 0.3134 inch
Exhaust**	0.3136 to 0.3142 inch

** Measured 1.26 inches from the top of the valve*
*** Measured 0.63 inches from the tip of the valve stem*

Valve spring	
Free length - minimum	1.770 inches
Valve stem installed height **(height above upper plane of cylinder head - dimension L in illustration 12.8)**	
B21F, B21FT and B23F	
Intake	0.6068 to 1.6146 inches
Exhaust	0.7053 to 0.7131 inches
B230F	
Intake	0.6063 to 0.6142 inches
Exhaust	0.7047 to 0.7126 inches
Valve guides inside diameter	
B21F, B21FT and B23F	
Intake	0.3152 to 0.3161 inches
Exhaust	0.3152 to 0.3161 inches
B230F	
Intake	0.3150 to 0.3158 inches
Exhaust	0.3150 to 0.3159 inches
Cam followers	
Diameter	
B21F, B21FT and B23F	1.4568 to 1.4576 inches
B230F	1.4557 to 1.4565 inches
Height	1.182 to 1.221 inches
Oil clearance (shim to cam follower)	
Standard	0.0004 to 0.0025 inch
Service limit	0.004 inch

Torque specifications*

	Ft-lbs (unless otherwise indicated)
Connecting rod bearing cap bolts	
B21F, B21FT and B23F	
New bolts	50
Old bolts	45
B230F	
First stage	15
Second stage	rotate an additional 1/4 turn (90 degrees)
Main bearing cap nuts	
B21F, B21FT and B23F	80
B230F	80
Intermediate shaft thrust plate bolts	12

** Refer to Part A for additional torque specifications.*

1 General information

Included in this portion of Chapter 2 are the general overhaul procedures for the cylinder head(s) and internal engine components.

The information ranges from advice concerning preparation for an overhaul and the purchase of replacement parts to detailed, step-by-step procedures covering removal and installation of internal engine components and the inspection of parts.

The following Sections have been written based on the assumption the engine has been removed from the vehicle. For information concerning in-vehicle engine repair, as well as removal and installation of the external components necessary for the overhaul, see Part A of this Chapter.

The Specifications included in this Part are only those necessary for the inspection and overhaul procedures which follow. Refer to Part A for additional Specifications.

The Volvo 240 models covered in this manual are equipped with either a B21F, B23F or B230F engine (fuel injected). Some Canadian models (1978 through 1984) are equipped with the B21A engine (carbureted) **(see Chapter 2A, illustration 1.4)**. The engine block is similar throughout all the models with slight variations in design. Here is a brief description of differences:

Note: *Each different model engine block is NOT interchangeable with any of the others.*

a) *The B230 crankshaft has eight counterweights while the B21 and B23 engines have four counterweights.*
a) *The B230 crankshaft has smaller bearing surfaces to reduce friction.*
c) *The B230 pistons are lighter and have thinner rings. Also, the pins are located higher in position in the piston to accommodate the longer connecting rods.*
d) *The B21 and B23 engines are non-interference designed. These engines are designed not to allow any valve/piston damage in the event the timing belt breaks or if the timing belt is installed improperly. The B230 engine is a interference design. Take special caution when performing any type of work on the timing belt.*
e) *The B21F-T and B230F engines are equipped with Stellite coating on the exhaust valves. These valve surfaces are much harder and wear less over a longer period of time. Do not grind these valves on any type of machinery but instead use valve lapping compound. B21F-T engines also use sodium filled exhaust valves that stay cooler at high engine temperatures. These type of valves must not be disposed with other metal because of their explosive properties.*

Note: *The specifications listed for the B21F are the same as the B21A engine for Canada.*

These four cylinder engines are equipped with a single overhead camshaft that controls two valves per cylinder (total 8-valves). The engine is also equipped with an aluminum cylinder head. An intermediate shaft turns at half crankshaft speed and controls the distributor and oil pump.

The pistons have two compression rings and one oil control ring. The semi-floating piston pins are press fitted into the small end of the connecting rod. The connecting rod big ends are also equipped with renewable insert-type plain bearings.

The engine is liquid-cooled, utilizing a centrifugal impeller-type pump, driven by a belt, to circulate coolant around the cylinders and combustion chambers and through the intake manifold.

Lubrication is handled by a gear-type oil pump mounted on the bottom of the engine. It is driven by a shaft that is gear driven by the intermediate shaft. The oil is filtered continuously by a cartridge-type filter mounted on the radiator side of the engine.

Engine application chart

B21F	1976 through 1982 USA
B21A	1976 through 1984 Canada
B21F-T	1978 through 1985 USA
B23F	1983 and 1984 USA
B230F	1985 through 1993 USA and Canada

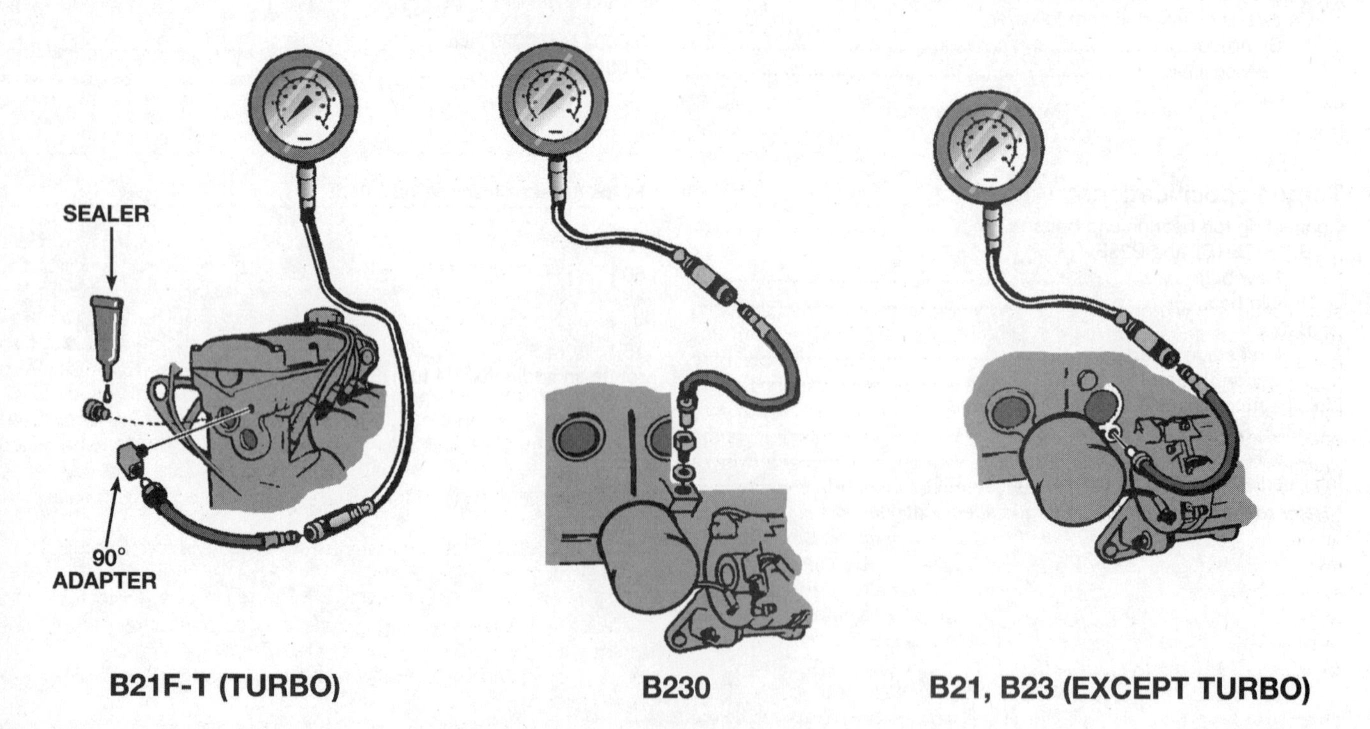

2.4a Oil pressure gauge mounting locations

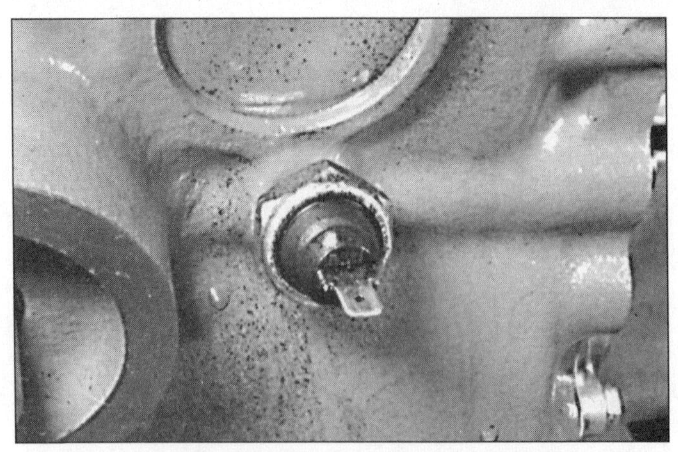

2.4b The oil pressure sending unit is located directly next to the oil filter on B21F and B23F

3.6 A compression gauge with a threaded fitting for the spark plug hole is preferred over the type that require hand pressure to maintain the seal

2 Engine overhaul - general information

Refer to illustrations 2.4a and 2.4b

It's not always easy to determine when, or if, an engine should be completely overhauled, as a number of factors must be considered.

High mileage isn't necessarily an indication an overhaul is needed, while low mileage doesn't preclude the need for an overhaul. Frequency of servicing is probably the most important consideration. An engine that's had regular and frequent oil and filter changes, as well as other required maintenance, will most likely give many thousands of miles of reliable service. Conversely, a neglected engine may require an overhaul very early in its life.

Excessive oil consumption is an indication that piston rings, valve seals and/or valve guides are in need of attention. Make sure oil leaks aren't responsible before deciding the rings and/or guides are bad. Perform a cylinder compression check to determine the extent of the work required (see Section 3).

Remove the oil pressure sending unit **(see illustrations)** and check the oil pressure with a gauge installed in its place. Compare the results to this Chapter's Specifications. As a general rule, engines should have ten psi of oil pressure for every 1,000 rpm. If the pressure is extremely low, the bearings and/or oil pump are probably worn out.

Loss of power, rough running, knocking or metallic engine noises, excessive valve train noise and high fuel consumption rates may also point to the need for an overhaul, especially if they're all present at the same time. If a complete tune-up doesn't remedy the situation, major mechanical work is the only solution.

An engine overhaul involves restoring the internal parts to the specifications of a new engine. During an overhaul, the piston rings are replaced and the cylinder walls are reconditioned (rebored and/or honed). If a rebore is done by an automotive machine shop, new oversize pistons will also be installed. The main bearings, connecting rod bearings and camshaft bearings are generally replaced with new ones and, if necessary, the crankshaft may be reground to restore the journals. Generally, the valves are serviced as well, since they're usually in less-than-perfect condition at this point. While the engine is being overhauled, other components, such as the starter and alternator, can be rebuilt as well. The end result should be a like-new engine that will give many trouble free miles. **Note:** *Critical cooling system components such as the hoses, drivebelts, thermostat and water pump MUST be replaced with new parts when an engine is overhauled. The radiator should be checked carefully to ensure it isn't clogged or leaking* (see Chapter 3).

Before beginning the engine overhaul, read through the entire chapter to familiarize yourself with the scope and requirements of the job. Overhauling an engine isn't particularly difficult, if you follow all of the instructions carefully, have the necessary tools and equipment and pay close attention to all specifications; however, it can be time consuming. Plan on the vehicle being tied up for a minimum of two weeks, especially if parts must be taken to an automotive machine shop for repair or reconditioning. Check on availability of parts and make sure any necessary special tools and equipment are obtained in advance. Most work can be done with typical hand tools, although a number of precision measuring tools are required for inspecting parts to determine if they must be replaced. Often an automotive machine shop will handle the inspection of parts and offer advice concerning reconditioning and replacement. **Note:** *Always wait until the engine has been completely disassembled and all components, especially the engine block, have been inspected before deciding what service and repair operations must be performed by an automotive machine shop. Since the block's condition will be the major factor to consider when determining whether to overhaul the original engine or buy a rebuilt one, never purchase parts or have machine work done on other components until the block has been thoroughly inspected. As a general rule, time is the primary cost of an overhaul, so it doesn't pay to install worn or substandard parts.*

As a final note, to ensure maximum life and minimum trouble from a rebuilt engine, everything must be assembled with care in a spotlessly clean environment.

3 Cylinder compression check

Refer to illustration 3.6

1 A compression check will tell you what mechanical condition the upper end (pistons, rings, valves, head gaskets) of the engine is in. Specifically, it can tell you if the compression is down due to leakage caused by worn piston rings, defective valves and seats or a blown head gasket. **Note:** *The engine must be at normal operating temperature and the battery must be fully charged for this check.*

2 Begin by cleaning the area around the spark plugs before you remove them. Compressed air should be used, if available, otherwise a small brush or even a bicycle tire pump will work. The idea is to prevent dirt from getting into the cylinders as the compression check is being done.

3 Remove all of the spark plugs from the engine (see Chapter 1).

4 Block the throttle wide open.

5 Disable the fuel system by removing the fuel pump fuse (see Chapter 4). Disable the ignition system by detaching the primary (low voltage) wires from the ignition coil (see Chapter 5).

6 Install the compression gauge in the number one spark plug hole **(see illustration)**.

7 Crank the engine over at least seven compression strokes and watch the gauge. The compression should build up quickly in a healthy engine. Low compression on the first stroke, followed by gradually increasing pressure on successive strokes, indicates worn piston

rings. A low compression reading on the first stroke, which doesn't build up during successive strokes, indicates leaking valves or a blown head gasket (a cracked head could also be the cause). Deposits on the undersides of the valve heads can also cause low compression. Record the highest gauge reading obtained.

8 Repeat the procedure for the remaining cylinders and compare the results to this Chapter's Specifications.

9 If the readings are below normal, add some engine oil (about three squirts from a plunger-type oil can) to each cylinder, through the spark plug hole, and repeat the test.

10 If the compression increases significantly after the oil is added, the piston rings are definitely worn. If the compression doesn't increase significantly, the leakage is occurring at the valves or head gasket. Leakage past the valves may be caused by burned valve seats and/or faces or warped, cracked or bent valves.

11 If two adjacent cylinders have equally low compression, there's a strong possibility the head gasket between them is blown. The appearance of coolant in the combustion chambers or the crankcase would verify this condition.

12 If one cylinder is about 20-percent lower than the others, and the engine has a slightly rough idle, a worn exhaust lobe on the camshaft could be the cause.

13 If the compression is unusually high, the combustion chambers are probably coated with carbon deposits. If that's the case, the cylinder head(s) should be removed and decarbonized.

14 If compression is way down or varies greatly between cylinders, it would be a good idea to have a leak-down test performed by an automotive repair shop. This test will pinpoint exactly where the leakage is occurring and how severe it is.

4 Vacuum gauge diagnostic checks

A vacuum gauge provides valuable information about what is going on in the engine at a low-cost. You can check for worn rings or cylinder walls, leaking head or intake manifold gaskets, incorrect carburetor adjustments, restricted exhaust, stuck or burned valves, weak valve springs, improper ignition or valve timing and ignition problems.

Unfortunately, vacuum gauge readings are easy to misinterpret, so they should be used in conjunction with other tests to confirm the diagnosis.

Both the gauge readings and the rate of needle movement are important for accurate interpretation. Most gauges measure vacuum in inches of mercury (in-Hg). As vacuum increases (or atmospheric pressure decreases), the reading will decrease. Also, for every 1,000 foot increase in elevation above approximately 2000 feet, the gauge readings will decrease about one inch of mercury.

Connect the vacuum gauge directly to intake manifold vacuum, not to ported (carburetor) vacuum. Be sure no hoses are left disconnected during the test or false readings will result.

Before you begin the test, allow the engine to warm up completely. Block the wheels and set the parking brake. With the transmission in neutral (or Park, on automatics), start the engine and allow it to run at normal idle speed. **Warning:** *Carefully inspect the fan blades for cracks or damage before starting the engine. Keep your hands and the vacuum tester clear of the fan and do not stand in front of the vehicle or in line with the fan when the engine is running.*

Read the vacuum gauge; a average, healthy engine should normally produce about 17 to 22 inches of vacuum with a fairly steady needle. Refer to the following vacuum gauge readings and what they indicate about the engine's condition:

1 A low steady reading usually indicates a leaking gasket between the intake manifold and carburetor or throttle body, a leaky vacuum hose, late ignition timing or incorrect camshaft timing. Check ignition timing with a timing light and eliminate all other possible causes, utilizing the tests provided in this Chapter before you remove the timing chain cover to check the timing marks.

2 If the reading is three to eight inches below normal and it fluctuates at that low reading, suspect an intake manifold gasket leak at an intake port or a faulty injector (on port-injected models only).

3 If the needle has regular drops of about two to four inches at a steady rate the valves are probably leaking. Perform a compression or leak-down test to confirm this.

4 An irregular drop or down-flick of the needle can be caused by a sticking valve or an ignition misfire. Perform a compression or leak-down test and read the spark plugs.

5 A rapid vibration of about four inches-Hg at idle combined with exhaust smoke indicates worn valve guides. Perform a leak-down test to confirm this. If the rapid vibration occurs with an increase in engine speed, check for a leaking intake manifold gasket or head gasket, weak valve springs, burned valves or ignition misfire.

6 A slight fluctuation, say one inch up and down, may mean ignition problems. Check all the usual tune-up items and, if necessary, run the engine on an ignition analyzer.

7 If there is a large fluctuation, perform a compression or leak-down test to look for a weak or dead cylinder or a blown head gasket.

8 If the needle moves slowly through a wide range, check for a clogged PCV system, incorrect idle fuel mixture, carburetor/throttle body or intake manifold gasket leaks.

9 Check for a slow return after revving the engine by quickly snapping the throttle open until the engine reaches about 2,500 rpm and let it shut. Normally the reading should drop to near zero, rise above normal idle reading (about 5 in.-Hg over) and then return to the previous idle reading. If the vacuum returns slowly and doesn't peak when the throttle is snapped shut, the rings may be worn. If there is a long delay, look for a restricted exhaust system (often the muffler or catalytic converter). An easy way to check this is to temporarily disconnect the exhaust ahead of the suspected part and redo the test.

5 Engine removal - methods and precautions

If you've decided the engine must be removed for overhaul or major repair work, several preliminary steps should be taken.

Locating a suitable place to work is extremely important. Adequate work space, along with storage space for the vehicle, will be needed. If a shop or garage isn't available, at the very least a flat, level, clean work surface made of concrete or asphalt is required.

Cleaning the engine compartment and engine before beginning the removal procedure will help keep tools clean and organized.

An engine hoist or A-frame will also be necessary. Make sure the equipment is rated in excess of the combined weight of the engine and its accessories. Safety is of primary importance, considering the potential hazards involved in lifting the engine out of the vehicle.

If the engine is being removed by a novice, a helper should be available. Advice and aid from someone more experienced would also be helpful. There are many instances when one person cannot simultaneously perform all of the operations required when lifting the engine out of the vehicle.

Plan the operation ahead of time. Arrange for or obtain all of the tools and equipment you'll need prior to beginning the job. Some of the equipment necessary to perform engine removal and installation safely and with relative ease are (in addition to an engine hoist) a heavy duty floor jack, complete sets of wrenches and sockets as described in the front of this manual, wooden blocks and plenty of rags and cleaning solvent for mopping up spilled oil, coolant and gasoline. If the hoist must be rented, be sure to arrange for it in advance and perform all of the operations possible without it beforehand. This will save you money and time.

Plan for the vehicle to be out of use for quite a while. A machine shop will be required to perform some of the work which the do-it-yourselfer can't accomplish without special equipment. These shops often have a busy schedule, so it would be a good idea to consult them before removing the engine in order to accurately estimate the amount of time required to rebuild or repair components that may need work.

Always be extremely careful when removing and installing the

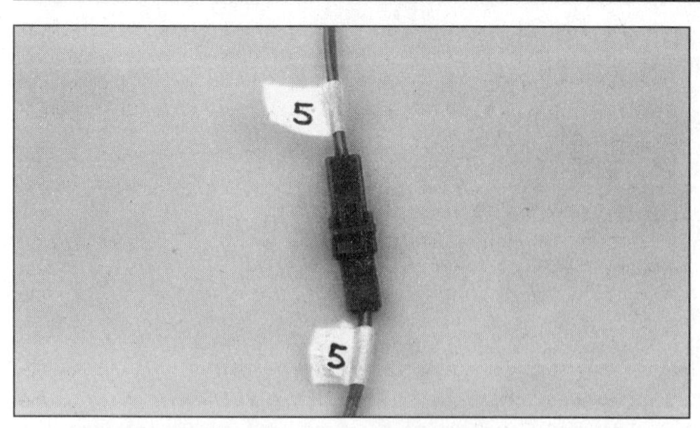

6.6 Label each wire before unplugging the connector

engine. Serious injury can result from careless actions. Plan ahead, take your time and a job of this nature, although major, can be accomplished successfully.

6 Engine - removal and installation

Refer to illustrations 6.6, 6.21, 6.24a, 6.24b, 6.25a, 6.25b and 6.30

Warning: *Gasoline is extremely flammable, so take extra precautions when you work on any part of the fuel system. Don't smoke or allow open flames or bare light bulbs near the work area, and don't work in a garage where a natural gas-type appliance (such as a water heater or a clothes dryer) with a pilot light is present. Since gasoline is carcinogenic, wear latex gloves when there's a possibility of being exposed to fuel, and, if you spill any fuel on your skin, rinse it off immediately with soap and water. Mop up any spills immediately and do not store fuel-soaked rags where they could ignite. The fuel system is under constant pressure, so, if any fuel lines are to be disconnected, the fuel pressure in the system must be relieved first (see Chapter 4 for more information). When you perform any kind of work on the fuel system, wear safety glasses and have a Class B type fire extinguisher on hand.*

Note: *Read through the following steps carefully and familiarize yourself with the procedure before beginning work. Also at this point it it'll helpful to use a penetrating fluid or spray on nuts and bolts that are difficult to remove, such as exhaust manifolds, bellhousing, engine mounts, etc.*

Removal

1 Disconnect the negative battery cable. **Caution:** *It is necessary to make sure the radio is turned OFF before disconnecting the battery cables to avoid damaging the microprocessor built into the radio.*

2 If the engine is fuel injected, refer to Chapter 4A and relieve the fuel system pressure.

3 Cover the fenders and cowl and remove the hood (see Chapter 11). Special pads are available to protect the fenders, but an old bedspread or blanket will also work.

4 Remove the air cleaner assembly (see Chapter 4A and 4B), and the air intake duct.

5 Disconnect the charcoal canister hose from the throttle body (see Chapter 6).

6 Label the vacuum lines, emissions system hoses, electrical connectors, ground straps and fuel lines to ensure correct reinstallation, then detach them. Pieces of masking tape with numbers or letters written on them work well **(see illustration)**. If there's any possibility of confusion, make a sketch of the engine compartment and clearly label the lines, hoses and wires.

7 Label and detach all coolant hoses from the engine.

8 Remove the coolant reservoir, cooling fan, shroud and radiator (see Chapter 3).

9 Remove the drivebelt(s) and idler, if equipped (see Chapter 1).

10 Disconnect the fuel lines running from the engine to the chassis (see Chapter 4A and 4B). Plug or cap all open fittings and lines.

11 Disconnect the accelerator linkage or cable, cruise control cable (if equipped) and kickdown cable (automatic transmission models) from the engine (see Chapters 4 and 7).

12 Unbolt the power steering pump and set it aside (see Chapter 10). Leave the lines/hoses attached and make sure the pump is kept in an upright position in the engine compartment.

13 Unbolt the air conditioning compressor (see Chapter 3) and set it aside. **Warning:** *Do not disconnect the hoses.*

14 Unbolt the alternator and mounting strap and set it aside (see Chapter 5).

15 Raise the vehicle and support it securely on jackstands. Drain the cooling system (see Chapter 1).

16 Drain the engine oil and remove the filter (see Chapter 1).

17 Remove the starter (see Chapter 5).

18 Remove the splash shields from the wheel wells and the underside of the engine compartment.

19 If the vehicle is equipped with an automatic transmission, remove the inspection cover from the bellhousing, which will give access to the torque converter (see Chapter 7).

20 Also on automatic transmission models, remove the crankshaft pulley and reinstall the bolt. Use a socket and a long breaker bar or ratchet to rotate the engine to position the torque converter bolts for removal. Remove the torque converter bolts.

21 Remove the oil dipstick tube **(see illustration)**.

22 Attach an engine sling or a length of chain to the lifting brackets on the engine.

23 Roll the hoist into position and connect the sling to it. Take up the slack in the sling or chain, but don't lift the engine. **Warning:** *DO NOT place any part of your body under the engine when it's supported only by a hoist or other lifting device.*

24 Remove the engine mount-to-crossmember bolts **(see illustration)**. Remove the crossmember **(see illustration)**.

2B

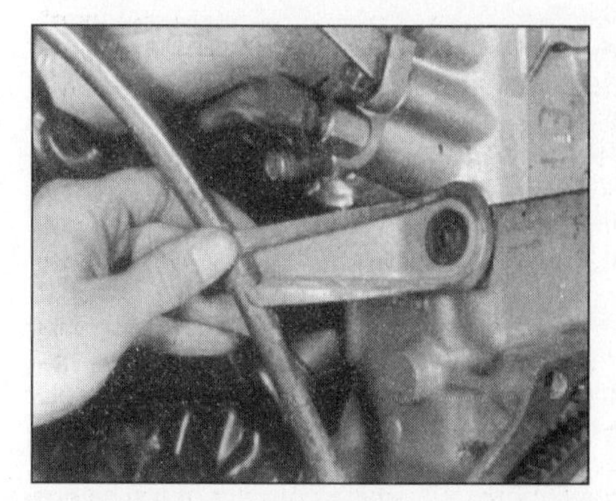

6.21 Remove the bolt from the oil dipstick tube brace, then carefully pull out the tube

6.24a Remove the engine mount bolts (arrows) from the crossmember

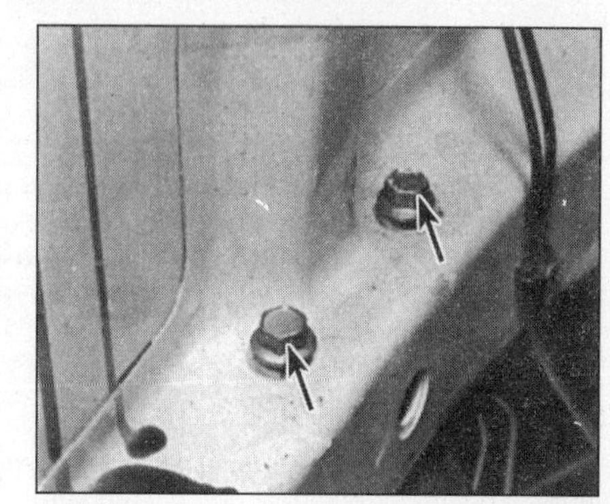

6.24b The crossmember-to-chassis bolts (arrows) are accessible from inside the engine compartment

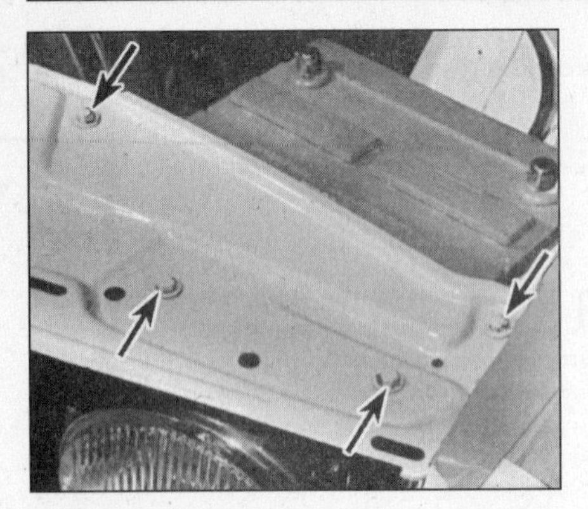

6.25a Remove the bolts (arrows) from the radiator support

6.25b Also remove the radiator support brace

6.30 Raise the engine and carefully guide it from the engine compartment

25 Remove the radiator support **(see illustration)** and the brace **(see illustration)**.
26 Disconnect the exhaust system from the engine (see Chapter 4).
27 Support the transmission with a jack (preferably a transmission jack). If you're not using a transmission jack, position a block of wood on the jack head to prevent damage to the transmission.
28 Remove the driveshaft(s) (see Chapter 8).
29 Recheck to be sure nothing is still connecting the engine to the vehicle. Disconnect anything still remaining.
30 Slowly raise the engine assembly out of the engine compartment, angling the front of the engine above the front bumper as necessary, for clearance. Also, slightly raise the jack supporting the transmission as the engine is raised. Check carefully to make sure nothing is hanging up as the hoist is raised **(see illustration)**.
31 Lower the engine assembly to the ground and support it with blocks of wood. Remove the clutch and flywheel or driveplate and mount the engine on an engine stand.

Installation

32 Check the engine and transmission mounts. If they're worn or damaged, replace them.
33 If you're working on a manual transmission vehicle, install the clutch and pressure plate (see Chapter 8). Now is a good time to install a new clutch. Apply a thin film of high-temperature grease to the input shaft.
34 Attach the hoist to the engine assembly and lower the assembly into the engine compartment.
35 The remainder of installation is the reverse of the removal steps.
36 Add coolant, oil, power steering and transmission fluid as needed (see Chapter 1).
37 Run the engine and check for leaks and proper operation of all accessories, then install the hood and test drive the vehicle.
38 If the air conditioning system was discharged, have it evacuated, recharged and leak tested by the shop that discharged it.

7 Engine rebuilding alternatives

The home mechanic is faced with a number of options when performing an engine overhaul. The decision to replace the engine block, piston/connecting rod assemblies and crankshaft depends on a number of factors, with the number one consideration being the condition of the block. Other considerations are cost, access to machine shop facilities, parts availability, time required to complete the project and the extent of prior mechanical experience.

Some of the rebuilding alternatives include:

Individual parts - If the inspection procedures reveal the engine block and most engine components are in reusable condition, purchasing individual parts may be the most economical alternative. The block, crankshaft and piston/connecting rod assemblies should all be inspected carefully. Even if the block shows little wear, the cylinder bores should be surface honed.

Short block - A short block consists of an engine block with a crankshaft and piston/connecting rod assemblies already installed. All new bearings are incorporated and all clearances will be correct. The existing camshaft, valve train components, cylinder head(s) and external parts can be bolted to the short block with little or no machine shop work necessary.

Long block - A long block consists of a short block plus an oil pump, oil pan, cylinder head, valve cover, camshaft and valve train components, timing sprockets and belt. All components are installed with new bearings, seals and gaskets incorporated throughout. The installation of manifolds and external parts is all that's necessary.

Give careful thought to which alternative is best for you and discuss the situation with local automotive machine shops, auto parts dealers and experienced rebuilders before ordering or purchasing replacement parts.

8 Engine overhaul - disassembly sequence

1 It's much easier to disassemble and work on the engine if it's mounted on a portable engine stand. A stand can often be rented quite cheaply from an equipment rental yard. Before it's mounted on a stand, the flywheel/driveplate should be removed from the engine.
2 If a stand isn't available, it's possible to disassemble the engine with it blocked up on the floor. Be extra careful not to tip or drop the engine when working without a stand.
3 If you're going to obtain a rebuilt engine, all external components must come off first, to be transferred to the replacement engine, just as they will if you're doing a complete engine overhaul yourself. These include:

Alternator and brackets
Power steering pump and brackets
Emissions control components
Distributor, spark plug wires and spark plugs
Thermostat and housing cover
Water pump and bypass pipe
Fuel injection components
Carburetor components
Intake/exhaust manifolds
Oil filter
Engine mounts
Clutch and flywheel or driveplate

Note: *When removing the external components from the engine, pay close attention to details that may be helpful or important during installation. Note the installed position of gaskets, seals, spacers, pins, brackets, washers, bolts, wiring and other small items.*
4 If you're obtaining a short block, which consists of the engine block, crankshaft, pistons and connecting rods all assembled, then the cylinder head(s), oil pan and oil pump will have to be removed as well. See *Engine rebuilding alternatives* for additional information regarding

9.2 A small plastic bag, with an appropriate label, can be used to store the valve train components so they can be kept together and reinstalled in the original position

9.3a Use a valve spring compressor to compress the springs, remove the keepers from the valve stem with a magnet or small needle-nose pliers . . .

9.3b . . . then lift the retainer and spring (and any shims that may be present) from the cylinder head

the different possibilities to be considered.

5 If you're planning a complete overhaul, the engine must be disassembled and the internal components removed in the following general order :

Intake and exhaust manifolds
Valve cover
Cylinder head with camshaft
Oil pan
Oil pick-up tube
Timing belt covers and bolts
Intermediate shaft
Bearing caps
Piston/connecting rod assemblies
Rear main oil seal retainer
Crankshaft and main bearings

6 Before beginning the disassembly and overhaul procedures, make sure the following items are available. Also, refer to *Engine overhaul - reassembly sequence* for a list of tools and materials needed for engine reassembly.

Common hand tools
Small cardboard boxes or plastic bags for storing parts
Gasket scraper
Ridge reamer
Vibration damper puller
Micrometers
Telescoping gauges
Dial indicator set
Valve spring compressor
Cylinder surfacing hone
Piston ring groove cleaning tool
Electric drill motor
Tap and die set
Wire brushes
Oil gallery brushes
Cleaning solvent

9 Cylinder head - disassembly

Refer to illustrations 9.2, 9.3a, 9.3b and 9.4

Note 1: *New and rebuilt cylinder heads are commonly available for most engines at dealer parts departments and auto parts stores. Due to the fact that some specialized tools are necessary for the disassembly and inspection procedures, and replacement parts aren't always readily available, it may be more practical and economical for the home mechanic to purchase replacement head(s) rather than taking the time to disassemble, inspect and recondition the original(s).*

Note 2: *The B21F-T and B230F engines are equipped with Stellite coating on the exhaust valves. These valve surfaces are much harder and wear less over a longer period of time. Do not grind these valves on any type of machinery but instead use valve lapping compound. B21F-T engines also use sodium-filled exhaust valves that stay cooler at high engine temperatures. These valves must not be disposed with other metal because of their explosive properties.*

1 Cylinder head disassembly involves removal of the intake and exhaust valves and related components. The camshaft and cam follower assemblies must be removed before beginning the cylinder head disassembly procedure (see Part A of this Chapter). Label the parts or store them separately so they can be reinstalled in their original locations.

2 Before the valves are removed, arrange to label and store them, along with their related components, so they can be kept separate and reinstalled in their original locations **(see illustration)**.

3 Compress the springs on the first valve with a spring compressor and remove the keepers **(see illustration)**. Carefully release the valve spring compressor and remove the retainer, the spring **(see illustration)** and the spring seat (if used).

4 Pull the valve out of the head, then remove the oil seal from the guide **(see illustration)**. If the valve binds in the guide (won't pull through), push it back into the head and deburr the area around the stem and the keeper groove with a fine file or whetstone.

5 Repeat the procedure for the remaining valves. Remember to keep all the parts for each valve together so they can be reinstalled in the same locations.

6 Once the valves and related components have been removed and stored in an organized manner, the head should be thoroughly cleaned and inspected. If a complete engine overhaul is being done, finish the engine disassembly procedures before beginning the cylinder head cleaning and inspection process.

9.4 Remove the valve stem seal from the valve guide

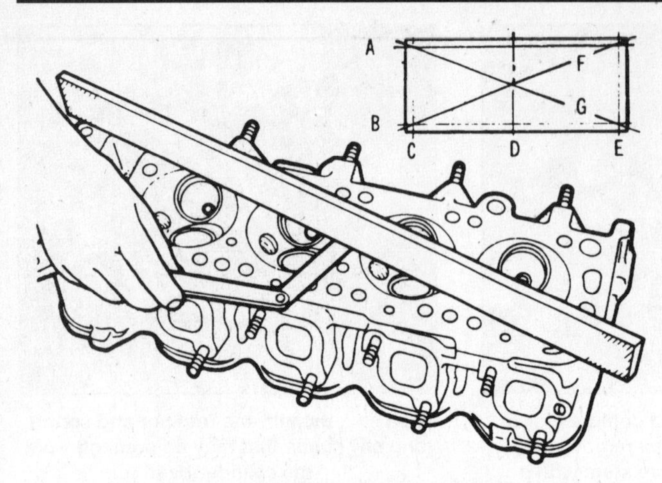

10.12 Check the cylinder head gasket surface for warpage by trying to slip a feeler gauge under the straightedge (see this Chapter's Specifications for the maximum warpage allowed and use a feeler gauge of that thickness)

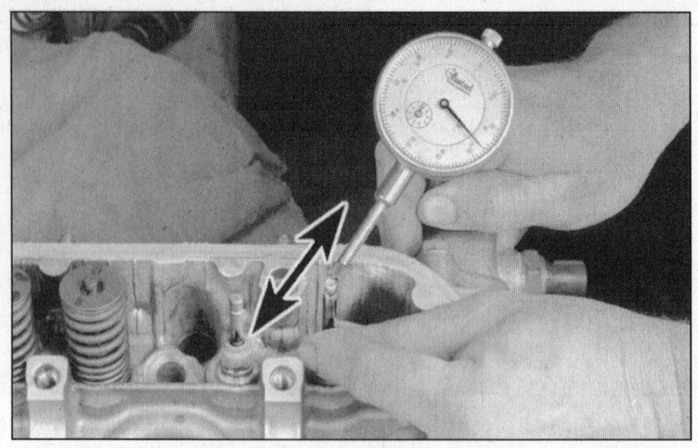

10.14 A dial indicator can be used to determine the valve stem-to-guide clearance (move the valve stem as indicated by the arrows)

10 Cylinder head - cleaning and inspection

Note: *The B21F-T and B230F engines are equipped with Stellite coating on the exhaust valves. These valve surfaces are much harder and wear less over a longer period of time. Do not grind these valves on any type of machinery but instead use valve lapping compound. B21F-T engines also use sodium filled exhaust valves that stay cooler at high engine temperatures. These valves must not be disposed with other metal because of their explosive properties.*

1 Thorough cleaning of the cylinder head(s) and related valvetrain components, followed by a detailed inspection, will enable you to decide how much valve service work must be done during the engine overhaul. **Note:** *If the engine was severely overheated, the cylinder head is probably warped.*

Cleaning

2 Scrape all traces of old gasket material and sealant off the head gasket, intake manifold and exhaust manifold mating surfaces. Be very careful not to gouge the cylinder head. Special gasket removal solvents that soften gaskets and make removal much easier are available at auto parts stores.

3 Remove all built-up scale from the coolant passages.

4 Run a stiff wire brush through the various holes to remove deposits that may have formed in them.

5 Run an appropriate size tap into each of the threaded holes to remove corrosion and thread sealant that may be present. If compressed air is available, use it to clear the holes of debris produced by this operation. **Warning:** *Wear eye protection when using compressed air!*

6 Clean the camshaft bearing cap bolt threads with a wire brush.

7 Clean the cylinder head with solvent and dry it thoroughly. Compressed air will speed the drying process and ensure that all holes and recessed areas are clean. **Note:** *Decarbonizing chemicals are available and may prove very useful when cleaning cylinder heads and valve train components. They're very caustic and should be used with caution. Be sure to follow the instructions on the container.*

8 Clean the rocker arms and bearing caps with solvent and dry them thoroughly (don't mix them up during the cleaning process). Compressed air will speed the drying process and can be used to clean out the oil passages.

9 Clean all the valve springs, spring seats, keepers and retainers with solvent and dry them thoroughly. Do the components from one valve at a time to avoid mixing up the parts.

10 Scrape off any heavy deposits that may have formed on the valves, then use a motorized wire brush to remove deposits from the valve heads and stems. Again, make sure the valves don't get mixed up.

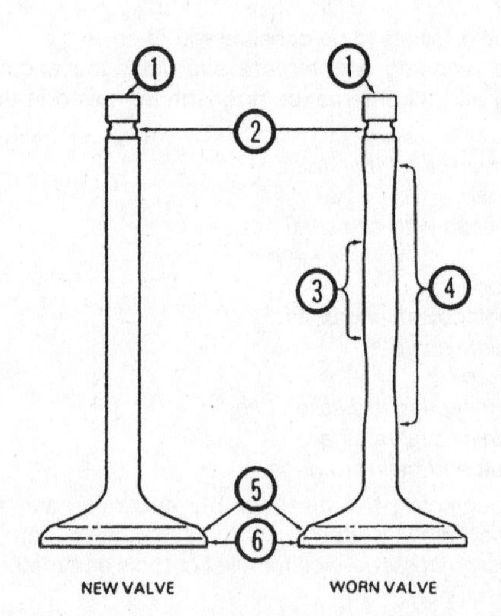

10.15 Check for valve wear at the points shown here

1 *Valve tip*
2 *Keeper groove*
3 *Stem (least worn area)*
4 *Stem (most worn area)*
5 *Valve face*
6 *Margin*

Inspection

Refer to illustrations 10.12, 10.14, 10.15, 10.16, 10.17 and 10.18

Note: *Be sure to perform all of the following inspection procedures before concluding machine shop work is required. Make a list of the items that need attention.*

Cylinder head

11 Inspect the head very carefully for cracks, evidence of coolant leakage and other damage. If cracks are found, check with an automotive machine shop concerning repair. If repair isn't possible, a new cylinder head should be obtained.

12 Using a straightedge and feeler gauge, check the head gasket mating surface for warpage **(see illustration)**. If the warpage exceeds the limit in this Chapter's Specifications, it can be resurfaced at an automotive machine shop.

13 Examine the valve seats in each of the combustion chambers. If they're pitted, cracked or burned, the head will require valve service that's beyond the scope of the home mechanic.

14 Check the valve stem-to-guide clearance by measuring the lateral movement of the valve stem with a dial indicator attached securely to

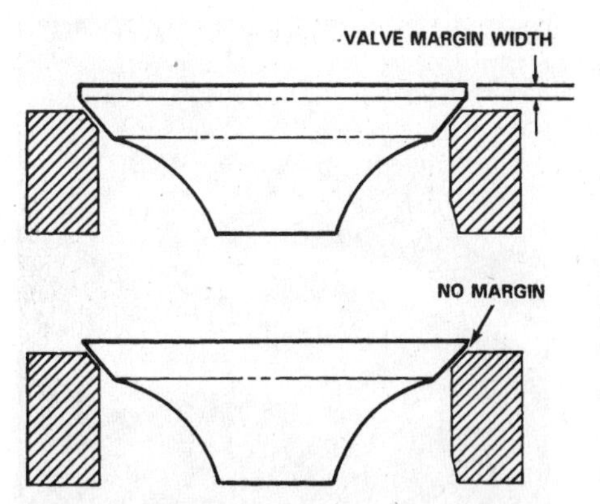

10.16 The margin width on each valve must be as specified (if no margin exists, the valve cannot be reused)

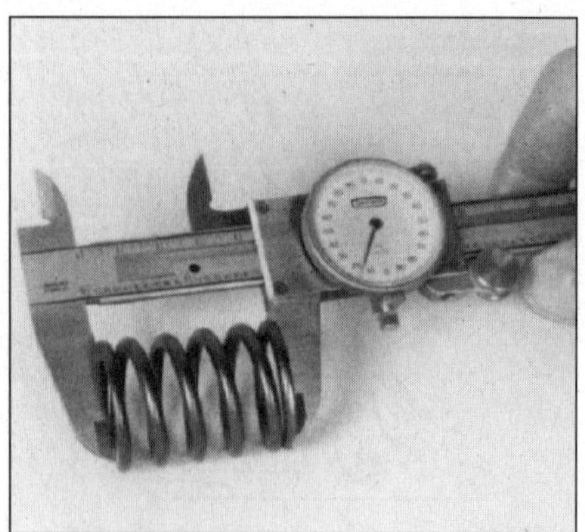

10.17 Measure the free length of each valve spring with a dial or vernier caliper

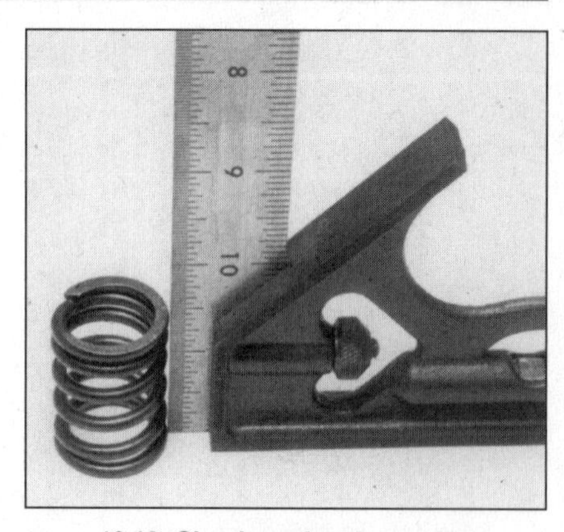

10.18 Check each valve spring for squareness

the head **(see illustration)**. The valve must be in the guide and approximately 1/16-inch off the seat. The total valve stem movement indicated by the gauge needle must be divided by two to obtain the actual clearance. After this is done, if there's still some doubt regarding the condition of the valve guides, they should be checked by an automotive machine shop (the cost should be minimal).

Valves

15 Carefully inspect each valve face for uneven wear, deformation, cracks, pits and burned areas. Check the valve stem for scuffing and galling and the neck for cracks. Rotate the valve and check for any obvious indication that it's bent. Look for pits and excessive wear on the end of the stem **(see illustration)**. The presence of any of these conditions indicates the need for valve service by an automotive machine shop.

16 Measure the margin width on each valve **(see illustration)**. Any valve with a margin narrower than 1/32 inch will have to be replaced with a new one.

Valve components

17 Check each valve spring for wear (on the ends) and pits. Measure the free length and compare it to this Chapter's Specifications **(see illustration)**. Any springs that are shorter than specified have sagged and shouldn't be reused. The tension of all springs should be checked with a special fixture before deciding they're suitable for use in a rebuilt engine (take the springs to an automotive machine shop for this check).

18 Stand each spring on a flat surface and check it for squareness **(see illustration)**. If any of the springs are distorted or sagged, replace all of them with new parts.

19 Check the spring retainers and keepers for obvious wear and cracks. Any questionable parts should be replaced with new ones, as extensive damage will occur if they fail during engine operation.

20 If the inspection process indicates the valve components are in generally poor condition and worn beyond the limits specified, which is usually the case in an engine that's being overhauled, reassemble the valves in the cylinder head and refer to Section 11 for valve servicing recommendations.

11 Valves - servicing

1 Because of the complex nature of the job and the special tools and equipment needed, servicing of the valves, the valve seats and the valve guides, commonly known as a valve job, should be done by a professional.

2 The home mechanic can remove and disassemble the head, do the initial cleaning and inspection, then reassemble and deliver it to a dealer service department or an automotive machine shop for the actual service work. Doing the inspection will enable you to see what condition the head and valvetrain components are in and will ensure that you know what work and new parts are required when dealing with an automotive machine shop.

3 The dealer service department, or automotive machine shop, will remove the valves and springs, recondition or replace the valves and valve seats, recondition the valve guides, check and replace the valve springs, rotators, spring retainers and keepers (as necessary), replace the valve seals with new ones, reassemble the valve components and make sure the installed height is correct. The cylinder head gasket surface will also be resurfaced if it's warped.

4 After the valve job has been performed by a professional, the head will be in like new condition. When the head is returned, be sure to clean it again before installation on the engine to remove any metal particles and abrasive grit that may still be present from the valve service or head resurfacing operations. Use compressed air, if available, to blow out all the oil holes and passages.

12 Cylinder head - reassembly

Refer to illustrations 12.4, 12.6 and 12.8

1 Regardless of whether or not the head was sent to an automotive repair shop for valve servicing, make sure it's clean before beginning reassembly.

2 If the head was sent out for valve servicing, the valves and related components will already be in place. Begin the reassembly procedure with Step 8.

3 Install the spring seats before the valve seals.

4 Install new seals on each of the valve guides. Using a hammer and a deep socket or seal installation tool, gently tap each seal into

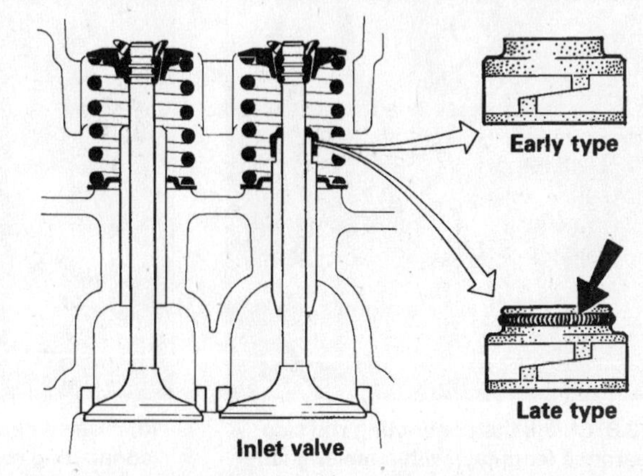

12.4 Be sure to replace the valve stem seals with the correct type

12.6 Apply a small dab of grease to each keeper as shown here before installation - it'll hold them in place on the valve stem as the spring is released

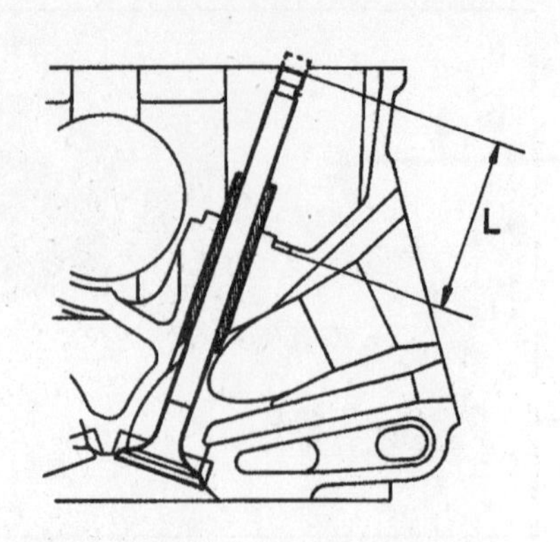

12.8 Be sure to check the valve stem installed height (the distance from the spring seat to the top of the valve stem)

13.1 A ridge reamer is required to remove the ridge from the top of each cylinder - do this before removing the pistons!

place until it's completely seated on the guide **(see illustrations)**. Don't twist or cock the seals during installation or they won't seal properly on the valve stems. **Note:** *Be sure to install the correct type valve stem seals. Early models use a rubber lip while later models use the spring around the lip.*

5 Beginning at one end of the head, lubricate and install the first valve. Apply moly-base grease or clean engine oil to the valve stem.

6 Position the valve springs (and shims, if used) over the valves. Compress the springs with a valve spring compressor and carefully install the keepers in the groove, then slowly release the compressor and make sure the keepers seat properly. Apply a small dab of grease to each keeper to hold it in place if necessary **(see illustration)**.

7 Repeat the procedure for the remaining valves. Be sure to return the components to their original locations - don't mix them up!

8 Check the installed valve stem height (dimension L) with a vernier or dial caliper **(see illustration)**. If the head was sent out for service work, the installed height should be correct (but don't automatically assume it is). The measurement is taken from the spring seat to the top of the valve stem. If the height is greater than listed in this Chapter's Specifications, shims can be added under the springs to correct it. **Caution:** *Do not, under any circumstances, shim the springs to the point where the installed height is less than specified.*

9 Apply camshaft assembly lube or moly-base grease to the rocker arm faces, the camshaft and the rocker shafts, then install the camshaft, rocker arms and shafts (refer to Chapter 2A).

13 Pistons and connecting rods - removal

Refer to illustrations 13.1, 13.3, 13.4 and 13.6

Note: *Prior to removing the piston/connecting rod assemblies, remove the cylinder head, the oil pan and the oil pump by referring to the appropriate Sections in Part A of Chapter 2.*

1 Use your fingernail to feel if a ridge has formed at the upper limit of ring travel (about 1/4-inch down from the top of each cylinder). If carbon deposits or cylinder wear have produced ridges, they must be completely removed with a special tool **(see illustration)**. Follow the manufacturer's instructions provided with the tool. Failure to remove the ridges before attempting to remove the piston/connecting rod assemblies may result in piston breakage.

2 After the cylinder ridges have been removed, turn the engine upside-down so the crankshaft is facing up.

3 Before the connecting rods are removed, check the connecting rod side clearance (endplay) with feeler gauges. Slide them between the first connecting rod and the crankshaft throw until the play is removed **(see illustration)**. The endplay is equal to the thickness of the feeler gauge(s). If the endplay exceeds the service limit, new connecting rods will be required. If new rods (or a new crankshaft) are installed, the endplay may fall under the minimum listed in this Chapter's Specifications (if it does, the rods will have to be machined to restore it - consult an automotive machine shop for advice if

13.3 Check the connecting rod side clearance (endplay) with a feeler gauge

13.4 Use a center-punch and mark each connecting rod on the bearing cap and then on the connecting rod itself (arrows)

13.6 To prevent damage to the crankshaft journals and cylinder walls, slip sections of rubber or plastic hose over the rod bolts before removing the pistons

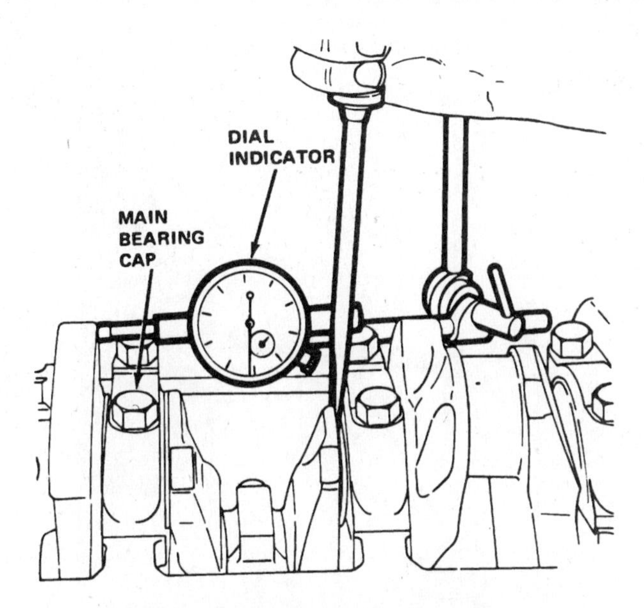

14.1 Checking crankshaft endplay with a dial indicator

necessary). Repeat the procedure for the remaining connecting rods.

4 Check the connecting rods and caps for identification marks. If they aren't plainly marked, use a small center-punch to make the appropriate number of indentations on each rod and cap (1, 2, 3, etc., depending on the engine type and cylinder they're associated with) **(see illustration)**.

5 Loosen each of the connecting rod cap nuts 1/2-turn at a time until they can be removed by hand. Remove the number one connecting rod cap and bearing insert. Don't drop the bearing insert out of the cap.

6 Slip a short length of plastic or rubber hose over each connecting rod cap bolt to protect the crankshaft journal and cylinder wall as the piston is removed **(see illustration)**.

7 Remove the bearing insert and push the connecting rod/piston assembly out through the top of the engine. Use a wooden or plastic hammer handle to push on the upper bearing surface in the connecting rod. If resistance is felt, double-check to make sure all of the ridge was removed from the cylinder.

8 Repeat the procedure for the remaining cylinders.

9 After removal, reassemble the connecting rod caps and bearing inserts in their respective connecting rods and install the cap nuts finger tight. Leaving the old bearing inserts in place until reassembly will help prevent the connecting rod bearing surfaces from being accidentally nicked or gouged.

10 Don't separate the pistons from the connecting rods (see Section 18 for additional information).

14 Crankshaft and intermediate shaft - removal

Note: *It's assumed the flywheel or driveplate, crankshaft pulley, timing belt, oil pan, oil pump and piston/connecting rod assemblies have already been removed. The rear main oil seal housing must be unbolted and separated from the block before proceeding with crankshaft removal.*

Crankshaft

Refer to illustrations 14.1 and 14.6

1 Before the crankshaft is removed, check the endplay. Mount a dial indicator with the stem in line with the crankshaft and touching one of the crank throws.

2 Push the crankshaft all the way to the rear and zero the dial indicator. Next, pry the crankshaft to the front as far as possible and check the reading on the dial indicator. The distance it moves is the endplay. If it's greater than specified in this Chapter's specifications, check the crankshaft thrust surfaces for wear. If no wear is evident, new main bearings should correct the endplay.

14.6 Lift the crankshaft from the engine block

3 If a dial indicator isn't available, feeler gauges can be used. Gently pry or push the crankshaft all the way to the front of the engine. Slip feeler gauges between the crankshaft and the front face of the thrust main bearing to determine the clearance.

4 Loosen the main bearing cap bolts 1/4-turn at a time each, until they can be removed by hand.

5 Gently tap each cap with a soft-face hammer, then separate the assembly from the engine block. If necessary, use a large screwdriver as a lever to remove the caps. Try not to drop the bearing inserts if they come out with the caps.

6 Carefully lift the crankshaft out of the engine **(see illustration)**. It may be a good idea to have an assistant available, since the crankshaft is quite heavy. With the bearing inserts in place in the engine block and main bearing caps, return the caps to their respective locations on the engine block and tighten the bolts finger tight.

Intermediate shaft

Refer to illustrations 14.9a, 14.9b and 14.10

7 Remove the sprocket from the intermediate shaft (see Chapter 2A). Remove the bolts from the intermediate shaft thrust plate.

8 Remove the distributor if not already done (see Chapter 5).

9 Remove the oil pump drive gear **(see illustrations)**.

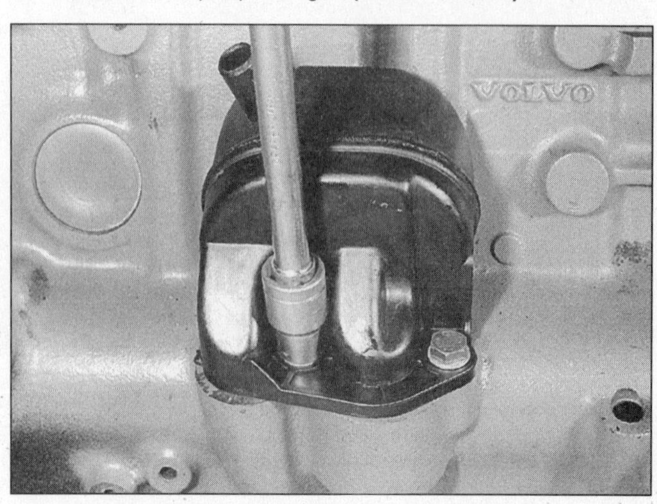

14.9a Remove the oil trap from the engine block

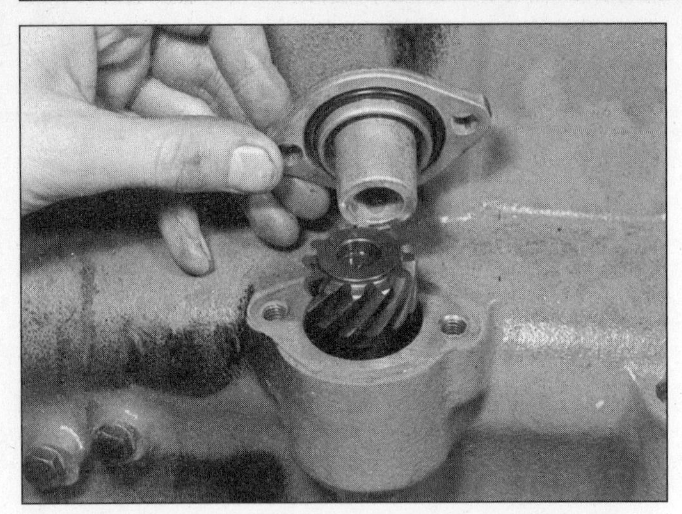

14.9b Remove the oil pump gear from the engine block

10 Remove the intermediate shaft assembly. **(see illustration). Caution:** *Pull the shafts straight out, taking care not to damage the bushings.*

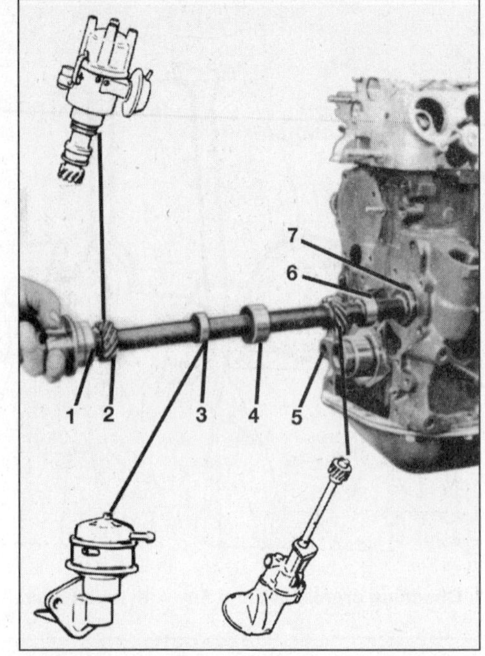

14.10 Removing the intermediate shaft

1 *Front bearing*
2 *Distributor gear*
3 *Fuel pump cam*
4 *Center bearing*
5 *Oil pump drive gear*
6 *Manufacturing support*
7 *Rear bearing*

15 Engine block - cleaning

Refer to illustrations 15.4a, 15.4b, 15.8 and 15.10

1 Remove the main bearing caps and separate the bearing inserts from the caps and the engine block. Tag the bearings, indicating which cylinder they were removed from and whether they were in the cap or the block, then set them aside.

2 Using a gasket scraper, remove all traces of gasket material from the engine block. Be very careful not to nick or gouge the gasket sealing surfaces.

3 Remove all of the covers and threaded oil gallery plugs from the block. The plugs are usually very tight - they may have to be drilled out and the holes retapped. Use new plugs when the engine is reassembled.

4 Remove the core plugs from the engine block. To do this, knock one side of the plug into the block with a hammer and a punch, then grasp them with large pliers and pull them out **(see illustrations).**

5 If the engine is extremely dirty, it should be taken to an automotive machine shop to be cleaned.

6 After the block is returned, clean all oil holes and oil galleries one more time. Brushes specifically designed for this purpose are available at most auto parts stores. Flush the passages with warm water until the water runs clear, dry the block thoroughly and wipe all machined surfaces with a light, rust preventive oil. If you have access to compressed air, use it to speed the drying process and blow out all the oil holes and galleries. **Warning:** *Wear eye protection when using compressed air!*

7 If the block isn't extremely dirty or sludged up, you can do an adequate cleaning job with hot soapy water and a stiff brush. Take plenty of time and do a thorough job. Regardless of the cleaning method used, be sure to clean all oil holes and galleries very thoroughly, dry the block completely and coat all machined surfaces with light oil.

8 The threaded holes in the block must be clean to ensure accurate torque readings during reassembly. Run the proper size tap into each of the holes to remove rust, corrosion, thread sealant or sludge and

15.4a A hammer and a large punch can be used to knock the core plugs sideways in their bores

15.4b Pull the core plugs from the block with pliers

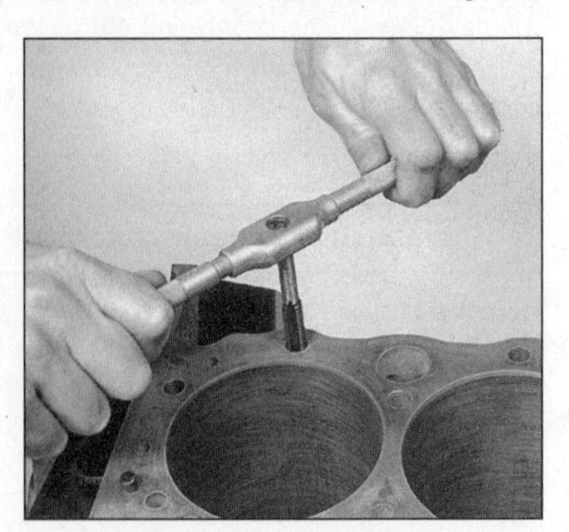

15.8 All bolt holes in the block - particularly the main bearing cap and head bolt holes - should be cleaned and restored with a tap (be sure to remove debris from the holes after this is done)

15.10 A large socket on an extension can be used to drive the new core plugs into the bores

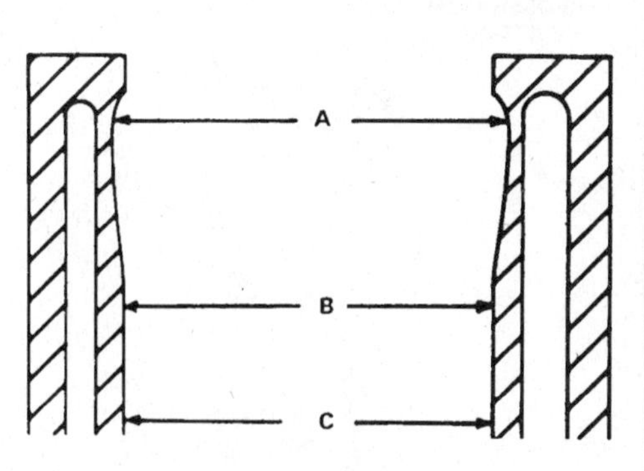

16.4a Measure the diameter of each cylinder just under the wear ridge (A), at the center (B) and at the bottom (C)

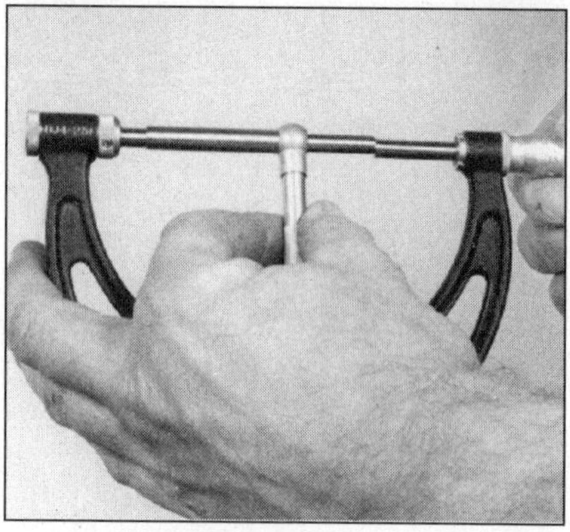

16.4b The gauge is then measured with a micrometer to determine the bore size

restore damaged threads **(see illustration)**. If possible, use compressed air to clear the holes of debris produced by this operation. Now is a good time to clean the threads on the head bolts and the main bearing cap bolts as well.

9 Reinstall the main bearing caps and tighten the bolts finger tight.

10 After coating the sealing surfaces of the new core plugs with Permatex no. 2 sealant (or equivalent), install them in the engine block **(see illustration)**. Make sure they're driven in straight and seated properly or leakage could result. Special tools are available for this purpose, but a large socket, with an outside diameter that will just slip into the core plug, a 1/2-inch drive extension and a hammer will work just as well.

11 Apply non-hardening sealant (such as Permatex no. 2 or Teflon pipe sealant) to the new oil gallery plugs and thread them into the holes in the block. Make sure they're tightened securely.

12 If the engine isn't going to be reassembled right away, cover it with a large plastic trash bag to keep it clean.

16 Engine block - inspection

Refer to illustrations 16.4a, 16.4b, 16.12a and 16.12b

1 Before the block is inspected, it should be cleaned as described in Section 15.

2 Visually check the block for cracks, rust and corrosion. Look for stripped threads in the threaded holes. It's also a good idea to have the block checked for hidden cracks by an automotive machine shop that has the special equipment to do this type of work. If defects are found, have the block repaired, if possible, or replaced.

3 Check the cylinder bores for scuffing and scoring.

4 Measure the diameter of each cylinder at the top (just under the ridge area), center and bottom of the cylinder bore, parallel to the crankshaft axis **(see illustrations)**. **Note:** *These measurements should not be made with the block mounted on an engine stand - the cylinders will be distorted and the measurements will be inaccurate.*

5 Next, measure each cylinder's diameter at the same three locations across the crankshaft axis. Compare the results to this Chapter's Specifications.

6 If the required precision measuring tools aren't available, the piston-to-cylinder clearances can be obtained, though not quite as accurately, using feeler gauge stock. Feeler gauge stock comes in 12-inch lengths and various thicknesses and is generally available at auto parts stores.

7 To check the clearance, select a feeler gauge and slip it into the cylinder along with the matching piston. The piston must be positioned exactly as it normally would be. The feeler gauge must be between the piston and cylinder on one of the thrust faces (90-degrees to the piston pin bore).

8 The piston should slip through the cylinder (with the feeler gauge in place) with moderate pressure.

9 If it falls through or slides through easily, the clearance is excessive and a new piston will be required. If the piston binds at the lower end of the cylinder and is loose toward the top, the cylinder is tapered. If tight spots are encountered as the piston/feeler gauge is rotated in the cylinder, the cylinder is out-of-round.

10 Repeat the procedure for the remaining pistons and cylinders.

11 If the cylinder walls are badly scuffed or scored, or if they're out-of-round or tapered beyond the limits given in this Chapter's Specifications, have the engine block rebored and honed at an automotive machine shop. If a rebore is done, oversize pistons and rings will be required.

12 Using a precision straightedge and feeler gauge, check the block deck (the surface that mates with the cylinder head) for distortion **(see illustrations)**.

2B

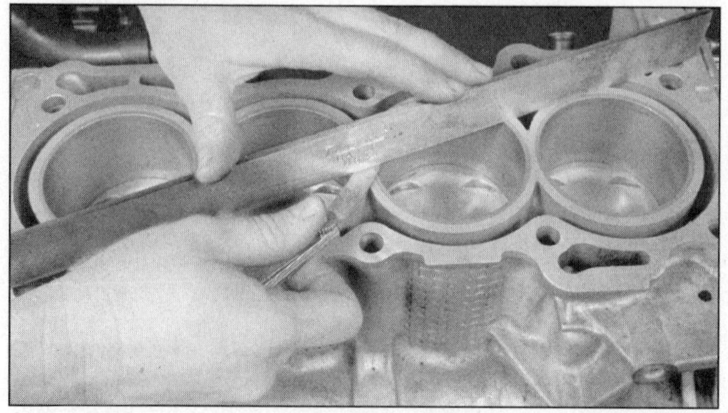

16.12a Check the block deck with a precision straightedge and feeler gauges

16.12b Lay the straightedge across the block, diagonally and from end-to-end when making the check

17.3a A "bottle brush" hone will produce better results if you've never honed cylinders before

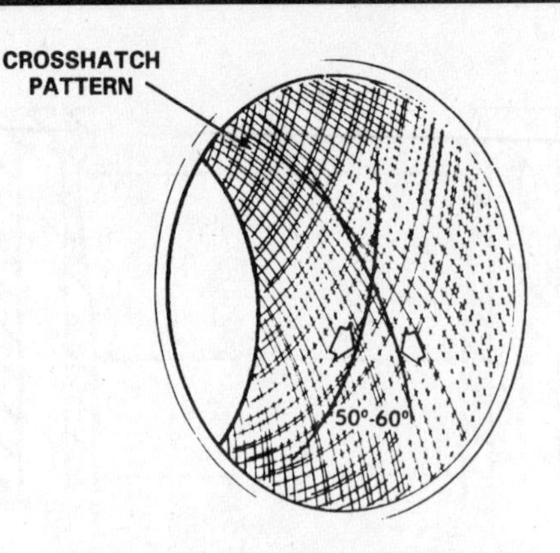

17.3b The cylinder hone should leave a smooth, cross-hatch pattern with the lines intersecting at approximately a 60-degree angle

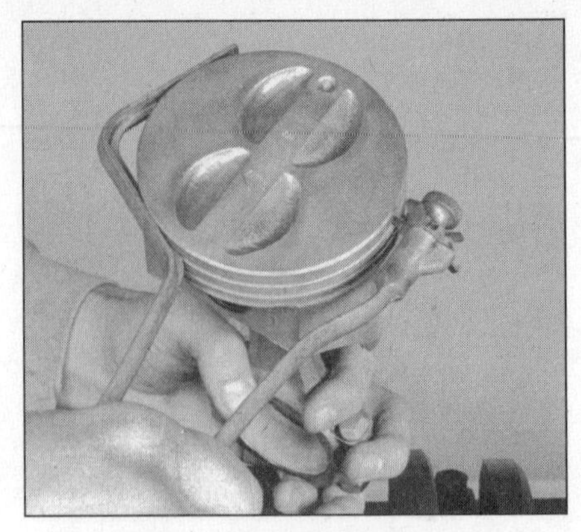

18.4a The piston ring grooves can be cleaned with a special tool, as shown here . . .

13 If the cylinders are in reasonably good condition and not worn to the outside of the limits, and if the piston-to-cylinder clearances can be maintained properly, they don't have to be rebored. Honing is all that's necessary (see Section 17).

17 Cylinder honing

Refer to illustrations 17.3a and 17.3b

1 Prior to engine reassembly, the cylinder bores must be honed so the new piston rings will seat correctly and provide the best possible combustion chamber seal. **Note:** *If you don't have the tools or don't want to tackle the honing operation, most automotive machine shops will do it for a reasonable fee.*

2 Before honing the cylinders, install the main bearing caps and tighten the bolts to the specified torque.

3 Two types of cylinder hones are commonly available - the flex hone or "bottle-brush" type and the more traditional surfacing hone with spring-loaded stones. Both will do the job, but for the less experienced mechanic the "bottle brush" hone will probably be easier to use. You'll also need some honing oil (kerosene will work if honing oil isn't available), rags and an electric drill motor. Proceed as follows:

a) *Mount the hone in the drill motor, compress the stones and slip it into the first cylinder* **(see illustration)**. *Be sure to wear safety goggles or a face shield!*
b) *Lubricate the cylinder with plenty of honing oil, turn on the drill and move the hone up-and-down in the cylinder at a pace that will produce a fine cross-hatch pattern on the cylinder walls. Ideally, the cross-hatch lines should intersect at approximately a 60-degree angle* **(see illustration)**. *Be sure to use plenty of lubricant and don't take off any more material than is absolutely necessary to produce the desired finish.* **Note:** *Piston ring manufacturers may specify a smaller cross-hatch angle than the traditional 60-degrees - read and follow any instructions included with the new rings.*
c) *Don't withdraw the hone from the cylinder while it's running. Instead, shut off the drill and continue moving the hone up-and-down in the cylinder until it comes to a complete stop, then compress the stones and withdraw the hone. If you're using a "bottle brush" type hone, stop the drill motor, then turn the chuck in the normal direction of rotation while withdrawing the hone from the cylinder.*
d) *Wipe the oil out of the cylinder and repeat the procedure for the remaining cylinders.*

4 After the honing job is complete, chamfer the top edges of the cylinder bores with a small file so the rings won't catch when the pistons are installed. Be very careful not to nick the cylinder walls with the end of the file.

5 The entire engine block must be washed again very thoroughly with warm, soapy water to remove all traces of the abrasive grit produced during the honing operation. **Note:** *The bores can be considered clean when a lint-free white cloth - dampened with clean engine oil - used to wipe them out doesn't pick up any more honing residue, which will show up as gray areas on the cloth. Be sure to run a brush through all oil holes and galleries and flush them with running water.*

6 After rinsing, dry the block and apply a coat of light rust preventive oil to all machined surfaces. Wrap the block in a plastic trash bag to keep it clean and set it aside until reassembly.

18 Pistons and connecting rods - inspection

Refer to illustrations 18.4a, 18,4b, 18.10 and 18.11

1 Before the inspection process can be carried out, the piston/connecting rod assemblies must be cleaned and the original piston rings removed from the pistons. **Note:** *Always use new piston rings when the engine is reassembled.*

2 Using a piston ring expander tool, carefully remove the rings from the pistons. Be careful not to nick or gouge the pistons in the process.

3 Scrape all traces of carbon from the top of the piston. A hand held wire brush or a piece of fine emery cloth can be used once the majority of the deposits have been scraped away. Do not, under any circumstances, use a wire brush mounted in a drill motor to remove deposits from the pistons. The piston material is soft and may be eroded away by the wire brush.

4 Use a piston ring groove cleaning tool to remove carbon deposits from the ring grooves. If a tool isn't available, a piece broken off the old ring will do the job. Be very careful to remove only the carbon deposits - don't remove any metal and do not nick or scratch the sides of the ring grooves **(see illustrations)**.

5 Once the deposits have been removed, clean the piston/rod assemblies with solvent and dry them with compressed air (if available). **Warning:** *Wear eye protection.* Make sure the oil return holes in the back sides of the ring grooves are clear.

6 If the pistons and cylinder walls aren't damaged or worn excessively, and if the engine block isn't rebored, new pistons won't be necessary. Normal piston wear appears as even vertical wear on the piston thrust surfaces and slight looseness of the top ring in its groove. New piston rings, however, should always be used when an engine is rebuilt.

7 Carefully inspect each piston for cracks around the skirt, at the pin bosses and at the ring lands.

8 Look for scoring and scuffing on the thrust faces of the skirt, holes in the piston crown and burned areas at the edge of the crown. If the skirt is scored or scuffed, the engine may have been suffering from

18.4b . . . or a section of a broken ring

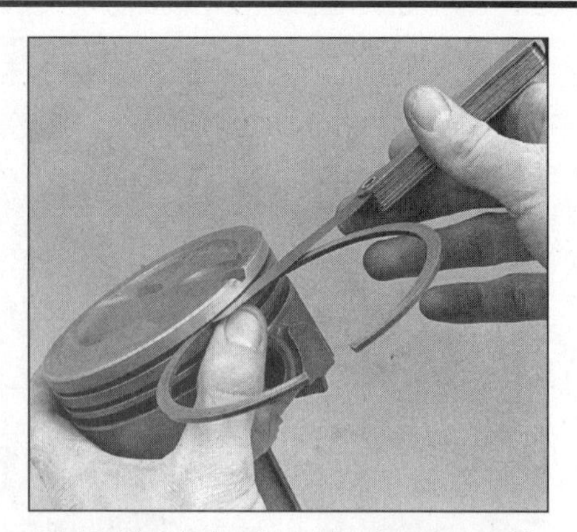

18.10 Check the ring side clearance with a feeler gauge at several points around the groove

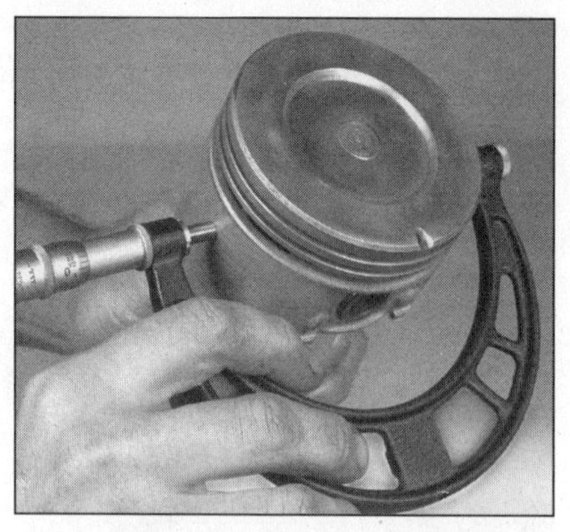

18.11 Measure the piston diameter at a 90-degree angle to the piston pin and in line with it

overheating and/or abnormal combustion, which caused excessively high operating temperatures. The cooling and lubrication systems should be checked thoroughly. A hole in the piston crown is an indication that abnormal combustion (pre-ignition) was occurring. Burned areas at the edge of the piston crown are usually evidence of spark knock (detonation). If any of the above problems exist, the causes must be corrected or the damage will occur again. The causes may include intake air leaks, incorrect fuel/air mixture, low octane fuel, ignition timing and EGR system malfunctions.

9 Corrosion of the piston, in the form of small pits, indicates coolant is leaking into the combustion chamber and/or the crankcase. Again, the cause must be corrected or the problem may persist in the rebuilt engine.

10 Measure the piston ring side clearance by laying a new piston ring in each ring groove and slipping a feeler gauge in beside it **(see illustration)**. Check the clearance at three or four locations around each groove. Be sure to use the correct ring for each groove - they are different. If the side clearance is greater than specified in this Chapter's specifications, new pistons will have to be used.

11 Check the piston-to-bore clearance by measuring the bore (see Section 16) and the piston diameter. Make sure the pistons and bores are correctly matched. Measure the piston across the skirt, at a 90-degree angle to the piston pin **(see illustration)** below the axis of the piston pin.

12 Subtract the piston diameter from the bore diameter to obtain the clearance. If it's greater than listed in this Chapter's Specifications, the block will have to be rebored and new pistons and rings installed.

13 Check the piston-to-rod clearance by twisting the piston and rod in opposite directions. Any noticeable play indicates excessive wear, which must be corrected. The piston/connecting rod assemblies should be taken to an automotive machine shop to have the pistons and rods re-sized and new pins installed.

14 If the pistons must be removed from the connecting rods for any reason, they should be taken to an automotive machine shop. While they are there have the connecting rods checked for bend and twist, since automotive machine shops have special equipment for this purpose. **Note:** *Unless new pistons and/or connecting rods must be installed, do not disassemble the pistons and connecting rods.*

15 Check the connecting rods for cracks and other damage. Temporarily remove the rod caps, lift out the old bearing inserts, wipe the rod and cap bearing surfaces clean and inspect them for nicks, gouges and scratches. After checking the rods, replace the old bearings, slip the caps into place and tighten the nuts finger tight. **Note:** *If the engine is being rebuilt because of a connecting rod knock, be sure to install new rods.*

19 Crankshaft and intermediate shaft - inspection

Crankshaft

Refer to illustrations 19.1, 19.2, 19.4, 19.6 and 19.8

1 Remove all burrs from the crankshaft oil holes with a stone, file or scraper **(see illustration)**.

2 Clean the crankshaft with solvent and dry it with compressed air (if available). **Warning:** *Wear eye protection when using compressed air.* Be sure to clean the oil holes with a stiff brush **(see illustration)** and flush them with solvent.

3 Check the main and connecting rod bearing journals for uneven wear, scoring, pits and cracks.

4 Rub a penny across each journal several times **(see illustration)**.

19.1 The oil holes should be chamfered so sharp edges don't gouge or scratch the new bearings

19.2 Use a wire or stiff plastic bristle brush to clean the oil passages in the crankshaft

19.4 Rubbing a penny lengthwise on each journal will reveal its condition - if copper rubs off and is embedded in the crankshaft, the journals should be reground

19.6 Measure the diameter of each crankshaft journal at several points to detect taper and out-of-round conditions

19.8 If the seals have worn grooves in the crankshaft journals, or if the seal contact surfaces are nicked or scratched, the new seals will leak

If a journal picks up copper from the penny, it's too rough and must be reground.

5 Check the rest of the crankshaft for cracks and other damage. It should be magnafluxed to reveal hidden cracks - an automotive machine shop will handle the procedure.

6 Using a micrometer, measure the diameter of the main and connecting rod journals and compare the results to this Chapter's Specifications **(see illustration)**. By measuring the diameter at a number of points around each journal's circumference, you'll be able to determine whether or not the journal is out-of-round. Take the measurement at each end of the journal, near the crank throws, to determine if the journal is tapered.

7 If the crankshaft journals are damaged, tapered, out-of-round or worn beyond the limits given in the Specifications, have the crankshaft reground by an automotive machine shop. Be sure to use the correct size bearing inserts if the crankshaft is reconditioned.

8 Check the oil seal journals at each end of the crankshaft for wear and damage. If the seal has worn a groove in the journal, or if it's nicked or scratched **(see illustration)**, the new seal may leak when the engine is reassembled. In some cases, an automotive machine shop may be able to repair the journal by pressing on a thin sleeve. If repair isn't feasible, a new or different crankshaft should be installed.

9 Refer to Section 20 and examine the main and rod bearing inserts.

Intermediate shaft

10 Check the journals of the intermediate shaft for signs of wear, scuffing and overheating (blue spots). Check the bushings in the cylinder block for the same conditions. If any undesirable conditions exist, the intermediate shaft and bushings must be replaced. Due to the special tools required to remove the intermediate shaft bushings and install the new ones, this job must be left to an automotive machine shop.

11 Using a micrometer, measure the diameter of the intermediate shaft journals and record these figures. Then measure the inside diameter of the corresponding bushings using a telescoping gauge and micrometer (similar to the technique shown in Section 16, step 4). Subtract the diameter of each intermediate shaft journal from the inside diameter of its corresponding bushing to calculate the oil clearance. Compare your findings with the values listed in this Chapter's Specifications. If any of the oil clearances are excessive, compare the measurements of the intermediate shaft journals with the values listed in this Chapter's Specifications. If the journal diameters are within the specified range, have new intermediate shaft bushings installed.

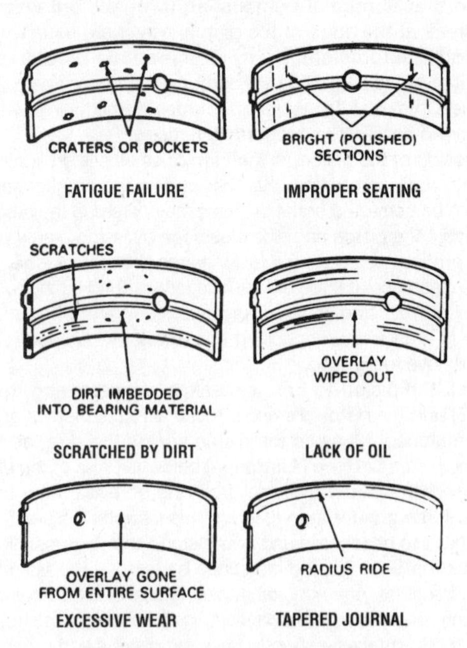

20.1 Typical bearing failures

20 Main and connecting rod bearings - inspection

Refer to illustration 20.1

1 Even though the main and connecting rod bearings should be replaced with new ones during the engine overhaul, the old bearings should be retained for close examination, as they may reveal valuable information about the condition of the engine **(see illustration)**.

2 Bearing failure occurs because of lack of lubrication, the presence of dirt or other foreign particles, overloading the engine and corrosion. Regardless of the cause of bearing failure, it must be corrected before the engine is reassembled to prevent it from happening again.

3 When examining the bearings, remove them from the engine block, the main bearing caps, the connecting rods and the rod caps and lay them out on a clean surface in the same general position as their location in the engine. This will enable you to match any bearing problems with the corresponding crankshaft journal.

4 Dirt and other foreign particles get into the engine in a variety of

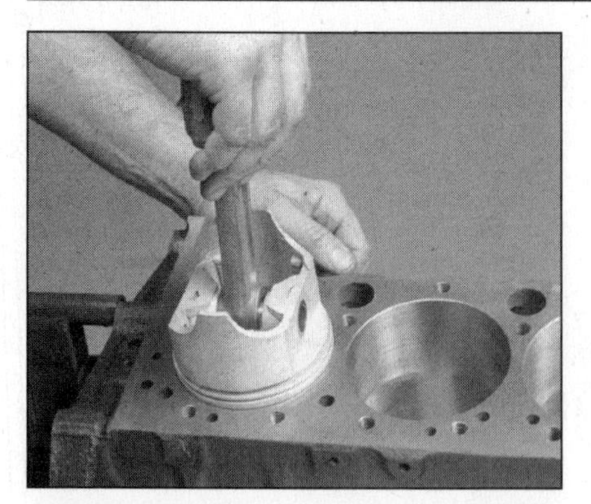

22.3 When checking piston ring end gap, the ring must be square in the cylinder bore (this is done by pushing the ring down with the top of a piston as shown)

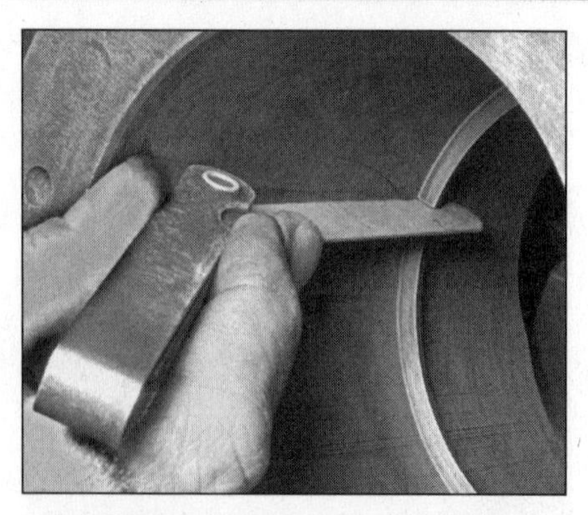

22.4 With the ring square in the cylinder, measure the end gap with a feeler gauge

22.5 If the end gap is too small, clamp a file in a vise and file the ring ends (from the outside in only) to enlarge the gap slightly

ways. It may be left in the engine during assembly, or it may pass through filters or the PCV system. It may get into the oil, and from there into the bearings. Metal chips from machining operations and normal engine wear are often present. Abrasives are sometimes left in engine components after reconditioning, especially when parts aren't thoroughly cleaned using the proper cleaning methods. Whatever the source, these foreign objects often end up embedded in the soft bearing material and are easily recognized. Large particles won't embed in the bearing and will score or gouge the bearing and journal. The best prevention for this cause of bearing failure is to clean all parts thoroughly and keep everything spotlessly clean during engine assembly. Frequent and regular engine oil and filter changes are also recommended.

5 Lack of lubrication (or lubrication breakdown) has a number of interrelated causes. Excessive heat (which thins the oil), overloading (which squeezes the oil from the bearing face) and oil leakage or throw off (from excessive bearing clearances, worn oil pump or high engine speeds) all contribute to lubrication breakdown. Blocked oil passages, which usually are the result of misaligned oil holes in a bearing shell, will also oil starve a bearing and destroy it. When lack of lubrication is the cause of bearing failure, the bearing material is wiped or extruded from the steel backing of the bearing. Temperatures may increase to the point where the steel backing turns blue from overheating.

6 Driving habits can have a definite effect on bearing life. Full throttle, low speed operation (lugging the engine) puts very high loads on bearings, which tends to squeeze out the oil film. These loads cause the bearings to flex, which produces fine cracks in the bearing face (fatigue failure). Eventually the bearing material will loosen in pieces and tear away from the steel backing. Short trip driving leads to corrosion of bearings because insufficient engine heat is produced to drive off the condensed water and corrosive gases. These products collect in the engine oil, forming acid and sludge. As the oil is carried to the engine bearings, the acid attacks and corrodes the bearing material.

7 Incorrect bearing installation during engine assembly will lead to bearing failure as well. Tight fitting bearings leave insufficient oil clearance and will result in oil starvation. Dirt or foreign particles trapped behind a bearing insert result in high spots on the bearing which lead to failure.

21 Engine overhaul - reassembly sequence

1 Before beginning engine reassembly, make sure you have all the necessary new parts, gaskets and seals as well as the following items on hand:

Common hand tools
Torque wrench (1/2-inch drive)
Piston ring installation tool
Piston ring compressor
Short lengths of rubber or plastic hose to fit over connecting rod bolts
Plastigage
Feeler gauges
Fine-tooth file
New engine oil
Engine assembly lube or moly-base grease
Gasket sealant
Thread locking compound

2 To save time and avoid problems, engine reassembly must be done in the following general order:

Crankshaft and main bearings
Rear main oil seal retainer
Piston/connecting rod assemblies
Intermediate shaft
Cylinder head, cam followers and camshaft
Timing belt and sprockets
Water pump
Oil pump
Oil pump pick-up
Oil pan
Intake and exhaust manifolds
Timing belt covers
Valve cover
Flywheel/driveplate

22 Piston rings - installation

Refer to illustrations 22.3, 22.4, 22.5, 22.9a, 22.9b and 22.11

1 Before installing the new piston rings, the ring end gaps must be checked. It's assumed the piston ring side clearance has been checked and verified correct (see Section 18).

2 Lay out the piston/connecting rod assemblies and the new ring sets so the ring sets will be matched with the same piston and cylinder during the end gap measurement and engine assembly.

3 Insert the top (number one) ring into the first cylinder and square it up with the cylinder walls by pushing it in with the top of the piston **(see illustration)**. The ring should be near the bottom of the cylinder, at the lower limit of ring travel.

4 To measure the end gap, slip feeler gauges between the ends of the ring until a gauge equal to the gap width is found **(see illustration)**. The feeler gauge should slide between the ring ends with a slight amount of drag. Compare the measurement to this Chapter's Specifications. If the gap is larger or smaller than specified, double-check to make sure you have the correct rings before proceeding. **Note:** *The ring end gap must be measured from 0.591 inches from the bottom of the cylinder.*

5 If the gap is too small, it must be enlarged or the ring ends may come in contact with each other during engine operation, which can

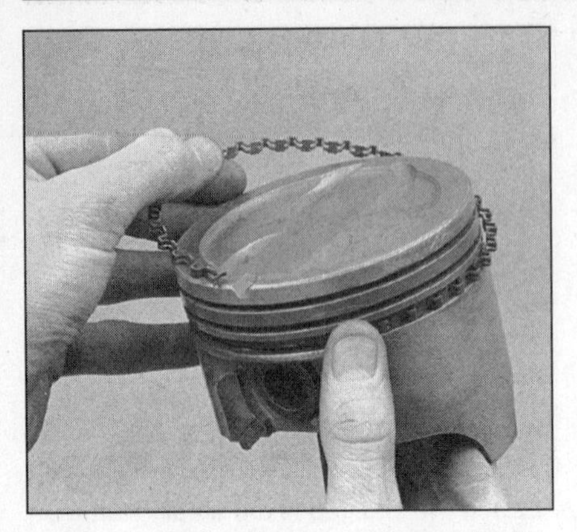

22.9a Installing the spacer/expander in the oil control ring groove

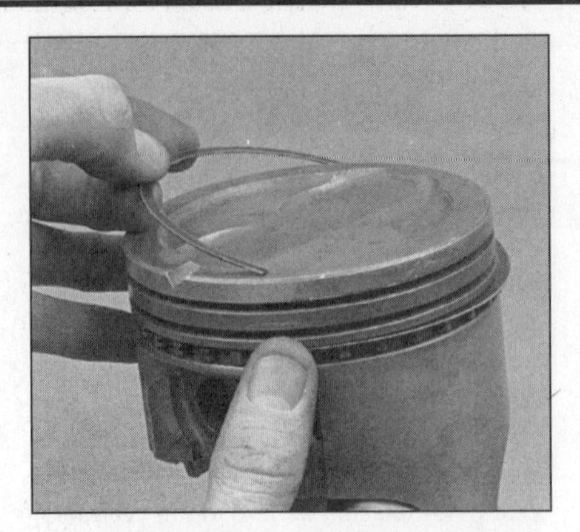

22.9b DO NOT use a piston ring installation tool when installing the oil ring side rails

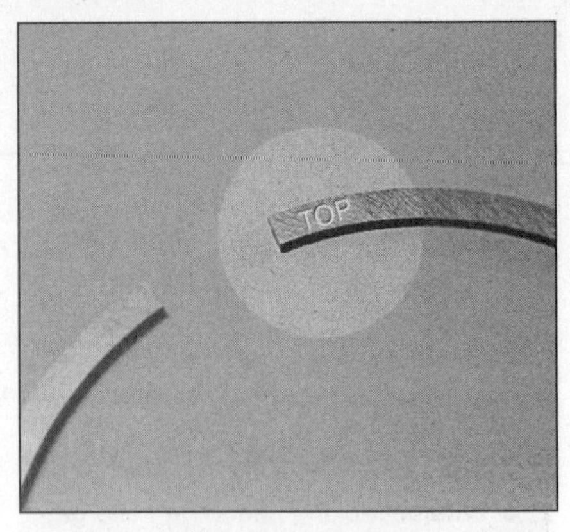

22.11 Installing the compression rings with a ring expander - the mark on the ring must face up

cause serious engine damage. The end gap can be increased by filing the ring ends very carefully with a fine file. Mount the file in a vise equipped with soft jaws, slip the ring over the file with the ends contacting the file teeth and slowly move the ring to remove material from the ends. When performing this operation, file only from the outside in **(see illustration)**.

6 Excess end gap isn't critical unless it's greater than 0.040-inch. Again, double-check to make sure you have the correct rings for the engine.

7 Repeat the procedure for each ring that will be installed in the first cylinder and for each ring in the remaining cylinders. Remember to keep rings, pistons and cylinders matched up.

8 Once the ring end gaps have been checked/corrected, the rings can be installed on the pistons.

9 The oil control ring (lowest one on the piston) is usually installed first. It's composed of three separate components. Slip the spacer/expander into the groove **(see illustration)**. If an anti-rotation tang is used, make sure it's inserted into the drilled hole in the ring groove. Next, install the lower side rail. Don't use a piston ring installation tool on the oil ring side rails, as they may be damaged. Instead, place one end of the side rail into the groove between the spacer/expander and the ring land, hold it firmly in place and slide a finger around the piston while pushing the rail into the groove **(see illustration)**. Next, install the upper side rail in the same manner.

10 After the three oil ring components have been installed, check to make sure both the upper and lower side rails can be turned smoothly in the ring groove.

11 The number two (middle) ring is installed next. It's usually stamped with a mark, which must face up, toward the top of the piston **(see illustration)**. **Note:** *Always follow the instructions printed on the ring package or box - different manufacturers may require different approaches. Don't mix up the top and middle rings, as they have different cross-sections.*

12 Use a piston ring installation tool and make sure the identification mark is facing the top of the piston, then slip the ring into the middle groove on the piston. Don't expand the ring any more than necessary to slide it over the piston.

13 Install the number one (top) ring in the same manner. Make sure the mark is facing up. Be careful not to confuse the number one and number two rings.

14 Repeat the procedure for the remaining pistons and rings.

23 Crankshaft and intermediate shaft - installation and main bearing oil clearance check

Refer to illustrations 23.5, 23.6, 23.11, 23.15, 23.27a and 23.27b

1 It's assumed at this point that the engine block and crankshaft have been cleaned, inspected and repaired or reconditioned.

2 Position the engine with the bottom facing up.

23.5 The last bearing shell (number 5) has thrust flanges on B21 and B23 engines

3 Remove the main bearing caps (if not already done) and make sure the original bearings are not in place in the block and caps.

4 If they're still in place, remove the original bearing inserts from the block and the main bearing caps. Wipe the bearing surfaces of the block and caps with a clean, lint-free cloth. They must be kept spotlessly clean.

Main bearing oil clearance check

Note: *Don't touch the faces of the new bearing inserts with your fingers. Oil and acids from your skin can etch the bearings.*

5 Clean the back sides of the new main bearing inserts and lay one in each main bearing saddle in the block. If one of the bearing inserts from each set has a large groove in it, make sure the grooved insert is installed in the block. Lay the other bearing from each set in the corresponding main bearing cap **(see illustration)**. Make sure the tab on the bearing insert fits into the recess in the block or cap. **Note:** *If the crankshaft has been "turned down" (lathed), use a set of oversize bearings from the local auto parts store. If the crankshaft is original and the bearings will be replaced, consult a qualified auto parts person to match the correct bearings.* **Caution:** *If the codes on the crankshaft are not visible, do not scrub them with a wire brush or scraper. Use only solvent or detergent.*

6 The flanged thrust washers must be installed in the number three cap and saddle (counting from the front of the engine) on B230 engines **(see illustration)**.

7 Clean the faces of the bearings in the block and the crankshaft main bearing journals with a clean, lint-free cloth.

8 Check or clean the oil holes in the crankshaft, as any dirt here can go only one way - straight through the new bearings.

9 Once you're certain the crankshaft is clean, carefully lay it in position in the main bearings.

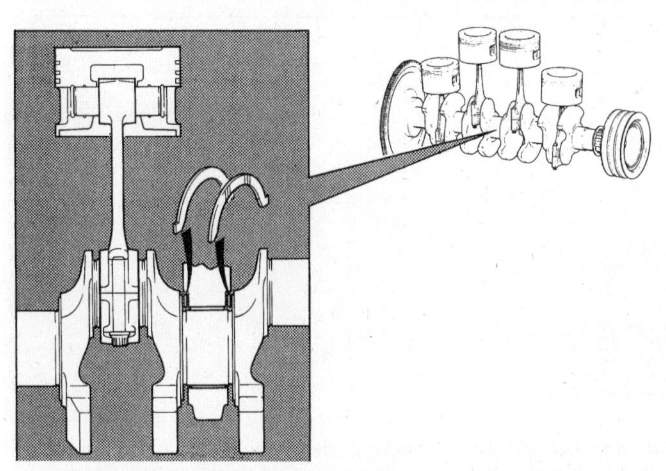

23.6 Location of the number 3 thrust bearings (arrows) on B230 engines. The grooved sides face OUT

10 Before the crankshaft can be permanently installed, the main bearing oil clearance must be checked.

11 Cut several pieces of the appropriate size Plastigage (they should be slightly shorter than the width of the main bearings) and place one piece on each crankshaft main bearing journal, parallel with the journal axis **(see illustration)**.

12 Clean the faces of the bearings in the main bearing caps and install the caps. Don't disturb the Plastigage.

13 Starting with the center main bearing caps and working out toward the ends, tighten the main bearings, in three steps, to the torque figure listed in this Chapter's Specifications. Don't rotate the crankshaft at any time during this operation.

14 Remove the bolts and carefully lift off the main bearing caps. Keep them in order. Don't disturb the Plastigage or rotate the crankshaft. If any of the main bearing caps are difficult to remove, tap them gently from side-to-side with a soft-face hammer to loosen them.

15 Compare the width of the crushed Plastigage on each journal to the scale printed on the Plastigage envelope to obtain the main bearing oil clearance **(see illustration)**. Check the Specifications at the beginning of this Chapter to make sure it's correct.

16 If the clearance is not as specified, the bearing inserts may be the wrong size (which means different ones will be required). Before deciding different inserts are needed, make sure no dirt or oil was between the bearing inserts and the caps or block when the clearance was measured. If the Plastigage was wider at one end than the other, the journal may be tapered (see Section 19).

17 Carefully scrape all traces of the Plastigage material off the main bearing journals and/or the bearing faces. Use your fingernail or the edge of a credit card - don't nick or scratch the bearing faces.

Final crankshaft installation

18 Carefully lift the crankshaft out of the engine.

19 Clean the bearing faces in the block, then apply a thin, uniform layer of moly-base grease or engine assembly lube to each of the bearing surfaces. Be sure to coat the thrust faces as well as the journal face of the thrust bearing.

20 Make sure the crankshaft journals are clean, then lay the crankshaft back in place in the block.

21 Clean the faces of the bearings in the caps, then apply lubricant to them.

22 Install the main bearing caps.

23 Install the bolts.

24 Tighten all main bearing cap bolts to the torque listed in this Chapter's Specifications, starting with the center main and working out toward the ends.

25 Rotate the crankshaft a number of times by hand to check for any obvious binding.

26 The final step is to check the crankshaft endplay with feeler

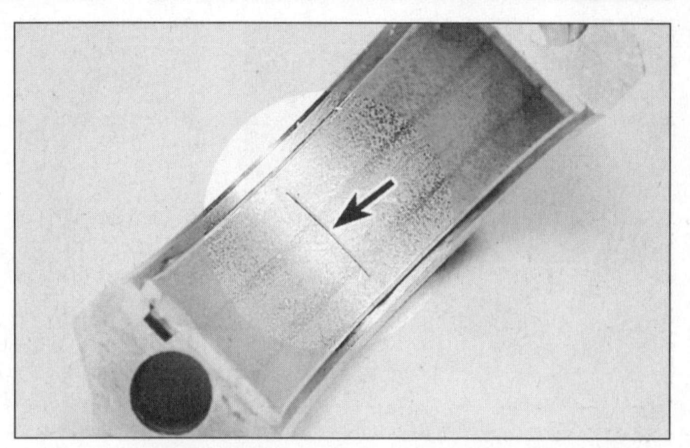

23.11 Lay the Plastigage strips (arrow) on the main bearing journals, parallel to the crankshaft centerline

23.15 Compare the width of the crushed Plastigage to the scale on the envelope to determine the main bearing oil clearance (always take the measurement at the widest point of the Plastigage); be sure to use the correct scale - standard and metric ones are included

23.27a Install the spacer over the woodruff key on the front section of the crankshaft

gauges or a dial indicator as described in Section 14. The endplay should be correct if the crankshaft thrust faces aren't worn or damaged and new bearings have been installed.

27 Install the front spacer over the woodruff key on the crankshaft **(see illustration)** and then the front bearing cover **(see illustration)**. Refer to Section 24 and install the new rear main oil seal.

23.27b Then, install the front seal housing

24.1 Support the retainer on wood blocks and drive out the old seal with a punch and hammer

24.2 Drive the new seal into the retainer with a wood block or a section of pipe, if you have one large enough - make sure you don't cock the seal in the retainer bore

Intermediate shaft installation

28 Make sure the intermediate shaft and the bushings in the block are clean, then lubricate the bushings and journals with engine assembly lube.

29 Carefully guide the intermediate shaft into its bores. **Refer to Section 14 for the illustrations.**

30 Install the thrust plate bolts on the front intermediate shaft and tighten them to the torque listed in this Chapter's Specifications **(see illustration 14.10)**. Rotate the intermediate shaft by hand to confirm smooth operation.

31 Install the timing belt and intermediate shaft belt following the procedure outlined in Chapter 2A.

24 Rear main oil seal - installation

Refer to illustrations 24.1, 24.2, 24.4 and 24.5

Note: *The crankshaft must be installed and the main bearing caps bolted in place before the new seal and retainer assembly can be bolted to the block.*

1 Remove the old seal from the retainer with a hammer and punch by driving it out from the back side **(see illustration)**. Be sure to note how far it's recessed into the retainer bore before removing it; the new seal will have to be recessed an equal amount. Be very careful not to scratch or otherwise damage the bore in the retainer or oil leaks could develop.

2 Make sure the retainer is clean, then apply a thin coat of engine oil to the outer edge of the new seal. The seal must be pressed squarely into the retainer bore, so hammering it into place isn't recommended. If you don't have access to a press, sandwich the retainer and seal between two smooth pieces of wood and press the seal into place with the jaws of a large vise. If you don't have a vise big enough, lay the retainer on a workbench and drive the seal into place with a block of wood and hammer **(see illustration)**. The piece of wood must be thick enough to distribute the force evenly around the entire circumference of the seal. Work slowly and make sure the seal enters the bore squarely. **Note**: Using a feeler gauge, confirm that the clearance between the seal and the retainer is equal all the way around.

3 Place a thin coat of RTV sealant to the entire edge of the retainer.

4 Lubricate the seal lips with multi-purpose grease or engine oil before you slip the seal/retainer over the crankshaft and bolt it to the block. Be sure to use a new gasket. **Note:** Apply a film of RTV sealant to both sides of the gasket before installation **(see illustration)**.

5 Tighten the retainer bolts, a little at a time, to the torque listed in the Chapter 2A Specifications. Trim the gasket flush with the oil pan gasket surface, being careful not to scratch it **(see illustration)**.

24.4 Apply a small amount of oil onto the rear main seal before sliding the assembly onto the crankshaft

24.5 Use a gasket scraper or utility knife to remove the excess gasket material from the seal retainer/engine block surface

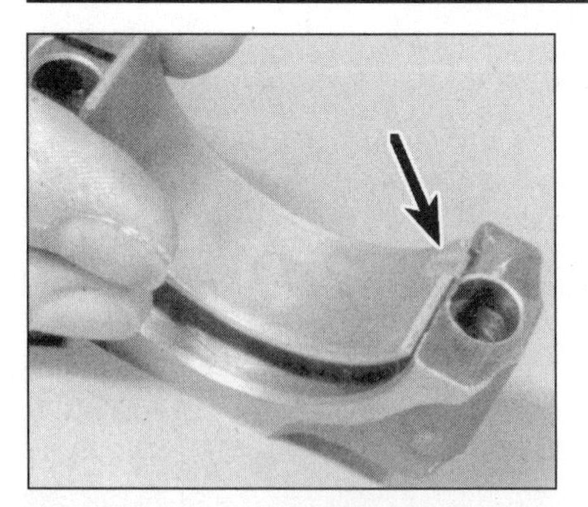

25.4 The tab on the bearing (arrow) must fit into the recess so that the bearing will seat properly

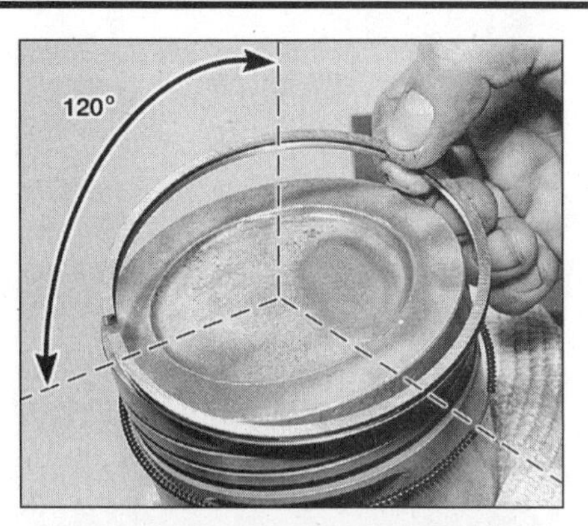

25.5 Position the ring gaps as shown before installing the piston/connecting rod assemblies into the engine

25.7 This type of ring compressing tool requires a speed wrench and extension to tighten the band adjustment (arrow) over the piston rings

25 Pistons and connecting rods - installation and rod bearing oil clearance check

Refer to illustrations 25.4, 25.5, 25.7, 25.9, 25.11, 25.13, 25.15a, 25.15b and 25.17

1 Before installing the piston/connecting rod assemblies, the cylinder walls must be perfectly clean, the top edge of each cylinder must be chamfered, and the crankshaft must be in place.

2 Remove the cap from the end of the number one connecting rod (check the marks made during removal). Remove the original bearing inserts and wipe the bearing surfaces of the connecting rod and cap with a clean, lint-free cloth. They must be kept spotlessly clean.

Connecting rod bearing oil clearance check

Note: *Don't touch the faces of the new bearing inserts with your fingers. Oil and acids from your skin can etch the bearings.*

3 Clean the back side of the new upper bearing insert, then lay it in place in the connecting rod. Make sure the tab on the bearing fits into the recess in the rod. Don't hammer the bearing insert into place and be very careful not to nick or gouge the bearing face. Don't lubricate the bearing at this time. **Note:** *If the crankshaft has been "turned down" (lathed), use a set of oversize bearings from the local auto parts store. If the crankshaft is original and the bearings will be replaced, consult an automotive parts person for the correct bearing selection.*

4 Clean the back side of the other bearing insert and install it in the rod cap. Again, make sure the tab on the bearing fits into the recess in the cap **(see illustration)**, and don't apply any lubricant. It's critically important that the mating surfaces of the bearing and connecting rod are perfectly clean and oil free when they're assembled.

5 Position the piston ring gaps at intervals around the piston **(see illustration)**.

6 Slip a section of plastic or rubber hose over each connecting rod cap bolt.

7 Lubricate the piston and rings with clean engine oil and attach a piston ring compressor to the piston. Leave the skirt protruding about 1/4-inch to guide the piston into the cylinder. The rings must be compressed until they're flush with the piston **(see illustration)**

8 Rotate the crankshaft until the number one connecting rod journal is at BDC (bottom dead center) and apply a coat of engine oil to the cylinder walls.

9 With the mark or notch on top of the piston facing the front of the engine **(see illustration)**, gently insert the piston/connecting rod assembly into the number one cylinder bore and rest the bottom edge of the ring compressor on the engine block. **Note:** *Make sure the connecting rod mark is facing the front of the engine also.*

10 Tap the top edge of the ring compressor to make sure it's contacting the block around its entire circumference.

11 Gently tap on the top of the piston with the end of a wooden or plastic hammer handle **(see illustration)** while guiding the end of the connecting rod into place on the crankshaft journal. The piston rings may try to pop out of the ring compressor just before entering the cylinder bore, so keep some pressure on the ring compressor. Work slowly, and if any resistance is felt as the piston enters the cylinder, stop immediately. Find out what's hanging up and fix it before proceeding. Do not, for any reason, force the piston into the cylinder - you might break a ring and/or the piston.

12 Once the piston/connecting rod assembly is installed, the connecting rod bearing oil clearance must be checked before the rod cap is permanently bolted in place.

13 Cut a piece of the appropriate size Plastigage slightly shorter than the width of the connecting rod bearing and lay it in place on the number one connecting rod journal, parallel with the journal axis **(see illustration)**.

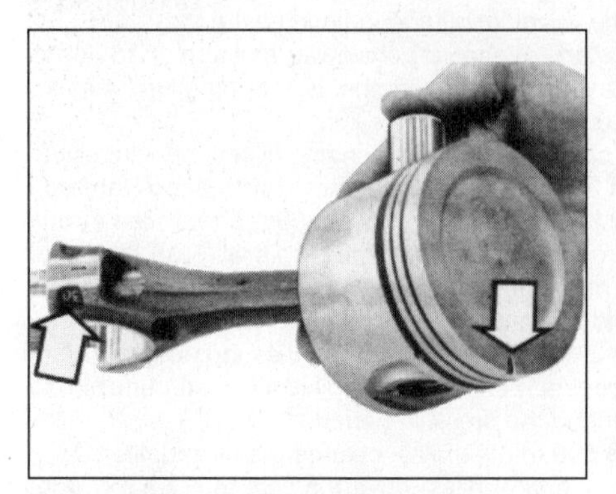

25.9 Remember the arrow points toward the front of the engine

25.11 Drive the piston gently into the cylinder bore with the end of a wooden or plastic hammer handle

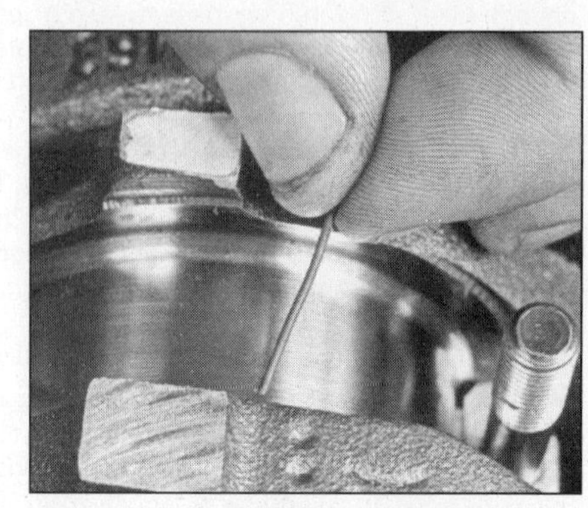

25.13 Lay the Plastigage strips on each rod bearing journal, parallel to the crankshaft centerline

25.15a Install the connecting rod bearing caps onto the connecting rods

25.15b Tighten the cap nuts to the torque listed in this Chapter's Specifications

25.17 Measuring the width of the crushed Plastigage to determine the rod bearing oil clearance (be sure to use the correct scale - standard and metric ones are included)

14 Clean the connecting rod cap bearing face, remove the protective hoses from the connecting rod bolts and install the rod cap. Make sure the mating mark on the cap is on the same side as the mark on the connecting rod.

15 Install the nuts and tighten them to the torque listed in this Chapter's Specifications **(see illustrations)**. Work up to it in three steps. **Note:** *Use a thin-wall socket to avoid erroneous torque readings that can result if the socket is wedged between the rod cap and nut. If the socket tends to wedge itself between the nut and the cap, lift up on it slightly until it no longer contacts the cap.* Do not rotate the crankshaft at any time during this operation.

16 Remove the nuts and detach the rod cap, being very careful not to disturb the Plastigage.

17 Compare the width of the crushed Plastigage to the scale printed on the Plastigage envelope to obtain the oil clearance **(see illustration)**. Compare it to this Chapter's Specifications to make sure the clearance is correct.

18 If the clearance is not as specified, the bearing inserts may be the wrong size (which means different ones will be required). Before deciding different inserts are needed, make sure no dirt or oil was between the bearing inserts and the connecting rod or cap when the clearance was measured. Also, recheck the journal diameter. If the Plastigage was wider at one end than the other, the journal may be tapered.

Final connecting rod installation

19 Carefully scrape all traces of the Plastigage material off the rod journal and/or bearing face. Be very careful not to scratch the bearing - use your fingernail or the edge of a credit card.

20 Make sure the bearing faces are perfectly clean, then apply a uniform layer of clean moly-base grease or engine assembly lube to both of them. You'll have to push the piston into the cylinder to expose the face of the bearing insert in the connecting rod - be sure to slip the protective hoses over the rod bolts first.

21 Slide the connecting rod back into place on the journal, remove the protective hoses from the rod cap bolts, install the rod cap and tighten the nuts to the torque listed in this Chapter's Specifications. Again, work up to the torque in three steps. **Note:** *Again, make sure the mating mark on the cap is on the same side as the mark on the connecting rod.*

22 Repeat the entire procedure for the remaining pistons/connecting rods.

23 The important points to remember are:

a) Keep the back sides of the bearing inserts and the insides of the connecting rods and caps perfectly clean when assembling them.

b) Make sure you have the correct piston/rod assembly for each cylinder.

c) The arrow or mark on the piston must face the front of the engine.

d) Lubricate the cylinder walls with clean oil.

e) Lubricate the bearing faces when installing the rod caps after the oil clearance has been checked.

24 After all the piston/connecting rod assemblies have been properly installed, rotate the crankshaft a number of times by hand to check for any obvious binding.

25 As a final step, the connecting rod endplay must be checked. Refer to Section 13 for this procedure.

26 Compare the measured endplay to this Chapter's Specifications to make sure it's correct. If it was correct before disassembly and the original crankshaft and rods were reinstalled, it should still be right. If new rods or a new crankshaft were installed, the endplay may be inadequate. If so, the rods will have to be removed and taken to an automotive machine shop for re-sizing.

26 Initial start-up and break-in after overhaul

Warning: *Have a fire extinguisher handy when starting the engine for the first time.*

1 Once the engine has been installed in the vehicle, double-check the engine oil and coolant levels.

2 With the spark plugs out of the engine and the ignition system disabled (see Section 3), crank the engine until oil pressure registers on the gauge or the light goes out.

3 Install the spark plugs, hook up the plug wires and restore the ignition system functions.

4 Start the engine. It may take a few moments for the fuel system to build up pressure, but the engine should start without a great deal of effort. **Note:** *If backfiring occurs through the throttle body, or carburetor, recheck the valve timing and ignition timing.*

5 After the engine starts, it should be allowed to warm up to normal operating temperature. While the engine is warming up, make a thorough check for fuel, oil and coolant leaks.

6 Shut the engine off and recheck the engine oil and coolant levels.

7 Drive the vehicle to an area with minimum traffic, accelerate from 30 to 50 mph, then allow the vehicle to slow to 30 mph with the throttle closed. Repeat the procedure 10 or 12 times. This will load the piston rings and cause them to seat properly against the cylinder walls. Check again for oil and coolant leaks.

8 Drive the vehicle gently for the first 500 miles (no sustained high speeds) and keep a constant check on the oil level. It isn't unusual for an engine to use oil during the break-in period.

9 At approximately 500 to 600 miles, change the oil and filter.

10 For the next few hundred miles, drive the vehicle normally. Don't pamper it or abuse it.

11 After 2000 miles, change the oil and filter again and consider the engine broken in.

Chapter 3
Cooling, heating and air conditioning systems

Contents

Specifications

General

Coolant capacity	
Automatic transmission	9.8 quarts
Manual transmission	10.0 quarts
Thermostat rating	
Opening temperature	190 degrees
Fully open at	212 degrees

General (continued)

Refrigerant capacity	
1990 and earlier	45.8 ounces
1991 and later	38.8 ounces
Refrigerant oil capacity	12.7 ounces
Oil to add when replacing air conditioning components	
Receiver-drier	0.35 ounces
Evaporator	1.4 ounces
Condenser	0.70 ounces
Hose assemblies	0.35 ounces

Torque specifications

	Ft-lbs (unless otherwise indicated)
Cooling fan bolts	
Center bolt (if equipped)	60
Fan clutch-to-hub bolts	15
Fixed fan bolts	7
Water pump bolts	16
Air conditioning compressor clutch center bolt	20
Air conditioning compressor rear housing bolts	15
Air conditioning compressor bottom cover	19
Air conditioning compressor oil drain plug	44 inch lbs.
Thermostat housing bolts	80 inch lbs.
Water pump pulley bolts	13
Air conditioning system hose unions	
Condenser	12
Compressor	26
Receiver/dryer	18
Expansion valve	22

1 General information

Engine cooling system

Refer to illustration 1.2

All vehicles covered by this manual employ a pressurized engine cooling system with thermostatically controlled coolant circulation.

An impeller-type water pump mounted on the front of the block pumps coolant through the engine **(see illustration)**. The coolant flows around each cylinder and toward the rear of the engine. Cast-in coolant passages direct coolant around the intake and exhaust ports, near the spark plug areas and in close proximity to the exhaust valve guides.

A wax-pellet-type thermostat is located in a housing near the front of the engine. During warm up, the closed thermostat prevents coolant from circulating through the radiator. As the engine nears normal operating temperature, the thermostat opens and allows hot coolant to travel through the radiator, where it's cooled before returning to the engine.

The cooling system is sealed by a pressure-type cap, which raises the boiling point of the coolant and increases the cooling efficiency of the radiator. If the system pressure exceeds the cap pressure relief value, the excess pressure in the system forces the spring-loaded valve inside the cap off its seat and allows the coolant to escape through the overflow tube.

The pressure cap is located on top of a translucent plastic expansion tank. The cap pressure rating is molded into the top of the cap. The pressure rating is between 9 to 12 psi. **Warning:** *Do not remove the pressure cap from the radiator or expansion tank until the engine has cooled completely and there's no pressure remaining in the cooling system.*

Heating system

The heating system consists of a blower fan and heater core located in the heater box, the hoses connecting the heater core to the engine cooling system and the heater/air conditioning control head on the dashboard. Hot engine coolant is circulated through the heater core. When the heater mode is activated, a flap door opens to expose the heater box to the passenger compartment. A fan switch on the control head activates the blower motor, which forces air through the core, heating the air.

Air conditioning system

The air conditioning system consists of a condenser mounted in front of the radiator, an evaporator mounted adjacent to the heater core, a compressor mounted on the engine, a filter-drier (receiver-drier) which contains a high pressure relief valve and the plumbing connecting all of the above components.

A blower fan forces the warmer air of the passenger compartment through the evaporator core (sort of a radiator-in-reverse), transferring the heat from the air to the refrigerant. The liquid refrigerant boils off into low pressure vapor, taking the heat with it when it leaves the evaporator.

2 Antifreeze - general information

Warning: *Do not allow antifreeze to come in contact with your skin or painted surfaces of the vehicle. Rinse off spills immediately with plenty of water. If consumed, antifreeze can be fatal; children and pets are attracted by its sweet taste, so wipe up garage floor and coolant spills immediately. Keep antifreeze containers covered and repair leaks in your cooling system as soon as they are noticed.*

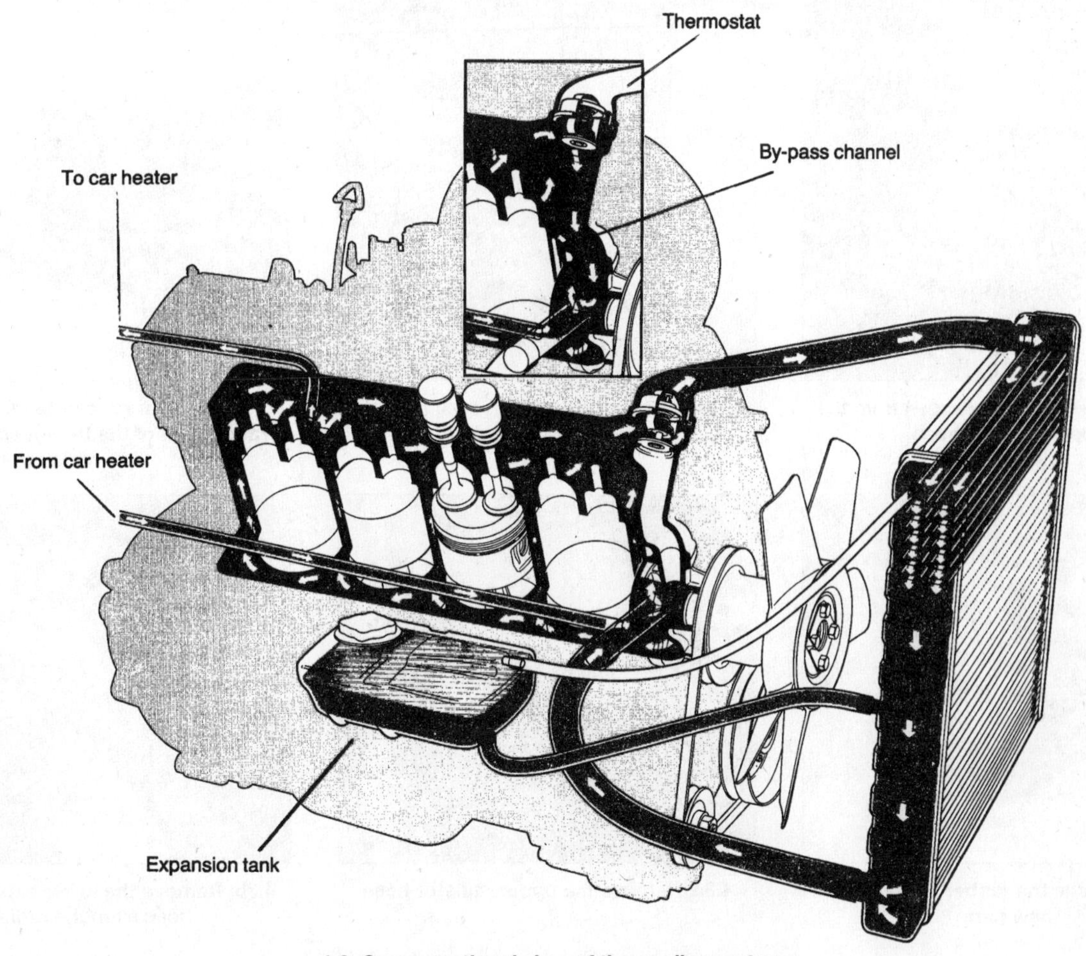

1.2 Cross-sectional view of the cooling system

The cooling system should be filled with a water/ethylene glycol based antifreeze solution, which will prevent freezing down to at least -20-degrees F, or lower if local climate requires it. It also provides protection against corrosion and increases the coolant boiling point.

The cooling system should be drained, flushed and refilled at the specified intervals (see Chapter 1). Old or contaminated antifreeze solutions are likely to cause damage and encourage the formation of rust and scale in the system. Use distilled water with the antifreeze.

Before adding antifreeze, check all hose connections, because antifreeze tends to search out and leak through very minute openings. Engines don't normally consume coolant, so if the level goes down, find the cause and correct it.

The exact mixture of antifreeze-to-water which you should use depends on the relative weather conditions. The mixture should contain at least 50 percent antifreeze, but should never contain more than 70 percent antifreeze. Consult the mixture ratio chart on the antifreeze container before adding coolant. Hydrometers are available at most auto parts stores to test the coolant. Use antifreeze which meets the vehicle manufacturer's specifications.

3 Thermostat - check and replacement

Note: *A thermostat that is stuck in the open position will cause the engine to warm up very slowly and run at a below normal operating temperature. A thermostat that is stuck closed, will cause the engine to overheat.*

Warning: *Do not remove the radiator cap, drain the coolant or replace the thermostat until the engine has cooled completely.*

Check

1 Before assuming the thermostat is to blame for a cooling system problem, check the coolant level, drivebelt tension (see Chapter 1) and temperature gauge (or light) operation.

2 If the engine seems to be taking a long time to warm up (based on heater output or temperature gauge operation), the thermostat is probably stuck open. Replace the thermostat with a new one.

3 If the engine runs hot, use your hand to check the temperature of the upper radiator hose. If the hose isn't hot, but the engine is, the thermostat is probably stuck closed, preventing the coolant inside the engine from escaping to the radiator. Replace the thermostat. **Caution:** *Don't drive the vehicle without a thermostat. The computer may stay in open loop and emissions and fuel economy will suffer.*

4 If the upper radiator hose is hot, it means that the coolant is flowing and the thermostat is open. Consult the *Troubleshooting* Section at the front of this manual for cooling system diagnosis.

Replacement

Refer to illustrations 3.8, 3.10, 3.11 and 3.12

5 Disconnect the cable from the negative battery terminal. **Caution:** *It is necessary to make sure the radio is turned OFF before disconnecting the battery cables to avoid damaging the microprocessor built into the radio.*

6 Drain the cooling system (see Chapter 1). If the coolant is relatively new or in good condition, save it and reuse it.

7 Follow the radiator hose to the engine to locate the thermostat housing.

8 Loosen the hose clamp, then detach the hose from the fitting **(see**

3.8 Remove the clamp (arrow) from the upper radiator hose

3.10 Remove the bolts (arrows) from the thermostat housing

3.11 Be sure to note the correct arrangement of the thermostat as it sits in the housing

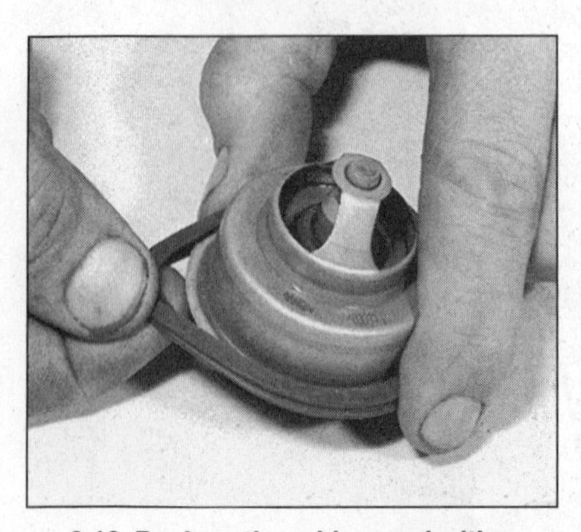

3.12 Replace the rubber seal with a new part

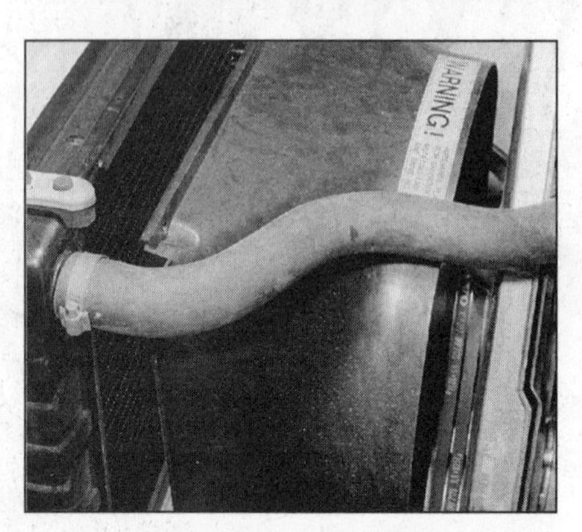

4.3a Remove the upper radiator hose

4.3b Remove the lower expansion tank hose from the radiator

illustration). If the hose clamps are the crimped type, grasp the clamp near the end with a pair of adjustable pliers and twist it to break the seal, then pull it off. If the clamps are the screw type, remove them by using a socket or a nut driver. If the hose is old or deteriorated, cut it off and install a new one.

9 If the outer surface of the fitting that mates with the hose is deteriorated (corroded, pitted, etc.), it may be damaged further by hose removal. If it is, the thermostat housing cover will have to be replaced.

10 Remove the bolts and detach the housing cover **(see illustration)**. If the cover is stuck, tap it with a soft-face hammer to jar it loose. Be prepared for some coolant to spill as the gasket seal is broken.

11 Note how it's installed (which end is facing out), then remove the thermostat **(see illustration)**.

12 Stuff a rag into the engine opening, then remove all traces of old gasket material, (if the gasket is the paper type material) or remove the rubber O-ring type gasket **(see illustration).** Remove any sealant from the housing and cover with a gasket scraper. Remove the rag from the opening and clean the gasket mating surfaces with lacquer thinner or acetone.

13 Install the new thermostat and gasket in the housing. Make sure the correct end faces out - the spring end is normally directed toward the engine **(see illustration 3.11)**.

14 Install the cover and bolts. Tighten the bolts to the torque listed in this Chapter's Specifications.

15 Reattach the hose to the thermostat assembly or housing and tighten the hose clamp(s) securely.

16 Refill the cooling system (see Chapter 1).

17 Start the engine and allow it to reach normal operating temperature, then check for leaks and proper thermostat operation (as described in Steps 2 through 4).

4 Radiator - removal and installation

Removal

Refer to illustrations 4.3a, 4.3b, 4.3c, 4.5, 4.8 and 4.9

Warning: *Wait until the engine is completely cool before beginning this procedure.*

1 Disconnect the cable from the negative battery terminal. **Caution:** *It is necessary to make sure the radio is turned OFF before disconnecting the battery cables to avoid damaging the microprocessor built into the radio.*

2 Drain the cooling system (see Chapter 1). If the coolant is relatively new or in good condition, save it and reuse it.

3 Loosen the hose clamps, then detach the radiator hoses from the fittings **(see illustrations)**. If they're stuck, grasp each hose near the end with a pair of adjustable pliers and twist it to break the seal, then pull it off - be careful not to distort the radiator fittings! If the hoses are old or deteriorated, cut them off and install new ones.

4 Disconnect the reservoir hose from the radiator.

5 Remove the screws that attach the shroud to the radiator and slide the shroud toward the engine **(see illustration)**.

6 If the vehicle is equipped with an automatic transmission, disconnect the cooler lines from the radiator. Use a drip pan to catch spilled fluid.

7 Plug the lines and fittings.

8 Remove the brackets that retain the radiator **(see illustration)**.

9 Carefully lift out the radiator **(see illustration)**. Don't spill coolant on the vehicle or scratch the paint.

10 With the radiator removed, it can be inspected for leaks and damage. If it needs repair, have a radiator shop or dealer service

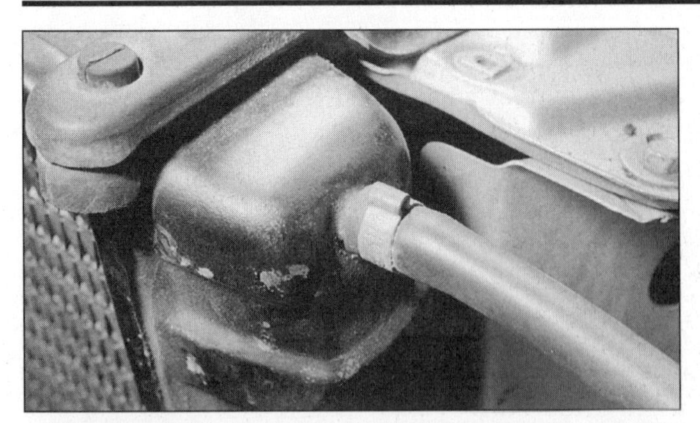

4.3c Remove the upper expansion tank hose from the radiator

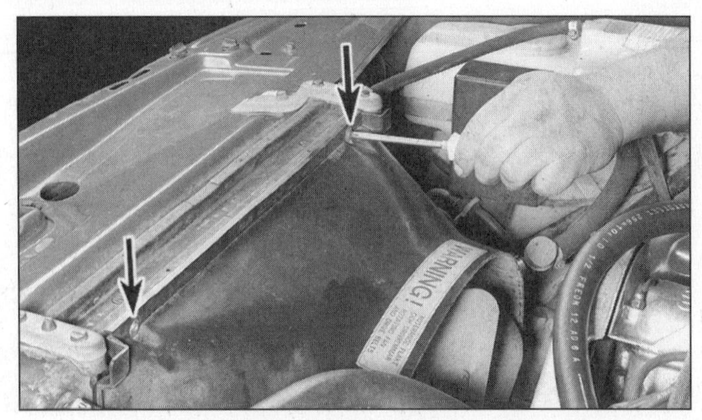

4.5 Remove the fan shroud screws (arrows)

4.8 Remove the radiator bracket bolts (arrows) from the engine compartment support

4.9 Lift the radiator from the engine compartment

department perform the work as special techniques are required.

11 Bugs and dirt can be removed from the radiator with compressed air and a soft brush. Don't bend the cooling fins as this is done.

12 Check the radiator mounts for deterioration and make sure there's nothing in them when the radiator is installed.

Installation

13 Installation is the reverse of the removal procedure.

14 After installation, fill the cooling system with the proper mixture of antifreeze and water. Refer to Chapter 1 if necessary.

15 Start the engine and check for leaks. Allow the engine to reach normal operating temperature, indicated by the upper radiator hose becoming hot. Recheck the coolant level and add more if required.

16 If you're working on an automatic transmission equipped vehicle, check and add fluid as needed.

5 Engine cooling fans - check and replacement

Warning: *To avoid possible injury or damage, DO NOT operate the engine with a damaged fan. Do not attempt to repair fan blades - replace a damaged fan with a new one.*

Mechanical fan with clutch

Refer to illustrations 5.6, 5.9a and 5.9b

Check

1 Disconnect the negative battery cable and rock the fan back and forth by hand to check for excessive bearing play. **Caution:** *It is necessary to make sure the radio is turned OFF before disconnecting the battery cables to avoid damaging the microprocessor built into the radio.*

2 With the engine cold, turn the fan blades by hand. The fan should turn on the hub with a slight resistance.

3 Visually inspect for substantial fluid leakage from the clutch assembly. If leakage is noted, replace the clutch assembly.

4 With the engine completely warmed up, turn off the ignition switch and disconnect the negative battery cable from the battery. Turn the fan by hand. Heavier resistance should be evident. If the fan turns easily, replace the fan clutch.

Replacement

5 Remove the fan shroud mounting screws and detach the shroud (see Section 4).

6 Using a box-end wrench, remove the bolts/nuts securing the fan/clutch assembly to the hub **(see illustration)**.

5.6 Remove the four bolts that mount the fan assembly to the water pump hub

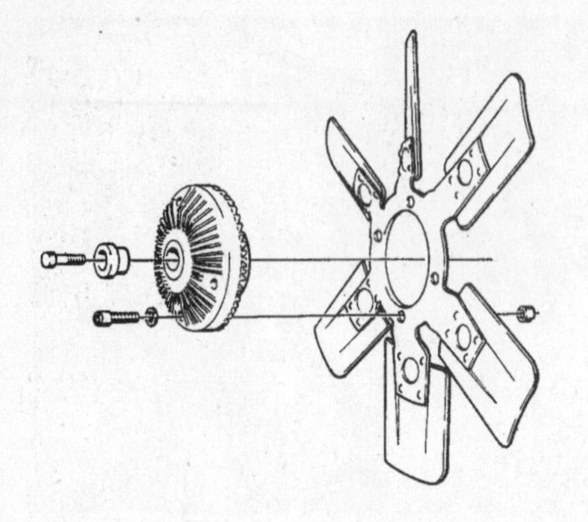

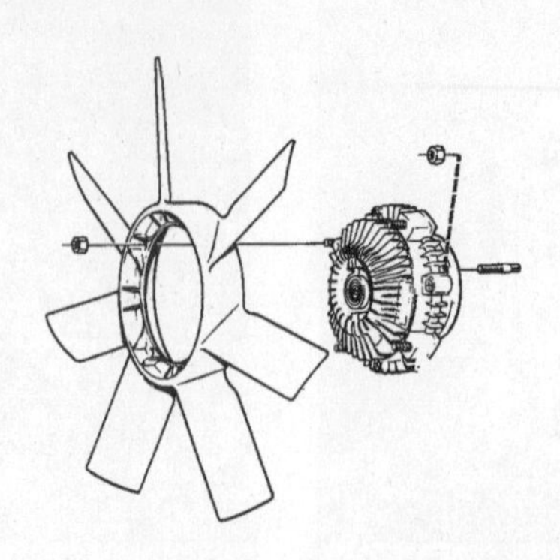

5.9a Three different types of clutches have been used, depending on year and model; on these two types the blades can be separated from the clutch . . .

5.9b . . . while this type of fan assembly is a single unit that cannot be separated

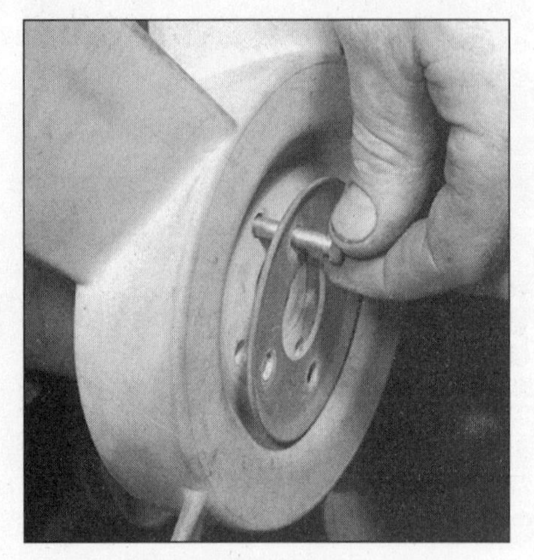

5.12 Remove the bolts from the fan assembly

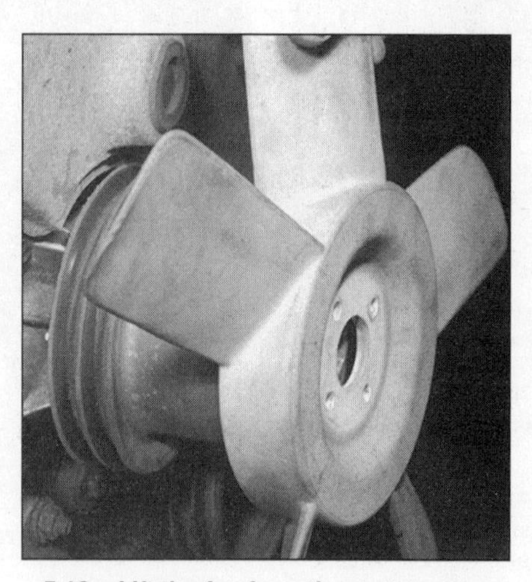

5.13a Lift the fan from the water pump

5.13b Remove the spacer from the water pump shaft

7 Lift the fan/clutch assembly (and shroud, if necessary) out of the engine compartment.

8 Carefully inspect the fan blades for damage and defects. Replace them if necessary.

9 At this point, the fan may be unbolted from the clutch, if necessary **(see illustrations)**. If the fan clutch is stored, position it with the radiator side facing down.

10 Installation is the reverse of removal.

5.13c Remove the pulley from the water pump hub

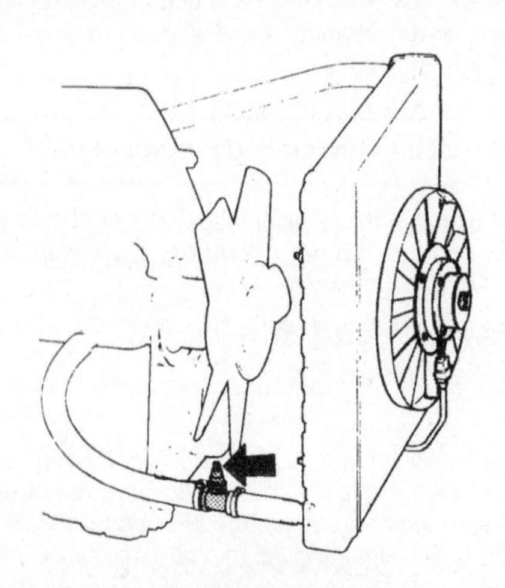

5.17 Location of the early model electric cooling fan thermal switch (arrow)

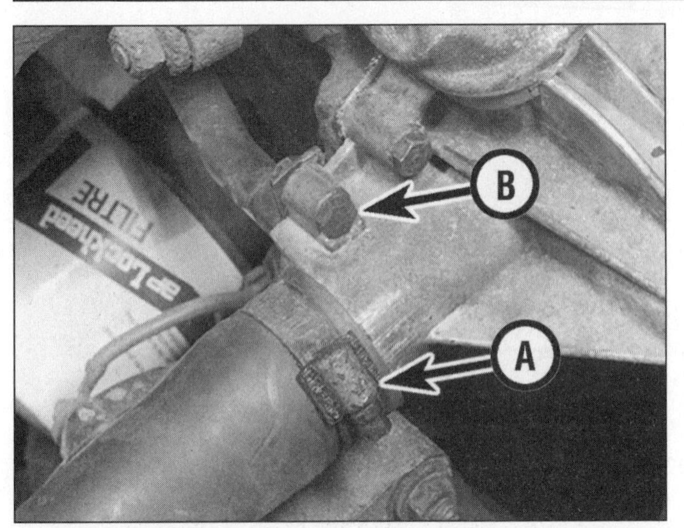

7.8a Remove the water pump bolts from the water pump discharge port

7.8b Remove the water pump bolts and nuts (arrows) (bottom bolt hidden from view)

Mechanical fan without clutch

Refer to illustrations 5.12, 5.13a, 5.13b and 5.13c

11 Disconnect the negative battery cable. Remove the fan shroud mounting screws and detach the shroud (see Section 4).

12 Use a box end wrench to remove the fan assembly **(see illustration).**

13 Lift the fan assembly **(see illustrations)** (and shroud, if necessary) out of the engine compartment.

14 Carefully inspect the fan blades for damage and defects. Replace it if necessary.

15 Installation is the reverse of removal.

Electric cooling fan

Note: *An auxiliary electric cooling fan was installed on turbocharged models with an intercooler, some early air-conditioned models and all 1991 and later air-conditioned models.*

Check

Refer to illustration 5.17

16 Place the transmission in Park (automatic) or Neutral (manual), set the parking brake and block the wheels. Start the engine, turn the air conditioning On and allow the engine to idle. The auxiliary electric cooling fan should come on when the engine temperature reaches 207 to 216 degrees F, on turbocharged and early air-conditioned models, or when the refrigerant pressure reaches the pre-determined value on 1991 and later air-conditioned models.

17 If the fan did not come on, stop the engine and locate the thermal switch electrical connector **(see illustration)**. **Note:** *For the location of the fan switch on 1991 and later models, see illustration 12.1b.* Disconnect the connector from the switch and place a jumper wire across the terminals of the connector.

18 Turn the ignition key to On. If the fan runs, there's a problem with the switch. The switch can be checked for continuity with an ohmmeter.

19 If the fan doesn't run, disconnect the electrical connector at the fan motor and connect a test light across the terminals at the connector. If the test light illuminates, indicating power is reaching the motor, the motor is probably defective. If the test light doesn't illuminate, the relay or wiring could be defective.

Replacement

20 Remove the front grille (see Chapter 11).

21 Disconnect the electrical connector from the fan motor.

22 Remove the screws retaining the fan assembly mounting brackets and remove the assembly from the vehicle.

23 Installation is the reverse of removal.

6 Water pump - check

1 A failure in the water pump can cause serious engine damage due to overheating.

2 There are two ways to check the operation of the water pump while it's installed on the engine. If the pump is defective, it should be replaced with a new or rebuilt unit.

3 Water pumps are equipped with vent holes. If a failure occurs in the pump seal, coolant will leak from the hole. In most cases you'll need a flashlight to find the hole on the water pump from underneath to check for leaks.

4 If the water pump shaft bearings fail there may be a howling sound at the front of the engine while it's running. Shaft wear can be felt if the water pump pulley is rocked up and down. Don't mistake drivebelt slippage, which causes a squealing sound, for water pump bearing failure.

7 Water pump - removal and installation

Removal

Refer to illustrations 7.8a, 7.8b, 7.9, 7.17 and 7.18

Warning: *Wait until the engine is completely cool before beginning this procedure.*

1 Disconnect the cable from the negative battery terminal. **Caution:** *It is necessary to make sure the radio is turned OFF before disconnecting the battery cables to avoid damaging the microprocessor built into the radio.*

2 Drain the cooling system (see Chapter 1). If the coolant is relatively new or in good condition, save it and reuse it.

3 Remove the cooling fan and shroud (see Section 5).

4 Remove the drivebelts (see Chapter 1).

5 Remove the pulley at the end of the water pump shaft (see Section 5).

6 Loosen the clamps and detach the hoses from the water pump. If they're stuck, grasp each hose near the end with a pair of adjustable pliers and twist it to break the seal, then pull it off. If the hoses are deteriorated, cut them off and install new ones.

7 Remove the timing belt covers from the engine (see Chapter 2A).

8 Remove the bolts that mount the water pump to the engine block **(see illustrations)**.

7.9 Remove the coolant bypass tube from the backside of the water pump

9 Remove the coolant bypass tube from the backside of the water pump **(see illustration)**.
10 Remove the water pump.

Installation

11 Clean the bolt threads and the threaded holes in the engine to remove corrosion and sealant.
12 Compare the new pump to the old one to make sure they're identical.
13 Remove all traces of old gasket material from the engine with a gasket scraper.
14 Clean the engine and new water pump mating surfaces with lacquer thinner or acetone.
15 Apply a thin coat of RTV sealant to the engine side of the new gasket.
16 Apply a thin layer of RTV sealant to the gasket mating surface of the new pump, then carefully mate the gasket and the pump.
17 Install the water pump-to-cylinder head seal **(see illustration)** and carefully attach the pump, gasket and seal to the engine. Secure the water pump with the two nuts, finger tight.
18 Pry the water pump up, toward the cylinder head **(see illustration)**, and install the remaining bolts (if they also hold an accessory bracket in place, be sure to reposition the bracket at this time). Tighten the bolts and nuts to the torque listed in this Chapter's Specifications in 1/4-turn increments. Don't over tighten them or the pump may be distorted.

7.17 Install a new water pump rubber seal

19 Reinstall all parts removed for access to the pump.
20 Refill the cooling system and check the drivebelt tension (see Chapter 1). Run the engine and check for leaks.

8 Coolant temperature sending unit - check and replacement

Warning: *Wait until the engine is completely cool before beginning this procedure.*
Refer to illustration 8.1
1 The coolant temperature indicator system is composed of a temperature gauge mounted in the instrument panel and a coolant temperature sending unit that's normally mounted in the cylinder head **(see illustration)**.
2 If an overheating indication occurs, check the coolant level in the system and then make sure the wiring between the gauge and the sending unit is secure and all fuses are intact.
3 Test the circuit by grounding the wire connected to the sending unit while the ignition is On (engine not running for safety). **Caution:** *Do not ground the wire for more than a second or two or damage to the gauge could occur.* If the gauge deflects full scale, replace the sending unit. If the gauge does not respond or does not indicate an overheating condition, the gauge or wiring to the gauge is faulty.
4 If the sending unit must be replaced, simply unscrew it from the

7.18 Pry the water pump up, toward the cylinder head while tightening the bolts

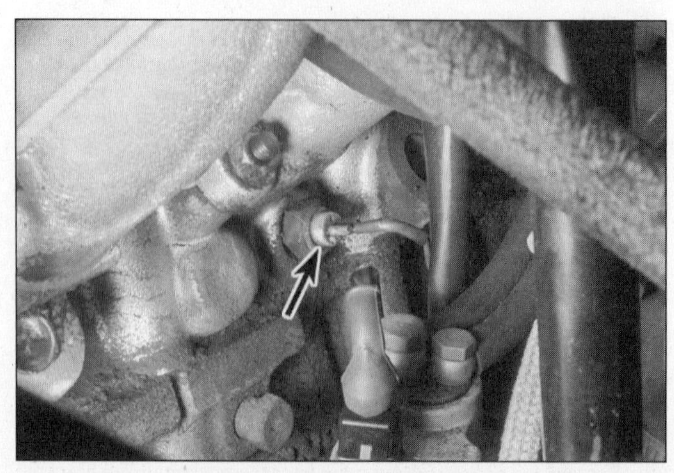

8.1 Location of the coolant temperature sensor

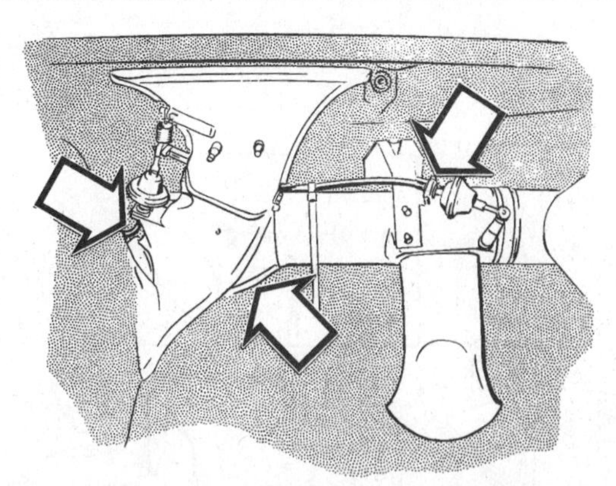

9.4 Remove the vacuum lines from the shutter actuators and slide the air ducts out of the engine compartment

9.8 Remove the center support screws

engine and install the replacement. Use sealant on the threads. Make sure the engine is cool before removing the defective sending unit. There will be some coolant loss as the unit is removed, so be prepared to catch it. Check the coolant level after the replacement unit has been installed.

9 Heater and air conditioning blower motor - removal and installation

Refer to illustrations 9.4, 9.8, 9.9, 9.11a and 9.11b

Note: *1976 through 1980 models are equipped with field wound fan motor while 1981 and later models are equipped with 4-speed permanent magnet motor. The early version has been superseded and in the event it needs to be replaced, additional parts and modifications will be necessary. Consult with a dealer parts department for the information and equipment.*

Removal

1 Disconnect the cable from the negative battery terminal. **Caution:** *It is necessary to make sure the radio is turned OFF before disconnecting the battery cables to avoid damaging the microprocessor built into the radio.*

2 The blower motor is contained within a combination unit that is mounted under the middle dash area. This combined unit houses the heater core and blower motor for the A/C system and heating system. Remove the screws attaching the control panel and the center console (see Chapter 11) and remove them from the passenger compartment.

3 Disconnect the control cables from the heater unit and from the heater and air conditioner control assembly (see Section 10).

4 Disconnect the vacuum lines to the shutter actuators and remove the air ducts **(see illustration)**.

5 Remove the radio (see Chapter 12).

6 Remove the glovebox (see Chapter 11).

7 Remove the steering column covers (see Chapter 11).

8 Disconnect the screws for the rear floor duct. Disconnect the lower heater unit screws, any vacuum lines, fan motor ground lead and the center support screws **(see illustration)**.

9 Disconnect the hoses to the heater valve **(see illustration)** and the heater core.

10 Remove the heater unit assembly. You can test the blower motor by applying battery voltage to the blower motor's terminals with fused jumper wires (be sure the fan cages won't hit anything when they rotate). If the blower motor spins the fan cages rapidly (this test simulates high-speed operation), the blower motor is OK. If the blower motor does not operate, or operates slowly or is noisy, replace it.

Note: *If the fan cages need to be removed, mark their relationship to the shaft. The cages are balanced during assembly and can cause excessive noise or shortened bearing life if the blower motor is not reinstalled in exactly the same position in relation to the shaft.*

11 Separate the two blower motor housing halves and remove the blower motor **(see illustrations)**.

Installation

12 Installation is the reverse of removal. **Note:** *Be sure to seal the case halves with silicon sealer to prevent any water from entering.* **Note:** *Spin the fan blades by hand before installing the unit back into the passenger compartment to make sure they are not binding or damaged in any way*

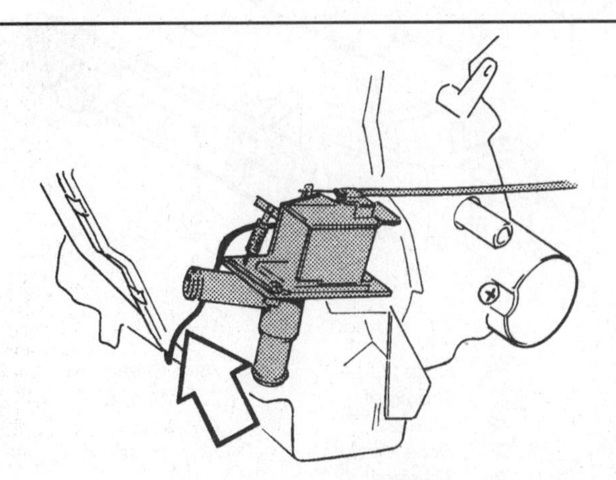

9.9 Separate the heater hoses from the heater valve and pull the capillary tube from the housing (arrow)

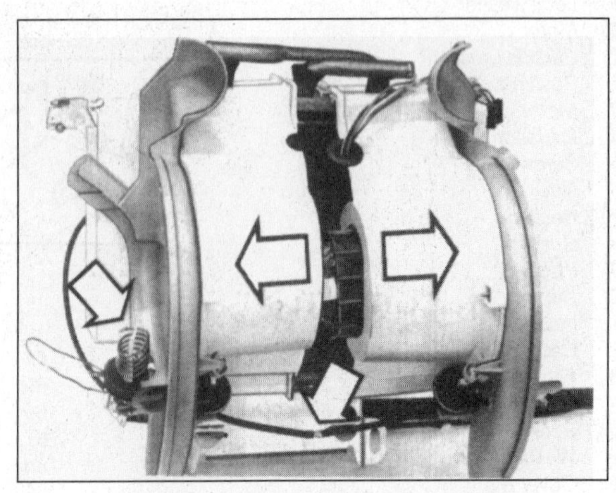

9.11a Separate the blower case halves

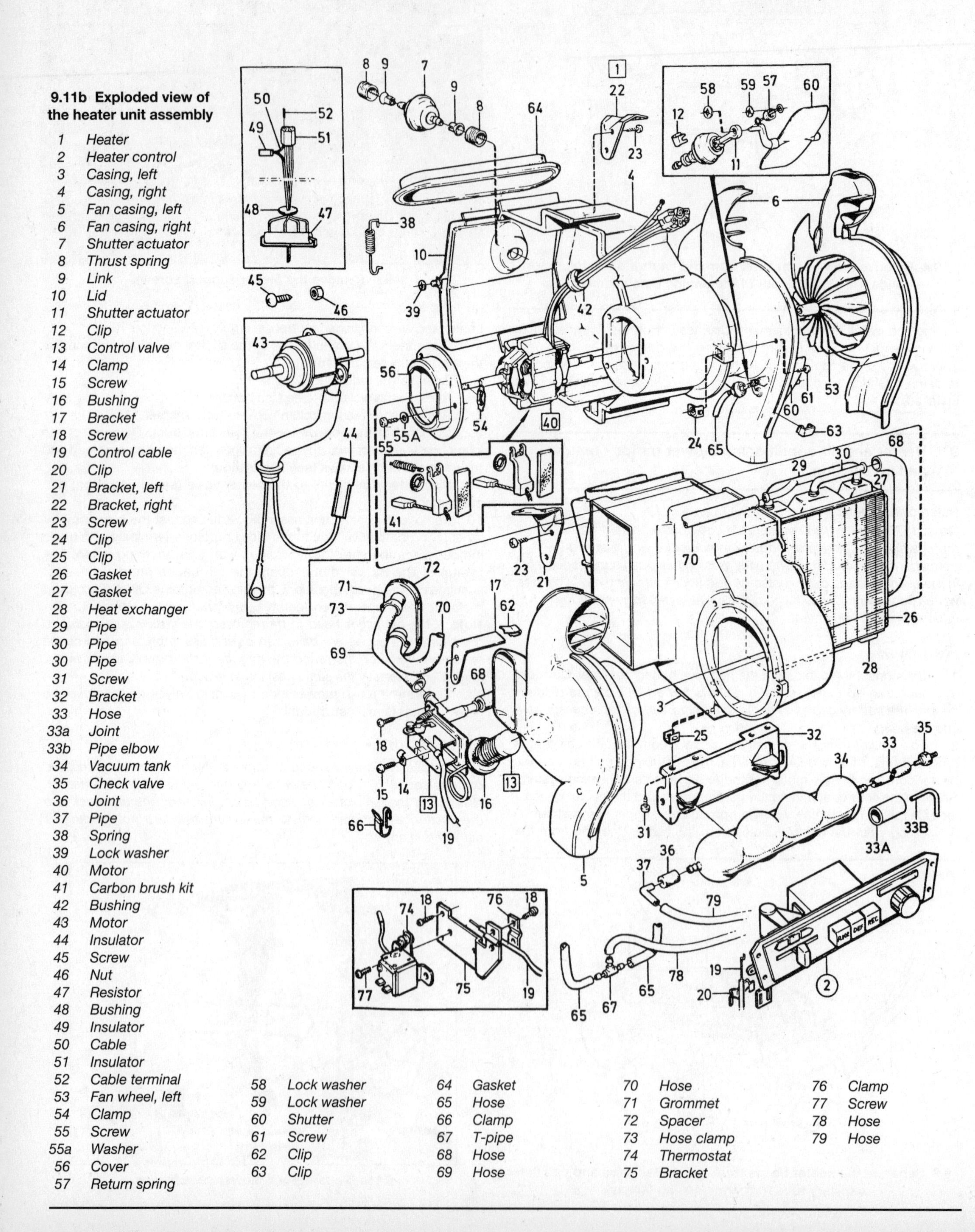

9.11b Exploded view of the heater unit assembly

1 Heater
2 Heater control
3 Casing, left
4 Casing, right
5 Fan casing, left
6 Fan casing, right
7 Shutter actuator
8 Thrust spring
9 Link
10 Lid
11 Shutter actuator
12 Clip
13 Control valve
14 Clamp
15 Screw
16 Bushing
17 Bracket
18 Screw
19 Control cable
20 Clip
21 Bracket, left
22 Bracket, right
23 Screw
24 Clip
25 Clip
26 Gasket
27 Gasket
28 Heat exchanger
29 Pipe
30 Pipe
30 Pipe
31 Screw
32 Bracket
33 Hose
33a Joint
33b Pipe elbow
34 Vacuum tank
35 Check valve
36 Joint
37 Pipe
38 Spring
39 Lock washer
40 Motor
41 Carbon brush kit
42 Bushing
43 Motor
44 Insulator
45 Screw
46 Nut
47 Resistor
48 Bushing
49 Insulator
50 Cable
51 Insulator
52 Cable terminal
53 Fan wheel, left
54 Clamp
55 Screw
55a Washer
56 Cover
57 Return spring
58 Lock washer
59 Lock washer
60 Shutter
61 Screw
62 Clip
63 Clip
64 Gasket
65 Hose
66 Clamp
67 T-pipe
68 Hose
69 Hose
70 Hose
71 Grommet
72 Spacer
73 Hose clamp
74 Thermostat
75 Bracket
76 Clamp
77 Screw
78 Hose
79 Hose

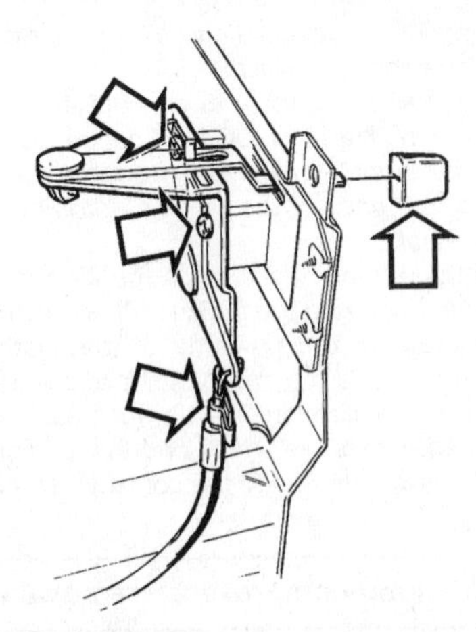

10.5 Disconnect the knobs and the heater cable connectors (arrows) from the control unit

10 Heater and air conditioner control assembly - removal and installation

Removal

Refer to illustrations 10.5

1 Disconnect the cable from the negative battery terminal. **Caution:** *It is necessary to make sure the radio is turned OFF before disconnecting the battery cables to avoid damaging the microprocessor built into the radio.*

2 Remove the center console and side trim pieces (see Chapter 11).

3 Remove the radio (see Chapter 12), then pull the knobs off the heater/air conditioning control levers.

4 Remove the heater trim panel to access the control cables.

5 Disconnect the cables, marking them for accurate reinstallation **(see illustration)**.

6 Disconnect the electrical connector.

7 Remove the control assembly from the dash.

Installation

8 Installation is the reverse of the removal procedure.

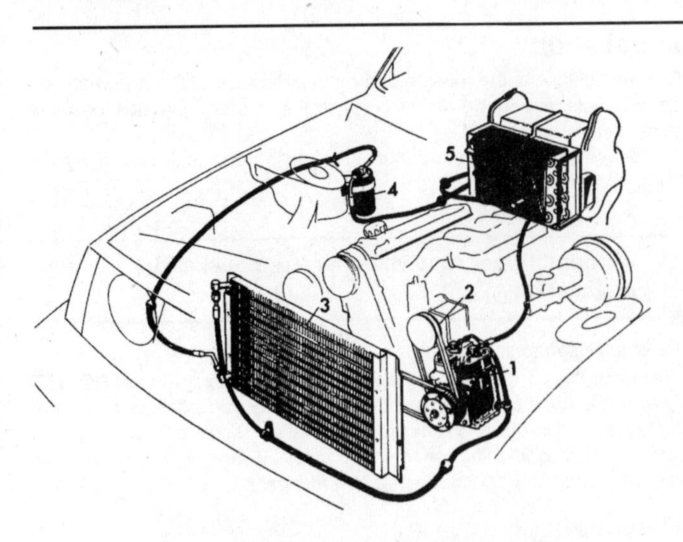

12.1a Early model air conditioning system

1 *Compressor*
2 *Steering pump*
3 *Condenser*
4 *Drier*
5 *Evaporator*

11 Heater core - removal and installation

Removal

1 Disconnect the cable from the negative battery terminal. **Caution:** *It is necessary to make sure the radio is turned OFF before disconnecting the battery cables to avoid damaging the microprocessor built into the radio.*

2 Drain the cooling system (see Chapter 1).

3 Remove the center console (see Chapter 11). Spread plastic sheeting on the front carpet in the center and on the passenger's side; this will prevent carpet stains if any residual coolant spills.

4 Remove the heater and air conditioner unit (see Section 9).

5 Separate the case halves and remove the heater core.

Installation

6 Installation is the reverse of removal. Refill the cooling system (see Chapter 1), run the engine with the heater on and check for leaks.

12 Air conditioning and heating system - check and maintenance

Warning: *The air conditioning system is under high pressure. DO NOT loosen any hose or line fittings or remove any components until after the system has been discharged. Air conditioning refrigerant should be properly discharged into an EPA-approved container at a dealer service department or an automotive air conditioning repair facility. Always wear eye protection when disconnecting air conditioning system fittings.*

Check

Refer to illustrations 12.1a, 12.1b and 12.1c

1 The following maintenance checks should be performed on a regular basis to ensure the air conditioner continues to operate at peak efficiency **(see illustrations)**.

a) *Check the drive-belt. If it's worn or deteriorated, replace it (see Chapter 1).*

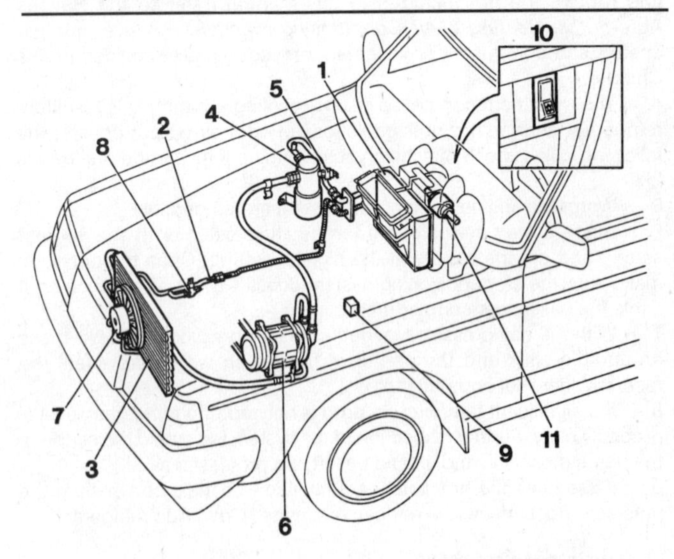

12.1b 1991 and later model air conditioning system

1 *Evaporator*
2 *Expansion pipe*
3 *Condenser*
4 *Accumulator*
5 *Pressure switch*
6 *Compressor*
7 *Electric cooling fan*
8 *Electric cooling fan switch*
9 *Electric cooling fan relay*
10 *Control switch*
11 *Blower motor*

3

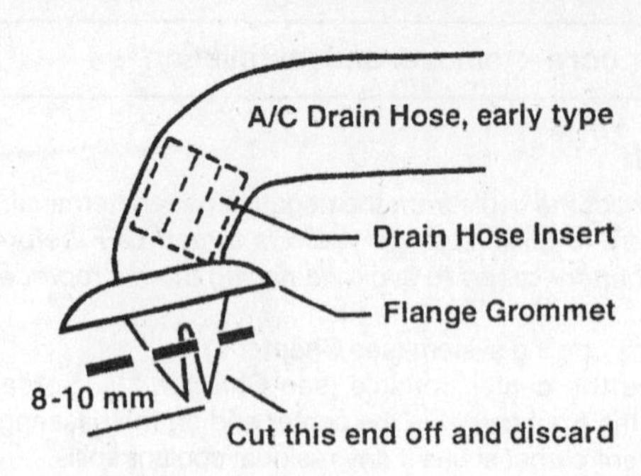

12.1c Install a special drain hose insert into the bottom section of the drain tube and cut a small portion of the tube rubber to prevent any collapsing of the hose

- b) *Check the system hoses. Look for cracks, bubbles, hard spots and deterioration. Inspect the hoses and all fittings for oil bubbles and seepage. If there's any evidence of wear, damage or leaks, replace the hose(s).*
- c) *Inspect the condenser fins for leaves, bugs and other debris. Use a "fin comb" or compressed air to clean the condenser.*
- d) *Make sure the system has the correct refrigerant charge.*
- e) *Check the air conditioning drain system. Tiny water droplets may spray inside the passenger compartment during air conditioning operation due to a clog in the evaporator drain. A back-flow of air through the plenum drain will occur during high fan speeds. To improve the situation, install a special tube insert to prevent the hose from collapsing* **(see illustration)**. *Remove the right side kick panel to gain access to the drain tube.*

2 It's a good idea to operate the system for about 10 minutes at least once a month, particularly during the winter. Long term non-use can cause hardening, and subsequent failure, of the seals.

3 Because of the complexity of the air conditioning system and the special equipment necessary to service it, in-depth troubleshooting and repairs are not included in this manual (refer to the *Haynes Automotive Heating & Air Conditioning Manual*). However, simple checks and component replacement procedures are provided in this Chapter.

4 The most common cause of poor cooling is simply a low system refrigerant charge. If a noticeable loss of cool air output occurs, the following quick check may help you determine if the refrigerant level is low.

5 Warm the engine up to normal operating temperature.

6 Place the air conditioning temperature selector at the coldest setting and put the blower at the highest setting. Open the doors (to make sure the air conditioning system doesn't cycle off as soon as it cools the passenger compartment).

7 With the compressor engaged - the compressor clutch will make an audible click and the center of the clutch will rotate - feel the receiver/drier inlet and outlet pipes.

8 If a significant temperature drop is noticed, the refrigerant level is probably okay. Further inspection of the system is beyond the scope of the home mechanic and should be left to a professional.

9 If the inlet line has frost accumulation or feels cooler than the receiver-drier surface, the refrigerant charge is low. Add refrigerant.

Adding refrigerant

Note: *Because of recent environmental legislation, 12-ounce cans of refrigerant may not be available in your area. If this is the case, you'll have to take the vehicle to a dealer service department or licensed air conditioning repair shop to have the air conditioning system recharged.*

10 Obtain a 12-ounce can of refrigerant, a tap valve and a short section of hose that can be attached between the tap valve and the system low-side service valve. Because one can of refrigerant may not be sufficient to bring the system charge up to the proper level, it's a good idea to obtain an additional can. **Warning:** *Never add more than two cans of refrigerant to the system.*

11 Use the tap valve and hose to connect the can to the system's low side service port. **Warning:** *DO NOT connect the charging kit hose to the system high side! If you have any doubts about how to hook up the system, obtain a copy of the Haynes Automotive Heating and Air Conditioning Manual.*

12 Warm up the engine and turn on the air conditioner. Keep the charging kit hose away from the fan and other moving parts.

13 Add refrigerant to the low side of the system until both the receiver-drier surface and the evaporator inlet line feel about the same temperature. Allow stabilization time between each addition.

14 Once the receiver-drier surface and the evaporator inlet line feel about the same temperature, add the contents remaining in the can.

13 Air conditioning compressor - removal and installation

Warning: *The air conditioning system is under high pressure. DO NOT disassemble any part of the system (hoses, compressor, line fittings, etc.) until after the system has been discharged by a dealer service department or licensed air conditioning repair shop.*

Note: *The receiver-drier or accumulator should be replaced whenever the compressor is replaced.*

Removal

1 Have the air conditioning system discharged (see Warning above).

2 Disconnect the cable from the negative battery terminal. **Caution:** *It is necessary to make sure the radio is turned OFF before disconnecting the battery cables to avoid damaging the microprocessor built into the radio.*

3 Disconnect the compressor clutch wiring harness.

4 Remove the drivebelt (see Chapter 1).

5 Disconnect the refrigerant lines from the rear of the compressor. Plug the open fittings to prevent entry of dirt and moisture.

6 Unbolt the compressor from the mounting brackets and lift it up and out of the vehicle.

7 If a new compressor is being installed, follow the directions with the compressor regarding the draining of excess oil prior to installation.

8 The clutch may have to be transferred from the original to the new compressor.

Installation

9 Installation is the reverse of removal. Replace all O-rings with new ones specifically made for air conditioning system use and lubricate them with refrigerant oil.

10 Have the system evacuated, recharged and leak tested by the shop that discharged it.

14 Air conditioner receiver-drier - removal and installation

Refer to illustrations 14.7 and 14.11

Warning: *The air conditioning system is under high pressure. DO NOT loosen any hose or line fittings or remove any components until after the system has been discharged by a dealer service department or licensed air conditioning repair shop. Always wear eye protection when disconnecting air conditioning system fittings.*

Removal

1 Have the system discharged (see Warning above).

2 Disconnect the cable from the negative battery terminal. **Caution:** *It is necessary to make sure the radio is turned OFF before disconnecting the battery cables to avoid damaging the microprocessor built into the radio.*

14.7 Location of the receiver/drier

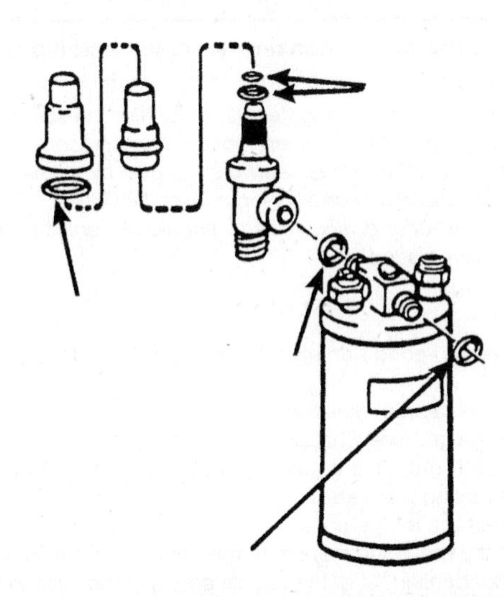
14.11 Locations of all the O-rings at the receiver/drier (arrows)

3 Remove the windshield washer fluid tank.
4 Disconnect the electrical connector from the pressure switch.
5 Disconnect the refrigerant lines from the receiver-drier.
6 Plug the open fittings to prevent the entry of dirt and moisture.
7 Remove the mounting screws and remove the receiver-drier **(see illustration)**.
8 If a new receiver-drier is being installed, pour in the amount of refrigerant oil indicated in the specifications at the front of this Chapter.
9 Remove the old refrigerant line O-rings and replace with new ones. This should be done regardless of whether a new receiver-drier is being installed or not.
10 If a new receiver-drier is being installed, unscrew the safety pressure switches and install them on the new unit before installation.
11 Lubricate O-rings with refrigerant oil before assembly **(see illustration)**.

Installation

12 Installation is the reverse of removal, but be sure to lubricate the O-rings on the fittings with refrigerant oil before connecting the fittings.
13 Have the system evacuated, recharged and leak tested by the shop that discharged it. **Note**: *The manufacturer recommends that all screwed -in fittings should be assembled with the use of a thread sealant. Specific sealant for the high or low pressure side of the system is available at an air conditioning repair shop or automotive parts store.*

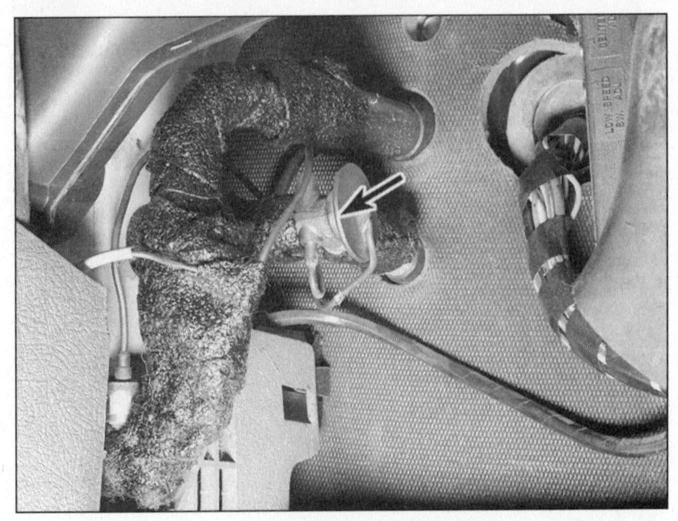
15.4 Expansion valve location

15 Air conditioning expansion valve - replacement

Refer to illustrations 15.4 and 15.6

Warning: *The air conditioning system is under high pressure. DO NOT loosen any hose or line fittings or remove any components until after the system has been discharged by a dealer service department or licensed air conditioning repair shop. Always wear eye protection when disconnecting air conditioning system fittings.*

1 Have the system discharged (see Warning above).
2 Disconnect the cable from the negative battery terminal. **Caution:** *It is necessary to make sure the radio is turned OFF before disconnecting the battery cables to avoid damaging the microprocessor built into the radio.*
3 Remove the side panels from the center console unit under the dash area (see Chapter 11).
4 Remove the insulation from around the expansion valve and the neighboring tubes **(see illustration)**.
5 Disconnect the expansion valve from the thermostat assembly.
6 Installation is the reverse of removal. Use new seals coated with compressor oil **(see illustration)**.
7 Have the system evacuated, recharged and leak tested by the shop that discharged it.

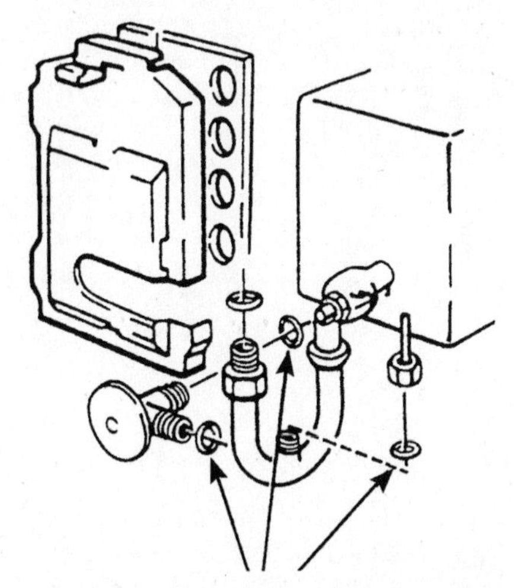
15.6 Locations of the O-rings at the expansion valve and evaporator (arrows)

16 Air conditioning condenser - removal and installation

Warning: *The air conditioning system is under high pressure. DO NOT disassemble any part of the system (hoses, compressor, line fittings, etc.) until after the system has been discharged by a dealer service department or licensed air conditioning service facility.*
Note: *The receiver-drier or accumulator should be replaced whenever the condenser is replaced.*

Removal

1 Have the air conditioning system discharged (see Warning above).
2 Remove the radiator (see Section 4).
3 Remove the grill (see Chapter 11).
4 If equipped, unbolt the auxiliary fan from the air conditioning condenser mounting brackets.
5 Disconnect the refrigerant lines from the condenser.
6 Remove the mounting bolts from the condenser brackets.
7 Lift the condenser out of the vehicle and plug the lines to keep dirt and moisture out.

Installation

8 If the original condenser will be reinstalled, store it with the line fittings on top to prevent oil from draining out.
9 If a new condenser is being installed, add new refrigerant oil to system according to the specifications at the beginning of this Chapter.
10 Reinstall the components in the reverse order of removal. Be sure the rubber pads are in place under the condenser.
11 Have the system evacuated, recharged and leak tested by the shop that discharged it.

17 Air conditioning evaporator core - removal and installation

Refer to illustration 17.7
Warning: *The air conditioning system is under high pressure. DO NOT loosen any hose or line fittings or remove any components until after the system has been discharged by a dealer service department or licensed air conditioning repair shop. Always wear eye protection when disconnecting air conditioning system fitting*

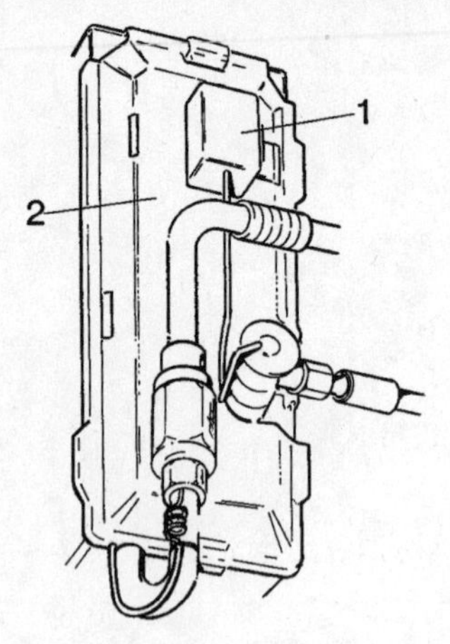

17.7 Evaporator cover

1 Thermostat 2 Evaporator cover

Removal

1 Have the air conditioning system discharged (see Warning above).
2 Remove the glove compartment (see Chapter 11)
3 Remove the panel beneath the glove compartment near the center console (see Chapter 11).
4 Remove the side panel next to the heater.
5 Remove the right side defroster vent and air duct.
6 Remove the expansion valve (see Section 15).
7 Remove the thermostat and the evaporator cover **(see illustration). Note:** *The thermostat is located in different areas depending upon the year of the vehicle. Early models (1976 through 1978) are located on the evaporator cover. Later models (1979 on), the thermostat is located on the control panel while the capillary tube runs into the evaporator. If it is necessary to change the thermostat, replace it with an updated version. Consult the dealer parts department for additional information and parts.*
8 Slide the evaporator from the housing.

Installation

9 Installation is the reverse of the removal procedure.
10 Have the system evacuated, recharged and leak tested by the shop that discharged it.

Chapter 4 Part A Fuel and exhaust systems - fuel-injected engines

Contents

Specifications

Fuel pressure checks

Mechanical injection systems

Delivery pressure	65 to 90 psi
System pressure	
Turbocharged engines	75 to 84 psi
Non-turbocharged engines	65 to 77 psi
Control pressure	
Typical cold	20 to 30 psi
Typical warm	40 to 58 psi

Control Pressure Regulator		463971-2 **004**	1219159-0 **014**	1219952-7 **021**	1276878-4 **079***	1276946-9 **082**
Engine type	**Model year**					
B 19	1977	X				
B 19 E Turbo	1982					X
B 21 E	1976	X				
B 21 E Turbo	1981					X
B 21 F	1976 through 1981	X	X	X		X
B 21 F-9	1981				X	
B 21 F Turbo	1981				X	
B 23 E	1979	X				

Note 1: *The type of control pressure regulator is stamped as the last three digits on the top of the unit.*
Note 2: *The 021 control pressure regulator is equipped with an altitude compensating device within the unit.*

Control pressure (cold)	
004	20 to 25 psi
014	20 to 25 psi
021	22 to 30 psi
079*	22 to 28 psi
082	20 to 25 psi
Control pressure (warm)	
004	50 to 54 psi
014	50 to 55 psi
021	35 to 60 psi
079*	50 to 55 psi
082	50 to 55 psi

** Some 1981 through 1983 240 models may have been retrofitted with a modified control pressure regulator (Bosch part no. 123, Volvo part no. 1336475-7) to correct an erratic idle or poor engine response condition after a cold start when the ambient temperature is below 60-degrees F. If your vehicle suffers from these symptoms and your engine is equipped with the early version control pressure regulator (Bosch part no. 079), install this modified control pressure regulator.*

Rest pressure	
Turbocharged engines	
1981 turbocharged engines	22 to 35 psi
1982 and later turbocharged engines	28 to 38 psi
Non-turbocharged engines	22 to 35 psi
Control pressure regulator resistance (CIS systems)	
004	20 to 30 ohms
014	20 to 30 ohms
021	20 to 30 ohms
079	10 to 20 ohms
082	20 to 30 ohms
123	
Below 53-degrees F	33.3 to 36.7 ohms
Above 63-degrees F	11.7 to 12.9 ohms
Electronic fuel injection systems	
Fuel system pressure (key ON, engine OFF, fuel system energized)	
LH 2.0 and 2.2	27 to 32 psi
LH 2.4 and 3.1	32 to 38 psi
Pressure regulator check (at idle)	
Vacuum hose attached	
LH 2.0 and 2.2	27 to 30 psi
LH 2.4 and 3.1	32 to 36 psi
Vacuum hose detached	
LH 2.0 and 2.2	35 to 38 psi
LH 2.4 and 3.1	41 to 44 psi
Fuel system residual pressure (minimum after 20 minutes)	14.5 psi
Fuel pump pressure (maximum) (deadhead pressure)	70 to 92 psi
Fuel pump hold pressure	80 psi
In-tank pump pressure	
1976 through 1982 CIS systems	2 to 4 psi
1982 through 1985 LH-Jetronic systems	2 to 3 psi
1986 on	4 to 5 psi
Cold start injector resistance	10 ohms
Injector resistance	16 ohms

Accelerator cable freeplay	3/64 to 3/32 inch

Torque specifications

Throttle body nuts/bolts	14 to 19 ft-lbs
Fuel rail mounting bolts	80 to 97 in-lbs
Turbocharger bolts	
Stage 1	12 in-lbs
Stage 2	33 ft-lbs
Stage 3	Turn an additional 45-degrees

1 General information

All US models are equipped with either mechanical fuel injection or electronic fuel injection systems. Canadian models also have the option of carburetors (some models between 1978 and 1984). The fuel injection system(s) consists of three basic sub-systems - the air intake system, the fuel delivery system and the electronic control system. Carbureted models are covered in Chapter 4, Part B.

Air intake system

Depending upon the type of fuel injection system installed, the air intake system consists of the air filter housing, the mass airflow sensor (LH-Jetronic systems), the throttle body and the intake manifold. All components except the intake manifold are covered in this Chapter; for information on removing and installing the intake manifold, refer to Chapter 2A.

The throttle valve inside the throttle body is actuated by the accelerator cable. When you depress the accelerator pedal, the throttle plate opens and airflow through the intake system increases. A throttle position switch (LH 2.0, 2.2 and 2.4) or throttle position sensor (LH 3.1) attached to the throttle shaft on the throttle body (LH-Jetronic systems) detects the angle of the flap (how much it's open) and converts this to a voltage signal which it sends to the computer.

Fuel system

An electric fuel pump supplies fuel under constant pressure to the fuel rail, which distributes fuel to the injectors (LH-Jetronic) or fuel distributor (CIS fuel injection). The electric fuel pump is located inside the fuel tank on newer models or adjacent to the fuel tank on early models. The early models are also equipped with a transfer pump located in the fuel tank. The transfer pump acts as an aid to the larger main pump for delivering the necessary pressure. A fuel pressure regulator controls the pressure in the fuel system. The fuel system is also equipped with a fuel pulsation damper located near the fuel filter. The damper reduces the pressure pulsations caused by fuel pump operation and the opening and closing of the injectors. The amount of fuel injected into the intake ports is precisely controlled by an Electronic Control Unit (ECU or computer). CIS turbocharged models are also equipped with a fuel accumulator. The accumulator absorbs fuel pressure surges and helps retain residual fuel pressure once the engine is shut off.

Electronic control system

Besides altering the injector opening duration as described above, the electronic control unit performs a number of other tasks related to fuel and emissions control. It accomplishes these tasks by using data relayed to it by a wide array of information sensors located throughout the engine compartment, comparing this information to its stored map and altering engine operation by controlling a number of different actuators. Since special equipment is required, most troubleshooting and repairing of the electronic control system is beyond the scope of the home mechanic. Additional information and testing procedures on the emissions system components (oxygen sensor, coolant temperature sensor, EVAP system, etc.) is contained in Chapter 6.

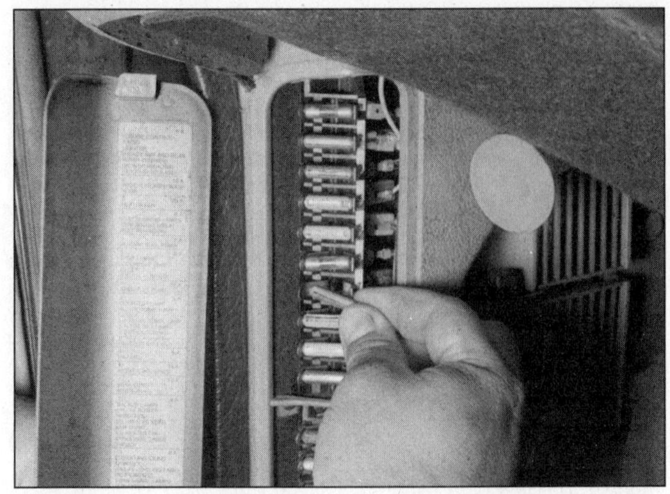

2.1 To relieve the fuel system pressure, remove the fuel pump fuse, start the engine and allow it to stall

Warranty information

These vehicles are covered by a Federally-mandated extended warranty (5 years or 50,000 miles at the time this manual was written) which covers nearly every fuel system component in this Chapter. Before working on the fuel system, check with a dealer service department for warranty terms on your particular vehicle.

2 Fuel pressure relief procedure

Refer to illustration 2.1

Warning: *Gasoline is extremely flammable, so take extra precautions when you work on any part of the fuel system. Don't smoke or allow open flames or bare light bulbs near the work area, and don't work in a garage where a natural gas-type appliance (such as a water heater or a clothes dryer) with a pilot light is present. Since gasoline is carcinogenic, wear latex gloves when there's a possibility of being exposed to fuel, and, if you spill any fuel on your skin, rinse it off immediately with soap and water. Mop up any spills immediately and do not store fuel-soaked rags where they could ignite. The fuel system is under constant pressure, so, if any fuel lines are to be disconnected, the fuel pressure in the system must be relieved first* (see Section 2 for more information). *When you perform any kind of work on the fuel system, wear safety glasses and have a Class B type fire extinguisher on hand.*

1 Remove the fuel pump fuse from the main fuse panel **(see illustration)**. **Note**: *Consult your owner's manual for the exact location of the fuel pump fuse if the information is not stamped onto the fuse cover.*

2 Start the engine and wait for the engine to stall, then turn off the ignition key.

3 Remove the fuel filler cap to relieve the fuel tank pressure.

4 The fuel system is now depressurized. **Note**: *Place a rag around the fuel line before removing the hose clamp to prevent any residual fuel from spilling onto the engine.*

5 Disconnect the cable from the negative terminal of the battery before working on any part of the system. **Caution:** *It is necessary to make sure the radio is turned OFF before disconnecting the battery cables to avoid damaging the microprocessor in the radio.*

3 Fuel pump/fuel pressure - check

Warning: *Gasoline is extremely flammable, so take extra precautions when you work on any part of the fuel system. Don't smoke or allow open flames or bare light bulbs near the work area, and don't work in a garage where a natural gas-type appliance (such as a water heater or a clothes dryer) with a pilot light is present. Since gasoline is carcinogenic, wear latex gloves when there's a possibility of being exposed to fuel, and, if you spill any fuel on your skin, rinse it off immediately with soap and water. Mop up any spills immediately and do not store fuel-soaked rags where they could ignite. The fuel system is under constant pressure, so, if any fuel lines are to be disconnected, the fuel pressure in the system must be relieved first* (see Section 2 for more information). *When you perform any kind of work on the fuel system, wear safety glasses and have a Class B type fire extinguisher on hand.*

Note 1: *The electric fuel pump is located outside the fuel tank (external) with an assisting fuel pump that primes the fuel system in the tank*

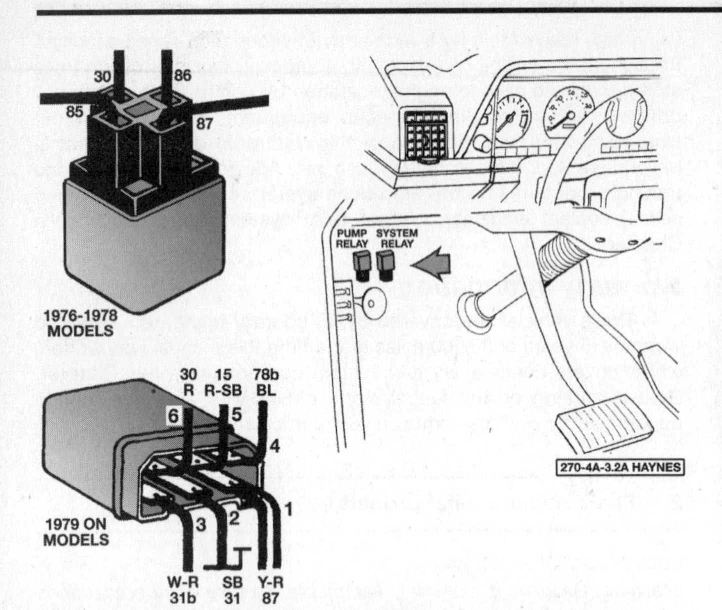

3.2a On CIS systems, the fuel pump relay is located under the driver's side kick panel. Remove the relay and jump the terminals that correspond to terminal numbers 30 and 87 to activate the fuel pump

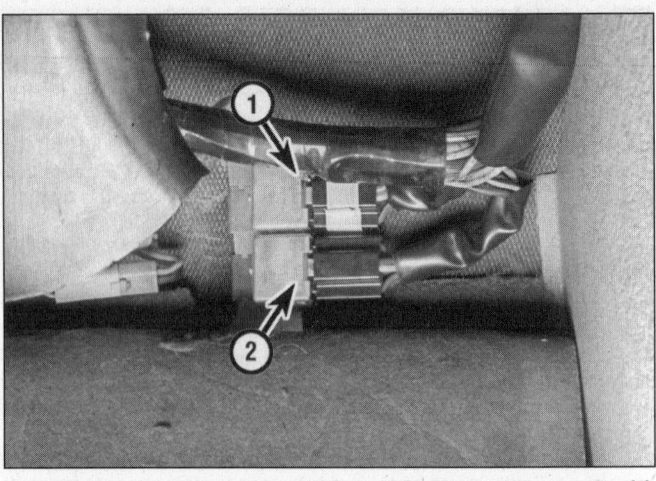

3.2b On LH-Jetronic systems before 1985, the fuel pump relay (1) is located next to the system relay (2) under the passenger's side kick panel

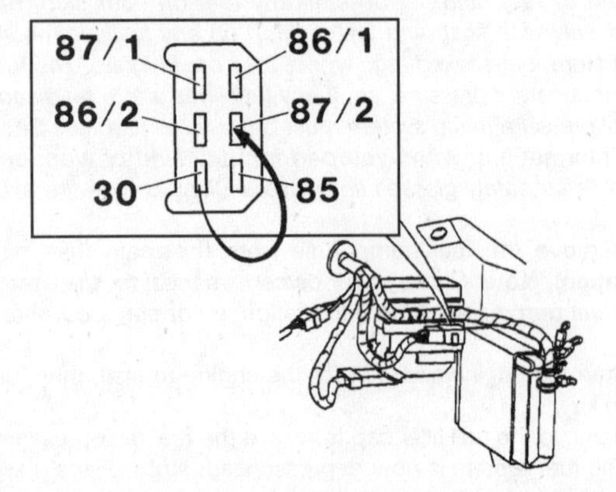

3.2c On LH 2.2, 2.4 and 3.1 systems, the fuel pump relay is located next to the ECU under the passenger side kick panel. Jump terminals 30 and 87/2 to activate the fuel pump

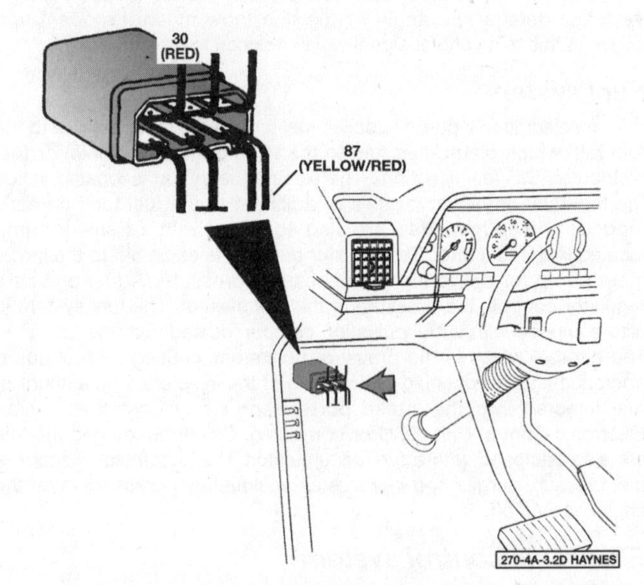

3.2d On turbocharged models, jump terminals 30 and 87

(internal). The internal pump feeds the main pump but can't generate the high pressure required by the system. There is a small percentage of 240 models that are not equipped with the in-tank fuel pump, depending on the model year and geographic location.

Note 2: *The fuel pump relay provides power to the fuel pumps (external and internal tank pumps) and the fuel injectors. The system relay provides power to the ECU for purposes of operating other system components.*

Note 3: *Some 1979 and later 240 series models use the buzzer relay as a power feed for the fuel pump relay. This isn't indicated on most Volvo factory wiring diagrams.*

Note 4: *The following checks assume the fuel filter is in good condition. If you doubt the condition of your fuel filter, replace it* (see Chapter 1).

Note 5: *If the engine is turbocharged and the fuel pressure readings are not correct, be sure to check the solenoid valve in the charge pressure control system. This valve can change fuel pressure in the event of excessive boost pressure. Refer to Section 16.*

Fuel pump/in-tank pump operational check

Refer to illustrations 3.2a through 3.2g

Note: *On engines equipped with mechanical fuel injection (CIS), first check the fuel pump operation (follow steps 1 and 2.). To check the working pressures for models with CIS, proceed to Section 12.*

1 First, should the fuel system fail to deliver fuel (no start) or the proper amount of fuel, inspect the system as follows. Remove the fuel filler cap. Have an assistant turn the ignition key ON (engine not running) while you listen at the fuel filler opening. You should hear a whirring sound that lasts for at least a couple of seconds. If you don't hear anything, inspect the fuel pump fuse (see Section 2). If the fuse is blown, replace it and see if it blows again. If it does, trace the fuel pump circuit for a short.

2 If the fuse is not blown, locate the fuel pump relay for the particular year and type of fuel system installed on the engine **(see illustrations).** Bridge the correct terminals **(see illustration)** with a suitable jumper wire. Have an assistant turn the ignition key to ON while you listen at the fuel tank (the main fuel pump is external to the fuel tank and forward). You should hear a whirring sound from the pump. **Note:** *Here is a quick method for activating the fuel pump without using the fuel pump relay on LH-Jetronic systems. Use a jumper wire across fuse terminals numbers 5 and 7* **(see illustration)** *on 1983 and 1984 models or numbers 4 and 6 on 1985 and later models. The fuel pump should activate. Refer to illustrations 13.1a through 13.1e for complete wiring diagrams of the various LH-Jetronic fuel injection systems.*

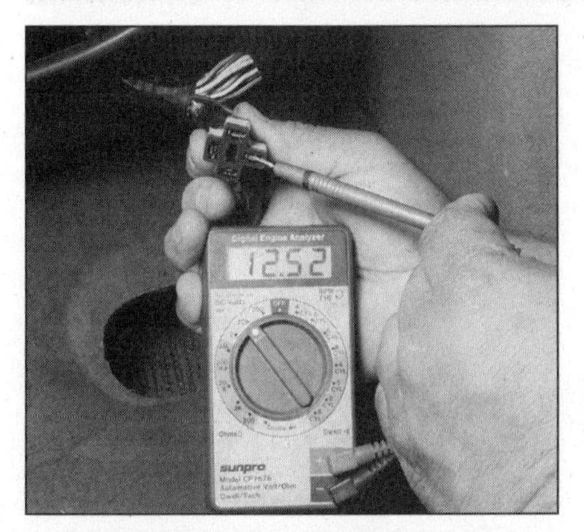

3.2e Check for battery voltage on terminal number 87 (1983 LH-Jetronic system shown)

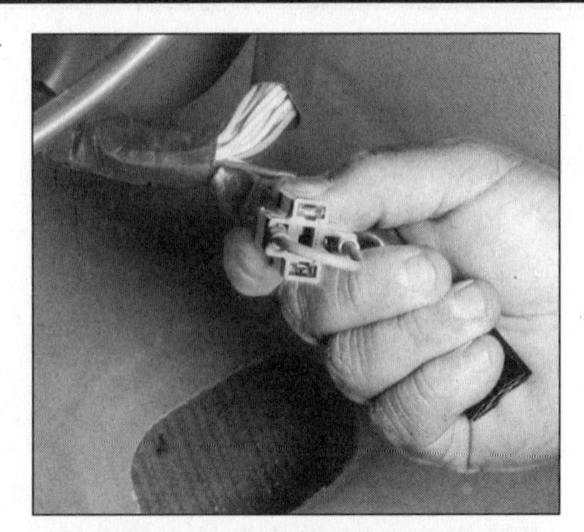

3.2f If battery voltage is present, install a jumper wire to activate the fuel pump (1983 LH-Jetronic system shown)

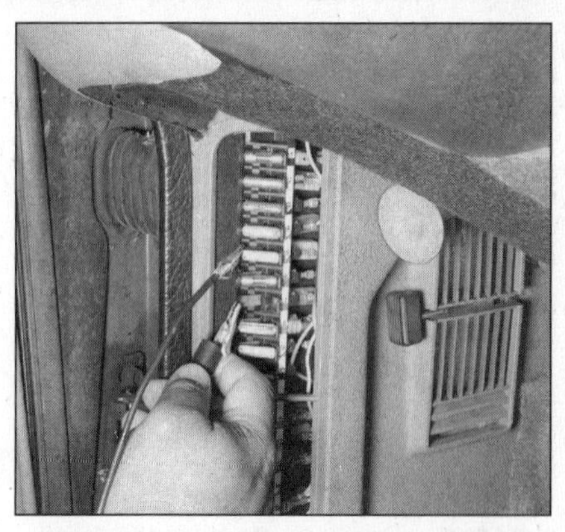

3.2g Another method to activate the fuel pump without the relay is to jump terminals number 5 and 7 on 1983 and 1984 LH-Jetronic systems directly at the fuse panel

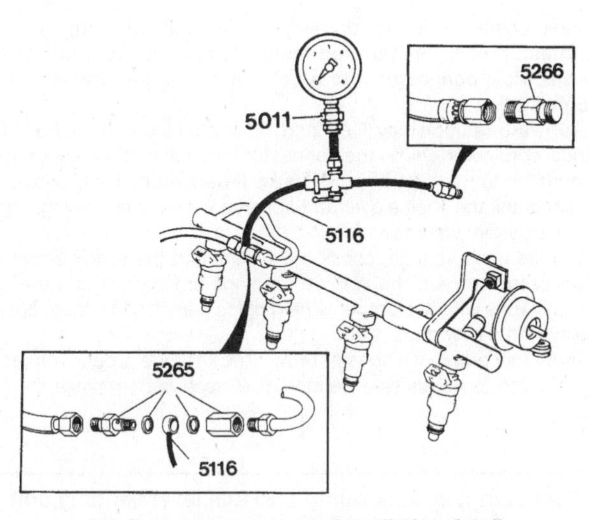

3.5 Fuel pressure gauge installation details

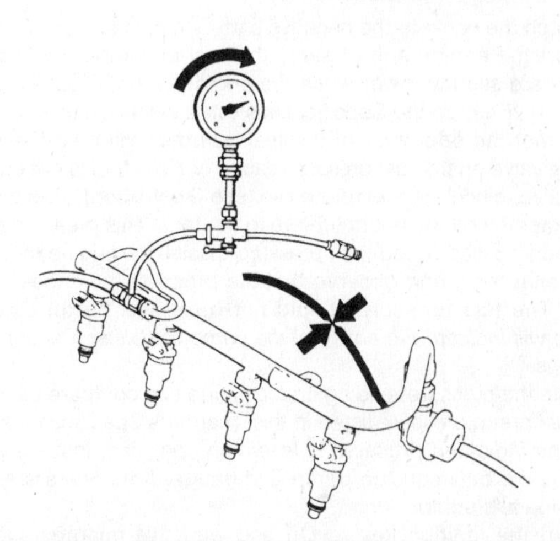

3.19 Pinch the return line and confirm that the fuel pressure increases. Raise the engine rpm and verify that the fuel pressure does not drop under load conditions

Fuel system pressure check

Refer to illustration 3.5

Note: *This procedure applies to LH-Jetronic systems only. For checking the working pressures of the CIS system, refer to Section 12.*

3 Relieve the system fuel pressure (see Section 2).

4 Detach the cable from the negative battery terminal. **Caution:** *It is necessary to make sure the radio is turned OFF before disconnecting the battery cables to avoid damaging the microprocessor in the radio.*

5 Detach the fuel feed line from the fuel rail **(see illustration)**.

6 The fuel pressure tests will require a special tool **(see illustration 3.5)** that will adapt the gauge to the particular thread size of the fuel lines. Obtain the fuel pressure gauge adapter at a specialty tool distributor. Using the special adapter, attach the gauge at the fuel rail.

7 Attach the cable to the negative battery terminal.

8 Bridge the terminals of the fuel pump relay using a jumper wire **(see illustrations 3.1a through 3.1d)**.

9 Turn the ignition switch to ON.

10 Note the fuel pressure and compare it with the pressure listed in this Chapter's Specifications.

11 If the system fuel pressure is less than specified:

a) *Inspect the system for a fuel leak. Repair any leaks and recheck the fuel pressure.*
b) *If there are no leaks, install a new fuel filter and recheck the fuel pressure.*
c) *If the pressure is still low, check the fuel pump pressure* (see below) *and the fuel pressure regulator* (see Section 13 for LH systems).

12 If the pressure is higher than specified, check the fuel return line for an obstruction. If the line is not obstructed, replace the fuel pressure regulator.

13 Turn the ignition switch to Off, wait five minutes and look at the gauge. Compare the reading with the system hold pressure listed in this Chapter's Specifications. If the hold pressure is less than specified:

a) *Inspect the fuel system for a fuel leak. Repair any leaks and recheck the fuel pressure.*
b) *Check the fuel pump pressure* (see below).
c) *Check the fuel pressure regulator* (see Section 13 for LH systems).
d) *Check the injectors* (see Section 13 for LH systems).

Fuel pump pressure check

Refer to illustration 3.19

14 Relieve the fuel pressure (see Section 2).

15 Detach the cable from the negative battery terminal. **Caution:** *It is necessary to make sure the radio is turned OFF before disconnecting the battery cables to avoid damaging the microprocessor in the radio.*

16 Detach the fuel feed line from the fuel rail on LH-Jetronic systems and connect a fuel pressure gauge using the special adapter (see Step 5).

3.26 To prepare for the transfer pump pressure check, detach the fuel line from the inlet side of the main fuel pump (arrow) and, using a T-fitting, install a pressure gauge

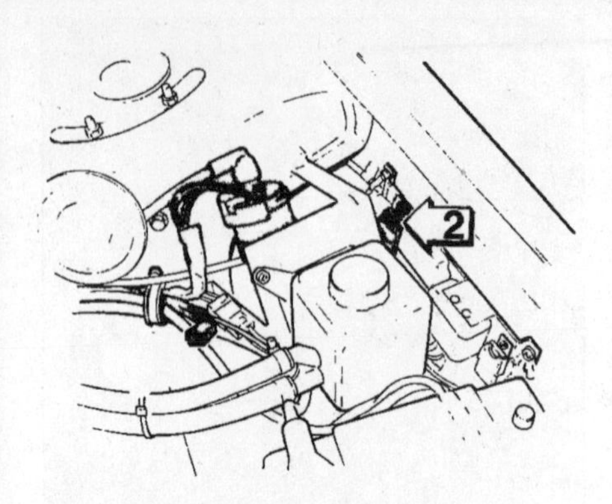

3.33 Be sure to check the fuse (2) that protects the fuel pump relay (LH-Jetronic shown)

4.4a Remove the three bolts from the fuel pump assembly (1982 LH-Jetronic system shown)

17 Attach the cable to the negative battery terminal.
18 Start the engine and observe the fuel pressure. Fuel pressure readings are slightly lower while the engine is running (see Steps 8 through 10). Refer to the Specifications listed in this Chapter.
19 To test the operation of the fuel pump, it will be necessary to close the valve on the fuel pressure gauge or if the tool is not equipped with a valve, pinch the fuel return line **(see illustration)**. This will allow fuel to enter the gauge but not return to the tank. This pressure reading is commonly called "deadhead pressure". Raise the engine rpm, let the throttle snap shut and observe that the pressure increases or stays steady. The fuel pressure should not decrease. **Note:** *Deadhead pressure will indicate the ability of the pump to deliver fuel under load conditions.*
20 Note the pressure reading on the gauge and compare the reading to the fuel pump pressure listed in this Chapter's Specifications.
21 If the indicated pressure is less than specified, inspect the fuel line for leaks between the pump and gauge. If no leaks are found, replace the fuel pump.
22 Turn the ignition key to Off and wait five minutes. Note the reading on the gauge and compare it to the fuel pump hold pressure listed in this Chapter's Specifications. If the hold pressure is less than specified, check the fuel lines between the pump and gauge for leaks. If no leaks are found, replace the fuel pump.
23 Relieve the fuel pressure (see Section 2). If the gauge is equipped with a bleeder valve, direct the fuel into an approved fuel container. Remove the gauge and reconnect the fuel lines.

Transfer pump pressure check

Refer to illustration 3.26

24 Relieve the system fuel pressure (see Section 2).
25 Detach the cable from the negative battery terminal. **Caution:** *It is necessary to make sure the radio is turned OFF before disconnecting the battery cables to avoid damaging the microprocessor in the radio.*
26 Detach the output hose on the in-tank pump and attach a fuel pressure gauge to the outlet pipe between the fuel pump and the in-tank pump **(see illustration)**. Use a T-fitting to install the fuel pressure gauge.
27 Attach the cable to the negative battery terminal.
28 Start the engine and check for leaks around the fuel pressure gauge fittings.
29 If there is no response from the in-tank pump, check the fuel pump relay and test for power to the pump (see Section 4).
30 Note the pressure reading on the gauge and compare the reading to the value listed in this Chapter's Specifications.
31 If the indicated pressure is less than specified, replace the transfer pump with a new one (see Section 4).

Fuel pump relay check

Refer to illustration 3.33

32 Turn the ignition key to the ON position.
33 First, check the fuel pump relay fuse **(see illustration)**. With the relay intact, probe the indicated terminals from the backside of the relay electrical connector with a voltmeter **(see illustrations 3.2a through 3.2d)**.
34 Turn the ignition key OFF and disconnect the relay from the electrical connector. Probe the connector terminal that corresponds to fuel pump relay pin number 87 **(see illustration 3.2d)**. Have an assistant crank the engine over and observe the voltage reading. There should be battery voltage.
35 If there is no voltage, check the fuse(s) and the wiring circuit for the fuel pump relay. If the voltage readings are correct and the fuel pump only runs with the jumper wire in place (see Step 1), then replace the relay with a new part.
36 If the fuel pump still does not run, check for the proper voltage at the fuel pump terminals (see Section 4). If necessary, replace the fuel pump.

4 Fuel pump, in-tank pump and fuel level sending unit - removal and installation

Warning: *Gasoline is extremely flammable, so take extra precautions when you work on any part of the fuel system. Don't smoke or allow open flames or bare light bulbs near the work area, and don't work in a garage where a natural gas-type appliance (such as a water heater or a clothes dryer) with a pilot light is present. Since gasoline is carcinogenic, wear latex gloves when there's a possibility of being exposed to fuel, and, if you spill any fuel on your skin, rinse it off immediately with soap and water. Mop up any spills immediately and do not store fuel-soaked rags where they could ignite. The fuel system is under constant pressure, so, if any fuel lines are to be disconnected, the fuel pressure in the system must be relieved first* (see Section 2 for more information). *When you perform any kind of work on the fuel system, wear safety glasses and have a Class B type fire extinguisher on hand.*
1 Relieve the system fuel pressure (see Section 2) and remove the fuel tank filler cap to relieve pressure in the tank.
2 Disconnect the cable from the negative battery terminal. **Caution:** *It is necessary to make sure the radio is turned OFF before disconnecting the battery cables to avoid damaging the microprocessor in the radio.*

Fuel pump (externally mounted)

Refer to illustrations 4.4a, 4.4b and 4.6

Note: *Refer to the fuel pressure checks in Section 3 to diagnose the fuel pump.*
3 Raise the vehicle and support it securely on jackstands.
4 Remove the bolts that secure the fuel pump cover, lower the

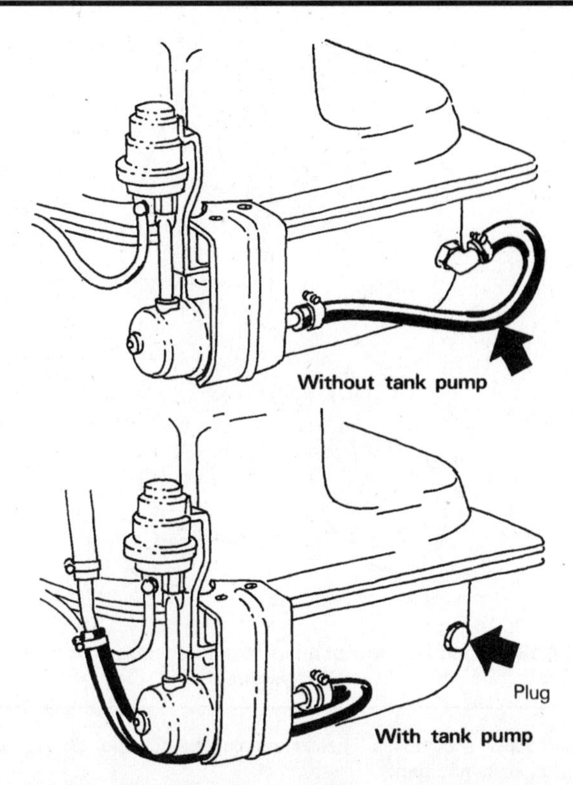

4.4b 1976 and 1977 fuel pump locations. Some are equipped with an in-tank pump (bottom) while others are not (top)

assembly and disconnect the wires from the pump **(see illustrations)**.

5 Use line clamps and pinch shut the fuel hoses on each side of the fuel pump. If you don't have line clamps, wrap the hoses with rags and clamp them shut with locking pliers, tightened just enough to prevent fuel from flowing out.

6 Loosen the clamps and disconnect the hoses from the pump **(see illustration)**.

7 Remove the fuel pump mounting screws and clamps and separate the fuel pump from the underside of the body.

8 Installation is the reverse of removal.

In-tank fuel pump

Refer to illustrations 4.9, 4.10, 4.13a, 4.13b, 4.13c and 4.13d

Note: *Some of the most common symptoms of a faulty in-tank fuel pump are hot starting problems, reduced top speed, vapor lock and engine hesitation and loud noise from the main fuel pump (overworked). If the transfer pump is not working, the engine usually starts and runs using only the main fuel pump.*

4.6 Mounting assembly tilted down to reveal the accumulator (1), fuel pump (2) and check valve (3)

9 On sedan models, access to the in-tank fuel pump is gained under the carpet in the trunk while on station wagons it will be necessary to remove the rear compartment lid to expose the access cover. **Note:** *To diagnose the in-tank fuel pump, follow the tests as described in Section 3. Another quick check of the system is to test for battery voltage with the ignition key ON (engine not running) directly at the electrical connector by the fuel tank* **(see illustration)**. *This will indicate a problem with the in-tank pump as opposed to electrical circuit problems.*

10 Remove the screws from the fuel pump access cover **(see illustration)**.

11 Remove the cover.

12 Locate the fuel pump and sending unit electrical connectors and unplug them. Also, disconnect the fuel inlet and return lines.

13 The assembly must be rotated counterclockwise to disengage the locking lugs from the fuel tank **(see illustrations)**. A special tool is available for this purpose, but you can use a brass punch and a hammer. **Warning:** *Don't use a steel punch or a screwdriver, as this could cause a spark.* Carefully lift the assembly from the fuel tank **(see illustrations)**. It may be necessary to slightly twist the assembly to get the float to clear the opening.

14 Remove the in-tank pump mounting screws and clamps and separate the transfer pump from the assembly.

15 Installation is the reverse of removal. If the gasket between the fuel pump and fuel tank is dried, cracked or damaged, replace it.

Fuel level sending unit - check and replacement

16 Remove the in-tank pump (see previous steps) along with the fuel level sending unit.

4A

4.9 Check for battery voltage at the electrical connector to the in-tank fuel pump

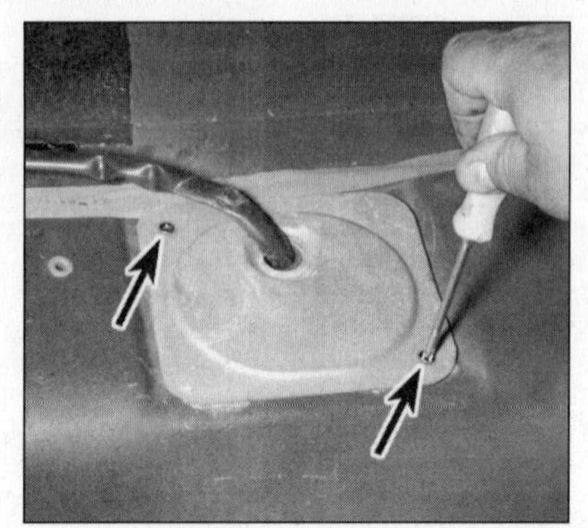

4.10 Remove the screws from the in-tank fuel pump access cover

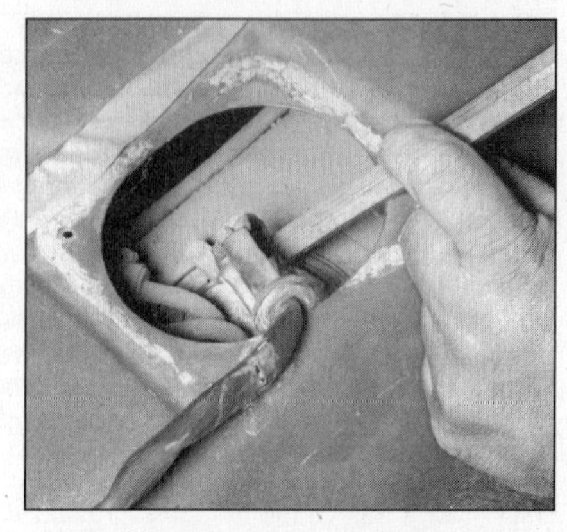

4.13a Use a brass punch to rotate the pump assembly

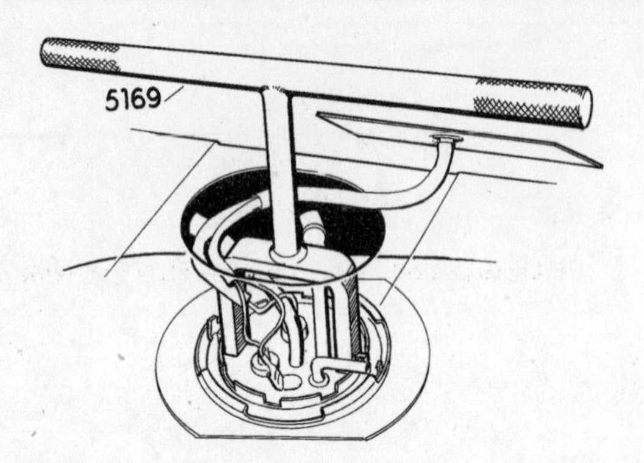

4.13b A special tool is available for removing the pump assembly from the tank

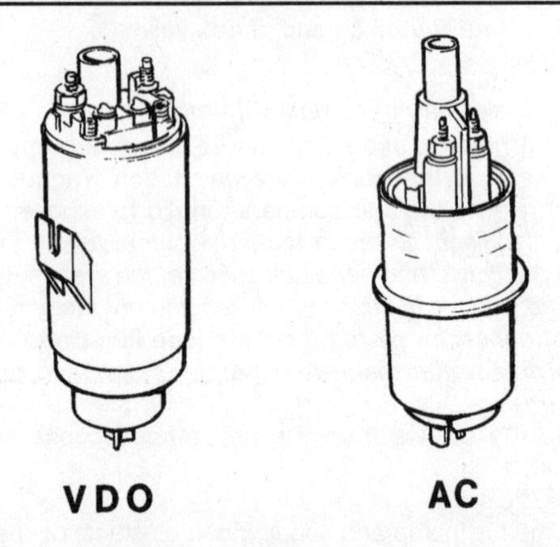

4.13d 1978 through 1981 models may be equipped with either an AC pump or a VDO pump. The VDO unit has been discontinued and must be replaced with an AC in-tank pump in the event of failure. The two pumps are interchangeable, but a separate suppressor must be installed to reduce radio interference.

17 Use an ohmmeter across the designated terminals and check for the correct resistance (see the *wiring diagrams* in Chapter 12). The resistance should decrease as the plunger rises.
18 If the resistance readings are incorrect, replace the sending unit with a new part.
19 Installation is the reverse of removal.

5 Fuel lines and fittings - repair and replacement

Warning: *Gasoline is extremely flammable, so take extra precautions when you work on any part of the fuel system. Don't smoke or allow open flames or bare light bulbs near the work area, and don't work in a garage where a natural gas-type appliance (such as a water heater or a clothes dryer) with a pilot light is present. Since gasoline is carcinogenic, wear latex gloves when there's a possibility of being exposed to fuel, and, if you spill any fuel on your skin, rinse it off immediately with soap and water. Mop up any spills immediately and do not store fuel-soaked rags where they could ignite. The fuel system is under constant pressure, so, if any fuel lines are to be disconnected, the fuel pressure in the system must be relieved first* (see Section 2 for more information). *When you perform any kind of work on the fuel system, wear safety glasses and have a Class B type fire extinguisher on hand.*

1 Always relieve the fuel pressure and disconnect the negative battery cable before servicing fuel lines or fittings (see Section 2).
2 The fuel feed, return and vapor lines extend from the fuel tank to the engine compartment. The lines are secured to the underbody with

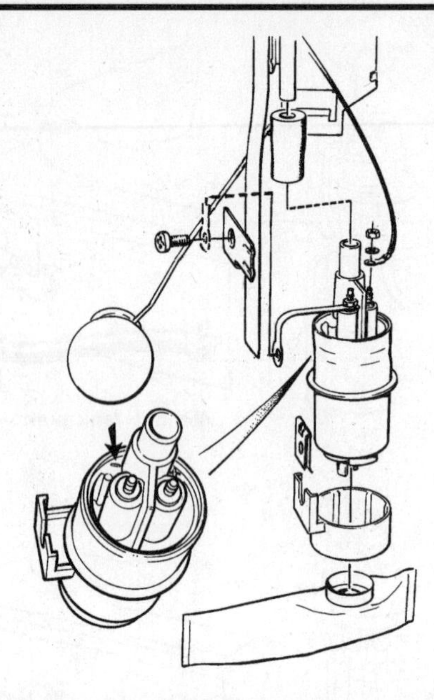

4.13c Exploded view of an in-tank fuel pump on a later CIS system

clip and screw assemblies. These lines must be occasionally inspected for leaks, kinks and dents.
3 If evidence of dirt is found in the system or fuel filter during disassembly, the line should be disconnected and blown out. Check the fuel strainer on the in-tank fuel pump for damage and deterioration.
4 Because fuel lines used on fuel-injected vehicles are under high pressure, they require special consideration. If replacement of a rigid fuel line or emission line is called for, use factory replacement parts to insure safety during engine operation. Don't use copper or aluminum tubing to replace steel tubing. These materials cannot withstand normal vehicle vibration.
5 When replacing fuel hose, use only hose approved for high-pressure fuel injection applications.

6 Fuel tank - removal and installation

Refer to illustrations 6.4a, 6.4b and 6.7

Warning: *Gasoline is extremely flammable, so take extra precautions when you work on any part of the fuel system. Don't smoke or allow open flames or bare light bulbs near the work area, and don't work in a garage where a natural gas-type appliance (such as a water heater or a clothes dryer) with a pilot light is present. Since gasoline is carcinogenic, wear latex gloves when there's a possibility of being exposed to fuel, and, if you spill any fuel on your skin, rinse it off immediately with soap and water. Mop up any spills immediately and do not store fuel-soaked rags where they could ignite. The fuel system is under constant pressure, so, if any fuel lines are to be disconnected, the fuel pressure in the system must be relieved first* (see Section 2 for more information). *When you perform any kind of work on the fuel system, wear safety glasses and have a Class B type fire extinguisher on hand.*

1 Remove the fuel tank filler cap to relieve fuel tank pressure.
2 Relieve the system fuel pressure (see Section 2).
3 Detach the cable from the negative battery terminal. **Caution:** *It is necessary to make sure the radio is turned OFF before disconnecting the battery cables to avoid damaging the microprocessor in the radio.*
4 Remove the drain plug **(see illustrations)** and drain the fuel into an approved gasoline container. If the tank doesn't have a drain plug, siphon the fuel into an approved fuel container. **Warning:** *Don't start the siphoning action by mouth - use a siphoning pump (available at most auto parts stores).*

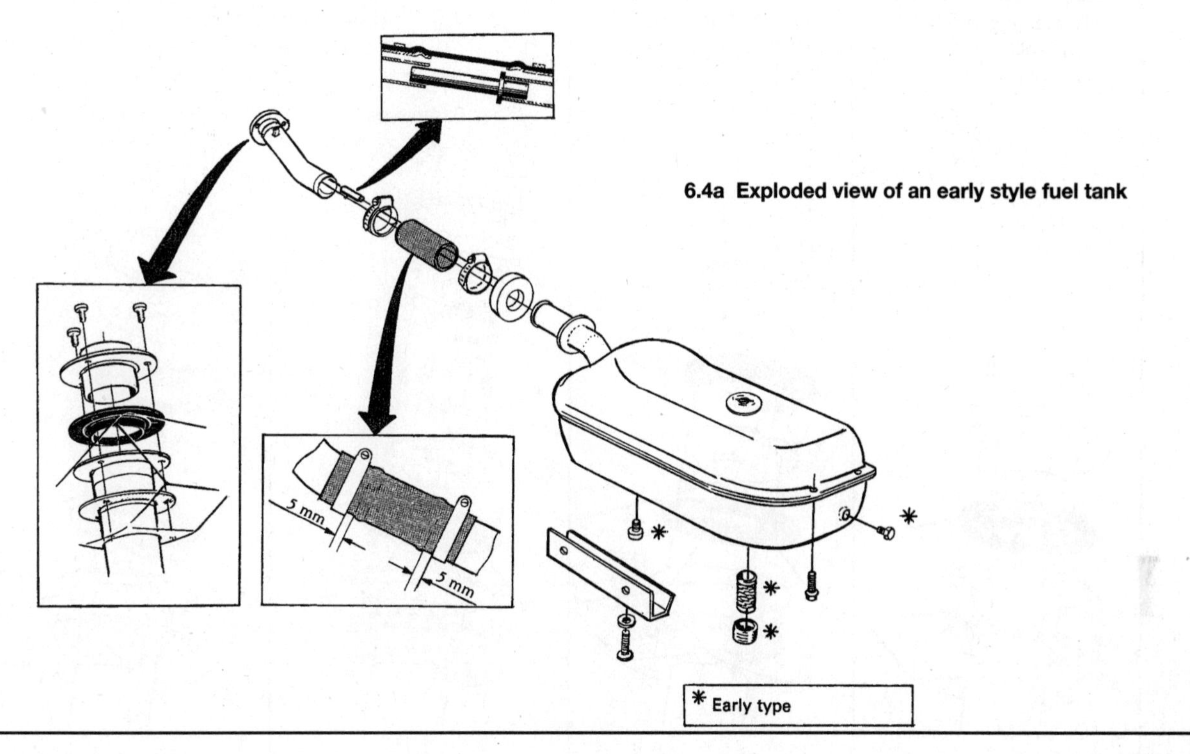

6.4a Exploded view of an early style fuel tank

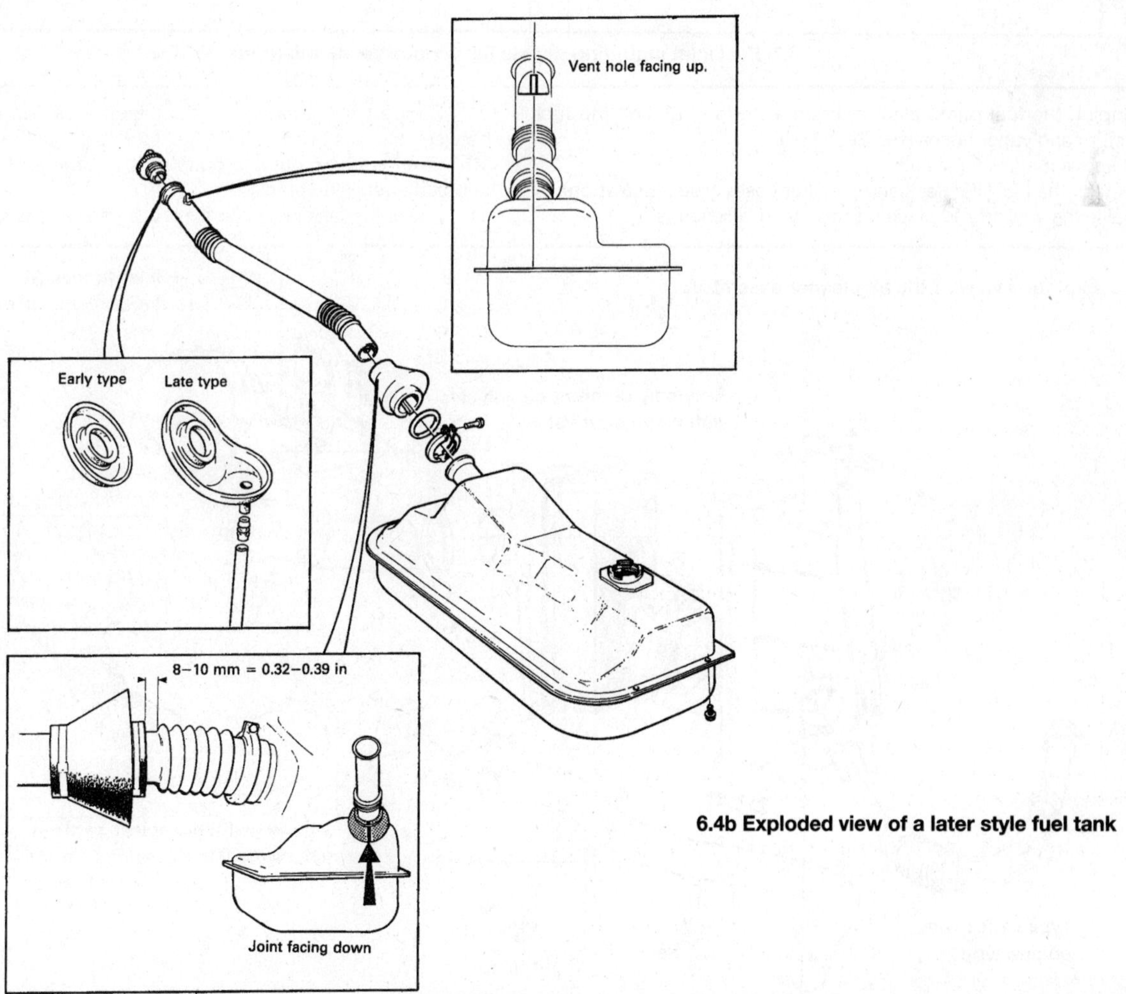

6.4b Exploded view of a later style fuel tank

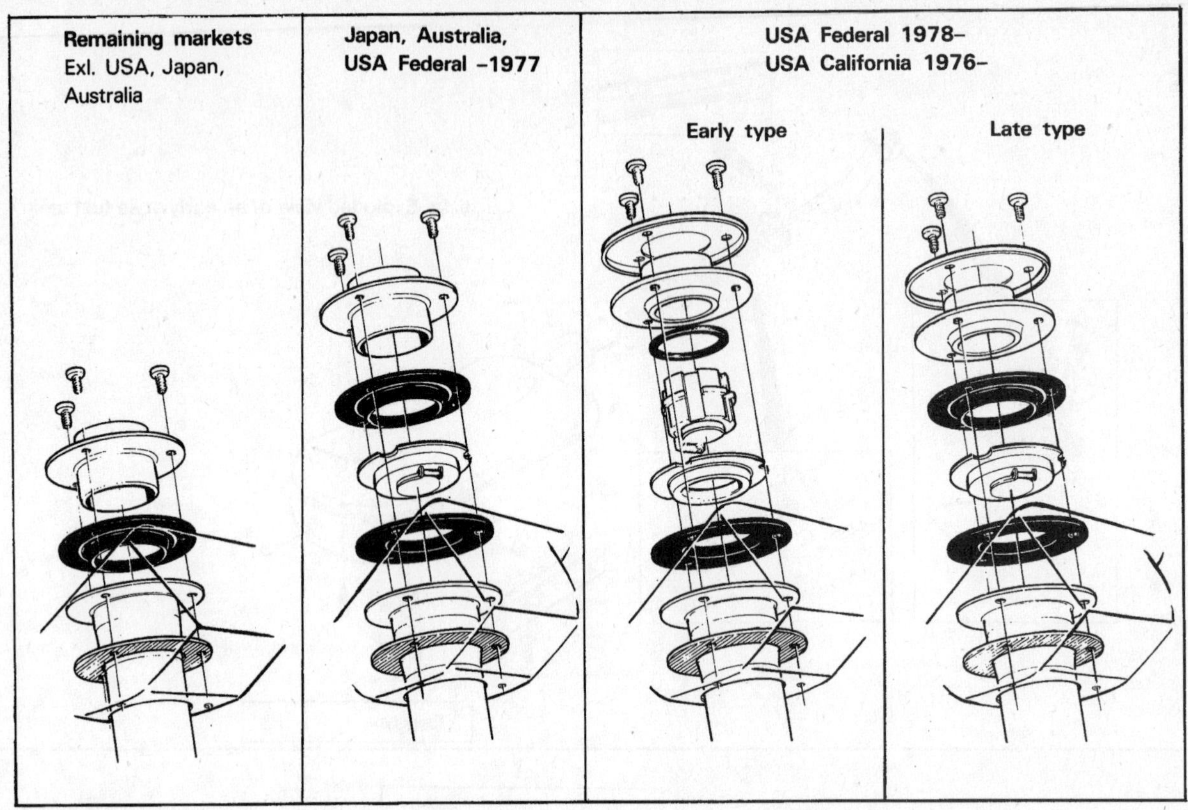

6.7 Fuel inlet mounting details for various years and areas

5 Unplug the fuel pump electrical connector and detach the fuel feed, return and vapor hoses (see Section 4).

6 Remove the fuel tank shield.

7 Detach the fuel filler neck and breather hoses **(see illustration)**.

8 Raise the vehicle and place it securely on jackstands.

9 Support the tank with a floor jack. Position a block of wood between the jack head and the fuel tank to protect the tank.

10 Remove the mounting bolts at the corners of the fuel tank and unbolt the retaining brackets.

11 Lower the tank just enough so you can see the top and make sure

8.2 Exploded view of the air cleaner assembly

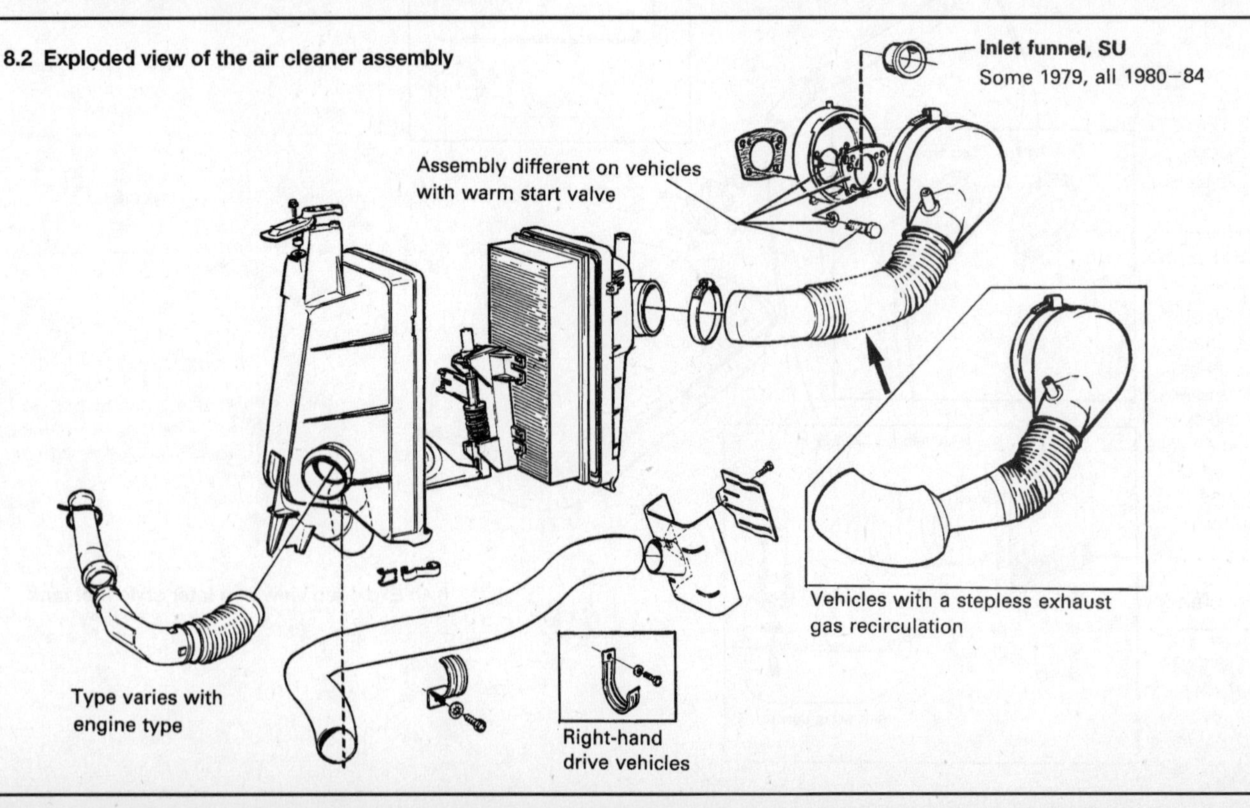

8.6a Remove the bolt (arrow) from the air cleaner bracket and lift the bracket from the body

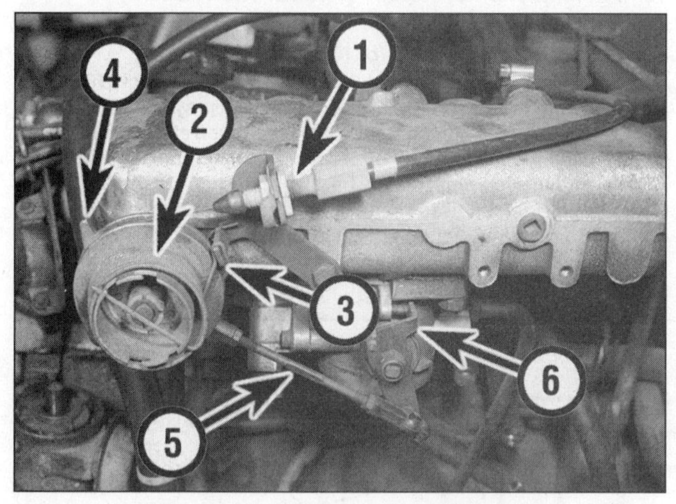

9.4a Accelerator control linkage details

1 *Accelerator cable adjusting nut A*
2 *Throttle pulley*
3 *Closed stop*
4 *Full open stop*
5 *Link rod*
6 *Throttle lever stop and adjusting screw*

you have detached everything. Finish lowering the tank and remove it from the vehicle.

12 Installation is the reverse of removal.

7 Fuel tank cleaning and repair - general information

1 All repairs to the fuel tank or filler neck should be carried out by a professional who has experience in this critical and potentially dangerous work. Even after cleaning and flushing of the fuel system, explosive fumes can remain and ignite during repair of the tank.

2 If the fuel tank is removed from the vehicle, it should not be placed in an area where sparks or open flames could ignite the fumes coming out of the tank. Be especially careful inside garages where a natural gas-type appliance is located, because the pilot light could cause an explosion.

8 Air cleaner assembly - removal and installation

Refer to illustrations 8.2, 8.6a and 8.6b

1 Detach the cable from the negative battery terminal. **Caution:** *It is necessary to make sure the radio is turned OFF before disconnecting the battery cables to avoid damaging the microprocessor in the radio.*

2 Detach the duct from the front of the air cleaner **(see illustration)**.

8.6b Pivot the air cleaner housing forward to allow the tabs to come out of the slots at the bottom

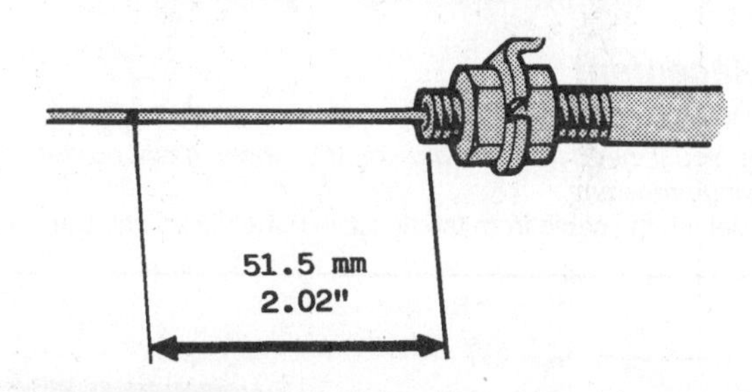

9.4b On cable assemblies equipped with the screw type adjusters, allow for sufficient play (2.0 inches)

3 Detach the duct between the air cleaner and the throttle body.

4 Remove the air filter (see Chapter 1).

5 Unplug the electrical connector from the air mass meter on LH-Jetronic systems. Remove the air mass meter from the engine compartment (see Chapter 6).

6 Remove the air cleaner mounting bolts **(see illustrations)** and lift the air cleaner assembly from the engine compartment.

7 Installation is the reverse of removal.

9 Accelerator cable - check, adjustment and replacement

Check

1 Have an assistant depress the accelerator pedal to the floor while you watch the throttle valve. It should move to the fully open (horizontal) position.

2 Release the accelerator pedal and make sure the throttle valve returns smoothly to the fully closed position. The throttle valve should not contact the body (intake manifold) at any time during its movement or else the unit must be replaced.

Adjustment

Refer to illustrations 9.4a and 9.4b

3 Warm the engine to normal operating temperature and turn it off. Depress the accelerator pedal to the floor twice, then check the cable freeplay at the throttle body. Compare it to the value listed in this Chapter's Specifications.

4 If the freeplay isn't within specifications, adjust it by turning nut A **(see illustrations)**.

5 Have an assistant help you verify the throttle valve is fully open when the accelerator pedal is depressed to the floor.

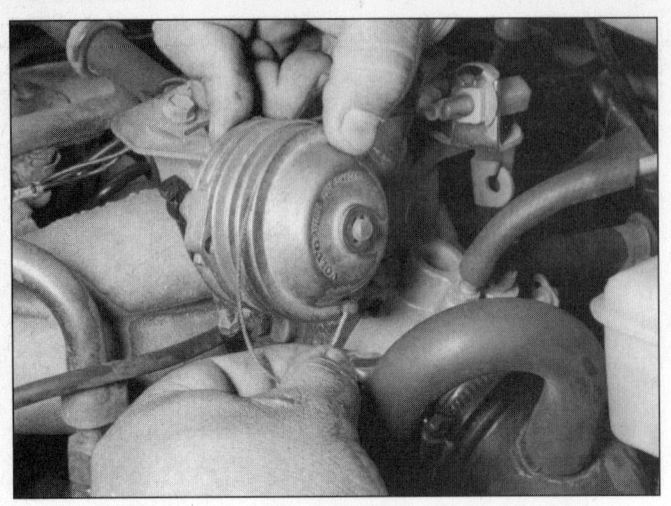

9.9 Rotate the throttle valve and remove the cable end from the slot

Replacement

Refer to illustration 9.9

Note: *You'll need a flashlight for the under-dash portion of the following procedure.*

6 Detach the cable from the negative battery terminal. **Caution:** *It is necessary to make sure the radio is turned OFF before disconnecting the battery cables to avoid damaging the microprocessor in the radio.*

7 Loosen the cable adjuster locknuts and detach the cable from its support bracket located on the intake manifold.

8 Pinch the plastic retainer with a pair of needle-nose pliers and push it out of the bracket.

9 Pull the cable down through the slot in the throttle bellcrank (lever) **(see illustration)**.

10 Working from underneath the driver's side of the dash, reach up and detach the throttle cable from the top of the accelerator pedal.

11 Pull the cable through the firewall, from the engine compartment side.

12 Installation is the reverse of removal. Be sure to adjust the cable as described in Steps 3 through 5.

10 Fuel injection systems - general information

Refer to illustrations 10.1a, 10.1b, 10.2a and 10.2b

Early United States model 240 series (1976 through 1982) are equipped with the Bosch K-Jetronic fuel injection system, also called the Continuous Injection System (CIS) **(see illustrations)**. The K-Jet system uses a constant flow of pressurized fuel that is injected into each cylinder simultaneously. The idle speed is controlled by the Constant Idle Speed system (also referred to as CIS, but not to be confused with the CIS fuel injection system) which consists of the ECU, Air Control Valve, TP switch, coolant temp sensor and fuel distributor. This system detects the position of the throttle valve and

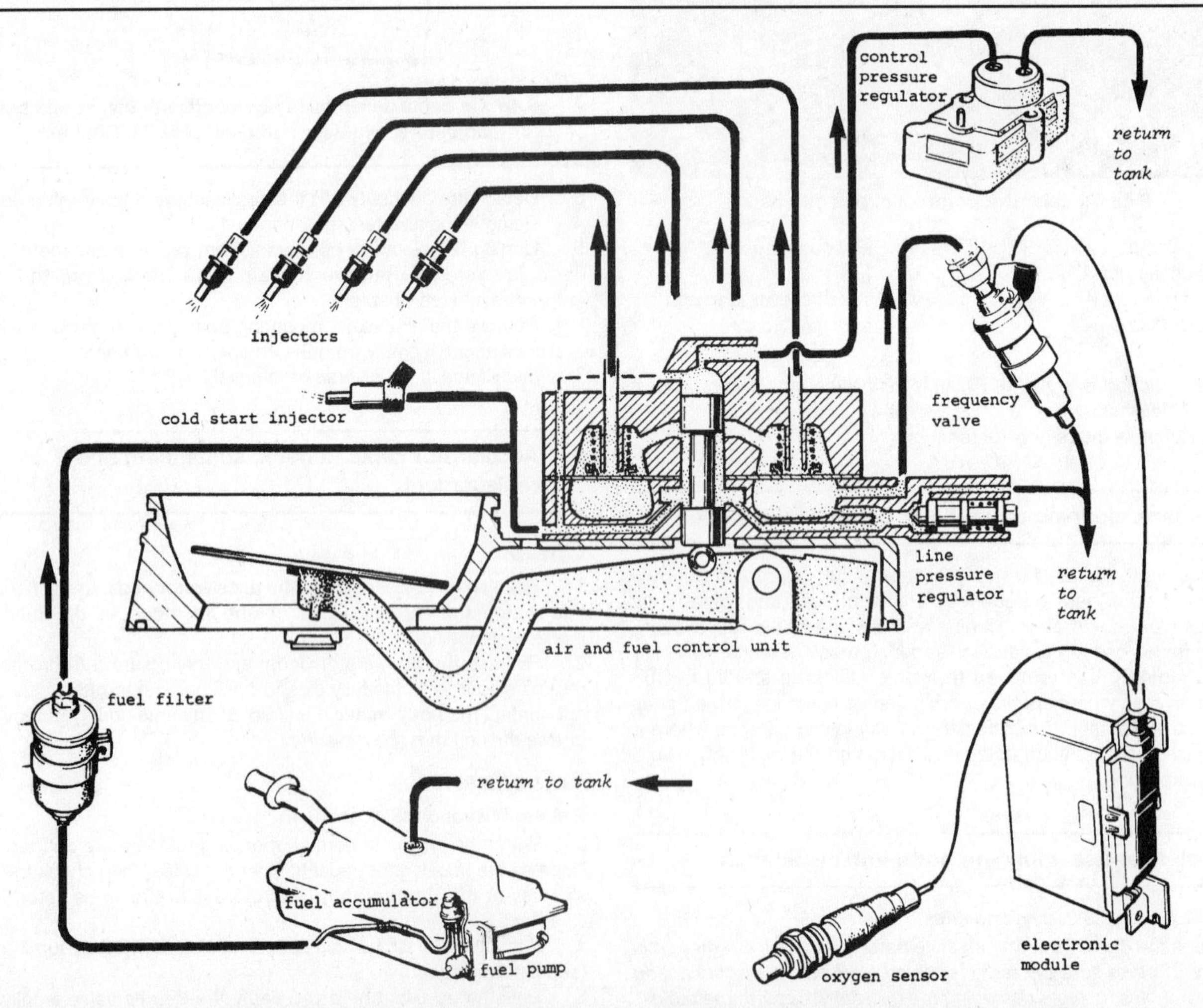

10.1a Typical CIS system with the oxygen sensor feedback system (also called Lambda-sond)

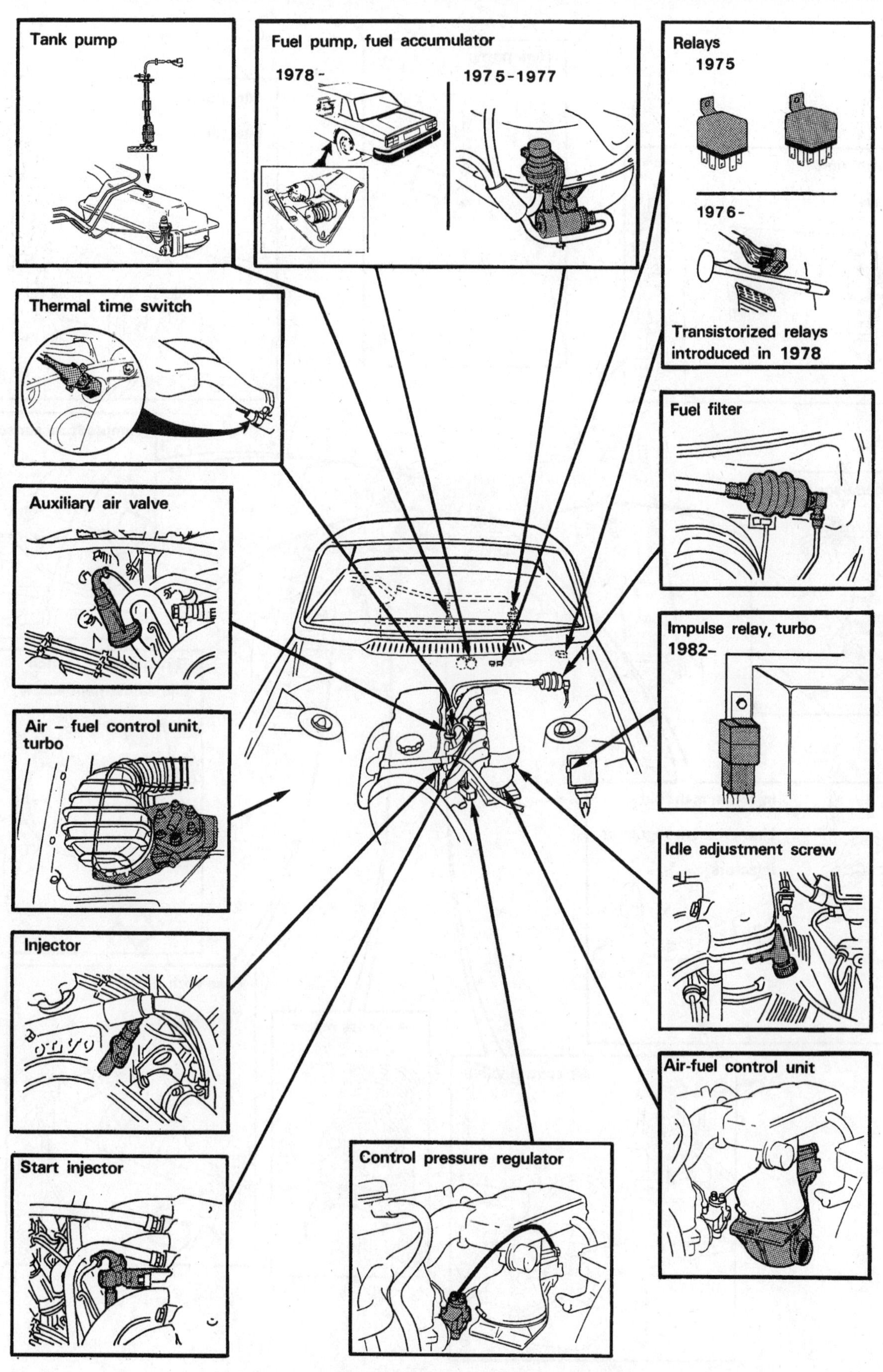

10.1b Schematic of a CIS system on a turbocharged model

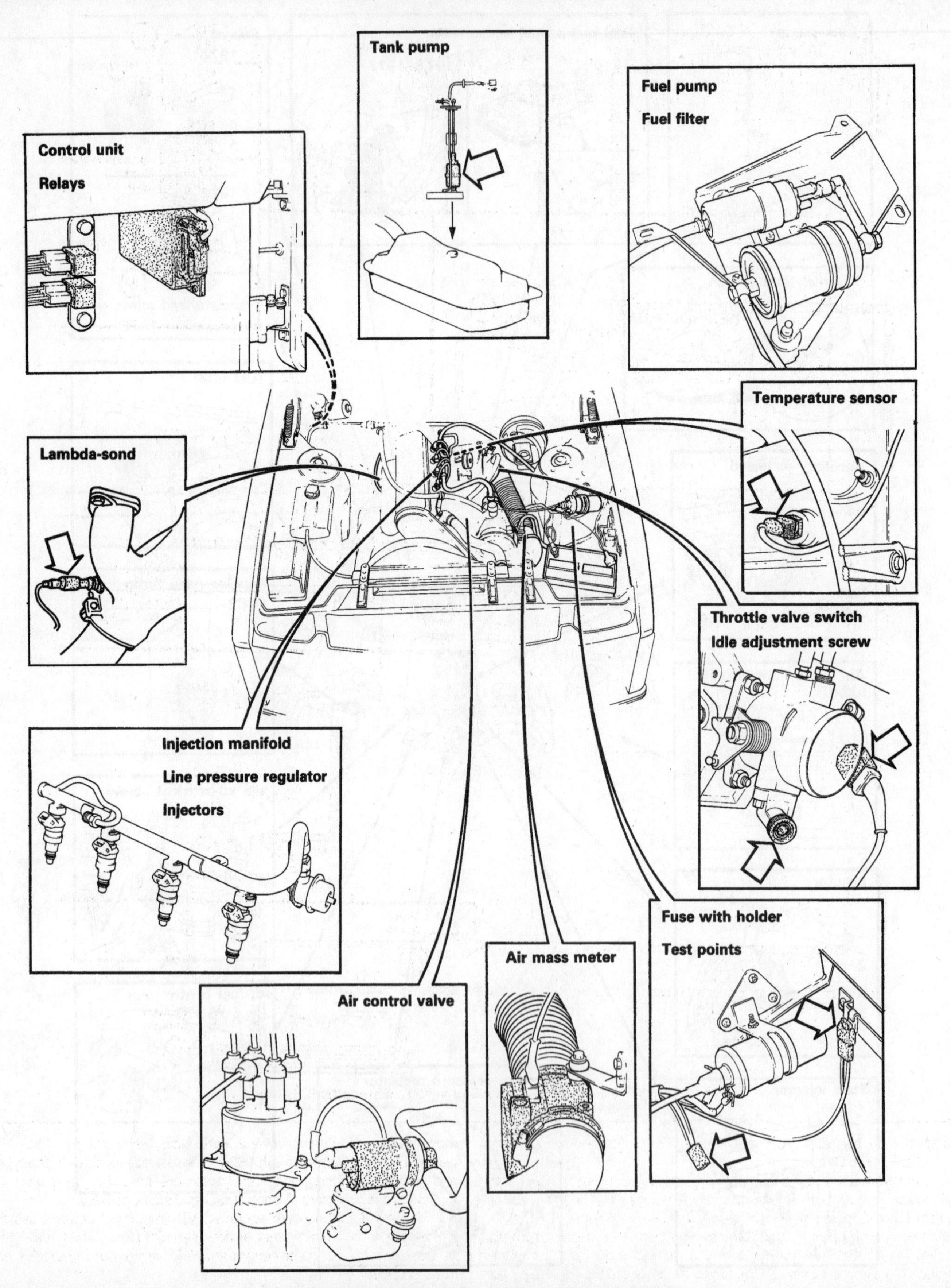

10.2a Component locations for a 2.0 LH-Jetronic system

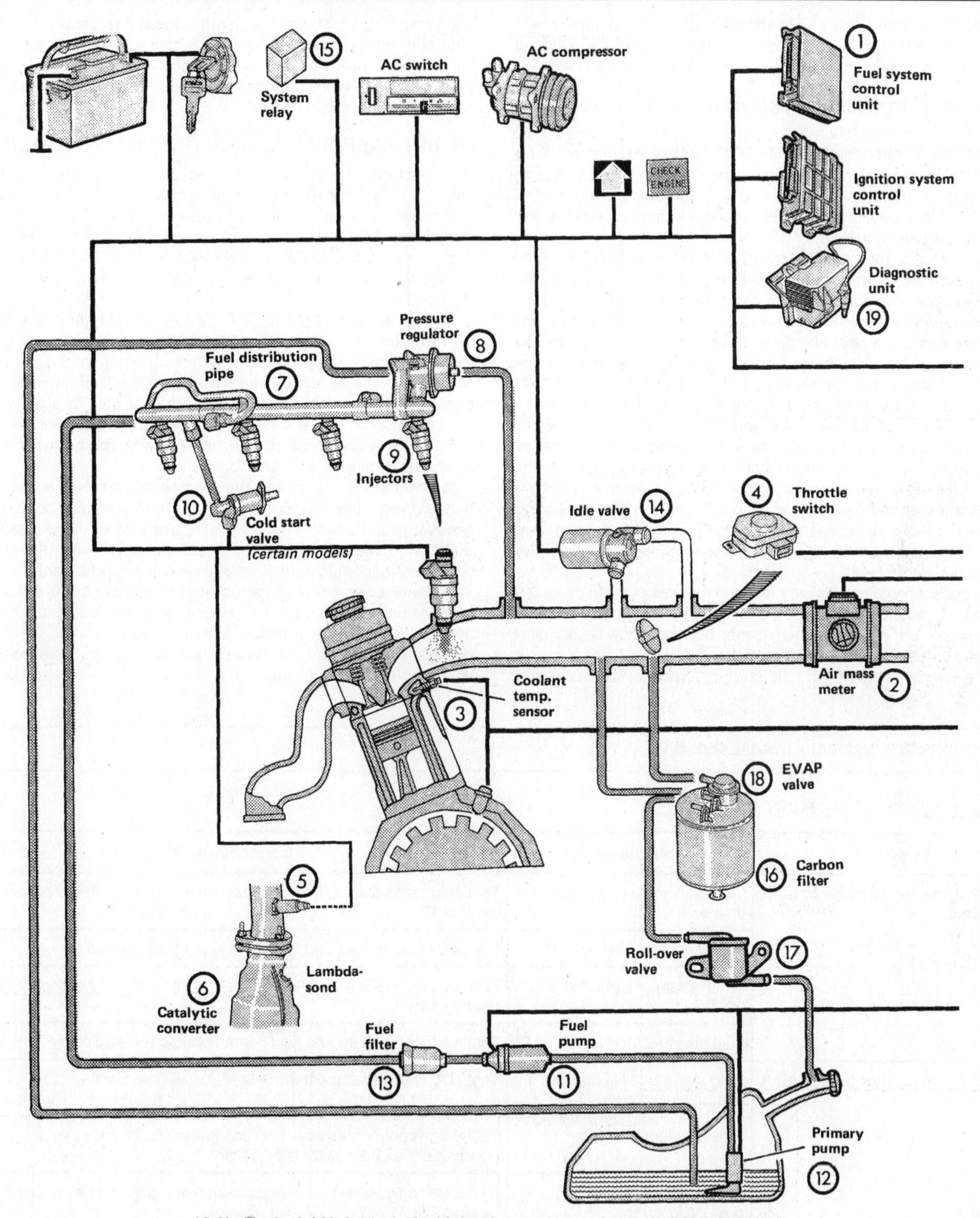

10.2b Typical LH-Jetronic fuel injection system schematic (LH 2.4 shown)

adjusts the idle speed accordingly. **Note:** *All turbocharged models use the CIS fuel injection system*

Later models (1982) are equipped with the LH-Jetronic system replaced by a modified version called LH-2.0 Jetronic (1983 and 1984) system **(see illustrations)**. Both systems use a "hot-wire" sensing device to calculate air mass (mass airflow sensor). The LH-Jetronic system (early) uses a vacuum switch to signal air intake pressure instead of a throttle switch (TPS). **Note:** *These systems are equipped with an idle air stabilizer valve (LH type fuel injection) instead of an auxiliary air valve (CIS type fuel injection).* Later models are slightly modified to include the LH-2.2 Jetronic (1985 through 1988) and LH-2.4 Jetronic (1989 through 1990). These systems differ in the types and number of sensors and output actuators. LH-2.4 Jetronic uses the ignition system sensors to control fuel and spark delivery.

Some 1991 models are equipped with the LH-3.1 Jetronic system which use a "hot-film" mass airflow sensor. **Note:** *1991 and 1992 models have the option of either the LH-2.4 system or the LH-3.1 fuel injection system.*

The self diagnosis system was first installed on 1989 models with LH-2.4 and continued to the 3.1 systems.

Canadian models from 1978 through 1984 are equipped with the B21A engine that is equipped with a carburetor. Refer to Chapter 4B for all the information concerning Canadian carbureted models.

Continuous Injection System (CIS) or K-Jetronic

The Continuous Injection System is found on 1976 through 1982 US models and all turbocharged engines covered by this manual. It is a well-proven system, with little to go wrong and no black boxes to worry about. As the name implies, fuel injection takes place continuously while the engine is running. The rate of injection is varied to suit the prevailing speed and load.

Fuel is drawn from the tank by an electric fuel pump. The pump pressurizes the system to approximately 75 psi. An accumulator next to the pump provides a reservoir of pressure to improve hot starting and to dampen the pulses from the pump. From the accumulator the fuel passes through a filter and then to the fuel distributor under the intake manifold. The turbocharged versions are equipped with the fuel distributor mounted near the fenderwell.

The fuel distributor looks a little like an ignition distributor, but it has fuel lines instead of spark plug wires. There is one fuel line per injector, with additional lines for the start injector and the control pressure regulator. The fuel distributor's main function is to regulate the fuel supply to the injectors in proportion to the incoming airflow. Incoming air deflects the airflow sensor plate, which moves the control plunger in the fuel distributor and so varies the supply to the injectors. The airflow sensor and the fuel distributor together are sometimes called the fuel control unit.

The control pressure regulator reduces the control pressure during warm-up and under condition of low manifold vacuum, and so enriches the mixture. A lower control pressure means that the airflow sensor plate is deflected further, and the quantity of fuel injected is increased.

An electrically-controlled cold start injector provides extra fuel during cold engine starting. A thermal time switch controls the duration of start injector operation when the engine is cold; on a hot engine an impulse relay provides a smaller quantity of extra fuel to be injected. An auxiliary air valve provides the extra air needed to maintain idle speed when the engine is cold.

Electronic Fuel Injection (EFI) or LH-Jetronic

The Bosch LH-Jetronic fuel injection system is used on 1982 and later US model engines. It is an electronically controlled fuel injection system that utilizes one solenoid-operated fuel injector per cylinder. The system is governed by an Electronic Control Unit (ECU) which processes information sent by various sensors, and in turn precisely meters the fuel to the cylinders by adjusting the amount of time that the injectors are open.

An electric fuel pump delivers fuel under high pressure to the injectors through the fuel feed line and an in-line filter. A pressure regulator keeps fuel available at an optimum pressure, allowing pressure to rise or fall depending on engine speed and load. Fuel in excess of injector needs is returned to the fuel tank by a separate line.

A sensor in the air intake duct constantly measures the mass of the incoming air, and through the ECU adjusts the fuel mixture to provide an optimum air/fuel ratio.

Another device, called the oxygen sensor, is mounted on the exhaust manifold and continually reads the oxygen content of the exhaust gas. This information is also used by the ECU to adjust the duration of injection, making it possible to meter the fuel very accurately to comply with strict emission control standards.

Other components incorporated in the system are the throttle valve (which controls airflow to the engine), the coolant temperature sensor, the throttle position switch, idle stabilizer valve (which bypasses air around the throttle plate to control idle speed) and associated relays and fuses.

11 Fuel injection system - troubleshooting

CIS fuel injection system

Symptom	Probable cause	Corrective action
Engine starts hard or fails to start when cold	Cold start valve or thermo-time switch faulty	Test cold start valve and thermo-time switch. Replace faulty parts (see Section 12)
	Fuel pump not running	Check fuel pump fuse and fuel pump relay (see Section 3)
	Air flow sensor plate rest position incorrect	Inspect air flow sensor plate rest position and adjust if necessary (see Section 12)
	Fuel pressure incorrect	Test system pressure and cold control pressure (see Section 12)
Engine starts hard or fails to start when warm	Cold start valve leaking or operating continuously	Test cold start valve and thermo-time switch (see Section 12)
	Fuel pressure incorrect	Test warm control pressure. Replace fuel pump check valve or fuel accumulator as necessary (see Section 12).
	Air flow sensor plate rest position incorrect	Inspect air flow sensor plate rest position and adjust if necessary (see Section 12)
	Insufficient residual fuel pressure	Test residual fuel pressure. Replace fuel pump check valve or fuel; accumulator as necessary (see Section 12)
	Fuel leak(s)	Inspect fuel lines and connections. Correct leaks as required (see Section 5 and Chapter 1)
	Turbocharged models: Pinched PCV hose	Crankcase ventilation hose is kinked at the lower connection to the intake manifold causing overpressure in crankcase and airflow sensor plate lift problems (see Chapter 6)

Symptom	Probable cause	Corrective action
Engine misses and hesitates under load	Fuel injector closed	Test fuel injectors. Check for clogged injector lines. Replace faulty injectors (see Section 12)
	Fuel pressure incorrect	Test system pressure and warm control pressure. Adjust system pressure regulator or replace control pressure regulator as necessary (see Section 12)
	Fuel leak(s)	Inspect fuel lines and connections. Correct leaks as required (see Chapter 1)
Engine starts but stalls at idle	Incorrect fuel pressure	Test system pressure and control pressure (see Section 12)
	Cold start valve leaking	Test and, if necessary, replace cold start valve (see Section 12)
	Auxiliary air regulator faulty	Test and, if necessary, replace auxiliary air regulator (see Section 12)
	Vacuum (intake air) leak	Inspect intake air components for leaking hoses, hose connections and cracks or other leaks. Repair as required (see Chapter 1)
Engine idles too fast	Accelerator pedal, cable or throttle valve binding	Inspect for worn or broken parts, kinked cable or other damage. Replace faulty injectors (see Section 12)
	Auxiliary air regulator faulty	Test and, if necessary, replace auxiliary air regulator (see Section 12)
	Air leaking past throttle valve	Inspect throttle valve and adjust or replace as required (see Section 12)
Hesitation on acceleration	Vacuum (intake air) leak	Inspect intake air components for leaking hoses, hose connections and cracks or other leaks. Repair as required
	Fuel injectors clogged	Test injector spray pattern and quantity. Replace faulty injectors (see Section 12)
	Cold start valve leaking	Test and, if necessary, replace cold start valve (see Section 12)
	Control plunger in fuel distributor binding or fuel distributor faulty	Check air flow sensor plate movement and, if necessary, replace fuel distributor (see Section 12)
	Air flow sensor plate out of adjustment	Inspect air flow sensor plate position and adjust if necessary (see Section 12)
	Fuel pressure incorrect	Test system pressure and warm control pressure. If necessary, replace control pressure regulator (see Section 12)
Poor fuel mileage	Idle speed, ignition timing and idle mixture	Check and adjust (see Chapter 1) (mixture adjustment must be performed by a dealer service department or other repair shop)
	Cold start valve leaking	Test and, if necessary, replace cold start valve (see Section 12)
	Fuel pressure incorrect	Test system pressure and warm control pressure. If necessary, replace control pressure regulator (see Section 12)
Engine continue to run (diesels) after ignition is turned off	Incorrect ignition timing or faulty ignition system	Check ignition timing (see Chapter 1)
	Engine overheated	Inspect cooling system (see Chapters 1 and 3)

LH-Jetronic fuel injection system

Note: *Begin troubleshooting by displaying any stored trouble codes* (see Chapter 6).

Symptom	Probable cause	Corrective action
Engine starts hard or fails to start when cold	Coolant temperature sensor faulty	Test coolant temperature sensor and replace if necessary (see Chapter 6)
	Fuel pump not running	Check fuel pump fuse and fuel pump relay (see Section 3)
	Fuel filter clogged	Check fuel filter (see Chapter 1)
	Vacuum (intake air) leak	Inspect intake air components for leaking hoses, hose connections and cracks or other leaks. Repair as required. Check for loose oil fill cap or dipstick

Symptom	Probable cause	Corrective action
Engine starts hard or fails to start when cold	Low fuel pressure	Test fuel pressure (see Section 3)
	Electronic control unit faulty	Have the system diagnosed by a dealer service department or other repair shop
Engine starts when cold but stalls at idle	Coolant temperature sensor faulty	Test coolant temperature sensor and replace if necessary (see Chapter 6)
	Electronic control unit faulty	Have the system diagnosed by a dealer service department or other repair shop
Engine idles rough or stalls (cold or warm)	Vacuum (intake air) leak	Inspect intake air components for leaking hoses, hose connections and cracks or other leaks. Repair as required. Check for loose oil fill cap or dipstick
	Air flow sensor flap binding or faulty	Check air flow sensor flap for binding. Have the system diagnosed by a dealer service department or other repair shop
	Inadequate fuel being delivered to engine	Test fuel pump (see Section 3)
	Blocked fuel filter	Replace fuel filter (see Chapter 1)
	Idle air control valve faulty	Test idle switch. Test air control valve (see Chapter 6)
	Low fuel pressure	Test fuel pressure (see Section 3)
	Electronic control unit faulty	Have the system diagnosed by a dealer service department or other repair shop
Engine misses, hesitates or stalls under load	Air flow sensor flap binding or faulty	Check air flow sensor flap. Have the system diagnosed by a dealer service department or other repair shop
	Intake air preheating system faulty	Test intake air preheating system and replace faulty components as required
	Vacuum (intake air) leak	Inspect intake air components for leaking hoses, hose connections and cracks or other leaks. Repair as required. Check for loose oil fill cap or dipstick.
	Low fuel pressure	Test fuel pressure (see Section 3)
Engine Idles too fast	Accelerator pedal, cable or throttle valve binding	Inspect for worn or broken parts, kinked cable or other damage. Replace faulty parts (see Section 9)
	Coolant temperature sensor wire disconnected or broken	Check wiring between control unit and sensor (see Chapter 12)
	Idle air stabilizer valve faulty	Test idle switch. Test idle air stabilizer valve (see Chapter 6)
Low power	Air intake restricted	Check air filter element, housing and preheating system (see Chapter 1)
	Air flow sensor flap not opening fully	Check movement of air flow sensor plate. Replace air flow sensor, if necessary (see Chapter 6)
	Throttle plate not opening fully	Check throttle cable adjustment, to make sure throttle is opening fully. Adjust cable if necessary (see Section 9)
	Full throttle switch faulty or incorrectly adjusted	Check throttle switch and adjust if necessary. Replace a faulty switch (see Section 13)
	Electronic control unit faulty	Have the system diagnosed by a dealer service department or other repair shop
Engine continues to run (diesels) after ignition Is turned off	Incorrect timing or faulty ignition system	Check ignition timing (see Chapter 1)
	Engine overheated	See Chapter 3

12 Continuous Injection System (CIS) - check

Warning: *Gasoline is extremely flammable, so take extra precautions when you work on any part of the fuel system. Don't smoke or allow open flames or bare light bulbs near the work area, and don't work in a garage where a natural gas-type appliance (such as a water heater or a clothes dryer) with a pilot light is present. Since gasoline is carcinogenic, wear latex gloves when there's a possibility of being exposed to fuel, and, if you spill any fuel on your skin, rinse it off immediately with soap*

12.10 To check system pressure, connect the fuel pressure gauge to the fuel distributor between the fuel distributor and the control pressure regulator and make sure the valve (arrow) is closed (no fuel allowed to the control pressure regulator

and water. Mop up any spills immediately and do not store fuel-soaked rags where they could ignite. The fuel system is under constant pressure, so, if any fuel lines are to be disconnected, the fuel pressure in the system must be relieved first (see Section 2 for more information). *When you perform any kind of work on the fuel system, wear safety glasses and have a Class B type fire extinguisher on hand.*

Preliminary checks

1 Check the ground wire connections on the intake manifold for tightness. Check all wiring harness connectors that are related to the system. Loose connectors and poor grounds can cause many problems that resemble more serious malfunctions.

2 Check to see that the battery is fully charged, as the control unit and sensors depend on an accurate supply voltage in order to properly meter the fuel.

3 Check the air filter element - a dirty or partially blocked filter will severely impede performance and economy.

4 Open the fuel filler cap and listen for fuel pump operation while an assistant cranks the engine. If no whirring noise is heard, check the fuel pump (see Section 3).

5 Check the fuses. If a blown fuse is found, replace it and see if it blows again. If it does, search for a grounded wire in the harness to the fuel pumps.

6 Check the air intake duct from the airflow sensor to the intake manifold for leaks, which will result in an excessively lean mixture. Also check the condition of all of the vacuum hoses connected to the intake manifold.

7 Remove the air intake duct from the throttle body and check for dirt, carbon or other residue build-up. If it's dirty, clean it with carburetor cleaner and a toothbrush.

Throttle switch

8 Turn the ignition switch to the On position and open the throttle lever by hand and listen for a click as soon as the throttle comes off its stop. This test will indicate that the idle switch is functioning. If no click is heard, proceed to Section 14 for adjustment (or if necessary, replacement) of the switch.

Fuel pressure checks

Refer to illustrations 12.10, 12.11a, 12.11b, 12.14, 12.15a, 12.15b and 12.17

Note: *On mechanical (CIS) fuel-injected engines, the fuel pump(s) delivery pressure must be checked before the actual CIS pressure tests can be performed. The CIS pressure tests only check the function of the fuel distributor, pressure regulator, injectors, control pressure regulator the fuel lines that flow from the fuel distributor to the injectors. In other words, it is necessary to make sure the fuel pump is getting the fuel to the fuel distributor from the fuel tank efficiently and at the correct pressure. For CIS systems, this will be referred to as the delivery fuel pressure. This is a very important distinction for separating the fuel pump and the delivery system versus the fuel distributor and the mechanical operation of the CIS fuel injection system. Follow the fuel pump relay checks* (see Section 3) *and the fuel pressure checks for the mechanical fuel injection systems to obtain the correct pressure. This will be referred to as delivery pressure. Once it is established that the fuel pump is working properly, go ahead with the CIS system checks.*

Note: *These fuel pressure tests will require a special fuel pressure gauge specially made for testing CIS systems.*

9 Relieve the fuel pressure (see Section 2).

10 Connect the fuel pressure gauge between the control pressure regulator and the fuel distributor with the valve side of the gauge toward the control pressure regulator **(see illustration)**.

11 Start the vehicle and observe the pressure reading. **Note:** *Disconnect the electrical connectors on the control pressure regulator and the auxiliary air valve* **(see illustrations)** *to prevent any change in the fuel pressure before testing the cold control pressure.* There are three significant fuel pressure values:

System pressure - the basic fuel pressure produced by the fuel pump and maintained by the pressure relief valve in the fuel distributor.

Control pressure - the difference between system pressure and lower chamber pressure in the fuel distributor as determined by the control pressure regulator. It is used to counter system pressure and regulate the movement of the control plunger.

Residual pressure - the amount of pressure which remains in the closed system after the engine and fuel pump are shut off.

12.11a First, disconnect the auxiliary air valve . . .

12.11b . . . then the control pressure regulator electrical connectors

12.14 To read control pressure (cold), make sure the valve (arrow) is now open to allow fuel flow

Checking system pressure

12 First check the system pressure. Close the valve on the pressure gauge (this prevents fuel from entering the control pressure regulator) and observe the reading. System pressure should be as listed in this Chapter's Specifications. If the system pressure is too low, look for leaks, a clogged fuel filter or a damaged fuel line blocking the fuel flow. If no other cause can be found, the pressure can be adjusted by adding shims to the pressure relief valve (see Step 34).

13 If the fuel pressure cannot be accurately adjusted, then the fuel distributor is faulty and must be replaced.

Checking control pressure

14 Next, check the control pressure. Turn the valve on the fuel pressure gauge to the open position **(see illustration)**. Make sure the vehicle is cold (68-degrees F) in order to obtain an accurate pressure reading. Connect the electrical connectors from the control pressure regulator and the auxiliary air valve. **Note:** *The cold control pressure and the warm control pressure will vary slightly with the different models of Bosch pressure regulators. Consult the Specifications at the beginning of this Chapter for the precise pressure specs.*

15 Start the vehicle and observe the gauge. The fuel pressure will increase as the temperature of the vehicle warms up **(see illustrations)**. The initial pressure value is called cold control pressure. The cold control pressure should be accurate according to the climate and altitude of the region.

16 If the cold control pressure is too high, check for a fuel line that is blocked or kinked. Also, check the fuel union at the control pressure regulator for a plugged filter screen. If no problems are found, replace the control pressure regulator. **Warning:** *Be sure to relieve the fuel pressure before disconnecting any fuel lines* (see Section 2).

17 To check the warm control pressure, run the engine until the control pressure is no longer increasing (approximately 2 minutes) and observe the gauge. The warm control pressure should be as listed in this Chapter's Specifications. Some models have control pressure regulators which compensate for changes in altitude. Refer to the chart to convert warm control pressure vs. altitude above sea level **(see illustration)**.

18 If the warm control pressure is too high, check for a blocked or kinked fuel line. Also, check the fuel union at the control pressure regulator for a plugged filter screen. If no problems are found, replace the control pressure regulator.

19 If warm control pressure is low, or takes more than 2 minutes to reach its highest value, test the resistance of the heating element and test for voltage reaching the electrical connector at the control pressure regulator.

Checking residual pressure

20 Finally, check the residual pressure. Check the residual pressure with the gauge connected as described in the previous fuel pressure tests.

21 When the engine is warm (control pressure 49 to 55 psi) shut the engine off and leave the gauge connected. Wait ten minutes and observe the gauge. The fuel pressure should not have dropped off below 22 psi.

22 If the pressure drops off excessively, check for leaks in the fuel lines, the fuel distributor, the injectors, the cold start valve and the oxygen sensor frequency valve. Also check residual pressure at the fuel supply line from the fuel pump. Disconnect the gauge from the fuel distributor and the control pressure regulator and reconnect those lines. Next, connect the gauge to the main supply line from the fuel pump and be sure to close the valve. Run the fuel pump with a jumper wire as described in Section 2 and pressurize the system until the gauge reads 49 to 55 psi. Once again, the pressure should not drop off below 22 psi within ten minutes.

23 If the pressure drops off excessively and there are no apparent leaks between the fuel pump and the gauge, pinch closed the fuel line between the tank and the fuel pump and observe the gauge.

24 If residual pressure now remains steady, then the check valve in the fuel pump is faulty.

25 If the residual pressure still drops off quickly, then the fuel accumulator is at fault.

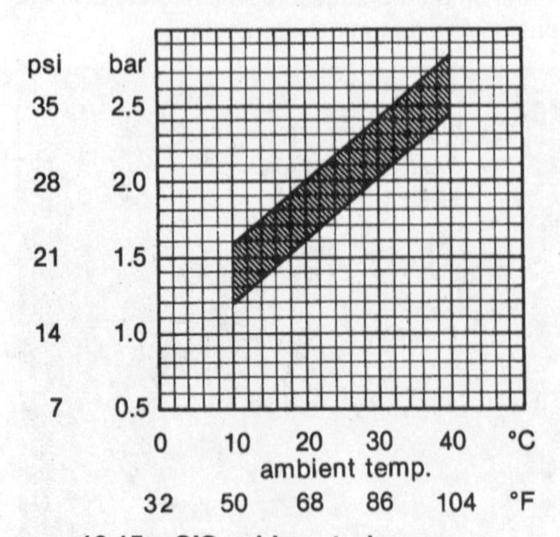

12.15a CIS cold control pressure graph for North American models (except California)

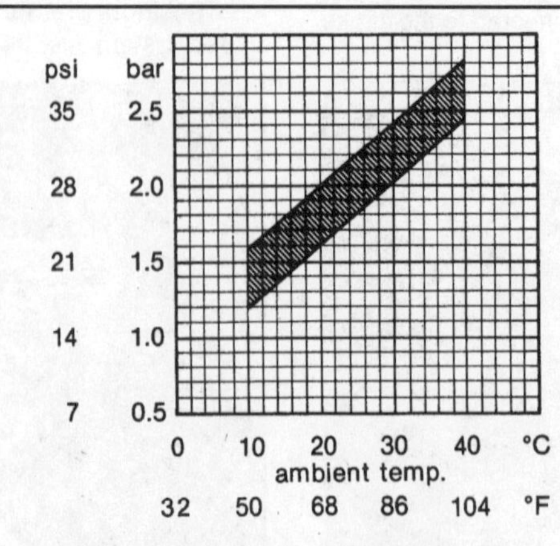

12.15b CIS cold control pressure graph for California

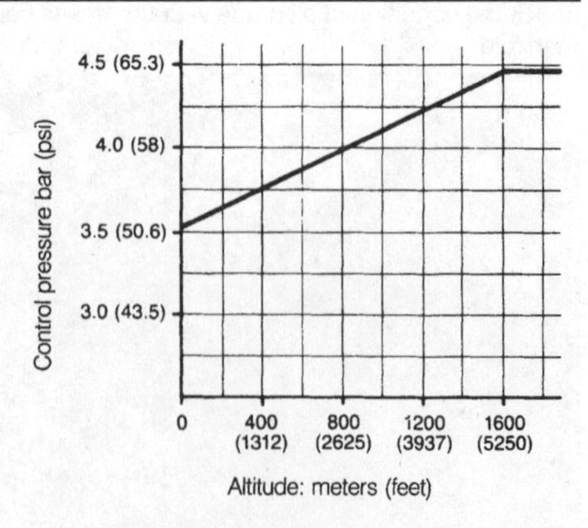

12.17 Graph of warm control pressure, as affected by altitude, for US cars except California - example: at 800 meters (2,625 feet) above sea level, warm control pressure should be approximately 58 psi

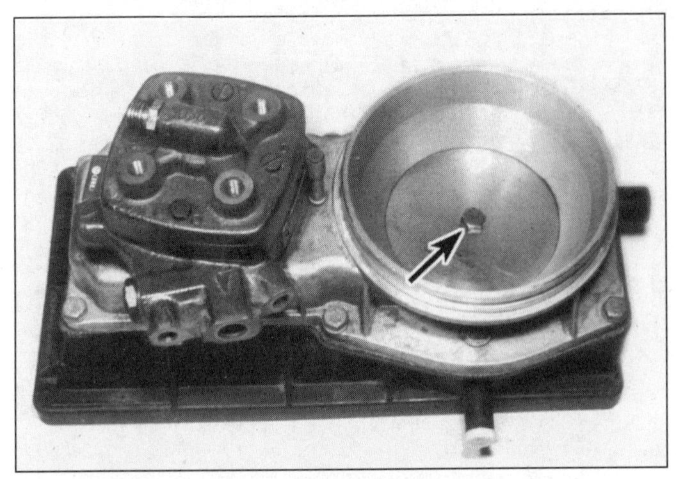

12.26 The sensor plate is located at the bottom of the venturi (arrow)

12.27 Be sure to clean the air intake housing and sensor plate before proceeding with the tolerance tests

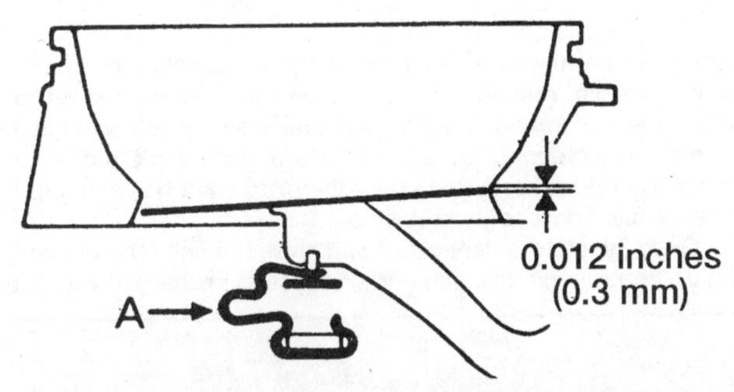

12.28a When the sensor plate is at rest (ignition off), the plate should be within 0.012 inches (0.3 mm) of the lower edge of the venturi

Airflow sensor - description, check and adjustment

Refer to illustrations 12.26, 12.27, 12.28a, 12.28b, 12.28c, 12.29a, 12.29b, 12.30a, 12.30b and 12.33

Description

26 The airflow sensor measures the air drawn in by the engine. As airflows past the airflow sensor plate, the plate is lifted, which lifts the control plunger in the fuel distributor to meter the fuel **(see illustration)**. The airflow sensor plate and its conical venturi are carefully machined to achieve an optimal ratio between air and fuel for every operating condition, from idle to full throttle. If the sensor plate is binding, or off-center, or the lever has too much resistance, the fuel distributor won't respond correctly to the airflow sensor plate.

Check

27 To check the position of the sensor plate it is necessary to remove the air intake casing, but before doing this, run the engine for a few minutes to build up pressure in the fuel lines. Loosen the clamp and take off the air intake duct. The sensor plate may now be seen **(see illustration)**. Check the position of the plate relative to the venturi. There must be a gap of 0.004 inch (0.10 mm) all around, between it and the venturi. The plate surface must also be even with the bottom of the air cone with the fuel line residual pressure is removed.

28 If the level is not correct then the plate should be lifted with a magnet or pliers, being careful not to scratch the bore. The clip underneath may be bent to adjust the level, using small pliers **(see illustrations)**. Pull the plate up as far as it will come and the job can be done without dismantling anything else. The tolerance is 0.012 inch (0.3 mm).

29 Centering the plate can be easy or difficult. Try the easy way first. Remove the center bolt - it is fairly stiff as it is held by thread locking compound. Take the bolt out and clean the threads. Now try to center the plate with the bolt loosely in position **(see illustrations)**. If this can be done then remove the bolt, put a drop of thread locking compound (a non-hardening type) on the threads and reinstall it holding the plate centralized. Tighten the bolt securely.

4A

12.28b Airflow sensor plate adjusting clip (arrow)

12.28c Lift the sensor plate and carefully move the clip to adjust the height of the plate

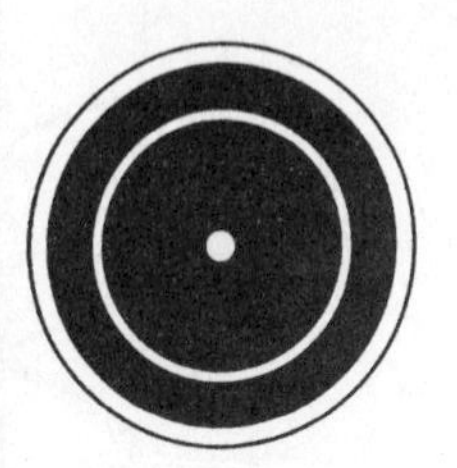

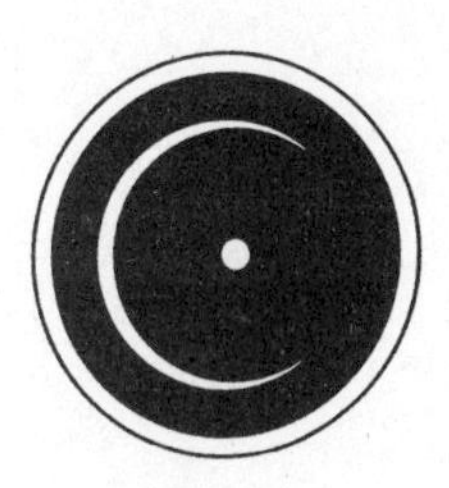

12.29a The sensor plate must be centered in the venturi

12.29b To center the plate, use three strips of typing paper and position them between the sensor plate and the walls of the air intake housing and check for any binding

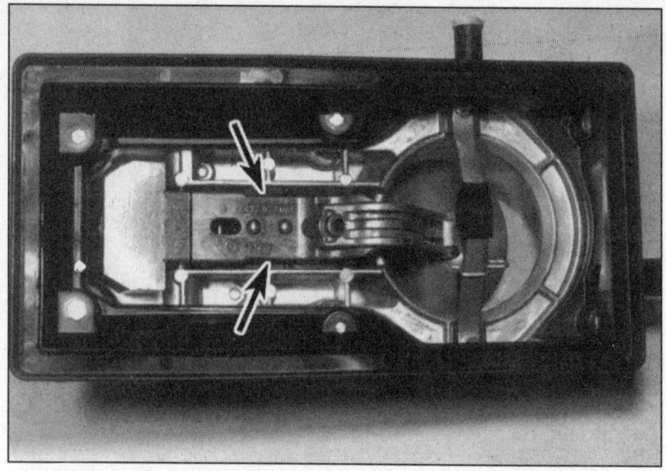

12.30a Bottom view of the airflow sensor unit - the clearance on either side of the sensor beam (arrows) should be even

30 If the plate will not center, then the sensor unit must be removed from the vehicle. It is probably easier to remove the mixture control unit from the sensor unit than to remove all the fuel lines, but be careful that the plunger doesn't drop out when you separate the units. Disconnect the sensor unit from the top of the air cleaner. Take the sensor unit out and turn it upside down. Now check that the sensor beam is central in its bearings **(see illustrations)**. If it is not, loosen the clamp bolt on the counterweight and it may be possible to center the beam in its bearings and at the same time center the sensor plate in the cone. If this is possible, remove the bolt, clean the threads, put a drop of thread locking compound on them and reinstall the bolt with the beam and plate in the correct positions. If this doesn't work, a new sensor unit must be purchased, because if the plate can't be centralized you will have major driveability problems.

31 Once the plate is centralized and level, the unit reassembled, the mixture control unit installed and the system recharged with fuel by

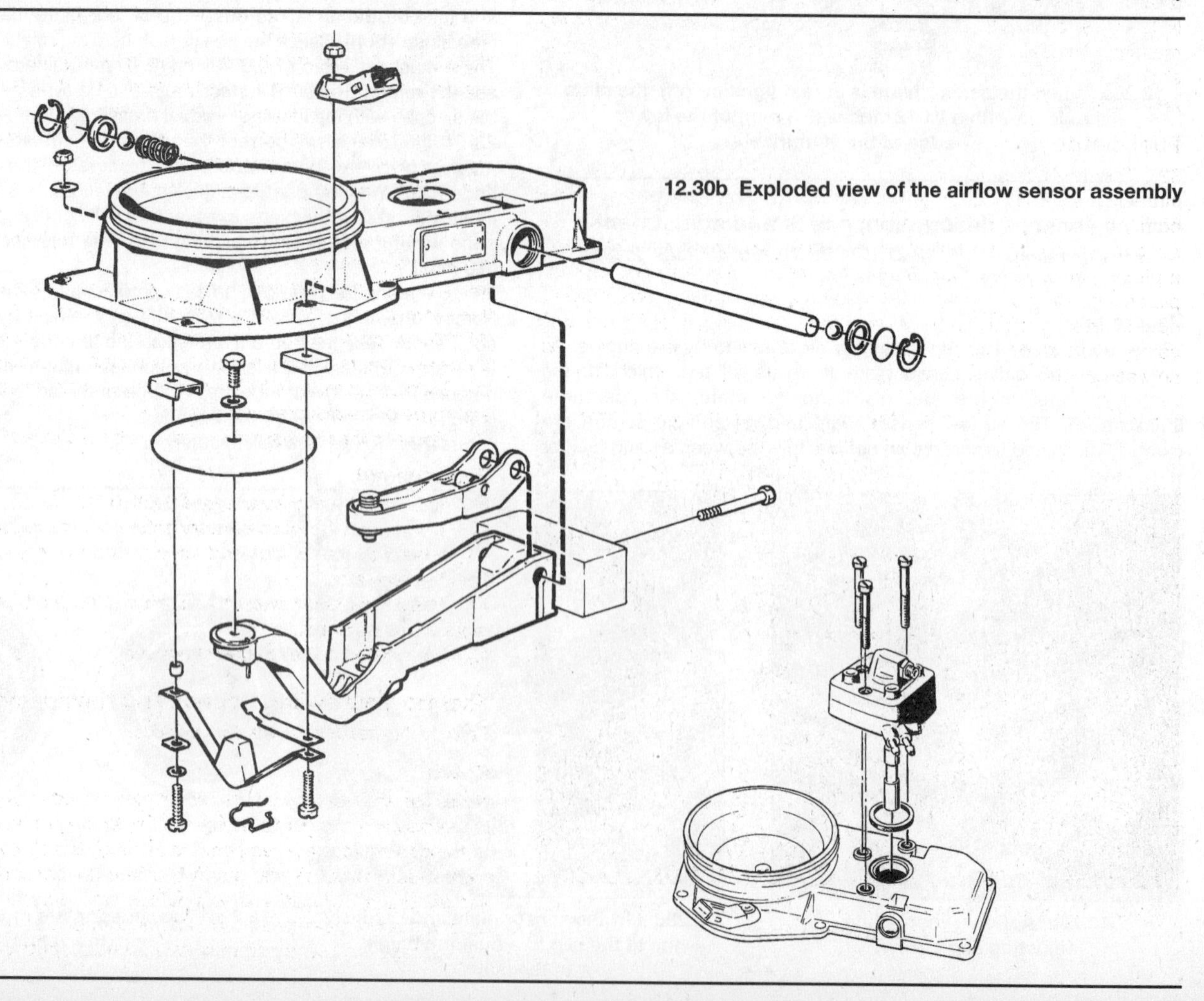

12.30b Exploded view of the airflow sensor assembly

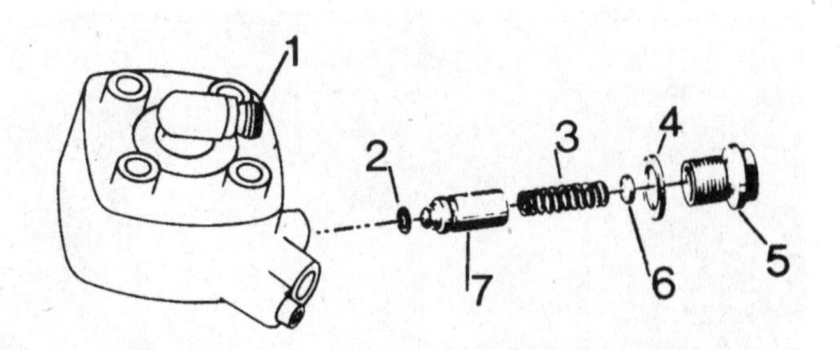

12.33 Exploded view of the pressure regulating valve on the fuel distributor body

1 *Fuel distributor body*
2 *Rubber ring*
3 *Spring*
4 *Copper washer*
5 *Plug*
6 *Shims for pressure adjustment*
7 *Valve piston*

turning the ignition on for a few seconds, it is possible to check the action of the airflow sensor. Turn the ignition off and, using a small magnet, lift the plate to the top of its movement. There must be a slight, but even, resistance, but no hard spots. Now depress the plate quickly. This time there should be no resistance to movement.

32 If there is resistance to movement, or hard spots in both directions, the plate might not be centralized, so check it again. If the resistance or hard spot happens only when lifting the plate, the problem is with the plunger of the fuel mixture unit. Remove the mixture unit from the sensor casing and carefully remove the plunger. Wash it with carburetor cleaner to remove any residue, reinstall it and try again. If this does not cure the problem then it is probable that a new mixture control unit is needed. DO NOT try to remove the hard spot with abrasives; this will only make matters worse. A visit to the dealer service department or other repair shop is indicated. They may be able to cure the problems but be prepared to purchase a new mixture control unit.

Fuel distributor - check and adjustment

33 The pressure regulating valve for the system pressure is included in the fuel distributor body **(see illustration)**. A hexagonal plug on the corner of the fuel distributor casing may be unscrewed and inside will be found a copper ring, shims for adjusting the pressure on the spring, a piston and a rubber ring. Be careful not to scratch the bore or the piston since these are mated on assembly and a new piston means a new distributor body. If the piston is stuck either blow it out with compressed air or work it out using a piece of soft wood. Always use new seals when refitting the plug.

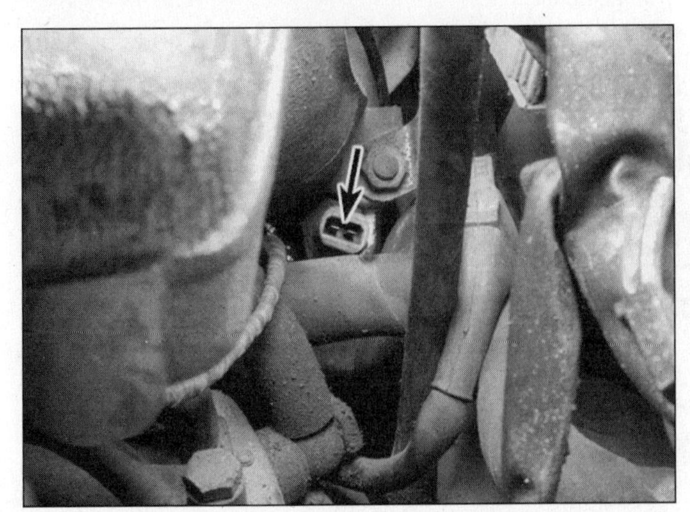

12.46a The thermo-time switch is located near the firewall, next to the intake manifold

34 Pressure can be adjusted by adding shims to the pressure relief valve. An additional 0.020-inch shim will increase system pressure by about 4 psi. A 0.040-inch shim will increase system pressure by about 8 psi.

35 If system pressure is too high, check for a blocked or damaged fuel return line. If the return line is good, the pressure can be lowered by reducing the thickness of the shims on the pressure relief valve. A reduction of a 0.020-inch shim thickness will decrease system pressure by about 4 psi. A reduction of 0.040-inch total shim thickness will decrease system pressure by 8 psi.

36 From the tests on the air sensor plate movement, the operation of the plunger will have been checked. If it is suspect then the fuel distributor body must be disconnected from the airflow sensor plate and lifted clear. Be careful that the plunger does not fall out and get damaged. Carefully extract the plunger and wash it in carburetor cleaner. When installing it, the small shoulder goes in first. Do not attempt to cure any hard spots by rubbing with abrasive. If washing it in carburetor cleaner does not cure the problem then a new assembly is required.

Control pressure regulator - check and replacement

Check

Note 1: *Bosch designed different control pressure regulators to function during adverse conditions and various applications; high altitude, cold weather, leaner emissions etc. Consult the charts in the Specifications at the beginning of this Chapter for complete information.*

Note 2: *The 0 438 140 079 and 0 438 140 123 control pressure regulators are equipped with a vacuum diaphragm and a thermostat valve that controls cold acceleration enrichment. In the event the fuel pressure increases upon acceleration* (see Steps 1 through 15) *then the system is not providing a rich mixture during cold acceleration. Have the control pressure regulator and the components diagnosed at a dealer service department.*

Note 3: *The 0 438 140 128 control pressure regulator is equipped with a compensator diaphragm that raises control pressure to lean the mixture out at higher elevations. This system cannot be tested directly. If the control pressure is within specifications, the control pressure regulator is functioning properly. If the unit passes the electrical tests but the control pressure is below specifications, then replace it with a new part.*

37 Disconnect the wiring from the control pressure (warm up) regulator and auxiliary air valve.

38 Connect a voltmeter across the electrical connectors and operate the starter briefly - there should be a minimum of 11.5 volts.

39 Connect an ohmmeter across the regulator heater element terminals - the resistance should be between 10 and 30 ohms. Refer to the Specifications in this Chapter for the exact resistance value for each type of control pressure regulator

40 Replace the regulator if necessary and reconnect the wiring.

Replacement

41 Relieve the fuel pressure (see Section 2).

42 Disconnect the electrical connector from the regulator.

43 Use a socket or box-end wrench and disconnect the fuel lines from the regulator.

44 Use a 6 mm Allen wrench and remove the two bolts that retain the regulator to the block.

45 Installation is the reverse of removal.

Thermo-time switch - check and replacement

Refer to illustrations 12.46a and 12.46b

Check

Note: *The thermo-time switch determines the operating time (duration of fuel spray) of the cold start injector depending upon the temperature of the coolant (coolant temperature sensor). Bosch has manufactured thermo-time switches that respond differently depending on the year and geographical location (cold weather, high altitude etc.). Be sure to replace the thermo-time switch with the correct part.*

46 Locate the thermo-time switch **(see illustrations)**. To test the

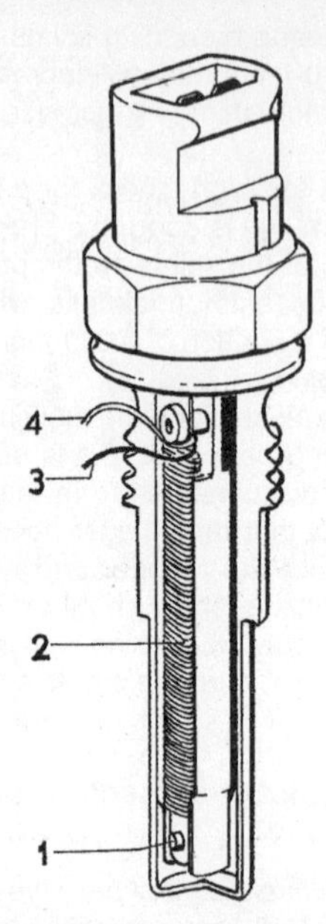

12.46b Cross-section of a thermo-time switch

1 *Contacts*
2 *Bi-metallic spring*
3 *Wire to starter motor*
4 *Wire to cold start injector*

switch, detach the electrical connector from the cold start valve and bridge the contacts in the electrical connector (harness side) with a test light or a voltmeter. The test must be done with a cold (coolant below 95-degrees F, 35-degrees C) engine.

47 Remove the coil wire from the center of the distributor and ground it with a jumper wire. Have an assistant operate the starter for ten seconds. Depending on the coolant temperature the bulb should light or the voltmeter register for a period of between three and ten seconds and then cease to register. If the circuit is not broken in ten seconds the thermo-time switch must be replaced. If the bulb does not light at all and you are sure the engine is cold, check that there is voltage supplied to the switch. If there is no voltage, the fuel pump relay must be checked.

Replacement

Warning: *The engine must be completely cool before replacing the thermo-time switch.*

48 Prepare the new thermo-time switch by wrapping the threads with Teflon tape.

49 Disconnect the electrical connector from the thermo-time switch.

50 Unscrew the thermo-time switch. Be prepared for coolant spillage.

51 Install the new switch as quickly as possible to prevent excessive coolant loss. Tighten the switch securely. Check the coolant level and add some, if necessary, to bring it to the appropriate level (see Chapter 1).

12.52a The cold start injector is located in the intake manifold (arrow)

Cold start valve - check and replacement

Check

Refer to illustrations 12.52a, 12.52b and 12.52c

52 Make sure engine coolant is below 86-degrees F. Preferably the engine should sit for several hours. Disconnect the electrical connector from the cold start valve **(see illustrations)** and move it aside, away from the work area - there will be fuel vapor present. Remove the two screws holding the valve to the intake chamber **(see illustration)** and take the valve out. The fuel line must be left connected to the valve. Wipe the nozzle of the valve. Pull the coil wire out of the center of the distributor and connect it to a good ground. Turn the ignition On and operate the fuel pump for one minute. There must be no fuel dripping from the nozzle. If there is, the valve is faulty and must be replaced. Switch off the ignition.

53 Now direct the stem of the valve into a metal can. Reconnect the plug to the valve. Unplug the electrical connector from the thermo-time switch and connect a jumper lead over the plug terminals. Have an assistant turn the ignition On and operate the starter. The valve should squirt a conical shaped spray into the jar. If the spray is correct, the valve is working properly. If the spray pattern is irregular the valve is damaged and should be replaced.

Replacement

54 Relieve the fuel pressure (see Section 2).

55 Use a box-end or socket wrench and remove the fuel line connected to the cold start valve.

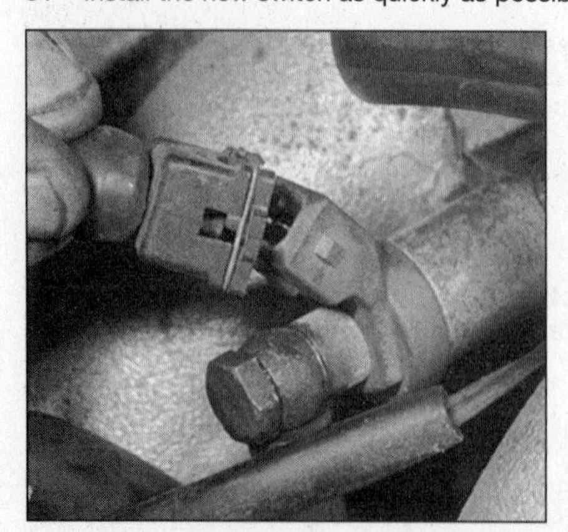

12.52b Disconnect the electrical connector from the cold start injector

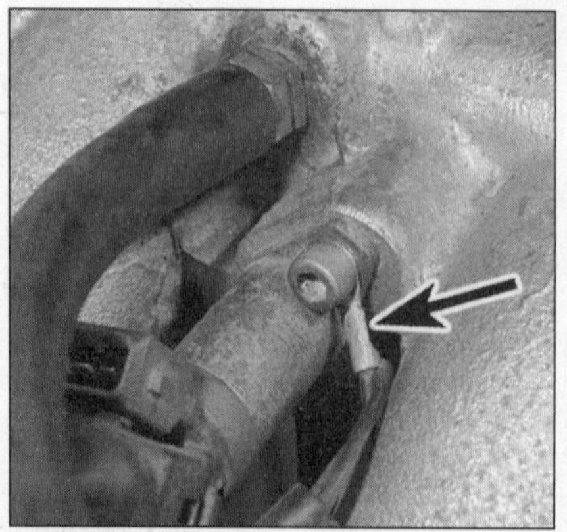

12.52c Be sure to not damage the ground strap when removing the bolts from the manifold

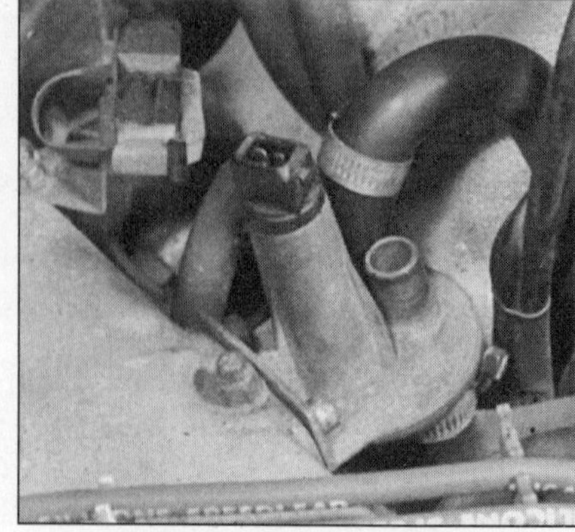

12.61 Auxiliary air regulator mounting details

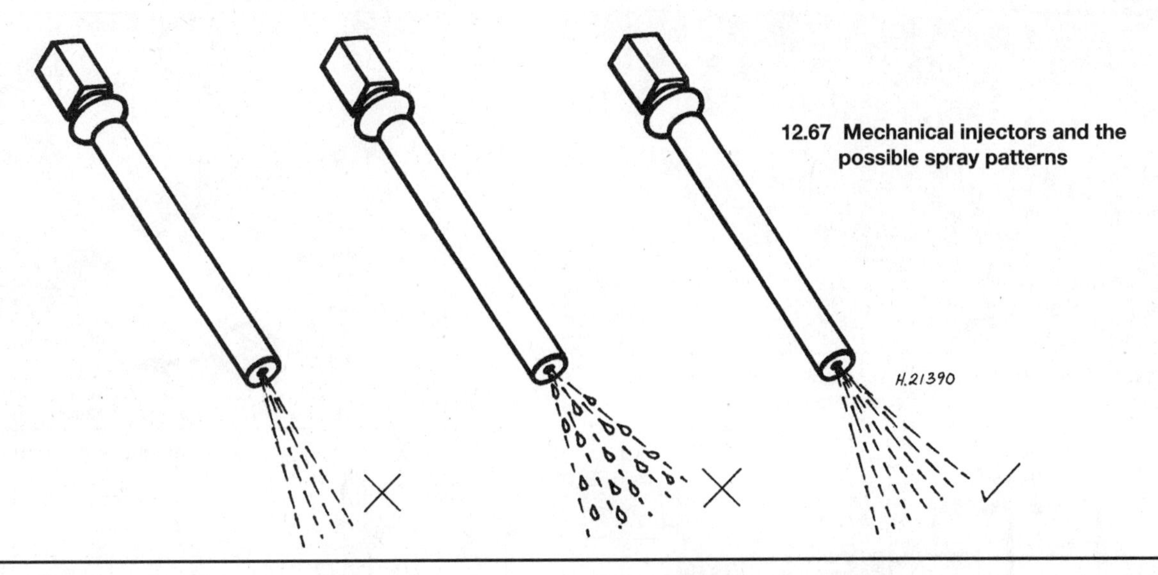

12.67 Mechanical injectors and the possible spray patterns

56 Remove the Allen bolts that retain the cold start valve to the air intake distributor and remove the valve.
57 Installation is the reverse of removal, but be sure to clean the mating surfaces and use a new gasket.

Auxiliary air regulator - check and replacement

Refer to illustration 12.61

Check

58 The auxiliary air regulator allows air to bypass the throttle plate while the engine is cold. When the ignition is switched On, the heater resistance causes a bi-metallic strip within the regulator to deform, slowly turning the rotating valve until the air passage is closed. It remains in this position during normal operation.
59 To check the operation of the regulator, remove it from the engine (see Step 4), disconnect the hoses and shine a flashlight into the port. If the unit is cold there must be a clear passage. Connect it to a 12-volt supply for five minutes and watch the operation through the inlet. At the end of the five minutes the valve should be closed. If it does not operate correctly check the resistance of the heater unit. This should be 30 ohms.
60 If the auxiliary air regulator resistance is in the correct range, disconnect the electrical connector from the control pressure regulator. Use a test light to determine if the battery voltage is reaching the heating element while the engine is running. If it is not, test the fuel pump relay.

Replacement

61 Disconnect both air hoses from the auxiliary air regulator to the intake air chamber **(see illustration)**.
62 Disconnect the electrical connector from the auxiliary air regulator.
63 Remove the mounting screws that retain the auxiliary air regulator to the intake air chamber.
64 Installation is the reverse of removal.

Fuel injectors - check and replacement

Refer to illustrations 12.67

Check

65 Each cylinder is equipped with one injector. They are pushed into bores in the cylinder head and secured by a bolt. After the bolt is removed they are pulled out quite easily. Inspect the rubber O-ring seal on each injector. If it is hardened or cracked, remove it and install a new one. Moisten the new seal with fuel before installing it. Tighten the bolt securely.
66 The injector may give trouble for one of four reasons. The spray pattern may be irregular in shape; the nozzle may not close when the engine is shut down, causing flooding when restarting; the nozzle filter may be clogged, giving less than the required ration of fuel, or the seal may be damaged allowing an air leak.
67 If the engine is running roughly and missing on one cylinder, allow it to idle and pull each spark plug wire off (use a pair of insulated pliers) and install it - one cylinder at a time (don't perform this check on electronic ignition models). If that cylinder is working properly this will have an even more adverse effect on the idle speed, when the wire is pulled off, which will promptly improve once the wire is reinstalled. If there is little difference when the spark plug wire is removed, then that is the cylinder giving trouble. Stop the engine and check and service the spark plug. Now have a look at the injector. Pull it out of the seal and hold it over a metal container. Start the engine and look at the shape of the spray. It should be of a symmetrical cone shape **(see illustration)**. If it is not, the injector must be changed because the vibrator pin is damaged or the spring is broken. Shut off the engine and wait for 15 seconds. There must be no leaks from the nozzle. If there is, the injector must be replaced, as leaking will cause flooding and difficult starting. If the spray is cone-shaped and no leak occurs then the fuel output should be checked.
68 The injector can't be disassembled for cleaning. If an injector is removed from the line, a new one should be installed and the union tightened to the torque listed in this Chapter's Specifications.

Removal

Refer to illustrations 12.69, 12.70 and 12.71

69 Remove the bolt that retains the injector assembly to the cylinder head **(see illustration)**.
70 Lift the injector and the assembly from the cylinder head **(see**

12.69 Injector retaining bolt

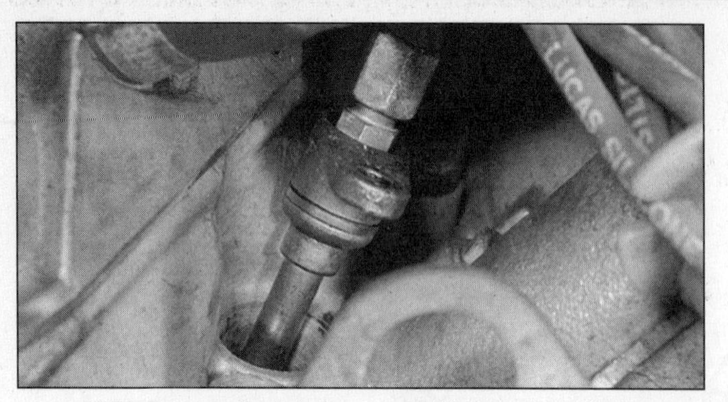

12.70 Remove the injector assembly from the cylinder head as a complete unit

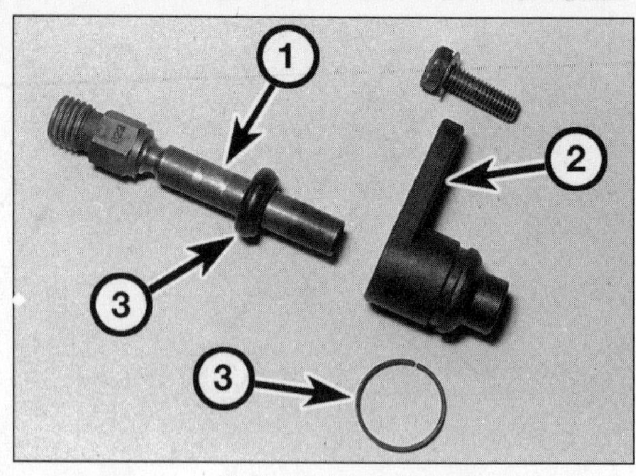

12.71 Mechanical injector assembly

1 Injector
2 Holder
3 Seals

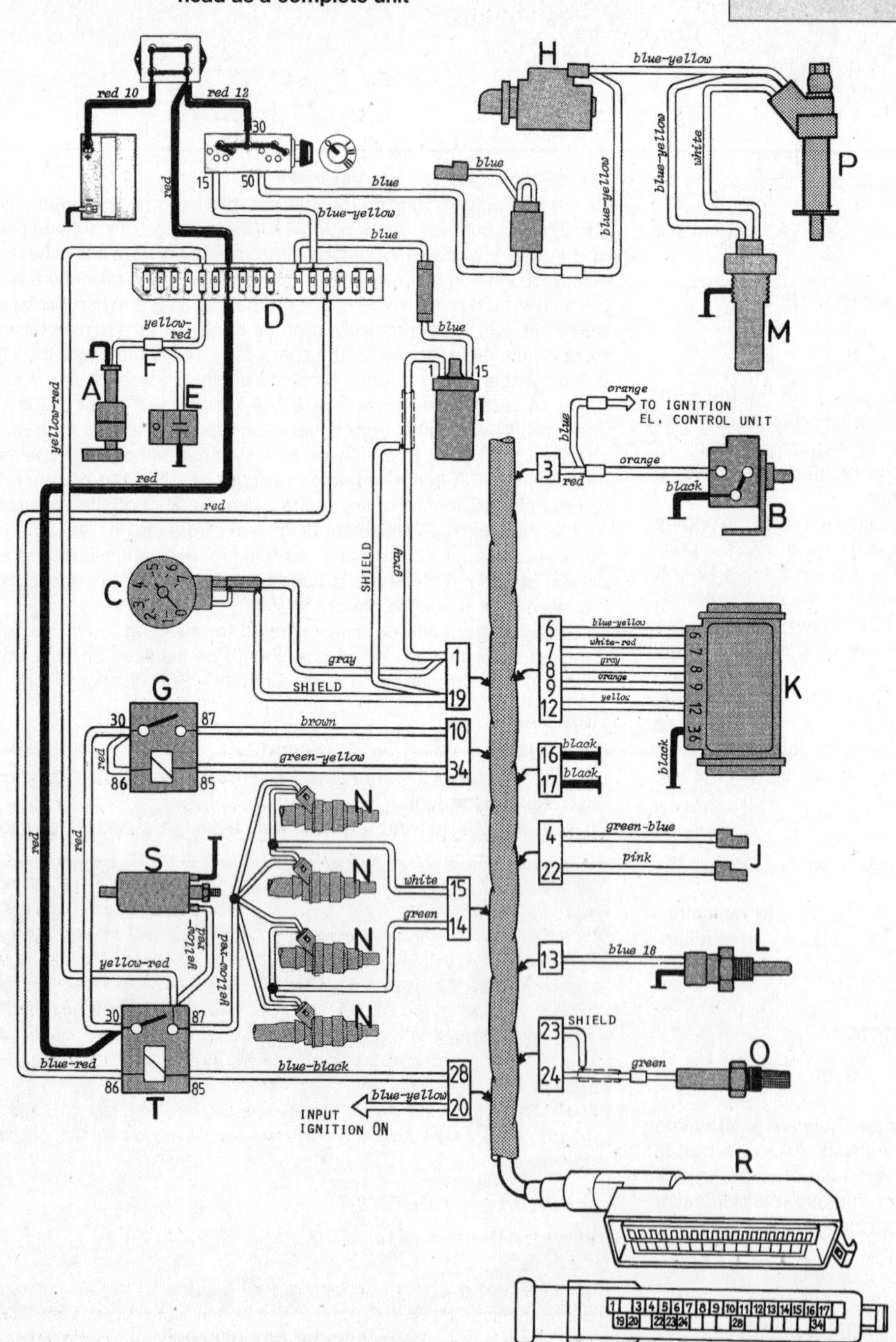

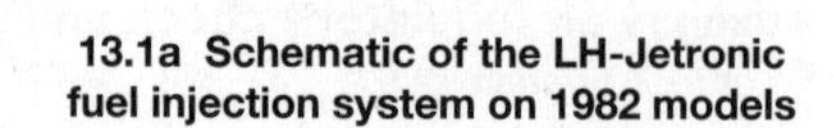

13.1a Schematic of the LH-Jetronic fuel injection system on 1982 models

A Fuel tank pump
B Vacuum switch
C Tachometer
D Fuse box
E Capacitor, tank pump
F Connector, tank pump
G System relay
H Starter motor
I Ignition coil
J Test pick-up connectors
K Air mass meter
L Temperature sensor
M Thermal time switch
N Injectors (four)
O Oxygen sensor
P Cold start injector
R Connector, Electronic Control Unit (ECU)
S Fuel pump (main pump)
T Fuel pump relay

Fuse number 5 (in-tank fuel pump)
Fuse number 7 (main fuel pump)
Fuse number 13 (system relay)

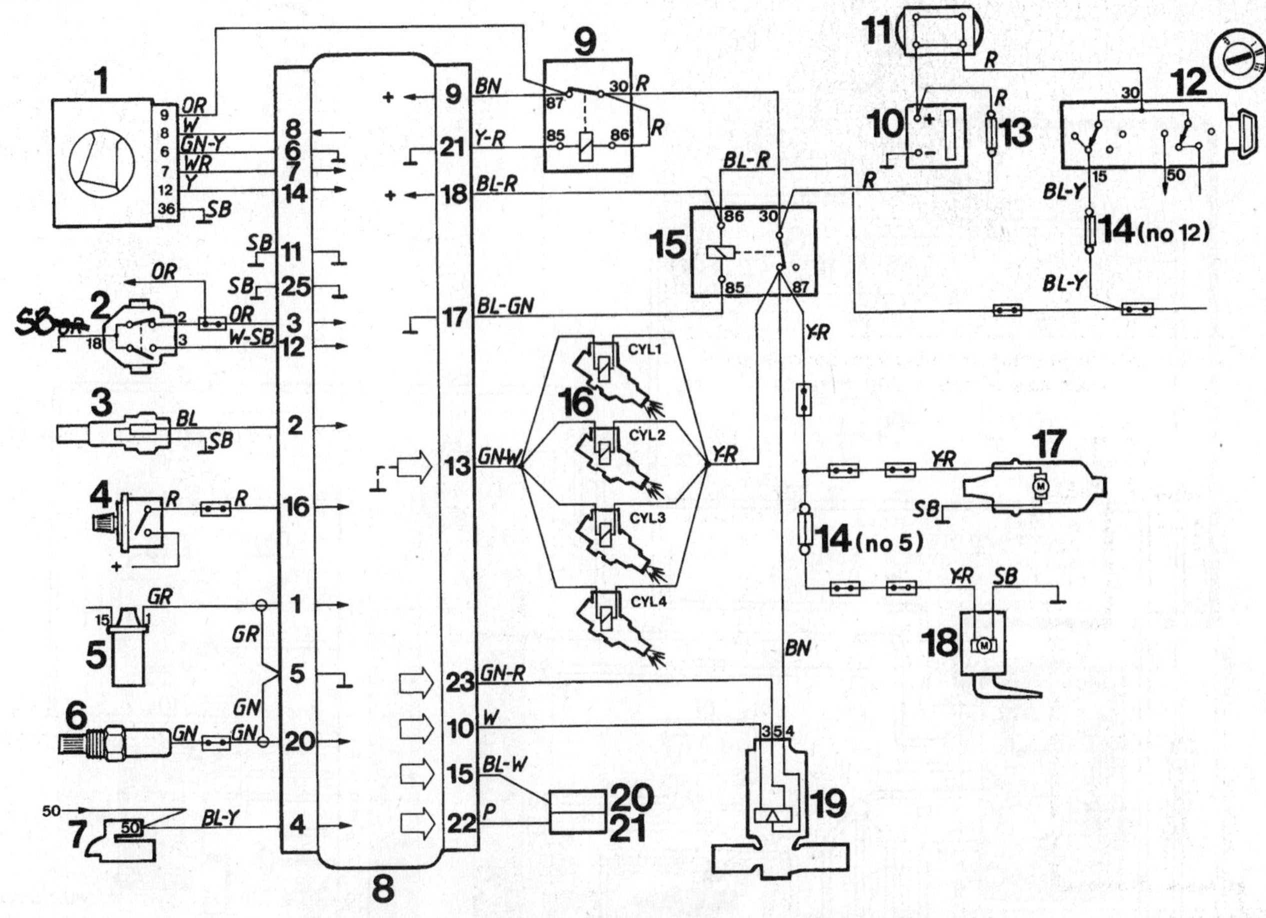

13.1b Schematic of the LH-2.0 Jetronic system on 1983 and 1984 models

1	Air mass meter	8	Control unit	15	Pump relay
2	Throttle valve switch	9	System relay	16	Injectors
3	Coolant temperature sensor	10	Battery	17	Fuel pump
4	Micro-switch, air conditioning	11	Junction box	18	Tank pump
5	Ignition coil	12	Ignition switch	19	Air control valve
6	Lambda-sond (oxygen sensor)	13	Fuse and holder	20	Test point, idle speed
7	Starter motor	14	Fuse box, fuses 5 and 12	21	Test point, Lambda-sond

illustration) and separate the injector from the fuel line using two open-end wrenches.

71 Separate the fuel injector from the holder **(see illustration)**. Be sure to replace all the seals with new ones.

72 Installation is the reverse of removal.

Fuel Accumulator

73 The accumulator on turbocharged engines is located next to the main fuel pump. The accumulator absorbs fuel pressure surges and helps retain residual fuel pressure once the engine is turned off. During running conditions, the fuel pressure actively depresses the accumulator piston against spring pressure and then once the engine is shut off, the spring closes the fuel line to allow residual pressure to accumulate.

74 Problems with the fuel accumulator usually show up as loss in residual pressure consequently causing hard, warm starting problems.

Removal

75 Remove the fuel accumulator along with the fuel pump (see Section 4).

76 Installation is the reverse of removal.

13 LH-Jetronic fuel injection system - check

Warning: *Gasoline is extremely flammable, so take extra precautions when you work on any part of the fuel system. Don't smoke or allow open flames or bare light bulbs near the work area, and don't work in a garage where a natural gas-type appliance (such as a water heater or a clothes dryer) with a pilot light is present. Since gasoline is carcinogenic, wear latex gloves when there's a possibility of being exposed to fuel, and, if you spill any fuel on your skin, rinse it off immediately with soap and water. Mop up any spills immediately and do not store fuel-soaked rags where they could ignite. The fuel system is under constant pressure, so, if any fuel lines are to be disconnected, the fuel pressure in the system must be relieved first* (see Section 2 for more information). *When you perform any kind of work on the fuel system, wear safety glasses and have a Class B type fire extinguisher on hand.*

Note: *Refer to Chapter 6 for additional information and testing procedures on the information sensors equipped on the LH-Jetronic systems*

Preliminary checks

Refer to illustrations 13.1a, 13.1b, 13.1c, 13.1d, 13.1e, 13.1f, 13.6, 3.7 and 13.8

1 Check all electrical connectors **(see illustrations)** that are related to the system. Check the ground wire connections **(see illustration)**. Loose connectors and poor grounds can cause many problems that resemble more serious malfunctions.

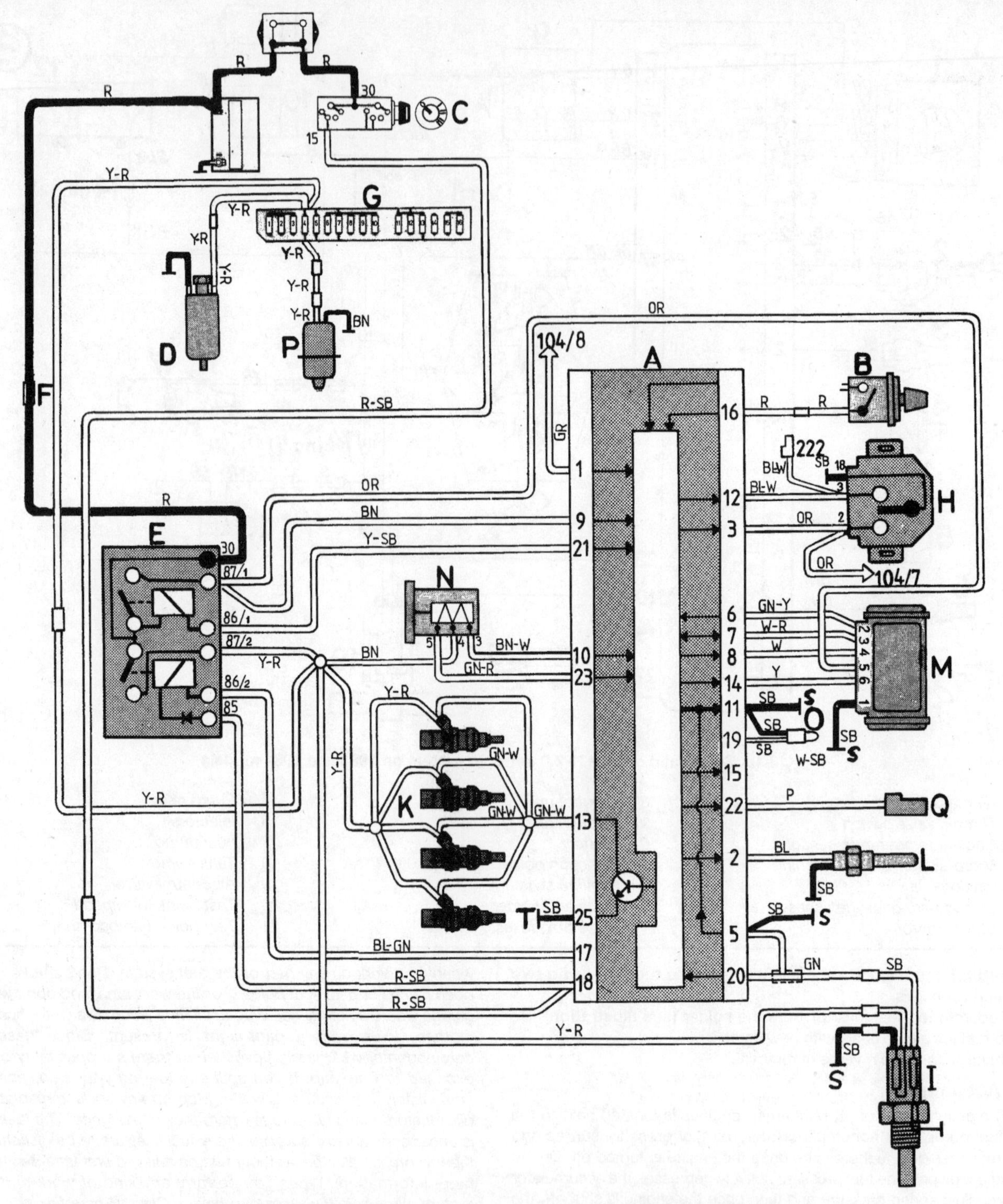

13.1c Schematic of the LH-2.2 Jetronic system on 1985 and 1986 models

A	Control unit	L	Coolant temperature sensor
B	Microswitch, air conditioning	M	Air mass meter
C	Ignition switch	N	Idle control valve
D	Fuel pump	O	Bridge connector
E	Main relay	P	Tank pump
F	In-line fuse	Q	Lambda-sond test point
G	Fusebox	S	Ground connection, intake manifold
H	Idle-full load switch	T	Ground connection, intake manifold
I	Lambda-sond (oxygen sensor)	104	Ignition system control unit
K	Injector	222	Idle speed connector

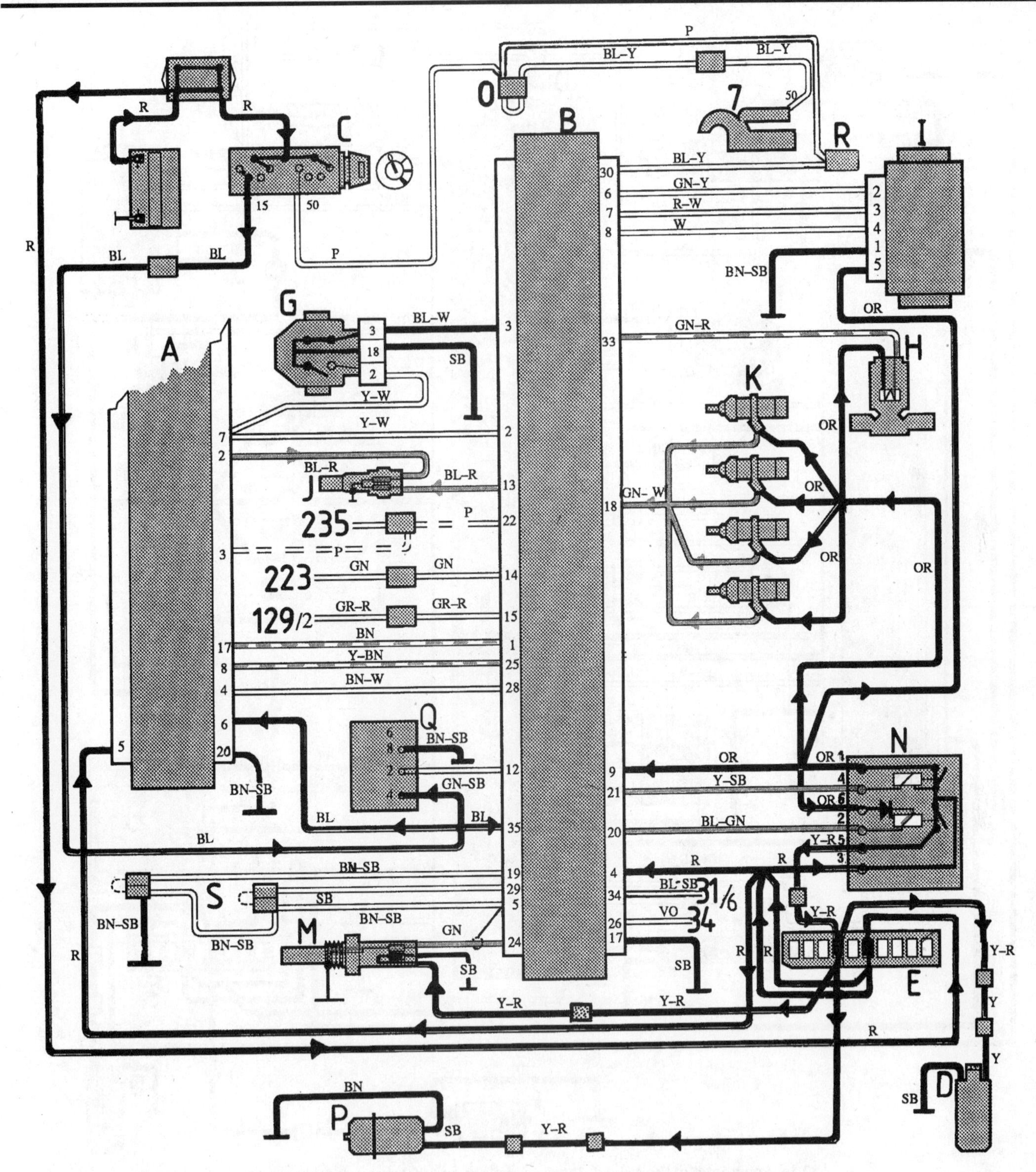

13.1d Schematic of the LH-2.4 Jetronic system on 1987 through 1991 models

A *Control module, EZ-116 K*
B *Control module, LH-2.4 Jetronic*
C *Ignition switch*
D *Fuel pump*
E *Fuse box*
G *Throttle switch*
H *Idle air control valve*
I *Mass airflow sensor*
J *Engine coolant temperature sensor*
K *Injectors*
L *Start Injectors*
M *Heated oxygen sensor*
N *Main relay*
O *Connector*
P *Tank pump*
Q *Data Link connector*
R *Service connection*
S *Bridge connection*
7 *Starter motor*
31 *Connector, instrument panel (Signal from speedometer)*
34 *Connector, instrument panel (Shift indicator light)*
129 *A/C Relay*
223 *Pressure sensor, A/C dryer*
235 *Connector, instrument panel (Malfunction Indicator lamp)*

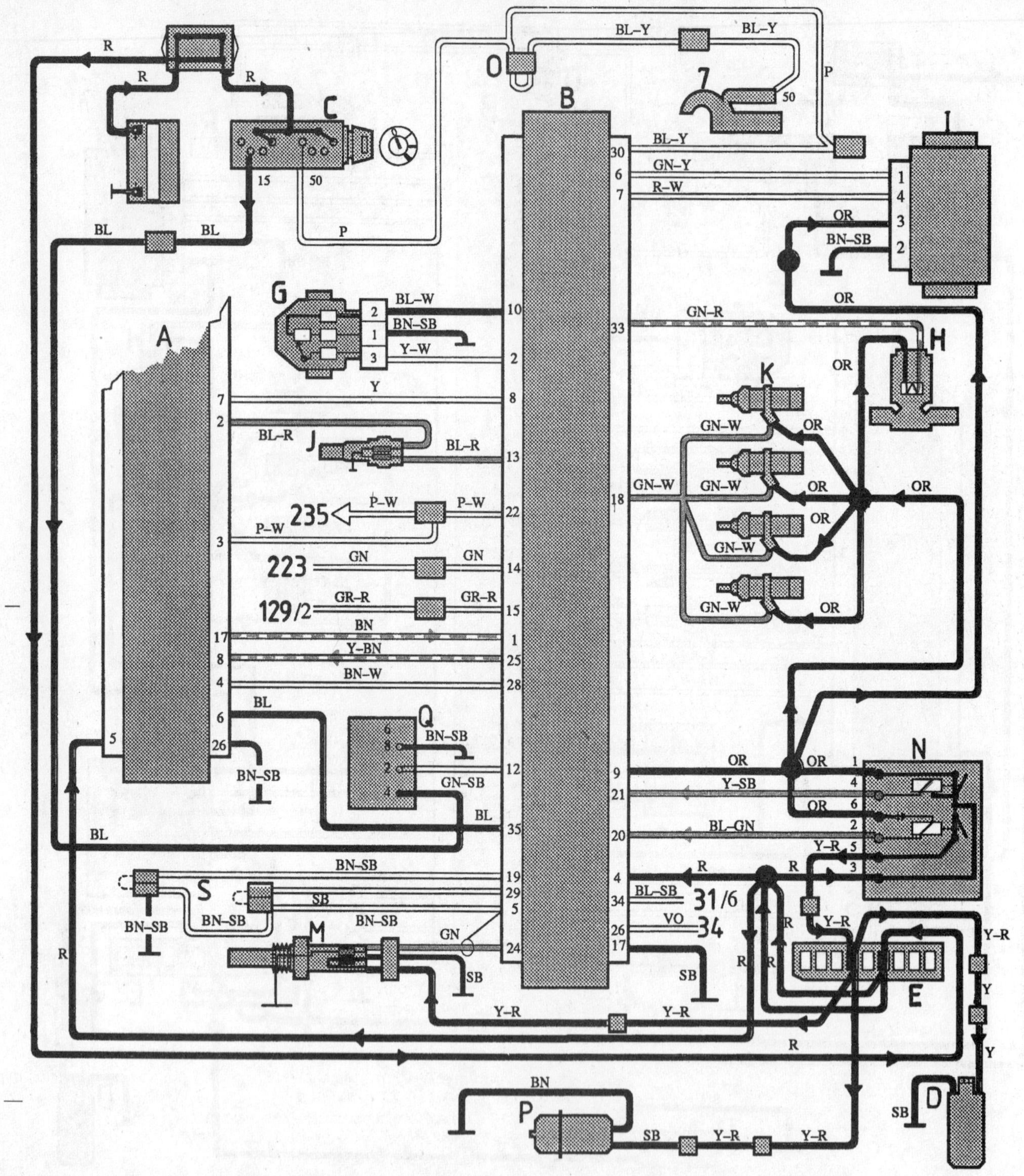

13.1e Schematic of the LH-3.1 Jetronic system on 1989 through 1993 models

- *A* *Control module, EZ-116 K ignition system*
- *B* *Control module, LH-3.1 Jetronic*
- *C* *Ignition switch*
- *D* *Fuel pump*
- *E* *Fuse box*
- *G* *Throttle position sensor*
- *H* *Sensor valve*
- *I* *Mass airflow sensor*
- *J* *Engine coolant temperature sensor*
- *K* *Injectors*
- *L* *Start injector*
- *M* *Heated oxygen sensor*
- *N* *Main relay*
- *O* *Connector*
- *P* *Tank pump*
- *Q* *Data link connector*
- *R* *Service connection*
- *S* *Bridge connection*
- *7* *Starter motor*
- *31* *Connector, instrument panel (Signal from speedometer)*
- *34* *Connector, instrument panel (shift gear indicator lamp)*
- *129* *A/C Relay*
- *223* *A/C dryer pressure sensor*
- *235* *Connector, instrument panel (Malfunction Indicator lamp)*

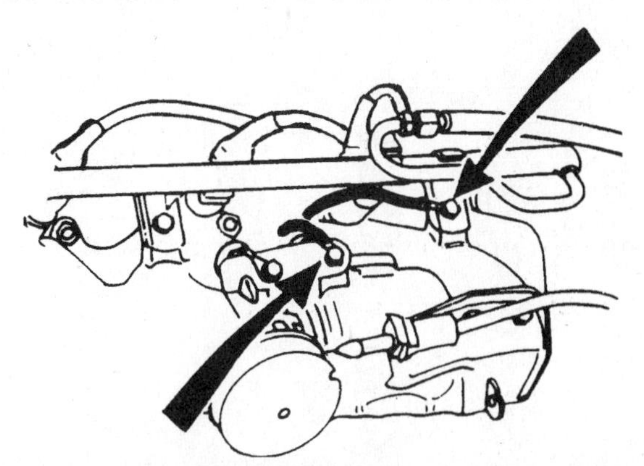

13.1f Make sure the electrical grounds (arrows) on the intake manifold are clean and secure

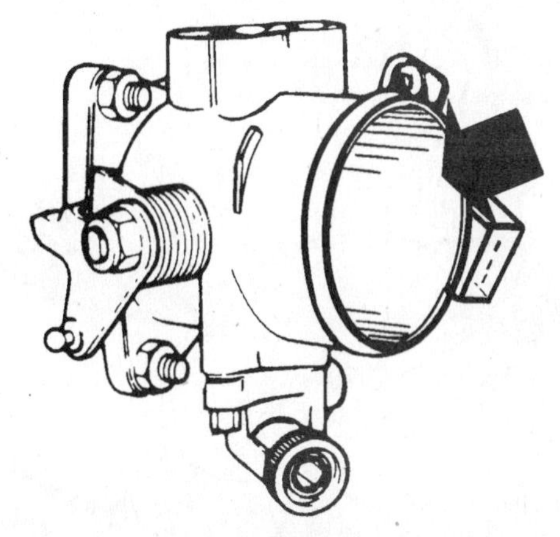

13.6 Check for any dirt or sludge deposits around the throttle valve area inside the throttle body bore

13.7 Use a stethoscope to determine if the injectors are working properly - they should make a steady clicking sound that rises and falls with engine speed changes

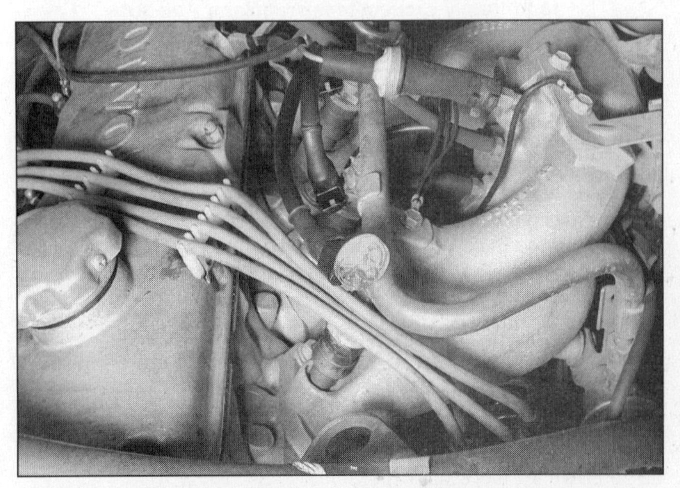

13.8 Install the "noid" light into the fuel injector harness and confirm that it blinks when the engine is running

2 Make sure the battery is fully charged, as the control unit and sensors depend on an accurate supply voltage in order to properly meter the fuel.

3 Check the air filter element - a dirty or partially blocked filter will severely impede performance and economy (see Chapter 1).

4 If a blown fuse is found, replace it and see if it blows again. If it does, search for a grounded wire in the harness related to the system.

5 Check the air intake duct from the air mass meter to the intake manifold for leaks, which will result in an excessively lean mixture. Also check the condition of the vacuum hoses connected to the intake manifold.

6 Remove the air intake duct from the throttle body and check for dirt, carbon and other residue build-up **(see illustration)**. If it's dirty, clean it with carburetor cleaner and a toothbrush.

7 With the engine running, place an automotive stethoscope against each injector, one at a time, and listen for a clicking sound, indicating operation **(see illustration)**. **Note:** *If you don't have a stethoscope, place the tip of a long screwdriver against the injector and listen through the handle.*

8 Install an injector test light (sometimes referred to as a "noid" light) into each injector electrical connector **(see illustration)** and make sure it flashes as the engine is running.

9 Check the fuel system pressure (see Section 3). Check for trouble codes stored in the ECU (see Chapter 6).

Throttle body - check, removal and installation

Check

10 Detach the air intake duct from the throttle body (see Section 8) and move the duct out of the way.

11 Have an assistant depress the accelerator pedal while you watch the throttle valve. Check that the throttle valve moves smoothly when the throttle is moved from closed (idle position) to fully open (wide open throttle).

12 If the throttle valve is not working properly, replace the throttle body unit.

Removal and installation

Warning: *Wait until the engine is completely cool before beginning this procedure.*

13 Detach the cable from the negative battery terminal. **Caution:** *It is necessary to make sure the radio is turned OFF before disconnecting the battery cables to avoid damaging the microprocessor in the radio.*

14 Detach the air intake duct from the throttle body and set it aside.

15 Detach the accelerator cable from the throttle body (see Section 9).

16 Detach the cruise control cable, if equipped.

17 Clearly label all electrical connectors (TPS, cold start injector, idle air stabilizer, etc.), then unplug them.

18 Clearly label all vacuum hoses, then detach them.

19 Unscrew the radiator or expansion tank cap to relieve any residual pressure in the cooling system, then reinstall it. Clamp shut the coolant hoses, then loosen the hose clamps and detach the hoses from the throttle body. Be prepared for some coolant leakage.

20 Remove the throttle body mounting nuts (upper) and bolts (lower) and detach the throttle body from the air intake plenum.

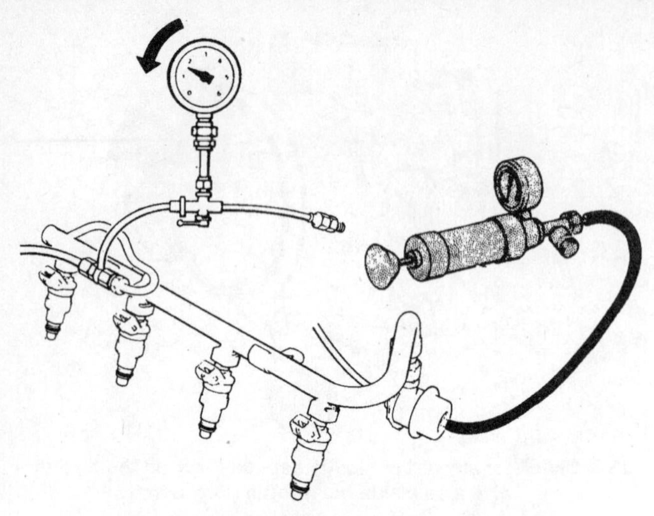

13.27 Apply vacuum to the regulator - the fuel pressure should decrease

21 Cover the air intake manifold opening with a clean cloth to prevent dust or dirt from entering while the throttle body is removed. Clean all traces of old gasket material from the mating surfaces of the throttle body and intake manifold.

22 Installation is the reverse of removal. Be sure to use a new gasket and tighten the throttle body mounting nuts and bolts to the torque listed in this Chapter's Specifications. Adjust the accelerator cable (see Section 9) when you're done.

Fuel pressure regulator - check and replacement

Check

Refer to illustration 13.27

23 Relieve the fuel system pressure (see Section 2).

24 Detach the cable from the negative battery terminal. **Caution:** *It is necessary to make sure the radio is turned OFF before disconnecting the battery cables to avoid damaging the microprocessor in the radio.*

25 Disconnect the fuel line and install a fuel pressure gauge (see Section 3). Reattach the cable to the battery.

26 Pressurize the fuel system by turning the ignition to the On position and check for leakage around the gauge connections. Start the engine.

27 Connect a vacuum pump to the fuel pressure regulator **(see illustration)**.

28 Read the fuel pressure gauge with vacuum applied to the pressure regulator and also with no vacuum applied. The fuel pressure should DECREASE as vacuum INCREASES. Compare your readings with the values listed in this Chapter's Specifications.

29 Reconnect the vacuum hose to the regulator and check the fuel pressure at idle, comparing your reading with the value listed in this Chapter's Specifications. Disconnect the vacuum hose and watch the gauge - the pressure should jump up to the maximum specified pressure as soon as the hose is disconnected.

30 If the fuel pressure is low, pinch the fuel return line shut and watch the gauge. If the pressure doesn't rise, the fuel pump is defective or there is a restriction in the fuel feed line. If the pressure rises sharply, replace the pressure regulator.

31 If the indicated fuel pressure is too high, turn off the engine and relieve the fuel system pressure. Disconnect the fuel return line and blow through it to check for a blockage. If there is no blockage, replace the fuel pressure regulator.

32 If the pressure doesn't fluctuate as described in Step 29, connect a vacuum gauge to the pressure regulator vacuum hose and check for vacuum.

33 If there is vacuum present, replace the fuel pressure regulator.

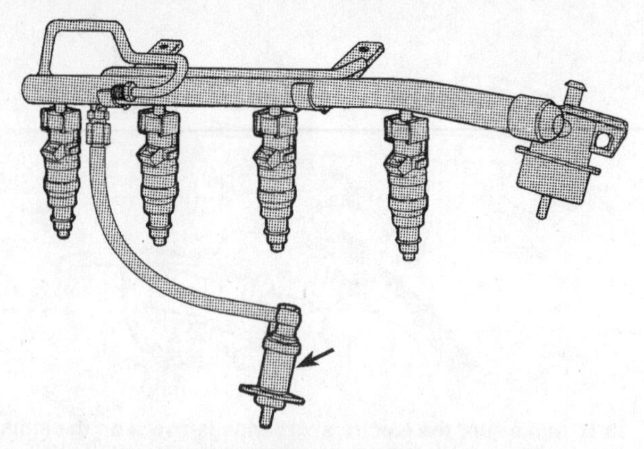

13.41 Remove the cold start injector (arrow) from the intake manifold but do not disconnect the fuel line

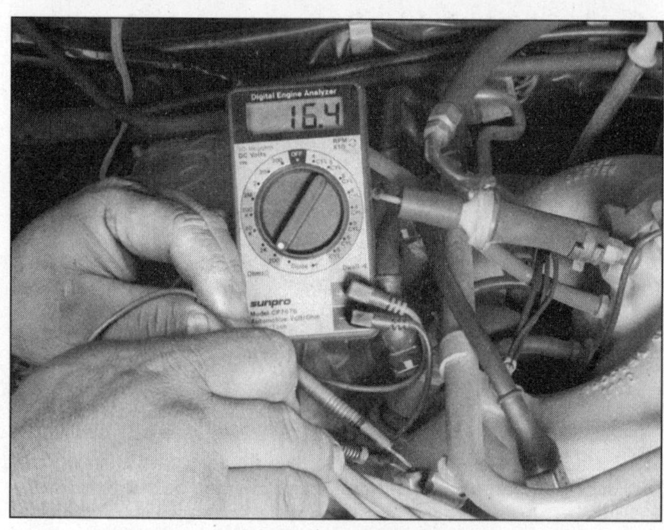

13.54 Use an ohmmeter and check the resistance of each fuel injector

34 If there isn't any reading on the gauge, check the hose and its port for a leak or a restriction.

Replacement

35 Relieve the system fuel pressure (see Section 2).

36 Detach the cable from the negative battery terminal. **Caution:** *It is necessary to make sure the radio is turned OFF before disconnecting the battery cables to avoid damaging the microprocessor in the radio.*

37 Detach the vacuum hose and fuel return hose from the pressure regulator, then unscrew the mounting bolts **(see illustration 13.57)**.

38 Remove the pressure regulator.

39 Installation is the reverse of removal. Be sure to use a new O-ring. Coat the O-ring with a light film of engine oil prior to installation.

40 Check for fuel leaks after installing the pressure regulator.

Cold start injector (LH 2.4 and 3.1) - check and replacement

Note: *The cold start injector injects extra fuel into the intake manifold during cranking when the temperature has dropped below 68-degrees F. The cold start injector is mounted in the underside of the intake manifold between the number 2 and 3 intake valves. The injector sprays fuel into the manifold until the cranking speed exceeds 900 rpm.*

Check

Refer to illustration 13.41

41 Make sure engine coolant is below 68-degrees F. Preferably, the

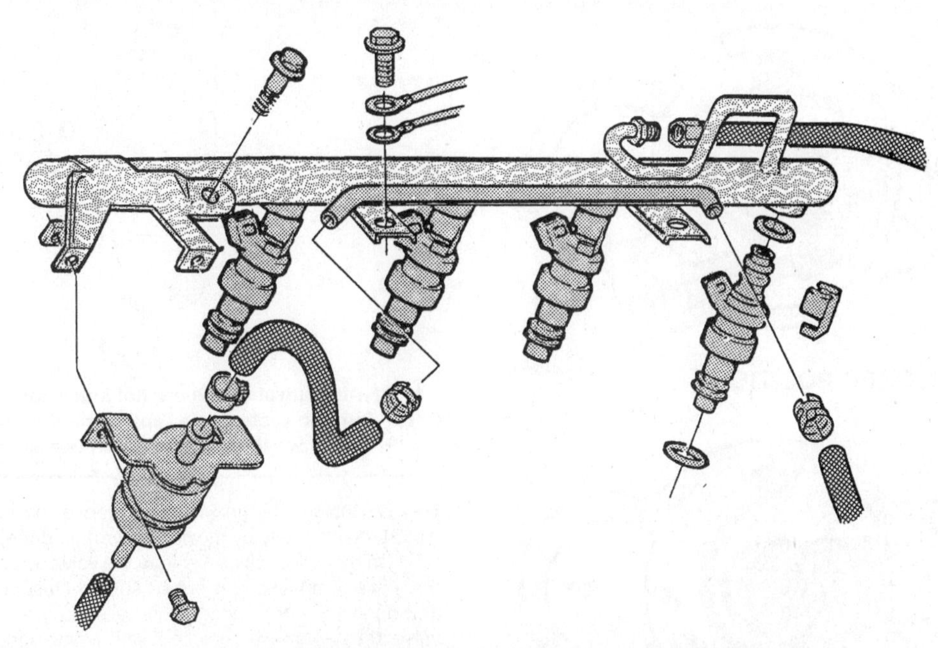

13.57 Exploded view of the fuel rail and injector assemblies

engine should sit for several hours before performing this check. Disconnect the electrical connector from the cold start valve and move it aside **(see illustration)**, away from the work area - there will be fuel vapor present. Remove the two screws holding the valve to the air intake manifold and take the valve out. The fuel line must be left connected to the valve. Wipe the nozzle of the valve. Disable the ignition system (see Chapter 2 Part B). Turn the ignition On and operate the fuel pump for one minute (see Section 3). There must be no fuel dripping from the nozzle. If there is, the valve is faulty and must be replaced. Switch off the ignition.

42 Now, direct the nozzle of the valve into a can or jar. Reconnect the electrical connector to the valve. Have an assistant turn the ignition On and operate the starter. The valve should squirt a conical shaped spray into the jar. If the spray pattern is good the valve is working properly. If the spray pattern is irregular, the valve is damaged and should be replaced.

43 If the cold start injector does not spray any fuel, check for a voltage signal at the electrical connector for the cold start valve. If there is no voltage, check the coolant temperature sensor and the wiring harness.

Replacement

44 Relieve the fuel pressure (see Section 2).

45 Disconnect the electrical connector from the cold start injector.

46 On models so equipped, use a box-end or socket wrench and remove the fuel line fitting connected to the cold start valve. On other models, simply loosen the hose clamp and detach the hose from the valve.

47 Remove the bolts that retain the cold start valve to the intake manifold and remove the valve.

48 Clean the mating surfaces and be sure to use a new gasket.

49 Installation is the reverse of removal.

Fuel injectors - check and replacement

Check

In-vehicle check

Refer to illustration 13.54

50 Using a mechanic's stethoscope (available at most auto parts stores), check for a clicking sound at each of the injectors while the engine is idling **(see illustration 13.7)**. The injectors should make a steady clicking sound if they are operating properly.

51 Increase the engine speed above 3,500 rpm. The clicking sound should rise with engine speed.

52 If you don't have a stethoscope, a screwdriver can be used. Place the tip of the screwdriver against the injector and press your ear against the handle.

53 If an injector isn't functioning (not clicking), purchase a special injector test light (sometimes called a "noid" light) and install it into the injector electrical connector **(see illustration 13.8)**. Start the engine and make sure the light flashes. If it does, the injector is receiving the proper voltage, so the injector itself must be faulty.

54 Unplug each injector connector and check the resistance of the injector **(see illustration)**. Check your readings with the values listed in this Chapter's Specifications. Replace any that do not have the correct amount of resistance.

Volume test

55 Because a special injection checker is required to test injector volume, this procedure is beyond the scope of the home mechanic. Have the injector volume test performed by a dealer service department or other repair shop.

Replacement

Refer to illustration 13.57

56 Relieve the fuel system pressure (see Section 2) then disconnect the cable from the negative terminal of the battery. Unplug the main electrical connector for the fuel injector wiring harness.

57 Detach the fuel hoses from the fuel rail and remove the fuel rail mounting bolts **(see illustration)**.

58 Lift the fuel rail/injector assembly from the intake manifold.

59 Unplug the electrical connectors from the fuel injectors. Detach the injectors from the fuel rail.

60 Installation is the reverse of removal. Be sure to replace all O-rings with new ones. Coat them with a light film of engine oil to prevent damage to the O-rings during installation. Be sure to pressurize the fuel system and check for leaks before starting the engine.

Air control valve - check and replacement

Refer to illustrations 13.61 and 13.62

61 The air control valve works to maintain engine idle speed within a 200 rpm range regardless of varying engine loads at idle. An electrically operated valve allows a small amount of air to flow past the

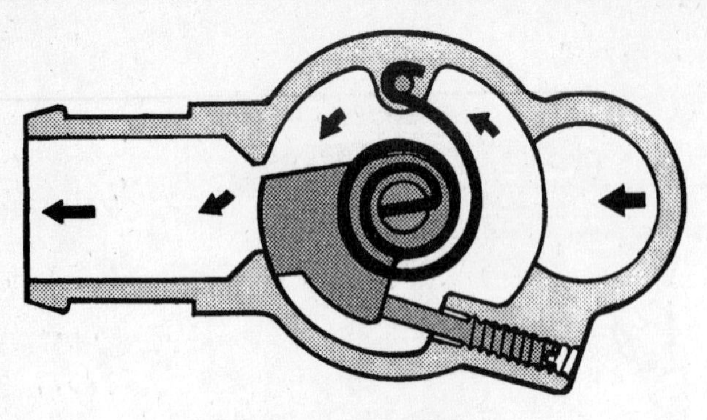

CLOSED POSITION

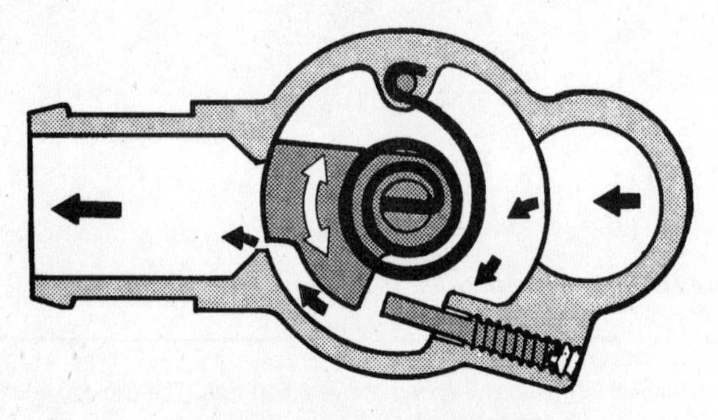

REGULATING POSITION

13.61 When the throttle switch is closed during idle conditions, the ECU receives a signal that controls the air valve electric motor to maintain a steady idle rpm

throttle plate to raise the idle speed whenever the idle speed drops below approximately 750 rpm **(see illustration)**.

62 The air control valve on LH 2.0 and 2.2 systems is located under the intake manifold near the distributor while on LH 2.4 and 3.1 systems, the air control valve is located under the throttle body. Different air control valves are used on the various types of fuel injection systems and they are not interchangeable **(see illustration)**. The valves on the LH 2.4 and 3.1 systems have a high idle speed screw that is preset at the factory and is not adjustable.

Preliminary check

63 Before performing any checks on the idle air stabilizer valve, make sure these criteria are met:

- *a) The engine has reached operating temperature (140-degrees F)*
- *b) Turn off all electrical accessories air conditioning, heater controls, headlights, auxiliary cooling fan, etc.)*
- *c) TPS sensor must be operating correctly* (see Chapter 6)
- *d) There must not be any exhaust leaks*
- *e) There must not be any vacuum leaks*
- *f) The oxygen sensor must be operating properly* (see Chapter 6)

64 The air control valve operates continuously when the ignition is on. Start the engine and make sure the valve is vibrating and humming slightly.

Check

65 With the ignition key OFF, disconnect the electrical connector from the air control valve. Using an ohmmeter, check the resistance of the valve.

Three-wire electrical connector	
Checked between outer terminals	40 ohms
Checked between center terminal to outer terminal	20 ohms
Two-wire electrical connector	8 ohms

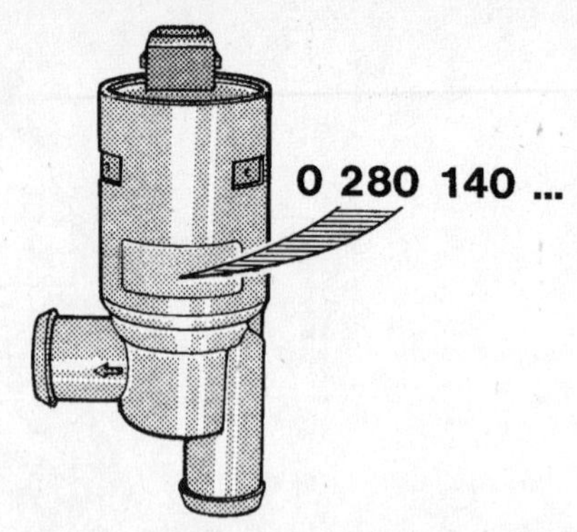

13.62 Air control valves are not interchangeable for the various LH-Jetronic systems - a replacement air control valve should have the same part number as the old one

66 On three-wire type valves, remove the part and check that the piston moves freely by moving and rotating the valve back and forth.

67 On two-wire valves, remove the valve and apply battery voltage to the valve terminals. The piston should close when voltage is applied and open when the voltage source is removed. If the piston inside the valve does not move, replace it with a new part.

Replacement

68 Turn the ignition key off and disconnect the electrical connector.

69 Pull the bypass hose from the air control valve, loosen the mounting bracket and remove the valve.

70 Installation is the reverse of removal. Be sure to replace any cracked or broken hoses with new ones.

14 Constant Idle Speed system - check, adjustment and component replacement

General information

Refer to illustration 14.1

1 The Constant Idle Speed system maintains idle speed within a 40 rpm range by regulating the amount of air that bypasses the closed throttle valve **(see illustration)**. **Note:** *The Constant Idle Speed system is installed on CIS fuel-injected models and 1982 LH-Jetronic systems only*

2 The Constant Idle Speed system is governed by an electronic control (ECU) that receives input signals from the throttle position switch, the coolant temperature sensor and tachometer and indicates to the computer the engine's temperature, cruising mode (throttle) and engine speed. The air control valve operates in three different modes:

Regulated flow - At idle, the throttle and throttle switch are closed. In this mode the valve varies bypass air to control idle speed by changing the position of the air valve several times per second. In this mode the frequency (amount the valve opens and closes) is great. Also, during warm-up while idling, engine rpm is increased to compensate for the cold engine and allow a quicker warm-up period and better driveability.

Low flow - During deceleration, the ignition coil provides engine speed to the computer to indicate the necessity to lower engine rpm. The computer also detects the throttle position from the TP switch and decreases flow through the air control valve.

High flow - During acceleration, the system detects the need to raise engine rpm to compensate for the additional load.

Note: *In 1982, the ECU is equipped with two extra terminals (numbers 7 and 10) that regulate the increase of idle air when the air conditioning is turned ON.*

Preliminary checks

Refer to illustrations 14.3 and 14.4

3 Be sure to first check the engine to make sure the basic operating

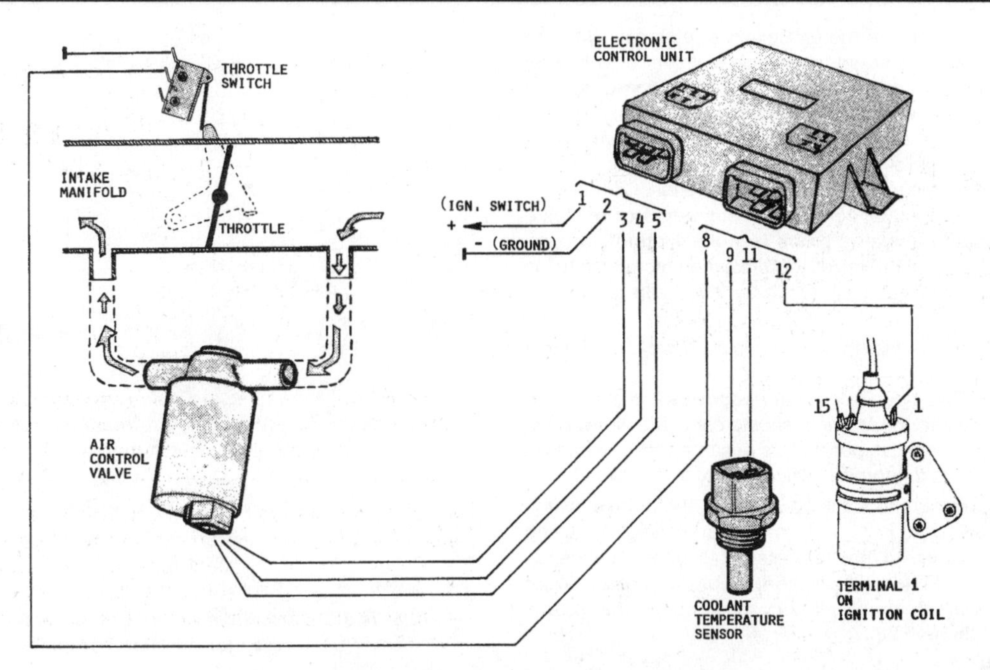

14.1 Schematic of the Constant Idle Speed system

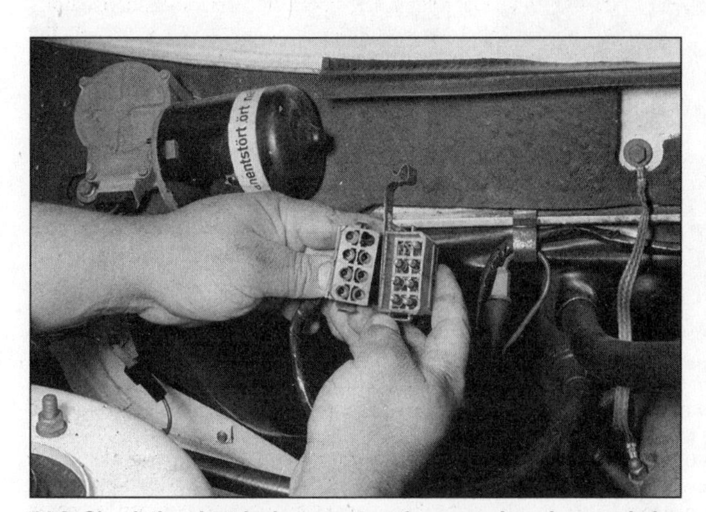

14.3 Check the electrical connectors for corrosion, damaged pins or terminals, broken wire harness etc.

14.4 Disconnect the electrical connector (arrow) from the air control valve

systems are working properly.

a) *Check the intake system for leaks, broken hoses, leaking manifold gaskets, damaged air filters or any other problem that would cause the engine to alter the normal vacuum supply to the combustion chambers. Also, check for sludge and deposit build-up around the throttle valve, inside the PCV hoses or inside any other vacuum lines near the crankcase.*
b) *Check the fuel system for the correct working pressure and volume* (see Section 3).
c) *Check the ignition system for the correct timing, the quality of the spark, the correct timing advance and any other problem within the ignition system that would cause the engine to misfire or run unevenly etc.*
d) *Check the battery and all electrical connectors that are directly related to the Constant Idle Speed system* **(see illustration)**.
e) *Check the Constant Idle Speed system fuse (number 13) on the fuse panel. If necessary, replace it with a new one.*

4 Next, check the system for electrical problems.

a) *Carefully listen for the buzzing sound from the air control valve while the engine is running. This will tell you quickly if the valve is operating. Another quick check for the air control valve is to disconnect the electrical connector* **(see illustration)** *from the valve and wait for a distinct drop in engine rpm. If there is no sound or rpm change, then start checking the Constant Idle Speed electrical system.*
b) *Check the throttle valve, throttle linkage or throttle switch adjustments to make sure they are within specifications. Many problems in the Constant Idle Speed system are a result of these components out of adjustment.*
c) *Test the coolant temperature sensor (see Chapter 6) to make sure the sensor is signaling the computer as the engine warms up and cools down.*

Component electrical tests

Refer to illustrations 14.5, 14.6, 14.7, 14.8, 14.9a, 14.9b, 14.10, 14.11a and 14.11b

5 Remove the panel from the right side (passenger side) kick panel to expose the Constant Idle Speed computer **(see illustration)**.

6 First, check the system for the proper voltage signal. With the ignition key ON (engine not running), connect a test light or voltmeter across the lower left pins (numbers 1 and 2) of the black connector and check for voltage **(see illustration).** There should be battery voltage available. If there is no voltage present, check the fuse (number 13) that governs the Constant Idle Speed system.

7 Next, check the throttle switch for correct activation. Connect the positive probe of an ohmmeter to terminal number 8 of the blue connector and ground the negative probe **(see illustration)**. Ground the test lamp to terminal number 1 of the black connector. With the ignition key ON (engine not running), touch the test lamp to terminal number 8 (blue connector).

a) *With the throttle closed, there should be no continuity and the test light should be OFF.*
b) *With the throttle valve open (accelerator depressed), there should be zero resistance and the test light should come on.* **Note:** *Have an assistant step on the gas pedal while observing the ohmmeter. These tests diagnose the throttle switch as well as the circuit.*

8 If the test is incorrect, check the operation of the throttle switch by itself. Simply turn the ignition key ON (engine not running) and using a test light, check the switch. With the engine at idle (throttle closed), the test light should be OFF **(see illustration)**. Depress the accelerator and the test light should come ON. If necessary, adjust the throttle switch (see Steps 12 through 26).

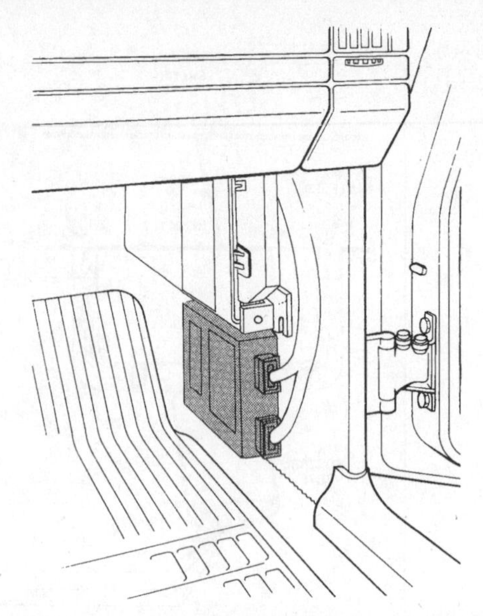

14.5 The Constant Idle Speed computer is located below the fuel injection computer

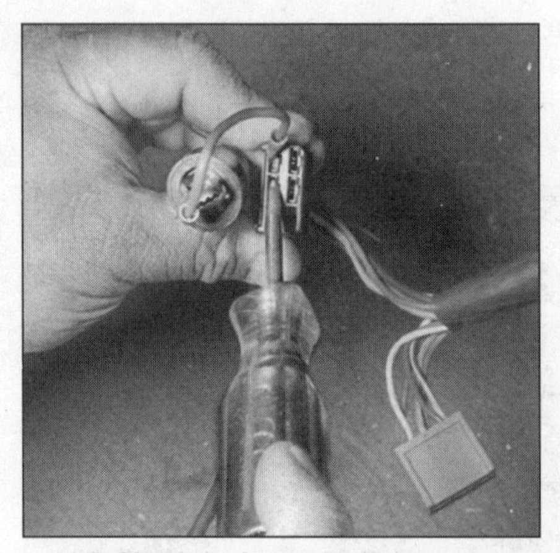

14.6 Check for battery voltage across terminals 1 (ground) and 2 (power)

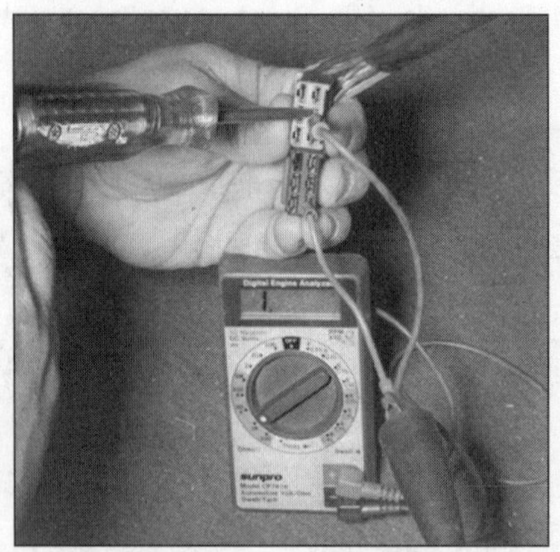

14.7 First connect an ohmmeter to terminal number 1 on the black connector and number 8 on the blue connector and then check for power on terminal number 8 using a test light

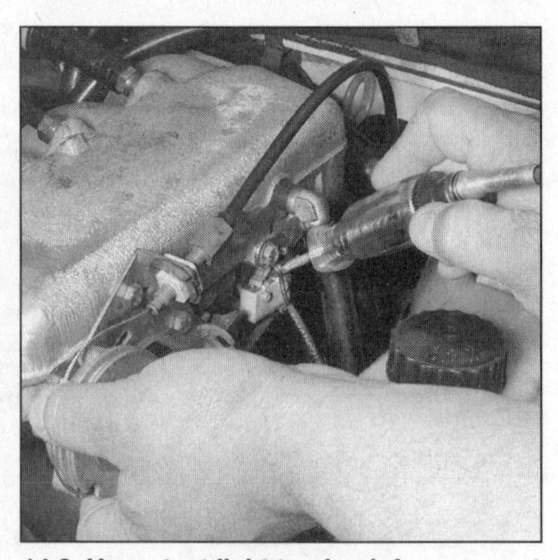

14.8 Use a test light to check for power at the throttle switch

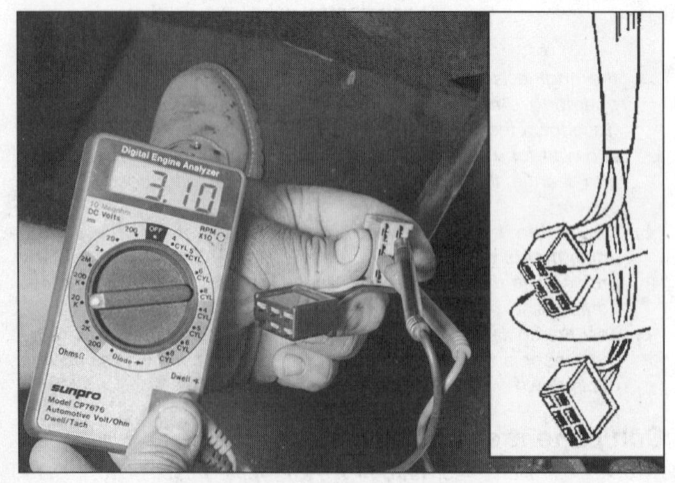

14.9a First check the resistance of the coolant temperature sensor at the Constant Idle Speed computer electrical connector . . .

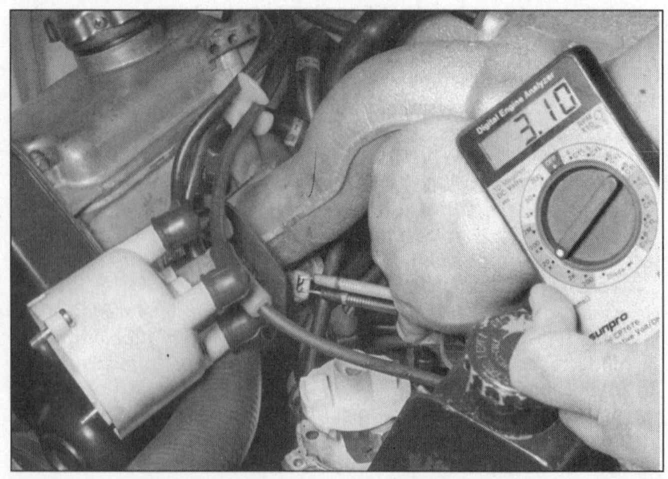

14.9b . . . then check the resistance of the coolant temperature sensor directly at the sensor

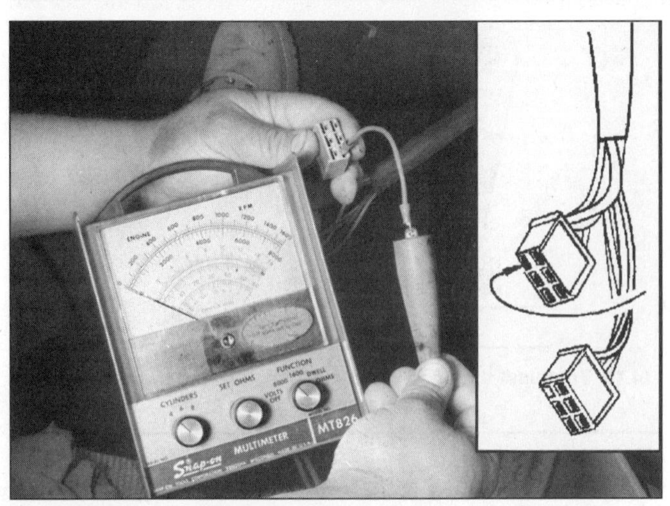

14.10 Use terminal number 12 on the blue connector to check the engine rpm

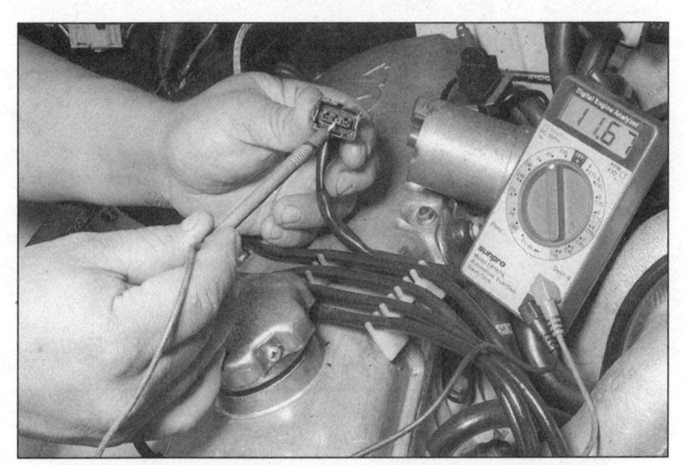

14.11b Always check to make sure the air control valve is receiving battery voltage to the electrical connector

9 Next, check the resistance of the coolant temperature sensor directly at the Constant Idle Speed computer electrical connector. Use an ohmmeter and check the resistance between terminals 9 and 11 of the blue connector **(see illustration)**. It should read exactly the same as the resistance value taken directly at the coolant temperature sensor **(see illustration)**.

10 Next, check the tachometer signal directly at the Constant Idle Speed computer electrical connector **(see illustration)**. With the engine running, the meter should indicate the engine rpm. Refer to Chapter 1 for the correct idle speed specifications.

11 Next, simulate a cold engine running condition to verify that the air control valve is operating properly. Jump terminals 4 and 1 on the black connector **(see illustration)** and another jumper wire across terminals 5 and 2. Start the engine and confirm that the engine speed increases to 1,600 to 2,400 rpm. If all the previous tests are correct and the engine speed does not increase to the proper range with the Constant Idle Speed computer disconnected, replace the air control valve **(see illustration)**. If all the previous tests are correct and the engine speed does not increase to the proper range with the Constant Idle Speed computer connected, replace the ECU.

Base idle speed adjustment

CIS engines

Refer to illustrations 14.12a, 14.12b, 14.17, 14.19 and 14.23

12 With the engine completely warmed up, disconnect the throttle rod from the throttle valve lever **(see illustrations)**.

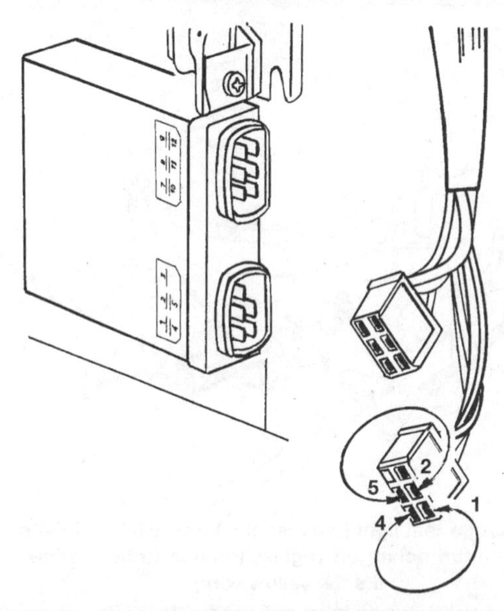

14.11a To simulate a cold running condition, jump terminals 1 and 4 and then jump terminals 5 and 2. The engine should run at 1,600 to 2,400 rpm

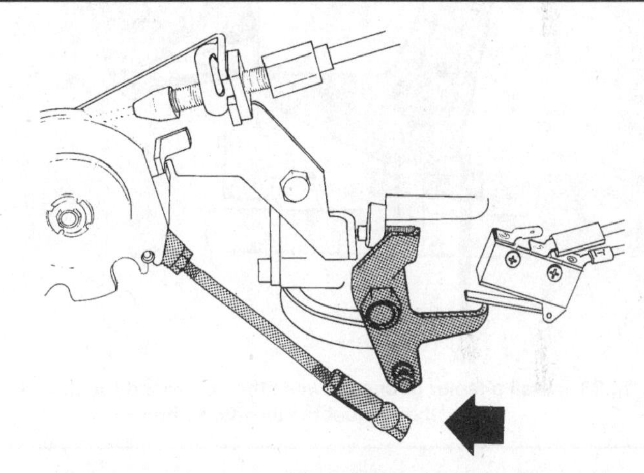

14.12a Disconnect the throttle linkage from the throttle valve (non-turbocharged engine)

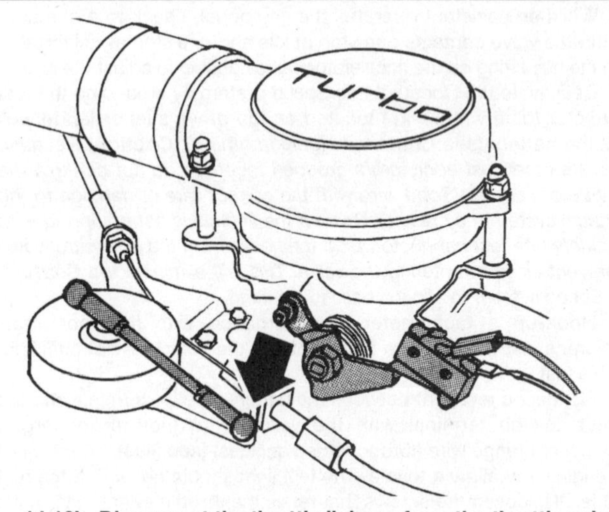

14.12b Disconnect the throttle linkage from the throttle valve (turbocharged engine)

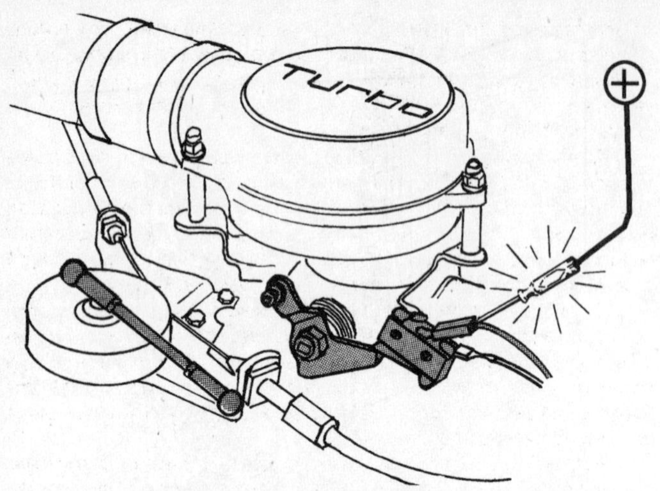

14.17 Connect a test light between the battery (+) and the orange wire on turbocharged engines (on non-turbo models it's the yellow wire)

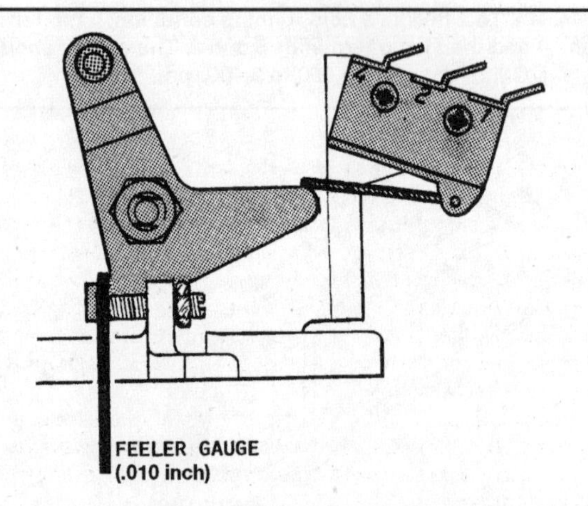

14.23 Install a feeler gauge between the screw and the throttle valve, then adjust the throttle switch

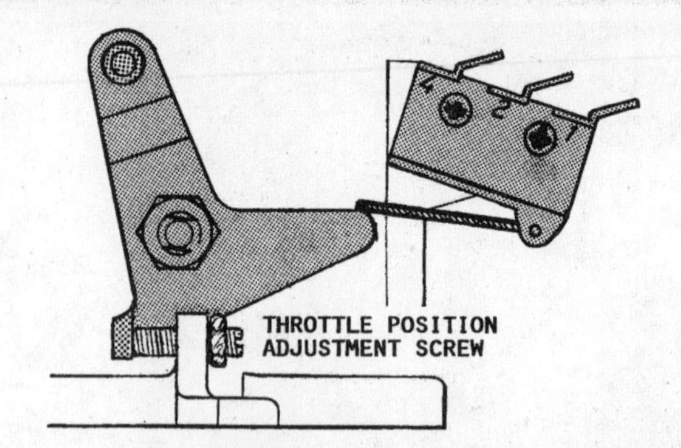

14.19 To adjust the baseline idle speed, turn the throttle position adjustment screw

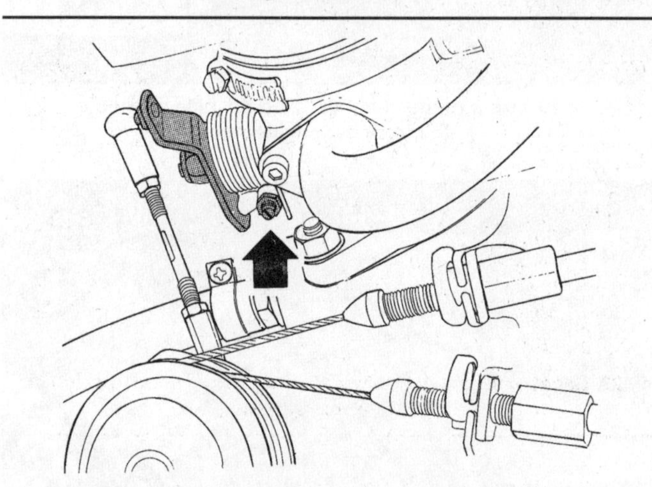

14.27 Location of the adjustment screw on LH-Jetronic systems

13 Check the accelerator cable to make sure it moves without binding. Also, check the throttle shaft and valve assembly to make sure it operates smoothly and it does not stick in any position.

14 While an assistant operates the gas pedal, check to make sure the throttle valve contacts one stop at idle and one stop at full throttle. Turn the hex fitting on the accelerator cable housing to adjust the cable.

15 Deactivate the Constant Idle Speed system by grounding the test connector (blue/white wire) located on the driver's inner fenderwell near the battery (see Chapter 1, *Ignition timing*). **Caution:** *Because there are other test connectors grouped together, do not confuse the red wire or Lambda-Sond wire with the correct wire or damage to the electrical system may result.* **Note:** *If the vehicle is not equipped with the blue/white test connector, plug the hoses from the air control valve to prevent air from entering the valve. This will eliminate the Constant Idle Speed system to allow a base idle setting.*

16 Hook-up a tachometer in accordance with the tool manufacturer's instructions. The tachometer in the vehicle is not sufficient for this test.

17 Connect a test light between the battery positive terminal and the throttle switch terminal with the yellow wire (non-turbocharged engines) or orange wire (turbocharged engines) **(see illustration)**. Start the engine and allow it to idle. The test light should be OFF. If the test light is ON, loosen the screws that retain the throttle switch and move the switch down until the test light goes OFF. Tighten the screws. This is a temporary adjustment.

18 Check the baseline idle speed. It should be between 850 and 900 rpm. **Note:** *The baseline idle speed is measured with the test connector grounded.*

19 If necessary, loosen the locknut and turn the throttle stop screw until the correct idle speed is attained **(see illustration)**. Tighten the locknut. **Note:** *The test light must not come on during this period of testing. If necessary readjust the throttle switch.*

20 Turn the engine key OFF and deactivate the Constant Idle Speed system by disconnecting the test terminal electrical connector.

21 Start the engine and observe that the engine idles slightly higher than the baseline idle speed - approximately 880 to 920 rpm.

22 Turn the ignition key OFF and reconnect the throttle linkage. Adjust the linkage to keep the idle speed within the allowable range (880 to 920 rpm).

23 Insert a feeler gauge (0.010 inch) between the throttle valve lever and the throttle stop screw **(see illustration)**.

24 With the ignition switch ON (engine not running), loosen the throttle switch screws and move the switch up until the test light comes ON and then down until the test light just goes OFF. Tighten the retaining screws and remove the feeler gauge.

25 Start the engine and confirm that the engine idles within the specified range. Disconnect the test light and the tachometer.

26 If applicable, adjust the automatic transmission kickdown cable (see Chapter 7B).

LH-Jetronic (1982 only)

Refer to illustrations 14.27, 14.29 and 14.30

27 Loosen the stop nut on the adjuster screw and back the adjustment screw out two turns **(see illustration)**.

28 Screw in the adjustment screw until it barely contacts the throttle lever, then turn it in 1/4-turn.

29 Disconnect the electrical connector from the vacuum switch **(see illustration)**.

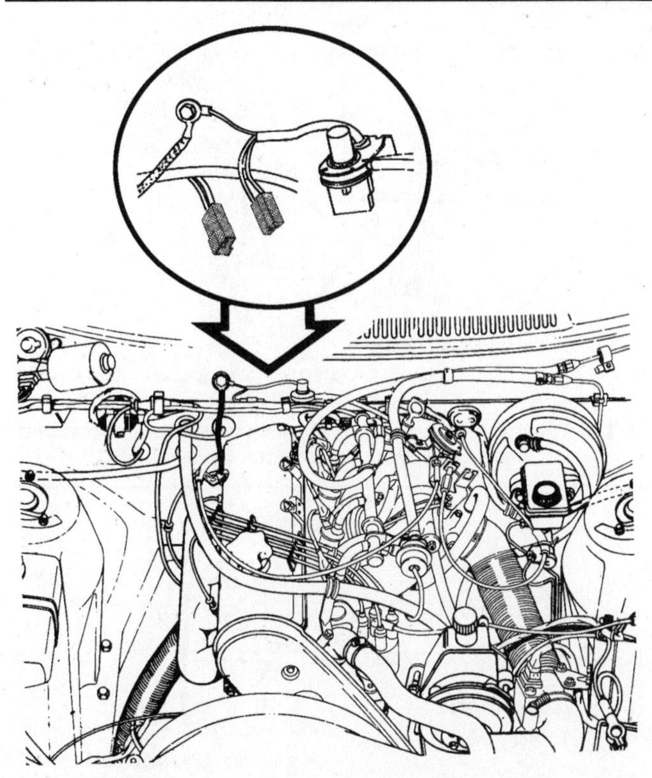

14.29 Location of the vacuum switch electrical connector on the LH-Jetronic system

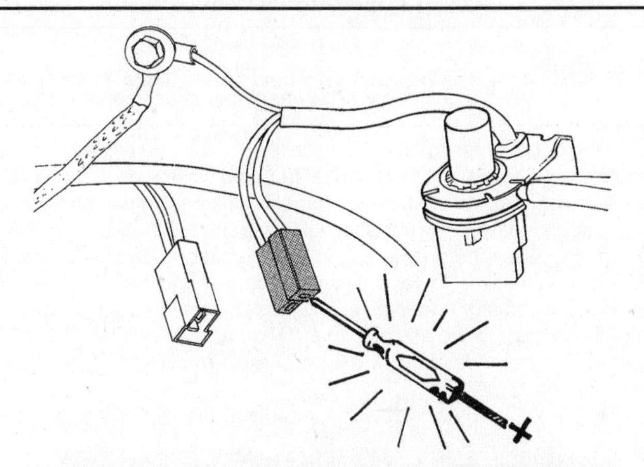

14.30 With the test light connected to the battery positive terminal, confirm that the light is ON at idle and OFF when throttled

30 Connect a test light from the battery positive (+) terminal to the orange wire on the electrical connector **(see illustration)**. With the engine idling, the test light should illuminate.
31 Open the throttle slightly - the test light should go out.
32 If the test light does not go out, replace the vacuum switch with a new part.

15 Turbocharger - general information

The turbocharger increases the efficiency of the engine by raising the pressure in the intake manifold above atmospheric pressure. Instead of the air/fuel mixture being simply sucked into the cylinders it is actively forced in.

Energy for the operation of the turbocharger comes from the exhaust gas. The gas flows through a specially shaped housing (the turbine housing) and in so doing spins the turbine wheel. The turbine wheel is attached to a shaft, at the other end of which is another vaned wheel known as the compressor wheel. The compressor wheel spins in its own housing and compresses the inducted air on the way to the intake manifold.

After leaving the turbocharger, the compressed air passes through an intercooler (on models so equipped), which is an air-to-air heat exchanger mounted in front of the radiator. Here the air gives up heat which it acquired when being compressed. This temperature reduction condenses the charge to an even further degree, which improves engine efficiency and reduces the risk of detonation.

Boost pressure (the pressure in the intake manifold) is limited by a wastegate, which diverts the exhaust gas away from the turbine wheel in response to a pressure-sensitive actuator. As a further precaution, a pressure-sensitive switch cuts out the fuel pump if boost pressure becomes excessive. Boost pressure is displayed to the driver by a gauge on the instrument panel.

The turbine shaft is pressure-lubricated by means of a feed pipe from the engine's main oil gallery. The shaft "floats" on a cushion of oil. A drain pipe returns the oil to the oil pan.

16 Turbocharger - check

General checks

1 While it is a relatively simple device, the turbocharger is also a precision component which can be severely damaged by an interrupted oil or coolant supply or loose or damaged ducts.
2 Due to the special techniques and equipment required, checking and diagnosis of suspected problems dealing with the turbocharger should be left to a dealer service department. The home mechanic can, however, check the connections and linkages for security, damage and other obvious problems. Also, the home mechanic can check components that govern the turbocharger such as the wastegate solenoid, bypass valve and wastegate actuator. Refer to the checks later in this section.
3 Because each turbocharger has its own distinctive sound, a change in the noise level can be a sign of potential problems.
4 A high-pitched or whistling sound is a symptom of an inlet air or exhaust gas leak.
5 If an unusual sound comes from the vicinity of the turbine, the turbocharger can be removed and the turbine wheel inspected. **Caution:** *All checks must be made with the engine off and cool to the touch and the turbocharger stopped, or personal injury could result. Operating the engine without all the turbocharger ducts and filters installed is also dangerous and can result in damage to the turbine wheel blades.*
6 With the engine turned off, reach inside the housing and turn the turbine wheel to make sure it spins freely. If it doesn't, it's possible the cooling oil has sludged or coked from overheating. Push in on the turbine wheel and check for binding. The turbine should rotate freely with no binding or rubbing on the housing. If it does, the turbine bearing is worn out.
7 Check the exhaust manifold for cracks and loose connections.
8 Because the turbine wheel rotates at speeds of up to 140,000 rpm, severe damage can result from the interruption of coolant or contamination of the oil supply to the turbine bearings. Check for leaks in the coolant and oil inlet lines and obstructions in the oil drain-back line, as this can cause severe oil loss through the turbocharger seals. Burned oil on the turbine housing is a sign of this. **Caution:** *Whenever a major engine bearing such as a main, connecting rod or camshaft bearing is replaced, the turbocharger should be flushed with clean oil.*

Component checks

Turbo boost pressure

Check

Refer to illustrations 16.9, 16.12a and 16.12b

9 Using a T-fitting, connect a pressure gauge (range approximately 0 to 15 psi) into the boost pressure hose near the pressure sensor

4A

16.9 Install the gauge into a T-fitting in the pressure line between the vacuum port on the intake manifold and the wastegate actuator - there should be enough hose to extend the gauge into the driver's compartment

16.12b Wastegate actuator rod and tamper seal

(wastegate actuator). Position the gauge so it can be read from inside the vehicle **(see illustration)**. If the hose on the gauge isn't long enough, have an assistant read the gauge for you. **Note:** *On engines equipped with an intercooler, disconnect the electrical connector from the charge air solenoid valve* **(see illustration 16.32)**.

Manual transmission

10 With the engine warmed up, read the gauge. Accelerate at full throttle from 1,500 rpm in 3rd gear, (apply the brakes for a few seconds if necessary to maintain the correct rpm range) to 3,000 rpm (intercooled models) or 4,000 rpm (non intercooled models). Maintain the specified rpm range while reading charge pressure. Note the highest pressure attained.

Without intercooler	6.0 to 6.8 psi
With intercooler	7.1 to 8.2 psi

Automatic transmission

11 With the engine warmed up, read the gauge. Accelerate at full throttle from 1,500 rpm in 2nd gear, (apply the brakes for a few seconds if necessary to maintain the correct rpm range) to 3,000 rpm (intercooled models) or 3,500 rpm (non intercooled models). Maintain the specified rpm range while reading charge pressure. Note the highest pressure attained.

Without intercooler	6.0 to 6.8 psi
With intercooler	7.1 to 8.2 psi

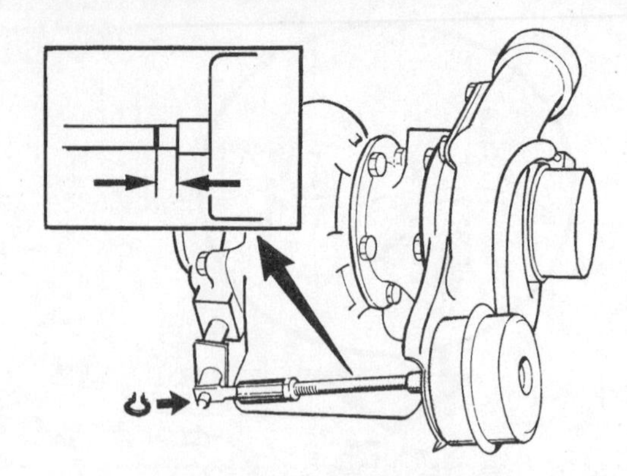

16.12a The wastegate actuator retraction distance is measured directly behind the actuator (arrows)

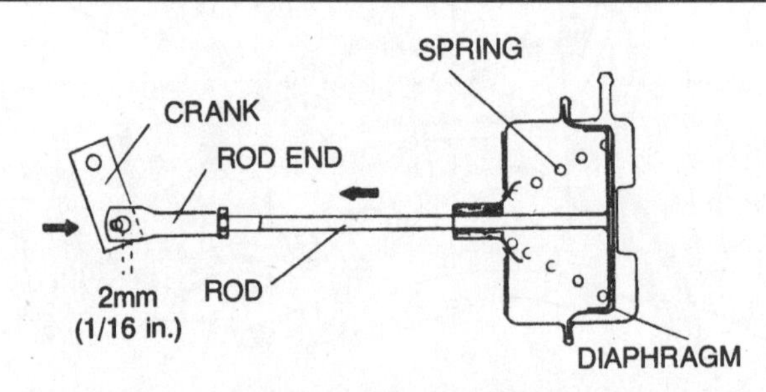

16.16 Once the actuator rod is released, it should retract only 1/16-inch or 2.0 mm

Adjustment

Caution: *Unskilled or improper adjustment can cause serious engine damage.*

12 If the boost pressure is not as specified, allow the engine to cool, then remove the wire and tamper seal from the wastegate actuator rod **(see illustrations)**. Unclip the rod from the wastegate lever and screw the threaded sleeve in or out after releasing the locknut. Lengthening the rod decreases the pressure, and vice-versa; one complete turn of the sleeve will raise or lower pressure by approximately 0.4 psi. **Note:** *The pressure actuator installed in early models cannot be adjusted. The sleeve is crimped to the actuator rod. If the pressure actuator must be replaced, it is possible to break the lead seal, turn the actuator rod and reseal the crimp (joint).*

13 Secure the actuator rod with a new retaining clip and repeat the test.

14 When adjustment is correct, install a new wire and tamper seal if required. Tighten the actuator rod locknut.

Turbo wastegate actuator - check and replacement

Refer to illustrations 16.16, 16.17 and 16.19

15 If the boost pressure is incorrect and cannot be adjusted, the wastegate actuator may be defective. Check as follows.

16 Break the seal and unclip the actuator rod from the wastegate lever. The rod should retract into the actuator a little way. Mark the position of the rod where it enters the actuator, then reconnect the rod to the wastegate and measure the distance moved by the mark **(see illustration)**. It must be 1/16-inch (2.0 mm); if not, replace the actuator.

17 To remove the actuator, unclip the link rod, remove the connecting hose and unscrew the two retaining nuts **(see illustration)**.

18 Begin installation by securing the actuator to its bracket using two new nuts, then perform a preliminary adjustment as follows.

19 Connect a pump and pressure gauge to the actuator **(see illustration)**. Apply 7 psi pressure.

20 With the pressure applied, adjust the link rod so it fits onto the wastegate lever (in the closed position) without being tight or slack. Tighten the adjuster locknut.

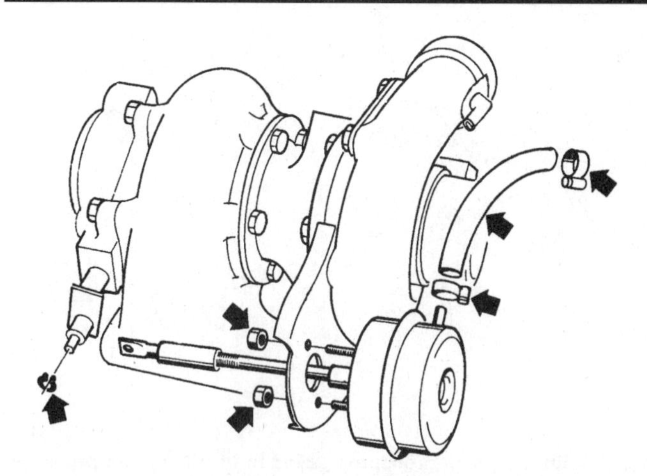

16.17 Remove the nuts from the actuator and separate the unit from the bracket

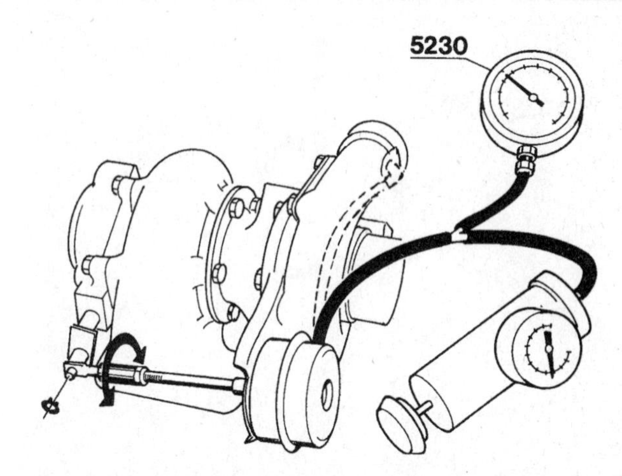

16.19 Apply 7 psi to the actuator and adjust the rod to fit without any slack while in the closed position

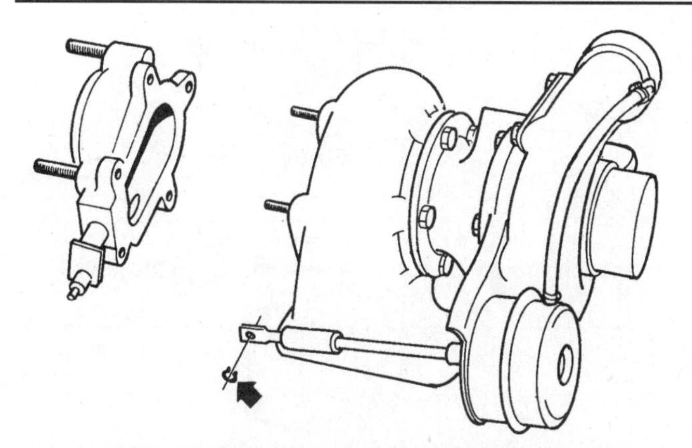

16.26 Remove the snap-ring (arrow) and separate the wastegate from the turbocharger

21 Secure the link rod to the wastegate lever using a new clip.
22 Remove the pump and gauge. Connect the actuator hose to the compressor housing.
23 Check the boost pressure.

Wastegate - check and replacement

Refer to illustration 16.26

24 Check the wastegate for binding, corrosion or exhaust leaks from the flange areas. In the event of problems, replace it with a new part.
25 Remove the nuts that retain the exhaust flange to the wastegate.
26 Remove the snap-ring from the wastegate actuator rod **(see illustration)**.
27 Remove the nuts from the wastegate and separate the wastegate assembly from the turbocharger.
28 Installation is the reverse of removal. Be sure to install a new gasket between the turbocharger and the wastegate to insure a tight seal.
29 Adjust the actuator rod (see Steps 19 through 23).

Charge pressure control system

Refer to illustrations 16.30, 16.32, 16.34, 16.36 and 16.38

Note: *On intercooled models, the charge air pressure control systems allows for a higher level of boost (turbocharged intake pressure) at engine speeds above 3,700 rpm. The system includes the solenoid valve, engine rpm delay and the charge air pressure switch.*

30 The charge pressure control system consists of the charge air pressure switch, solenoid valve and the relays that govern the engine cooling fan, engine rpm and air conditioning disengagement **(see illustration)**.

4A

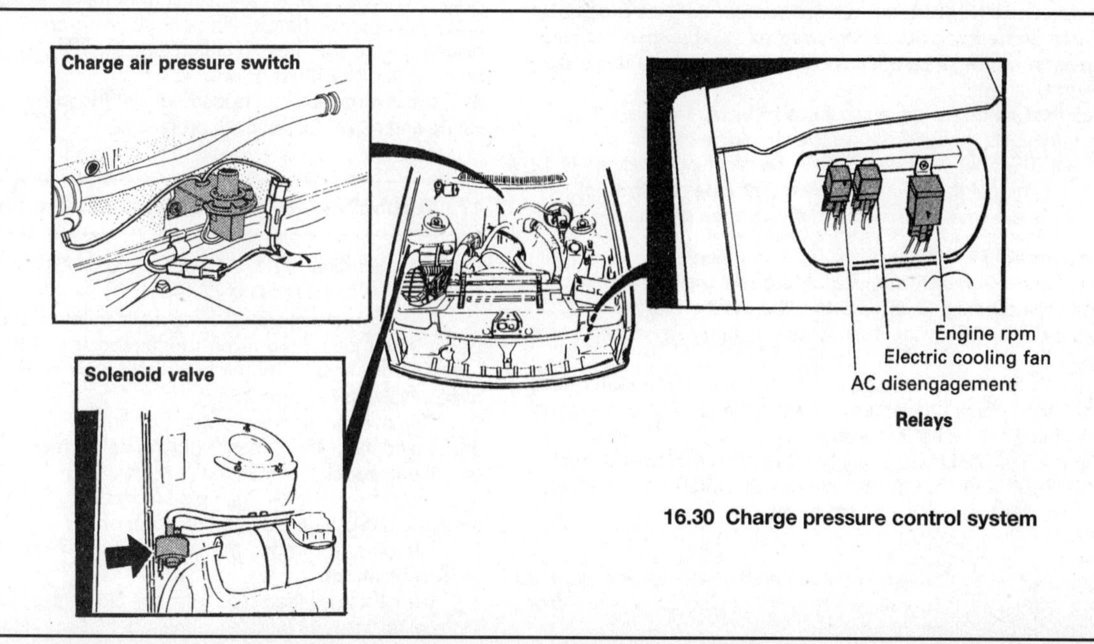

16.30 Charge pressure control system

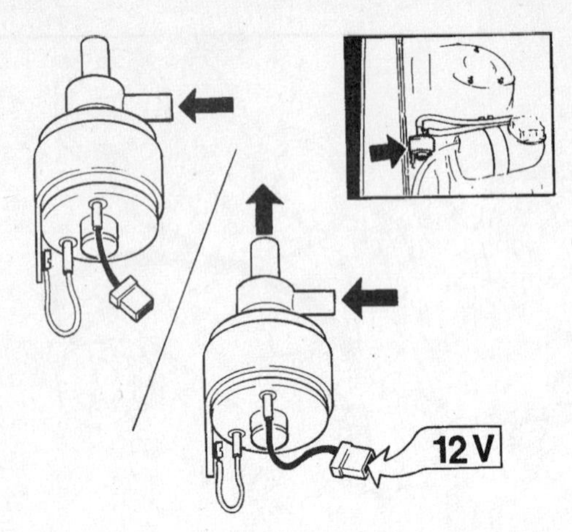

16.32 First check the solenoid valve without battery voltage applied and then with voltage applied (the solenoid valve should open and allow air to pass through when voltage is applied)

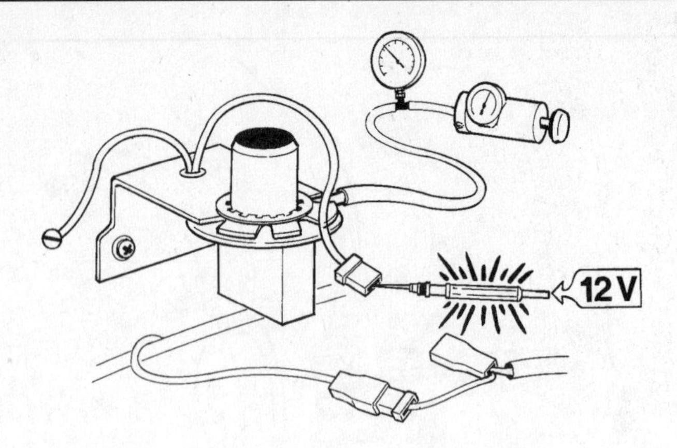

16.34 With 2.9 psi of pressure applied to the charge air pressure switch, make sure the test lamp comes on as the contacts close

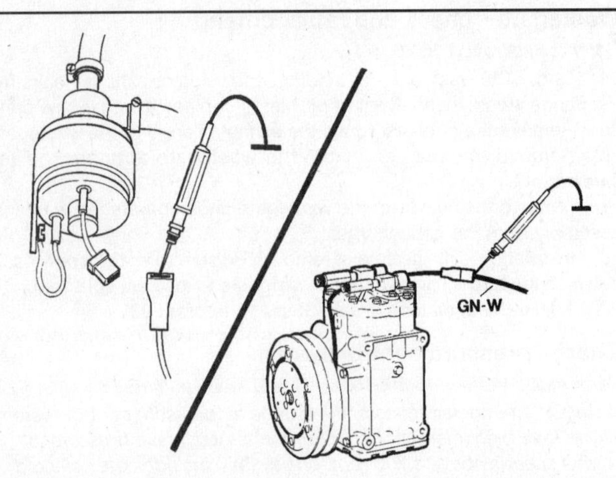

16.36 First connect two test lights; one to the solenoid valve and the other to the air conditioning compressor

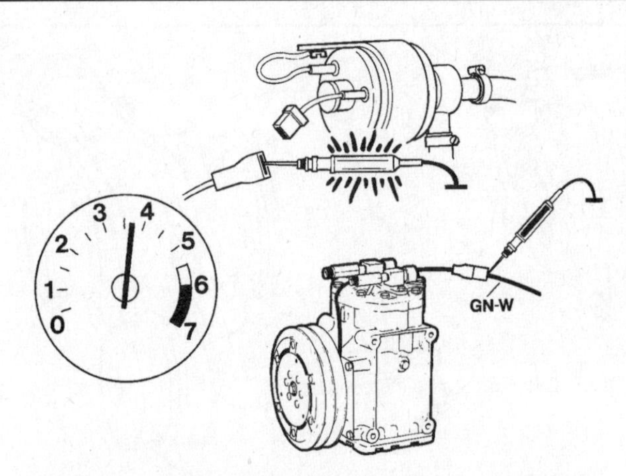

16.38 Next, confirm that when the engine runs at high rpm (3,700 rpm) the solenoid valve opens (light ON) and the air conditioning compressor shuts off (test light off)

Solenoid valve

Note: *The solenoid valve (overpressure switch) acts as a back-up system to the wastegate. If boost pressure exceeds a specified level, then the solenoid valve stops the fuel pump circuit and the fuel supply to the injectors.*

31 Check the function of the solenoid valve by blowing air through it with no battery voltage applied. The valve should be closed.

32 Connect a jumper wire from the B+ terminal (approximately 12 volts) and confirm that the solenoid valve opens **(see illustration)**.

33 If the valve does not respond correctly, replace it with a new part.

Charge air pressure switch

Note: *The charge air pressure switch signals the oxygen sensor feedback computer (see Chapter 6) to enrichen the fuel mixture. On intercooled models, the switch also signals the rpm relay during boost conditions.*

34 Connect a hand-held vacuum/pressure pump to the switch **(see illustration)** and connect a test light between the battery (B+) terminal and the electrical connector for the switch.

35 Using the hand-held pump, apply 2.9 psi of pressure and confirm that the test light comes ON. If the charge air switch does not close (light ON) then replace the switch with a new part.

Relays

36 Connect a test light between the solenoid connector and ground and then another test light between the air conditioning compressor wire and ground **(see illustration)**.

37 Start the engine and turn the air conditioning to MAX COOL.

38 Raise the engine speed to 3,700 rpm and check the engine rpm relay is energized when the test light at the solenoid valve comes ON. Also, at the same time the air conditioning compressor relay should disengage the compressor clutch (test light OFF) at the air conditioning compressor **(see illustration)**.

39 If this simulated test does not get the correct results, check the wiring and relays for possible problems.

17 Turbocharger - removal and installation

Refer to illustrations 17.1, 17.3, 17.4, 17.5, 17.6, 17.7, 17.13a, 17.13b, 17.14a, 17.14b, 17.14c and 17.14d

1 The turbocharger and exhaust manifold are removed together. Begin by removing the turbo-to-intercooler and the airflow meter-to-turbocharger hoses. The airflow meter hose is also connected to the bypass valve **(see illustration)**.

2 Remove the air cleaner warm air duct.

3 Disconnect the exhaust downpipe from the turbocharger outlet **(see illustration)**. Remove the heat shield.

4 Disconnect the oil drain pipe from the turbocharger. Be prepared for some oil spillage **(see illustration)**.

5 Unbolt and remove the stiffener plate from below the manifold **(see illustration)**.

6 Unbolt the oil feed pipe from the block **(see illustration)**. Discard the sealing washers (new ones must be used during installation).

17.1 Turbocharger hose connections

17.3 Turbocharger-to-downpipe bolts (third bolt hidden from view)

17.4 Turbocharger oil drain pipe connection

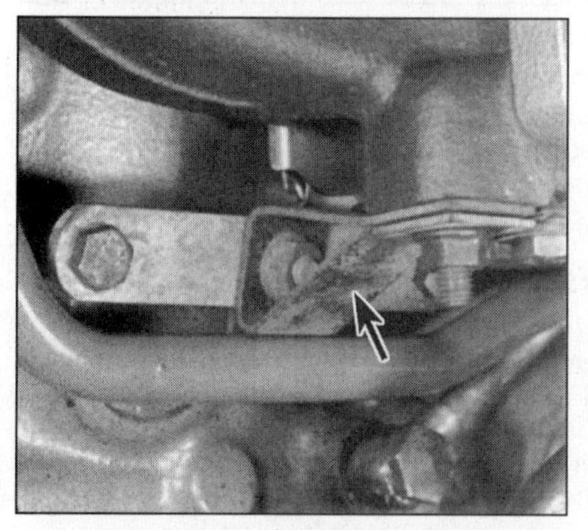

17.5 Exhaust manifold stiffener plate (arrow)

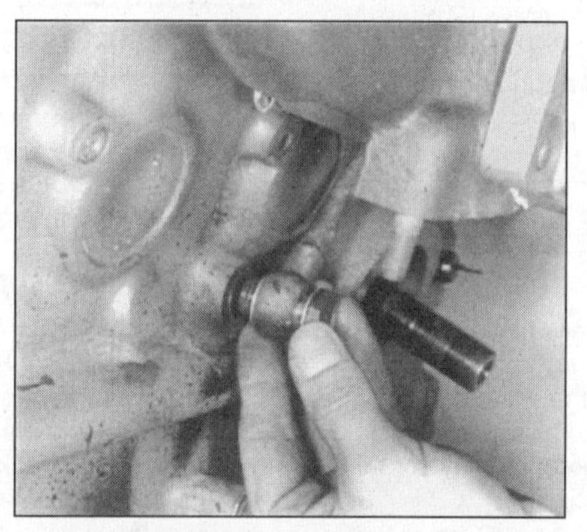

17.6 Turbo oil feed connection - note sealing washers

17.7 Remove the exhaust manifold nuts

7 Remove the eight nuts which secure the exhaust manifold to the cylinder head. Note that one of the nuts secures a lifting eye **(see illustration)**.

8 Lift off the exhaust manifold and turbocharger. Remove the exhaust port gaskets.

9 Remove the oil feed pipe and remove the gasket.

10 Remove the locking plates from the four bolts which secure the turbocharger to the manifold. Bend up the plate tabs with a chisel and pry or drive them off the bolts. New plates will be needed for reassembly.

11 Clamp the manifold in a vise and remove the four bolts. Lift the turbocharger off the manifold. Remove the other halves of the locking plates.

12 Install the turbocharger to the manifold and secure it with the fou bolts, applying anti-seize compound to their threads. Remember tc install the new locking plate sections.

13 Tighten the bolts to the torque listed in this Chapter's Specifi cations, following the correct sequence **(see illustrations)**. Make up template for the final angular tightening.

14 Install the outer halves of the locking plates. Drive them over th bolt heads with a hammer and a tube, crimp the tabs with pliers and finally knock the tabs down with a hammer **(see illustrations)**.

15 The remainder of installation is a reversal of the removal procedure. Use new gaskets, oil pipe sealing washers, etc.

16 Before starting the engine, disable the ignition system by disconnecting the primary wires from the ignition coil and crank the engine

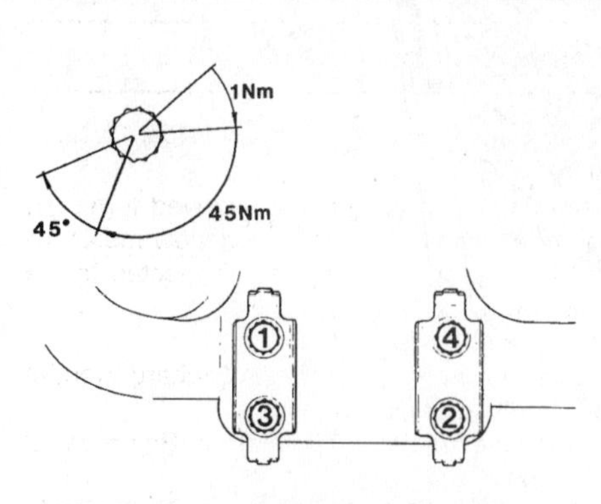

17.13a Turbocharger bolt tightening sequence and stages

17.13b A simple angle-tightening gauge can be fabricated using a piece of cardboard and a protractor

17.14a Install the lockplate outer half and . . .

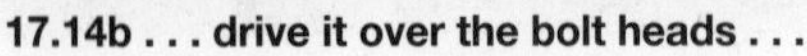

17.14b . . . drive it over the bolt heads . . .

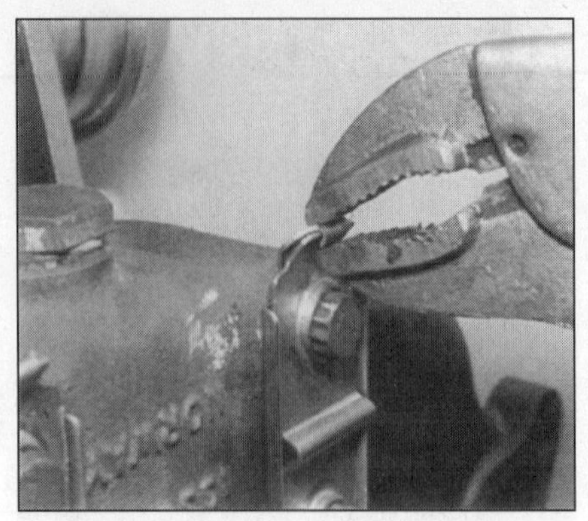

17.14c . . . crimp over the tabs . . .

17.14d . . . and finally knock the tabs down

with the starter for six 10-second bursts. This will prime the turbocharger with oil.
17 Reconnect the ignition coil, run the engine and verify that there are no oil leaks.

18 Intercooler - removal and installation

Refer to illustration 18.2

1 Remove the radiator top mounts and carefully position the radiator out of the way.
2 Loosen the hose clamps, disconnect the hoses from the intercooler and lift it out **(see illustration)**.
3 If turbocharger failure has occurred, the intercooler may contain a substantial quantity of oil. A drain plug is provided.
4 Install by reversing the removal operations.

19 Exhaust system servicing - general information

Warning: *Inspect or repair exhaust system components only after enough time has elapsed after driving the vehicle to allow the system components to cool completely. Also, when working under the vehicle make sure it is securely supported by jackstands.*

Muffler and pipes

Refer to illustrations 19.1a, 19.1b, 19.1c, 19.1d, 19.1e

1 The exhaust system consists of the exhaust manifold, catalytic converter, mufflers and all connecting pipes, brackets, hangers **(see illustrations)** and clamps. The exhaust system is attached to the body with mounting brackets and rubber hangers. If any of the parts are improperly installed, excessive noise and vibration may be transmitted to the body.

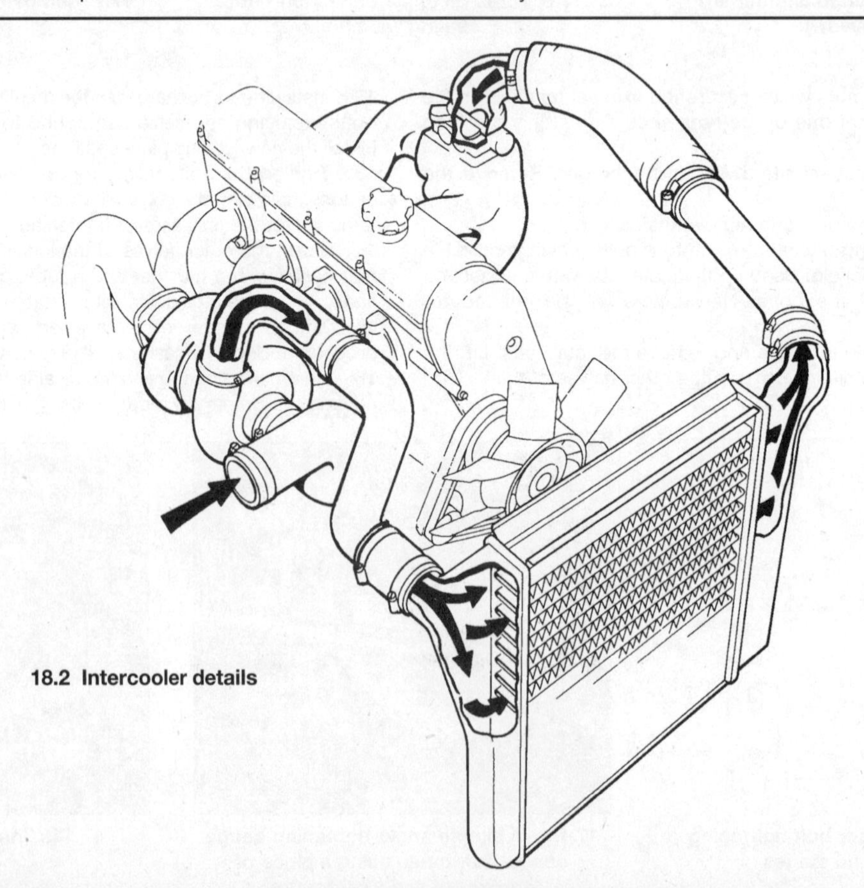

18.2 Intercooler details

19.1a Layout of the muffler and hangers on a 240 sedan

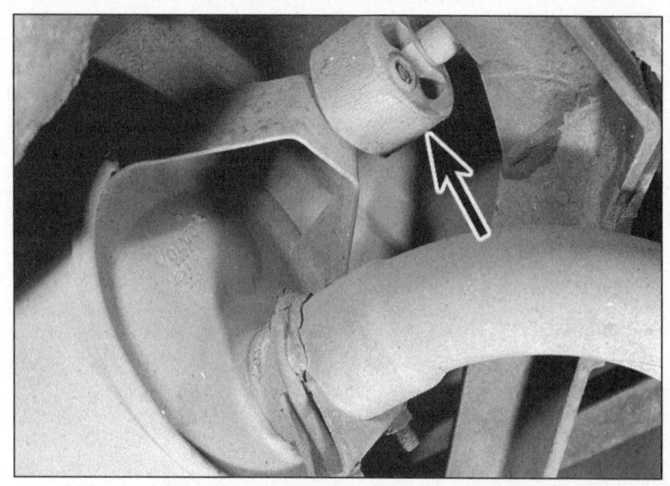

19.1b The hanger is made out of rubber and can usually be removed with a prybar or large screwdriver

19.1c Remove the bolts to release the clamp from the muffler pipe

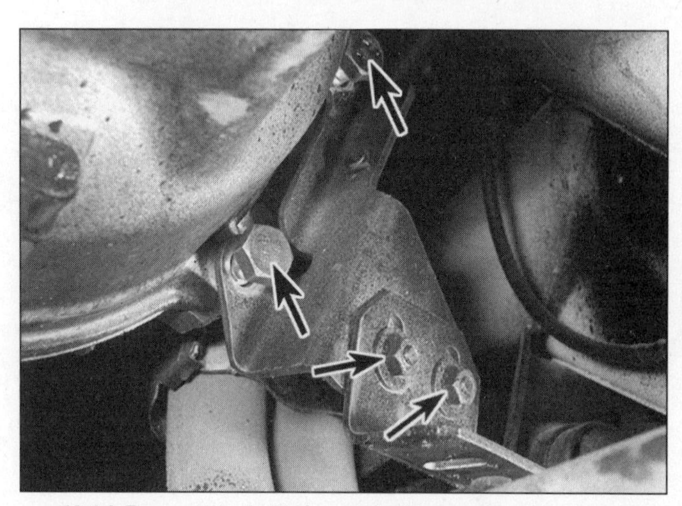

19.1d Remove the bolts (arrows) and separate the muffler hanger from the bellhousing

2 Inspect the exhaust system regularly to keep it safe and quiet. Look for any damaged or bent parts, open seams, holes, loose connections, excessive corrosion or other defects which could allow exhaust fumes to enter the vehicle. Deteriorated exhaust system components should not be repaired; they should be replaced with new parts.

19.1e Check the heat shield for cracks or damage

3 If the exhaust system components are extremely corroded or rusted together, welding equipment will probably be required to remove them. The convenient way to accomplish this is to have a muffler repair shop remove the corroded sections with a cutting torch. If, however, you want to save money by doing it yourself (and you don't have a welding outfit with a cutting torch), cut off the old components with a hacksaw. If you have compressed air, a special pneumatic cutting chisel can also be used. If you decide to tackle the job at home, be sure to wear safety goggles to protect your eyes from metal chips and use work gloves to protect your hands.

4 Here are some simple guidelines to follow when repairing the exhaust system:

- *a) Work from the back to the front of the vehicle when removing exhaust system components.*
- *b) Apply penetrating oil to the exhaust system fasteners to make them easier to remove.*
- *c) Use new gaskets, hangers and clamps when installing exhaust system components.*
- *d) Apply anti-seize compound to the threads of all exhaust system fasteners during reassembly.*
- *e) Be sure to allow sufficient clearance between newly installed parts and all points on the under body to avoid overheating the floor pan and possibly damaging the interior carpet and insulation. Pay particularly close attention to the catalytic converters and heat shields.*

Catalytic converters

5 Although the catalytic converters are emissions-related components, they are discussed here because, physically, they're integral parts of the exhaust system. Always check the converters whenever you raise the vehicle to inspect or service the exhaust system.
6 Raise the vehicle and place it securely on jackstands.
7 Visually inspect all catalytic converters on the vehicle for cracks or damage.
8 Check all converters for tightness.
9 Check the insulation covers welded onto the catalytic converters for damage or a loose fit.
10 Start the engine and run it at idle speed. Check all converter connections for exhaust gas leakage.

Chapter 4 Part B
Fuel and exhaust systems - carbureted engines

Contents

Specifications

Fuel pressure	2 to 5 psi
Idle speed	
SU-HIF6	
1976 and 1977	850 rpm
1978 through 1981	900 rpm
Solex Zenith 175 CD	
1976 and 1977	850 rpm
1978 through 1984	900 rpm
Pierburg (DVG) 175CDUS	900 rpm
Fast idle speed (all)	1250 to 1350 rpm
Damper oil type	Automatic transmission fluid

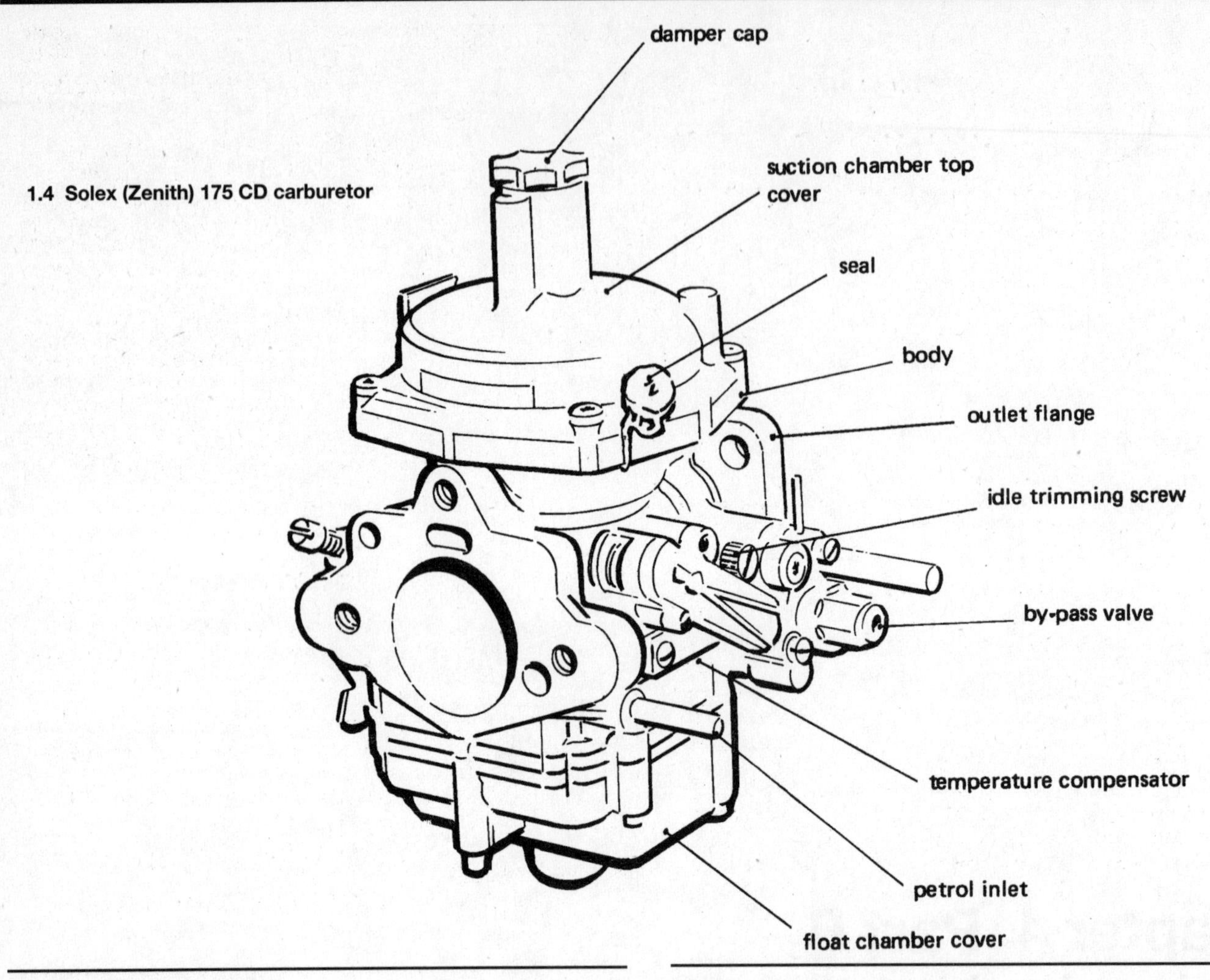

1.4 Solex (Zenith) 175 CD carburetor

1 General Information

General information

1 Canadian 240 Volvo models are equipped with three different types of carburetors. 1978 through 1984 models are equipped with the Solex (Zenith) carburetor or the Pierburg (DVG) 175 CDUS carburetor. The SU-HIF6 covers 1978 through 1981 models only. Adjustments and tune-up specifications may vary slightly depending upon the model and the year of production.

2 All carbureted engines are equipped with a mechanical fuel pump which is mounted on the lower left section of the engine block.

3 All carbureted engines use a manual choke system that must be actuated with a cable.

Solex (Zenith) 175 CD

Refer to illustration 1.4

4 Originally known as a Stromberg **(see illustration)**, this carburetor is similar in many respects to the SU-HIF6 carburetor, being a constant depression, side-draft type carburetor.

5 The main difference is in the suction chamber, where the piston is attached to a rubber diaphragm, and not sealed directly against the sides of the chamber.

6 As the depression (vacuum) in the upper part of the chamber changes with engine speed and load, the piston moves up and down in response, moving the fuel metering needle which is fixed to the piston, in and out of the fuel jet.

7 The carburetor is equipped with a temperature compensator, which allows more air into the idle circuit in relation to temperature. As the compensator warms up, more air is admitted. This reduces variations in idle speed relative to temperature.

8 There is also a warm start valve, which offsets the effects of fuel vapor from the float bowl during high under-hood temperatures, which can lead to difficulty in starting.

Pierburg (DVG) 175 CDUS

9 The Pierburg (DVG) carburetor is similar in principle and construction to the Solex (Zenith) 175 CD carburetor.

10 The main difference is in the temperature compensation of the main jet, which is achieved by using a floating jet which is spring-loaded against a bi-metallic washer. As temperature increases, the bi-metallic washer expands and moves the jet up against the spring pressure, thus leaning the mixture.

11 A manual barrel (cold start device) is equipped, and from 1980 an additional vacuum valve is installed, which gives increased control of the mixture when the barrel is in operation.

12 On 1981 and later B23 models, a modified damper reservoir and piston are installed, which automatically keeps the damper fluid at the correct level, provided the reservoir is kept topped up.

SU-HIF6

13 The variable barrel SU carburetor is a relatively simple instrument. It differs from most other carburetors in that instead of having a number of various fixed jets for different conditions, only one variable jet is used to deal with all possible conditions.

14 Air passing rapidly through the carburetor draws fuel from the jet, forming the fuel/air mixture. The amount of fuel drawn from the jet depends on the position of the tapered carburetor needle, which moves up and down the jet orifice according to the engine load and throttle opening, effectively altering the size of jet so that exactly the right amount of fuel is metered for all driving conditions.

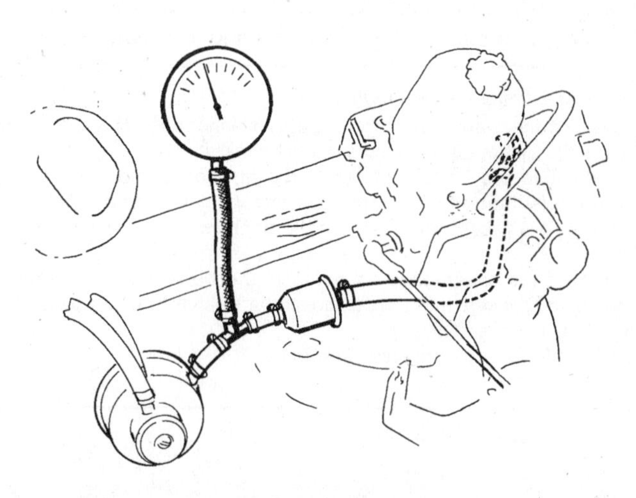

2.1 Install a fuel pressure gauge between the fuel pump and the carburetor using a T-fitting

15 The position of the tapered needle in the jet is determined by engine vacuum. The shank of the needle is held at its top end in a piston which slides up and down the dashpot in response to the degree of manifold vacuum.

16 With the throttle fully open, the full effect of intake manifold vacuum is felt by the piston which has an air bleed into the barrel tube on the outside of the throttle. This causes the piston to rise fully, bringing the needle with it. With the gas pedal partially closed, only slight intake manifold vacuum is felt by the piston (although, of course, on the engine side of the throttle the vacuum is greater), and the piston only rises a little, blocking most of the jet orifice with the metering needle.

17 To prevent the piston fluttering and giving a richer mixture when the gas pedal is suddenly depressed, an oil damper and light spring are fitted inside the dashpot.

18 The only portion of the piston assembly to come into contact with the piston chamber or dashpot is the actual central piston rod. All the other parts of the piston assembly, including the lower barrel portion, have sufficient clearance to prevent any direct metal to metal contact which is essential if the carburetor is to function correctly.

19 From 1978, a bypass idling system is used.

2 Fuel pump/fuel pressure check

Refer to illustration 2.1

Warning: *Gasoline is extremely flammable, so take extra precautions when you work on any part of the fuel system. Don't smoke or allow open flames or bare light bulbs near the work area, and don't work in a garage where a natural gas-type appliance (such as a water heater or a clothes dryer) with a pilot light is present. Since gasoline is carcinogenic, wear latex gloves when there's a possibility of being exposed to fuel, and, if you spill any fuel on your skin, rinse it off immediately with soap and water. Mop up any spills immediately and do not store fuel-soaked rags where they could ignite. The fuel system is under constant pressure, so, if any fuel lines are to be disconnected, be sure to use shop rags to catch any residual fuel. When you perform any kind of work on the fuel system, wear safety glasses and have a Class B type fire extinguisher on hand.*

Note: *It is a good idea to first check the fuel pump for any obvious damage or fuel leakage around the gaskets or diaphragm. It is possible that the fuel pump diaphragm has ruptured internally and is leaking fuel into the crankcase. In the event of excess fuel consumption or noticeable fuel vapor smell, check the condition of the engine oil for any signs of fuel mixing with the engine oil.*

1 Disconnect the fuel line from the carburetor and install a T-fitting. Connect a fuel pressure gauge to the T-fitting with a section of fuel hose that is no longer than six inches **(see illustration)**.

2 Disconnect the gauge from the end of the fuel hose and direct the end of the hose into a metal container. Operate the starter until fuel spurts out of the hose, to bleed the hose (this eliminates any air in the hose, which could affect the pressure reading). Reattach the gauge to the fuel hose.

3 Start the engine and allow it to idle. The pressure on the gauge should be as listed in this Chapter's Specifications, remain constant and return to zero slowly when the engine is shut off.

4 An instant pressure drop indicates a faulty outlet valve, This will require replacement of the fuel pump.

5 If the pressure is too high, check the air vent to see if it is plugged before replacing the pump.

6 If the pressure is too low, be sure the fuel inlet line is not plugged before replacing the pump.

3.3 The fuel pump is mounted on the cylinder block

3 Fuel pump - removal and installation

Refer to illustration 3.3

Warning: *Gasoline is extremely flammable, so take extra precautions when you work on any part of the fuel system. Don't smoke or allow open flames or bare light bulbs near the work area, and don't work in a garage where a natural gas-type appliance (such as a water heater or a clothes dryer) with a pilot light is present. Since gasoline is carcinogenic, wear latex gloves when there's a possibility of being exposed to fuel, and, if you spill any fuel on your skin, rinse it off immediately with soap and water. Mop up any spills immediately and do not store fuel-soaked rags where they could ignite. The fuel system is under constant pressure, so, if any fuel lines are to be disconnected, the fuel pressure in the system must be relieved first* (see Chapter 4A for more information). *When you perform any kind of work on the fuel system, wear safety glasses and have a Class B type fire extinguisher on hand.*

1 The fuel pump is mounted near the front of the engine near the timing cover on the left side (driver's) of the engine block.

2 Place rags under the fuel pump to catch any gasoline which is spilled during removal.

3 Loosen the fuel line clamps and detach the lines from the pump **(see illustration)**.

4 Unbolt and remove the fuel pump. Remove all traces of old gasket material.

5 Before installation, coat both sides of the gasket surface with RTV sealant, position the gasket and fuel pump against the block and install the bolts, tightening them securely.

6 Attach the lines to the pump and tighten the fittings securely.

7 Run the engine and check for leaks.

4.2a Remove the air filter housing bolts

4 Air cleaner assembly - removal and installation

Refer to illustrations 4.2a through 4.2f

1 Disconnect any hoses from the air cleaner assembly and mark them with paint or tape to insure correct installation.

2 Remove the bolts that retain the air cleaner housing to the carburetor(s) **(see illustrations)**. On rectangular shaped air cleaner housings, it is necessary to remove the air filters first to gain access to the bolts.

3 Place the assemblies on a bench and be sure not to drop or damage the gaskets that are positioned between the carburetor(s) and the air cleaner assembly.

4 Installation is the reverse of removal.

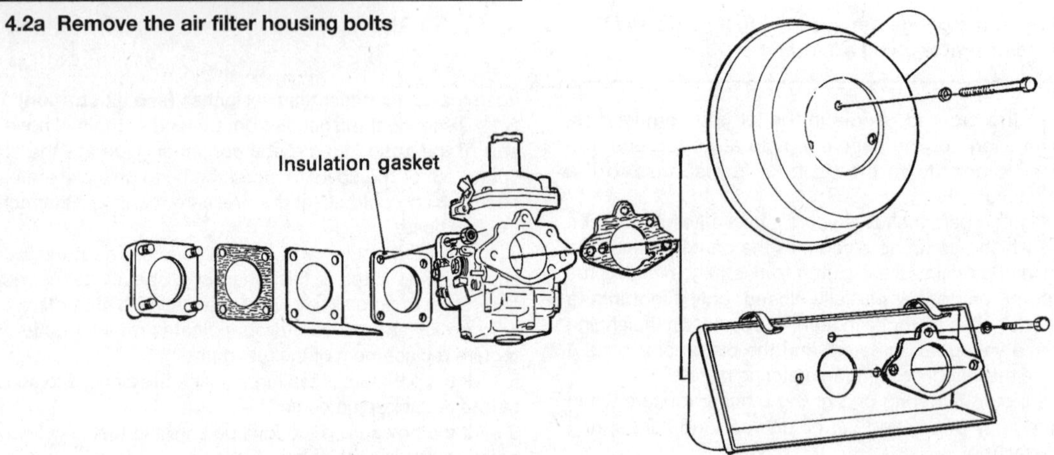

4.2b Installation details of the Solex (Zenith) 175 CD carburetor and air cleaner assembly on the B20A engine

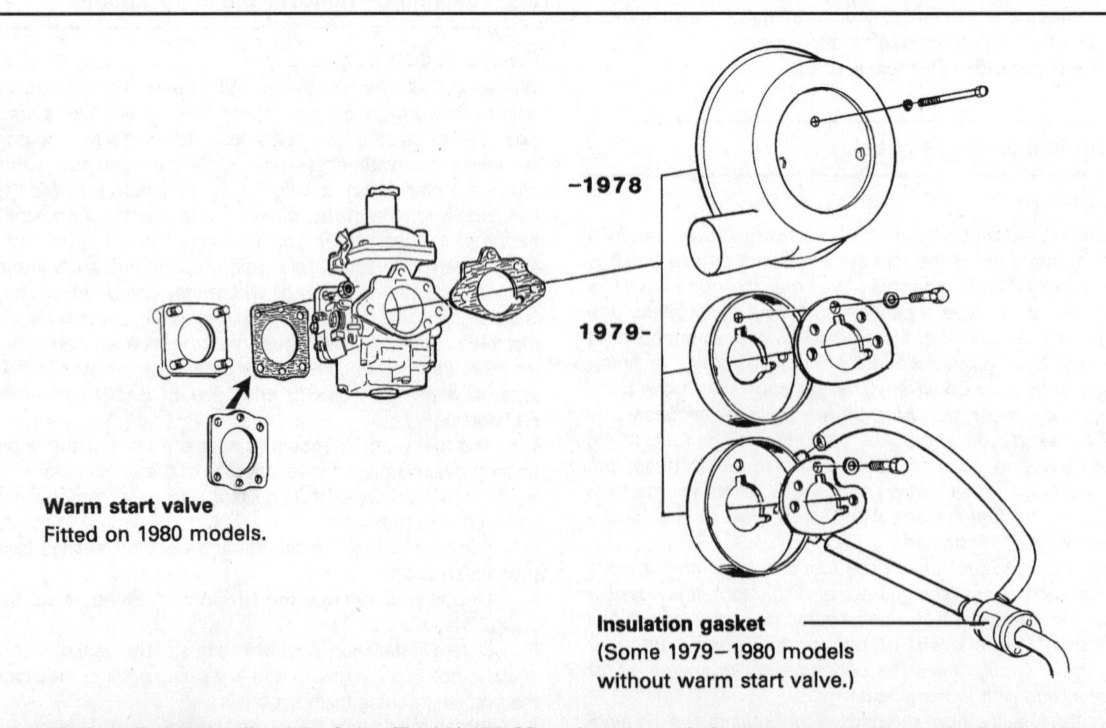

4.2c Installation details of the Solex (Zenith) 175 CD carburetor and air cleaner assembly on the B21A engine

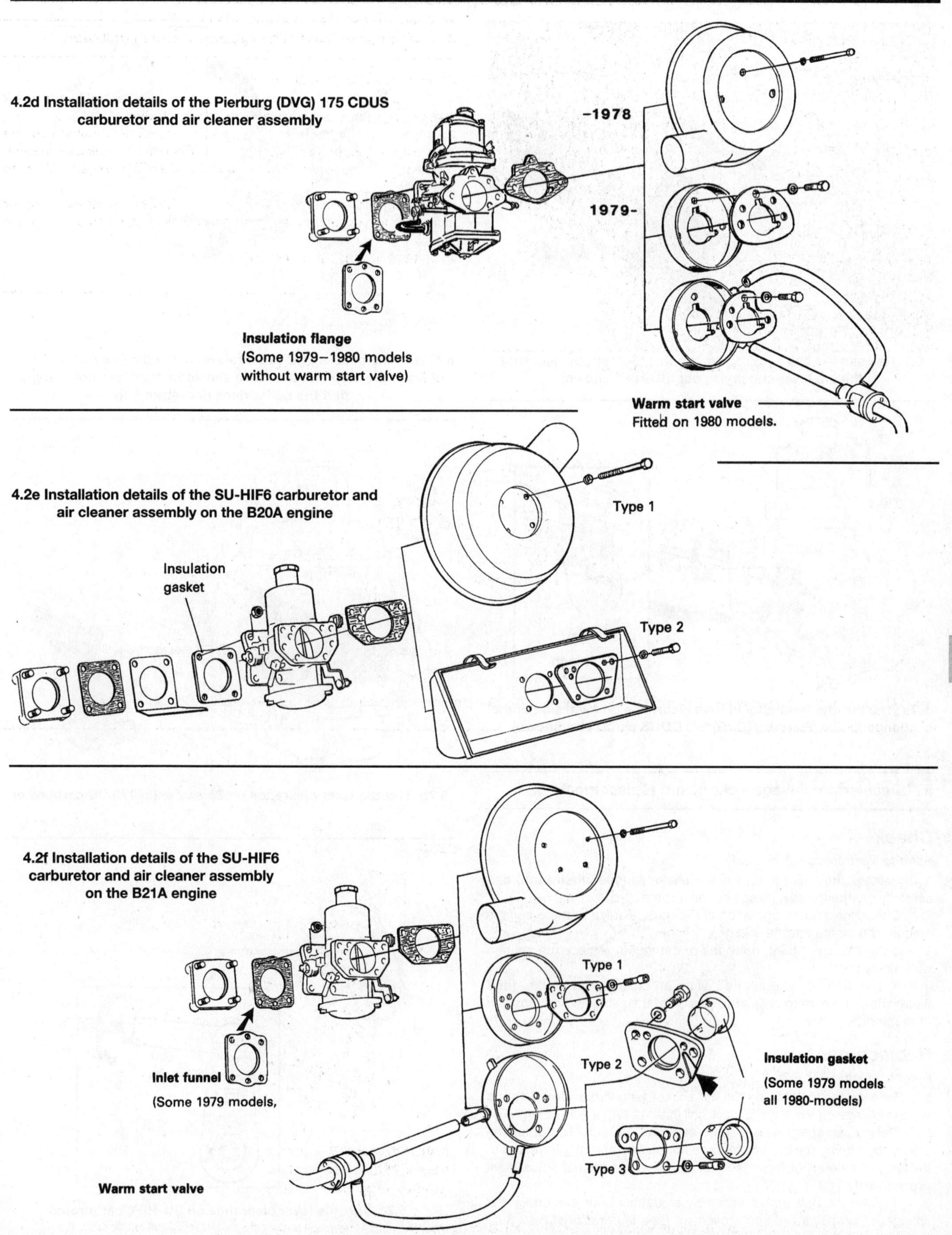

4.2d Installation details of the Pierburg (DVG) 175 CDUS carburetor and air cleaner assembly

4.2e Installation details of the SU-HIF6 carburetor and air cleaner assembly on the B20A engine

4.2f Installation details of the SU-HIF6 carburetor and air cleaner assembly on the B21A engine

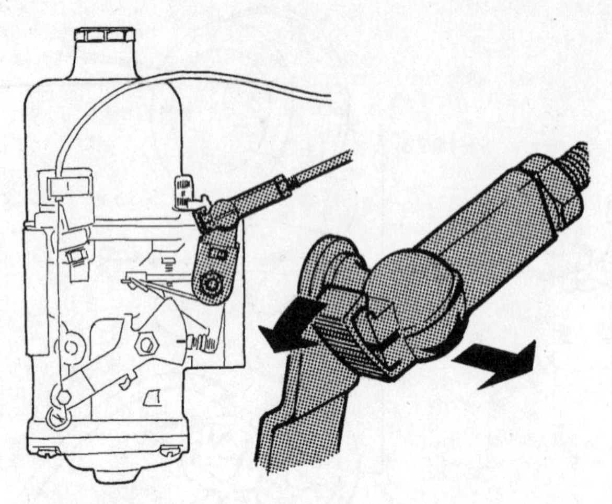
5.1 Release the ball socket from the throttle lever by releasing the tab lever and prying out (SU-HIF6 shown)

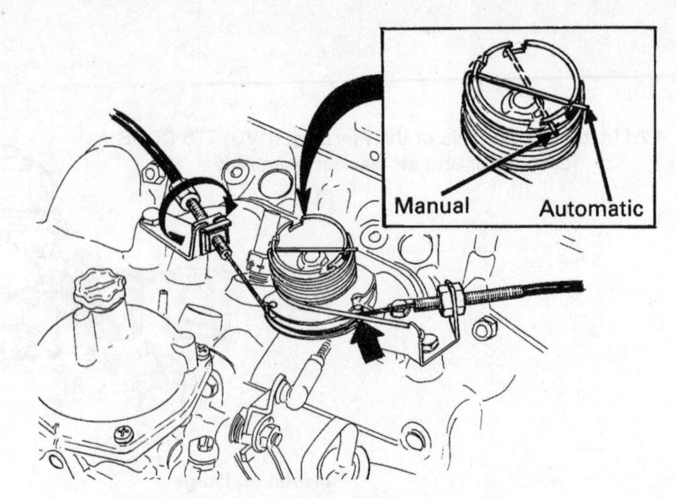

5.4 With the accelerator cable released, the throttle pulley should contact the stop and the cable should be tight, but not so tight that the pulley does not return fully

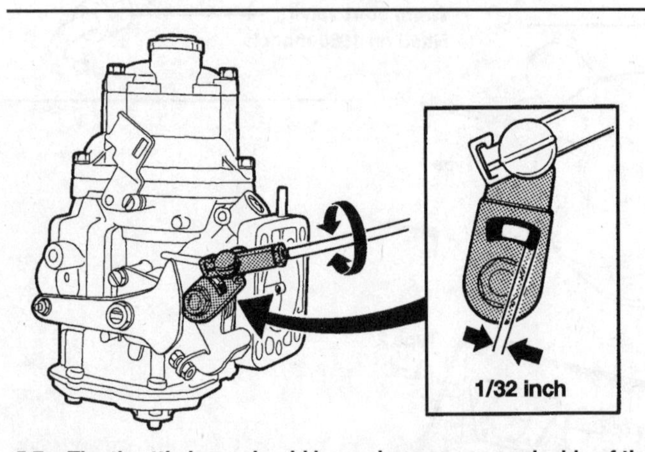

5.7a The throttle lever should have clearance on each side of the flange at idle (Pierburg [DVG] 175 CDUS carburetor shown)

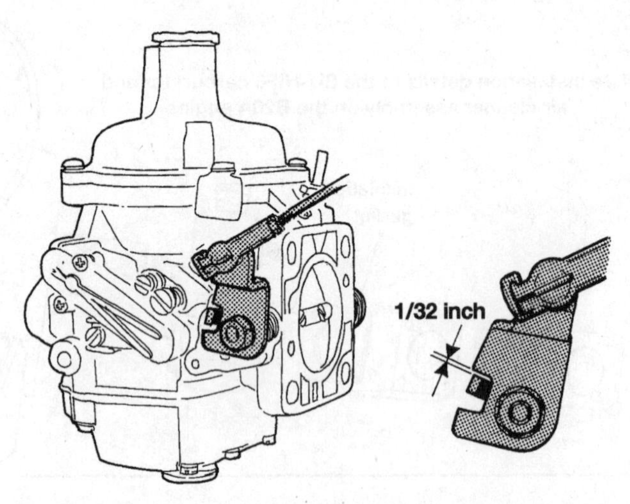

5.7b Throttle lever clearance on Solex Zenith 175 CD carburetor

5 Accelerator linkage - check and replacement

Check

Refer to illustrations 5.1 and 5.4

1 Release the lock tab on the ball socket(s) **(see illustration)** and carefully pry the throttle linkage off the throttle lever (ball hitch).

2 Check for smooth operation of the throttle valve and the spindle mechanism on the throttle linkage.

3 If the linkage is bent, corroded or damaged, replace the section with a new part.

4 If the engine is equipped with an accelerator cable **(see illustration)**, be sure to check the tension of the cable and the position of the throttle pulley.

Replacement

Refer to illustrations 5.7a, 5.7b and 5.7c

5 Release the lock tab on the ball socket **(see illustration 5.1)** and pry the linkage off the throttle valve ball (hitch) using a screwdriver.

6 Before replacing new linkage sections, be sure to adjust the linkage by turning the ball socket assembly on the end of the link rod to the proper amount of turns (see Section 9 for additional adjustment procedures).

7 Tighten the ball socket assembly adjustment nut and check the linkage for proper travel and movement **(see illustrations)** (see Section 9).

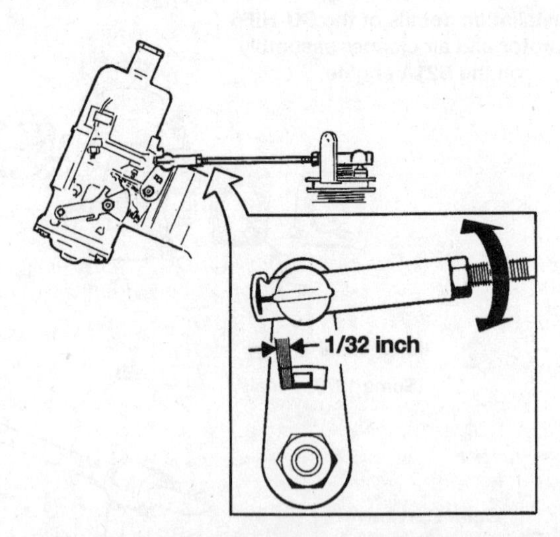

5.7c Throttle lever clearance on SU-HIF6 carburetor

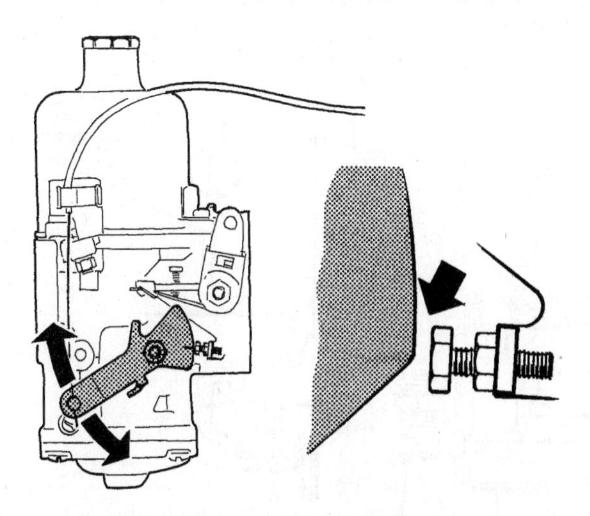

6.1 Be sure the choke cable allows complete movement on the fast idle cam from the cold (fast idle) to the warm position (arrow) (SU-HIF6 shown)

6 Choke cable - check and replacement

Refer to illustration 6.1

Check

1 With the choke cable fully drawn out (CHOKE ON), check to make sure there is complete movement of the lever **(see illustration)**. Refer to the adjustments in Section 9 for additional information.

2 Press the choke using one finger to make sure that it has bottomed at the lower stop position and the fast idle adjustment screw does not touch the lever.

Replacement

3 Disconnect the choke cable end from the throttle lever and withdraw the cable from the carburetor brackets.

4 Remove the choke assembly from the dash and pull the choke cable through.

5 Installation is the reverse of removal.

7 Carburetor - diagnosis and overhaul

Warning: *Gasoline is extremely flammable, so take extra precautions when you work on any part of the fuel system. Don't smoke or allow open flames or bare light bulbs near the work area, and don't work in a garage where a natural gas-type appliance (such as a water heater or a clothes dryer) with a pilot light is present. Since gasoline is carcinogenic, wear latex gloves when there's a possibility of being exposed to fuel, and, if you spill any fuel on your skin, rinse it off immediately with soap and water. Mop up any spills immediately and do not store fuel-soaked rags where they could ignite. The fuel system is under constant pressure, so, if any fuel lines are to be disconnected, the fuel pressure in the system must be relieved first* (see Chapter 4A for more information). *When you perform any kind of work on the fuel system, wear safety glasses and have a Class B type fire extinguisher on hand.*

Diagnosis

1 A thorough road test and check of carburetor adjustments should be done before any major carburetor service work. Specifications for some adjustments are listed on the Vehicle Emissions Control Information (VECI) label found in the engine compartment.

2 Carburetor problems usually show up as flooding, hard starting, stalling, severe backfiring and poor acceleration. A carburetor that's leaking fuel and/or covered with wet looking deposits definitely needs attention.

3 Some performance complaints directed at the carburetor are actually a result of loose, out-of-adjustment or malfunctioning engine or electrical components. Others develop when vacuum hoses leak, are disconnected or are incorrectly routed. The proper approach to analyzing carburetor problems should include the following items:

a) *Inspect all vacuum hoses and actuators for leaks and correct installation.*
b) *Tighten the intake manifold and carburetor mounting nuts/bolts evenly and securely.*
c) *Perform a cylinder compression test* (see Chapter 2).
d) *Clean or replace the spark plugs as necessary* (see Chapter 1).
e) *Check the spark plug wires* (see Chapter 1).
f) *Inspect the ignition primary wires.*
g) *Check the ignition timing (follow the instructions printed on the Emissions Control Information label)* (see Chapter 1).
h) *Check the fuel pump pressure/volume* (see Section 2).
i) *Check the heat control valve in the air cleaner for proper operation* (see Chapter 1).
j) *Check/replace the air filter element* (see Chapter 1).
k) *Check the PCV system* (see Chapter 6).
l) *Check/replace the fuel filter* (see Chapter 1). *Also, the strainer in the tank could be restricted.*
m) *Check for a plugged exhaust system.*
n) *Check EGR valve operation* (see Chapter 6).
o) *Check the choke - it should be completely open at normal engine operating temperature* (see Chapter 1).
p) *Check for fuel leaks and kinked or dented fuel lines* (see Chapters 1 and 4A)
q) *Check for incorrect fuel or bad gasoline.*
r) *Check the valve clearances (if applicable) and camshaft lobe lift* (see Chapters 1 and 2).
s) *Have a dealer service department or repair shop check the electronic engine and carburetor controls.*

4 Diagnosing carburetor problems may require that the engine be started and run with the air cleaner off. While running the engine without the air cleaner, backfires are possible. This situation is likely to occur if the carburetor is malfunctioning, but just the removal of the air cleaner can lean the fuel/air mixture enough to produce an engine backfire. **Warning:** *Do not position any part of your body, especially your face, directly over the throat of the carburetor during inspection and servicing procedures. Wear eye protection!*

4B

Overhaul

Refer to illustrations 7.7a through 7.7f

5 Once it's determined that the carburetor needs an overhaul, several options are available. If you're going to attempt to overhaul the carburetor yourself, first obtain a good quality carburetor rebuild kit (which will include all necessary gaskets, internal parts, instructions and a parts list). You'll also need some special solvent and a means of blowing out the internal passages of the carburetor with air.

6 An alternative is to obtain a new or rebuilt carburetor. They are readily available from dealers and auto parts stores. Make absolutely sure the exchange carburetor is identical to the original. A tag is usually attached to the top of the carburetor or a number is stamped on the float bowl. It will help determine the exact type of carburetor you have. When obtaining a rebuilt carburetor or a rebuild kit, make sure the kit or carburetor matches your application exactly. Seemingly insignificant differences can make a large difference in engine performance.

7 If you choose to overhaul your own carburetor, allow enough time to disassemble it carefully, soak the necessary parts in the cleaning solvent (usually for at least one-half day or according to the instructions listed on the carburetor cleaner) and reassemble it, which will usually take much longer than disassembly **(see illustrations)**. When disassembling the carburetor, match each part with the illustration in the carburetor kit and lay the parts out in order on a clean work surface. Overhauls by inexperienced mechanics can result in an engine which runs poorly or not at all. To avoid this, use care and patience when disassembling the carburetor so you can reassemble it correctly.

8 Because carburetor designs are constantly modified by the

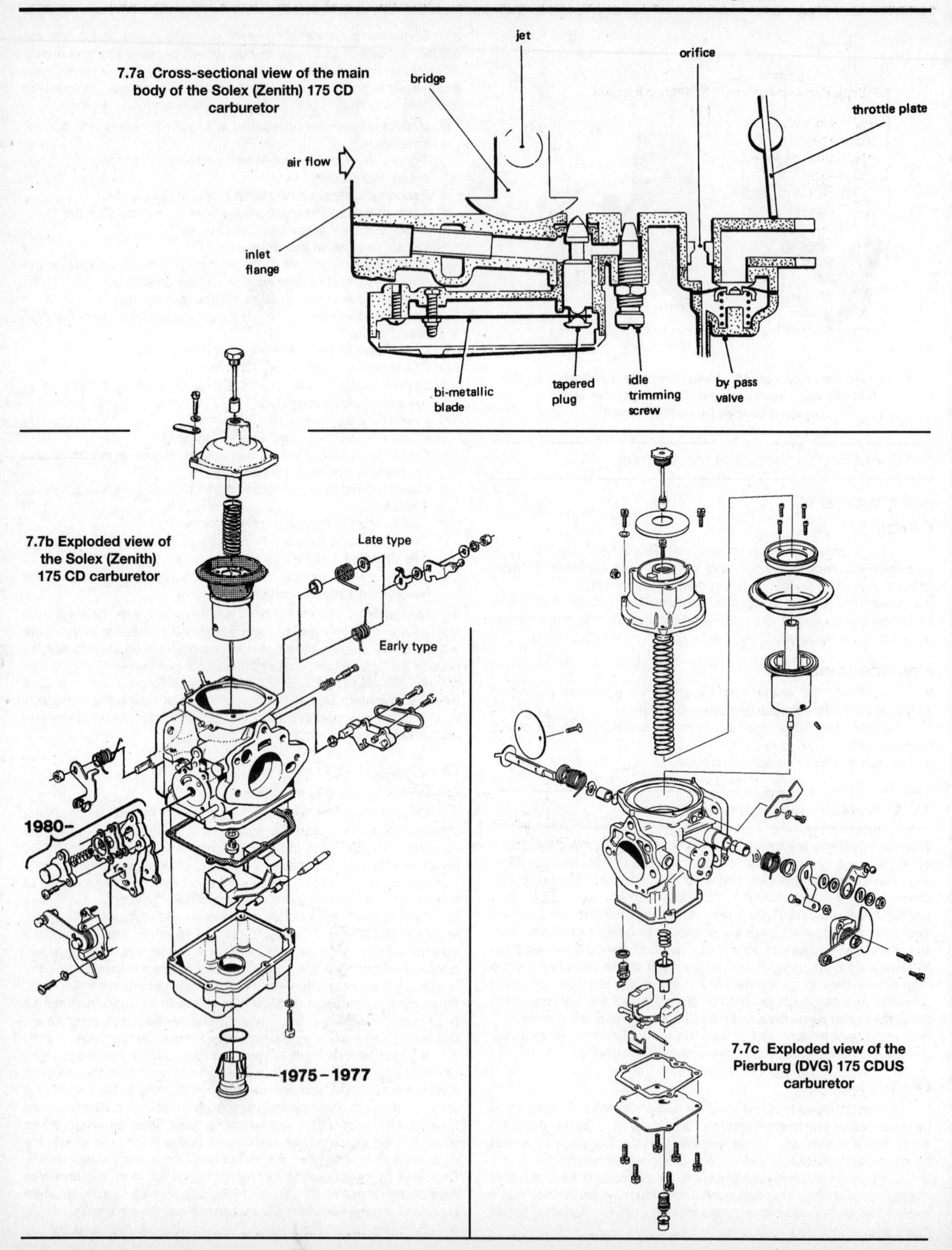

7.7a Cross-sectional view of the main body of the Solex (Zenith) 175 CD carburetor

7.7b Exploded view of the Solex (Zenith) 175 CD carburetor

7.7c Exploded view of the Pierburg (DVG) 175 CDUS carburetor

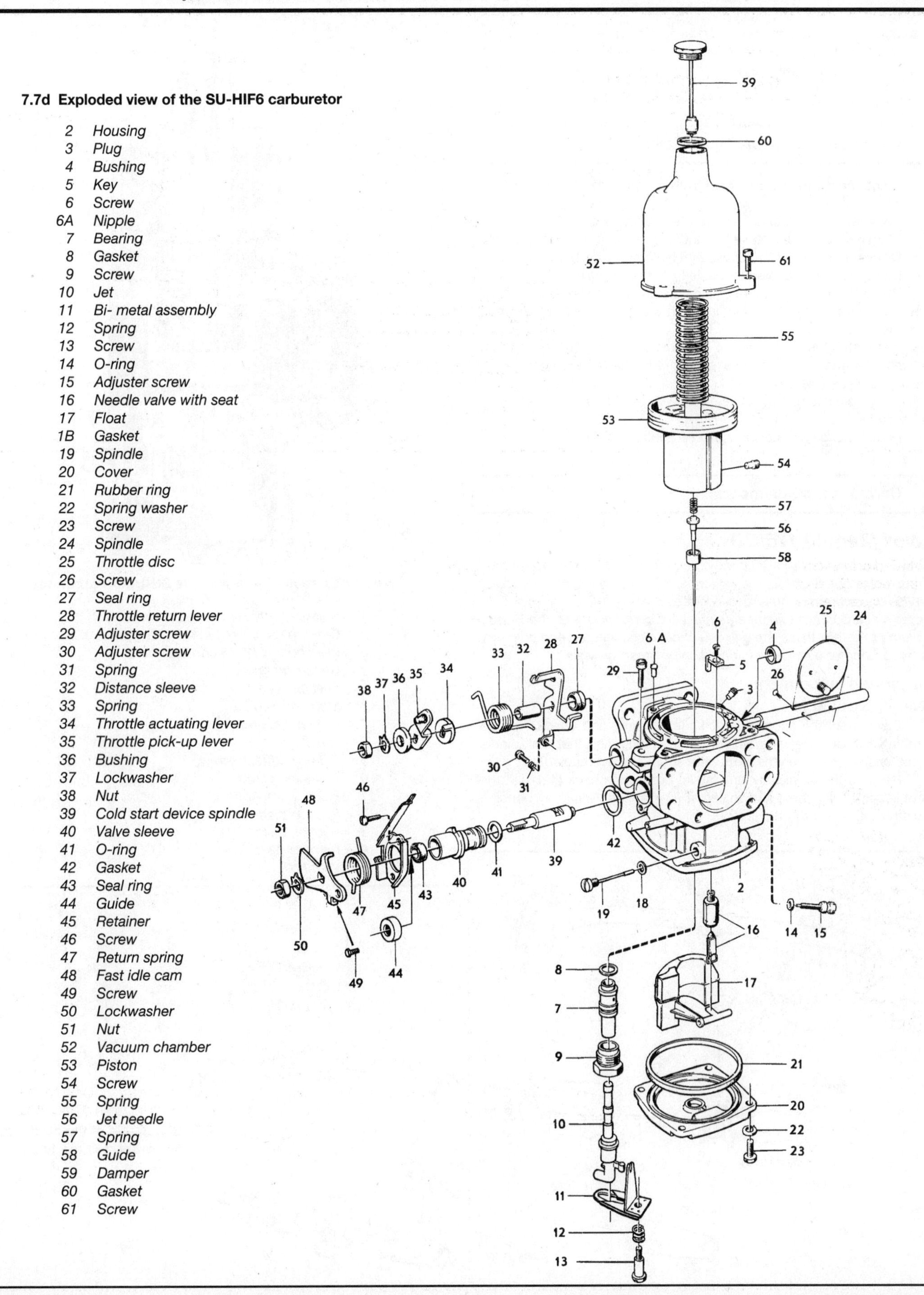

7.7d Exploded view of the SU-HIF6 carburetor

2 *Housing*
3 *Plug*
4 *Bushing*
5 *Key*
6 *Screw*
6A *Nipple*
7 *Bearing*
8 *Gasket*
9 *Screw*
10 *Jet*
11 *Bi- metal assembly*
12 *Spring*
13 *Screw*
14 *O-ring*
15 *Adjuster screw*
16 *Needle valve with seat*
17 *Float*
1B *Gasket*
19 *Spindle*
20 *Cover*
21 *Rubber ring*
22 *Spring washer*
23 *Screw*
24 *Spindle*
25 *Throttle disc*
26 *Screw*
27 *Seal ring*
28 *Throttle return lever*
29 *Adjuster screw*
30 *Adjuster screw*
31 *Spring*
32 *Distance sleeve*
33 *Spring*
34 *Throttle actuating lever*
35 *Throttle pick-up lever*
36 *Bushing*
37 *Lockwasher*
38 *Nut*
39 *Cold start device spindle*
40 *Valve sleeve*
41 *O-ring*
42 *Gasket*
43 *Seal ring*
44 *Guide*
45 *Retainer*
46 *Screw*
47 *Return spring*
48 *Fast idle cam*
49 *Screw*
50 *Lockwasher*
51 *Nut*
52 *Vacuum chamber*
53 *Piston*
54 *Screw*
55 *Spring*
56 *Jet needle*
57 *Spring*
58 *Guide*
59 *Damper*
60 *Gasket*
61 *Screw*

4B

manufacturer in order to meet increasingly more stringent emissions regulations, it isn't feasible to include a step-by-step overhaul of each type. You'll most likely receive a detailed, well illustrated set of instructions with the carburetor overhaul kit.

8 Carburetor - removal and installation

1 Disconnect the battery cable from the negative terminal of the battery.
2 Remove the air cleaner assembly (see Section 4).
3 Disconnect the fuel intake line and plug the end.
4 Disconnect the distributor vacuum hose at the carburetor.
5 Disconnect the accelerator cable from the carburetor.
6 Remove the carburetor retaining nuts and lift off the carburetor(s). Cover the intake manifold opening with shop rags to keep dirt and dampness out.
7 Installation is the reversal of removal, using a new gasket between the carburetor and intake manifold, making sure all traces of the old gasket are removed.
8 After installation, adjust the idle speed, fast idle etc. as described in Section 9.
9 Adjust the accelerator linkage (see Section 5).

9 Carburetor adjustments

Solex (Zenith) 175 CD

Note: *Adjustment of the CO content requires the use of a special tool. If this tool is not available, the adjustment should be left to your dealer service department or other qualified repair facility*
Note: *All carburetor adjustments must be performed with the Pulsair system or Air Injection system (if equipped) disconnected and plugged. Refer to Chapter 6 for more information on these systems.*

Idle speed adjustment

Refer to illustration 9.9

1 Disconnect the accelerator linkage ball end, and check that the throttle valve operates smoothly without binding, and that the spindle is not loose. If it is, then the carburetor should be overhauled.
2 Remove the air cleaner and intake duct, and check that with the throttle valve fully open, it is angled at 90-degrees. Adjust by bending the stop lug on the throttle spindle.

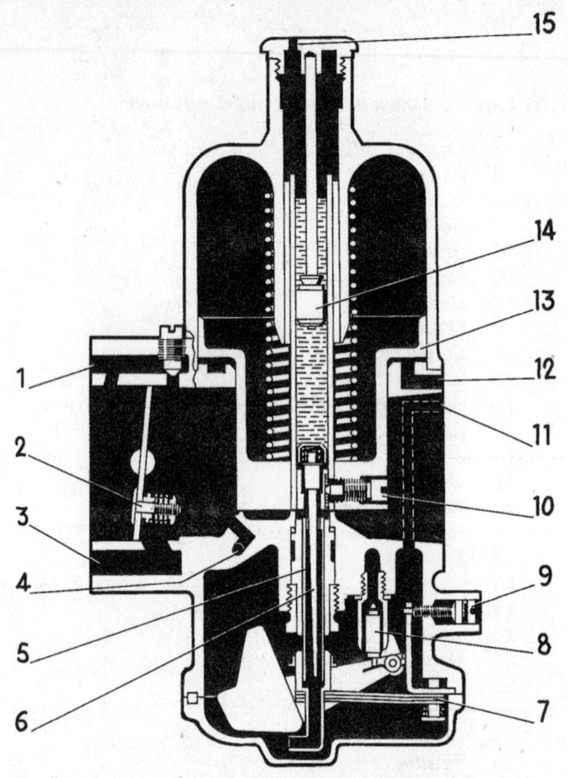

7.7e Cross-sectional view of the SU-HIF6 carburetor

1 *Solenoid valve (idle bypass channel)*
2 *Overrun valve (limited to certain models)*
3 *Idle channel (1978 and later models)*
4 *Choke channel*
5 *Fuel jet*
6 *Metering needle*
7 *Bi-metal spring*
8 *Needle valve (float chamber)*
9 *CO adjustment screw*
10 *Retaining screw*
11 *Float chamber vent*
12 *Suction piston vent*
13 *Suction piston*
14 *Damper piston*
15 *Vent hole*

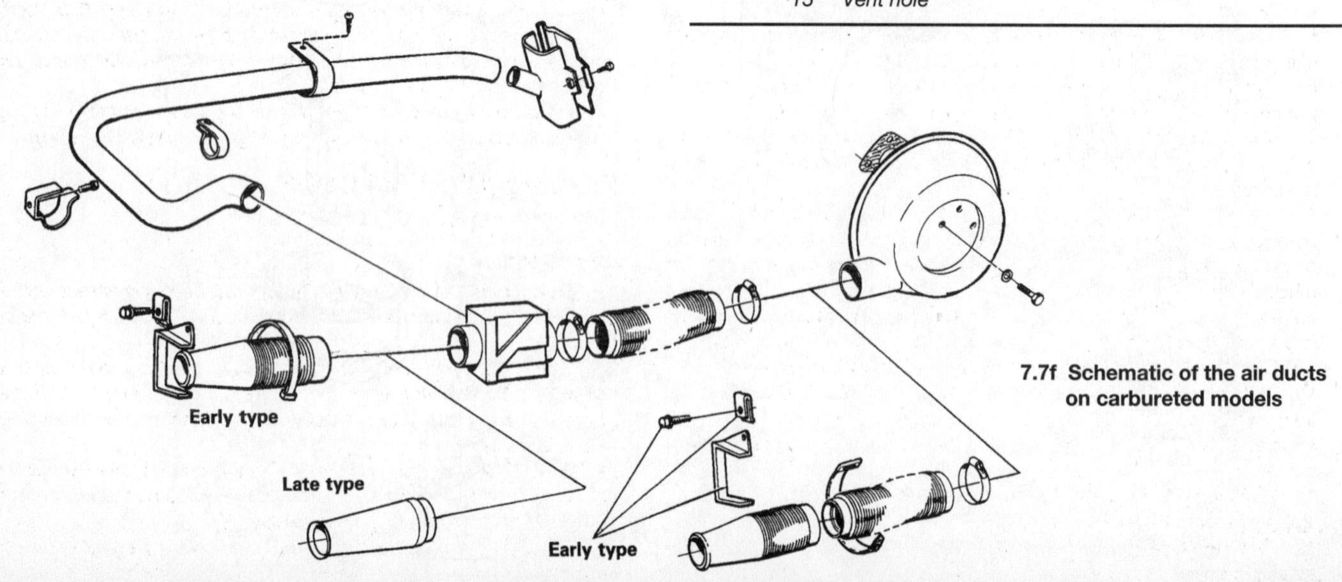

7.7f Schematic of the air ducts on carbureted models

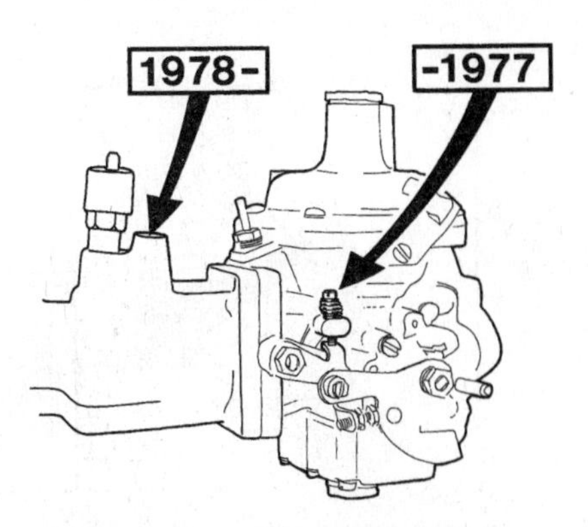

9.9 Solex (Zenith) 175 CD idle adjusting screws

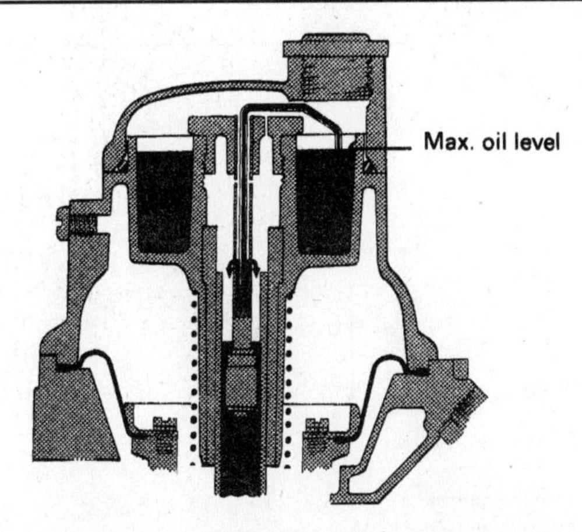

9.21 Correct damper fluid level on the Solex (Zenith)

3 If the throttle valve adjustment has been disturbed, set it as follows. Remove the locknut on the throttle valve adjusting screw and unscrew the adjuster until the throttle valve is fully closed. Screw the adjuster in to obtain the initial opening. On completion, tighten the adjuster locknut and seal it with paint. There should be no need to repeat this adjustment.
4 Install the air cleaner or intake duct and reconnect the accelerator linkage rod.
5 Check that the choke cable is pushed fully in, and that the fast idle cam is clear of the fast idle adjusting screw.
6 Check that the oil level in the damper chamber is correct, topping-up as necessary.
7 Connect a tachometer to the manufacturer's specifications.
8 Start the engine and allow it to reach normal operating temperature.
9 Set the engine idle speed as follows. On models without an idling by-pass system (see Section 10), adjust the throttle adjusting screw in or out to obtain the specified idling speed **(see illustration)**. On models with an idling bypass system, adjust the idling bypass system adjusting screw to obtain the specified idling speed. **Note:** *Idle speed is adjusted on the throttle valve adjusting screw on models before 1977, and on later models on the idle bypass adjustment screw.*
10 If the correct idle speed cannot be obtained on the bypass adjusting screw, check the throttle valve setting as follows. **Note:** *This method should also be used where dieseling (run-on) has been a problem.*
11 Screw the idle bypass adjusting screw in until it bottoms, then out again by four complete turns.
12 Set the engine idle speed to 1,100 to 1,200 rpm using the throttle adjusting screw.

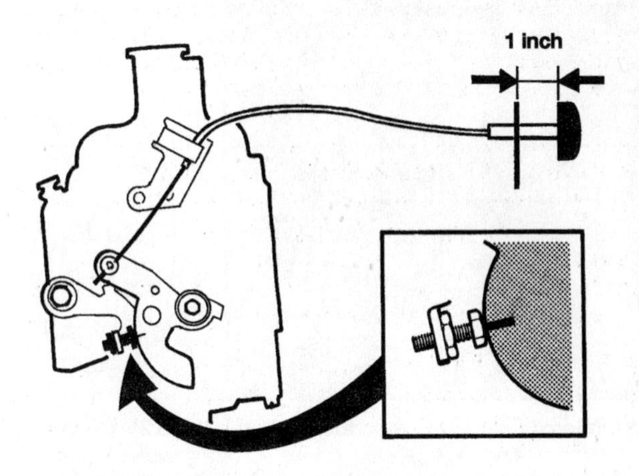

9.16 Fast idle speed adjustment on the Solex Zenith 175 CD carburetor

13 Reduce the engine idle speed to that specified by screwing in the idle bypass adjusting screw.
14 On completion, tighten all adjuster screw locknuts and remove the tachometer. Seal the throttle adjusting screw locknut with paint if it was disturbed.

CO adjustment

15 This procedure requires the use of a special exhaust gas analyzer tool. This apparatus analyzes the CO, HC, CO2 and NOx compounds in the exhaust system out of the tailpipe. Take the vehicle to an emissions repair shop or other qualified repair facility that is equipped to analyze emissions levels.
Note: *On models up to 1977, the main jet is adjusted. This requires the use of press tools, and should be done by a Volvo dealer or other specialist. On 1977 and later models, the air valve metering needle is adjusted. The special tool required is readily available from most auto parts stores and accessory shops.*

Fast idle adjustment

Refer to illustration 9.16
Note: *Adjust the idle speed before attempting fast idle adjustment.*
16 Pull out the choke cable lever approximately 1.0 inch (25.0 mm) so that the index mark on the choke lever cam is opposite the fast idle adjustment screw **(see illustration)**.
17 Connect a tachometer. Start the engine and check that the fast idle is as specified. If not, loosen the locknut on the adjuster screw and turn the screw until the specified speed is reached. On completion, tighten the locknut.
18 Push the choke lever cable fully in, and check that there is a gap between the choke lever cam and the head of the adjuster screw.

Pierburg (DVG) 175 CDUS

Refer to illustrations 9.21, 9.24 and 9.25

Idle speed

19 Disconnect the accelerator linkage rod from the carburetor, and check that the throttle valve and spindle operate smoothly. Reconnect the rod.
20 Check the operation of the choke lever, ensuring that it does not contact the throttle stop when the choke is pushed fully in, and that it opens fully when the choke is pulled fully out. Push the choke fully in for this adjustment.
21 Ensure the damper/reservoir fluid level is correct **(see illustration)**.
22 If equipped, disconnect the Pulsair system hose at the air cleaner, and plug the end of the hose (see Chapter 6).
23 Start the engine and allow it to reach normal operating temperature.

9.24 Idle adjusting screw on the Pierburg DVG 175 CDUS

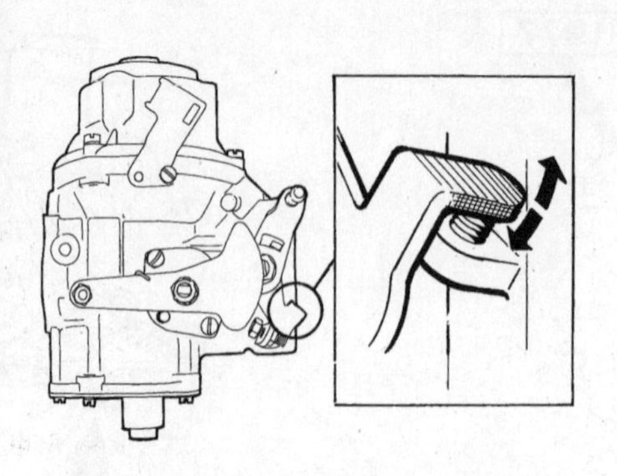

9.25 On the Pierburg, be sure the throttle valve is closed and the adjustment screw barely touches the stop (initial setting)

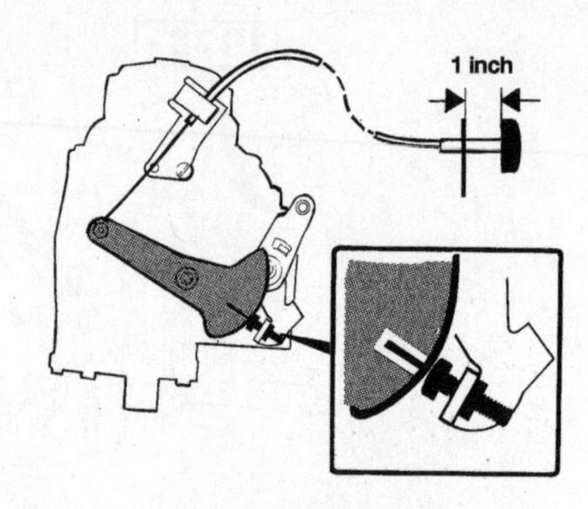

9.27 Fast idle speed adjustment on the Pierburg DVG 175 CDUS

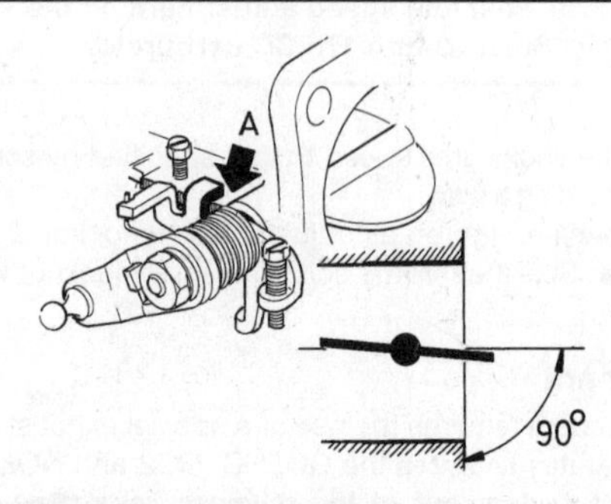

9.29 Bend the tang in the event the throttle does not open fully (90-degrees)

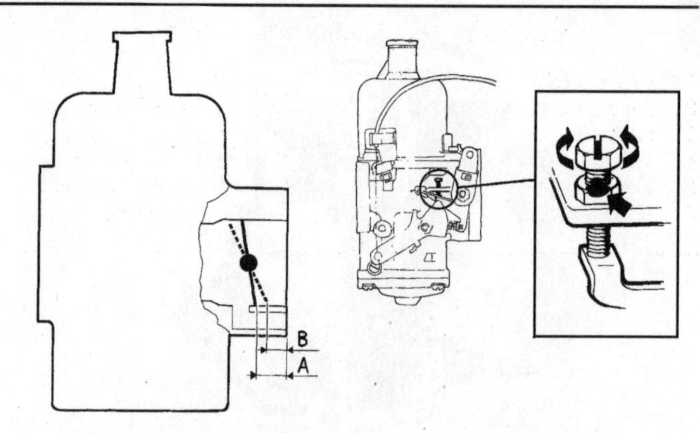

9.30 Throttle valve basic setting on the SU-HIF6 carburetor

24 Adjust the idle speed to the Specifications listed in this Chapter, using the idle bypass adjustment screw **(see illustration)**.
25 If the correct idle speed cannot be achieved using the idle bypass screw, check the basic setting of the throttle valve (see Step 3) **(see illustration)**.

CO content

26 This procedure requires the use of a special exhaust gas analyzer tool. This apparatus analyzes the CO, HC, CO_2 and NOx compounds in the exhaust system out of the tailpipe. Take the vehicle to a emissions repair shop or other qualified repair facility that is equipped to analyze emissions levels.

Fast idle adjustment

Refer to illustration 9.27

27 Follow the fast idle adjustment for the Solex (Zenith) 175 CD carburetor (see Steps 16 through 18), but use the accompanying illustration **(see illustration)**.

SU-HIF6

Idle speed adjustment

Refer to illustrations 9.29, 9.30, 9.33 and 9.36

28 Disconnect the accelerator linkage ball end, and check that the throttle valve operates smoothly without binding, and that the spindle is not loose. If it is, then the carburetor should be overhauled.
29 Remove the air cleaner or intake duct, and check that with the throttle valve fully open, it is angled 90-degrees. Adjust by bending the stop lug on the throttle spindles **(see illustration)**.
30 On 1978 to 1981 models, if the throttle valve adjustment has been disturbed, set it as follows. Remove the locknut on the throttle valve adjusting screw, and unscrew the adjuster until the throttle valve is fully closed. Screw the adjuster in to obtain the initial opening **(see illustration)**. On completion, tighten the adjuster locknut and seal it with paint. There should be no need to repeat this adjustment.

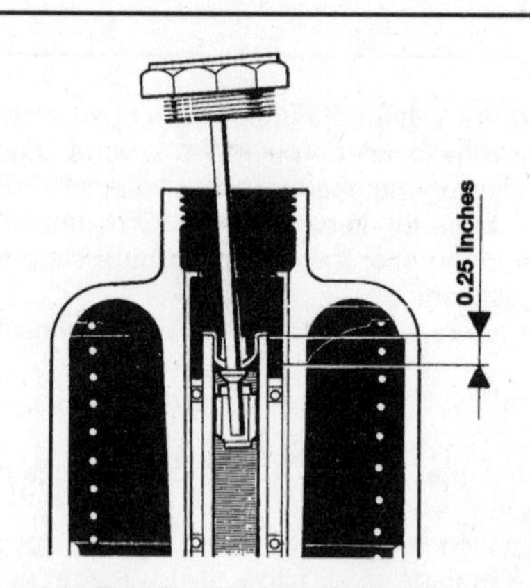

9.33 Damper fluid level on the SU-HIF6 carburetor

31 Install the air cleaner or intake duct and reconnect the accelerator linkage rod.
32 Check that the choke cable is pushed fully in, and that the fast idle cam is clear of the fast idle adjusting screw.
33 Check that the oil level in the damper chamber is correct, topping-up as necessary **(see illustration)**.
34 Connect up a tachometer.
35 Start the engine and allow it to reach normal operating temperature.
36 Set the engine idle speed as follows. On models without an idling by-pass system (see Section 10), adjust the throttle adjusting screw in or out to obtain the specified idling speed **(see illustration)**. On models with an idling bypass system, adjust the idling bypass system adjusting screw to obtain the specified idling speed.

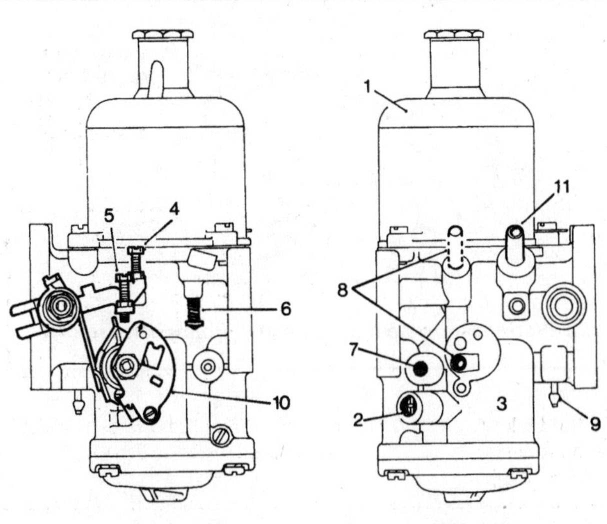

9.36 Adjustment points on the SU-HIF6

1	*Piston/suction chamber*	7	*Fuel inlet*
2	*Jet adjusting screw*	8	*Vent tube (alternative positions)*
3	*Float chamber*	9	*Auto ignition connection*
4	*Throttle adjusting screw*	10	*Cold start enrichment lever*
5	*Fast idle adjusting screw*	11	*Crankcase ventilation tube*
6	*Piston lifting pin*		

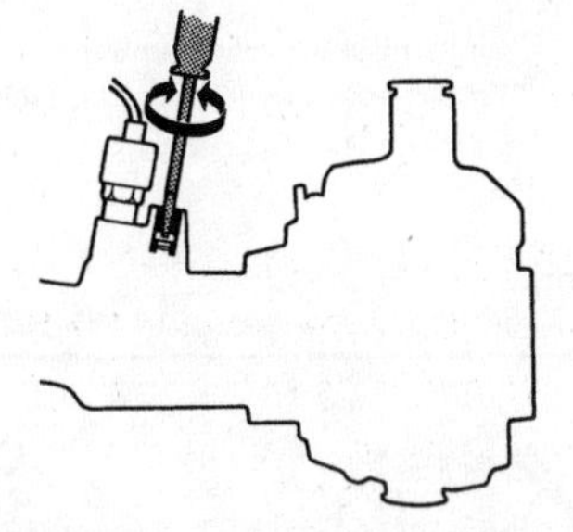

10.5 Use the adjustment screw in the bypass channel to adjust the idle speed on carburetors equipped with the idling air bypass system

37 If the correct idle speed cannot be obtained on the bypass adjusting screw, check the throttle valve setting as follows. **Note:** *This method should also be used where dieseling (running-on) has been a problem.*

38 Screw the idle bypass adjusting screw in until it bottoms, then out again by four complete turns.

39 Set the engine idle speed to 1,100 to 1,200 rpm using the throttle adjusting screw.

40 Reduce the engine idle speed to that specified by screwing in the idle bypass adjusting screw.

41 On completion, tighten all adjuster screw locknuts and remove the tachometer. Seal the throttle adjusting screw locknut with paint if it was disturbed.

CO adjustment

42 This procedure requires the use of a special exhaust gas analyzer tool. This apparatus analyzes the CO, HC, CO2 and NOx compounds in the exhaust system out of the tailpipe. Take the vehicle to a emissions repair shop or other qualified repair facility that is equipped to analyze emissions levels.

Fast idle adjustment

Refer to illustration 9.43

Note: *Adjust the idle speed before attempting fast idle adjustment.*

43 Pull out the choke cable approximately 1.0 inch (25.0 mm) so that the index mark on the fast cam is opposite the fast idle adjustment screw **(see illustration).**

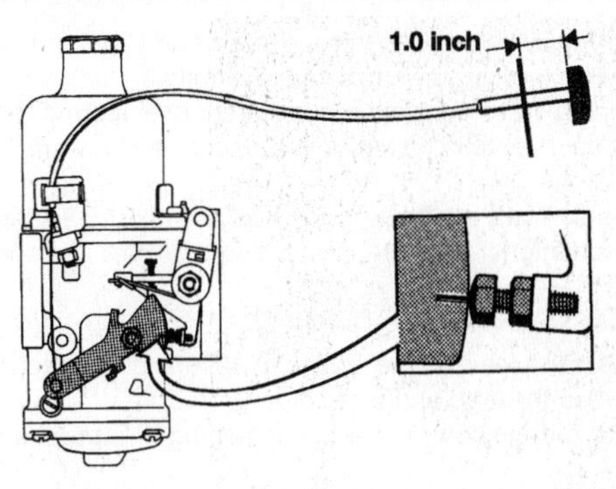

9.43 Fast idle adjustment on the SU-HIF6

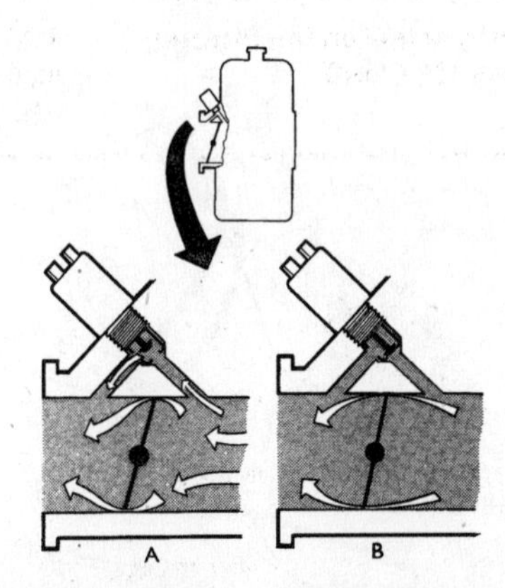

10.7 The additional idling solenoid increases airflow on models equipped with air conditioning

44 Connect a tachometer. Start the engine and check that the fast idle is as specified. If not, loosen the locknut on the adjuster screw and turn the screw until the specified speed is reached. On completion, tighten the locknut.

45 Push the choke cable fully in, and check that there is a gap between the fast idle cam and the head of the adjuster screw.

10 Idling bypass system

Idle speed

Refer to illustrations 10.5 and 10.7

1 The idling bypass system is equipped from 1976 through 1984 on certain engines.

2 It consists of a solenoid valve, screwed into the intake manifold, and connected electrically to the ignition switch.

3 The intake manifold and carburetor have channels bored in them, through which a quantity of the fuel/air mixture can flow, bypassing the throttle valve.

4 When the ignition is switched on, the solenoid valve operates and opens the bypass channel, and closes when the ignition is switched off, giving greater control of the idling system and preventing running-on.

5 Idle speed on carburetors with a bypass system should be set by using the adjuster screw in the bypass channel **(see illustration)**, and not by the throttle valve adjuster screw.

6 Where the idle speed cannot be set satisfactorily using the bypass adjustment screw, the basic setting of the carburetor main jet, needle and throttle valve should be checked and adjusted.

Idle speed - vehicles with air conditioning

7 On some vehicles equipped with air conditioning a solenoid valve

(see illustration), which opens when the air conditioning is switched off, is installed on the carburetor or intake manifold.

8 The solenoid valve controls a bypass channel around the throttle valve, similar to the idle bypass system, but there is no adjusting screw.

9 The solenoid valve allows more fuel/air mixture to the engine when the air conditioner is in use, preventing a drop in idle speed.

Idling bypass valves - testing

10 Disconnect the lead to the solenoid valve and connect a 12 volt test light between the lead and a good ground.

11 Switch on the ignition (and air conditioning if appropriate), when the test light should light. If it does not, check the supply to the valve, starting at fuse No. 13 (refer to the *wiring diagrams* in Chapter 12).

12 Connect an ohmmeter between the connection on top of the valve and the valve body. Resistance should be approximately 30 ohms.

13 Similarly, a reading of 30 ohms should be obtained between the connector and intake manifold.

14 Remove the valve from the intake manifold, and clean off any carbon deposits from the valve seat, using a stiff wire brush.

Note: *Carbon deposits are caused by:*

(a) Blocked or defective crankcase ventilation system
(b) Dirty or incorrectly installed air cleaner or air intake system
(c) Poor quality fuel

15 These defects should be rectified before installation the valve.

16 Smear the threads of the solenoid valve with grease or anti-seize compound before installation.

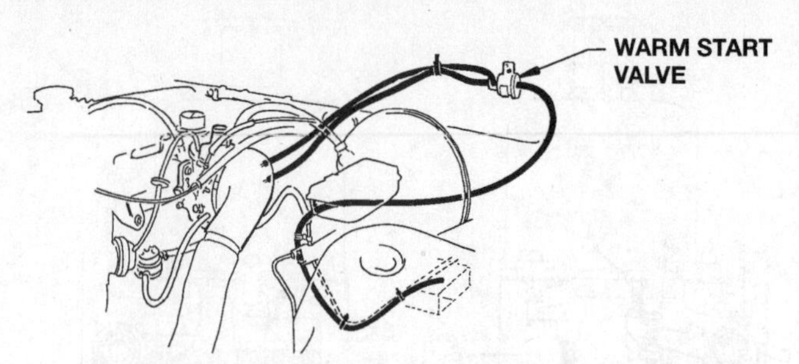

11.2a Warm start system (Solex and Pierburg carburetors)

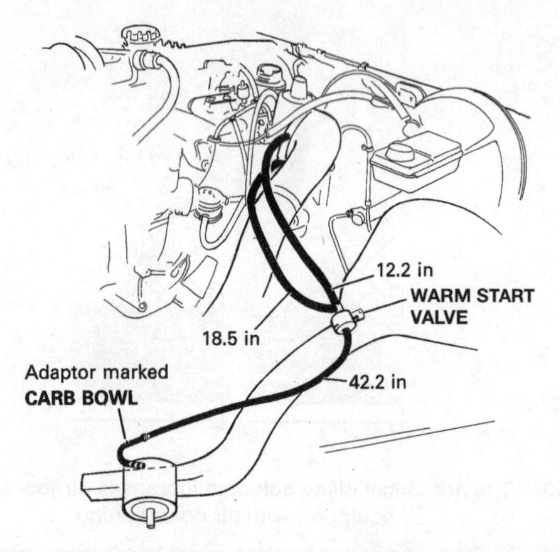

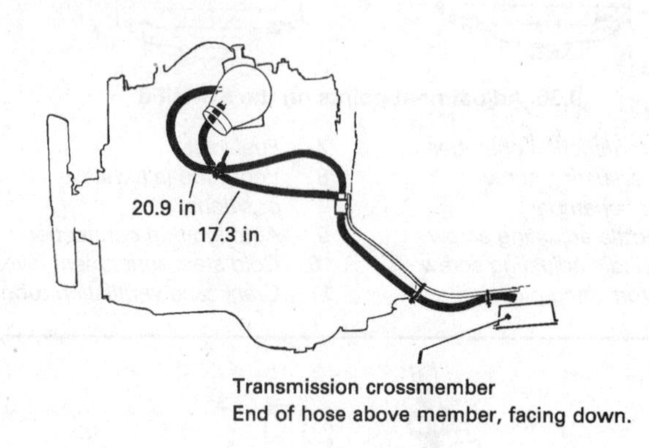

11.2b Warm start system (SU-HIF6 carburetor)

11.7 With the ignition key OFF, blow through the valve from the carb side and confirm that air passes through the valve, then with the ignition key ON, that air does not pass through the valve

11 Warm start valve

General information

Refer to illustrations 11.2a and 11.2b

1 The warm start valve is installed on certain models from mid-1980 on.

2 On early Solex and Pierburg carburetors, it is mounted on the engine firewall **(see illustration)**, and on SU and later versions of the above, it is mounted on the inner wheel well **(see illustration)**. From 1984 it is mounted directly on to the carburetor.

3 The valve is connected to the carburetor float bowl by tubing (via your dealer for fitting the warm start valve to 1979 vehicles.

6 Malfunction of the warm start valve can cause difficult warm starting, high fuel consumption, high CO level at idle and rough running.

Warm start valve - testing

Refer to illustration 11.7

7 With the ignition switched off, it should be possible to blow through the valve from the carburetor side to the vent tube side **(see illustration)**.

8 With the ignition switched on, the valve should close and it should not be possible to blow through it.

9 If the valve fails to operate, check the electrical supply to it, and its ground before replacing the valve.

Chapter 5
Engine electrical systems

Contents

Specifications

General

Coil primary resistance	
Contact breaker systems	
1976 through 1978 models	2.7 to 3.0 ohms
1979 through 1984 models	1.8 to 2.0 ohms
Breakerless systems	1.8 to 2.0 ohms
Computerized systems	1.1 to 1.3 ohms
EZ-116K systems	0.6 to 1.0 ohms
Coil secondary resistance	
Contact breaker systems	
1976 through 1978 models	7.0 to 12.0K ohms
1979 through 1984 models	8.0 to 11.0K ohms
Breakerless systems	8 to 10K ohms
Computerized systems	9.6 to 11.6K ohms
EZ-116K systems	6.5 to 8.5K ohms
Ballast resistor resistance	
Contact breaker systems	
1976 through 1978 models	0.9 ohms
1979 on	1.3 ohms
Breakerless systems	0.9 ohms
Computerized systems	Not applicable

Distributor

Air gap	0.010 inch (0.25 mm)

Ignition timing

USA models	
1976 models	15 degrees BTDC at 800 rpm
1977 ** and 1978 *** models	12 degrees BTDC at 800 rpm
1979 *** and 1980 *** **** models	8 degrees BTDC at 800 rpm
1981 and later models *	
Turbocharged engines	12 degrees BTDC at 900 rpm
All others ****	10 to 14 degrees BTDC at 750 rpm
Canadian models	
1976 and 1977 models	15 degrees BTDC at 800 rpm
1978 models *	12 degrees BTDC at 800 rpm
1979 and 1980 models	10 degrees BTDC at 800 rpm
1981 through 1984 models *	8 degrees BTDC at 800 rpm
1985 and later models	Refer to the VECI label in the engine compartment

** Distributor vacuum hose disconnected and plugged at distributor, if equipped*
*** On 1977 models with Exhaust Gas Recirculation (EGR), disconnect and plug the EGR valve vacuum hose*
**** On 1978 through 1980 models with air injection (AIR), disconnect the air injection hose*
***** On 1980 and 1981 models with AIR , disconnect the air injection hose and turn the A/C off*

1 General information

The engine electrical systems include all ignition, charging and starting components. Because of their engine-related functions, these components are discussed separately from chassis electrical devices such as the lights, the instruments, etc. (which are included in Chapter 12).

Always observe the following precautions when working on the electrical systems:

a) *Be extremely careful when servicing engine electrical components. They are easily damaged if checked, connected or handled improperly.*
b) *Never leave the ignition switch on for long periods of time with the engine off.*
c) *Don't disconnect the battery cables while the engine is running.*
d) *Observe the rules when jump-starting your vehicle. Read the precautions at the front of this manual.*
e) *Always disconnect the negative cable first and hook it up last or the battery may be shorted by the tool being used to loosen the cable clamps.*
f) *Don't charge the battery with the cables connected to the terminals.*

It's also a good idea to review the safety-related information regarding the engine electrical systems located in the *Safety first* section near the front of this manual before beginning any operation included in this Chapter. **Caution:** *It is necessary to make sure the radio is turned OFF before disconnecting the battery cables to avoid damaging the microprocessor built into the radio*

2 Battery - emergency jump starting

Refer to the *Booster battery (jump) starting* procedure at the front of this manual.

3 Battery - removal and installation

1 Disconnect the cable from the negative terminal of the battery.

2 Detach the cable from the positive terminal. **Caution:** *It is necessary to make sure the radio is turned OFF before disconnecting the battery cables to avoid damaging the microprocessor built into the radio.*

3 Remove the battery hold-down bracket and lift out the battery. Be careful - it's heavy.

4 While the battery is out, inspect the carrier (tray) for corrosion (see Chapter 1).

5 If you are replacing the battery, make sure that you get one that's identical, with the same dimensions, amperage rating, cold cranking rating, etc.

6 Installation is the reverse of removal.

4 Battery cables - check and replacement

1 Periodically inspect the entire length of each battery cable for damage, cracked or burned insulation and corrosion. Poor battery cable connections can cause starting problems and decreased engine performance. **Caution:** *It is necessary to make sure the radio is turned OFF before disconnecting the battery cables to avoid damaging the microprocessor built into the radio*

2 Check the cable-to-terminal connections at the ends of the cables for cracks, loose wire strands and corrosion. The presence of white, fluffy deposits under the insulation at the cable terminal connection is a sign that the cable is corroded and should be replaced. Check the terminals for distortion, missing mounting bolts and corrosion.

3 When removing the cables, always disconnect the negative cable first and hook it up last or the battery may be shorted by the tool used to loosen the cable clamps. Even if only the positive cable is being replaced, be sure to disconnect the negative cable from the battery first (see Chapter 1 for further information regarding battery cable removal).

4 Disconnect the old cables from the battery, then trace each of them to their opposite ends and detach them from the starter solenoid and ground terminals. Note the routing of each cable to ensure correct installation.

5 If you are replacing either or both of the old cables, take them with you when buying new cables. It is vitally important that you replace the cables with identical parts. Cables have characteristics

that make them easy to identify: positive cables are usually red, larger in cross-section and have a larger diameter battery post clamp; ground cables are usually black, smaller in cross-section and have a slightly smaller diameter clamp for the negative post.

6 Clean the threads of the solenoid or ground connection with a wire brush to remove rust and corrosion. Apply a light coat of battery terminal corrosion inhibitor, or petroleum jelly, to the threads to prevent future corrosion.

7 Attach the cable to the solenoid or ground connection and tighten the mounting nut/bolt securely.

8 Before connecting a new cable to the battery, make sure that it reaches the battery post without having to be stretched.

9 Connect the positive cable first, followed by the negative cable.

5 Ignition system - general information and precautions

General information

The ignition system includes the ignition switch, the battery, the distributor, the primary (low voltage) and secondary (high voltage) wiring circuits, the spark plugs and the spark plug wires. Early models and carbureted Canadian models are equipped with the contact breaker points type ignition system. Middle years are equipped with electronic breakerless ignition systems. Newer models are equipped with the EZ 116K electronic ignition system.

When working on the ignition system, take the following precautions:

- a) *If the engine won't start, don't keep the ignition switch on for more than 10 seconds.*
- b) *Never allow an ignition coil terminal to contact ground. Grounding the ignition coil can damage the igniter and/or the coil itself.*
- c) *Don't disconnect the battery when the engine is running.*
- d) *Make sure the igniter is properly grounded.*

Contact breaker points systems

This ignition system controls the ignition spark using the conventional points and condenser arrangement and timing is controlled by mechanical advance components. The ballast resistor limits the voltage to the coil during engine operation. Its main purpose is to prevent current overload of the coil during low speed operation. While the engine is cranking with the starter, a higher voltage is supplied to the coil to improve the starting capabilities.

Breakerless ignition systems

The breakerless ignition systems use either Bosch electronic distributors (early breakerless system) or Volvo/Chrysler electronic distributors (computerized). The early systems (1976 through 1982 USA/1981 and later turbo models) will be referred to as "breakerless ignitions" and the later systems (1983 through 1988 non-turbo) will be referred to as "computerized ignitions". Most of these type ignition systems are equipped with four major components; the pick-up coil, the ignition control unit (ignition module), the ignition coil and the spark plugs. The pick-up coil (impulse generator) provides a timing signal for the ignition system. Equivalent to cam-actuated breaker points in a standard distributor, the pick-up coil creates an A/C voltage signal every time the trigger wheel tabs pass the impulse generator tabs. When the ignition control unit (ignition module) receives the voltage signal, it triggers a spark discharge from the coil by interrupting the primary coil circuit. The ignition dwell (coil charging time) is adjusted by the ignition control unit for the most intense spark. **Note:** *The air gap (distance between the impulse generator and trigger wheel tabs) can be adjusted (see Sec-tion 12).*

Later models are equipped with a slightly more sophisticated breakerless ignition system or computerized ignition system. This system, co-manufactured by Volvo and Chrysler, includes a knock sensor to monitor and control ignition knock or ping. The control unit receives inputs from the distributor Hall Effect switch, the throttle switch and the knock sensor. The control unit senses engine load (manifold vacuum) from a hose connected to the intake manifold. All of these sensors help to control the ignition timing for all the engine's operating conditions. This computerized ignition system controls ignition timing only. The later system (EZ-116K) controls fuel characteristics as well as ignition timing.

These systems also include a ballast resistor. The ballast resistor limits the voltage to the coil during engine operation. Its main purpose is to prevent current overload of the coil during low speed operation. While the engine is cranking with the starter, a higher voltage is supplied to the coil to improve the starting capabilities.

EZ-116K ignition systems

Refer to illustration 5.7

This Bosch system **(see illustration on next page)** works together with the LH-2.4 and LH-3.1 fuel injection systems to control ignition timing and fuel delivery. The EZ-116K control unit analyzes the best ignition timing based on inputs the ECU (control unit) receives for engine load, engine speed, ignition quality, coolant temperature and air intake temperature. The distributor responds to distribute high voltage signals to the individual spark plugs.

The EZ-116K system uses adaptive knock control to adjust the ignition timing for the individual cylinders. If compression knock is detected, the timing is retarded for the particular cylinder. If the knock continues, fuel delivery to that particular cylinder is increased by the computer to reduce the combustion temperature. Once the knock is controlled, the ignition timing is slowly returned to the original setting by the computer.

The EZ-116K computerized ignition system delivers only the high tension voltage to the spark plugs. The EZ-116K system is included with a diagnostic circuit that monitors the operation of the complete system. This self diagnostic system detects and stores fault codes (see Chapter 6).

Ignition timing is electronically controlled and should not receive adjustment in normal service. Once the engine is running, the ignition timing is continually changing based on the various input signals to the ECU. Engine speed is signaled by an inductive speed pick-up.

Precautions:

Certain precautions must be observed when working on a transistorized ignition system.

- a) *Do not disconnect the battery cables when the engine is running*
- b) *Make sure the ignition control unit is always well grounded (see Section 10 and 11).*
- c) *Keep water away from the distributor*
- d) *If a tachometer is to be connected to the engine, always connect the tachometer positive (+) lead to the ignition coil negative terminal (-) and never to the distributor.*
- e) *Do not allow the coil terminals to be grounded, as the impulse generator or coil could be damaged.*
- f) *Do not leave the ignition switch on for more than ten minutes if the engine isn't running or will not start.*

6 Ignition system - check

Caution: *Always disconnect the battery before disconnecting the electrical connectors from the module or electronic control unit.*

All ignition systems

Refer to illustrations 6.1 and 6.6

Warning: *Because of the high voltage generated by the ignition system, extreme care should be taken whenever an operation is performed involving ignition components. This not only includes the igniter (electronic ignition), coil, distributor and spark plug wires, but related components such as spark plug connectors, tachometer and other test equipment.*

1 If the engine will not start even though it turns over, check for spark at the spark plug by installing a calibrated ignition system tester

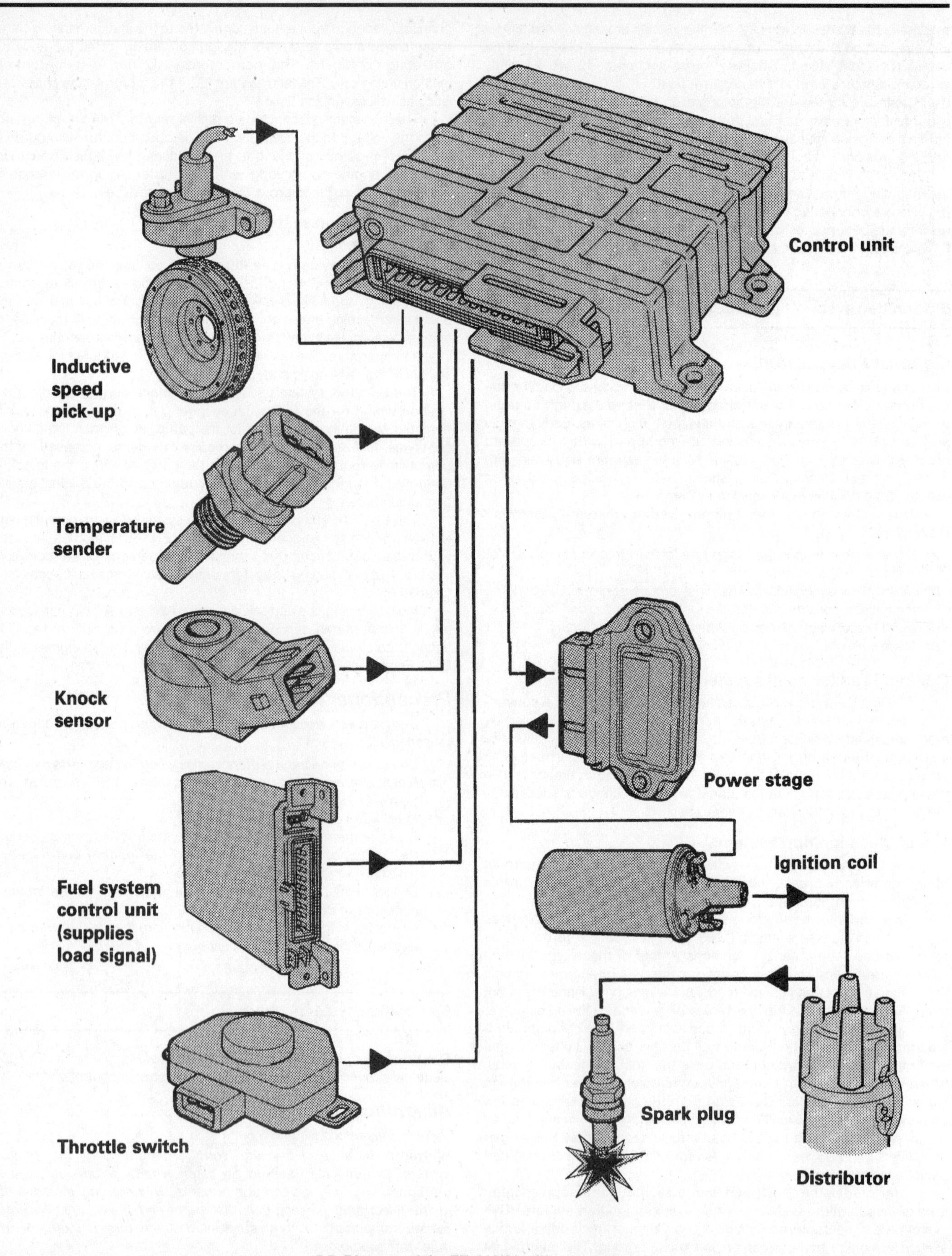

5.7 Diagram of the EZ-116K ignition system

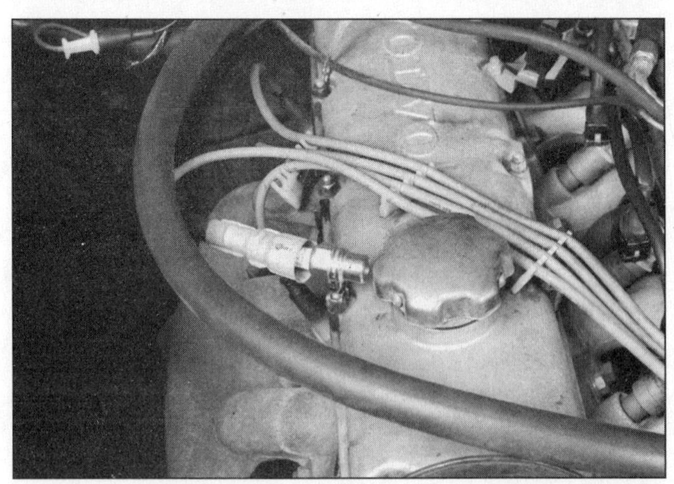

6.1 To use a calibrated ignition tester, simply disconnect a spark plug wire, clip the tester to a convenient ground (like a valve cover bolt) and operate the starter - if there is enough power to fire the plug, sparks will be visible between the electrode tip and the tester body

to the end of the plug wire **(see illustration)**. The tool is available at most auto parts stores. Be sure to order the correct tool for your particular ignition system (breaker type [points] or breakerless type [electronic]).

2 Connect the clip on the tester to a ground such as a metal bracket or valve cover bolt, crank the engine and watch the end of the tester for a bright blue, well defined spark.

3 If sparks occur, sufficient voltage is reaching the plugs to fire the engine. However the plugs themselves may be fouled, so remove and check them as described in Chapter 1 or replace them with new ones.

4 If no spark occurs, remove the distributor cap and check the cap and rotor as described in Chapter 1. If moisture is present, use WD-40 or something similar to dry out the cap and rotor, then reinstall the cap and repeat the spark test.

5 If there is still no spark, the tester should be attached to the wire from the coil and the test repeated again.

6 If no spark occurs, check the primary wire connections at the coil

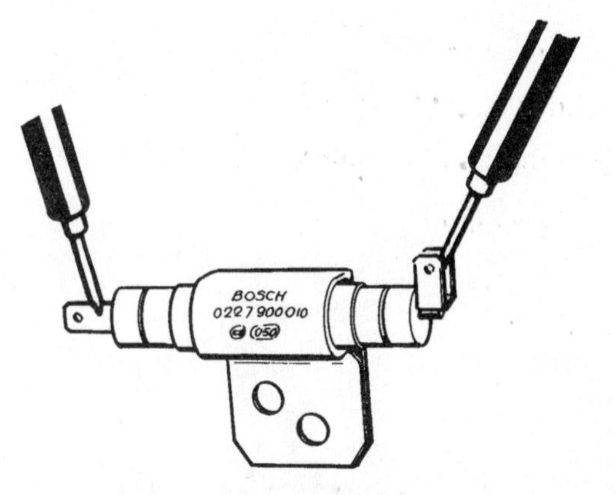

6.6 Check the ballast resistor with an ohmmeter

to make sure they are clean and tight. Make any necessary repairs, then repeat the check again. Also, check the ballast resistor **(see illustration)**. Refer to the specifications listed in this Chapter for the correct resistance values.

7 If sparks now occur, the distributor cap, rotor, plug wires or spark plugs may be defective. If there is still no spark, the coil to cap wire may be bad. Remove the wire from the distributor and have an assistant hold it about 1/4-inch away from the engine block with insulated pliers. Crank the engine over (be careful of moving engine parts) and check for spark at the gap between the secondary wire terminal and the engine block.

Breaker ignition systems (points)

Refer to illustrations 6.9a and 6.9b

Note: *Some of the following check(s) may require a voltmeter, ohmmeter and/or a jumper cable.*

8 If there is still no spark (see Step 1), check the ignition points (refer to Chapter 1). If the points appear to be in good condition (no pits or burned spots on the point surface) and the primary wires are hooked up correctly, adjust the points as described in Chapter 1.

9 Measure the voltage at the points with a voltmeter **(see**

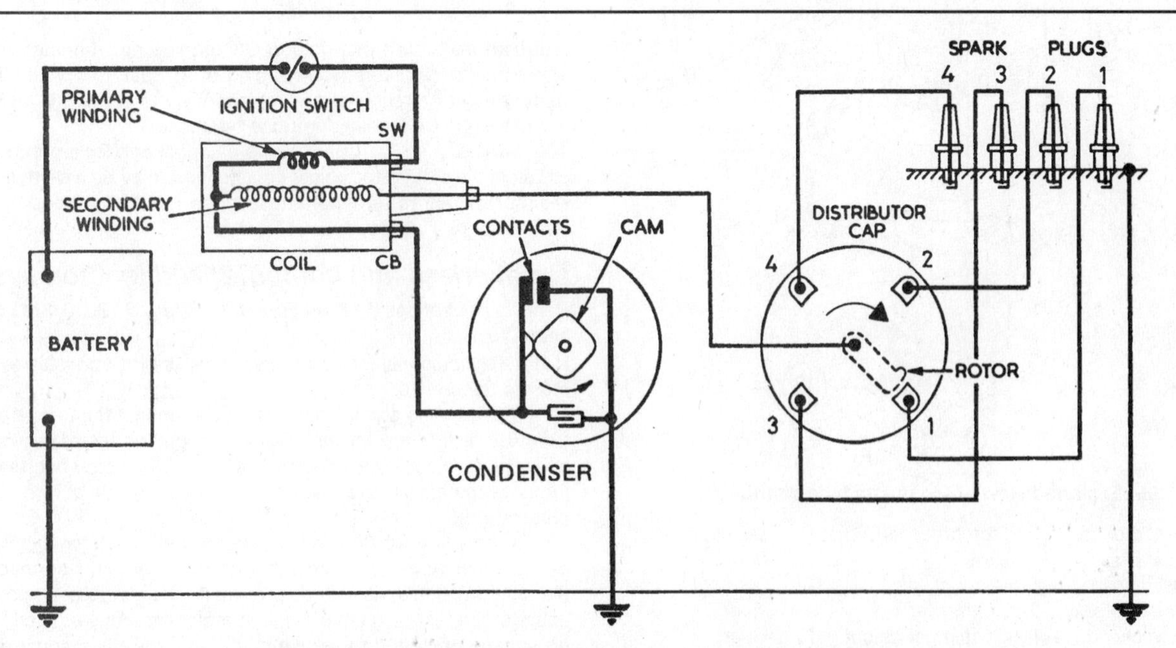

6.9a Schematic of the contact breaker type ignition system

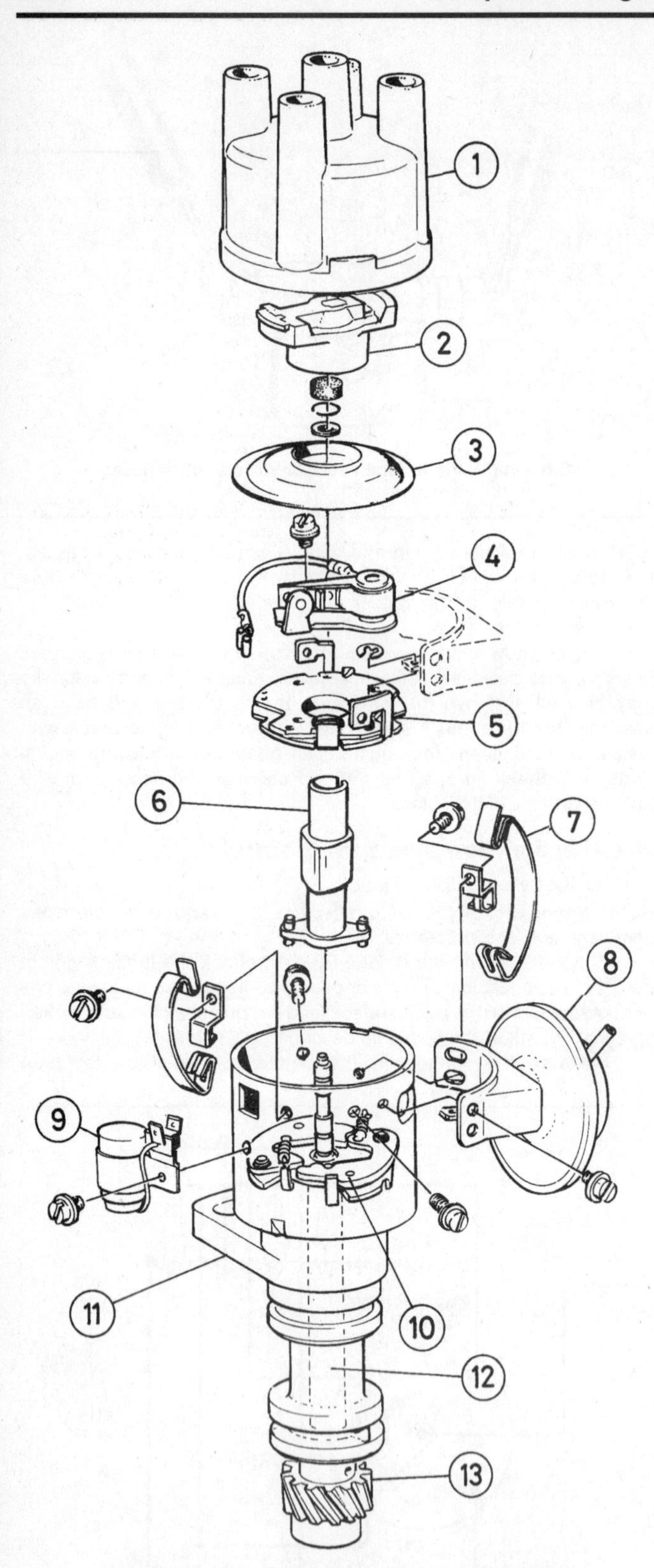

6.9b Exploded view of the points type distributor

1 Distributor cap
2 Rotor arm
3 Plastic cover
4 Contact breaker assembly
5 Baseplate
6 Cam lobe
7 Distributor cap clips
8 Vacuum unit
9 Condenser
10 Centrifugal weights assembly
11 Distributor body
12 Distributor drive shaft
13 Gear

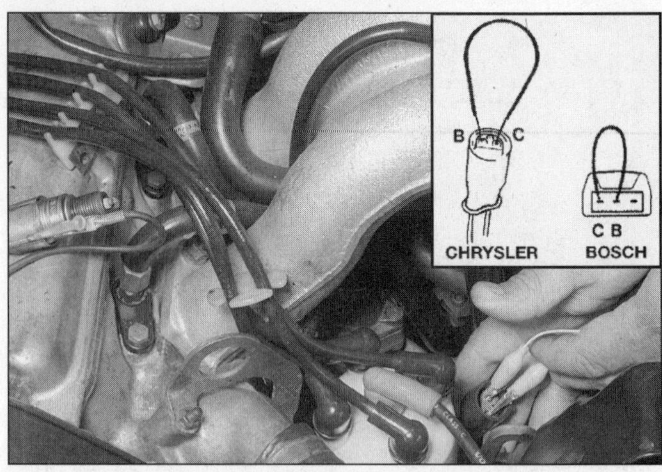

6.11 Jump terminals B and C and observe the spark at the plug. Be sure to ground the test plug or use a calibrated ignition tester (Chrysler ignition system shown)

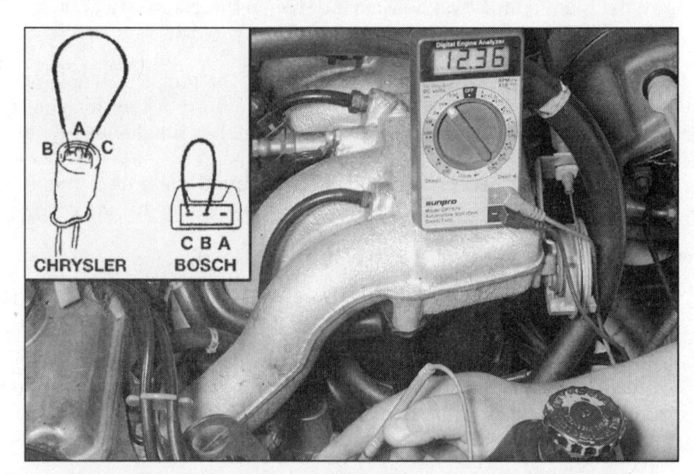

6.12a Check for battery voltage (B+) on terminal A of the distributor electrical connector

illustrations). With the ignition ON (engine not running), the points should produce a voltmeter reading of at least 10.5 volts. If not, the battery must be recharged or replaced. If the reading is 10.5 volts or more, record the reading for future reference.

10 If there is still no spark at the plugs, check for a ground or open circuit in the distributor/points circuit. There may be a damaged ballast resistor or points terminal causing the ignition voltage to become shorted or diminished.

Breakerless and computerized ignition systems

Refer to illustrations 6.11, 6.12a, 6.12b, 6.12c, 6.12d, 6.12e , 6.14 and 6.15

Note: *For additional ignition system checks on turbocharged engines, refer to Section 10*

11 If there is no spark (see Step 1), disconnect the 3-pole connector from the distributor. Install a calibrated ignition tester into one of the spark plug wires. Turn the ignition switch ON (engine not running) and jump terminals B and C and observe a spark at the plug **(see illustration)**.

12 If there is a spark, use a voltmeter and check for battery voltage on terminal A of the 3-pole distributor electrical connector **(see illustration)**. If there is voltage present (to the distributor) and there is a spark at the test plug (step 11) then replace the pick-up coil. If there is no voltage present (to the distributor) and there is a spark at the test plug (step 11), check the electrical circuit to the distributor **(see illustrations)**. If necessary, replace the ignition module or control unit.

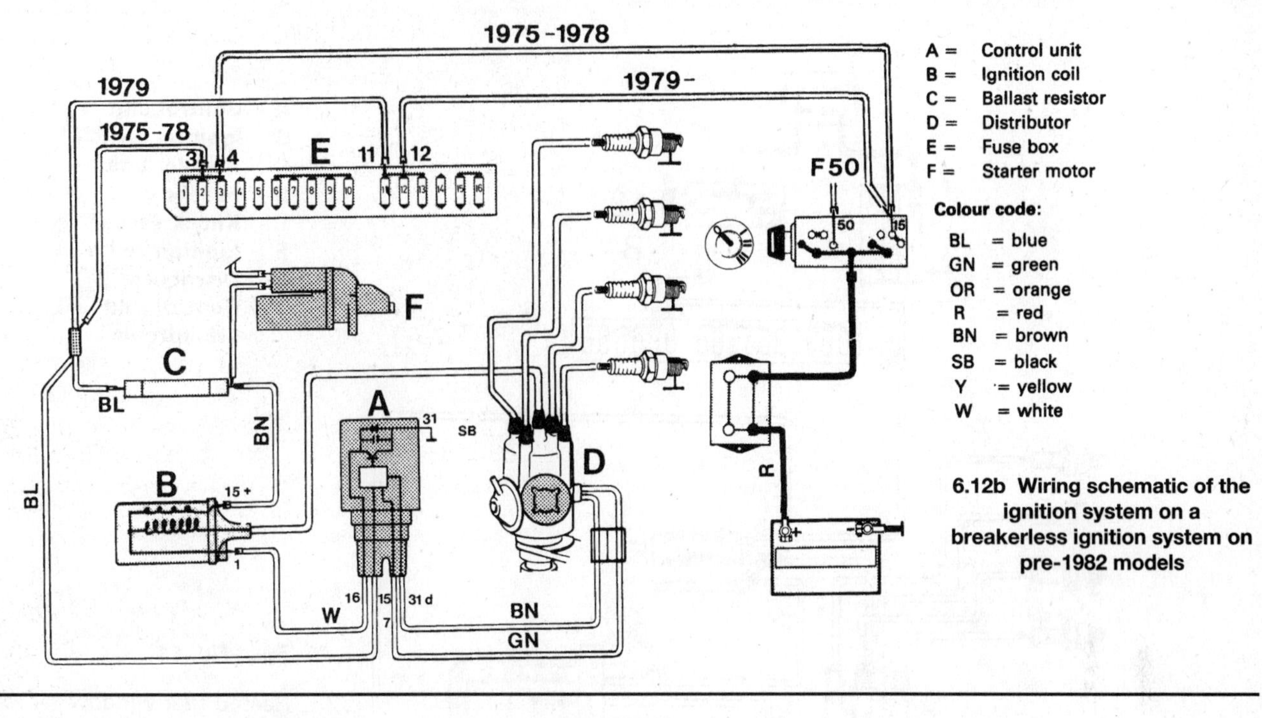

6.12b Wiring schematic of the ignition system on a breakerless ignition system on pre-1982 models

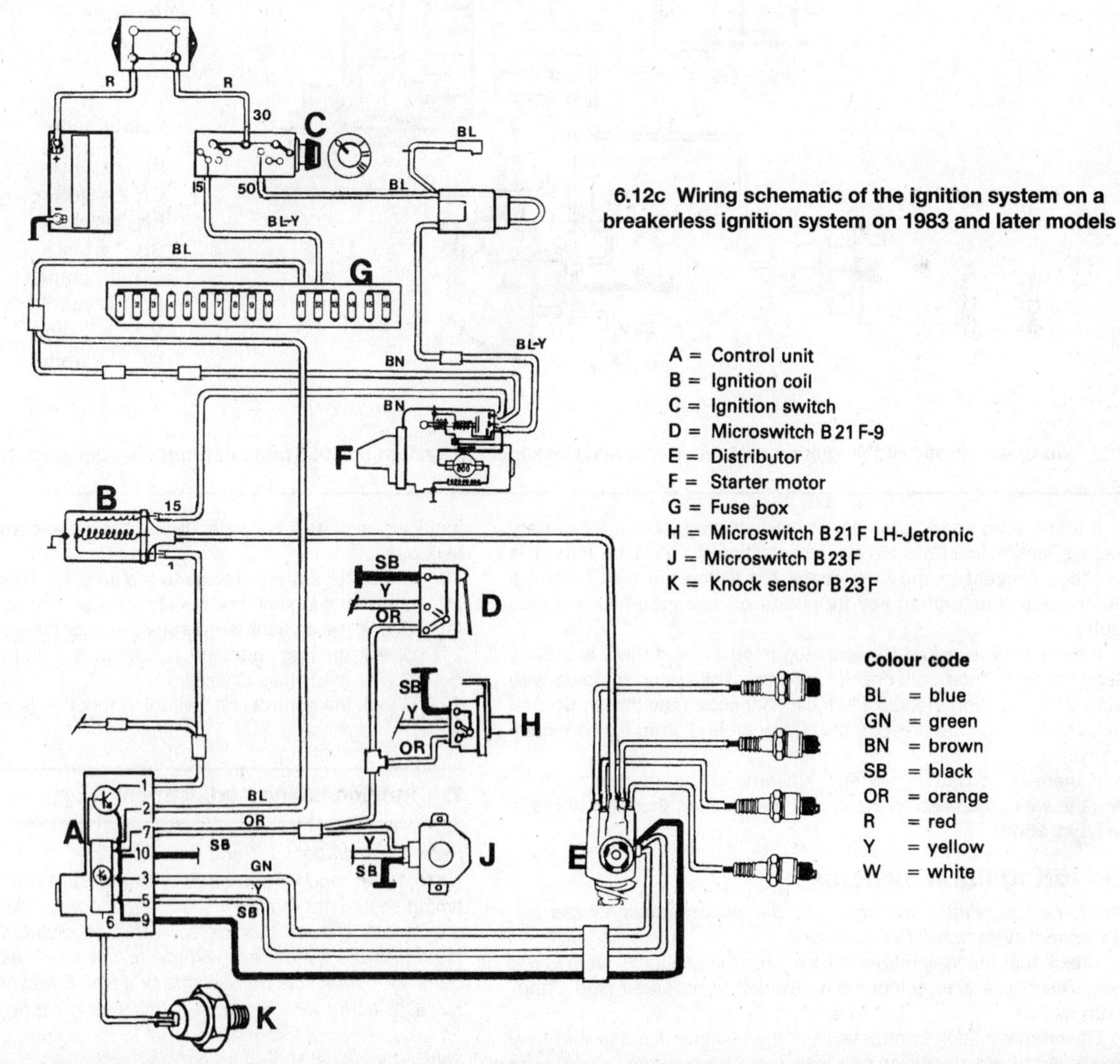

6.12c Wiring schematic of the ignition system on a breakerless ignition system on 1983 and later models

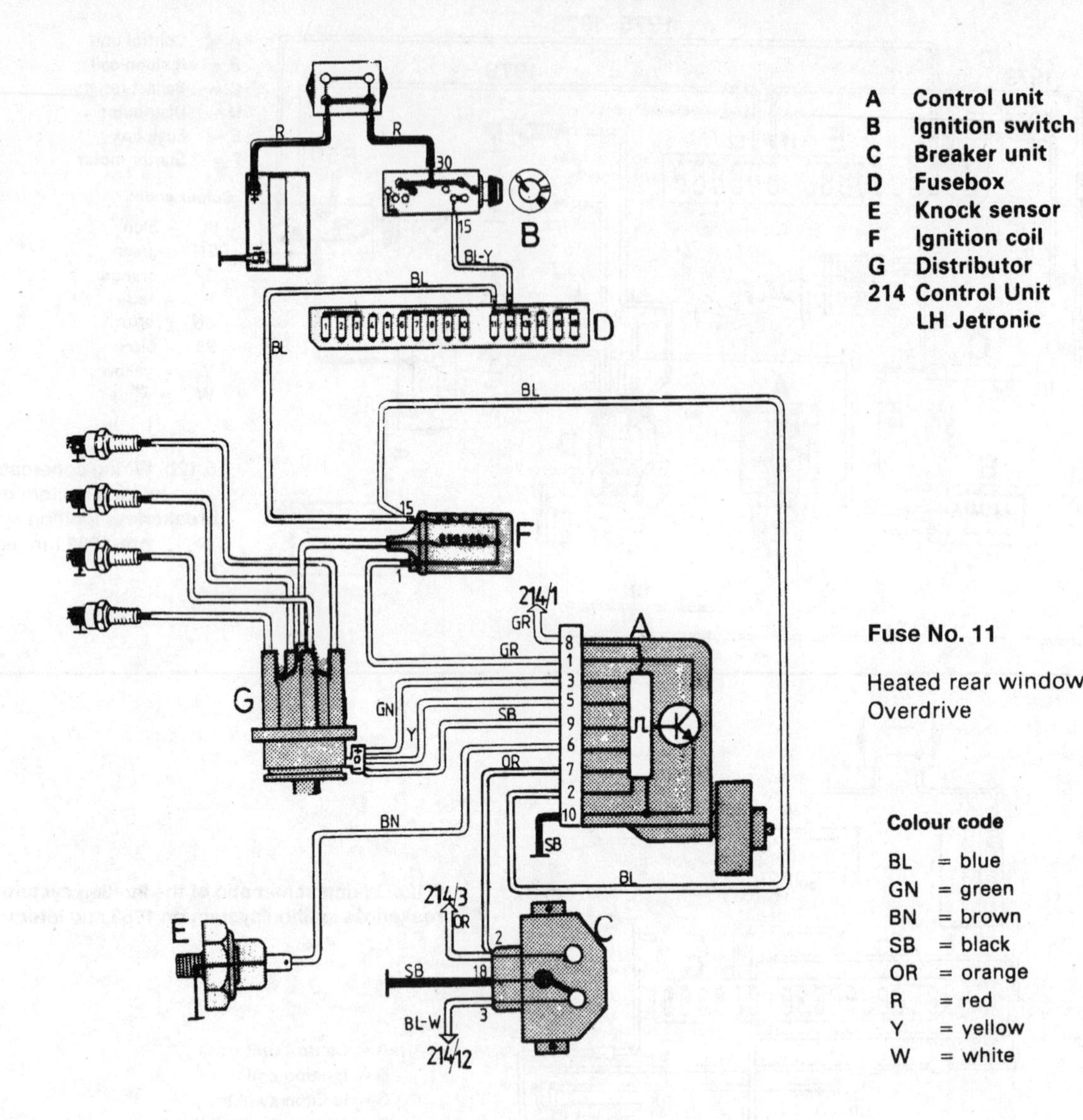

6.12d Wiring schematic of the ignition system on a breakerless ignition system on 1985 and later models equipped with the B230 engine

13 If there is no spark (step 11) at the test plug, check for battery voltage at terminal number 15 (+) on the ignition coil terminal. If there is no voltage present at the coil, check the wiring harness from the ignition coil to the ignition key for shorts or damaged harness (see Chapter 12).

14 If there is no spark at the test plug (step 11) and there is battery voltage present at the coil, check for battery voltage on the blue wire directly at the ignition module electrical connector **(see illustration)**. If there is battery voltage present, check the ground strap for continuity or damage.

15 If there is no battery voltage available at the ignition module, check the wiring harness to the ignition module for shorts or damage **(see illustration)**.

EZ-116K ignition systems

Note: *For information concerning the diagnostic codes for the EZ-116K ignition system, refer to Chapter 6.*

16 Check that the in-line fuse that governs the ignition system is not blown. The in-line fuse is located on the left front fender (see Chapter 12).

17 Check the ignition control unit for any obvious damage that may have resulted from a collision or a leak in the body of the vehicle. The ignition control unit is located below the glove compartment behind the kick panel.

18 Check for battery voltage to the amplifier (see Section 11).

19 Check the ignition coil resistance (see Section 9).

20 Check the coolant temperature sensor (see Chapter 6).

21 Check the engine speed sensor to make sure the rpm are being detected properly (see Chapter 6).

22 Check the control unit switching function (see Section 11).

7 Ignition timing - adjustment

Refer to illustrations 7.1 and 7.8

Note 1: *On models equipped with the EZ-116K ignition system, the timing should not require adjustment in normal service, although it may require adjustment if tampered with. The distributor functions only to distribute the secondary voltage to the spark plugs. The distributor does not contain counterweights or a Hall-Effect switch.*

Note 2: *If the information in this Section differs from the Vehicle Emission Control Information label in the engine compartment of your vehicle, the label should be considered correct.*

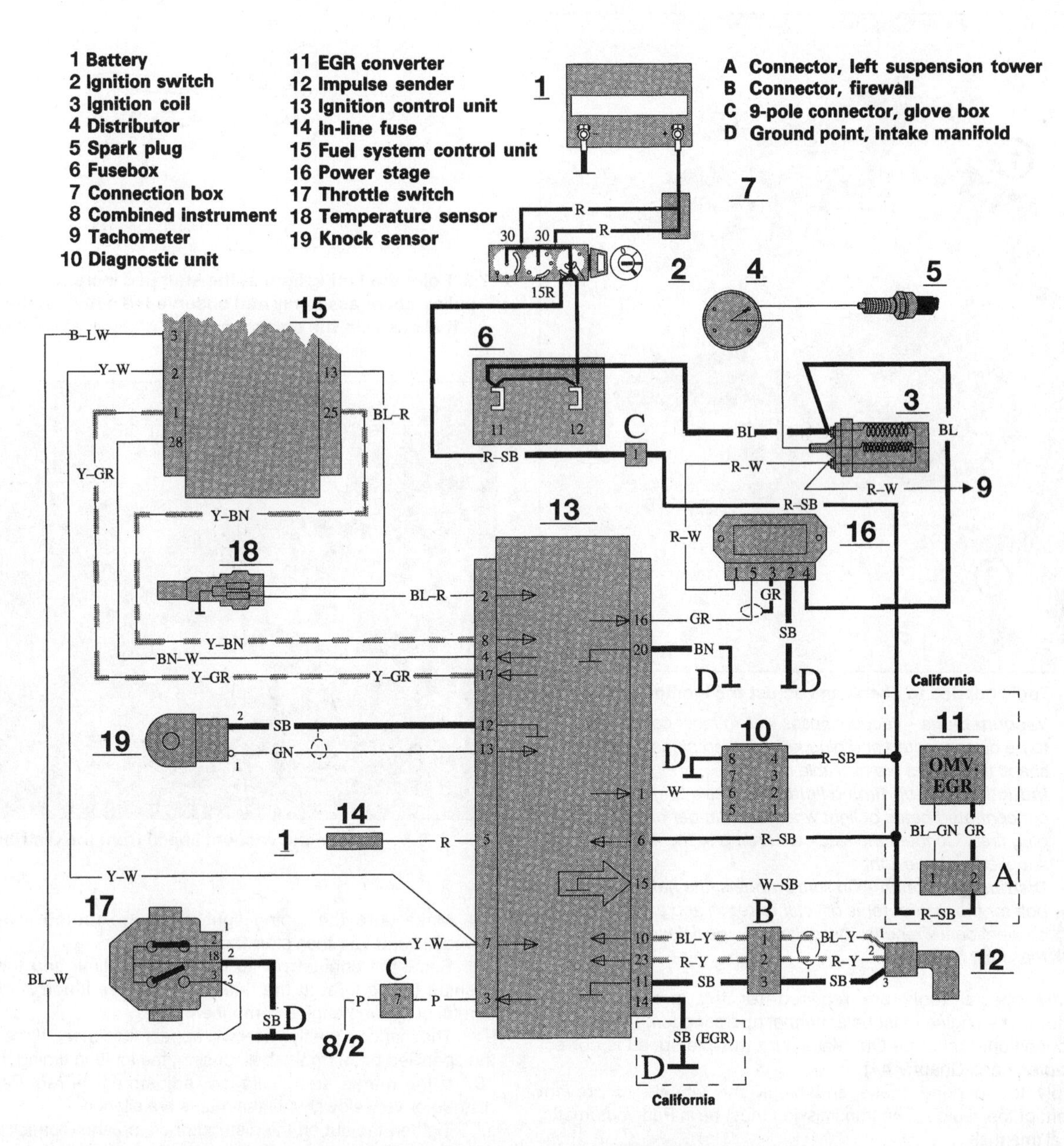

6.12e Wiring schematic of the EZ-116K ignition system on the B230F engine

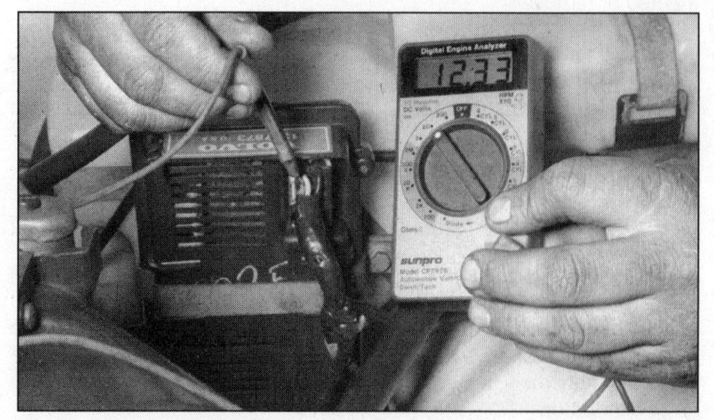

6.14 Check for battery voltage on the blue wire on the ignition module connector

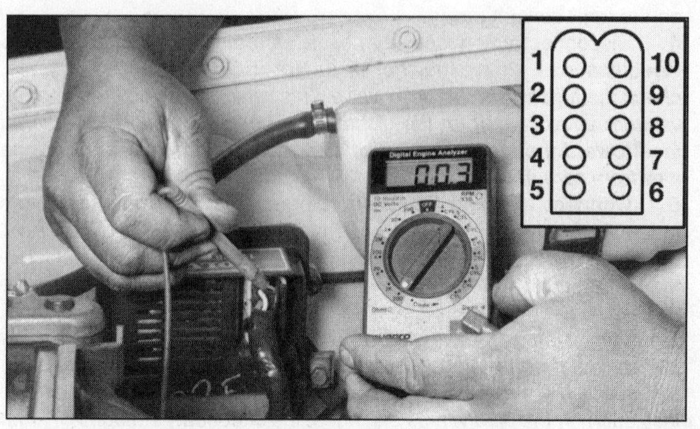

6.15 Check the ground circuit on terminal number 10 (top right). It should be less than 0.5 ohms resistance

5

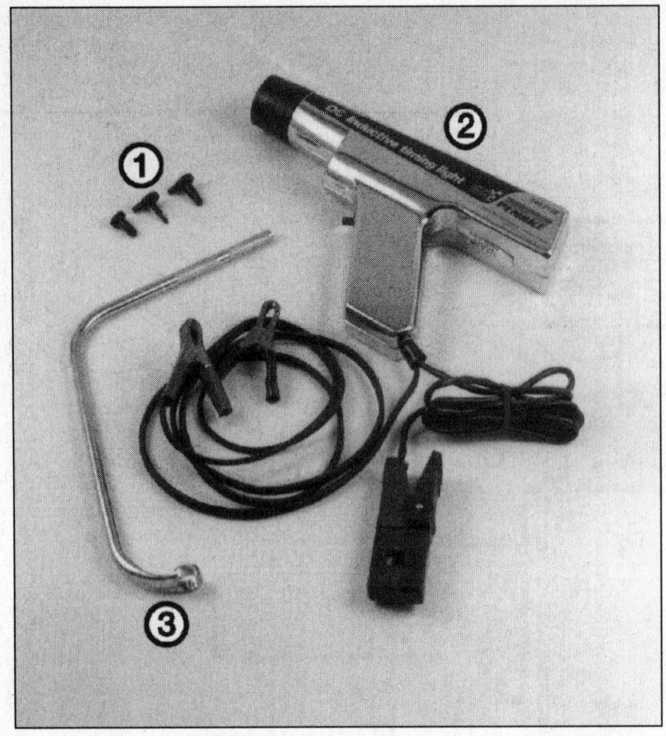

7.1 Tools needed to check and adjust the ignition timing

1 ***Vacuum plugs*** *- Vacuum hoses will, in most cases, have to be disconnected and plugged. Molded plugs in various shapes and sizes are available for this*

2 ***Inductive pick-up timing light*** *- Flashes a bright, concentrated beam of light when the number one spark plug fires. Connect the leads according to the instructions supplied with the light*

3 ***Distributor wrench*** *- On some models, the hold-down bolt for the distributor is difficult to reach and turn with conventional wrenches or sockets. A special wrench like this must be used*

1 Some special tools are required for this procedure **(see illustration)**. The engine must be at normal operating temperature and the air conditioner must be Off. Make sure the idle speed is correct (see Chapter 1 and Chapter 4A).

2 Apply the parking brake and block the wheels to prevent movement of the vehicle. The transmission must be in Park (automatic) or Neutral (manual).

3 The timing marks on these models are located on the front pulley and are viewed by observing the timing increments that are mounted on the front cover housing.

4 Disconnect any necessary hoses, as indicated in this Chapter's Specifications, then plug the hose. **Note:** *On carbureted engines, check the function of the delay valve. This valve is located in-line to the vacuum advance unit on the distributor. The designation* **DIST** *should face toward the distributor. Make sure that it is possible to blow one way from carburetor to vacuum advance unit only.*

5 Connect a timing light according to the tool manufacturer's instructions (an inductive pick-up timing light is preferred). Generally, the power leads are attached to the battery terminals and the pick-up lead is attached to the number one spark plug wire. The number one spark plug is the one closest to the drivebelt end of the engine. **Caution:** *If an inductive pick-up timing light isn't available, don't puncture the spark plug wire to attach the timing light pick-up lead. Instead, use an adapter between the spark plug and plug wire. If the insulation on the plug wire is damaged, the secondary voltage will jump to ground at the damaged point and the engine will misfire.*

6 With the ignition OFF, loosen the distributor clamp nut just enough to allow the distributor to pivot without any slipping.

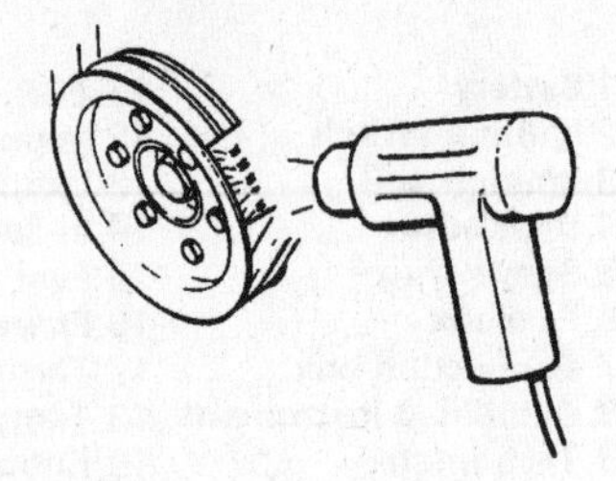

7.8 Point the timing light at the stamped increments on the front pulley/cover assembly and observe the notch on the pulley as it aligns with the correct number of degrees on the timing belt cover

8.5 Remove the vacuum line(s) from the distributor

7 Make sure the timing light wires are routed away from the drivebelts and fan, then start the engine.

8 Raise the engine rpm to the specified limit and then point the flashing timing light at the timing marks **(see illustration)** - be very careful of moving engine components.

9 The mark on the flywheel will appear stationary. If it's aligned with the specified point on the bellhousing, the ignition timing is correct.

10 If the marks aren't aligned, adjustment is required. Turn the distributor very slowly until the marks are aligned.

11 Tighten the nut on the distributor clamp and recheck the timing.

12 Turn off the engine and remove the timing light (and adapter, if used). Reconnect and install any components which were disconnected or removed.

8 Distributor - removal and installation

Removal

Refer to illustrations 8.5, 8.7a and 8.7b

1 After carefully marking them, remove the coil wire and spark plug wires from the distributor cap (see Chapter 1).

2 Remove the number one spark plug (the one nearest you when you are standing in front of the engine).

3 Manually rotate the engine to Top Dead Center (TDC) on the compression stroke for number one piston (see Chapter 2A)

4 Carefully mark the vacuum hoses if more than one is present on your distributor.

5 Disconnect the vacuum hose(s) **(see illustration)**.

6 Disconnect the primary wires from the distributor.

7 Mark the relationship of the rotor tip to the distributor housing

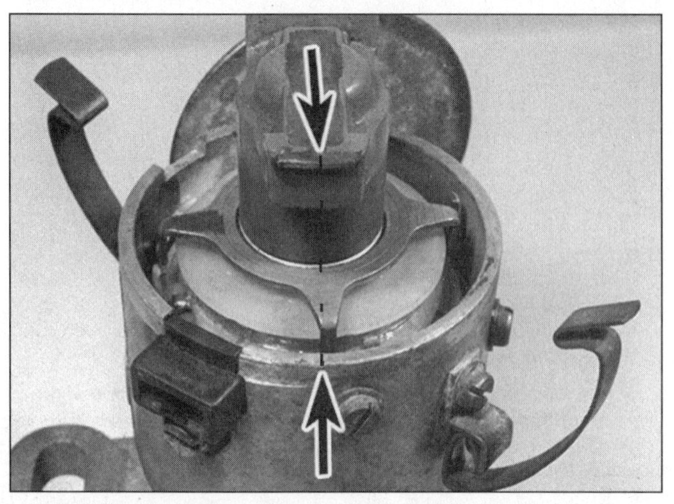

8.7a Align the rotor with the notch in the distributor housing

8.7b Before removing the distributor, make sure you paint alignment marks on the engine block and distributor

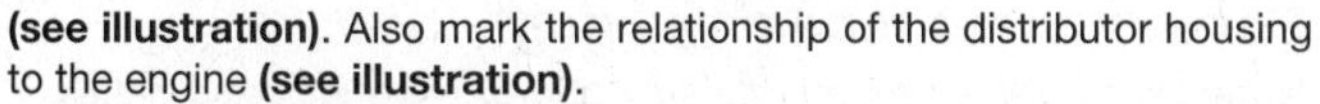

(see illustration). Also mark the relationship of the distributor housing to the engine **(see illustration)**.

8 Remove the hold-down bolt and clamp.

9 Remove the distributor. **Note:** *Do not rotate the engine with the distributor out.*

Installation

10 Before installing the distributor, make certain the number one piston is still at TDC on the compression stroke.

11 Insert the distributor into the engine with the adjusting clamp centered over the hold-down hole. Make sure the gear does not turn as the distributor is inserted.

12 Install the hold-down bolt. The marks previously made on the distributor housing and on the rotor and engine should line up before the bolt is tightened.

13 Install the distributor cap.

14 Connect the wiring for the distributor.

15 Install the spark plug wires.

16 Install the vacuum hoses as previously marked.

17 Adjust the ignition timing (see Section 7).

9 Ignition coil - check and replacement

Refer to illustrations 9.4 and 9.5

Caution: *If the coil terminals touch a ground source, the coil and/or impulse generator could be damaged.*

1 Mark the wires and terminals with pieces of numbered tape, then remove the primary wires and the high-tension wire from the coil.

2 Remove the coil assembly from its mount, clean the outer case and check it for cracks and other damage.

3 Inspect the coil primary terminals and the coil tower terminal for corrosion. Clean them with a wire brush if any corrosion is found.

4 Check the coil primary resistance by attaching the leads of an ohmmeter to the primary terminals **(see illustration)**. Compare the measured resistance to the Specifications listed in this Chapter.

5 Check the coil secondary resistance by hooking one of the ohmmeter leads to one of the primary terminals and the other ohmmeter lead to the coil high-tension terminal **(see illustration)**. Compare the measured resistance to the Specifications listed in this Chapter.

6 If the measured resistances are not as specified, the coil is defective and should be replaced with a new one.

7 It is essential for proper ignition system operation that all coil terminals and wire leads to be kept clean and dry.

8 Install the coil in its mount and hook up the wires. Installation is the reverse of removal.

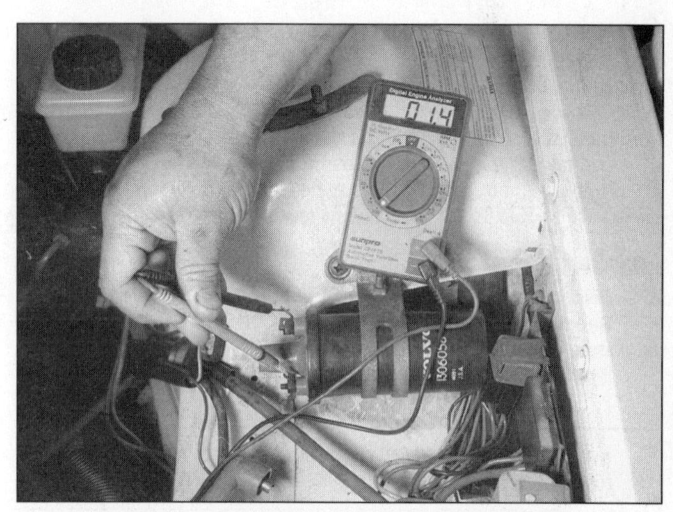

9.4 Checking ignition coil primary resistance

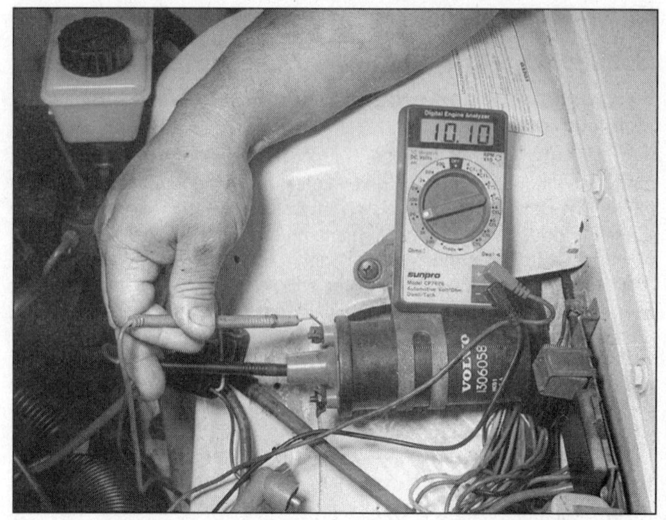

9.5 Checking ignition coil secondary resistance

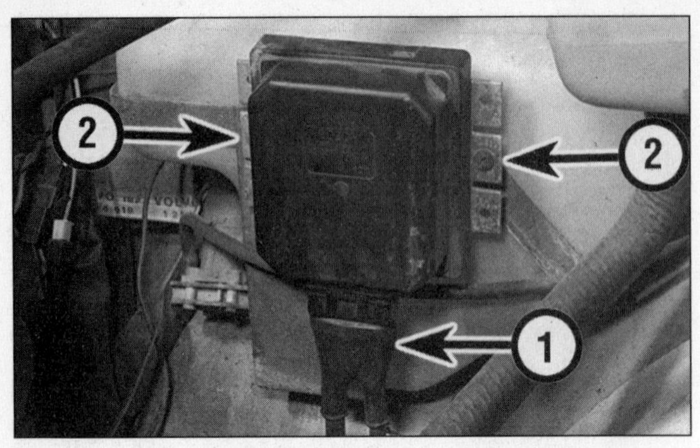

11.2 Ignition module on early breakerless ignition systems

1 Multi-plug 2 Securing screws

10 Ballast resistor - check and replacement

Check

1 The ballast resistor limits voltage to the coil during low speed operation but allows it to increase as the engine speed increases. While the engine is cranking, the ballast resistor is by-passed to ensure adequate voltage through the coil.

2 Disconnect the electrical leads from the ballast resistor and using an ohmmeter, check the resistance of the ballast resistor. The resistance should be 0.8 to 1.0 ohms **(see illustration 6.6)**.

Replacement

3 Disconnect the electrical leads from the ballast resistor.

4 Remove the screws from the ballast resistor and lift it from the engine compartment.

5 Installation is the reverse of removal.

11 Impulse generator and ignition control unit - check and replacement

1 The impulse generator and ignition control unit need to be tested in the event there is no spark at the spark plugs. Make sure the plug wires, ignition coil and spark plugs are working properly (see Section 6 and Section 9).

Breakerless ignition systems - checks

Voltage supply and ground to ignition control unit

Refer to illustration 11.2

2 With the ignition key ON, check for continuity on the ground circuit on the ignition control unit (ignition module). Refer to the ignition wiring diagrams 6.12b through 6.12d. Backprobe the black wire on the ignition module **(see illustration)** and chassis ground and measure the resistance of the ground circuit. It should not exceed 0.5 ohms.

3 Turn the ignition key ON. There should be battery voltage present at the ignition module. If there is no voltage, check the wiring harness for an open circuit. Backprobe the green wire on the ignition module **(see illustration 6.12b)**. It should be approximately 12 volts. **Note:** *If the ignition control unit passes all the tests and you still suspect there is a problem (possibly an intermittent problem), have the unit tested by a dealer service department.*

Impulse generator check

Refer to illustration 11.4

4 Using an ohmmeter, check the resistance of the impulse generator (pick-up coil) on breakerless type ignition systems. Install the probes of the ohmmeter onto the terminals of the electrical connector

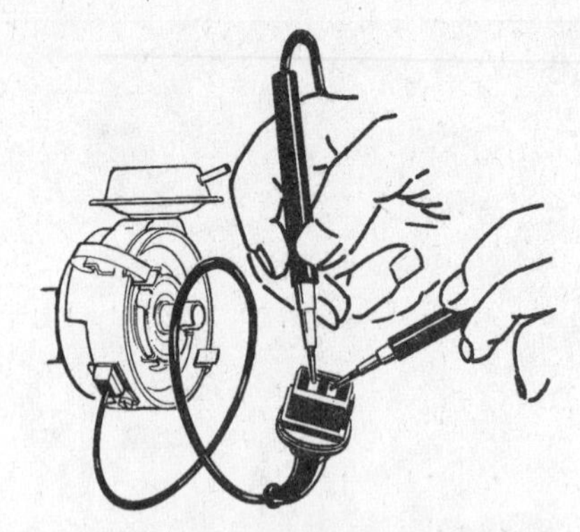

11.4 Check the resistance of the impulse generator using an ohmmeter at the harness connector. It should be 950 to 1,250 ohms

(see illustration) and measure the resistance. It should be 950 to 1,250 ohms.

Computerized ignition system - checks

Note: *On models equipped with the Volvo/Chrysler ignition systems, do not disconnect the harness connector from the control unit during the testing process. The connector uses special one-way sleeves that are damaged when disconnected. If the harness connector is removed, then the complete harness will need to be replaced.*

Voltage supply and ground to ignition control unit

5 With the ignition key ON, check for continuity on the ground circuit on the ignition control unit (ignition module). Refer to the ignition wiring diagrams 6.12b through 6.12d. Backprobe the black wire (terminal number 10 [top right]) **(see illustration 6.15)** and chassis ground and measure the resistance of the ground circuit. It should not exceed 0.5 ohms.

6 Turn the ignition key ON. There should be battery voltage present at the ignition module. If there is no voltage, check the wiring harness for an open circuit. Backprobe the blue wire (terminal number 2 [2nd from top left]) **(see illustration 6.14)**. It should be approximately 12 volts.

Impulse generator check

7 Check the switching function of the Hall-Effect switch in the distributor. Disconnect the 3-pole connector from the distributor. Install a calibrated ignition tester into one of the spark plug wires. Turn the ignition switch ON (engine not running) and jump terminals B and C and observe a spark at the plug **(see illustration 6.11)**.

8 Next, check for the presence of battery voltage at the distributor. Use a voltmeter and check for battery voltage on terminal A of the 3-pole distributor electrical connector **(see illustration 6.12a)**. If there is voltage present (to the distributor) and there is a spark at the test plug (step 5) then replace the pick-up coil. If there is no voltage present (to the distributor) and there is a spark at the test plug (step 5), check the electrical circuit to the distributor. If necessary, replace the ignition module (control unit).

9 Check the throttle switch. With the ignition key OFF, open the throttle slowly and listen for a click just as the throttle comes to a stop. If the click is not immediate, adjust the throttle switch by loosening the screws and rotating the switch clockwise and turning it counter-clockwise until the switch contacts the stop. Tighten the retaining screws and recheck the adjustment.

10 Check the knock sensor signal. Refer to the procedure for checking the knock sensor in Chapter 6.

EZ-116K ignition systems

Voltage supply and switching signal to amplifier

Refer to illustration 11.11

Note: *The amplifier is located on the left front fender, behind the*

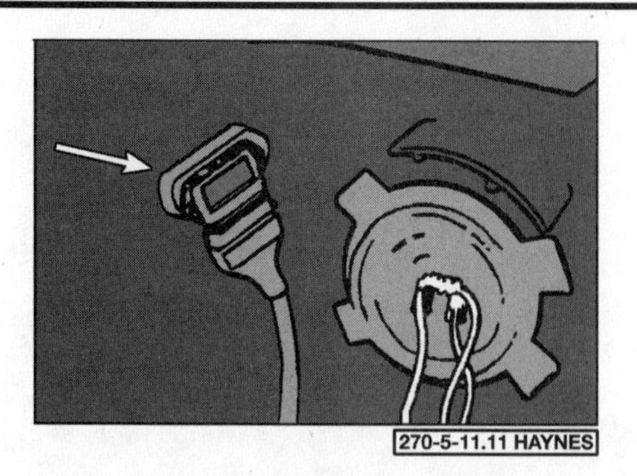

11.11 The EZ-116 ignition amplifier is located in the corner of the engine compartment (arrow)

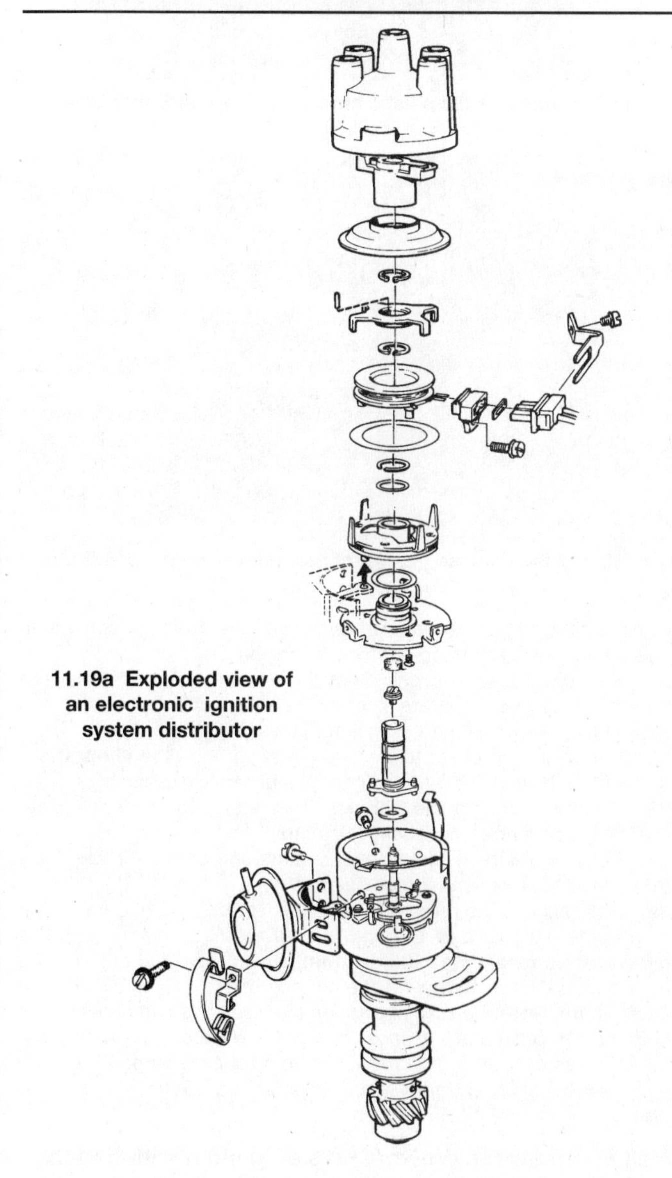

11.19a Exploded view of an electronic ignition system distributor

headlight assembly.

11 With the ignition key ON, peel back the boot on the amplifier electrical connector and check for voltage **(see illustration)** on terminals number 4 and number 2 **(see illustration 6.12e)**. There should be battery voltage present. If not, check the wiring harness for a short or open circuit.

12 With the ignition key OFF, disconnect the electrical connector from the amplifier and connect a voltmeter between terminal number 5 and ground **(see illustration 6.12e).** Crank the starter over and observe that the voltmeter fluctuates between 0 and 2 volts. If the test is correct, replace the amplifier. If the test is incorrect, have the ECU diagnosed at a dealer service department.

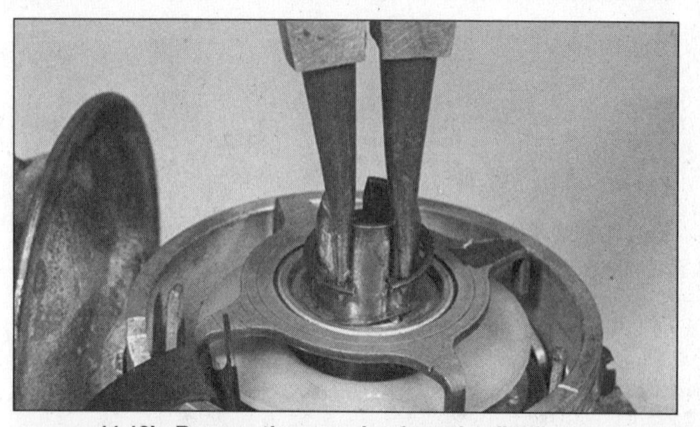

11.19b Remove the snap ring from the distributor

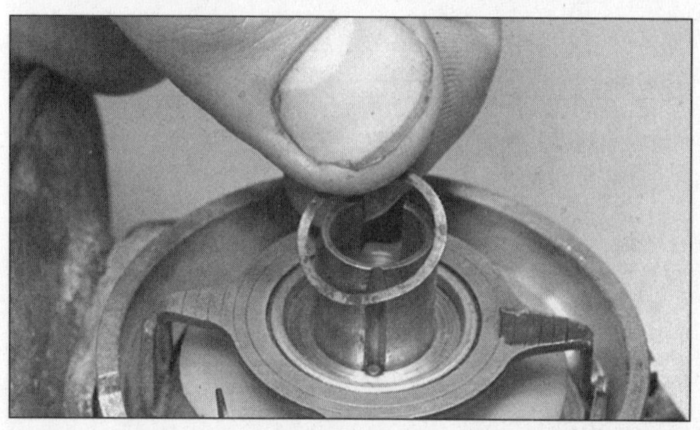

11.19c Also, remove the washer from the trigger wheel assembly

Replacement

Note: *On vehicles equipped with Volvo/Chrysler ignition systems, do not disconnect the harness connector from the ignition control unit. This connector uses special one-way sleeves that will become damaged during removal. The harness will have to be replaced in the event the connector is tampered with. Perform the tests on the module by carefully backprobing the electrical connector.*

Ignition control unit or amplifier

13 Make sure the ignition key is OFF.

14 Disconnect the electrical connector(s) from the control unit.

15 Remove the mounting screws from the control unit and lift it from the engine compartment.

16 Installation is the reverse of removal. **Note:** *On Bosch amplifiers, a special dielectric grease is used between the heat sink and the back side of the control unit. In the event the two are separated (replacement or testing) the old grease must be removed and the heat sink cleaned off using 180-grit sandpaper. Apply Curil K2 (Bosch part number 81 22 9 243). A silicon dielectric compound can be used as a substitute. This treatment is very important for the long life of these expensive ignition parts.*

Impulse generator (breakerless ignition)

Refer to illustrations 11.19a, 11.19b, 11.19c, 11.21a, 11.21b, 11.22, 11.23, 11.24, 11.25a and 11.25b

17 Disconnect the cable from the negative battery terminal.

18 Remove the distributor from the engine (see Section 8).

19 Using a pair of snap-ring pliers, remove the snap-ring retaining the trigger wheel **(see illustrations)**.

11.21a Remove the screws (arrow) from the vacuum advance unit

11.21b Remove the plastic block from the distributor body

11.22 Remove the snap ring that retains the impulse generator assembly

11.23 Lift the impulse generator assembly from the distributor

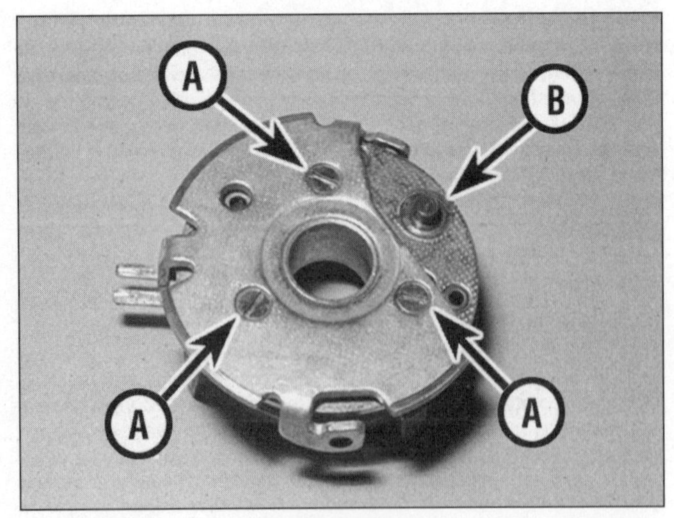

11.24 Remove the screws (A) that retain the impulse generator to the distributor plate. The vacuum advance arm is attached the post (B)

20 Use two flat-bladed screwdrivers positioned at opposite sides of the trigger wheel and carefully pry it up. **Note:** *Push the screwdrivers in as far as possible without bending the trigger wheel. Pry only on the strongest, center portion of the trigger wheel. In the event the trigger wheel is bent, it must be replaced with a new one.* **Note:** *Be sure not to lose the roll pin when lifting out the trigger wheel.*

21 Remove the two screws from the vacuum advance unit **(see illustrations)** and separate it from the distributor by moving the assembly down while unhooking it from the base plate pin.

22 Use snap ring pliers to remove the snap ring **(see illustration)** that retains the impulse generator and the base plate assembly.

23 Carefully remove the impulse generator and the base plate assembly as a single unit **(see illustration)**.

24 Remove the three screws and separate the base plate assembly from the impulse generator **(see illustration).**

25 Installation is the reverse of removal. **Note:** *Be sure to check the centrifugal weights and the springs before installing the impulse generator assembly* **(see illustration)**. **Note:** *Drive the pin into the assembly until it is flush with the surface of the trigger wheel* **(see illustration)**. **Note:** *Be sure to position the insulating ring between the coil and the base plate. It must be centered before tightening the mounting screws. Also, it is necessary to adjust the air gap once the trigger wheel has been removed or tampered with to the point that the clearance is incorrect* (see Section 12).

Hall Effect switch (Volvo/Chrysler ignition with Bosch distributor)

Refer to illustration 11.28

26 Disconnect the cable from the negative battery terminal.

27 Remove the distributor from the engine (see Section 8).

28 Using a pair of snap-ring pliers, remove the snap-ring retaining the trigger wheel **(see illustration)**.

29 Use two flat-bladed screwdrivers positioned at opposite sides of the trigger wheel and carefully pry it up. **Note:** *Push the screwdrivers in*

11.25a Make sure the centrifugal weights and the springs are in place and properly attached

11.25b Drive the pin into the trigger wheel assembly until it is flush with the surface

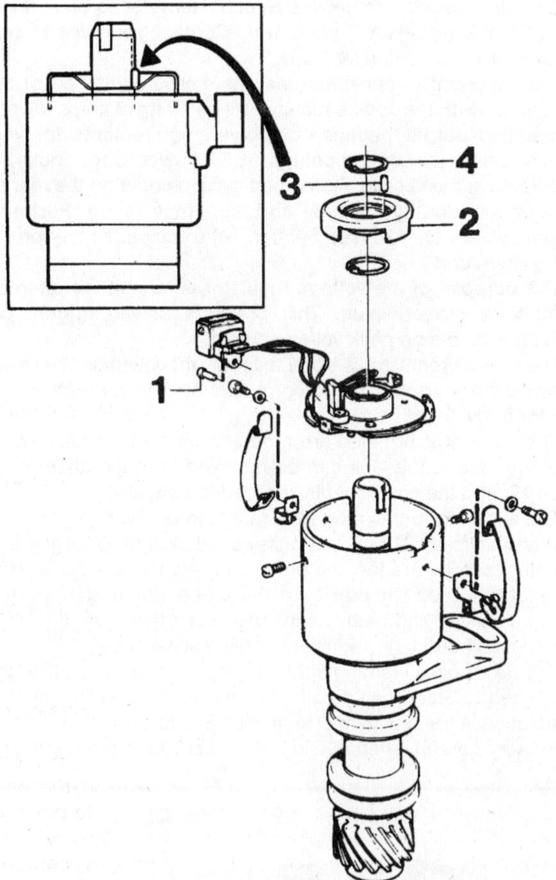

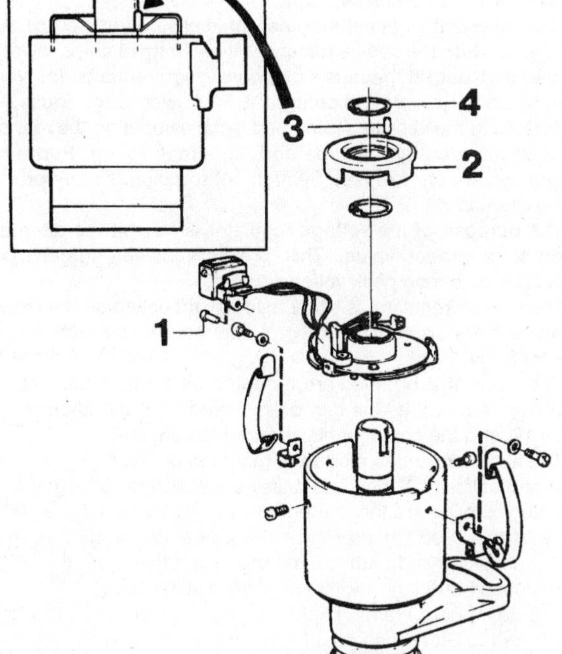

11.28 Exploded view of the Volvo/Chrysler ignition system in a Bosch distributor

1 Screw
2 Impulse generator
3 Pin
4 Snap-ring

as far as possible without bending the trigger wheel. Pry only on the strongest, center portion of the trigger wheel. In the event the trigger wheel is bent, it must be replaced with a new one. **Note:** *Be sure not to lose the roll pin when lifting out the trigger wheel.*

30 Remove the snap-ring and the spacer from above the Hall-Effect switch.

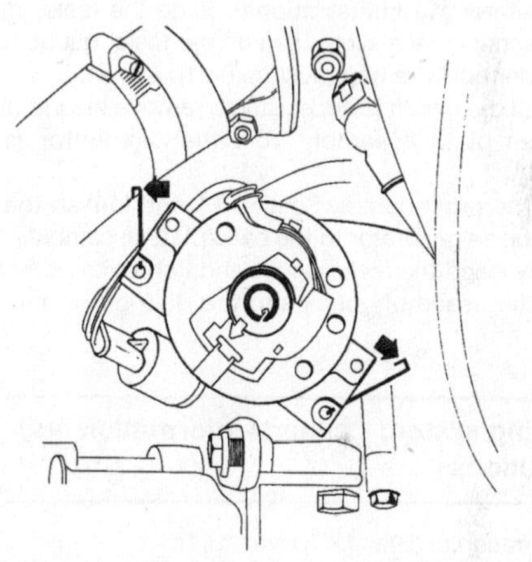

11.35 Press in the retaining clip to install the Hall Effect Switch in the distributor

31 Carefully pry the plastic connector from the side of the distributor body and lift the switch from the distributor. **Note:** *Make sure the spacer below the Hall Effect switch does not come out with the assembly.*

32 Installation is the reverse of removal.

Hall Effect switch (Volvo/Chrysler ignition with Chrysler distributor)

Refer to illustration 11.35

33 Disconnect the cable from the negative battery terminal.

34 Remove the distributor from the engine (see Section 8).

35 Pry open the spring clips holding the sender to the distributor body **(see illustration)**.

36 Install the new Hall Effect switch into the body of the distributor and press in the retaining clips. Make sure the wiring is installed properly in such a way as to not interfere with the rotation of the rotor and shaft.

37 Installation is the reverse of removal.

12 Air gap - check and adjustment

Refer to illustrations 12.2a and 12.2b

1 Disconnect the cable from the negative terminal on the battery. **Caution:** *It is necessary to make sure the radio is turned OFF before disconnecting the battery cables to avoid damaging the microprocessor built into the radio.*

5

12.2a Use a feeler gauge to check the air gap (be sure the gauge rubs lightly against the trigger wheel as well as the locating pin for the correct adjustment

2 Insert a brass feeler gauge between the trigger wheel tab and the impulse generator **(see illustrations)**. Slide the feeler gauge up and down - you should feel a slight drag on the feeler gauge as it is moved if the gap is correct. The gap must be 0.010 inches.

3 To adjust the gap, it is necessary to remove the impulse generator and the base plate assembly from the distributor **(see illustration 11.24)**.

4 Follow the procedure in Section 11 and loosen the screws that retain the impulse generator to the base plate assembly.

5 Carefully insert the feeler gauge and tighten the screws.

6 Install the assembly back into the distributor and recheck the adjustment.

13 Charging system - general information and precautions

Refer to illustrations 13.1a, 13.1b and 13.1c

This alternator has an output amperage rating between 12 and 90 amps depending on the load and the engine rpm. There are two different types of alternators **(see illustrations)** installed on these models; Bosch and Marchel. Also, there are different amperage ratings available:

1976 and 1977	Bosch 55 amps
1978 through 1981	Bosch 55 amps (different terminal arrangement)
1978 through 1981	Marchel SEV 55 amps
1981 (B21)	Bosch 70 amps
1981 (B23 and B230)	Bosch 70 amps (different output ratings)
1981 turbocharged	Bosch 55 amps (less output at high rpm)
1982 on	Bosch 55 amps
1982 on	Bosch 70 amps
1982 on	Bosch 80 amps

12.2b Top view of the air gap adjustment

Note: *It is possible to install a special kit that will adapt the wiring and the alternator brackets to allow a Bosch alternator to be installed in place of the Marchel SEV alternator. Contact your dealer service department for the completer kit.*

It is important to remember that the measured amperage output at the battery with the engine running will be 10 to 15 amps lower than the specified output because of power requirements for various electrical units (air conditioning, ABS, power door locks, etc.) incorporated in the vehicle. A stamped serial number on the rear of the alternator will identify the type and amperage rating. Perform the charging system checks (see Section 14) to diagnose any problems with the alternator.

The purpose of the voltage regulator is to limit the alternator's voltage to a preset value. This prevents power surges, circuit overloads, etc., during peak voltage output.

The voltage regulator is either external, mounting on the firewall in the engine compartment, or integral, mounted to the outside of the alternator body.

The alternator brushes are mounted as a single assembly. On Bosch alternators, this unit can be removed from the alternator (see Section 17) and the components serviced individually.

The alternator on all models is mounted on the right, front of the engine and utilizes a V-belt and pulley drive system. Drivebelt tension and battery service are the two primary maintenance requirements for these systems. See Chapter 1 for the procedures regarding engine drivebelt checking and battery servicing. Other than that, the charging system doesn't ordinarily require periodic maintenance.

The dashboard warning light should come on when the ignition key is turned to Start, then go off immediately. If it remains on, there is a malfunction in the charging system (see Section 14).

Be very careful when making electrical circuit connections to a

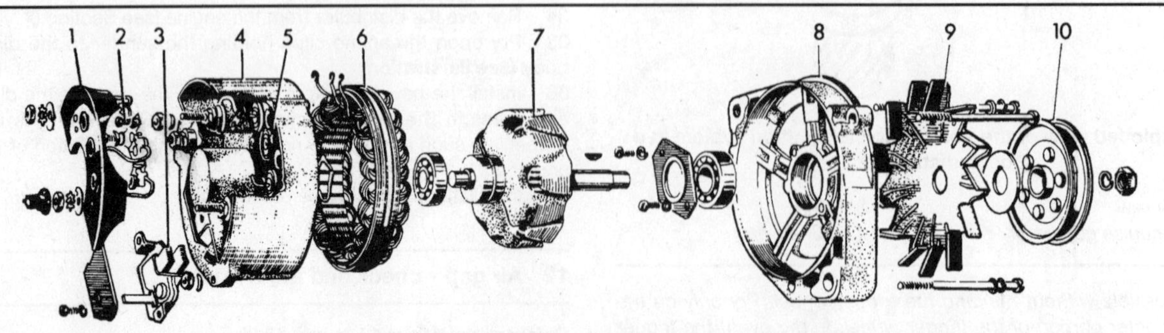

13.1a Exploded view of an early Bosch alternator

1 Rectifier (plus diode plate)
2 Magnetizing rectifier
3 Brush holder
4 Slip ring end shield
5 Rectifier (negative diodes)
6 Stator
7 Rotor
8 Drive end shield
9 Fan
10 Pulley

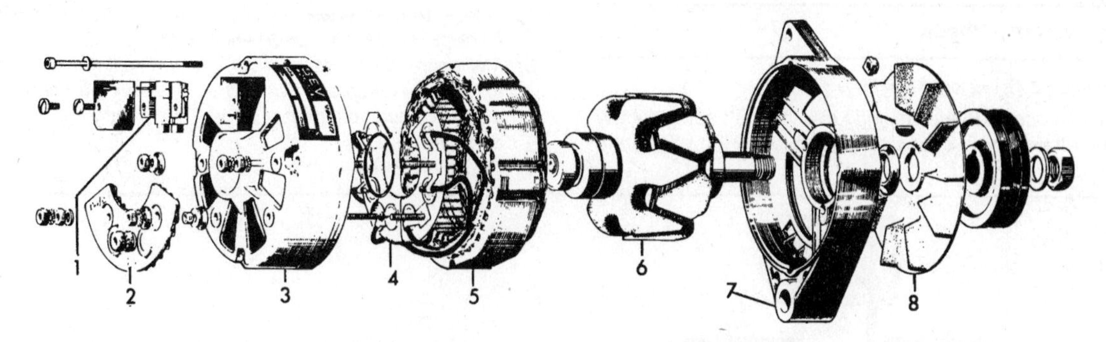

13.1b Exploded view of the Marchel SEV alternator

1 *Brush holder*
2 *Isolation diodes with holder*
3 *Slip ring end shield*
4 *Rectifier (silicon diodes)*
5 *Stator*
6 *Rotor*
7 *Drive end shield*
8 *Fan*

vehicle equipped with an alternator and note the following:

a) *When reconnecting wires to the alternator from the battery, be sure to note the polarity.*
b) *Before using arc welding equipment to repair any part of the vehicle, disconnect the wires from the alternator and the battery terminals.*
c) *Never start the engine with a battery charger connected.*
d) *Always disconnect both battery cables before using a battery charger.* **Caution:** *It is necessary to make sure the radio is turned OFF before disconnecting the battery cables to avoid damaging the microprocessor built into the radio.*
e) *The alternator is turned by an engine drivebelt which could cause serious injury if your hands, hair or clothes become entangled in it with the engine running.*
f) *Because the alternator is connected directly to the battery, it could arc or cause a fire if overloaded or shorted out.*
g) *Wrap a plastic bag over the alternator and secure it with rubber bands before steam cleaning the engine.*

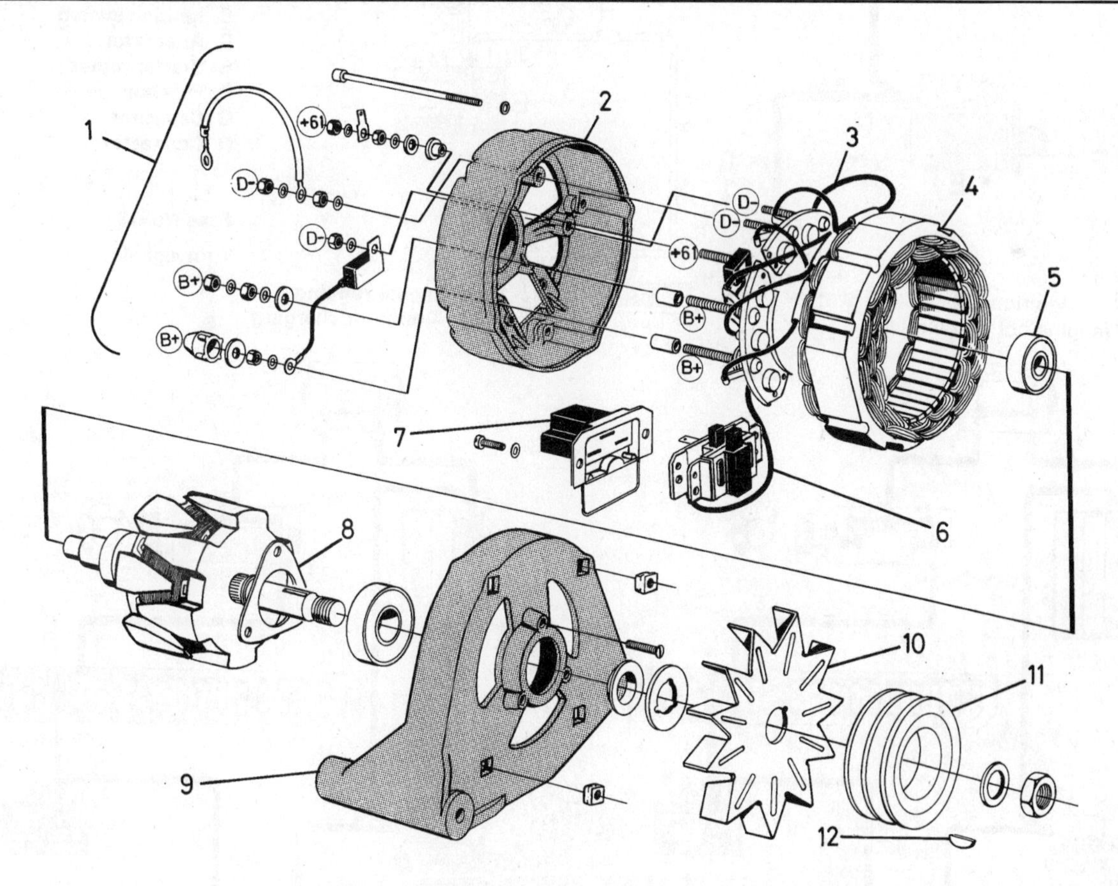

13.1c Exploded view of the late model Marchel SEV alternator

1 *Connections*
2 *End cover*
3 *Rectifier*
4 *Stator*
5 *Bearing*
6 *Brush holder*
7 *Voltage regulator*
8 *Rotor*
9 *Drive end shield*
10 *Fan*
11 *Pulley*
12 *Key*

14 Charging system - check

Refer to illustrations 14.1a, 14.1b, 14.1c and 14.d

1 If a malfunction occurs in the charging circuit, don't automatically assume that the alternator is causing the problem. First check the following items:

a) Check the drivebelt tension and condition (see Chapter 1). Replace it if it's worn or deteriorated.

b) Make sure the alternator mounting and adjustment bolts are tight.

c) Inspect the alternator wiring harness and the connectors at the alternator and voltage regulator. They must be in good condition and tight.

d) Check the fuses.

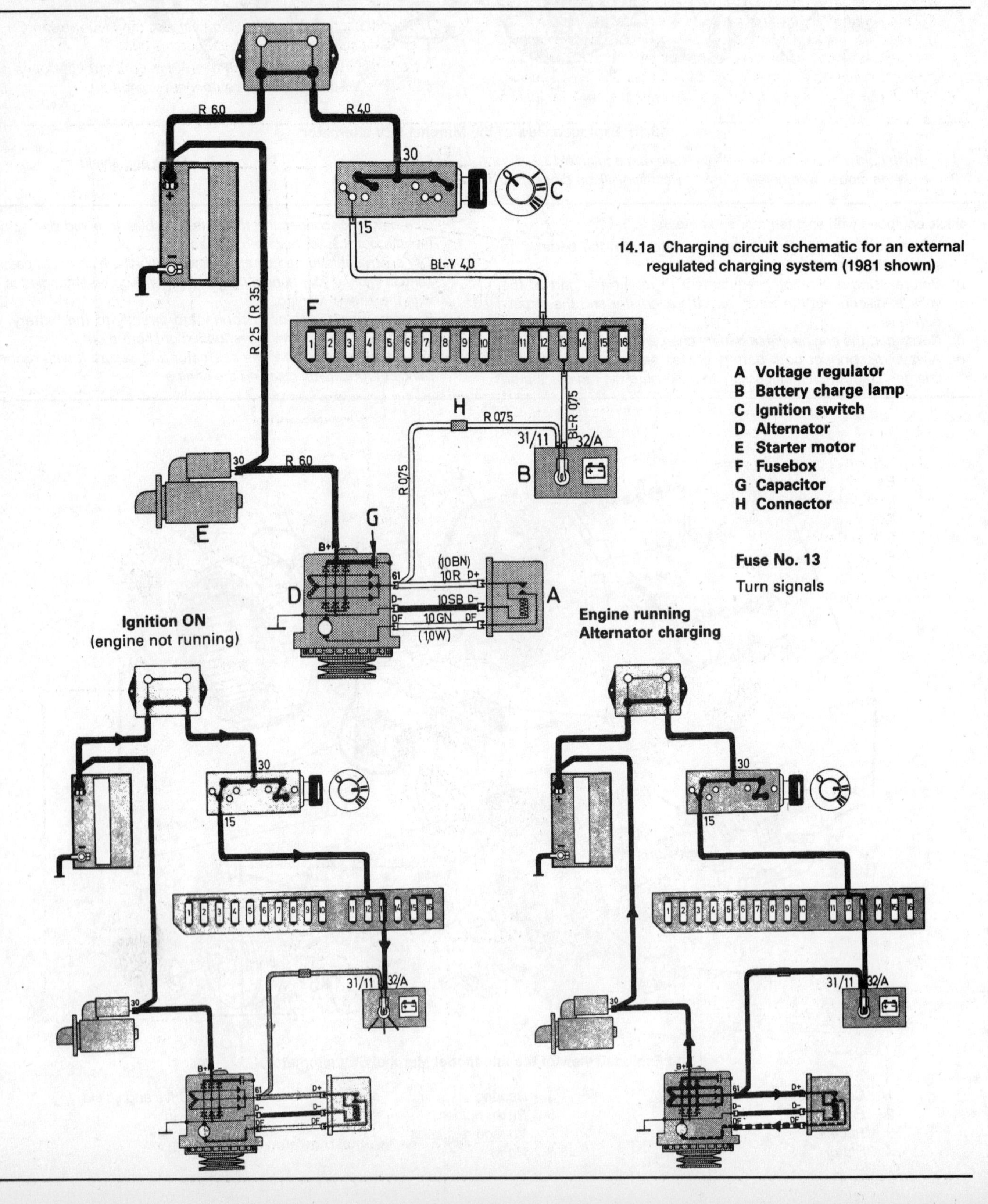

14.1a Charging circuit schematic for an external regulated charging system (1981 shown)

e) Start the engine and check the alternator for abnormal noises (a shrieking or squealing sound indicates a bad bearing).
f) Check the specific gravity of the battery electrolyte. If it's low, charge the battery (doesn't apply to maintenance-free batteries).
g) Make sure the battery is fully charged (one bad cell in a battery can cause overcharging by the alternator).
h) Disconnect the battery cables (negative first, then positive). Inspect the battery posts and the cable clamps for corrosion. Clean them thoroughly if necessary (see Chapter 1). **Caution:** It is necessary to make sure the radio is turned OFF before disconnecting the battery cables to avoid damaging the microprocessor built into the radio. Reconnect the cable to the positive terminal.
i) With the key off, connect a test light between the negative battery post and the disconnected negative cable clamp.
1) If the test light does not come on, reattach the clamp and proceed to Step 3.
2) If the test light comes on, there is a short (drain) in the electrical system of the vehicle. The short must be repaired before the charging system can be checked. **Note:** Accessories which are always on (such as the clock) must be disconnected before performing this check.
3) Disconnect the alternator wiring harness **(see illustrations)**.
   a) If the light goes out, the alternator is bad.
   b) If the light stays on, pull each fuse until the light goes out (this will tell you which component is shorted).

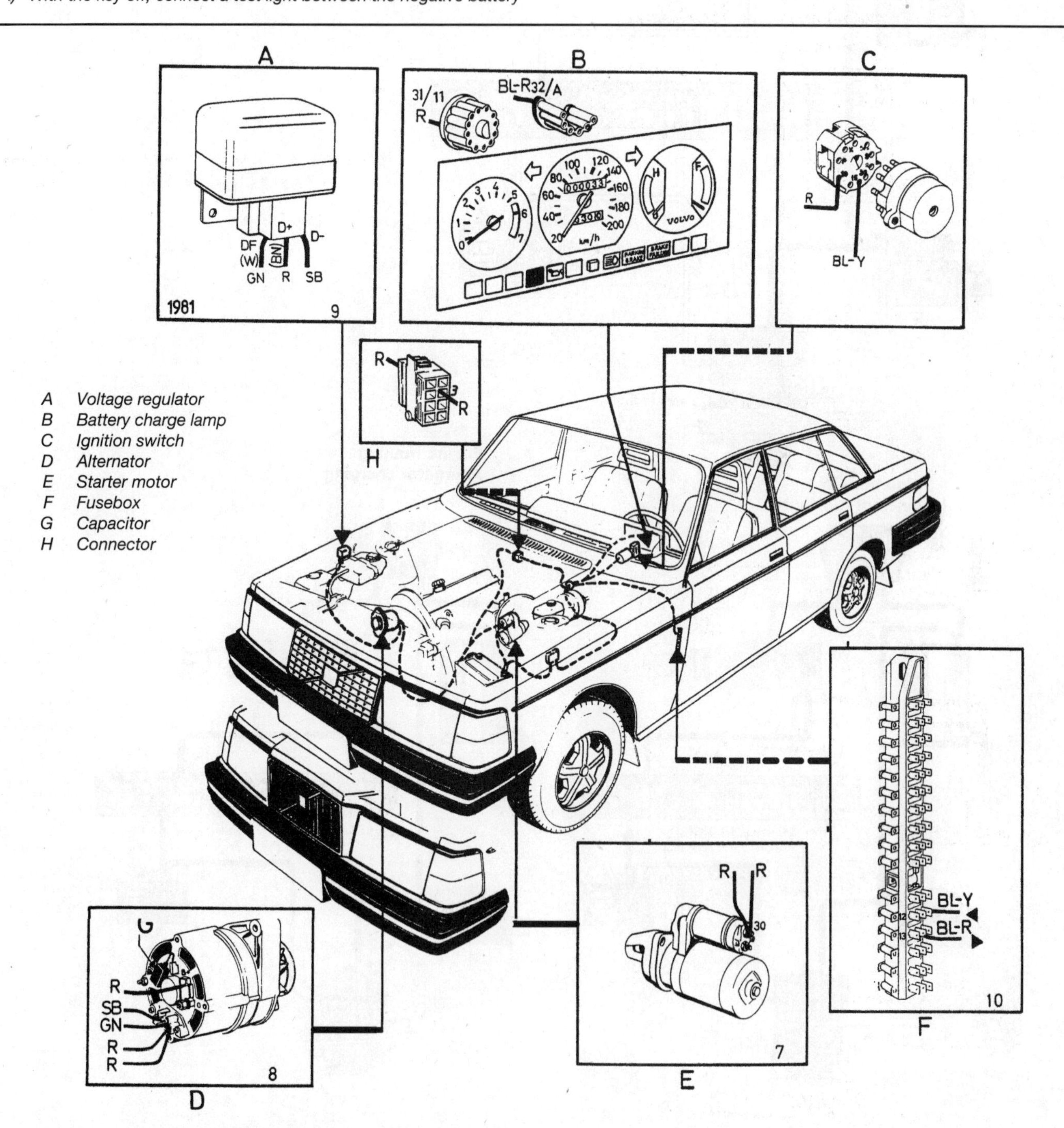

14.1b Component locations for the charging system on an external regulated charging system

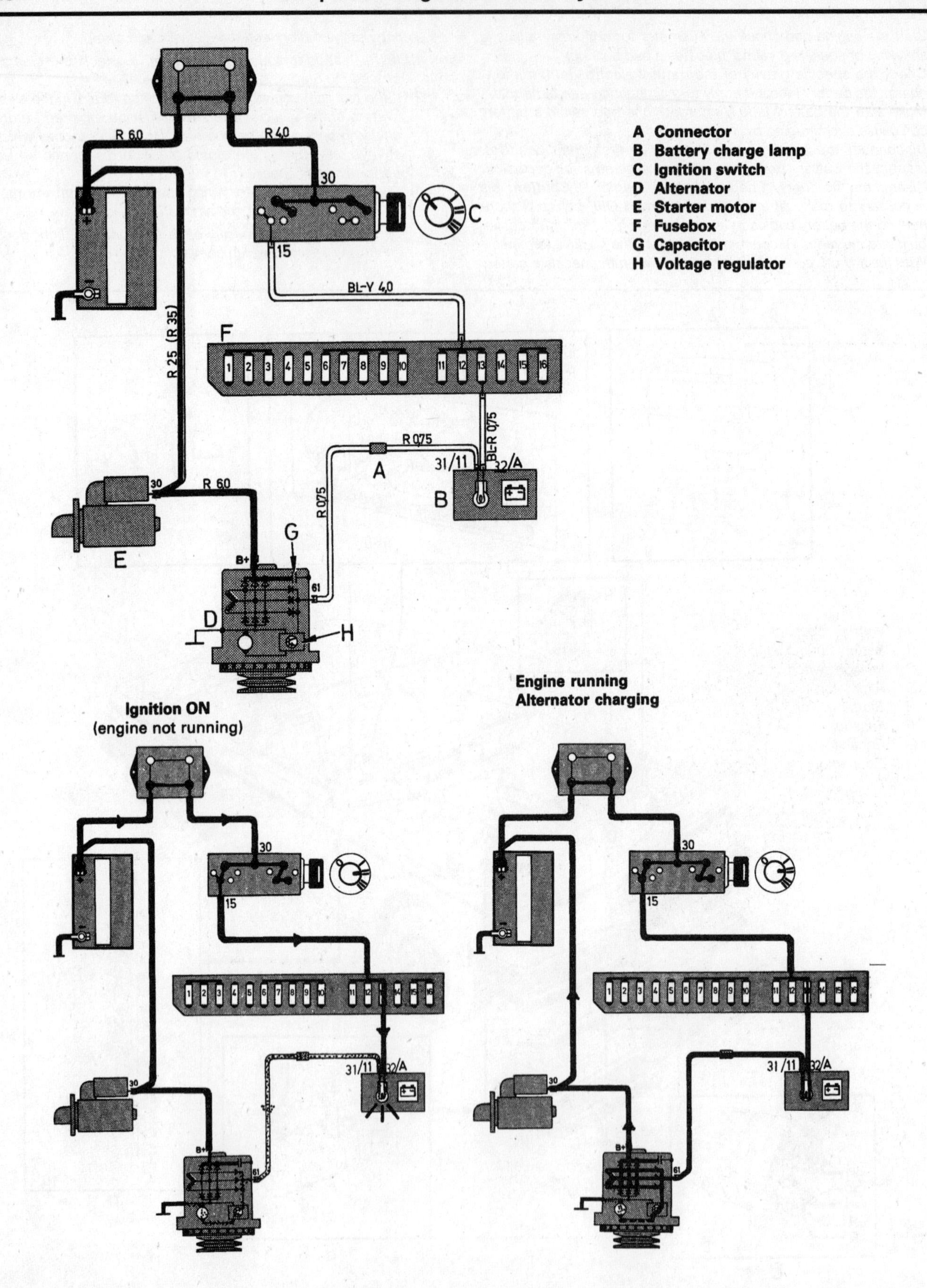

14.1c Charging circuit schematic for an integral regulated charging system (1983 shown)

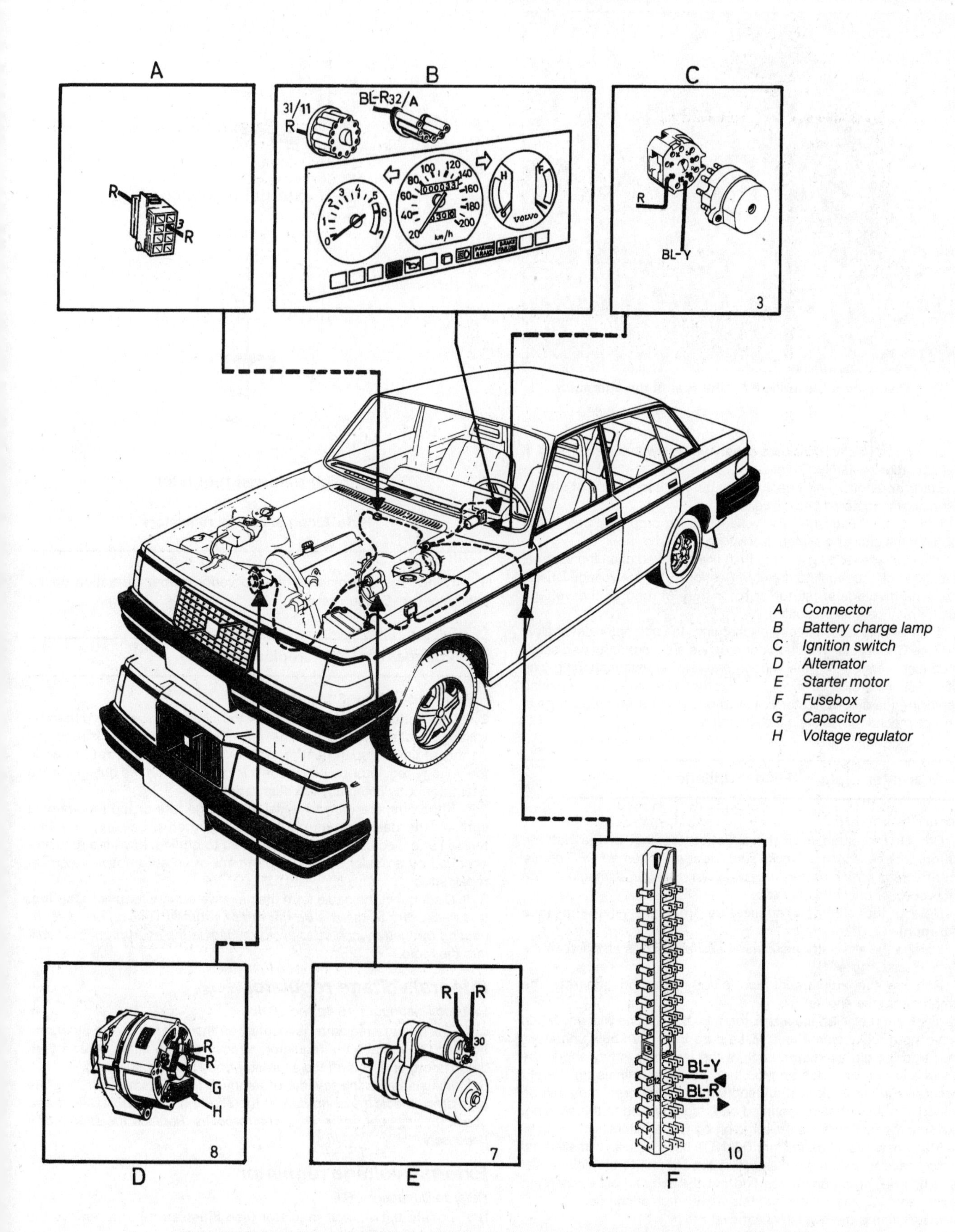

14.1d Component locations for the charging system on an integral regulated charging system

15.2 Disconnect the leads from the rear of the alternator

2 Using a voltmeter, check the battery voltage with the engine off. If should be approximately 12-volts.
3 Start the engine and check the battery voltage again. It should now be approximately 14-to-15 volts.
4 Turn on the headlights. The voltage should drop, and then come back up, if the charging system is working properly.
5 If the voltage reading is more than the specified charging voltage, replace the voltage regulator (refer to Section 16). If the voltage is less, the alternator diode(s), stator or rotor may be bad or the voltage regulator may be malfunctioning.
6 If the battery is constantly discharging, the alternator drivebelt is loose (see Chapter 1), the alternator brushes are worn, dirty or disconnected (see Section 17), the voltage regulator is malfunctioning (see Section 16) or the diodes, stator coil or rotor coil is defective. Repairing or replacing the diodes, stator coil or rotor coil is beyond the scope of the home mechanic. Replace the alternator.

15 Alternator - removal and installation

Refer to illustration 15.2

1 Detach the cable from the negative terminal of the battery. **Caution:** *It is necessary to make sure the radio is turned OFF before disconnecting the battery cables to avoid damaging the microprocessor built into the radio.*
2 Detach the electrical connectors from the alternator **(see illustration)**.
3 Loosen the alternator adjustment and pivot bolts and detach the drivebelt (see Chapter 1).
4 Remove the adjustment and pivot bolts and separate the alternator from the engine.
5 If you are replacing the alternator, take the old one with you when purchasing a replacement unit. Make sure the new/rebuilt unit looks identical to the old alternator. Look at the terminals - they should be the same in number, size and location as the terminals on the old alternator. Finally, look at the identification numbers - they will be stamped into the housing or printed on a tag attached to the housing. Make sure the numbers are the same on both alternators.
6 Many new/rebuilt alternators DO NOT have a pulley installed, so you may have to switch the pulley from the old unit to the new/rebuilt one. When buying an alternator, find out the shop's policy regarding pulleys - some shops will perform this service free of charge.
7 Installation is the reverse of removal.
8 After the alternator is installed, adjust the drivebelt tension (see Chapter 1).

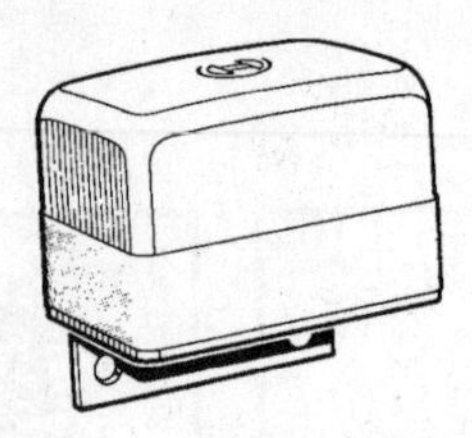

SEV external regulator

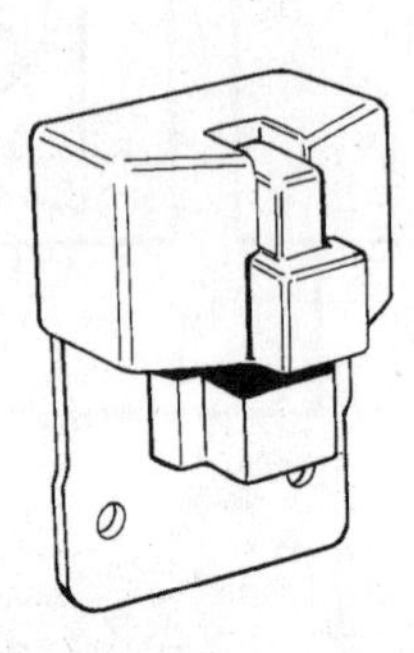

Bosch external regulator

16.1a External voltage regulators

9 Check the charging voltage to verify proper operation of the alternator (see Section 14).

16 Voltage regulator - replacement

Refer to illustrations 16.1a and 16.1b

1 The voltage regulator controls the charging system voltage by limiting the alternator output. The regulator on early models is mounted on the engine compartment firewall (external) **(see illustration)** while the voltage regulator on late models is contained on the outside of the alternator body (integral) **(see illustration).**
2 If the ammeter fails to register a charge rate or the red warning light on the dash comes on and the alternator, battery, drivebelt tension and electrical connections seem to be fine, have the regulator checked by a dealer service department or an automotive electrical repair shop.
3 Disconnect the cable from the negative battery terminal. **Caution:** *It is necessary to make sure the radio is turned OFF before disconnecting the battery cables to avoid damaging the microprocessor built into the radio.*

Integral voltage regulator

Refer to illustrations 16.4a and 16.4b

4 The voltage regulator is located on the exterior of the alternator housing. To replace the regulator, remove the mounting screws **(see illustration)** and lift it off the alternator body **(see illustration)**.
5 Installation is the reverse of removal. **Note:** *Before installing the regulator, check the condition of the slip rings. Use a flashlight and check for any scoring or deep wear grooves. Replace the alternator if necessary.*

External voltage regulator

Refer to illustration 16.6

6 Locate the voltage regulator **(see illustration)** and remove the electrical connector from the regulator.
7 Remove the screws that retain it to the firewall and lift the

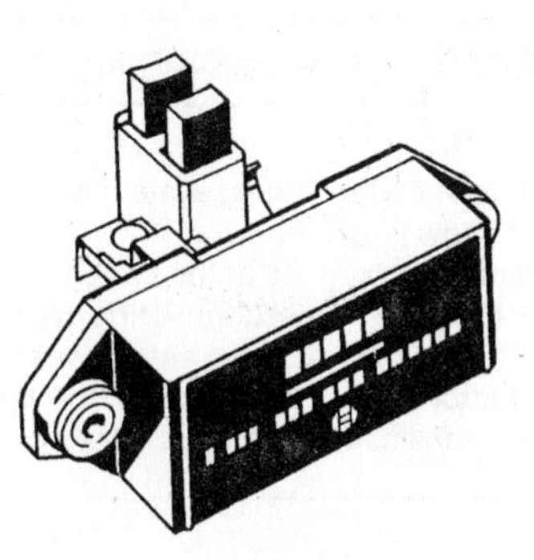

Integral regulator

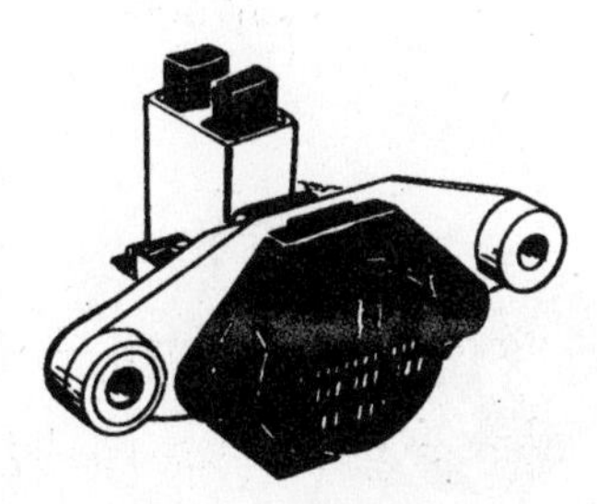

16.1b Integral voltage regulators

16.4b Lift the voltage regulator assembly from the alternator body

regulator out of the engine compartment.
8 Installation is the reverse of removal.

17 Alternator brushes - check and replacement

Refer to illustration 17.3

1 Disconnect the negative cable from the battery. **Caution:** *It is necessary to make sure the radio is turned OFF before disconnecting the battery cables to avoid damaging the microprocessor built into the radio.*

16.4a Remove the voltage regulator/brush holder assembly

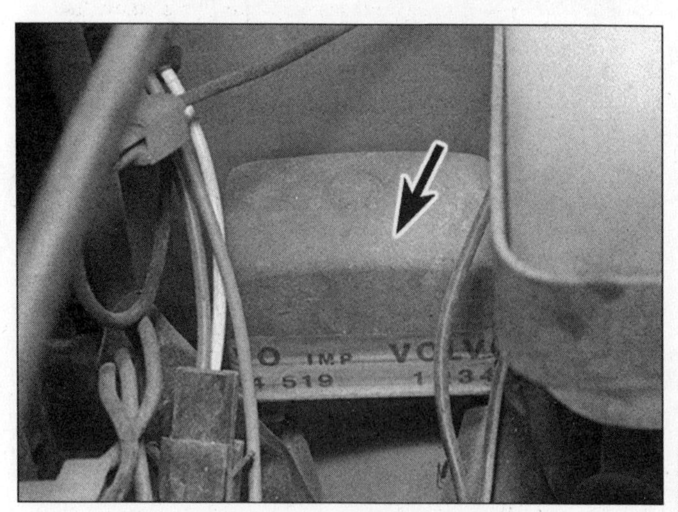

16.6 Location of the voltage regulator (arrow)

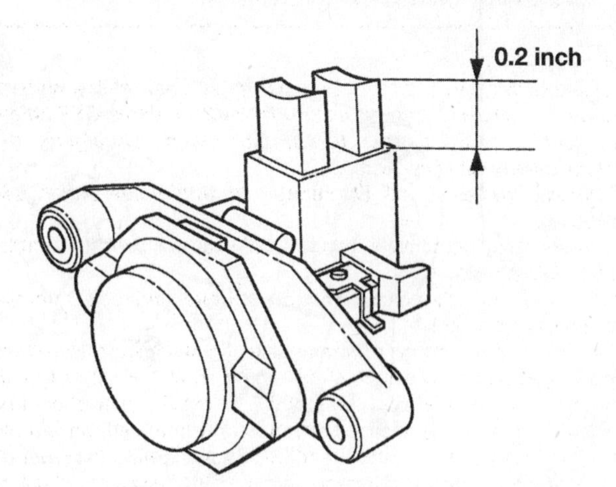

17.3 The brush length should not be less than 0.2 inches

2 Remove the voltage regulator from the back of the alternator (see Section 16).
3 Measure the length of the brushes **(see illustration)**. They should not be less than 0.2-inch (0.5 mm). If they are worn past this point, replace them with a new set.

4 Also, check for excessively worn slip rings.

5 The brushes are retained by either set screws or by solder. **Note:** *Be careful not to apply heat to the solder joint for more than 5 seconds. If necessary, install a heat sink to capture the excess heat. This can be accomplished by clamping a pair of needle-nose pliers next to the solder joint.*

6 On the screw type, hold the assembly in place and install the screws. Tighten them evenly, a little at a time, so the holder isn't distorted.

7 Install the regulator assembly onto the alternator.

8 Reconnect the negative battery cable.

18 Starting system - general information and precautions

Refer to illustration 18.2

The sole function of the starting system is to turn over the engine quickly enough to allow it to start.

The starting system consists of the battery, the starter motor, the starter solenoid and the wires connecting them. The solenoid is mounted directly on the starter motor **(see illustration)**. The starter/solenoid motor assembly is installed on the lower part of the engine, next to the transmission bellhousing.

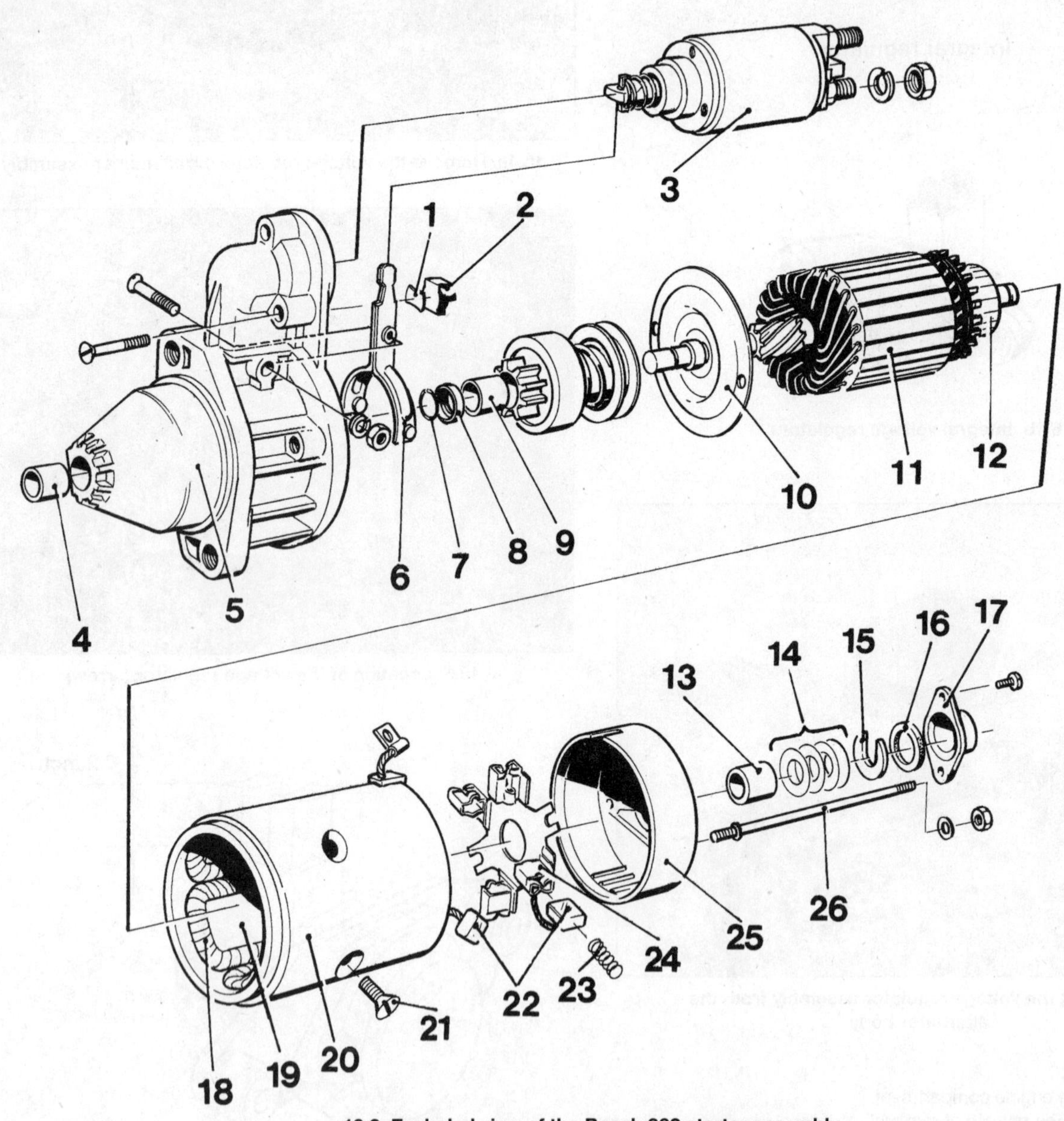

18.2 Exploded view of the Bosch 362 starter assembly

1 *Steel washer*
2 *Rubber washer*
3 *Solenoid*
4 *Drive end bearing housing*
5 *Bushing*
6 *Shift arm*
7 *Lock ring*
8 *Stop ring*
9 *Bushing*
10 *Center bearing*
11 *Armature*
12 *Commutator*
13 *Bushing*
14 *Shims*
15 *Lock washer*
16 *Sealing ring*
17 *Bushing cap*
18 *Field coil*
19 *Pole shoe*
20 *Starter body*
21 *Screw*
22 *Brush*
23 *Brush spring*
24 *Brush holder*
25 *Bushing cover*
26 *Rod*

20.3 Remove the electrical connectors from the starter (arrows)

20.4a Remove the starter bolts from the bellhousing (arrows)

20.4b Remove the bolts that retain the starter bracket to the engine block

When the ignition key is turned to the Start position, the starter solenoid is actuated through the starter control circuit. The starter solenoid then connects the battery to the starter. The battery supplies the electrical energy to the starter motor, which does the actual work of cranking the engine.

The starter motor on a vehicle equipped with a manual transmission can only be operated when the clutch pedal is depressed; the starter on a vehicle equipped with an automatic transmission can only be operated when the transmission selector lever is in Park or Neutral.

Always observe the following precautions when working on the starting system:

a) *Excessive cranking of the starter motor can overheat it and cause serious damage. Never operate the starter motor for more than 10-seconds at a time without pausing to allow it to cool for at least two minutes.*
b) *The starter is connected directly to the battery and could arc or cause a fire if mishandled, overloaded or shorted out.*
c) *Always detach the cable from the negative terminal of the battery before working on the starting system.*

19 Starter motor - in-vehicle check

Note: *Before diagnosing starter problems, make sure the battery is fully charged.*

1 If the starter motor does not turn at all when the switch is operated, make sure that the shift lever is in Neutral or Park (automatic transmission) or that the clutch pedal is depressed (manual transmission).

2 Make sure that the battery is charged and that all cables, both at the battery and starter solenoid terminals, are clean and secure.

3 If the starter motor spins but the engine is not cranking, the overrunning clutch in the starter motor is slipping and the starter motor must be replaced.

4 If, when the switch is actuated, the starter motor does not operate at all but the solenoid clicks, then the problem lies with either the battery, the main solenoid contacts or the starter motor itself (or the engine is seized).

5 If the solenoid plunger cannot be heard when the switch is actuated, the battery is bad, the fusible link is burned (the circuit is open) or the solenoid itself is defective.

6 To check the solenoid, connect a jumper lead between the battery (+) and the ignition switch wire terminal (the small terminal) on the solenoid. If the starter motor now operates, the solenoid is OK and the problem is in the ignition switch, neutral start switch or the wiring.

7 If the starter motor still does not operate, remove the starter/solenoid assembly for disassembly, testing and repair.

8 If the starter motor cranks the engine at an abnormally slow speed, first make sure that the battery is charged and that all terminal connections are tight. If the engine is partially seized, or has the wrong viscosity oil in it, it will crank slowly.

9 Run the engine until normal operating temperature is reached, then disconnect the coil wire from the distributor cap and ground it on the engine.

10 Connect a voltmeter positive lead to the positive battery post and connect the negative lead to the negative post.

11 Crank the engine and take the voltmeter readings as soon as a steady figure is indicated. Do not allow the starter motor to turn for more than 10 seconds at a time. A reading of 9 volts or more, with the starter motor turning at normal cranking speed, is normal. If the reading is 9 volts or more but the cranking speed is slow, the motor is faulty. If the reading is less than 9 volts and the cranking speed is slow, the solenoid contacts are probably burned, the starter motor is bad, the battery is discharged or there is a bad connection.

20 Starter motor - removal and installation

Refer to illustrations 20.3, 20.4a and 20.4b

1 Detach the cable from the negative terminal of the battery. **Caution:** *It is necessary to make sure the radio is turned OFF before disconnecting the battery cables to avoid damaging the microprocessor built into the radio.*

2 Raise the vehicle and support it securely on jackstands.

3 Clearly label, then disconnect the wires from the terminals on the starter motor and solenoid **(see illustration)**. Carefully label any hoses or components that need to be removed from the engine compartment to avoid confusion when reassembling.

4 Remove the mounting bolts **(see illustrations)** and detach the starter.

5 Installation is the reverse of removal.

21 Starter solenoid - removal and installation

1 Disconnect the cable from the negative terminal of the battery. **Caution:** *It is necessary to make sure the radio is turned OFF before disconnecting the battery cables to avoid damaging the microprocessor built into the radio.*
2 Remove the starter motor (see Section 20).
3 Disconnect the strap from the solenoid to the starter motor terminal.
4 Remove the screws which secure the solenoid to the starter motor.
5 Detach the solenoid from the starter body.
6 Remove the plunger and plunger spring.
7 Installation is the reverse of removal.

Chapter 6
Emissions and engine control systems

Contents

1 General information

Refer to illustration 1.6

To prevent pollution of the atmosphere from incompletely burned or evaporating gases, and to maintain good driveability and fuel economy, a number of emission control systems are used on these vehicles. They include the:

Catalytic converter (CAT) system
Evaporative emission control (EVAP) system
Positive crankcase ventilation (PCV) system
Pulsair system
Oxygen sensor feedback system
Air Injection (AI) system
Electronic engine controls

Note: *It is important to remember that many of the systems have different mounting and hose arrangements depending upon the type of fuel system they are installed on carburetors, CIS or EFI fuel injection. Some of the systems are installed on carbureted engines (Pulsair) while others are specific to CIS (AIR). If necessary, purchase the exact model and year vacuum hose routing label from the dealer parts department to give you the exact information on all the required emission control systems that were originally installed on your vehicle.*

The Sections in this Chapter include general descriptions and checking procedures within the scope of the home mechanic and component replacement procedures (when possible) for each of the systems listed above.

Before assuming that an emissions control system is malfunctioning, check the fuel and ignition systems carefully (see Chapters 4 and 5). The diagnosis of some emission control devices requires specialized tools, equipment and training. If checking and servicing become too difficult or if a procedure is beyond your ability, consult a dealer service department or other repair shop. Remember - the most frequent cause of emissions problems is simply a loose or broken vacuum hose or wire, so always check the hose and wiring connections first.

This doesn't mean, however, that emission control systems are

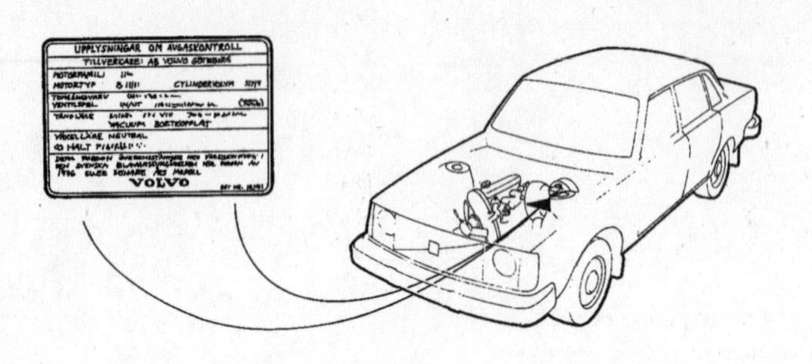

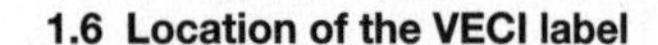

1.6 Location of the VECI label

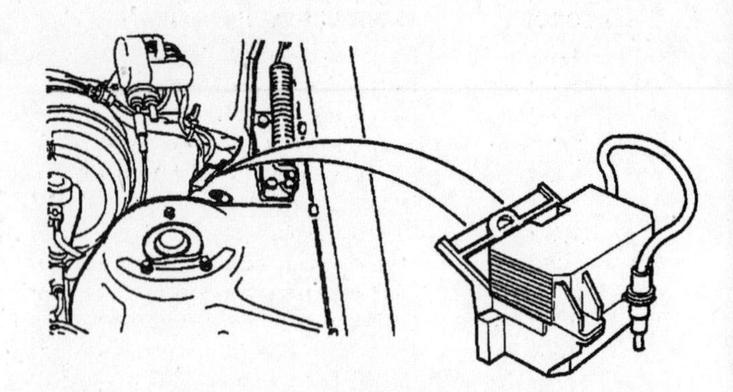

2.3a The diagnostic socket is located in the engine compartment on the left side behind the strut tower

particularly difficult to maintain and repair. You can quickly and easily perform many checks and do most of the regular maintenance at home with common tune-up and hand tools. **Note**: *Because of a Federally mandated extended warranty which covers the emission control system components, check with your dealer about warranty coverage before working on any emissions-related systems. Once the warranty has expired, you may wish to perform some of the component checks and/or replacement procedures in this Chapter to save money.*

Pay close attention to any special precautions outlined in this Chapter. It should be noted that the illustrations of the various systems may not exactly match the system installed on your vehicle because of changes made by the manufacturer during production or from year to year.

A Vehicle Emissions Control Information (VECI) label is located in the engine compartment **(see illustration)**. This label contains important emissions specifications and adjustment information. When servicing the engine or emissions systems, the VECI label in your particular vehicle should always be checked for up-to-date information.

2.3b To obtain diagnostic codes, install the test lead into socket number 2 and press the button (arrow) for 1 second but not more than 3 seconds to activate the LED

2 EFI system self-diagnosis capability

Refer to illustrations 2.3a and 2.3b

The EFI system control unit (computer) on all LH 2.4 and 3.1 Jetronic (1989 through 1993) systems has a built-in self-diagnosis system which detects malfunctions in the system sensors and actuators and alerts the driver by illuminating a CHECK ENGINE warning light in the instrument panel. If a malfunction is detected the computer stores a diagnostic code in its memory. The warning light goes out automatically after the malfunction has been repaired, however the diagnostic code will remain in memory until the diagnostic system is cleared (see below).

When the system is operating normally the CHECK ENGINE warning light will come on when the ignition switch is placed in the On position. When the engine is started, the warning light should go out. The light will remain ON (with the engine running) if the diagnostic system has detected a malfunction or abnormality in the system.

Communication with the diagnostic system is carried out through the diagnostic socket, which is located in the engine compartment to the rear of the left strut tower **(see illustration)**. To obtain a stored diagnostic code, Open the cover to the diagnostic socket and insert the test lead into the number 2 socket **(see illustration).** Push the diagnostic unit push-button for one second but not more than three seconds and wait for any stored codes to be displayed, via a series of flashes, from the LED on the unit. **Note:** *The ECU will activate the CHECK ENGINE light and keep it on for emission systems related diagnostic codes only. Consult the code chart for a list of the codes that hold the light ON. Do not press the diagnostic button more than the prescribed duration (time) or the computer will give faulty information as result of the incorrect commands. If the battery cable is removed or power is interrupted to the diagnostic computer, all stored diagnostic codes will be erased.*

The diagnostic code is the number of flashes indicated on the LED. If any malfunctions have been detected, the light will blink the digits of the code. For example, code 233 (Air control valve malfunction) will blink two flashes, there will be a pause, and then it will blink three flashes, there will be a pause and then it will blink three more flashes. Any other codes that are stored will then be flashed. There will be a slightly longer pause between separate codes. Recheck the codes by depressing the button again. If the same code(s) are repeated, then there are no additional codes stored.

To clear any diagnostic codes, install the test lead into socket number 2 and turn the key to ON (engine not running). Press the diagnostic button for at least 5 seconds and release. The LED should come on. Once the LED is activated, depress the button again for at least 5 seconds and wait for the LED to cancel. Now, press the button a third time for less than 3 seconds and wait for the LED to flash code 1-1-1. This code indicates that the memory has been erased and cleared. Turn the ignition key OFF, remove the test lead from socket number 2 and install the cover on the diagnostic socket.

The accompanying tables explain the code that will be flashed for each of the malfunctions. The accompanying charts indicate the diagnostic code - in blinks - along with the system, diagnosis and specific areas of trouble.

LH 2.4 and 3.1 Jetronic DIAGNOSTIC CODES

CODE	PROBABLE CAUSE	CHECK ENGINE light ON
1-1-1	No codes detected	No
1-1-2	Fault in Electronic Control Unit (ECU)	Yes
1-1-3	Fault in fuel injectors	Yes
1-2-1	Fault in signal for air mass meter	Yes
1-2-3	Fault in signal for coolant temperature sensor	Yes
1-3-1	Engine rpm signal missing	No
1-3-2	Battery voltage high or low	No
1-3-3	LH 2.4 only - Throttle switch idle contacts shorted or out of adjustment	No
2-1-2	Faulty oxygen sensor signal	Yes
2-1-3	LH 2.4 only - throttle switch full-throttle contacts misadjusted or shorted	No
2-2-1	Fuel system compensating for extremely rich or extremely lean running condition at cruise	Yes
2-2-3	Fault in signal for air control valve	No
2-3-1	Fuel system compensating for rich or lean mixture at cruise	No
2-3-2	Fuel system compensating for rich or lean mixture at idle	No
2-3-3	Air control valve closed to compensate for possible air leak	No
3-1-1	Vehicle speed signal missing	No
3-1-2	Anti-knock signal missing	No
3-2-2	LH 2.4 Air mass meter hot wire burn-off not functioning	No
4-1-1	LH 3.1 Throttle sensor signal missing or faulty	No

Check the indicated system or component or take the vehicle to a dealer service department to have the malfunction repaired. Be sure to cancel all the diagnostic codes once the repairs have been completed.

The EZ-116K ignition system features a self-diagnosis system that detects and stores diagnostic codes in a similar manner as the LH 2.4 and 3.1 fuel injection system codes described previously. These codes can also be displayed by flashes of the diagnostic LED light. The test lead is installed into socket number 6 to gain access to the EZ 116K codes. The diagnostic unit can store up to 3 of the possible 7 codes.

Remove the cover from the diagnostic unit and insert the test lead into socket number 6. Turn the ignition key ON (engine not running) and press the button for 1 second but not more than 3 seconds. Follow the procedure described previously to interpret the LED flashes and write down all the trouble codes stored.

The accompanying tables explain the code that will be flashed for each of the malfunctions. The accompanying charts indicate the diagnostic code - in blinks - along with the system, diagnosis and specific areas.

EZ 116K ignition system DIAGNOSTIC CODES

CODE	PROBABLE CAUSE
1-1-1	No codes detected
1-4-2	Fault in Electronic Control Unit (ECU)
1-4-3	Knock sensor signal is missing. Timing is retarded 10 degrees
1-4-4	Load signal missing from LH control unit. Computer assumes the engine is at full load
2-1-4	Speed sender signal missing or sender faulty
2-2-4	Coolant temperature sensor is faulty
2-3-4	Throttle control faulty. Engine runs rough at idle. Timing is retarded 10 degrees
2-4-1	EGR system faulty (California 1991 and later)
4-3-1	EGR temperature sender signal missing or faulty (California 1991 and later)

To clear any diagnostic trouble codes, install the test lead into socket number 6 and turn the key to ON (engine not running). Press the diagnostic button for at least 5 seconds and release. The LED should come on. Once the LED is activated, depress the button again for at least 5 seconds and wait for the LED to cancel. Now, press the button a third time for less than 3 seconds and wait for the LED to flash code 1-1-1. This code indicates that the memory has been erased and cleared. Turn the ignition key OFF, remove the test lead from socket number 6 and install the cover on the diagnostic unit.

3 Electronic Control Unit (ECU) - replacement

Refer to illustrations 3.6a and 3.6b

1 The Electronic Control Unit (ECU) is located inside the passenger compartment under the dashboard (right side) behind a kick panel.

2 Disconnect the negative battery cable from the battery. **Caution:** *It is necessary to make sure the radio is turned OFF before disconnecting the battery cables to avoid damaging the microprocessor built into the radio.*

3 Remove the lower trim panel that runs along the door ledge.

4 Remove the kick panel from the right side of the engine compartment to locate the ECU.

5 Unplug the electrical connectors from the ECU.

6 Remove the retaining nut from the ECU bracket **(see illustrations on next page)**.

7 Carefully remove the ECU. **Note:** *Avoid any static electricity damage to the computer by using gloves and a special anti-Static pad to store the ECU on once it is removed.*

4 Information sensors

Note: *Refer to Chapters 4 and 5 for additional information on the location and diagnosis of the output actuators that are not directly covered in this section.*

6

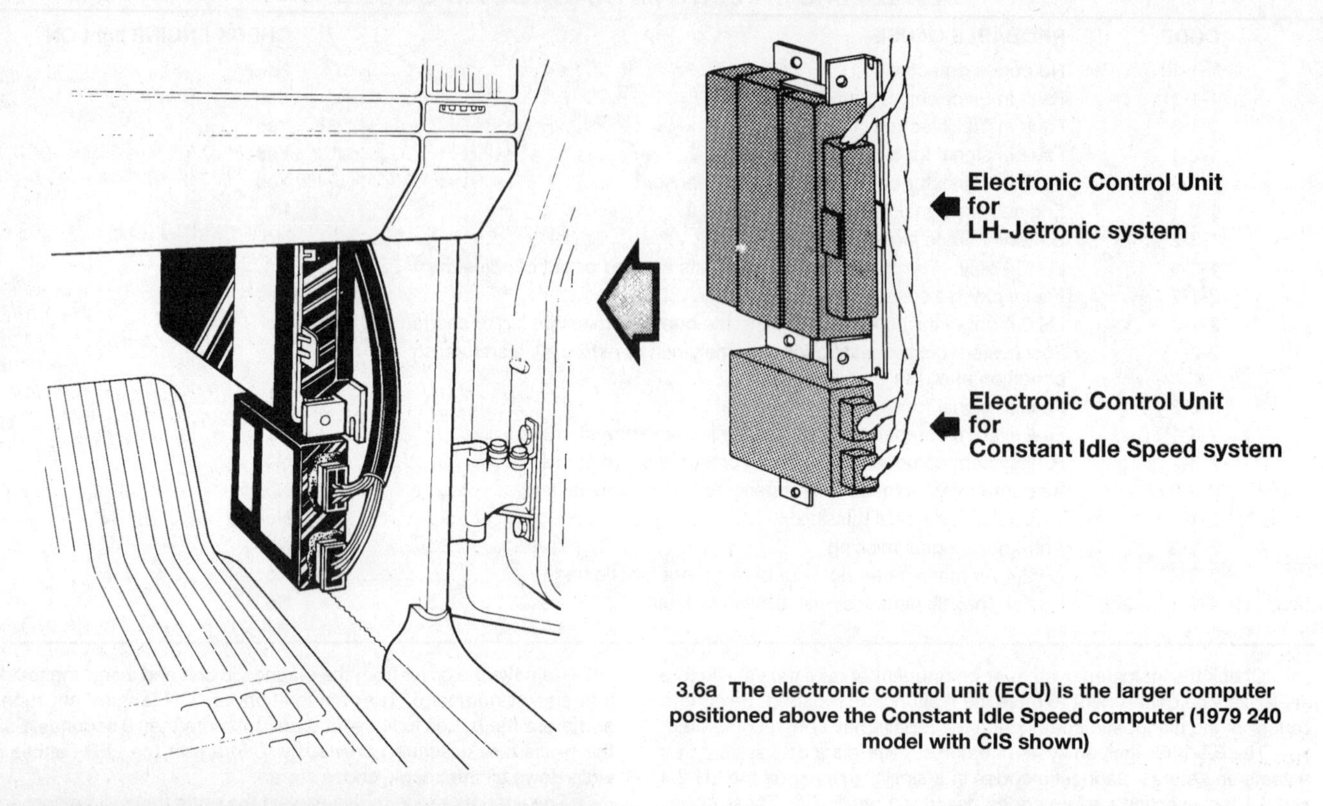

3.6a The electronic control unit (ECU) is the larger computer positioned above the Constant Idle Speed computer (1979 240 model with CIS shown)

3.6b Aftermarket ECU on a LH Jetronic equipped vehicle (1983 240 DL shown)

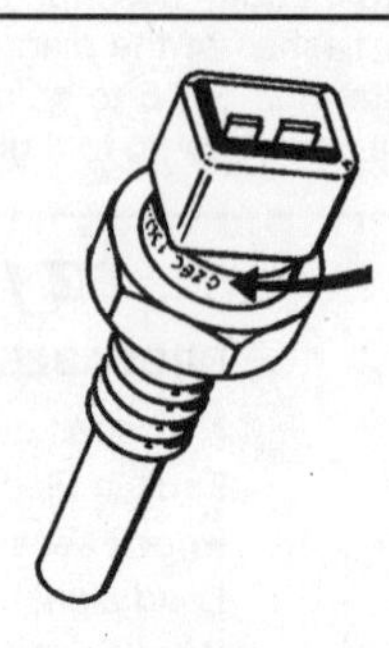

4.1 The coolant temperature sensor is mounted in the left side of the cylinder head below the number 3 intake port

Coolant temperature sensor

Refer to illustration 4.1

General description

1 The coolant temperature sensor **(see illustration)** is a thermistor (a resistor which varies its resistance value in accordance with temperature changes). The change in the resistance value regulates the amount of voltage that can pass through the sensor. As the sensor temperature DECREASES, the resistance values will INCREASE. As the sensor temperature INCREASES, the resistance values will DECREASE. A failure in this sensor circuit should set a Code 1-2-3 (2.4 and 3.1 LH codes) or a code 2-2-4 (EZ 116K codes). This code indicates a failure in the coolant temperature sensor circuit, so in most cases the appropriate solution to the problem will be either repair of a wire or replacement of the sensor.

Check

2 To check the sensor, check the resistance value of the coolant temperature sensor while it is completely cold (50 to 80-degrees F = 2,100 to 2,900 ohms). Next, start the engine and warm it up until it reaches operating temperature. The resistance of the coolant temperature sensor should be lower (180 to 200 degrees F = 250 to 300 ohms). **Note**: *If the lack of access to the coolant temperature sensor makes it difficult to position electrical probes on the terminals, remove the sensor and perform the tests in a pan of heated water to simulate the conditions.* **Note**: *In the case of extremely cold temperatures (about 15-degrees F) the sensor should read 8,000 to 11,000 ohms.*

Replacement

Warning: *Wait until the engine is completely cool before beginning this procedure.*

3 Before installing the new sensor, wrap the threads with Teflon

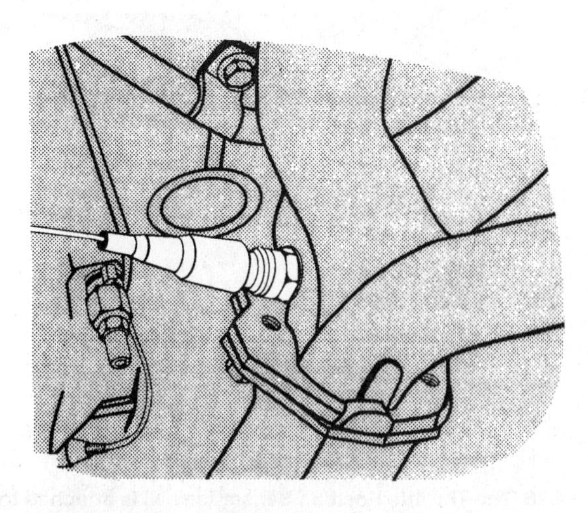

4.6 The oxygen sensor is located in the lower section of the exhaust manifold on CIS and early LH Jetronic systems

tape to prevent leakage and thread corrosion.

4 To remove the sensor, depress the spring lock, unplug the electrical connector, then carefully unscrew the sensor. Be prepared for some coolant spillage and install the new sensor as quickly as possible. **Caution:** *Handle the coolant sensor with care. Damage to this sensor will affect the operation of the entire fuel injection system.* **Note:** *It may be necessary to drain a small amount of the coolant from the radiator before removing the sensor.*

5 Installation is the reverse of removal.

Oxygen sensor

General description

Refer to illustrations 4.6 and 4.17

Note: *Most oxygen sensors are located in the exhaust pipe, downstream from the exhaust manifold.*

6 The oxygen sensor, which is located in the exhaust system **(see illustration)**, monitors the oxygen content of the exhaust gas stream. The oxygen content in the exhaust reacts with the oxygen sensor to produce a voltage output which varies from 0.1-volt (high oxygen, lean mixture) to 0.9-volts (low oxygen, rich mixture). The ECU constantly monitors this variable voltage output to determine the ratio of oxygen to fuel in the mixture. The ECU alters the air/fuel mixture ratio by controlling the pulse width (open time) of the fuel injectors. A mixture ratio of 14.7 parts air to 1 part fuel is the ideal mixture ratio for minimizing exhaust emissions, thus allowing the catalytic converter to operate at maximum efficiency. It is this ratio of 14.7 to 1 which the ECU and the oxygen sensor attempt to maintain at all times.

7 The oxygen sensor produces no voltage when it is below its normal operating temperature of about 600-degrees F. During this initial period before warm-up, the ECU operates in open loop mode.

8 If the engine reaches normal operating temperature and/or has been running for two or more minutes, and if the oxygen sensor is producing a steady signal voltage below 0.45-volts at 1,500 rpm or greater, the ECU will set a Code 2-1-2 (LH 2.4 and 3.1).

9 When there is a problem with the oxygen sensor or its circuit, the ECU operates in the open loop mode - that is, it controls fuel delivery in accordance with a programmed default value instead of feedback information from the oxygen sensor.

10 The proper operation of the oxygen sensor depends on four conditions:

a) ***Electrical*** - *The low voltages generated by the sensor depend upon good, clean connections which should be checked whenever a malfunction of the sensor is suspected or indicated.*

b) ***Outside air supply*** - *The sensor is designed to allow air circulation to the internal portion of the sensor. Whenever the sensor is removed and installed or replaced, make sure the air passages are not restricted.*

c) ***Proper operating temperature*** - *The ECU will not react to the sensor signal until the sensor reaches approximately 600-degrees F. This factor must be taken into consideration when evaluating the performance of the sensor.*

d) ***Unleaded fuel*** - *The use of unleaded fuel is essential for proper operation of the sensor. Make sure the fuel you are using is of this type.*

11 In addition to observing the above conditions, special care must be taken whenever the sensor is serviced.

a) *The oxygen sensor has a permanently attached pigtail and electrical connector which should not be removed from the sensor. Damage or removal of the pigtail or electrical connector can adversely affect operation of the sensor.*

b) *Grease, dirt and other contaminants should be kept away from the electrical connector and the louvered end of the sensor.*

c) *Do not use cleaning solvents of any kind on the oxygen sensor.*

d) *Do not drop or roughly handle the sensor.*

e) *The silicone boot must be installed in the correct position to prevent the boot from being melted and to allow the sensor to operate properly.*

Check

12 Warm up the engine and let it run at idle. Disconnect the oxygen sensor electrical connector and connect the positive probe of a voltmeter to the oxygen sensor connector terminal (output signal) and the negative probe to ground.

Note 1: *Engines equipped with CIS, LH 2.0 and LH 2.2 systems use a single wire oxygen sensor located in the exhaust manifold. Engines equipped with LH 2.4 and 3.1 systems use a three wire oxygen sensor and heater located farther down near the catalytic converter.*

Note 2: *Most oxygen sensor electrical connectors are located on the firewall in the engine compartment. Look for a large rubber boot attached to a thick green wire harness (early) or a single wire along with a sensor heater two wire arrangement.*

13 Increase and then decrease the engine speed and monitor the voltage.

14 When the speed is increased, the voltage should increase to 0.5 to 1.0 volts. When the speed is decreased, the voltage should decrease to about 0 to 0.4 volts.

15 Allow the engine to warm up and enter "closed loop" operation. This period of time is usually 2 to 3 minutes. **Note:***Typically once in "closed loop", the meter will respond with a fluctuating millivolt reading (0.1 to 0.9 volts) when connected properly. If the oxygen sensor is slow to respond in the "closed loop" mode, the sensor is not operating efficiently and is termed "lazy". Keep watching the voltmeter. If the oxygen sensor fails to respond after "closed loop" mode has been obtained (3 to 4 minutes total), replace the oxygen sensor. Be certain the engine is completely warmed up and actually operating in "closed loop" mode and there is not a problem with the thermostat or cooling system. "Lazy" oxygen sensors will quite often fail emissions testing and if there is any doubt, replace it with a new one.*

16 Watch very carefully as the voltage oscillates. The display will flash values ranging from 100 mV to 900 mV (0.100 to 0.900 V). The numbers will flash very quickly, so be observant. Record the high and low values over a period of one minute. The response of the oxygen sensor is very important in determining the condition of the sensor. Start the test when the engine is cold (open loop) and observe that the oxygen sensor is steady at approximately 0.5 to 0.9 volts. As the sensor warms up (closed loop) it will switch suddenly back and forth between 0.1 and 0.9 volts. These signals should be constant and within this range or the sensor is defective. Also, if the engine warms up and the sensor delays before entering into closed loop readings (fluctuating between 0.1 and 0.9 volts), the sensor is defective. Also, if the sensor does not output a voltage signal greater than 0.5 volts, replace it.

17 If the engine is equipped with the three wire type oxygen sensor with a heater, test the heating capabilities. With the ignition system

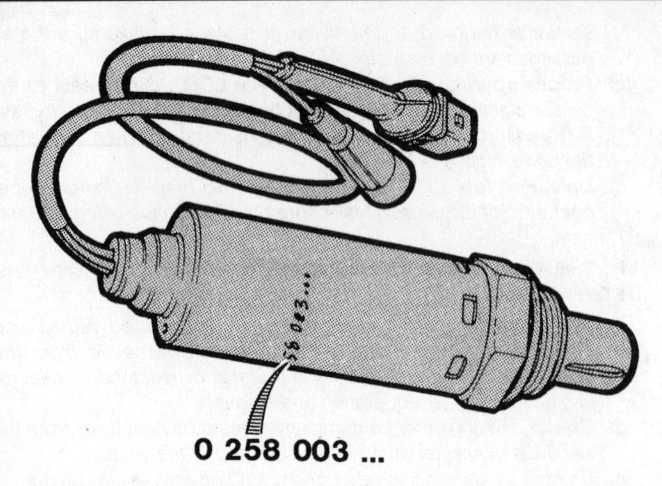

4.17 The oxygen sensor heater resistance should be 13 ohms when warm and 3 ohms when cold

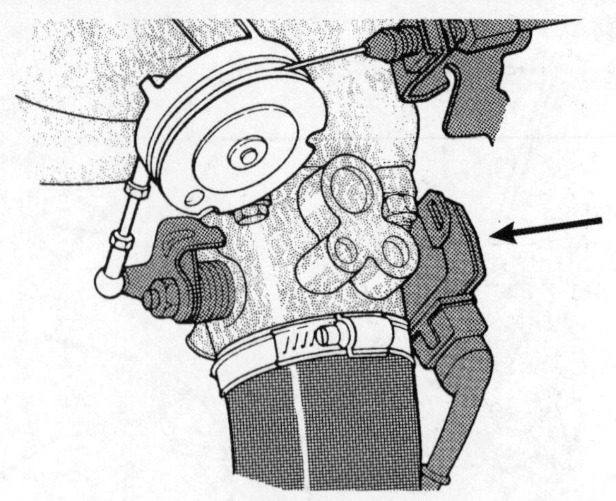

4.26 The Throttle Position Sensor (arrow) is attached to the throttle body

OFF, disconnect the two wire electrical connector from the sensor **(see illustration)**. Using an ohmmeter, check the resistance of the sensor warm (13 ohms hot) then let it cool down and recheck (3 ohms cold). If the resistance is incorrect, replace the oxygen sensor.

Replacement

Note: *Because it is installed in the exhaust manifold, converter or pipe, which contracts when cool, the oxygen sensor may be very difficult to loosen when the engine is cold. Rather than risk damage to the sensor (assuming you are planning to reuse it in another manifold or pipe), start and run the engine for a minute or two, then shut it off. Be careful not to burn yourself during the following procedure.*

18 Disconnect the cable from the negative terminal of the battery. **Caution:** *It is necessary to make sure the radio is turned OFF before disconnecting the battery cables to avoid damaging the microprocessor built into the radio*

19 Raise the vehicle and place it securely on jackstands.

20 Disconnect the electrical connector from the sensor.

21 Carefully unscrew the sensor. **Caution**: *Excessive force may damage the threads.*

22 Anti-Seize compound must be used on the threads of the sensor to facilitate future removal. The threads of new sensors will already be coated with this compound, but if an old sensor is removed and reinstalled, recoat the threads.

23 Install the sensor and tighten it securely.

24 Reconnect the electrical connector of the pigtail lead to the main engine wiring harness.

25 Lower the vehicle and reconnect the cable to the negative terminal of the battery.

Throttle Position Sensor (TPS) (LH 3.1 only)

Refer to illustrations 4.26 and 4.28

General description

26 The Throttle Position Sensor (TPS) is located on the end of the throttle shaft on the throttle body **(see illustration)**. By monitoring the output voltage from the Throttle Position Sensor (TPS), the ECU can determine fuel delivery based on throttle valve angle (driver demand). A broken or loose TPS can cause intermittent bursts of fuel from the injector and an unstable idle because the ECU thinks the throttle is moving.

Check

27 To check the TPS, turn the ignition switch to ON (engine not running) and install the probes of the volt-ohmmeter into the ground wire (terminal 1) and signal wire (terminal 3) on the backside of the electrical connector. This test checks for the proper signal voltage from

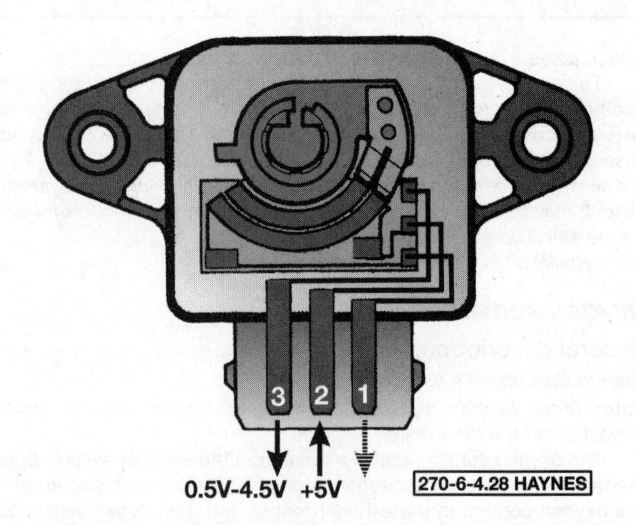

4.28 Check for a voltage signal on terminals 1 and 3. It should be 0.25 to 4.8 volts from idle to wide-open-throttle (WOT)

the TPS. **Note:** *The ground wire is a black wire on all models.*

28 The sensor should read 0.25 to 0.98-volts at idle. Have an assistant depress the accelerator pedal to simulate full throttle and the sensor should increase voltage from 4.0 to 4.8-volts **(see illustration)**. If the TPS voltage readings are incorrect, replace it with a new unit.

29 Also, check the TPS reference voltage. With the ignition key ON (engine not running), install the positive (+) probe of the ohmmeter **(see illustration 4.28)** onto the 2 terminal. There should be approximately 5.0 volts sent from the ECU to the TPS.

Replacement

30 Disconnect the cable from the negative battery terminal.

31 Disconnect the electrical harness, remove the retaining screws and separate the sensor from the throttle body.

32 Installation is the reverse of removal. There is no adjustment required for the new TPS.

Throttle switch (all models except LH 3.1)

Refer to illustrations 4.34 and 4.38

General description

33 The Throttle Switch is located on the end of the throttle shaft on the throttle body. One set of throttle valve switch contacts is closed

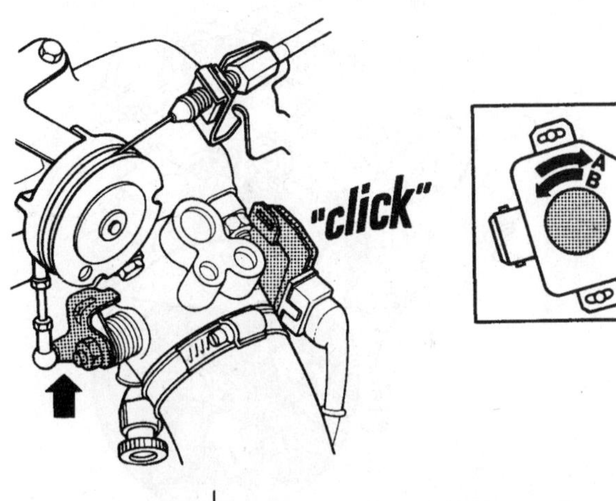

4.34 The throttle switch should "click" immediately upon opening the throttle

4.38 Turn the base idle screw all the way out (arrow) until the engine reaches slowest idle

(continuity) only at idle. A second set of contacts closes as the engine approaches full throttle. Both sets of contacts are open (no continuity) between these positions. A broken or loose TPS can cause intermittent bursts of fuel from the injector and an unstable idle because the ECU thinks the throttle is moving. The self diagnosis system will set a code 2-1-3 or 4-1-1 (LH 2.4 and 3.1) or 2-3-4 (EZ 116K) in the event of throttle switch malfunction.

Check

34 With the ignition key OFF, open the throttle and listen for the sound of a "click" **(see illustration)** immediately as the throttle opens. If necessary adjust the switch (see Steps 37 through 42).

35 Disconnect the wiring harness from the throttle switch and check for continuity between terminals number 2 (orange wire - LH 2.0, 2.2) or (yellow/white wire LH 2.4) and ground. The meter should indicate continuity at closed throttle and open during acceleration.

36 Connect an ohmmeter between switch terminal number 3 (white/black wire) and ground. The meter should indicate continuity at wide open throttle and open at idle. If necessary, adjust the throttle switch.

Adjustment

37 Turn the ignition key off and disconnect the throttle control rod from the throttle pulley **(see illustration 4.34)**. Loosen the screws that retain the throttle switch to the throttle body and turn the switch clockwise.

38 Loosen the locknut and backout the throttle stop screw **(see illustration)** until there is clearance between the end of the screw and the throttle valve lever.

39 Turn the throttle stop screw clockwise (IN) an additional 1/4 to 1/2 turn and tighten the locknut.

40 Rotate the throttle switch counterclockwise until it stops (do not open the throttle valve). Tighten the retaining screws and test the switch. The switch should "click" just as the throttle is opened.

41 Adjust the throttle linkage (see Chapter 4).

42 Adjust the idle speed (see Chapter 1).

Air Mass meter

General description

Note : *A no-start condition is the prime symptom of a defective air mass sensor. Follow the test procedures very carefully to determine the condition of the unit before replacing it with a new part. Air mass meters (also called Mass Airflow Sensor) are very expensive electronic parts that are often not returnable.*

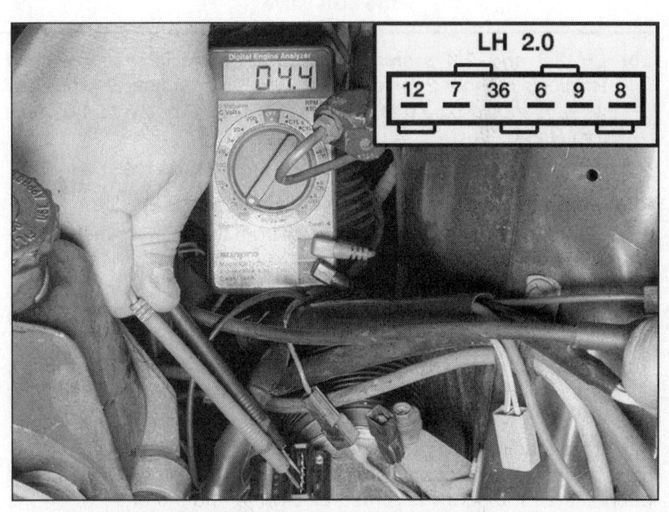

4.45 On LH 2.0 systems, check the resistance on terminals numbers 6 and 7 to test the heated wire

6

43 The air mass meter measures the mass of the intake air by monitoring the air temperature (resistance) using a electrically heated platinum wire suspended in the airstream. As intake air cools the platinum wire, the variations in electrical current needed to maintain the base temperature of the wire are converted into an air mass signal that is sent to the ECU. The air mass meter is also equipped with a burn-off control which rids the platinum wire of any contaminants that have collected on the surface. Each time the engine is shut down, the wire is momentarily heated to a temperature over 1,832 degrees F. **Note :** *LH 3.1 systems use a "hot film" instead of a heated platinum wire. The hot film consists of four resistors which react to the airflow into the engine. This type of air mass meter does not require a burn-off system. Refer to Chapter 4 for additional information on the different series of LH-Jetronic fuel injection systems to properly identify the system that is installed on your model and year*

Checks

Heated wire check (LH 2.0, 2.2 and 2.4)

Refer to illustration 4.45

44 Disconnect the electrical connector from the air mass meter.

45 Using an ohmmeter:

a) *LH 2.0 systems, measure the resistance between terminals number 6 (green/yellow wire) and number 7 (white/red wire)* **(see illustration)**. *It should be 3.5 to 4.0 ohms.*

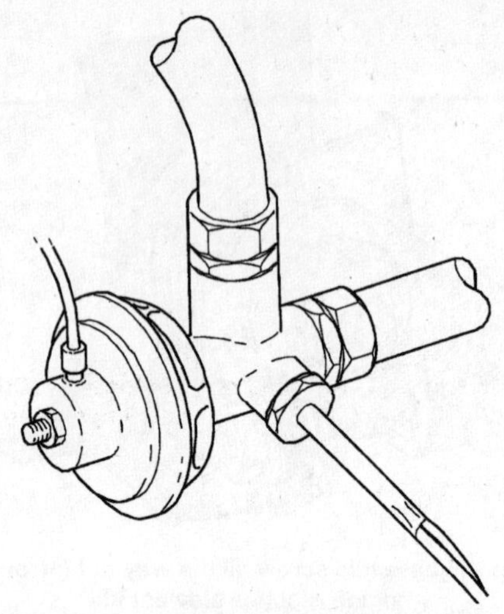

4.61 The EGR gas temperature sensor is located directly behind the EGR valve

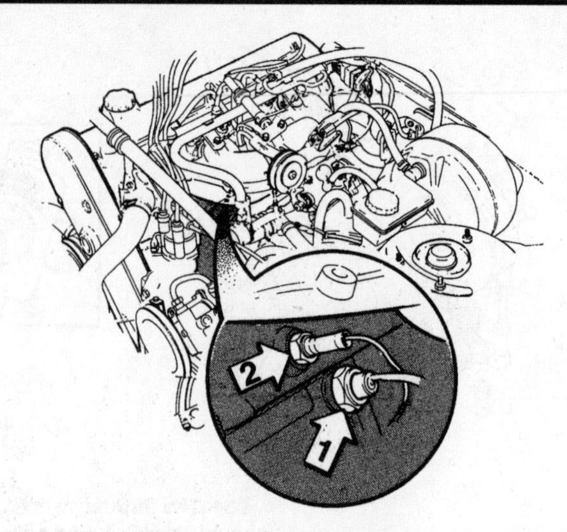

4.64 Location of the knock sensor (1) and coolant temperature sensor (2)

b) *LH 2.2 and 2.4 systems, measure the resistance between terminals number 2 (green/yellow wire) and number 3 (white/red wire). It should be 3.5 to 4.0 ohms for LH 2.2 or 2.5 to 4.0 ohms for LH 2.4.*

46 If the readings are incorrect, replace the air mass meter.

Idle mixture potentiometer check (LH 2.0 and 2.2)

47 Disconnect the electrical connector from the air mass meter.

48 Using an ohmmeter:

a) *LH 2.0 systems, measure the resistance between terminals number 6 (green/yellow wire) and number 12 (yellow wire)* **(see illustration 4.45)**. *It should be 0 to 1,000 ohms.*

b) *LH 2.2 systems, measure the resistance between terminals number 2 (green/yellow wire) and number 6 (yellow wire). It should be 0 to 1,000 ohms.*

49 If the readings are incorrect, replace the air mass meter.

50 Turn the idle mixture screw and observe the ohmmeter. It should decrease as the screw is turned clockwise and increase when the screw is turned counterclockwise. If the resistance does not change, replace the air mass meter. If the resistance changes properly, set the idle mixture screw at the exact same original resistance value.

Burn-off function check (LH 2.0, 2.2 and 2.4)

51 Run the engine until it reaches normal operating temperature.

52 Pull back the rubber boot on the air mass meter electrical connector and carefully backprobe the connector wires:

a) *LH 2.0 systems, measure the voltage between terminals number 8 (white wire) and ground* **(see illustration 4.48).**

b) *LH 2.2 and 2.4 systems, measure the voltage between terminals number 4 (white wire) and ground.*

53 Increase the engine speed to 2,500 rpm then allow it to return to idle, then turn the ignition OFF. After four seconds, the voltmeter should show a voltage reading for approximately one second. Look quick. This is the burn-off signal.

54 If there is no result, check the system relay (LH 2.0) or main relay (LH 2.2 and 2.4)

Heated film check (LH 3.1)

55 Disconnect the electrical connector from the air mass meter.

56 Using an ohmmeter, measure the resistance between terminals number 1 (green/yellow wire) and number 4 (red/white wire). It should be 108 ohms.

Replacement

57 Disconnect the electrical connector from the air mass meter.

58 Remove the air cleaner assembly (see Chapter 4).

59 Remove the nuts and lift the air mass meter from the engine compartment or from the air cleaner assembly.

60 Installation is the reverse of removal.

EGR Gas temperature sensor

Refer to illustration 4.61

61 The EGR gas temperature sensor is installed on 1991and later models **(see illustration)**. The sensor monitors the temperature of the exhaust gasses and relays the information to the computer which in turn adjusts the amount of exhaust gasses that are allowed to recirculate. Problems with the sensor will cause the CHECK ENGINE light to come on and a diagnostic code will be set (see Section 2).

62 To test the sensor, disconnect the sensor electrical connector and measure the resistance across the terminals. The resistance should be between 500 to 1,000 ohms. **Note:** *This sensor is a positive coefficient (PTC) electrical sensor. The resistance will increase with an increase in the temperature.*

63 Installation is the reverse of removal. Be sure to apply a small amount of anti-seize compound on the threads of the sensor before installation.

Knock sensor

General description

Refer to illustration 4.64

64 The knock sensor **(see illustration)** is located in the engine block next to the coolant temperature sensor (LH 2.4 and 3.1). The knock sensor detects abnormal vibration in the engine. The sensor produces an AC output voltage which increases with the severity of the knock. The signal is fed into the ECM and the timing is retarded up to 10 degrees to compensate for the severe "rattling" or "heavy" detonation.

Check

65 With the ignition key ON (engine not running) backprobe the sensor electrical connector and check for voltage. It should be between 1.5 and 5.0 volts. If the voltage is higher or lower, there is a short or open in the sensor circuit.

66 Connect a timing light in accordance with the manufacturer's instructions; then use a wrench to rap on the intake manifold (not too hard or you may damage the manifold!) or cylinder block near the sensor while the engine is idling. Never strike the sensor directly. Observe the timing mark with the timing light. The vibration from the wrench will produce enough of a shock to cause the knock sensor to signal the computer to back off the timing. The timing should retard momentarily. If nothing happens, check the wiring, electrical connector or computer for any obvious shorts or problems. If the wiring and connectors are OK, the sensor is probably faulty.

67 Also check for a loose knock sensor. Tighten with an open-end wrench if necessary.

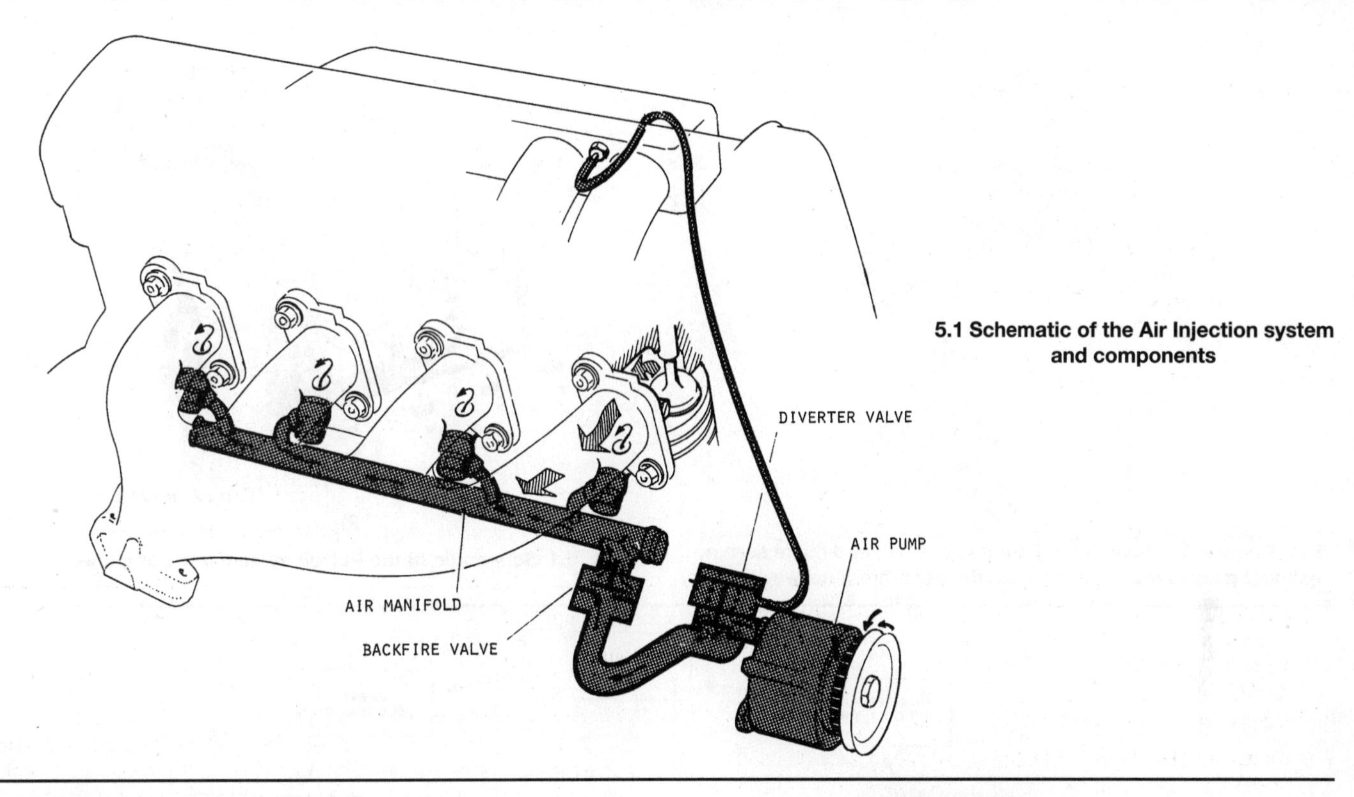

5.1 Schematic of the Air Injection system and components

Replacement

68 Unplug the electrical connector from the sensor.
69 Unscrew the sensor from the block (Volvo/Chrysler type ignition) or remove the bolt and detach the sensor from the block (EZ-116K ignition).
70 Installation is the reverse of removal. On models with the EZ-116K ignition system, tighten the mounting bolt to 15ft-lbs.

5 Air Injection (AI) system

Refer to illustrations 5.1, 5.9 and 5.11

General description

1 The air injection system is employed to reduce carbon monoxide and hydrocarbon emissions. The system consists of an air supply pipe, a diverter valve, a backfire valve, injection tubes and connecting hoses **(see illustration).**
2 The air injection system promotes combustion of unburned gases after they leave the combustion chamber by injecting fresh air into the hot exhaust stream leaving the exhaust ports. At this point, the fresh air mixes with hot exhaust gases to increase oxidation of both hydrocarbons and carbon monoxide, thereby reducing their concentration and converting some of them into harmless carbon dioxide and water. During some modes of operation, such as higher engine speeds, the air is dumped into the atmosphere by the air control valve to prevent overheating of the exhaust system.
3 There are two types of air control valves used in Volvo air injection systems:

- a) *The diverter valve has two functions. It regulates the air pump pressure and also shuts off the air delivery when the engine vacuum is high.*
- b) *The backfire valve admits air into the exhaust manifold but prevents the return of exhaust gas to the air pump so that in the event of a backfire or air pump problem, damage to the pump or drivebelt will not occur.*

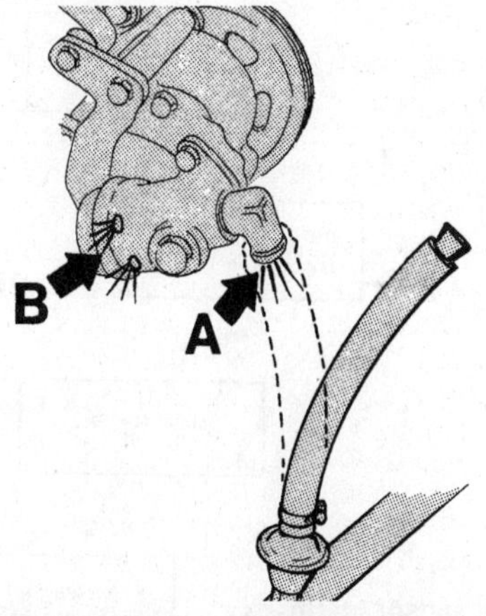

5.9 Fresh air is pumped from outlet (A). There should not be any excessive flow of air out of the relief ports (B). This condition would indicate a leaking diverter valve

Checks

Air supply pump

4 Check and adjust the drivebelt tension, if necessary (refer to Chapter 1).
5 Disconnect the air supply hose at the diverter valve inlet.
6 The pump is operating satisfactorily if airflow is felt at the pump outlet with the engine running at idle, increasing as the engine speed is increased.
7 If the air pump does not successfully pass the above tests, replace it with a new or rebuilt unit.

Diverter valve

8 Failure of the relief valve or diverter valve will cause excessive noise and air pump output at the valve.
9 There should be little or no air escaping from the silencer of the valve body at engine speed **(see illustration)**. If the valve itself becomes faulty, fresh air mixed with over-rich exhaust vapors would

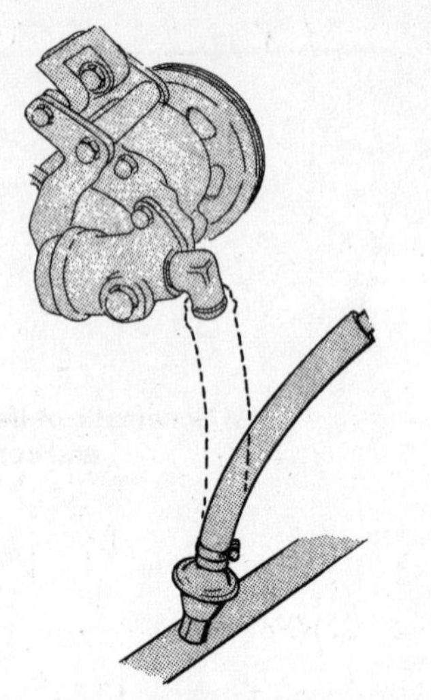

5.11 Remove the hose from the backfire valve and make sure no exhaust gasses are escaping past the back-fire (one-way) valve

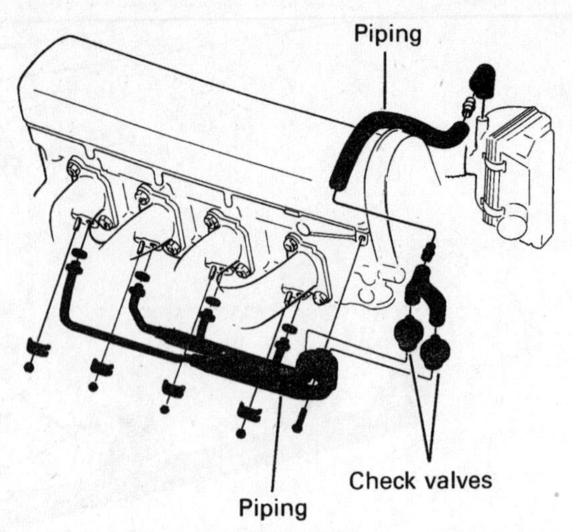

6.1 Schematic of the Pulsair system and components

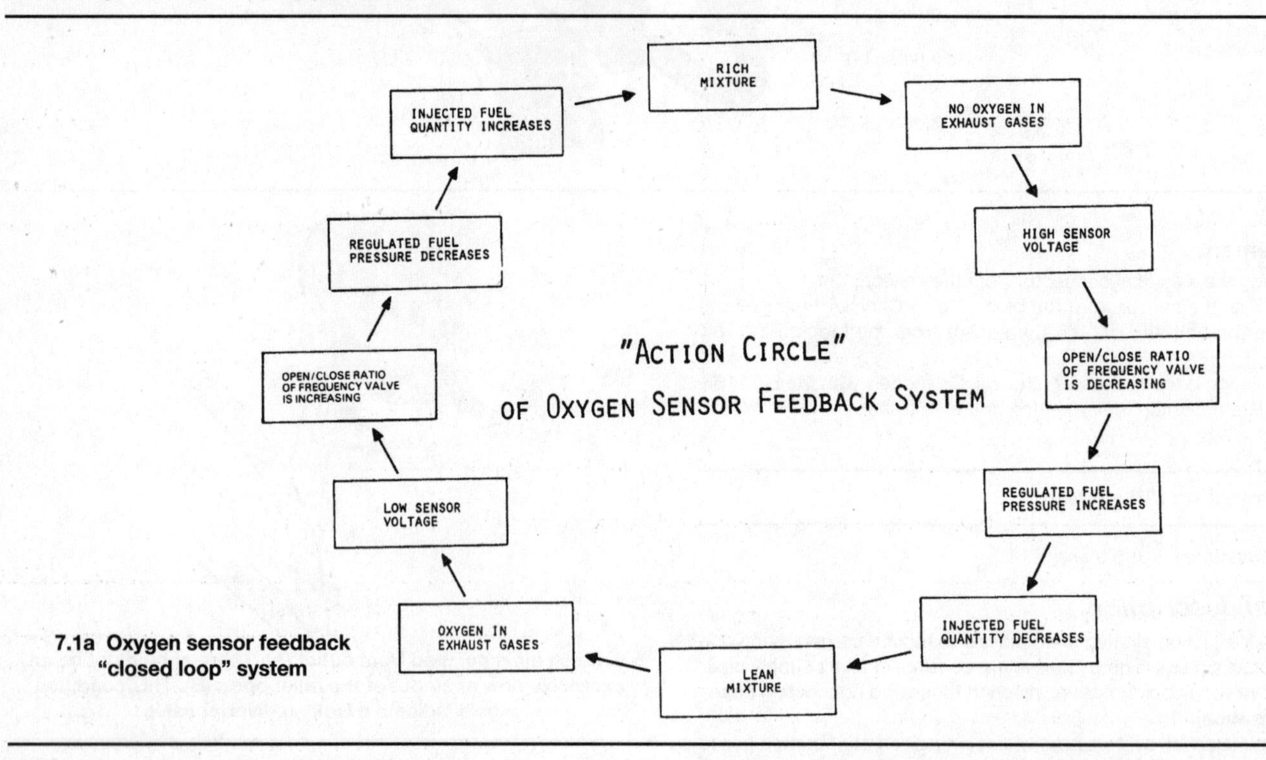

7.1a Oxygen sensor feedback "closed loop" system

leak into the exhaust system and cause back-fire.

10 If these tests indicate a faulty diverter valve, the unit should be replaced.

Backfire valve

11 The back-fire valve, located on the injection tube assembly, can be tested by removing the air hose from the valve inlet tube **(see illustration)**.

12 Start the engine and check for exhaust gas escaping from the valve inlet tube.

13 If there is an exhaust leak, the check valve is faulty and should be replaced.

Component replacement

14 The check valve and air control valve may be replaced by disconnecting the hoses leading to them (be sure to label the hoses as they are disconnected to facilitate reconnection), replacing the faulty element with a new one and reconnecting the hoses to the proper ports. Be sure to use new gaskets and remove all traces of the old gasket, where applicable. Make sure that the hoses are in good condition. If not, replace them with new ones.

15 To replace the air supply pump, first loosen the appropriate engine drivebelt(s) (refer to Chapter 1), then remove the faulty pump from the mounting bracket, labeling all wires and hoses as they are removed to facilitate installation of the new unit.

16 After the new pump is installed, adjust the drivebelt(s) to the specified tension (refer to Chapter 1).

6 Pulsair system (Canada only)

General information

Refer to illustration 6.1

1 The Pulsair system **(see illustration)** replaced the Air Injection

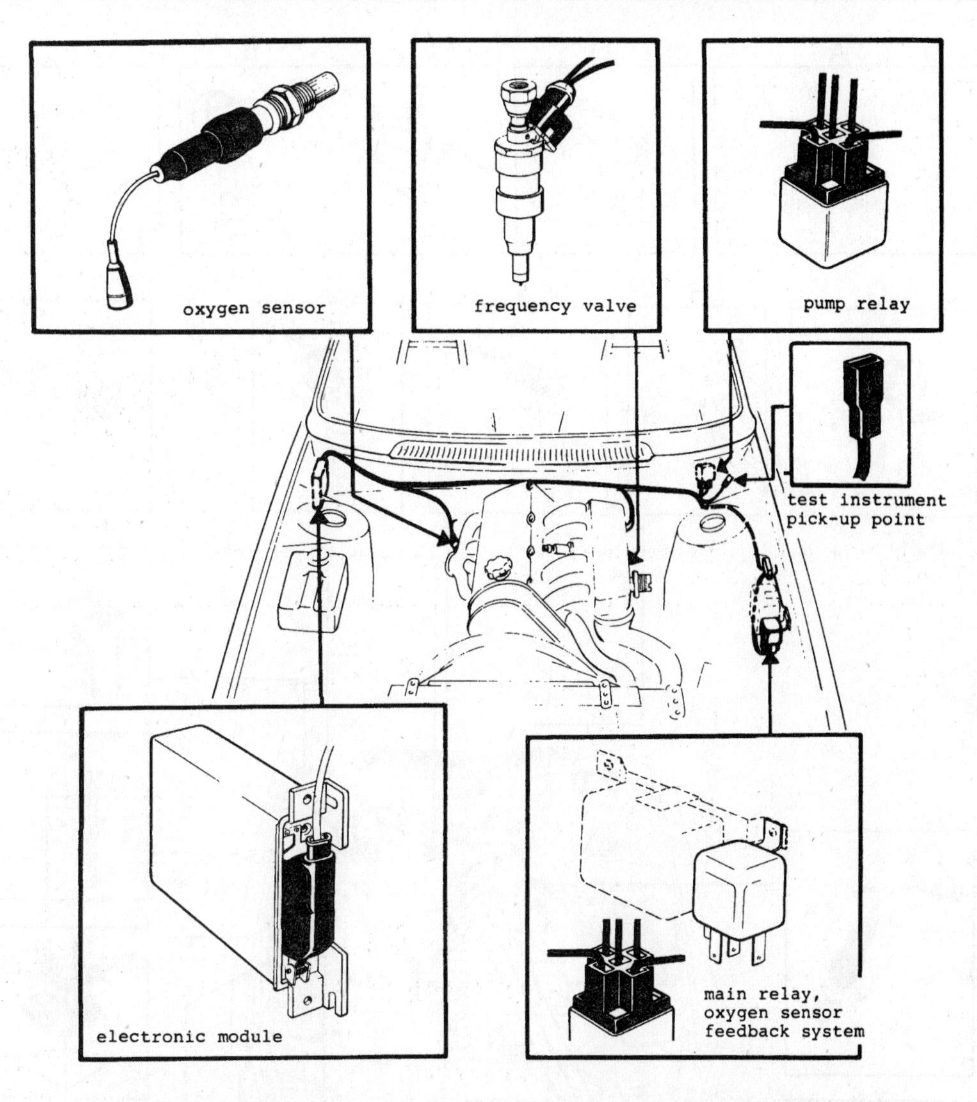

7.1b Typical early oxygen sensor feedback system schematic and components

system on some models produced for Canada. When the engine is running, pressure pulses in the exhaust manifold are closely followed by vacuum pulses. The alternating pressure and vacuum pulsation created in the exhaust system allows the special one-way valves to draw air from the air cleaner and enter the exhaust manifold. This additional air further aids in the burning of exhaust gasses in the exhaust system thereby lowering the emissions levels.

Check

2 Because of the complexity of the Pulsair system, it is difficult to diagnose a malfunctioning system at home. Here is a simple check to give the home mechanic an idea of the possible problem.

3 Inspect all hoses, vacuum lines and wires. Be sure they are in good condition and all connections are clean and tight.

4 If there is still a malfunction in the system, take the vehicle to a dealer service department or a certified emissions control repair shop for testing.

Replacement

5 Refer to illustration 6.1 for an exploded view of the separate parts. It is a good idea to spray penetrating lubricant onto the various connections before attempting to remove them. These parts are usually very tight due to the extreme heat and corrosion.

7 Oxygen sensor feedback system

Note: *Before attempting to diagnose the oxygen sensor feedback system, be sure to check the oxygen sensor for proper operation (see Section 4). If the oxygen sensor feedback system has been checked and all the components are working properly, adjust the CO mixture as described in Section 8.*

General information

Refer to illustrations 7.1a, 7.1b, 7.1c and 7.1d

1 The oxygen sensor feedback system **(see illustrations)** or commonly called the "Lambda-sond" is designed to reduce emissions and improve fuel economy on fuel injected engines. On LH-Jetronic engines, the Lambda-sond is fully integrated into the LH fuel injection system. On CIS engines, the Lambda-sond is independent of the CIS fuel injection system. The oxygen sensor feedback system consists basically of the oxygen sensor, ECU, frequency valves, relays and the wiring harness. This system continuously adjusts the air/fuel ratio to provide optimum conditions for combustion and cleaner exhaust emissions. **Note:** *The illustrations provided do not cover all the years and versions of the Lambda- sond system. Refer to the fuel injection wiring diagrams in Chapter 4 for additional detailed component locations and wiring schematics.* The ECU receives a feedback signal

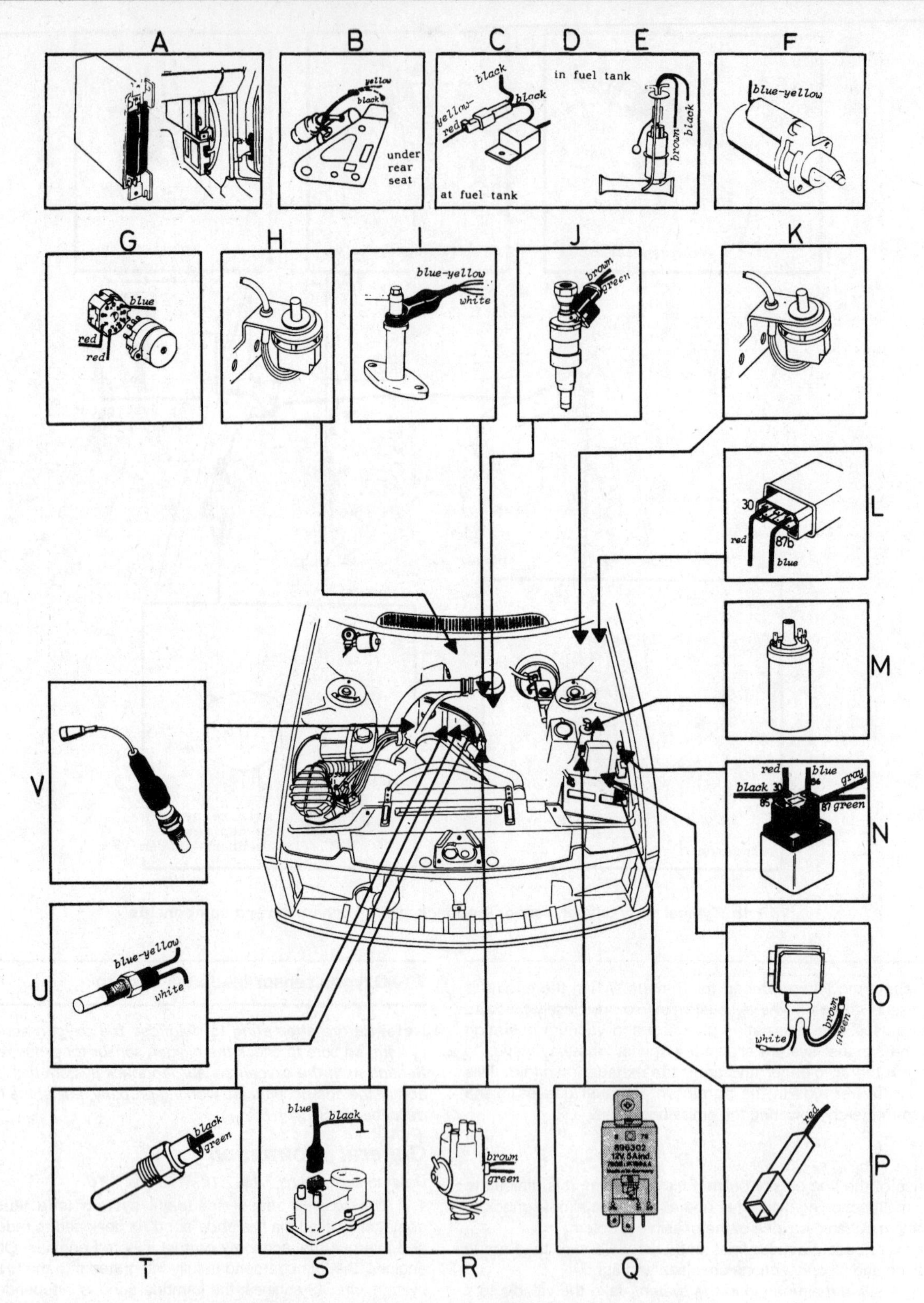

7.1c Oxygen sensor system for the 1983 B21F turbocharged engine

A Electronic Control Unit (ECU)
B Fuel pump assembly
C Capacitor
E Fuel pump
F Starter
G Ignition switch
H Pressure switch, overload protection
I Cold start injector
J Frequency valve
K Pressure switch, acceleration enrichment
L Fuel pump relay
M Coil
N Relay, oxygen sensor system
O Ignition module
P Test connector
Q Relay
R Distributor
S Control pressure regulator
T Temperature switch
U Thermal time switch
V Oxygen sensor

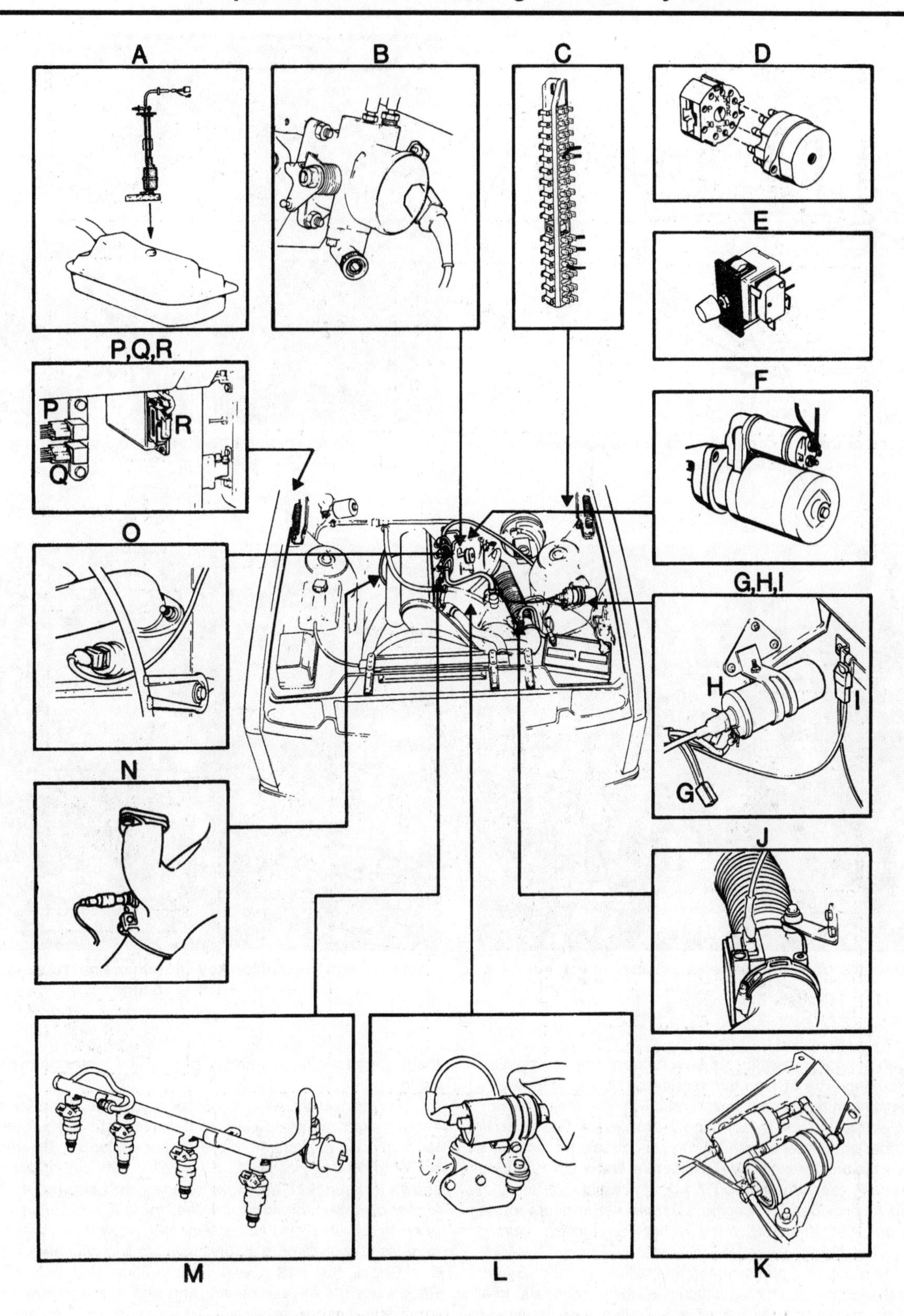

7.1d Oxygen sensor system for the 1983 and 1984 B23F LH-2.2 Jetronic system

A Fuel tank pump
B Throttle body assembly
C Fuses
D Ignition switch
E Air conditioning switch
F Starter
G Test connector
H Ignition coil
I Fuse (25 amp)
J Air mass meter
K Fuel pump assembly
L Air control valve
M Fuel injectors
N Oxygen sensor
O Temperature sensor
P System relay
Q Fuel pump relay
R ECU

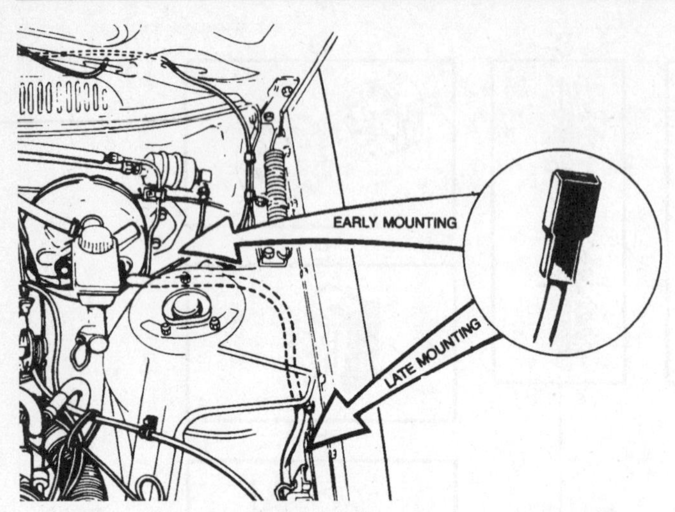

7.3a Location of the test connectors for the oxygen sensor feedback system

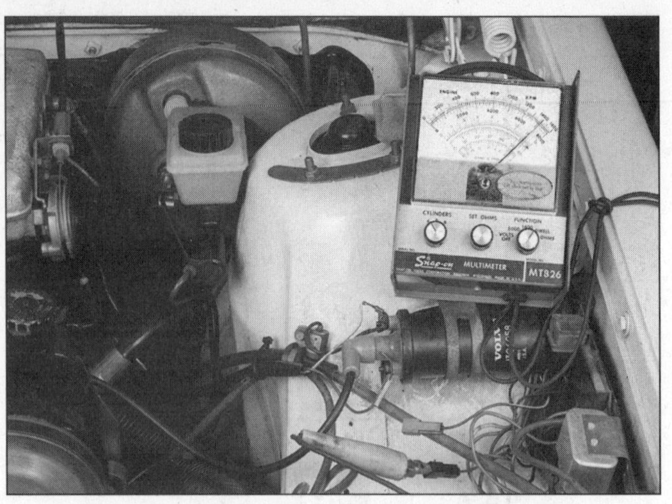

7.3b Dwellmeter hooked-up to a test connector

7.6a Disconnect the frequency valve electrical connector and . . .

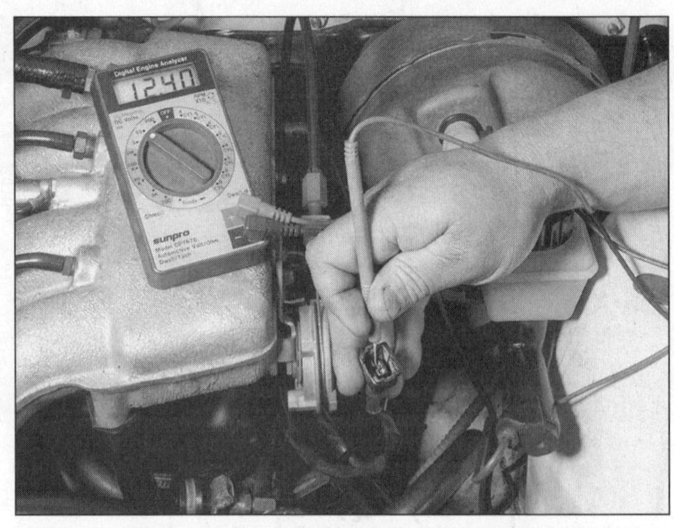

7.6b . . . with the ignition key ON (engine not running), check for battery voltage

from the oxygen sensor that indicates how efficient the combustion ratios are performing. The ECU in turn regulates the frequency valve (duty cycle) to maintain the correct air/fuel ratio. On CIS, the frequency valve alters the air/fuel mixture by adjusting the fuel system pressure in the lower chamber of the fuel distributor. On LH-Jetronic systems, the computer varies the fuel through the fuel injectors. **Note:** *On 1984 and 1985 turbo models* **(see illustration 7.1c)**, *a pressure differential switch is used to provide cold engine acceleration enrichment. A Lambda pressure switch allows for a richer fuel mixture during acceleration.*

2 Some of the most common symptoms of failure with the oxygen sensor feedback system are starting difficulties when hot, erratic idle, poor engine performance and low fuel mileage. To find out right away if the system is working, listen for any buzzing sounds coming from the frequency valve located in the fuel distributor. If no sounds are heard while the engine is running, there is a problem with the system.

Checks - CIS models

Refer to illustrations 7.3a, 7.3b, 7.6a, 7.6b, 7.7, 7.8, 7.11 and 7.12

3 Connect a dwellmeter to the special test adapter (red wire) near the battery **(see illustrations)** and set the meter on the four-cylinder scale.

4 Disconnect the oxygen sensor connector at the firewall, start the engine and read the dwell. Do not allow the oxygen sensor wire to touch ground or damage to the sensor will occur. This reading should be 42 to 48 degrees (47 to 53 percent duty cycle). **Note:** *The dwell shows frequency of the valve opening and closing. If the reading is higher or lower, this indicates that there is a problem with the valve opening (duration) too long or not long enough. A dwell reading of 45 degrees is equivalent to a duty cycle reading of 50 percent.*

5 Check the idle speed and ignition timing (see Chapter 1). Reconnect the oxygen sensor and warm up the engine completely. Verify the dwell steadies out between 41 and 44 degrees (46 to 49 percent duty cycle).

No buzzing sound from the frequency valve

6 Disconnect the frequency valve electrical connector **(see illustration)** and check for battery voltage **(see illustration)**.

7 Check the resistance of the frequency valve **(see illustration)**. It should be 2 to 3 ohms.

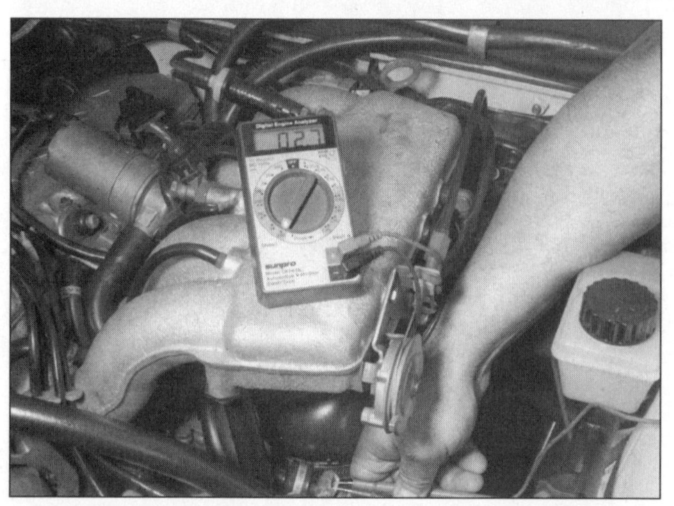

7.7 The frequency valve resistance should be 2 to 3 ohms

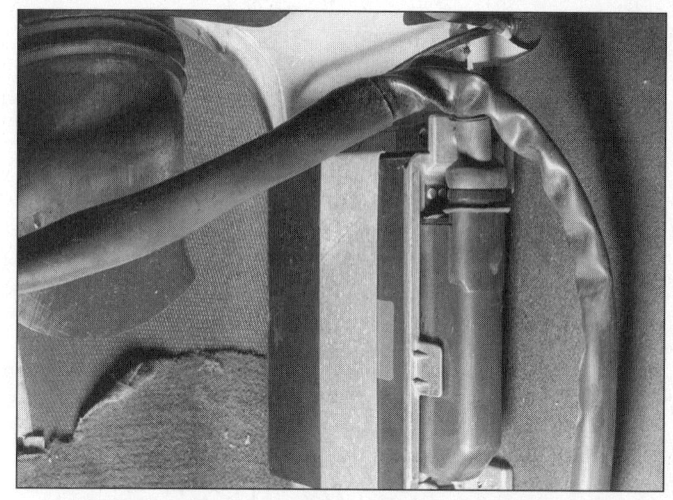

7.8 The computer is located under the passenger's side kick-panel

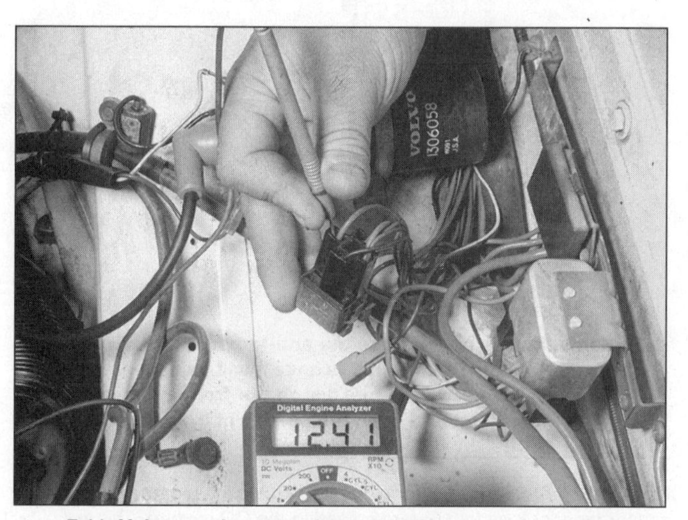

7.11 Using a voltmeter, check for battery voltage at the relay connector

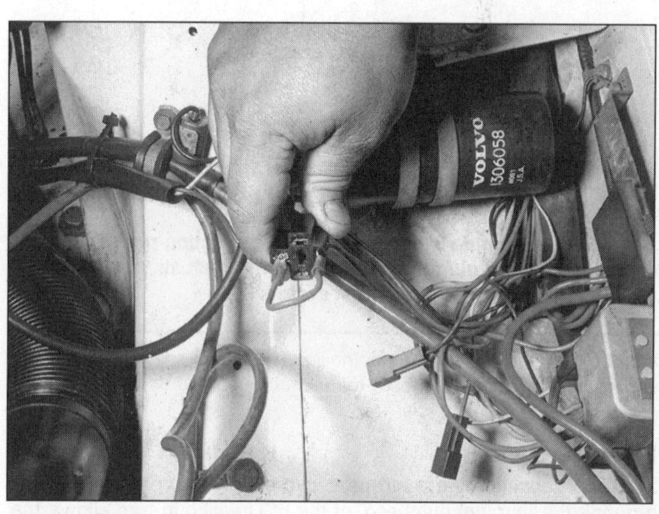

7.12 If battery voltage is available, jump the relay connector to activate the frequency valve

8 Locate the ECU for the oxygen sensor feedback system behind the driver's side kick-panel **(see illustration)**. Disconnect the electrical connector and check for battery voltage on terminal number 15 (three pins from the bottom).

9 If the frequency valve is receiving battery voltage, replace it with a new part. If there is no voltage present, diagnose the wiring harness for shorts or open circuits.

10 If the ECU is receiving battery voltage, have it checked by a dealer service department. If it is not receiving battery voltage, diagnose the wiring harness for shorts or open circuits.

Relay check

11 The Lambda-sond receives power from a relay located in the left side (driver's) inner fender. This relay (system relay) is activated by power from the fuel pump relay when the starter motor cranks or the engine is running. Use a voltmeter and check for battery voltage at the relay electrical connector **(see illustration)**.

12 If battery voltage exists, remove the relay and jump the terminals to activate the frequency valve **(see illustration)**. Listen for the buzzing sound from the frequency valve.

13 If the frequency valve buzzes, then the relay is faulty. If there is no battery voltage present at the relay connector, diagnose a short or open circuit in the wiring harness.

Replacement

Frequency valve

14 Relieve the fuel pressure (see Chapter 4).

15 Remove the fuel lines from the frequency valve.

16 Disconnect the electrical connector.

17 Remove the clamp and the frequency valve as a single unit.

Checks - Turbocharged engines

Thermal cut-out switch

18 The thermal cutout switch **(see illustration 7.1c)** provides a ground for terminal number 7 on the ECU which in turn allows the ECU to hold the signal to the frequency valve fixed (64 to 70 degrees) for a richer warm-up mixture. The thermal cut-out switch is threaded into the cylinder block below the intake manifold.

19 Make sure the coolant temperature is below 59 degrees F and disconnect the harness connector from the thermal cut-out switch. Check for continuity between the switch and ground. There should be continuity.

20 Run the engine until normal operating temperature is reached, recheck for continuity to ground (see Step 19). There should be no continuity. Replace the switch if the tests are incorrect. **Note:***To doublecheck the thermal cut-out switch, using the dwellmeter (above*

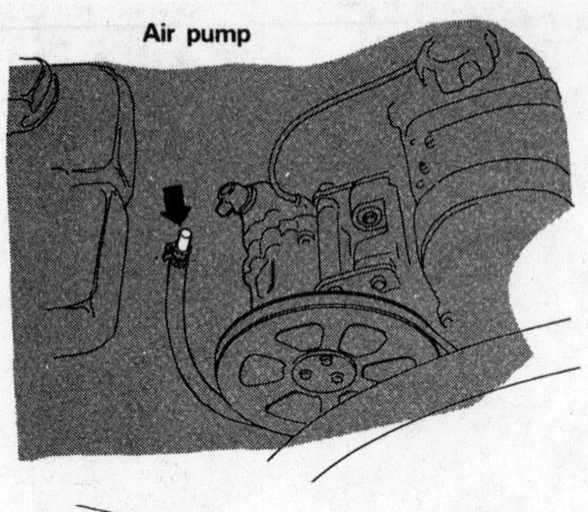

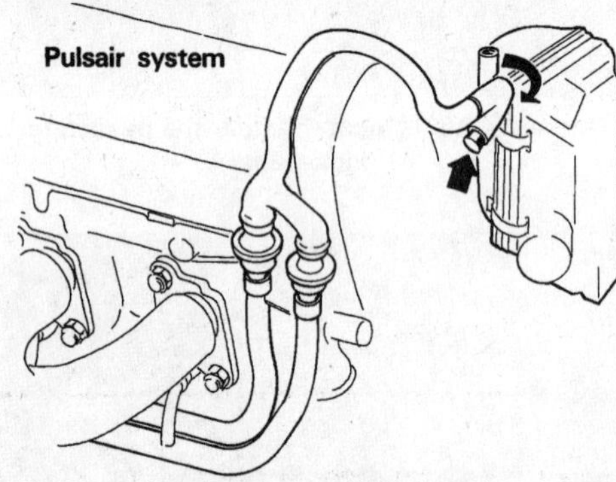

8.2 If the engine is equipped with Air Injection or Pulsair, disconnect and plug the hose to prevent fresh air to enter the exhaust system

method), ground the switch electrical connector to ground against the engine block. The dwell should read a rich condition 64 to 70 degrees (71 to 77 percent duty cycle).

Pressure switch

21 Upon accelerating a turbocharged engine, the pressure switch grounds the terminal number 7 of the ECU which in turn allows the ECU to hold the signal to the frequency valve fixed (64 to 70 degrees dwell) for a richer accelerating fuel/air mixture. The pressure switch is mounted on the firewall in the middle of the engine compartment **(see illustration 7.1c).**

22 With the ignition key OFF, disconnect the electrical connector from the pressure switch and using an ohmmeter, check the resistance across the terminals. There should be no continuity across the terminals.

23 Connect a hand-held vacuum/pressure pump to the pressure switch (use a T-fitting) and apply pressure to the switch while observing the ohmmeter. The switch should close (continuity) when the pressure in the switch exceeds 2.9 psi. **Note:***To doublecheck the pressure switch, using the dwellmeter (as described above), apply pressure (more than 2.9 psi) to the switch using the hand held vacuum pump. The dwell should read a rich condition 64 to 70 degrees (71 to 77 percent duty cycle).*

24 If the test results are incorrect, replace the pressure switch.

8 CO adjustment (CIS engines only)

Refer to illustrations 8.2, 8.7a and 8.7b

Note 1: *Adjusting the CO mixture or air/fuel mixture is not a regular maintenance item. This setting is pre-set at the factory and not usually changed unless someone has tampered with the CIS fuel injection system, a frequency valve or fuel distributor has been replaced or the engine has failed its state-required emissions test.*

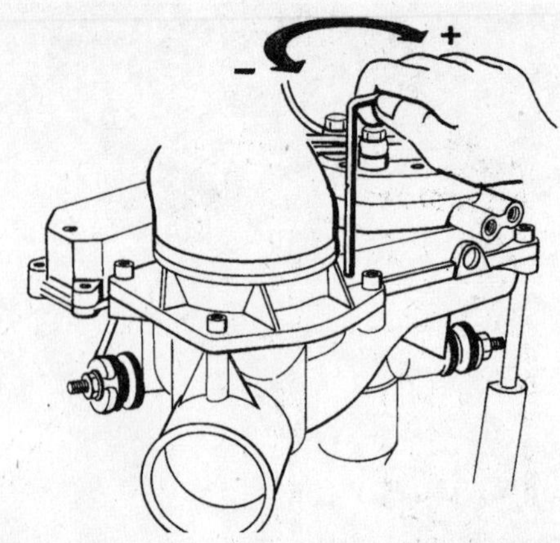

8.7a Turn the 3mm wrench clockwise for a richer mixture or counterclockwise for a leaner mixture

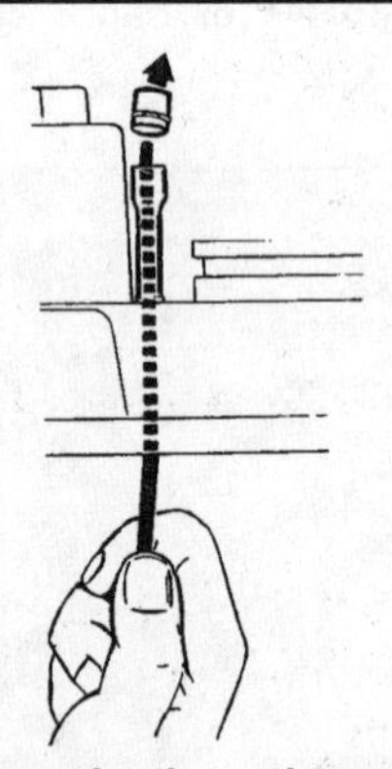

8.7b Some models may require the anti-tamper plug be removed from the fuel distributor. In this case it will be necessary to completely remove the fuel distributor from the engine compartment and push the plug from the bottom

Note 2: *Refer to Section 7 for a complete checkover of the oxygen sensor feedback system before adjusting the CO mixture. This will insure a precise adjustment and working knowledge of the entire system.*

Note 3: *Turbocharged engines typically run slightly richer air/fuel mixture at idle. The adjustment procedure in this section will suffice for all CIS systems*

1 With the ignition key OFF, hook up a tachometer according to the manufacturer's instructions.

2 If the engine is equipped with Air Injection or a Pulsair system, disconnect and plug the hose to prevent it from allowing fresh air into the exhaust system **(see illustration).**

3 Connect a dwellmeter to the test connector (red wire) located near the strut tower on the driver's side **(see illustrations 7.3a and 7.3b)** and select the 4 cylinder scale.

4 Start the engine and run it to approximately 1,500 rpm for at least 5 minutes after it reaches normal operating temperatures and then allow it to idle (parking brake ON, transmission in Neutral).

5 Disconnect the wire harness from the oxygen sensor electrical connector but do not allow the connector to touch ground or damage to the oxygen sensor will occur.

6 Observe the dwellmeter reading at idle. It should be between 42 and 48 degrees with the oxygen sensor disconnected (open loop).

7 If the dwell is incorrect, remove the anti-tamper plug from the mixture control unit and insert an extra long, 3mm wrench down into the access hole **(see illustrations)** in the fuel distributor and turn the adjustment screw until the correct dwell is attained.

a) *Counterclockwise reduces the CO content (low dwell reading).*
b) *Clockwise increases the CO content (high dwell reading).*

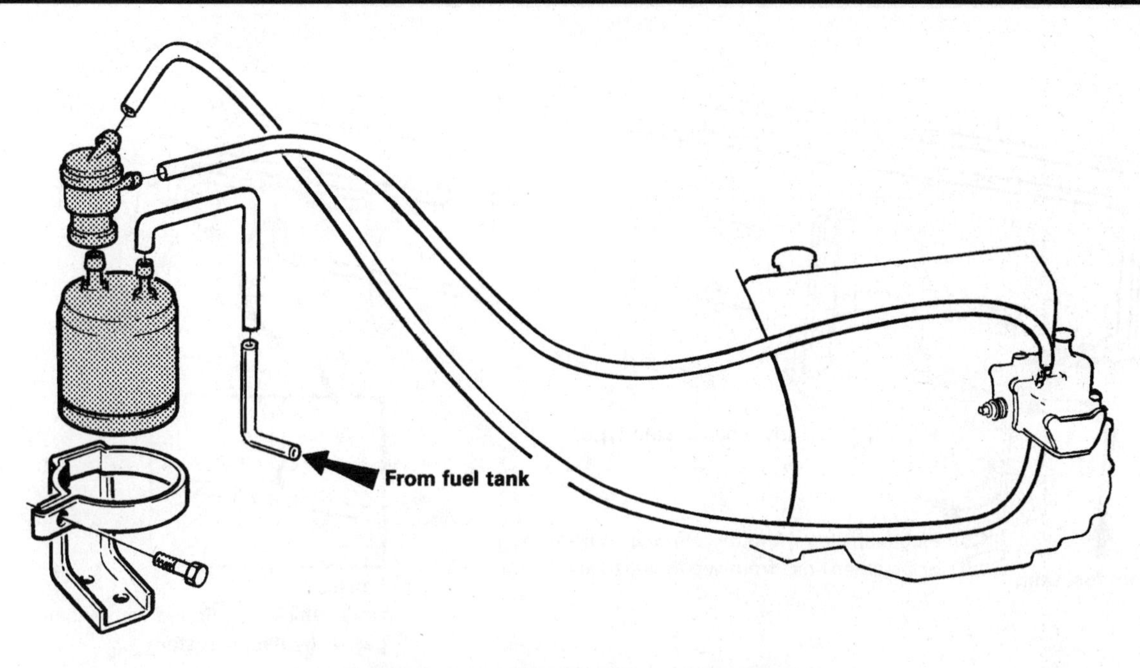

9.1a EVAP system - 1976 and 1977 carbureted engines

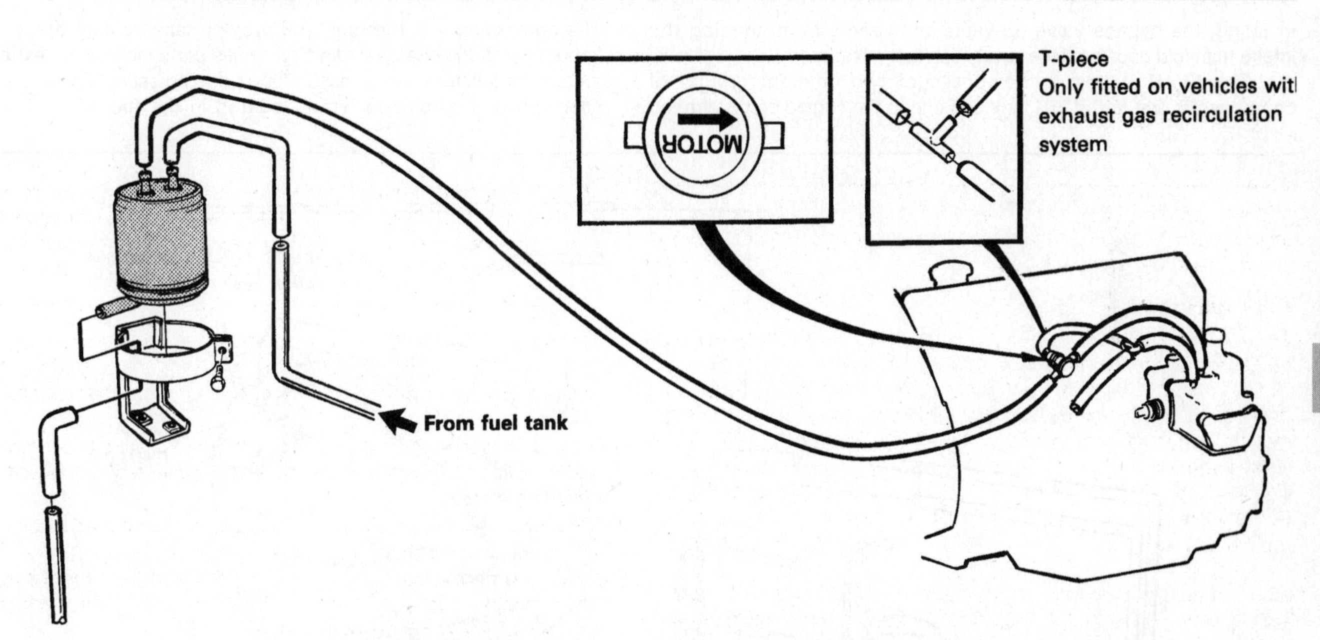

9.1b EVAP system - 1978 and 1979 carbureted engines

8 Ground the disconnected wire from the oxygen sensor on the harness side (green wire) and read the dwellmeter. It should exceed 68 degrees (rich condition). If the system does not respond properly, check the oxygen sensor feedback system (see Section 7).

9 Turn the ignition key off and reconnect the oxygen sensor wire to the oxygen sensor and start the engine. The dwell should read 41 to 44 degrees. This reading should fluctuate between these two numbers as the oxygen sensor feeds information and corrects the rich/lean condition. For additional information on the oxygen sensor operation, refer to Section 4. If there is no response from the oxygen sensor, check the oxygen sensor function (see Section 4).

10 If the dwellmeter reads correctly, the oxygen sensor is working properly and the CO mixture is correct.

9 Evaporative emissions control (EVAP) system

General description

Refer to illustrations 9.1a through 9.1h and 9.2

1 The evaporative emissions control system **(see illustrations)** stores fuel vapors generated in the fuel tank in a charcoal canister located in the engine compartment when the engine isn't running. When the engine is started, the fuel vapors are drawn into the intake manifold and burned. The crankcase emission control system works like this: When the engine is cruising, the purge control valve (bypass valve) is opened slightly and a small amount of blow-by gas is drawn into the intake manifold and burned. When the engine is starting cold

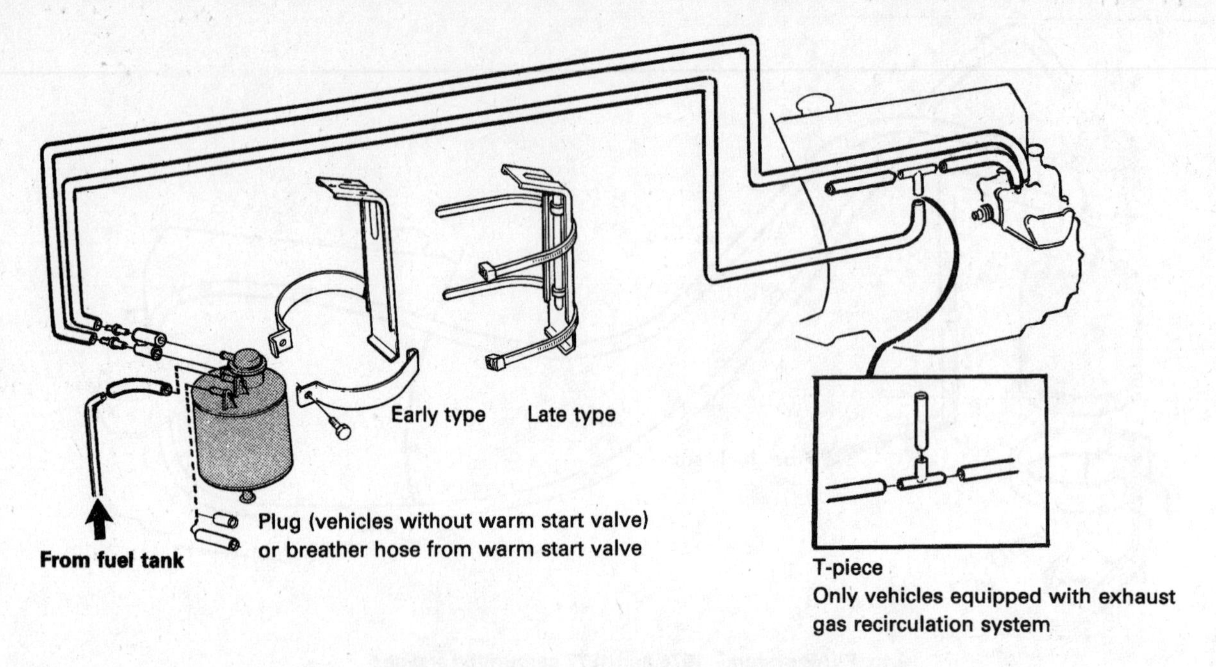

9.1c EVAP system - 1980 and 1981 carbureted engines

or idling, the bypass valve prevents any vapors from entering the intake manifold causing excessively rich fuel mixture.

2 The EVAP system is composed of two different groups of components; one at the fuel tank and one at the engine compartment. The components at the fuel tank provide tank venting and provide vapor flow to the charcoal canister. These parts include the expansion area in the fuel tank, overpressure and underpressure valves in the fuel filler cap and a roll-over valve located in the fuel filler pipe **(see**

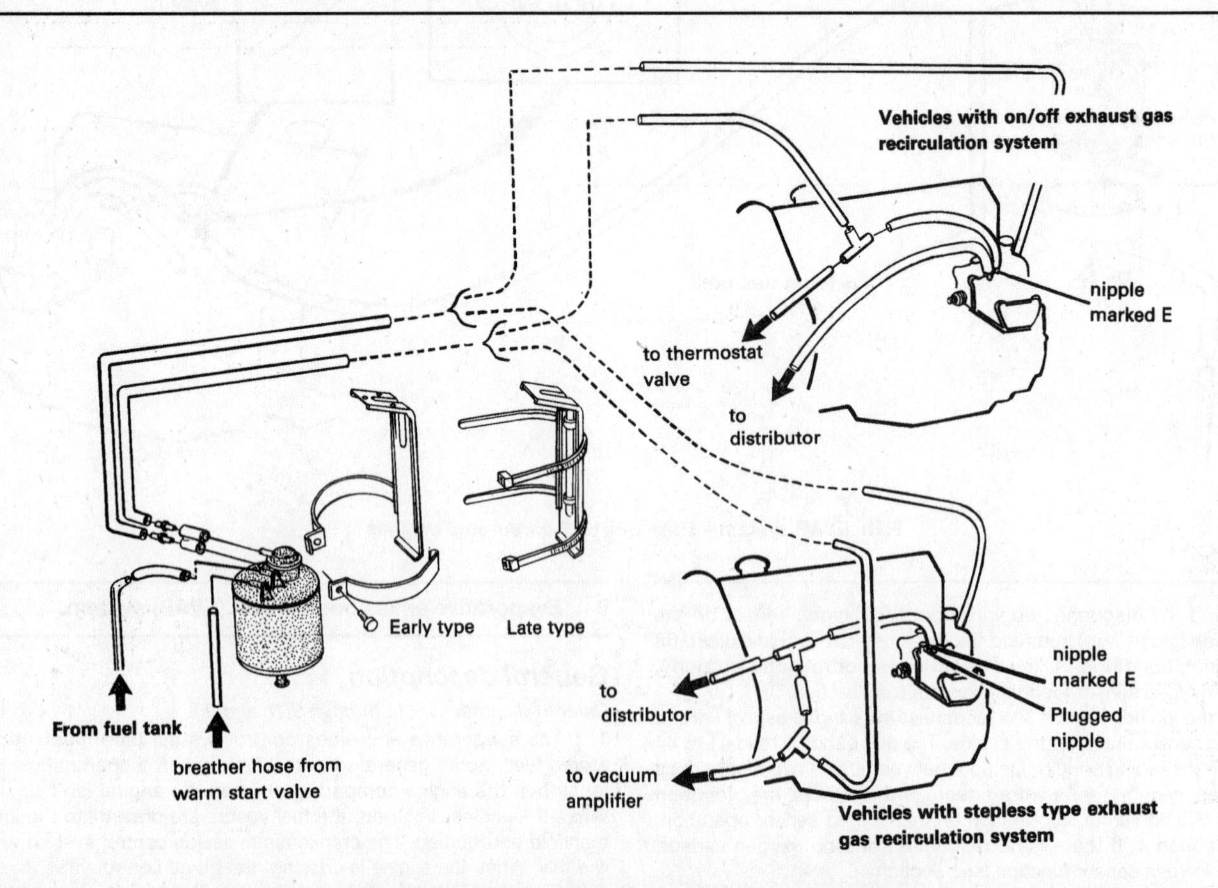

9.1d EVAP system - 1982 through 1984 carbureted engines

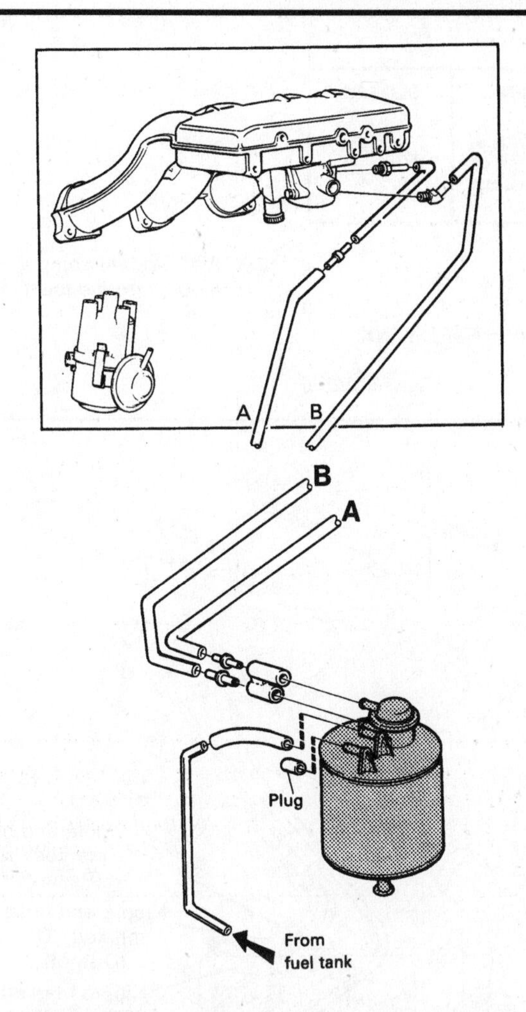

9.1e EVAP system - 1981 and 1982 B21F engines with CIS

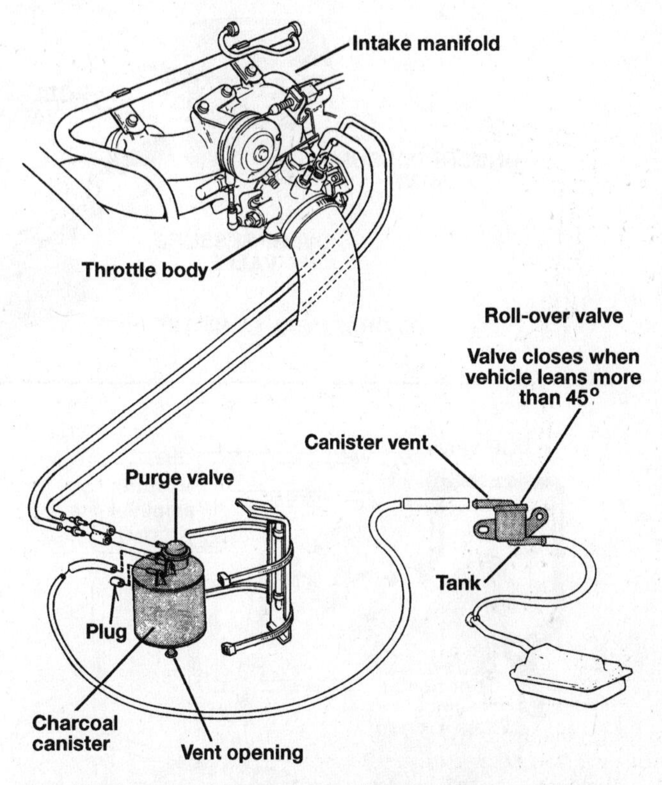

9.1f EVAP system - 1983 and later B23F engines and 1985 and later B230F engines with LH Jetronic (EFI)

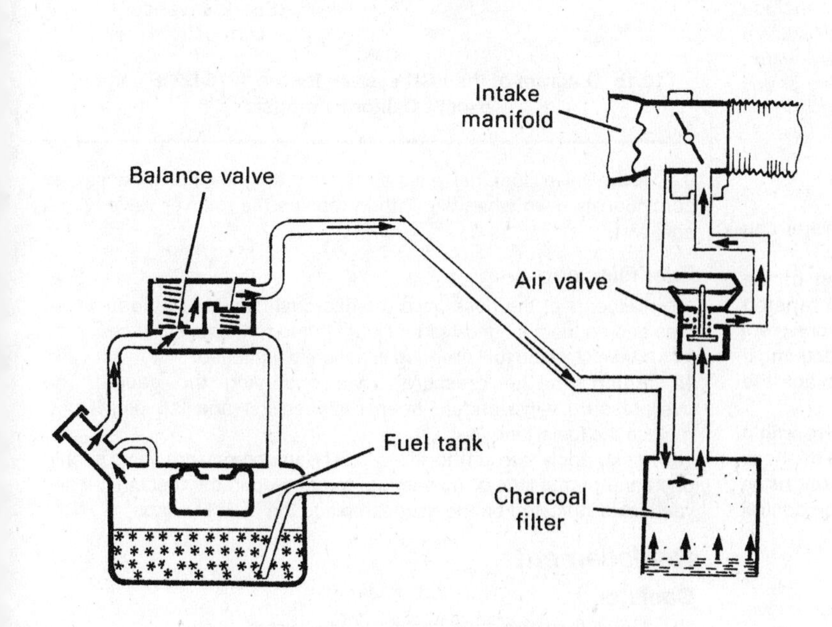

9.1g EVAP system - 1976 through 1980 B21F engines with CIS

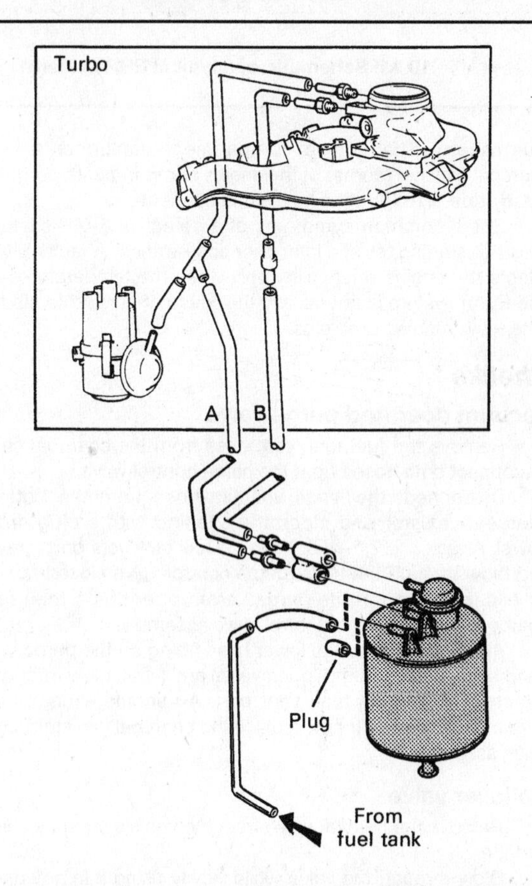

9.1h EVAP system - 1982 B21F turbocharged engines

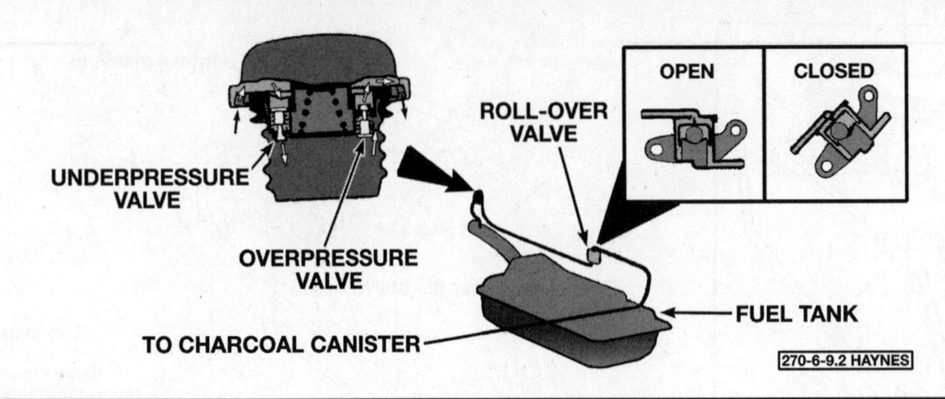

9.2 EVAP system components located in/around fuel tank

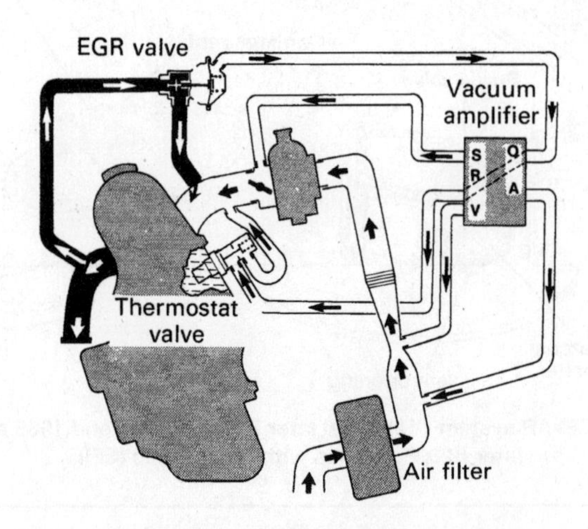

10.1a Schematic of a typical EGR system

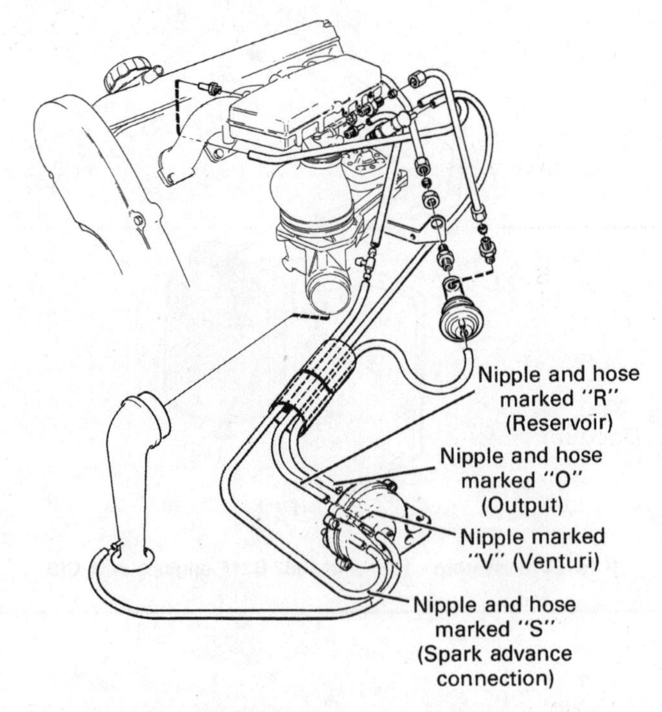

10.1b Diagram of the EGR system for the 1976 B21F engine, except California models

illustration). The engine compartment components include the charcoal canister (corner of the engine compartment), purge valve **(see illustration 9.1b)** and the hoses and connections.

3 Some common symptoms of a defective EVAP system include hard-hot starting, erratic idle, poor acceleration. A faulty EVAP system affects the engine driveability only when the temperatures are warm. The EVAP system is not usually the cause of hard, cold starting or any other cold running problems.

Checks

Vacuum operated purge valve

4 Remove the fuel tank vent hose from the charcoal canister and disconnect both hoses from the purge control valve.

5 Disconnect the small vacuum hose from the bottom of the charcoal canister and block the opening with a plug or a tapered dowel. Attach a length of hose to the fuel tank vent port (previous step) and blow through the hose (purge control valve closed/no vacuum). If air escapes through the purge valve upper fitting then replace the charcoal canister/purge control valve assembly.

6 Apply vacuum to the lower hose fitting on the purge valve with a hand held vacuum pump (open valve) and again blow through the hose attached to the fuel tank vent port. Air should escape through the purge control valve. If not, replace the charcoal canister/purge control valve assembly.

Rollover valve

7 Remove the rollover valve from the vehicle and place it on a level surface.

8 Blow through the valve while slowly tilting it in any direction. Air should pass freely through the valve until it becomes offset 45 degrees or more, then air should not pass through.

9 If the valve does not pass air when sitting level or it passes air continuously even when tilted, then replace the rollover valve with a new part.

Fuel filler cap

10 Disconnect the hose from the charcoal canister at the fuel filler pipe and connect a hand-held vacuum pump to the fuel filler cap.

11 Make sure the fuel filler cap is securely tightened.

12 Pump up the pressure while observing the gauge. The overpressure valve should open between 2.1 and 3.6 psi. If not, replace the fuel filler cap.

13 Next, apply vacuum to the cap. The underpressure valve should open and permit little or no vacuum to build-up in the fuel tank. If any vacuum is indicated on the gauge, replace the fuel filler cap.

Replacement

Canister

14 Label, then detach all hoses to the canister.

15 Slide the canister out of its mounting clip.

16 Visually examine the canister for leakage or damage.

17 Replace the canister if you find evidence of damage or leakage.

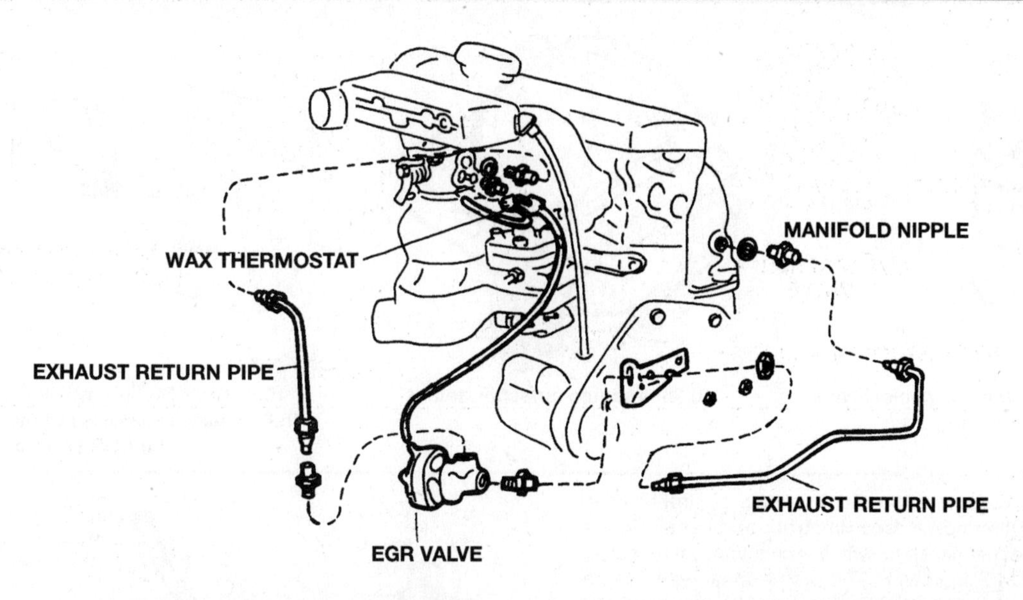

10.1c Diagram of the EGR system for the B21F engine, except California models without the vacuum amplifier system

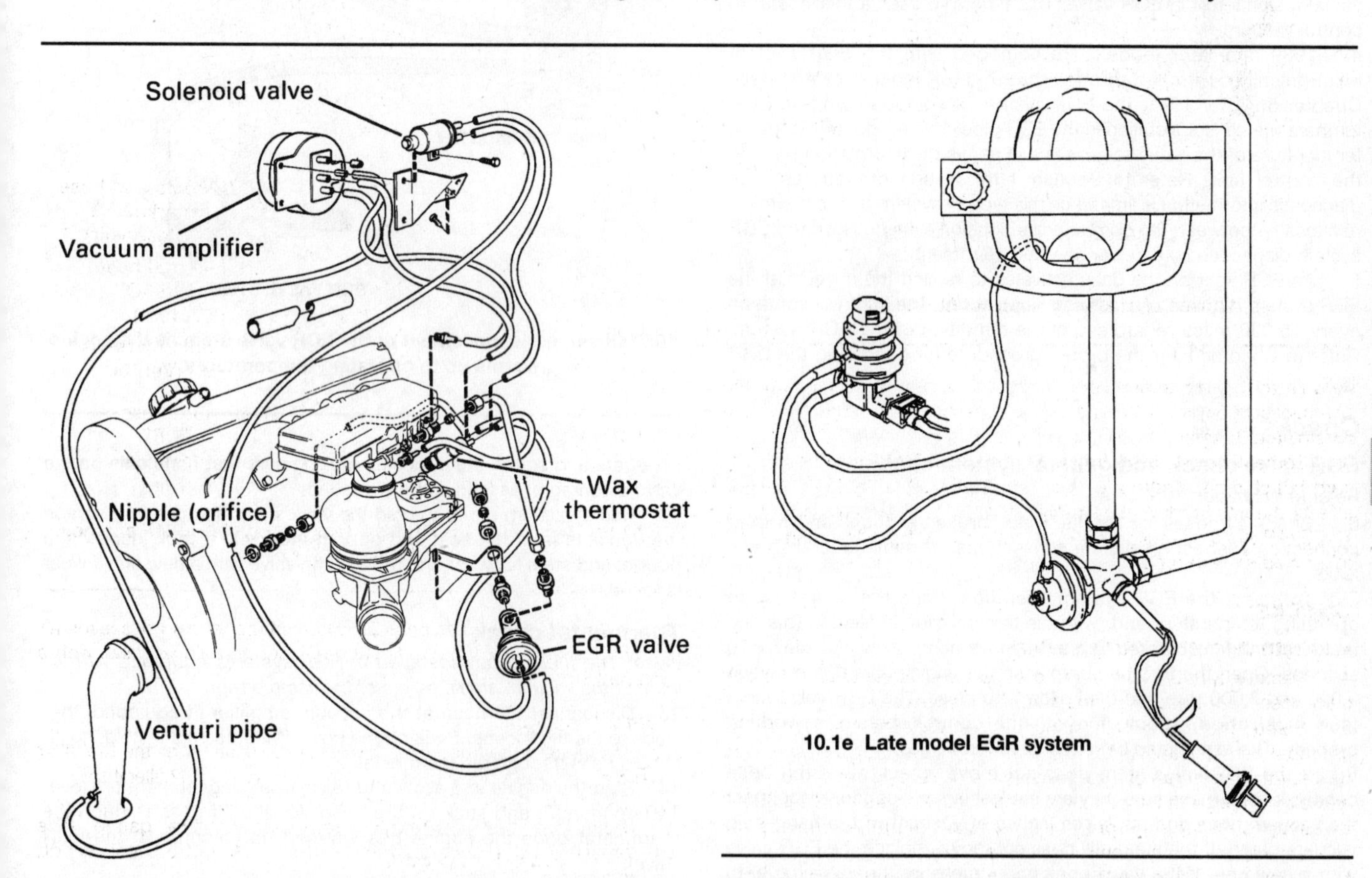

10.1e Late model EGR system

10.1d Diagram of the EGR system for the 1976 through 1978 B21F engine, California models

10 Exhaust Gas Recirculation (EGR) system

General description

Refer to illustrations 10.1a through 10.1e, 10.3a, 10.3b and 10.5

1 This system recirculates a portion of the exhaust gases into the intake manifold in order to reduce the combustion temperatures and decrease the amount of nitrogen oxide (NOx) produced **(see illustrations)**. The main component in the system is the EGR valve.

2 There are two types of EGR systems; the mechanical type which uses thermostats and solenoid valves to control the vacuum to the EGR valve or the electronic type which uses a control unit (computer) to control the actions of the EGR valve.

3 There are basically two different types of mechanical EGR systems. Depending upon the state or country (USA or Canada), early CIS engines use either the ON/OFF type or the Proportional type EGR system. Both types vary the EGR valve using vacuum. Each type

10.3a Thermostat for the mechanical type EGR systems

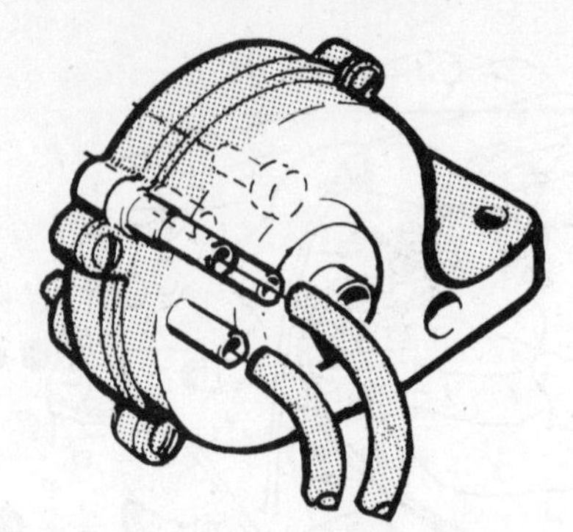

10.3b Vacuum amplifier details

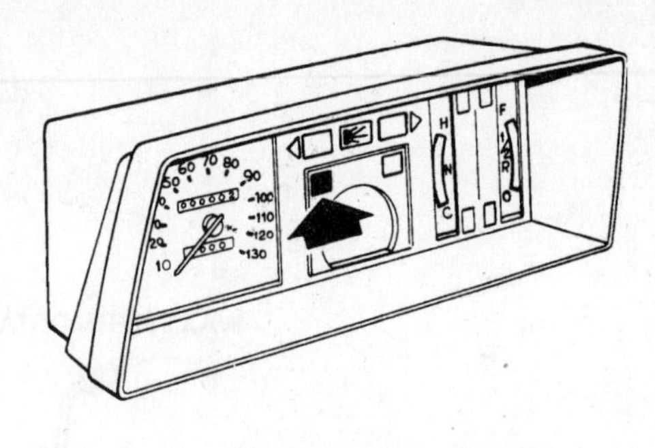

10.5 The EGR light will illuminate every 15,000 miles regardless of the condition of the EGR system

differs in the way the vacuum to the EGR valve is controlled. The simple versions use a thermostat **(see illustration)** mounted in the coolant system near the thermostat to switch vacuum once the coolant temperature reaches 140 degrees F. The proportional type uses a vacuum amplifier **(see illustration)** to increase or decrease the vacuum signal to the EGR valve. This type also uses a thermostat to control vacuum.

4 1989 and later models are equipped with the EGR system electronically controlled through the EZ-116K ignition system (see Chapter 5). After 1991, the EGR system is equipped with an EGR temperature sensor installed in the EGR pipe. This sensor monitors the temperature of the exhaust gasses and relays the information back to the control unit. Refer to Section 4 for testing procedures. The diagnostic information is limited on this type of system due to the inter-relationship between the computerized ignition system. Have the EGR system diagnosed by a dealer service department.

5 An EGR light on the dash activates to remind the driver that the EGR system requires service **(see illustration)**. The light will come on every 15,000 miles regardless of the condition of the EGR system. Refer to Chapter 1 for the proper procedure for cancelling the EGR light.

WAX THERMOSTAT

S

E

EGR VALVE STEM

10.7 Observe the movement of the EGR valve stem as the engine warms up to operating temperatures

Checks

EGR valve check and general system check

Refer to illustration 10.7

6 Check all hoses for cracks, kinks, broken sections and proper connection. Inspect all system connections for damage, cracks and leaks.

7 To check the EGR system operation, bring the engine up to operating temperature and, with the transmission in Neutral (parking brake set and tires blocked to prevent movement), allow it to idle for 70 seconds. Open the throttle abruptly so the engine speed is between 2,000 and 3,000 rpm and then allow it to close. The EGR valve stem **(see illustration)** should move if the control system is working properly. The test should be repeated several times.

8 If the EGR valve stem does not move, check all of the hose connections to make sure they are not leaking or clogged. Disconnect the vacuum hose and apply ten inches of vacuum with a hand-held vacuum pump. If the stem still does not move, replace the EGR valve with a new one. If the valve does open, measure the valve travel to make sure it is approximately 1/8-inch. Also, the engine should run roughly when the valve is open. If it doesn't, the passages are probably clogged.

9 Apply vacuum with the pump and then clamp the hose shut. The valve should stay open for 30 seconds or longer. If it does not, the diaphragm is leaking and the valve should be replaced with a new one.

10 If the engine idles roughly and it is suspected the EGR valve is not closing, remove the EGR valve and inspect the poppet and seat area for deposits.

11 If the deposits are more than a thin film of carbon, the valve should be cleaned. To clean the valve, apply solvent and allow it to penetrate and soften the deposits, making sure that none gets on the valve diaphragm, as it could be damaged.

12 Use a vacuum pump to hold the valve open and carefully scrape the deposits from the seat and poppet area with a tool. Inspect the poppet and stem for wear and replace the valve with a new one if wear is found.

Thermostat check

Note: *The thermostats installed in EGR systems equipped with or without the vacuum amplifier, are of the same design.*

13 Disconnect the hose at the vacuum amplifier (if equipped) that leads to the thermostat. Suck on the hose. With the engine cold, no air should pass through the thermostat.

14 Start the engine and allow it to reach operating temperature (over 140 degrees F) and suck on the hose. Air should pass through the thermostat once the engine has warmed up (vacuum to the EGR valve).

15 If the thermostat does not operate properly, replace it with a new part.

Vacuum amplifier check

Refer to illustration 10.19

16 Locate the vacuum amplifier (if equipped) and with the engine idling, remove the lower hose and check for the presence of vacuum. Turn the ignition key OFF.

17 Disconnect the wiring harness connector from the vacuum amplifier and measure the resistance across the terminals of the vacuum amplifier. The resistance should be 75 to 95 ohms.

18 Another check on the vacuum amplifier is to check for battery voltage on the blue wire to make sure power is reaching the vacuum amplifier.

19 If there is no power available (approximately 12 volts) then locate the open wire or shorted circuit in wiring harness. Also, check the performance of the microswitch **(see illustration)** and if necessary adjust the switch to the correct specification (see Chapter 4, Section 14).

Component replacement

EGR valve

Refer to illustration 10.20

20 Unscrew the fitting attaching the EGR valve tube **(see illustration)** to the intake manifold (if equipped). Remove the two nuts retaining the valve and remove the assembly.

21 Check the tube for signs of leaking or cracks.

22 Installation is the reverse of removal.

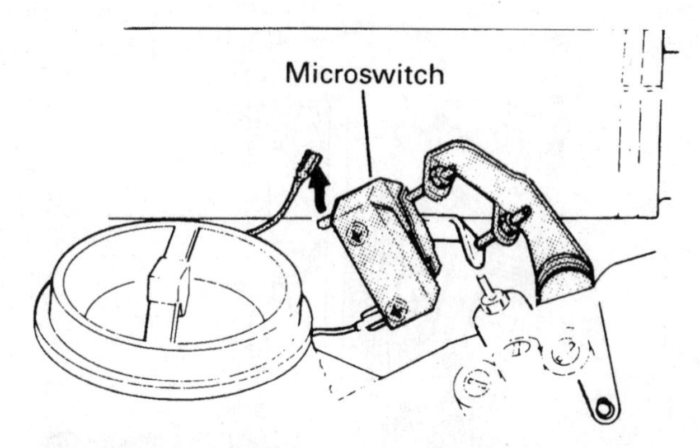

10.19 The microswitch is located on the throttle body housing

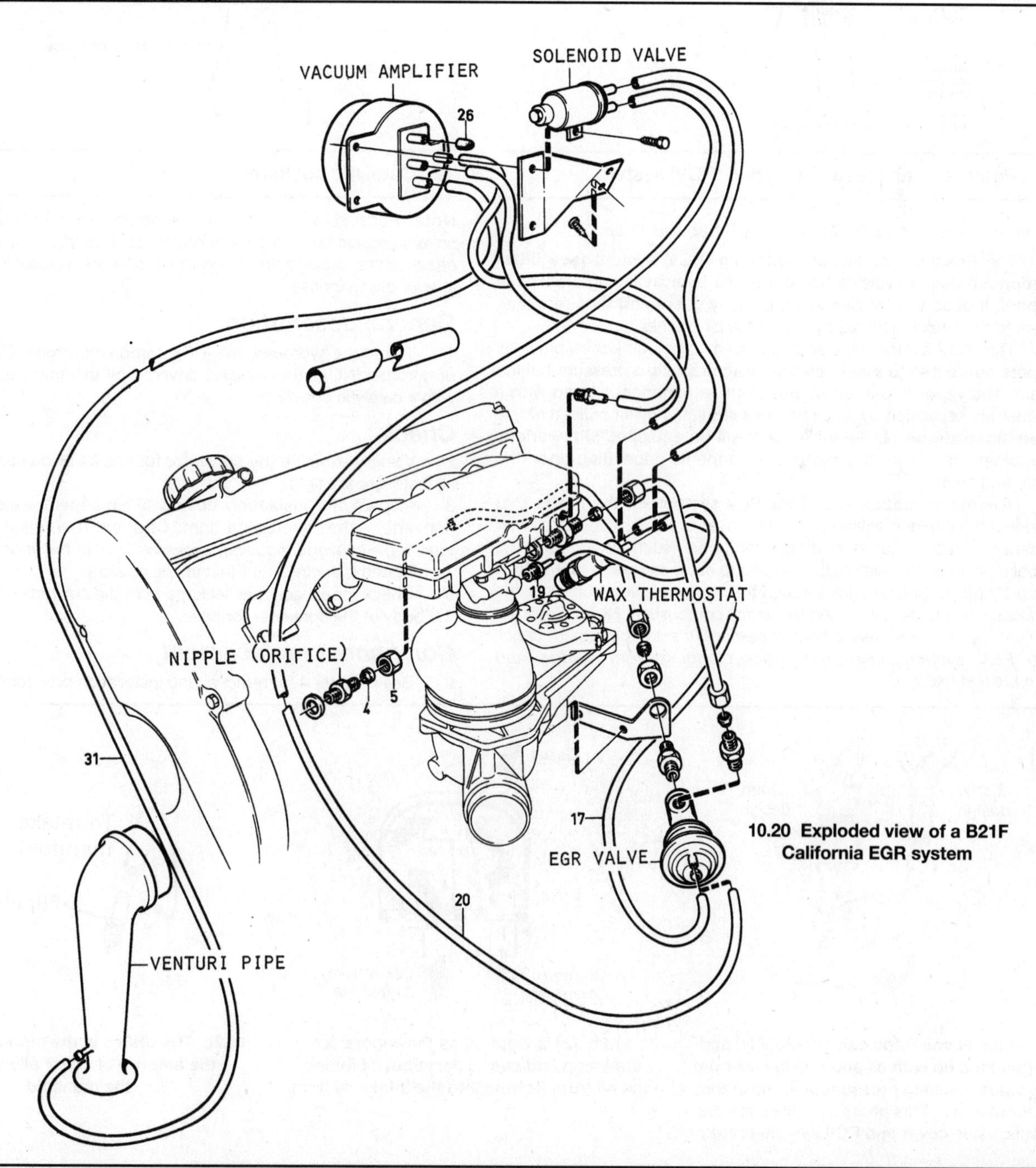

10.20 Exploded view of a B21F California EGR system

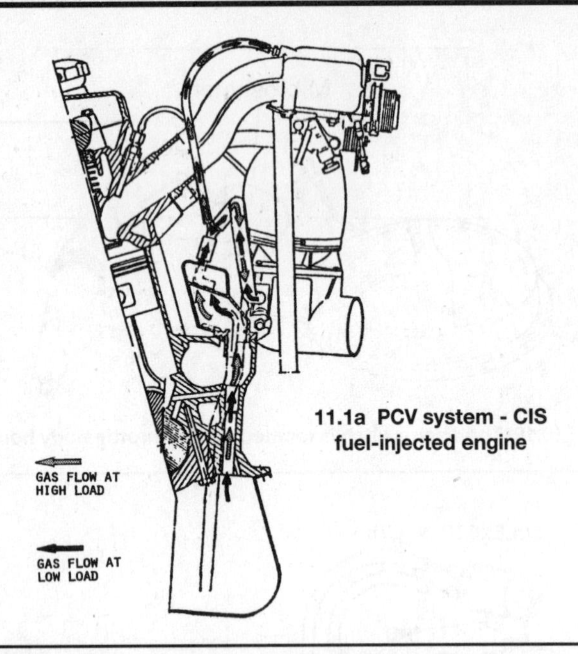

11.1a PCV system - CIS fuel-injected engine

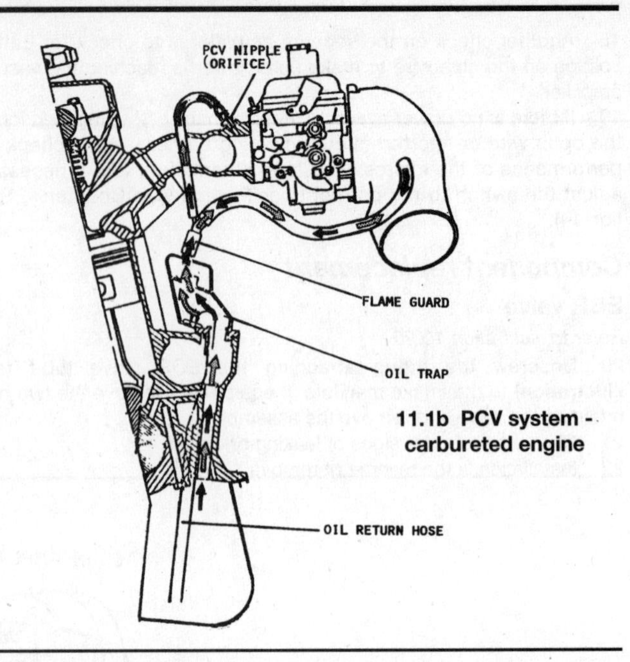

11.1b PCV system - carbureted engine

11 Positive Crankcase Ventilation (PCV) system

Refer to illustrations 11.1a, 11.1b, 11.2a, 11.2b and 11.2c

1 The Positive Crankcase Ventilation (PCV) system **(see illustrations)** reduces hydrocarbon emissions by scavenging crankcase vapors. It does this by circulating blow-by gases and then rerouting them to the intake manifold by way of the air cleaner.

2 This PCV system is a sealed system. The crankcase blow-by vapors are routed to the air cleaner with crankcase pressure behind them. The vapor is purged by one of three methods; filtered with a flame trap, separated by an oil trap or a regulated by an orificed nipple **(see illustrations).** Different models will be equipped with various combinations of the PCV system components depending upon the state and country.

3 The main components of the PCV system are the hoses that connect the intake manifold and valve cover to the throttle body or air cleaner. If abnormal operating conditions (such as piston ring problems) arise, the system is designed to allow excessive amounts of blow-by gases to flow back through the crankcase vent tube into the intake system to be consumed by normal combustion. **Note**: *Since the orificed nipples don't use a filtering element, it's a good idea to check the PCV system passageways for clogging from sludge and combustion residue.*

12 Catalytic converter

Note: *Because of a Federally mandated extended warranty which covers emissions-related components such as the catalytic converter, check with a dealer service department before replacing the converter at your own expense.*

General description

1 To reduce hydrocarbons (HC), carbon monoxide (CO) and oxides of nitrogen (NOx), the vehicles covered by this manual are equipped with a catalytic converter.

Check

2 Visually examine the converter for cracks or damage. Make sure all fasteners are tight.

3 Inspect the insulation covers (if equipped) welded onto the converter - they should be tight. **Caution**: *If an insulation cover is touching a converter housing, excessive heat at the floor may result.*

4 Start the engine and run it at idle speed.

5 Check for exhaust gas leakage from the converter flanges. Check the body of the converter for holes.

Component replacement

6 See Chapter 4 for removal and installation procedures.

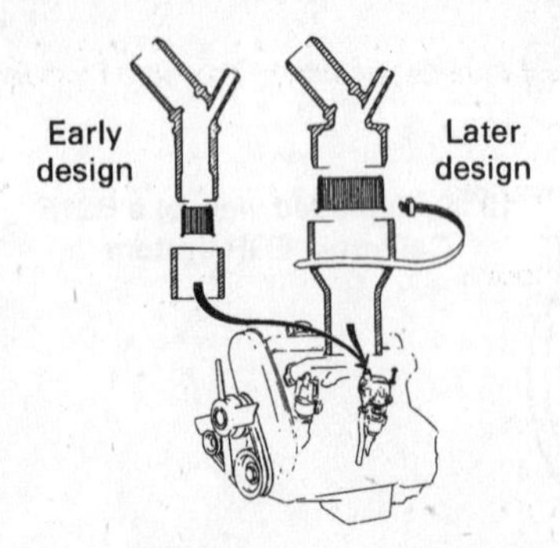

11.2a Flame traps can get clogged and gummed up with oil and residue over the years, causing pressure build-up in the crankcase. This pressure will cause the seals, valve cover and PCV system to leak oil

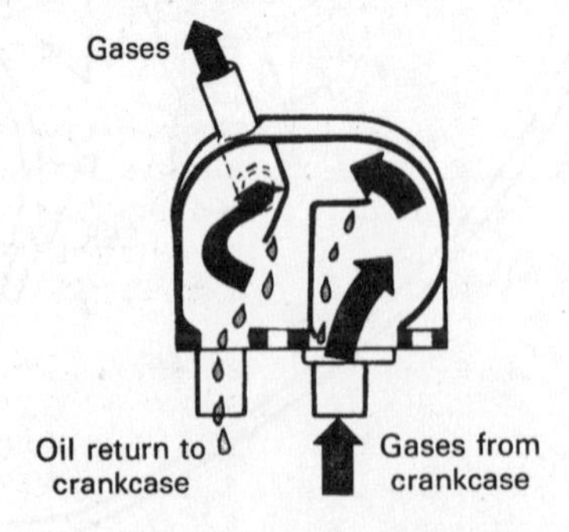

11.2b Oil is trapped as the vapors are drawn up and over a partition, inhibiting the oil from flowing into the intake system

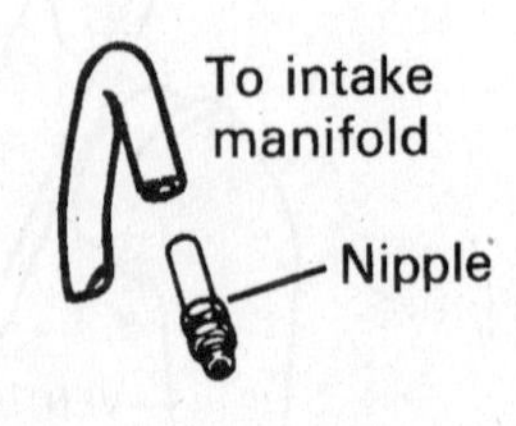

11.2c The orifice in the nipple regulates the amount of vapor allowed into the manifold

Chapter 7 Part A
Manual transmission and overdrive

Contents

Specifications

General

Type	Four or five forward speeds and reverse; overdrive on some models
Identification:	
M40	4-speed, early models (to 1976)
M41	4-speed with overdrive, early models
M45	4-speed, later models (to 1984)
M46	4-speed with overdrive, later models
M47	5-speed (1984 on)

Ratios (typical)	M40/41	M45/46	M47
First	3.41 : 1	3.71 : 1	4.03 : 1
Second	1.99 : 1	2.16 : 1	2.16 : 1
Third	1.36 : 1	1.37 : 1	1.37 : 1
Fourth	1.0 : 1	1.0 : 1	1.0 : 1
Fifth			0.83 : 1
Reverse	3.25 : 1	3.68 : 1	3.68 : 1

Overdrive

Ratio	0.8 : 1
Lubricant type/capacity	See Chapter 1

Torque specifications

	Ft-lbs
Transmission	
Bearing retainer plate bolts (M47)	11 to 18
Bellhousing bolts (M45/6/7)	26 to 37
Countershaft bolt (M47)	26 to 33
Drain plug	20 to 30
Fifth gear synchro nut (later type M47)	89
Output flange nut (M40)	70 to 77
Output flange retaining nut (M45/47):	
M16	52 to 66
M20	66 to 81
Rear cover bolts	26 to 37
Transmission top cover bolts (M45/6/7)	11 to 18
Overdrive	
Oil pan bolts	7
Output flange retaining nut	130
Overdrive main-to-rear case nuts	9
Overdrive solenoid	30 to 40
Overdrive-to-intermediate flange nuts	9
Pressure filter plug	16
Solenoid valve	37

1 General information

Transmission

There are five different versions of manual transmission in use on the 240 series. All are derived from the same basic unit, and are as follows:

M40 - 4-speed
M41 - 4-speed with overdrive
M45 - modified 4-speed
M46 - modified 4-speed with overdrive
M47 - 5-speed

The M40/41 types were installed to early models, and were replaced with the modified M45/46 in 1976. The M45/46 was again modified in 1979, a simpler version being introduced. The five-speed M47 was introduced in 1984, and is basically the M45 with a fifth gear housed in a separate unit on the rear of the transmission. Some modifications to the fifth gear components took place in 1986.

All units have synchromesh on all forward gears, gear changing being by floor-mounted remote shift lever via mechanical linkage.

The overdrive unit is electro-hydraulic, controlled by a switch in the shift lever, and operates on high gear only. On later models, the overdrive is automatically disengaged when changing down to third gear.

Overdrive

The overdrive is essentially an extra transmission, driven by the output shaft of the main transmission and producing on its own output shaft a step-up ratio of 0.797:1. The upshift to overdrive is controlled hydraulically; the hydraulic control valve is operated by a solenoid. A switch on the cover of the main transmission ensures that overdrive can only be engaged when the vehicle is in high gear. The activating switch for the overdrive system is mounted in the top of the shift lever knob.

2 Shift lever - removal, overhaul and installation

Refer to illustrations 2.1a, 2.1b, 2.2, 2.6, 2.7, 2.8, 2.9, 2.10 and 2.15

Note: *This procedure applies specifically to 1979 and later models installed with the M45, M46 and M47 transmissions, but the procedure is similar for earlier models.*

1 Unclip the shift lever boot from the console and slide the boot up the shift lever **(see illustrations)**.

2 Drive out the roll pin **(see illustration)** at the lower end of the shift lever, using a pry bar to protect the linkage from bending.

3 On models with overdrive, remove the console facia panels (see Chapter 11) and disconnect the overdrive wiring. Use a length of cord tied to the cable to help you re-route the cable during installation **(see illustration 2.2)**.

4 On models with overdrive, pry out the top of the shift lever knob and disconnect the overdrive switch,

5 Remove the shift lever.

6 Grip the shift lever in a soft-jawed vise, then use a soft mallet to tap off the knob **(see illustration)**.

7 Remove the screw (steel pullrod only). Withdraw the reverse detent knob **(see illustration)**.

8 Remove the pullrod, spring and interlock sleeve **(see illustration)**.

9 Remove the screw and separate the sleeve from the pullrod. Note that the sleeve has been modified **(see illustration)**, and the later type should always be fitted, together with a 0.08 inch thick, 0.63 inch diameter washer beneath the shift lever knob, to ensure that the sleeve clears the detent bracket.

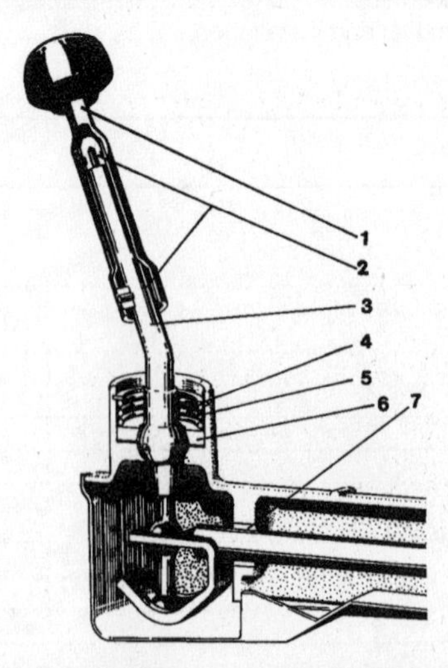

2.1a Cutaway view of a typical early type shift lever assembly

1 Shift lever (upper part)
2 Rubber bushings
3 Shift lever (lower part)
4 Circlip
5 Spring
6 Washer
7 Selector rod

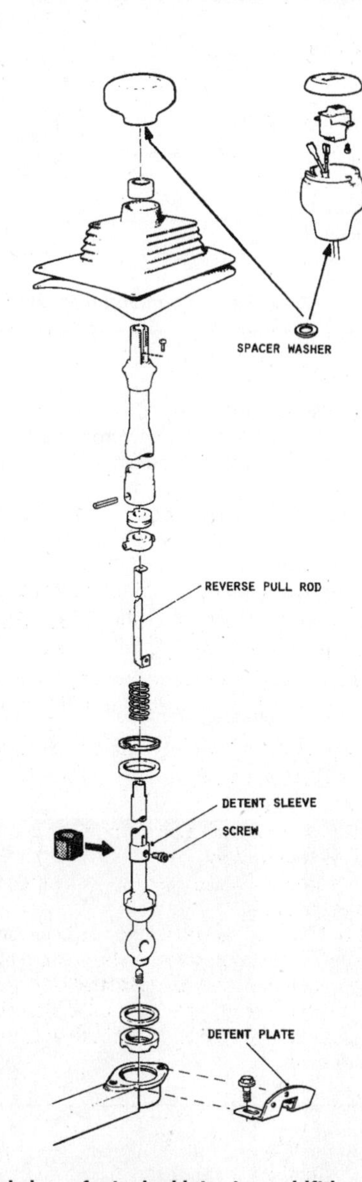

2.1b Exploded view of a typical later type shift lever assembly

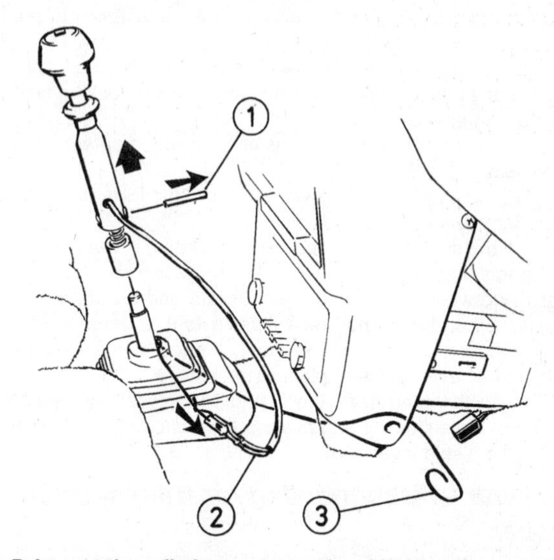

2.2 Drive out the roll pin to remove the shift lever (to reroute the overdrive wiring harness, use a length of cord as shown)

1 *Pin*
2 *Overdrive cable*
3 *Cord*

2.6 To remove the shift lever knob, put the shift lever in a vise and carefully tap the knob loose with a brass or plastic mallet

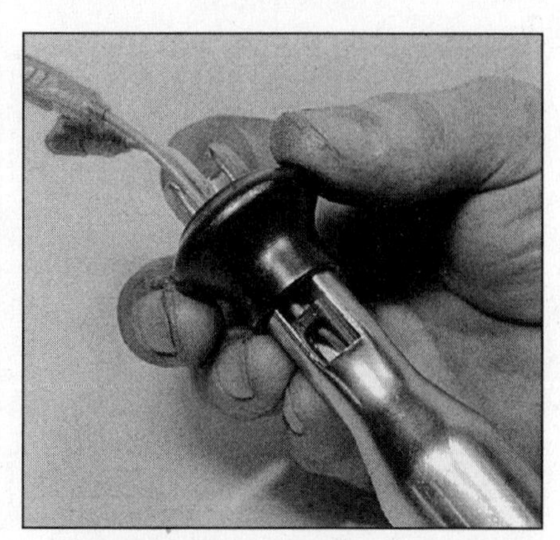

2.7 Withdrawing the reverse detent knob; it doesn't need to be completely removed once the rod is released (plastic pullrod type)

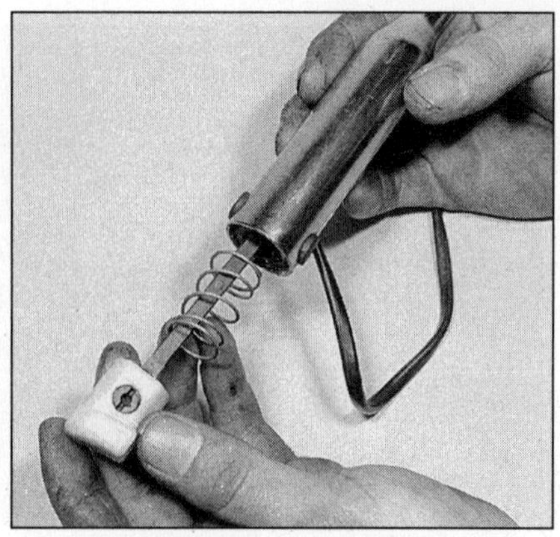

2.8 Removing the plastic type pullrod, spring and interlock sleeve

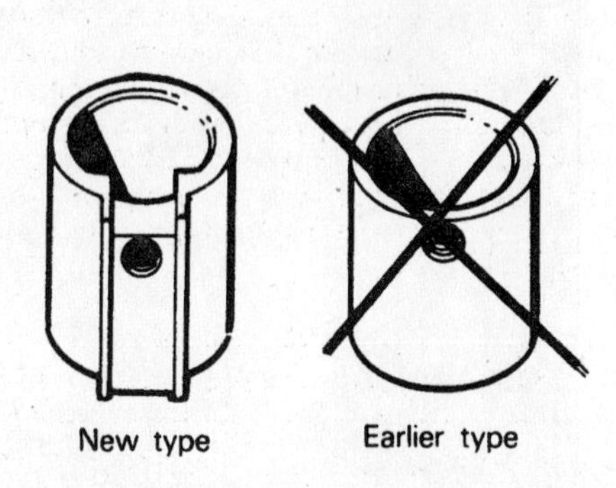

2.9 Note that there are two different types of pullrod interlock sleeves - if you're replacing the sleeve, always use the newer type

10 Bend the end of the steel pullrod **(see illustration)** to prevent it rattling.
11 Soak the plastic pullrod in water for one hour before fitting. Install the spring and interlock sleeve to the pullrod, apply locking fluid to the screw then tighten it.
12 Grease the steel pullrod and install the reverse detent knob to the rod and secure it with the screw.
13 Install the pullrod to the lever, then grip the lever in a soft-jawed vise and tap the knob into position,
14 Install the shift lever and secure it by tapping in the pin.
15 Engage first gear and use a feeler gauge to check that the clearance between the reverse detent plate and the detent screw is between 0.02 and 0.06 inch **(see illustration)**. Engage second gear and verify that the clearance is the same. If necessary, loosen the bolts and adjust the reverse detent plate as necessary.
16 Connect the overdrive cable (if equipped), connecting the switch at the top of the shift lever and pushing the switch back into the knob.
17 Install the shift lever boot.

Shift lever - disconnecting for transmission removal

18 On models with overdrive, disconnect the overdrive cable as described in Section 8

M40/41 type

19 Remove the shift lever dust boot.
20 Using snap-ring pliers, remove the snap-ring from the shift lever ball end housing.
21 Lift out the spring and washer.
22 Lift out the shift lever.

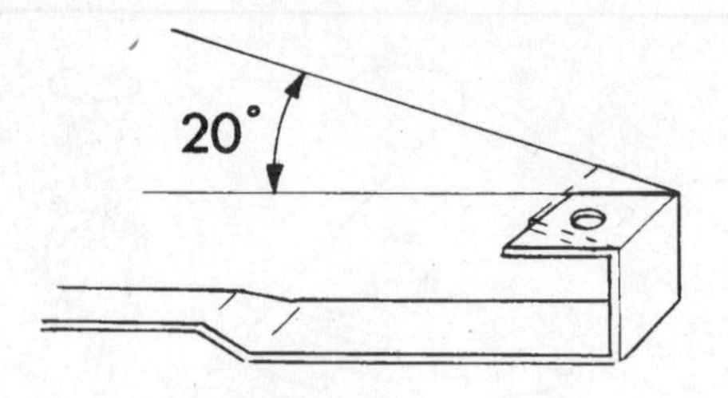

2.10 Pullrod bending diagram

M45, 46 and 47 type

Refer to illustrations 2.23 and 2.25

23 Remove the 4 mm Allen set screw from the selector rod fork end underneath the vehicle, push out the pin and disconnect the fork **(see illustration)**.
24 Remove the shift lever dust boot.
25 Remove the bolts from the reverse detent bracket and lift out the bracket **(see illustration)**.
26 Remove the snap-ring and lift out the shift lever.
Note: *On some models, the back-up light switch is mounted on the shift lever carrier, and must be removed prior to lowering the transmission.*

All types

27 Installation is the reverse of removal.
28 After you're done, check and, if necessary, adjust the reverse detent bracket (see Step 15).

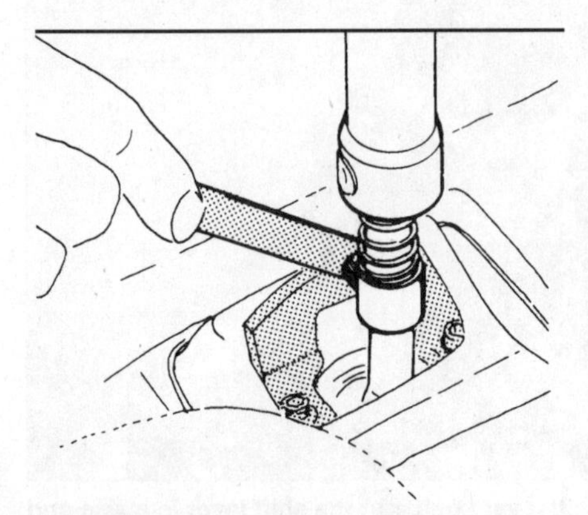

2.15 Use a feeler gauge to check reverse detent clearance

2.23 Shift lever setscrew (arrow) on M45/46/47 transmission

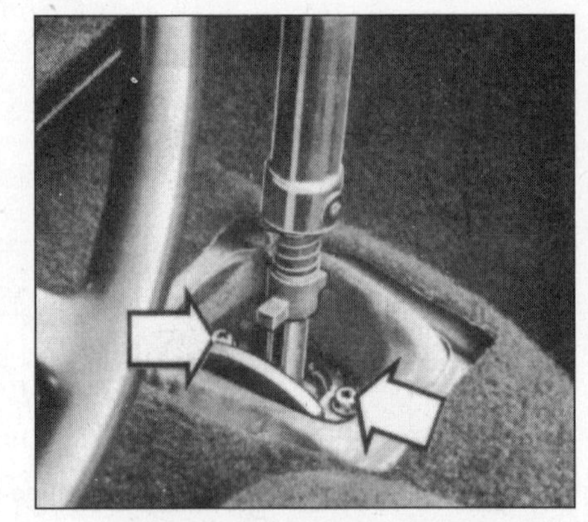

2.25 Reverse detent bracket bolts (arrows)

3.2a Be sure to mark the relationship of the front U-joint to the drive flange to ensure that the driveshaft maintains its dynamic balance after reassembly

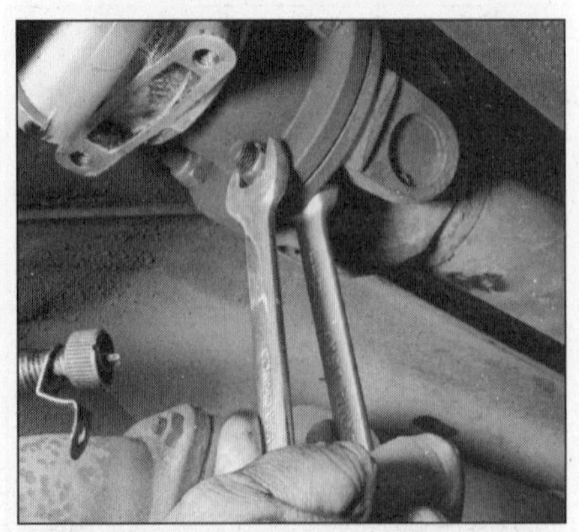

3.2b Using a backup wrench, remove the nuts and bolts that attach the front U-joint to the drive flange

3.3a Before you can remove the drive flange, you'll have to remove the retaining nut (arrow); this nut is very tight, so you'll need some means of holding the flange while you break the nut loose

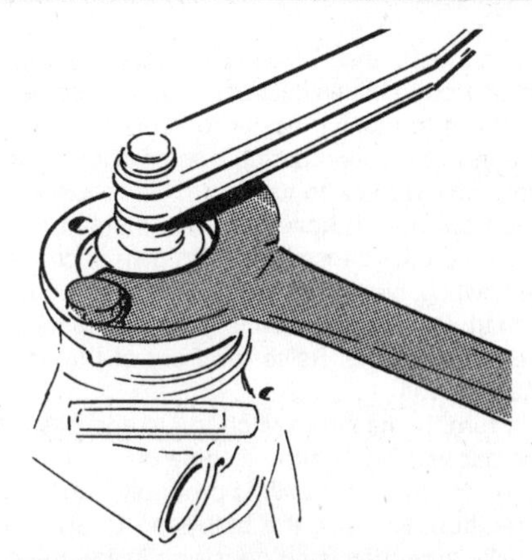

3.3b The best way to hold the output flange while you break the flange retaining nut loose is to use Volvo's special tool (shown), but there are aftermarket pin-type spanners available that will do the job

3.5 Pry out the extension housing seal with a seal removal tool or a screwdriver - make sure you don't damage the seal bore

3.6 Install a new extension housing seal with a seal installer, a large socket or a short section of pipe; if you use a socket or pipe, make sure its outside diameter is slightly smaller than the outside diameter of the seal itself

3 Oil seal replacement

Extension housing seal

Refer to illustrations 3.2a, 3.2b, 3.3a, 3.3b, 3.5 and 3.6

1 Raise the vehicle and place it securely on jackstands.

2 Mark the relationship of the driveshaft to the drive flange and remove the bolts and nuts attaching the driveshaft to the flange **(see illustrations)**. Detach the driveshaft from the flange and move the driveshaft aside.

3 Hold the flange and remove the flange retaining nut **(see illustrations)**. Depending on the model, this nut is torqued on at somewhere between 50 and 80 ft-lbs. If you don't have the special Volvo tool, an aftermarket universal pin spanner or a suitable equivalent tool, put the transmission in gear and hold the flange with a pair of large water pump pliers.

4 Place a drain pan underneath the flange. Pull out the flange. Use a puller if necessary. Do not try to hammer it off.

5 Pry out the old oil seal **(see illustration)**. Clean up the seal bore. Inspect the seal friction surface on the flange hub. You can remove small imperfections with crocus or emery cloth. If it's permanently scored, replace the flange.

6 Lubricate the new seal and install it using a piece of tubing or a large socket to drive it into place **(see illustration)**. On M47 transmissions, the seal should be recessed by 0.1 inch.

7 On transmissions with overdrive, apply locking compound to the output shaft splines. Be careful not to contaminate the seal.

8 Install the flange and tighten the flange retaining nut to the torque listed in this Chapter's Specifications.

9 Install the driveshaft (see Chapter 8).

10 Top-off the transmission oil (see Chapter 1).

11 Remove the jackstands and lower the vehicle.

12 Test drive the vehicle and check for leaks.

Speedometer drive oil seal

Refer to illustrations 3.14 and 3.17

13 Raise the vehicle and place it securely on jackstands.

14 Disconnect the speedometer drive cable by removing the locking plate **(see illustration)** and unscrewing the knurled end of the cable, then pulling it from the housing.

15 Using a wrench on the machined flats of the bearing housing, unscrew the housing and lift out the bearing, housing and drive pinion.

16 The drive pinion and bearing can now be removed from the housing.

17 Pry the O-ring from the housing **(see illustration)** and clean the housing in solvent.

18 Coat all new parts in clean transmission oil, then install the new O-ring.

19 Push a new seal into the housing, followed by the bearing and drive pinion.

20 Install the housing.

21 Installation of the speedometer cable is the reverse of removal.

22 Check and top-off the transmission oil level.

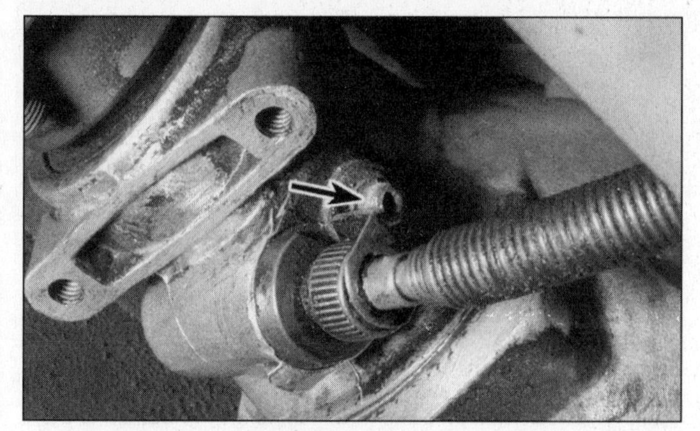

3.14 To remove the speedometer driven gear lock plate, remove this Allen screw (arrow)

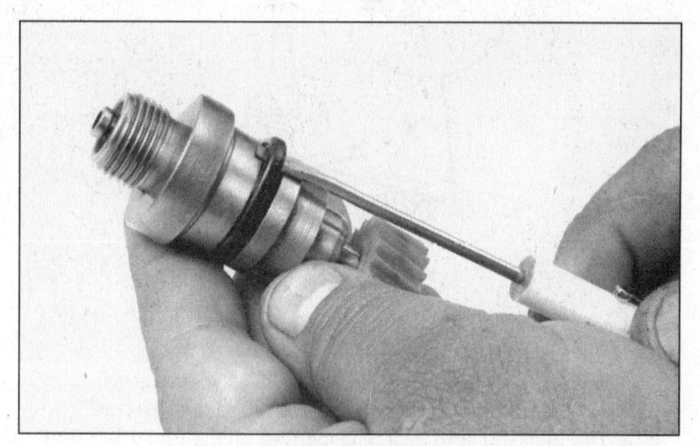

3.17 Pry off the speedometer driven gear O-ring with a small screwdriver

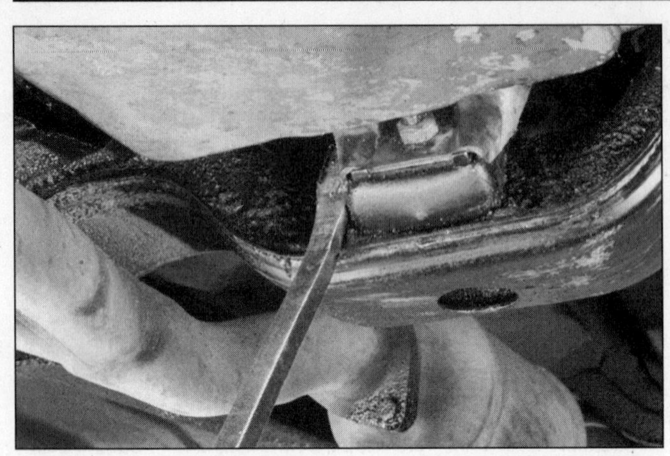

4.1 Insert a large screwdriver or pry bar into the space between the transmission extension housing and the crossmember and try to pry the transmission up slightly

4 Transmission mount - check and replacement

Refer to illustration 4.1

1 Insert a large screwdriver or pry bar into the space between the transmission extension housing and the crossmember **(see illustration)**. Try to pry the transmission up slightly.

2 The transmission should not move away from the transmission mount very much.

3 To replace the mount, remove the nuts attaching the mount to the crossmember and the bolts attaching the mount to the transmission.

4 Raise the transmission slightly with a floor jack and remove the transmission mount.

5 Installation is the reverse of removal. Be sure to tighten the nuts/bolts securely.

5 Back-up light switch - check and replacement

Check

Refer to illustrations 5.1a and 5.1b

1 The back-up light switch is mounted near the top left rear corner of the transmission housing on most transmissions **(see illustration)**. The switch was temporarily relocated to the shift lever support on some late 70's vehicles **(see illustration)** to make it easier to service. However, the switch was vulnerable in this location during transmission removal and installation, so it was moved back to its original location. If you have to replace the switch on one of these vehicles, install the new switch at the original location on the transmission. To gain access to a shift-lever carrier mounted switch, simply unclip the shift lever boot and slide it up the shift lever shaft. On all other vehicles, the switch must be accessed from underneath the vehicle. Raise the front of the vehicle and place it securely on jackstands.

2 To check the switch, turn the ignition switch to On and place the shift lever in Reverse. The back-up lights should come on.

3 If only one light comes on, the other bulb is probably burned out, loose or has a bad connection. Remove the bulb (see Chapter 12), check the filament and make sure there's no corrosion inside the bulb holder. Replace as necessary and retest the back-up lights.

4 If neither of the lights comes on, check both bulbs and replace as necessary (see Chapter 12).

5 If the bulbs are both okay, check the fuse (see Chapter 12).

6 If the fuse is okay, check the wire from the fuse to the electrical connector for the back-up light switch for voltage (see the Wiring Diagrams at the end of Chapter 12). If that wire is okay, check the wire from the switch to the left back-up light for voltage. If that wire is okay, check the wire from the left light to the right light. If that wire is also okay, the switch is bad. Replace it.

Replacement

M40/41 type

7 The back-up light switch is mounted on the top left rear part of the transmission case.

8 Raise the vehicle and place it securely on jackstands. Disconnect the negative battery cable and the electrical cables to the switch.

9 Unscrew the switch from the transmission case.

10 Installation is the reverse of removal. Be sure to use a new sealing washer under the switch.

M45/46 type (up to 1980)

Refer to illustrations 5.13a, 5.13b and 5.18

11 Gain access to the rear of the center console and unplug the electrical connector for the back-up light.

12 Remove the shift lever dust boot.

13 Bend a piece of stiff wire into a hook **(see illustration)**, and insert it down between the left-hand side soundproofing and tunnel **(see**

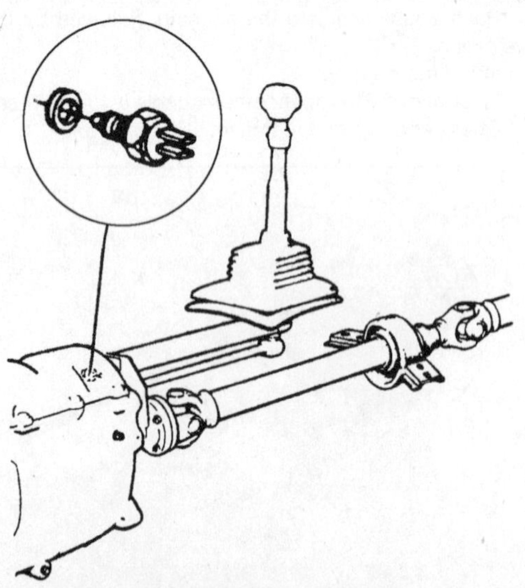

5.1a The back-up light switch is located on the upper left rear part of the transmission on most vehicles

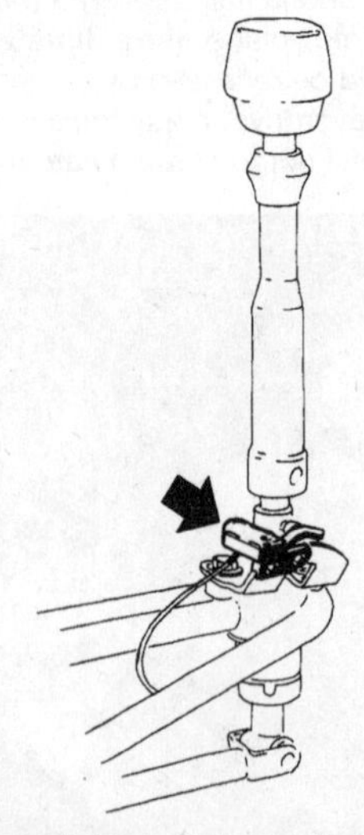

5.1b The back-up light switch was temporarily relocated to the shift lever carrier on some vehicles during the late Seventies

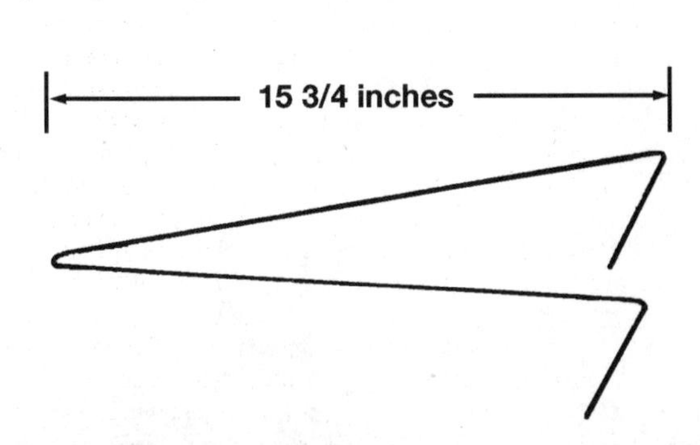

5.13a In order to thread the new back-up light harness back up through the space between the transmission and the tunnel, you'll need to fabricate a hook shaped like this

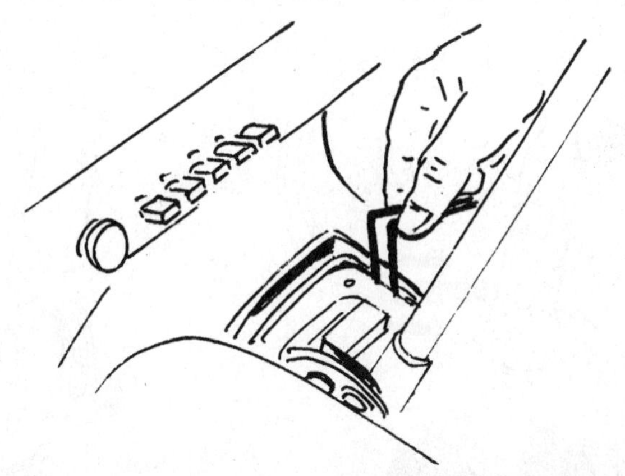

5.13b Insert your hooked tool down through the space between the shift lever base and the edge of the hole in the floor

illustration) so you can pull the cable back through when installing.

14 Raise the vehicle and place it securely on jackstands. Remove the bolts from the right side of the transmission crossmember (see Section 10, if necessary).

15 Feed the back-up light wire harness down through the side of the tunnel and soundproofing to the underside of the vehicle. Free the harness from the shift lever carrier.

16 Unscrew and remove the back-up light switch from the transmission housing (you'll need a 22mm or a 7/8-inch crow's foot).

17 Install the new switch. Be sure to use a new sealing washer under the switch.

18 Hook the wire harness through the wire **(see illustration)**, and pull the harness through to the inside of the vehicle. Secure the harness to the shift lever carrier.

19 Install the transmission crossmember bolts and tighten them securely. Remove the jackstands and lower the vehicle.

20 Plug in the electrical connector and check the operation of the back-up light, then install the console side panel and shift lever dust boot.

M45/46 (1980 on) and M47 types

21 Raise the vehicle and place it securely on jackstands. Position a floorjack under the transmission and raise the jack head just enough to take the weight of the transmission off the crossmember.

22 Remove the transmission crossmember bolts.

23 Unhook the rubber hangers supporting the exhaust muffler.

24 Lower the transmission far enough to allow room to unscrew the back-up light switch, working from under the vehicle or inside the engine bay.

25 Unplug the electrical leads, and unscrew the switch from the transmission case.

26 Installation is the reverse of removal. Be sure to use a new sealing washer under the switch.

27 Remove the jackstands and lower the vehicle.

28 Check the operation of the back-up light.

6 Overdrive wiring harness modification

Refer to illustrations 6.2, 6.4, 6.5 and 6.7

Note: *The following procedure applies to 1976, 1977 and some 1978 models, which were equipped with an overdrive wiring harness that was too hard and too short. These harnesses were prone to breakage. (Later models are equipped with a longer, softer, better-insulated harness.) If you have one of these early vehicles with the older wiring harness - and it hasn't yet been modified - alter it as follows:*

1 Slide the shift lever dust boot up and out of the way.

2 Cut the overdrive wires about 1-1/2 inches from the shift lever **(see illustration)**.

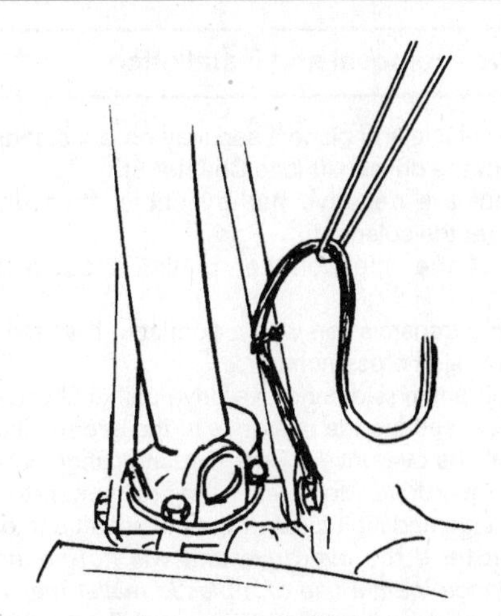

5.18 Anchor the lower end of the back-up light switch harness to a selector rod with a cable tie, then pull the other end up through the space between the transmission and the floorpan with your hooked tool

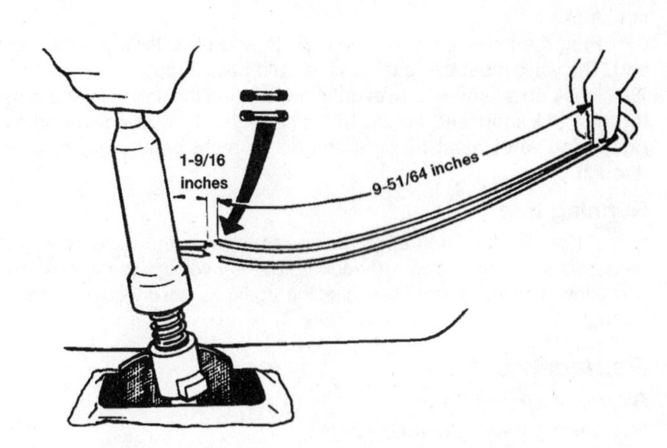

6.2 To modify the older type wiring harness for the overdrive unit, cut the wires about 1-1/2 inches from the shift lever and splice on a pair of 10-inch long, 14-gauge wires as shown

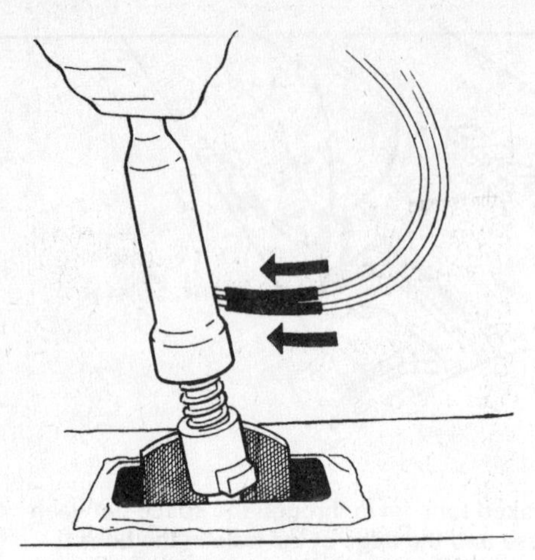

6.4 Push a section of 2-inch insulation tubing onto each splice

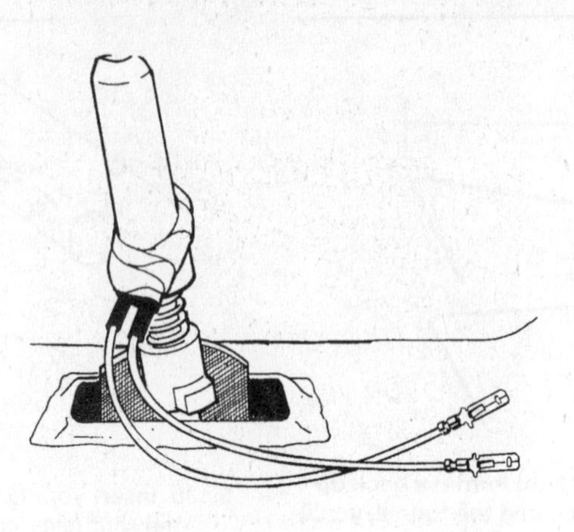

6.5 Attach the wiring harness to the shift lever as shown with duct tape and attach a couple of spade connectors to the ends of the new leads

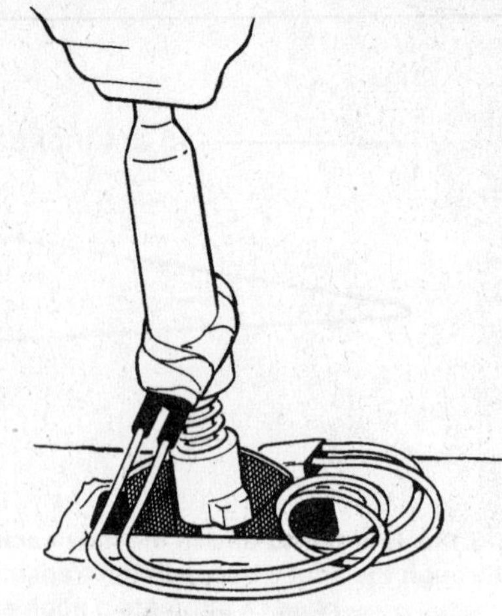

6.7 Loop the wires as shown, to allow plenty of flexibility, and plug in the spade terminals to the overdrive connector

3 Splice on a pair of 10-inch long, 14-gauge wires to the harness.
4 Push on two 2-inch sections of insulation tubing **(see illustration)**.
5 Using duct tape, attach the wire harness to the shift lever as shown **(see illustration)**.
6 Attach two spade connectors to the wire ends.
7 Loop the wiring harness to permit maximum flexibility **(see illustration)**. Plug the spade terminals into the electrical connector for the overdrive unit.
8 Install the shift lever dust boot.

7 Overdrive solenoid - check and replacement

Check

Electrical

1 Raise the vehicle and place it securely on jackstands. Locate the solenoid valve, which is mounted on the left side of the overdrive unit.
2 Turn the ignition to On, shift into fourth gear and engage Overdrive.
3 Verify that there's voltage in the yellow wire at the solenoid.

Mechanical

4 Remove the solenoid.
5 Using a flashlight, inspect the oil passages and verify that they're not blocked.
6 Plug the holes between the O-rings and blow through the short end. The valve must be tight; no air should pass through.
7 Hook up a jumper between the battery and the solenoid and blow through the short end again, this time without covering the holes. Again, the valve must be tight and no air must be allowed to pass through.

Running test

8 If the overdrive unit operates correctly when the transmission is cold, but not when it's warm, hook up battery voltage to the solenoid and allow it to warm up. Then check it again as described in Steps 1 through 7.

Replacement

Refer to illustration 7.10

9 Unplug the electrical connector.
10 Attach a 25mm or 1-inch crows-foot wrench **(see illustration)**. Using an extension and ratchet wrench, remove the solenoid.
11 Dip the new O-rings into fresh ATF.
12 Installation is the reverse of removal. Be sure to tighten the solenoid to the torque listed in this Chapter's Specifications.

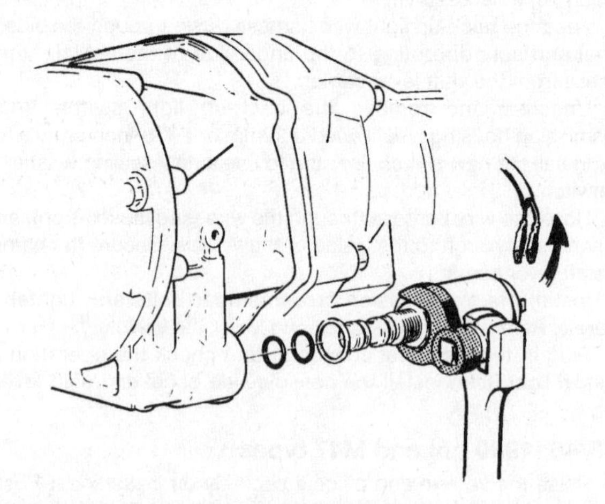

7.10 An exploded view of the solenoid assembly and Volvo's special crow's-foot wrench (any 25mm or 1-inch crow's foot will work)

8 Overdrive - removal and installation

1 Raise the vehicle and place it securely on jackstands.
2 Disconnect the driveshaft (see Chapter 8).
3 Disconnect the negative battery cable, then disconnect the electrical leads at the solenoid.
4 Disconnect the speedometer cable, if equipped, from the overdrive unit.
5 Support the transmission with a floorjack, then remove the bolts from the transmission crossmember.
6 Lower the transmission and overdrive unit until you can reach the uppermost bolts securing the overdrive to the intermediate flange.
7 Remove all the overdrive-to-intermediate flange nuts or bolts.
8 Pull the overdrive unit rearward to separate it from the intermediate flange and lift it clear of the mainshaft and out from under the vehicle. **Note:** *If the overdrive unit will not separate from the intermediate flange, gentle use of a plastic mallet may loosen it, or a slide hammer could be bolted to the output flange. If none of these measures work, do not force the unit from the flange, but dismantle the unit in sequence, working from the rear, until it can be removed.*
9 Installation is the reverse of removal. Be sure to use a new gasket between the overdrive and the intermediate flange, and tighten the

9.11a Transmission crossmember retaining bolts (two at each end)

9.11b Transmission center support block bolt (later models)

9.11c Support block-to-transmission bolts (later models)12.64 Measuring the distance between the front face of the input shaft bearing and surface of the transmission case

flange nuts or bolts to the torque listed in this Chapter's Specifications.
10 Top-off the transmission oil (see Chapter 1), road test the vehicle, then check the oil level again.

9 Transmission - removal and installation

Note: *The M40/41 transmission can be removed from the bellhousing, leaving the bellhousing attached to the engine. On all other types the transmission and bellhousing are removed together. The procedure given here for the M40/41 is for removal, leaving the bellhousing in position. If the bellhousing is to be removed with the transmission, the procedure is similar to that described for M45/46 transmissions.*

Transmission support crossmember - removal and installation

1 The transmission support crossmember is bolted to the floorpan on either side of the transmission tunnel, underneath the transmission.
2 Raise the vehicle and support it securely on jackstands. Position a floor jack under the transmission and raise it just enough to take the weight of the transmission off the crossmember. Remove the bolts at either end of the crossmember.
3 Remove the bolt(s) from the center support block (this has been modified several times, and varies between models).
4 Lift the crossmember over the exhaust and remove it from the vehicle.
5 Installation is the reverse of the removal procedure.

Transmission removal

M40/41 type

Refer to illustrations 9.11a, 9.11b and 9.11c

6 Disconnect the negative battery cable.
7 Remove the shift lever (see Section 2).
8 Raise the vehicle and support it securely on jackstands.
9 Drain the transmission lubricant (see Chapter 1).
10 Position a floor jack under the transmission and raise it just enough to take the weight of the transmission off the mount.
11 Remove the transmission support crossmember **(see illustrations)**.
12 Remove the engine rear mount (see Chapter 2).
13 Disconnect the exhaust bracket (see Chapter 4).
14 Disconnect the driveshaft from the output flange and hang it out of the way (see Chapter 8).
15 Disconnect the speedometer cable.
16 Place a piece of wood between the rear of the engine and the firewall. Lower the transmission on the jack until the upper transmission bolts can be reached.
17 Disconnect any electrical wiring still attached to the transmission.
18 Remove the right upper and lower left transmission retaining bolts. These must be replaced with guide pins, which could be made up from old bolts with the heads cut off and screwdriver slots cut in to the shank.
19 Remove the remaining two bolts, then slide the transmission rearward off the guide pins, and lower the transmission to the ground. Do not allow the weight of the transmission to hang on the input shaft.

M45/46 and 47 types

Note: *On these transmissions, the clutch bellhousing and transmission are removed together. The procedure is similar to that given for M40/41 transmissions with the following additional points.*
20 Disconnect the clutch cable or slave cylinder (see Chapter 8).
21 Disconnect the rubber support rings from the forward end of the muffler (see Chapter 4).
22 Remove the starter motor (see Chapter 5).
23 Remove the engine-to-bellhousing bolts, and lower the transmission to the ground.

Installation

24 Installation is the reverse of removal. Note the following points:
- a) *Apply a coat of molybdenum-based grease to the input shaft splines.*
- b) *Make sure that the clutch driven plate is properly centered, and that the clutch release components have been installed in the bellhousing.*
- c) *Adjust the clutch cable (if equipped).*
- d) *Refill or top-off the transmission lubricant* (see Chapter 1).

10 Transmission - overhaul

Overhauling a manual transmission is a difficult job for the do-it-yourselfer. It involves the disassembly and reassembly of many small parts. Numerous clearances must be precisely measured and, if necessary, changed with select fit spacers and snap-rings. As a result, if transmission problems arise, it can be removed and installed by a competent do-it-yourselfer, but overhaul should be left to a transmission repair shop. Rebuilt transmissions may be available - check with your dealer parts department and auto parts stores. At any rate, the time and money involved in an overhaul is almost sure to exceed the cost of a rebuilt unit.

Nevertheless, it's not impossible for an inexperienced mechanic to rebuild a transmission if the special tools are available and the job is done in a deliberate step-by-step manner so nothing is overlooked.

The tools necessary for an overhaul include internal and external snap-ring pliers, a bearing puller, a slide hammer, a set of pin punches, a dial indicator and possibly a hydraulic press. In addition, a large, sturdy workbench and a vise or transmission stand will be required. During disassembly of the transmission, make careful notes of how each piece comes off, where it fits in relation to other pieces and what holds it in place.

Before taking the transmission apart for repair, it will help if you have some idea what area of the transmission is malfunctioning. Certain problems can be closely tied to specific areas in the transmission, which can make component examination and replacement easier. Refer to the *Troubleshooting* section at the front of this manual for information regarding possible sources of trouble.

Chapter 7 Part B
Automatic transmission

Contents

Specifications

General

Fluid type/specification	See Chapter 1
Fluid capacity	See Chapter 1

Torque specifications*

	Ft-lbs (unless otherwise indicated)
Brake band adjuster locknut	30 to 40
Converter housing to transmission case	96 to 156 in-lbs
Converter to driveplate	25 to 30
Oil cooler connection nipple	144 to 180 in-lbs
Oil cooler connection nut	120 to 144 in-lbs
Oil drain plug	108 to 144 in-lbs
Speedometer gear locknut	48 to 84 in-lbs
Neutral starter switch contact	72 to 96 in-lbs
Kickdown cable connection at transmission case	96 to 108 in-lbs

***Note:** *Torque specifications apply to all units unless otherwise noted.*

Torque specifications (continued)

	Ft-lbs (unless otherwise indicated)
BW55	
Converter housing cover:	
M6 bolts	48 to 72 in-lbs
M8 bolts	156 to 228 in-lbs
Converter housing-to-engine	26 to 37
Drain plug	108 to 156 in-lbs
Driveplate to converter bolts:	
M10 bolts	30 to 37
M12 bolts	41 to 66
Oil cooler pipe union	15 to 22
Oil filler pipe nut	48 to 52
Output flange bolts	30 to 37
AW70/AW71*	
Drain plug	156 to 204 in-lbs
Driveplate to torque converter bolts	30 to 36

**Note:* Other torque specifications same as BW55*

1 General information

Two types of automatic transmission may be installed, depending on model. They are as follows:

1) *Borg Warner 55: A three-speed unit, installed on models equipped with a B21A, B21E, B23A or B23E engine.*
2) *Aisan Warner 70 and 71: A three-speed unit with a lock-up fourth gear overdrive, derived from the BW55. Installed on later models equipped with a B23A or B23E engine, and on models with a B200E, B200K, B230A, B230E or B230K engine.*

Drive from the engine is transmitted via a torque converter, which provides a fluid coupling and a variable torque multiplication ratio. The planetary transmission is hydraulically controlled through a valve assembly, which automatically selects the correct gear for engine speed and load.

An oil cooler, which is an integral part of the radiator (it's located inside the right tank), is used to cool the transmission fluid. The oil cooler and transmission are connected by a pair of lines, one to route hot oil to the cooler, another to return the cooled oil back to the transmission. On some models, an auxiliary cooler, mounted in front of the radiator, provides extra cooling. The hoses for the auxiliary cooler are attached to a special junction at the radiator outlet for the regular cooler's return line. In other words, oil flowing out of the regular cooler is then routed to the auxiliary cooler, then back to the junction, before returning to the transmission.

2 Diagnosis - general

Note: *Automatic transmission malfunctions may be caused by five general conditions: poor engine performance, improper adjustments, hydraulic malfunctions, mechanical malfunctions or malfunctions in the computer or its signal network. Diagnosis of these problems should always begin with a check of the easily repaired items: fluid level and condition (see Chapter 1), shift linkage adjustment and throttle rod linkage adjustment. Next, perform a road test to determine if the problem has been corrected or if more diagnosis is necessary. If the problem persists after the preliminary tests and corrections are completed, additional diagnosis should be done by a dealer service department or transmission repair shop. Refer to the Troubleshooting Section at the front of this manual for information on symptoms of transmission problems.*

Preliminary checks

1 Drive the vehicle to warm the transmission to normal operating temperature.

2 Check the fluid level as described in Chapter 1:

a) *If the fluid level is unusually low, add enough fluid to bring the level within the designated area of the dipstick, then check for external leaks (see below).*
b) *If the fluid level is abnormally high, drain off the excess, then check the drained fluid for contamination by coolant. The presence of engine coolant in the automatic transmission fluid indicates that a failure has occurred in the internal radiator walls that separate the coolant from the transmission fluid (see Chapter 3).*
c) *If the fluid is foaming, drain it and refill the transmission, then check for coolant in the fluid or a high fluid level.*

3 Check the engine idle speed. **Note:** *If the engine is malfunctioning, do not proceed with the preliminary checks until it has been repaired and runs normally.*

4 Check the throttle linkage for freedom of movement. Adjust it if necessary (see Section 5). **Note:** The throttle rod may function properly when the engine is shut off and cold, but it may malfunction once the engine is hot. Check it cold and at normal engine operating temperature.

5 Inspect the shift linkage (see Section 3). Make sure it's properly adjusted and that the linkage operates smoothly.

Fluid leak diagnosis

6 Most fluid leaks are easy to locate visually. Repair usually consists of replacing a seal or gasket. If a leak is difficult to find, the following procedure may help.

7 Identify the fluid. Make sure it's transmission fluid and not engine oil or brake fluid (automatic transmission fluid is a deep red color).

8 Try to pinpoint the source of the leak. Drive the vehicle several miles, then park it over a large sheet of cardboard. After a minute or two, you should be able to locate the leak by determining the source of the fluid dripping onto the cardboard.

9 Make a careful visual inspection of the suspected component and the area immediately around it. Pay particular attention to gasket mating surfaces. A mirror is often helpful for finding leaks in areas that are hard to see.

10 If the leak still cannot be found, clean the suspected area thoroughly with a degreaser or solvent, then dry it.

11 Drive the vehicle for several miles at normal operating temperature and varying speeds. After driving the vehicle, visually inspect the suspected component again.

12 Once the leak has been located, the cause must be determined before it can be properly repaired. If a gasket is replaced but the sealing flange is bent, the new gasket will not stop the leak. The bent flange must be straightened.

13 Before attempting to repair a leak, check to make sure that the following conditions are corrected or they may cause another leak. **Note:** *Some of the following conditions cannot be fixed without highly specialized tools and expertise. Such problems must be referred to a transmission repair shop or a dealer service department.*

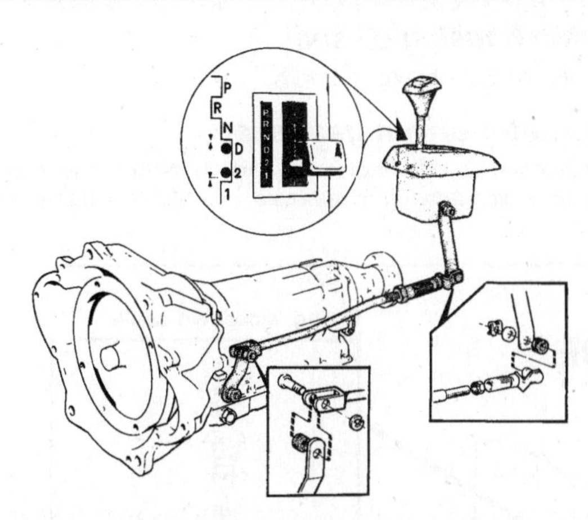

3.2 To check the shift lever positions, engage D and move the shift lever against the gate. The clearance should be the same or greater than the clearance in position 2

Gasket leaks

14 Check the pan periodically. Make sure the bolts are tight, no bolts are missing, the gasket is in good condition and the pan is flat (dents in the pan may indicate damage to the valve body inside).

15 If the pan gasket is leaking, the fluid level or the fluid pressure may be too high, the vent may be plugged, the pan bolts may be too tight, the pan sealing flange may be warped, the sealing surface of the transmission housing may be damaged, the gasket may be damaged or the transmission casting may be cracked or porous. If sealant instead of gasket material has been used to form a seal between the pan and the transmission housing, it may be the wrong sealant.

Seal leaks

16 If a transmission seal is leaking, the fluid level or pressure may be too high, the vent may be plugged, the seal bore may be damaged, the seal itself may be damaged or improperly installed, the surface of the shaft protruding through the seal may be damaged or a loose bearing may be causing excessive shaft movement.

17 Make sure the dipstick tube seal is in good condition and the tube is properly seated. Periodically check the area around the speedometer gear or sensor for leakage. If transmission fluid is evident, check the O-ring for damage.

Case leaks

18 If the case itself appears to be leaking, the casting is porous and will have to be repaired or replaced.

19 Make sure the oil cooler hose fittings are tight and in good condition.

Fluid comes out vent pipe or fill tube

20 If this condition occurs, the transmission is overfilled, there is coolant in the fluid, the case is porous, the dipstick is incorrect, the vent is plugged or the drain-back holes are plugged.

3 Shift linkage - adjustment

Refer to illustrations 3.2 and 3.3

1 First verify that all bushings in the linkage are in good condition. Replace any which are worn.

2 Engage 'D' and move the shift lever towards the stop in the gate. Measure the clearance between the lever and the stop. This clearance should be equal to, or greater than, the same measurement between the lever and stop with the lever in position 2 **(see illustration)**.

3 The clearance can be adjusted from under the vehicle by disconnecting the linkage clevis **(see illustration)**, which may be at either end of the rod, depending on the model. For a "ballpark" adjustment, screw the clevis in or out. For a finer adjustment, turn the knurled sleeve in or out. When you're done, the visible threads on the link rod must not exceed 1.38 inches in length. ***Note:*** *Increasing the link rod length reduces the clearance at D, and increases clearance at position 2.*

4 After adjustment, engage position 1, then engage position P, then repeat the check in Step 2.

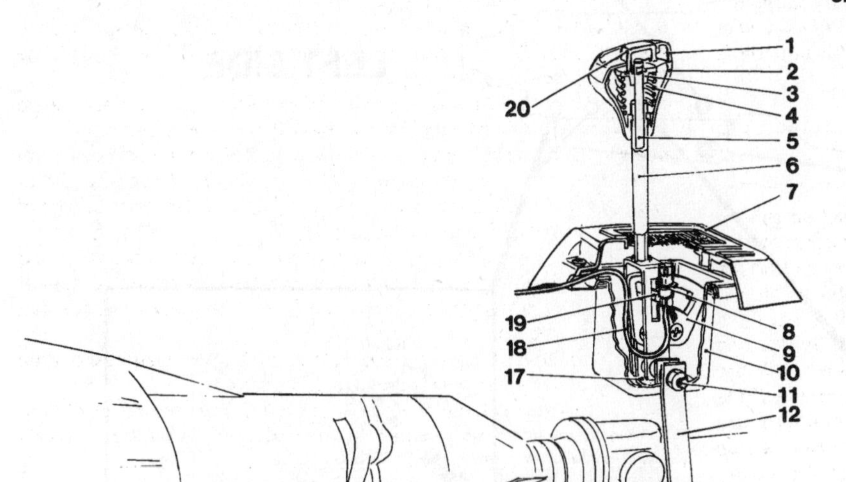

3.3 Installation details of a typical shift lever assembly

1 *Shift lever knob, upper section*
2 *Shift lever knob, lower section*
3 *Washer*
4 *Spring*
5 *Pushrod*
6 *Shift lever*
7 *Shift positions cover*
8 *Shift position light*
9 *Inhibitor plate*
10 *Housing*
11 *Shaft*
12 *Lever*
13 *Gearshift rod adjuster*
14 *Locknut*
15 *Control rod*
16 *Lever*
17 *Bracket*
18 *Cable, shift positions light*
19 *Detent*
20 *Button*
21 *Visible thread on linkage rod (35 mm max)*

4 Oil seals - replacement

1 Oil leaks can occur because of wear at the extension housing seal, the speedometer drive gear seal or O-ring, etc.. Replacing these seals is relatively easy, since the repairs can usually be performed without removing the transmission from the vehicle.

Extension housing oil seal

2 See Section 3 in Chapter 7, Part A.

Speedometer driven gear seal

3 See Section 3 in Chapter 7, Part A. **Note:** *Any time the speedometer pinion adapter is removed, you MUST install a new O-ring.*

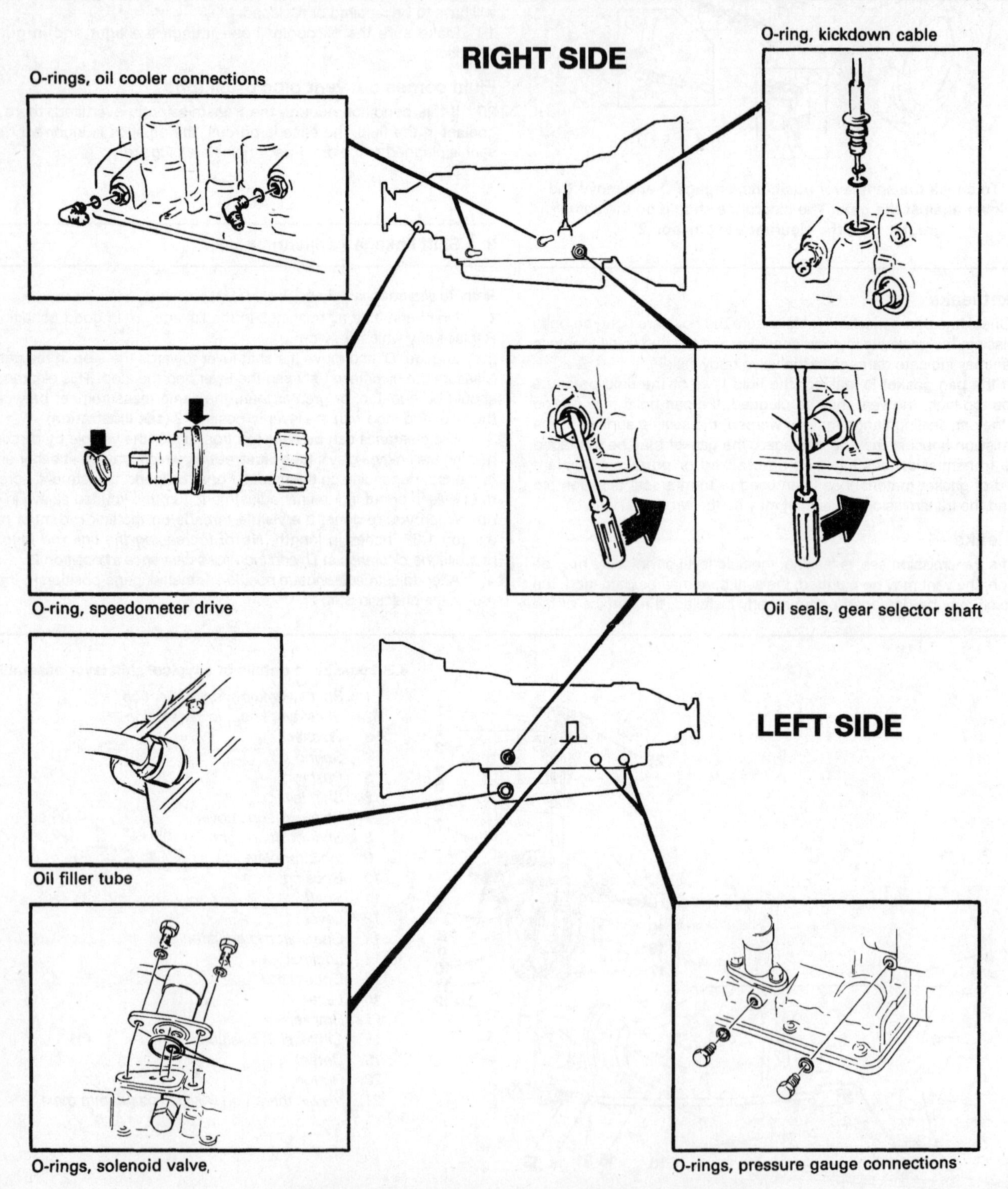

4.6 Automatic transmission oil seal and O-ring location chart

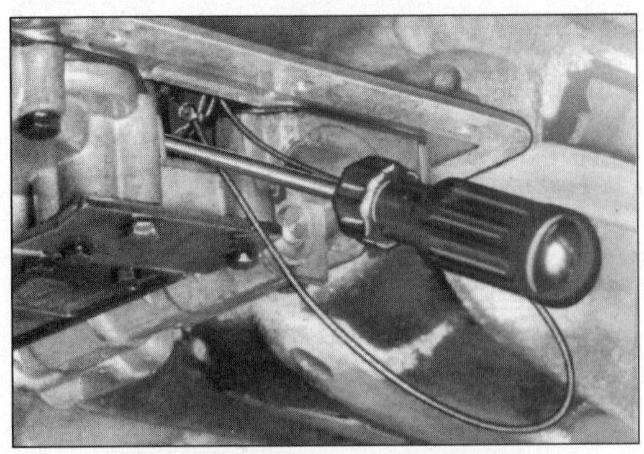

5.4 Retaining the kickdown cable cam

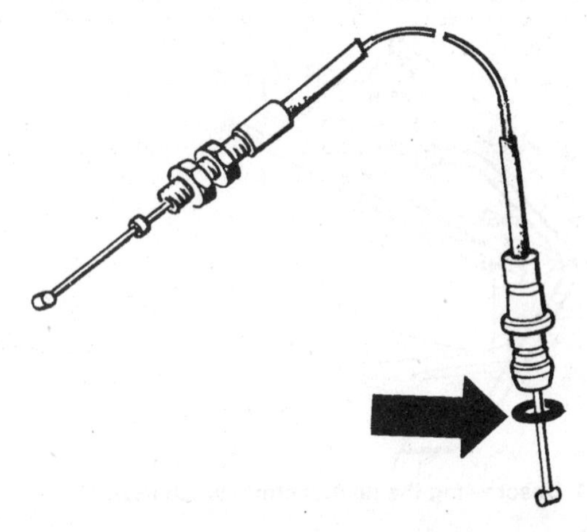

5.7 Kickdown cable O-ring seal (arrow)

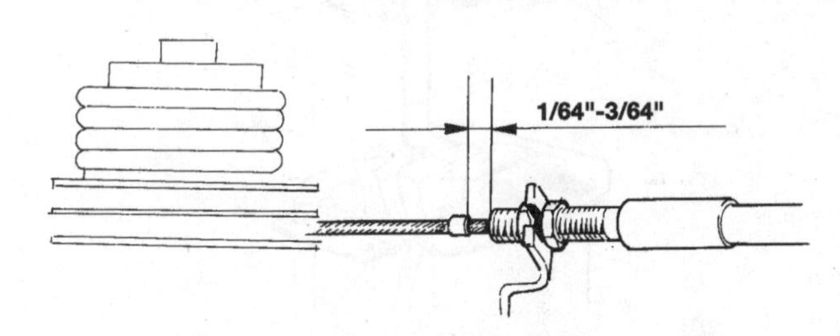

5.10 Kickdown cable stop adjustment (accelerator pedal released)

5.11 Kickdown cable stop adjustment (accelerator pedal depressed)
A 1.986 to 2.072 inches

Oil cooler connection O-rings

Refer to illustration 4.6

4 These fittings are located on the right side of the transmission.
5 Drain the transmission oil (see Chapter 1).
6 Remove the fittings **(see illustration)**, discard the old O-rings and install new ones.
7 Installation is the reverse of removal.

Kickdown cable O-ring

8 See Section 5.

Gear selector shaft seals

9 The gear selector mechanism is inside the front lower part of the transmission, but the ends of the selector shaft protrude through the walls of the transmission case. There are seals located at either end of this shaft **(see illustration 4.6)**.
10 To replace either seal, simply remove the linkage, dig out the old seal with an awl and tap the new seal into place with a small deep socket, then attach the linkage.

Pressure gauge connection seals

11 A pair of bolts on the lower left side of the transmission **(see illustration 4.6)** plug the test ports used by transmission technicians for hooking up pressure gauges. Unless these bolts have been repeatedly unscrewed, it's unlikely the O-rings will ever leak, but should they ever do so, remove the bolt(s), discard the old O-ring(s), install the new O-ring(s), install the bolt(s) and top-off the transmission fluid.

Overdrive solenoid O-rings

12 See Section 7.

5 Kickdown cable - replacement and adjustment

Caution: *Cable replacement involves removal of the transmission oil pan. Scrupulous cleanliness must be observed.*

Replacement

Refer to illustrations 5.4, 5.7, 5.10 and 5.11

1 Drain the transmission fluid and remove the oil pan (see Chapter 1).
2 Cut off the cable at the engine end, between the crimped stop and the cable adjuster. Remove and discard the cable end from the throttle pulley.
3 Release the cable adjuster from the cable mounting bracket.
4 Using a pair of long-nosed pliers, pull the inner cable through the sheath from the transmission end, forming a loop **(see illustration)**. Use the cable to pull the throttle cam of the kickdown valve around so that the cable end is visible. Wedge the cam in this position with a screwdriver.
5 Disconnect the cable from the throttle cam and pull it out of the sheath.
6 Pry the outer sheath from the housing on the side of the transmission case.
7 Install a new O-ring seal on the cable **(see illustration)** and push the sheath into the transmission housing.
8 Attach the cable to the kickdown cam and remove the screwdriver.
9 Install the cable adjuster loosely in the mounting bracket at the engine end, and attach the cable to the throttle pulley.
10 Pull on the cable (not the sheath) from the engine end until slight resistance is felt. Crimp the stop to the cable in this position so that it is the specified distance from the sheath **(see illustration)**. Note that when the cable inner is released, the stop will be hard against the sheath.
11 Have an assistant depress the accelerator pedal fully. Adjust the sheath as necessary so that the distance between the sheath and the stop is as shown **(see illustration)**. It should be possible to pull the kickdown cable inner out by a further 0.08 inch.

7B

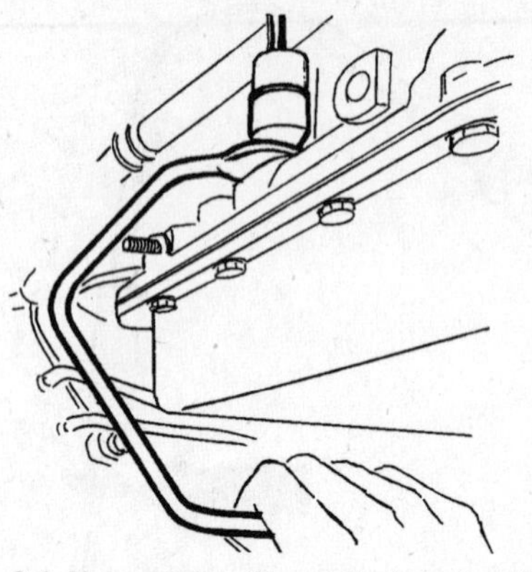

6.4 Unscrewing the neutral start switch (BW35)

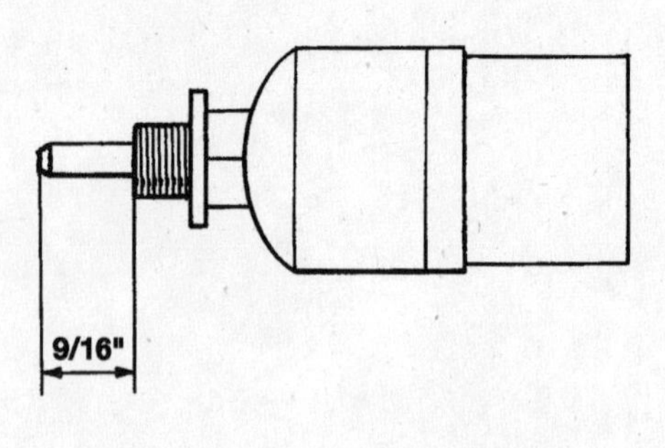

6.5 Neutral-start switch pin protrusion

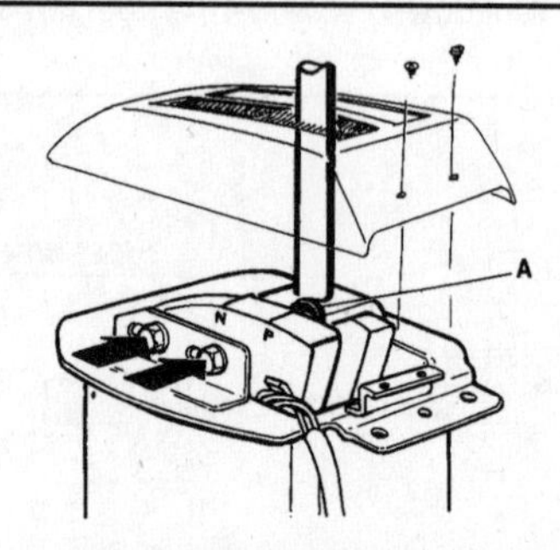

6.10 Neutral-start switch adjustment bolts (arrows)

A Shift lever

12 Tighten the adjuster locknuts when adjustment is correct.

13 Install the oil pan, using a new gasket, and refill the transmission with the specified fluid (see Chapter 1).

Adjustment

14 Adjustment is as described above (see Steps 10 to 12).

15 After adjustment, there should be no play in the cable with the throttle closed, and in the fully open position, it should be possible to pull out the cable by a further 0.08 inch.

6 Neutral start switch - check and replacement

BW35

Refer to illustrations 6.4 and 6.5

1 The switch is located on the left side of the transmission unit.

2 To remove the switch, first place the gear selector in 'P'. Raise the vehicle and support it securely on jackstands.

3 Disconnect the electrical leads to the switch.

4 Remove the switch by unscrewing it from the transmission case **(see illustration).**

5 Verify that the operating pin projects by 9/16 inch **(see illustration). Note:** *Do this check on new switches too.*

6 If the protrusion is less than the specified length, press the switch pin in and out and measure the protrusion again. Each time the switch is pressed in and released, the pin will protrude a little more. Be careful not to let the pin come out too far.

7 If the pin protrusion is greater than the specified length, install a new switch.

8 Installation is the reverse of removal. Be sure to use a new sealing washer under the switch.

9 On completion, verify that the engine can only be started in 'P' or 'N', and that the back-up light comes on when 'R' is selected.

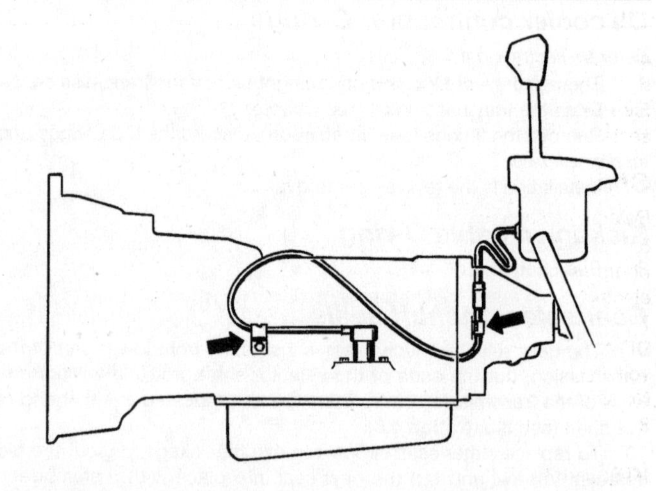

7.3 The electrical connector (arrow) for the overdrive solenoid valve is located on the left side of the transmission; and so is the cable clip (arrow) securing the harness to the transmission case

BW55, AW70 and AW71

Refer to illustration 6.10

10 Remove the shift lever cover **(see illustration)** and verify that the neutral start switch lever aligns correctly with the 'N' and 'P' marks. If not, loosen the bolts, reposition the switch then tighten the bolts.

11 Move the shift lever through all positions, and verify that the contact pin does not slide out of the switch lever. Verify that the engine can only be started with the shift lever in positions 'P' and 'N', and that the back-up lights come on in position 'R'. If the back-up light flashes when the vehicle is reversed, move the switch forward 0.04 inch, but verify that the engine can only be started in positions 'P' and 'N'.

7 Overdrive solenoid valve (AW70/AW71) - check and replacement

Removal

Refer to illustrations 7.3 and 7.5

1 The overdrive solenoid is located on the left-hand side of the transmission unit.

2 Raise the vehicle and support it securely on jackstands.

3 Unplug the electrical connector on the left side of the transmission **(see illustration)**, and remove the cable clip securing the cable to the case.

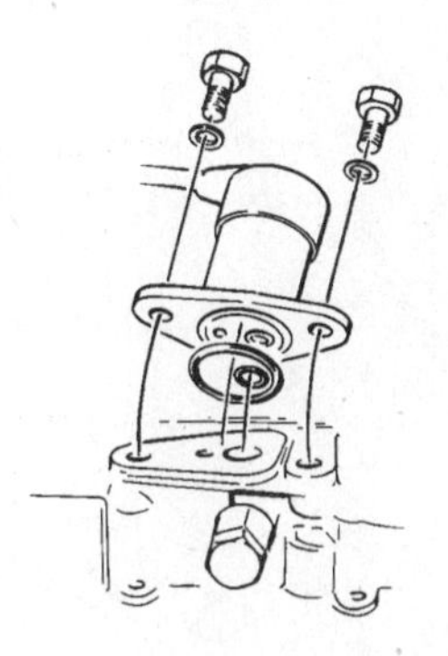
7.5 Typical overdrive solenoid valve assembly details

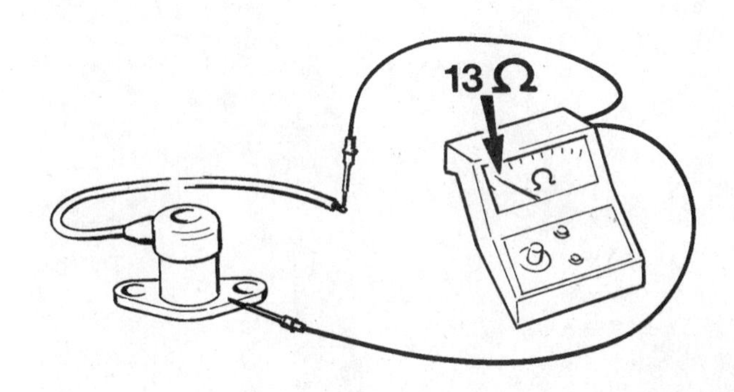

7.6 To test an overdrive solenoid valve, remove it and check the resistance as shown; the resistance should be 13 ohms

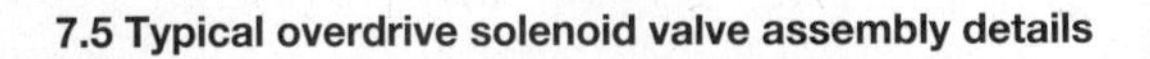

4 Clean off the area around the solenoid valve.
5 Remove the two bolts securing the valve to the transmission case **(see illustration)**, and lift out the solenoid, retrieving the two 0-ring seals.

Check

Refer to illustrations 7.6 and 7.7

6 The resistance of the solenoid can be checked using an ohmmeter connected between the valve body and the end of the electrical lead **(see illustration)**. Resistance should be 13 ohms.
7 The operation of the valve can be checked by applying 12 volts DC to the valve (positive to the lead, negative to the body). With voltage applied, air should pass through the valve **(see illustration)**. No air should pass with the voltage removed.
8 If the solenoid fails either of these tests, replace it.

Installation

9 Installation is the reverse of removal. Be sure to use new O-rings, lightly greased with petroleum jelly.

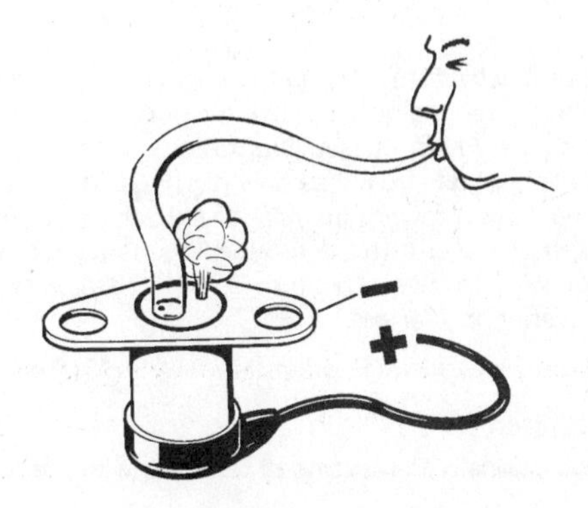

7.7 Using a piece of plastic hose or tubing as shown, apply battery voltage to the solenoid and verify that air passes through the valve when you blow into the tube; now disconnect the voltage source and verify that no air will pass through the valve when you blow into the tube

8 Transmission - removal and installation

Removal

Refer to illustrations 8.15a and 8.15b

1 Disconnect the negative battery cable. Raise the vehicle and support it securely on jackstands.
2 Drain the transmission fluid (see Chapter 1).
3 Move the shift lever to position 2.
4 Disconnect the kickdown cable (see Section 5).
5 Remove the transmission crossmember and the transmission mount (see Section 10 in Chapter 7, Part A).
6 Disconnect the driveshaft from the transmission output flange (see Chapter 8).
7 Disconnect the speedometer cable (if equipped).
8 Disconnect the shift lever linkage at the lever end (see Section 3).
9 Disconnect the oil cooler lines at the transmission end, and plug the open ends.
10 Disconnect the overdrive solenoid on AW70/71 units (see Section 7).
11 Remove the starter motor (see Chapter 5).
12 Disconnect and remove the oil dipstick tube.
13 Disconnect the exhaust bracket on the right-hand side of the converter housing (if equipped). Remove the exhaust pipe, if it's in the way (see Chapter 4).
14 Support the transmission unit securely on a floor jack, then remove all but the upper transmission housing-to-engine bolts.
15 Remove the torque converter cover plate and (if equipped) the

8.15a Removing a torque converter cooling grille

8.15b Removing a torque converter-to-driveplate bolt

8.19 Transmission oil cooler return line fitting (arrow)

cooling grilles **(see illustration)**. Turning the engine with a screwdriver levered against the starter ring gear, remove the bolts which secure the torque converter to the driveplate **(see illustration)**.

16 Remove the upper bolts from the converter housing.

17 Pull the transmission toward the rear of the vehicle to release it from the engine, then lower it to the ground. It's heavy, so have an assistant help you with this Step. **Caution:** *Do not tilt the unit forwards or the torque converter may fall out.*

Installation

Refer to illustration 8.19

18 Installation is basically the reverse of removal, with the following additions:

(a) *Make sure the torque converter is pushed fully to the rear so that the oil pump dogs are fully engaged.*

(b) *Make sure the two dowel location pins are in position on the engine block.*

(c) *Lightly grease the torque converter guide and female end.*

(d) *There are two different lengths of torque converter retaining bolts, 0.55 inch and 0.63 inch. Substitute the short bolts for the long bolts, which can shear in the torque converter. Tighten the bolts to the torque listed in this Chapter's Specifications.*

(e) *Make all connections, then adjust the accelerator cable (see Chapter 4), shift linkage (see Section 3) and kickdown cable (see Section 5).*

(f) *Fill the transmission unit with the specified fluid (see Chapter 1).*

19 Flush the oil cooling system when you're done:

(a) *Overfill the transmission by about 0.3 quart.*

(b) *Disconnect the oil return line from the rear of the transmission* ***(see illustration)*** *and position a container beneath it.*

(c) *With position P selected and the parking brake applied, have an assistant start the engine and let it idle.*

(d) *Switch off the engine when clean fluid comes out, then reconnect the return line and top-off the fluid level (see Chapter 1).*

(e) *To clean the auxiliary cooler (if equipped), disconnect the lines at the standard cooler, then use a pump to force new fluid through the cooler until it emerges from the return line.*

(f) *Connect the lines, then top-off the fluid level (see Chapter 1).*

Chapter 8 Clutch and driveline

Contents

Specifications

Clutch

General

Clutch type	Single dry-plate, diaphragm-spring type
Actuation	Cable or hydraulic
Hydraulic fluid type/specification	See Chapter 1

Adjustment

Hydraulic type	Automatic adjustment
Cable type (freeplay at release arm)	0.04 to 0.12 inch
Clutch pedal travel	6-5/16 inches

Pressure plate

Warpage (maximum)	0.008 inch

Driveshaft

General

Type	Tubular, two-section with sliding center joint and support bearing, Hardy-Spicer universal joints at center bearing and at front and rear flanges, with rubber front joint on some later models

Rear axle

Lubricant capacity	See Chapter 1
Lubricant type	See Chapter 1

Pinion bearing preload (models with compression sleeve)

New bearings	22 to 31 in-lbs
Used bearings	13 to 21 in-lbs

Torque specifications

	Ft-lbs (unless otherwise indicated)
Pinion nut	
Without compression sleeve	148 to 184
With compression sleeve	133 to 207
Clutch pressure plate bolts	17

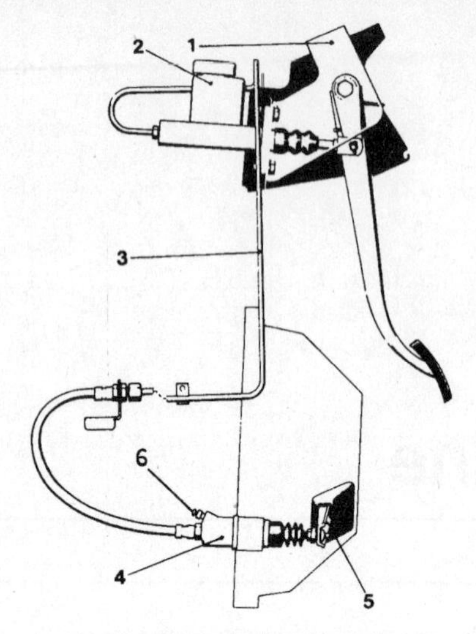

2.1 Clutch hydraulic system

1 Pedal bracket
2 Master cylinder
3 Hydraulic line
4 Release cylinder
5 Release arm
6 Bleed screw

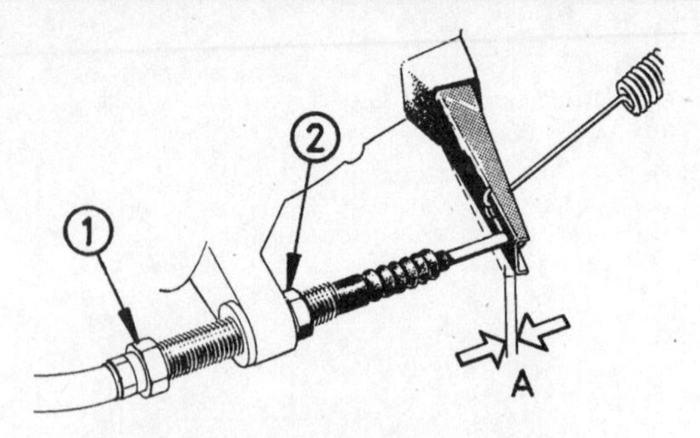

3.2 Release arm freeplay adjustment

1 Adjuster
2 Locknut

A = 0.04 to 0.12 in (1.0 to 3.0 mm)

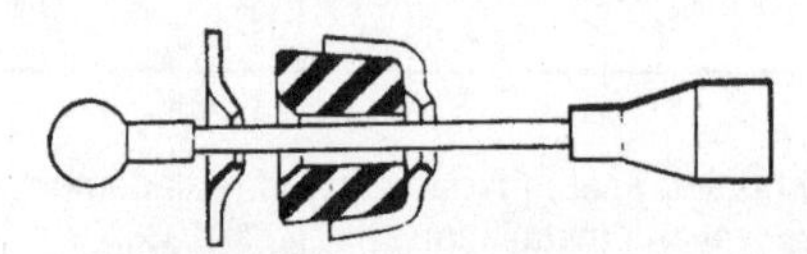

3.7 Correct fitting of the rubber buffer in the release arm

1 General information

The information in this Chapter deals with the components from the rear of the engine to the rear wheels, except for the transmission, which is dealt with in the previous Chapter. For the purposes of this Chapter, these components are grouped into three categories: clutch, driveshaft and axles. Separate Sections within this Chapter offer general descriptions and checking procedures for components in each of the three groups.

Since nearly all the procedures covered in this Chapter involve working under the vehicle, make sure it's securely supported on sturdy jackstands or on a hoist where the vehicle can be easily raised and lowered.

2 Clutch - description and check

Refer to illustration 2.1

Warning: *The clutch friction linings may contain asbestos. Any dust in or around the pressure plate, friction disc and bellhousing should be assumed to contain asbestos. Wash off all traces of this dust with brake system cleaner.*

1 The clutch is a single dry-plate, diaphragm-spring type, cable or hydraulically-operated **(see illustration)** via a release arm and bearing.

2 The main components of the clutch are the pressure plate (also called the cover), which incorporates the diaphragm spring, the driven plate (or friction plate or disc) and the release arm and bearing.

3 The pressure plate is bolted to the flywheel, the driven plate is sandwiched between them, but free to move axially on the splines of the transmission input shaft.

4 With the engine running and the clutch pedal released, the diaphragm spring forces the driven plate against the flywheel, thus transmitting drive from engine to transmission.

5 When the clutch pedal is depressed, the release arm causes the release bearing to press upon the diaphragm spring, which relieves the pressure on the driven plate and interrupts the drive.

6 Hydraulically-operated clutches need no adjustment during service, but cable-operated clutches must be adjusted periodically to compensate for wear in the friction linings on the driven plate.

7 On early models, the clutch hydraulic system has its own reservoir. On 1984 and 1985 models, the system shares the brake fluid reservoir. Later models have separate reservoirs.

8 Terminology can be a problem when discussing the clutch components because common names are in some cases different from those used by the manufacturer. For example, the driven plate is also called the clutch plate or disc, the clutch release bearing is sometimes called a throwout bearing, the release fork is also called a release lever, and so on.

9 Before replacing any components with obvious damage, some preliminary checks should be performed to diagnose clutch problems.

a) *The first check should be "clutch spin-down time." Run the engine at normal idle speed with the transmission in Neutral (clutch pedal up - engaged). Disengage the clutch (pedal down), wait several seconds and shift the transmission into Reverse. No grinding noise should be heard. A grinding noise would most likely indicate a problem in the pressure plate or the clutch disc.*

b) *To check for complete clutch release, run the engine (with the parking brake applied to prevent movement) and hold the clutch pedal approximately 1/2-inch from the floor. Shift the transmission between First gear and Reverse several times. If the shift is hard or the transmission grinds, component failure is indicated. Make sure clutch pedal freeplay is adjusted correctly (see Chapter 1).*

c) *Visually inspect the pivot bushing at the top of the clutch pedal to make sure there is no binding or excessive play.*

d) *Crawl under the vehicle and make sure the clutch release lever is solidly mounted on the ball stud.*

3 Clutch cable - removal, installation and adjustment

Refer to illustrations 3.2 and 3.7

1 Raise the vehicle and place it securely on jackstands.

2 Loosen the adjuster locknut **(see illustration)** and screw the adjuster in to release all tension in the cable.

3 Unhook the return spring from the release arm, then release the cable end from the arm.

4 Gain access to the clutch pedal, then remove the cotter pin from the clevis pin, remove the clevis pin and disconnect the cable end.

5 Pull the cable through the engine firewall, into the engine compartment.

6 Reinstall the cable in the reverse order of removal, routing the cable to the right-hand side of the steering shaft.

7 Make sure the rubber buffer in the release arm is installed correctly **(see illustration)**.

8 Screw in the adjuster to give the specified freeplay in the release arm (see Specifications), tightening the adjuster locknut on completion.

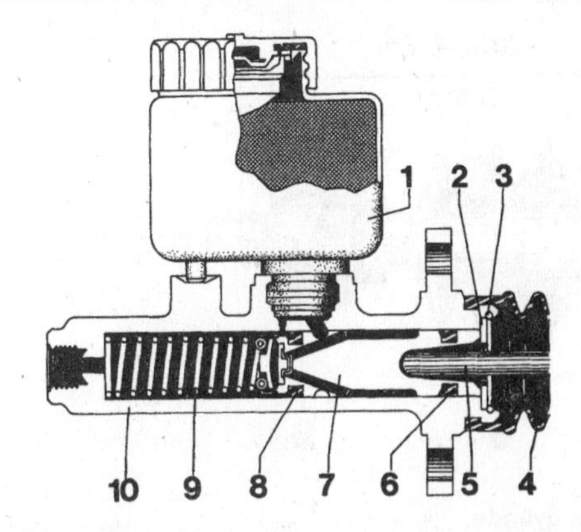

4.6 Cutaway view of the clutch master cylinder

1	*Fluid reservoir*	6	*Piston seal*
2	*Washer*	7	*Piston*
3	*Snap-ring*	8	*Piston seal*
4	*Dust cover*	9	*Spring*
5	*Pushrod*	10	*Cylinder body*

4 Clutch master cylinder - removal, overhaul and installation

Refer to illustration 4.6

Removal

1 Attach a suitable length of tubing to the bleed screw of the release cylinder and place the other end in a jar. Loosen the bleed screw one turn and pump the clutch pedal up and down until the clutch fluid reservoir is empty of fluid. Remove the tube and tighten the bleed screw.
2 Disconnect the fluid outlet line from the master cylinder. On 1984 and 1985 models, which use a single reservoir for the brakes and the clutch, disconnect the line which connects the master cylinder to the fluid reservoir and plug it. **Caution:** *Protect the paint against fluid spillage; wash off spilled fluid immediately.*
3 Remove the panel under the dashboard (see Chapter 11), then disconnect the pushrod from the clutch pedal by removing the clevis pin.
4 Unscrew and remove the bolts from the master cylinder mounting flange and remove the master cylinder.
5 Clean away all external dirt before taking the master cylinder to a totally clean working area.

Overhaul

6 Pull off the dust boot **(see illustration)** and remove the pushrod.
7 Extract the snap-ring, then remove the washer, piston, seals and return spring.
8 Inspect the surfaces of the piston and the cylinder bore for rust, scoring or bright wear areas. If such conditions are evident, replace the master cylinder assembly.
9 If these components are in good condition, discard the rubber seals and wash each part with brake system cleaner.
10 Obtain a repair kit with the necessary seals and other replaceable items.
11 Dip the new seals in clean brake fluid and position them on the piston using the fingers only.
12 Insert the spring, piston, seals and washer carefully into the master cylinder and secure them with the snap-ring.
13 Insert the pushrod and install the dust boot.

Installation

14 Installation of the master cylinder is the reverse of removal.
15 Verify that there's a clearance of 3/64-inch between the pushrod and the piston. If there isn't, adjust the clearance by turning the adjusting nuts on each side of the pushrod yoke that connects to the pedal.
16 Install the pedal mounting cover panel under the dashboard.
17 Fill the master cylinder reservoir with brake fluid (see Chapter 1) and bleed the system (see Section 6).

5 Clutch release cylinder - removal, overhaul and installation

Refer to illustrations 5.10a, 5.10b, 5.10c and 5.11

Removal

1 Disconnect the flexible hose from the rigid hydraulic line by unscrewing the nut from the latter.
2 Disconnect the hose from its bracket and plug the open end of the line to avoid loss of fluid.
3 On early models, remove the snap-ring and remove the release cylinder from the clutch housing. On later models, unbolt and remove the release cylinder.

Overhaul

4 Clean away all external dirt from the release cylinder and remove the rubber dust boot and pushrod.
5 Remove the snap-ring and pull out the piston assembly and spring.
6 Inspect the surfaces of the piston and cylinder bores for rust, scoring or bright wear areas. If these are evident, the complete slave cylinder must be replaced.
7 If the components are in good condition, wash them in clean hydraulic fluid or brake system cleaner. **Caution:** *Do not use any other fluid.*
8 Obtain a repair kit with the new seal and other replaceable items.
9 Dip the new piston seal in clean hydraulic fluid and manipulate it into position on the piston with your fingers.
10 Install the spring and piston assembly into the cylinder and secure them with the snap-ring. Install the pushrod and dust boot **(see illustrations)**.

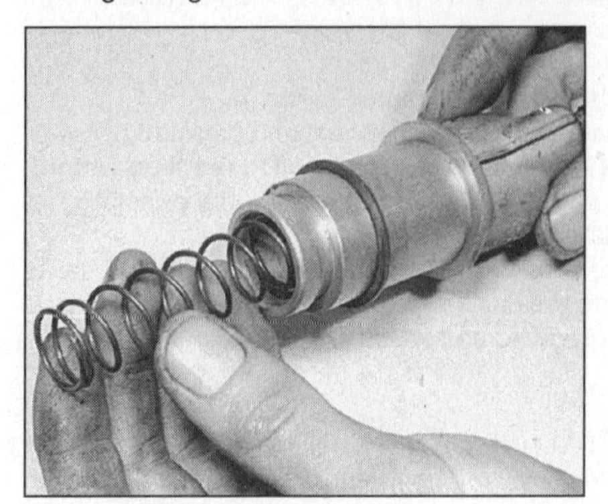

5.10a Insert the spring . . .

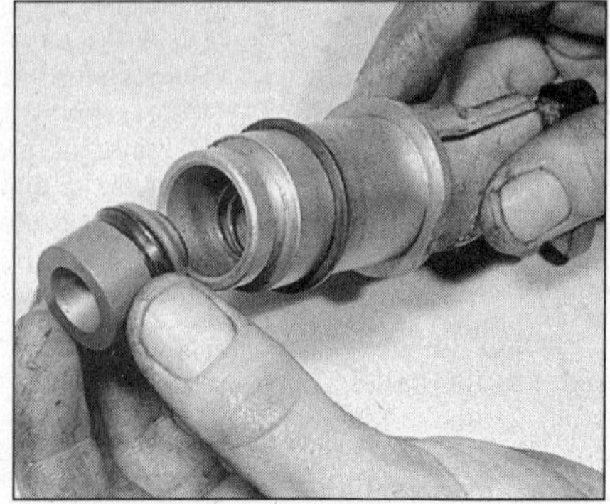

5.10b . . . then insert the piston - make sure the piston is facing the right way, O-ring end in first

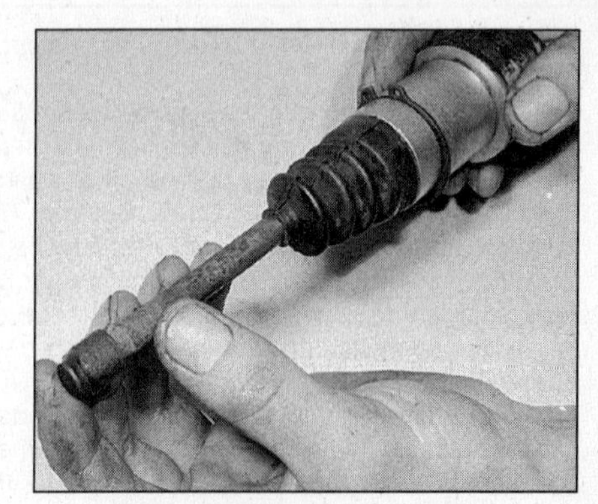

5.10c Installing the dust boot and pushrod

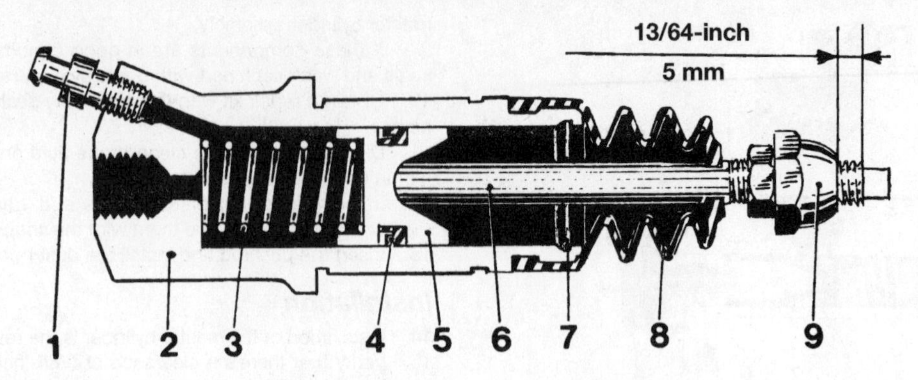

5.11 Cutaway view of a typical release cylinder

1 Bleed screw
2 Cylinder body
3 Spring
4 Seal
5 Piston
6 Pushrod
7 Stop ring
8 Dust boot
9 Domed nut (early models only)

7.3 Five of the pressure plate-to-flywheel bolts (arrows) - sixth bolt, at top, is not visible in this photo

7.6 Flywheel friction surface, showing light scoring and a small crack

11 On early models, verify that the domed nut is located on the pushrod as shown **(see illustration)**. If necessary, loosen the locknut, re-position the domed nut, then tighten the locknut.

Installation

12 Installation is the reverse of removal. Top-off the system with the specified brake fluid (see Chapter 1), then bleed the system (see Section 6).

6 Clutch hydraulic system - bleeding

Note: *The use of one-man bleeding equipment is described in Chapter 9.*

1 Fill the reservoir with the specified brake fluid (see Chapter 1).

2 Raise the vehicle and place it securely on jackstands. Loosen the bleed screw on the release cylinder, then push the end of a piece of plastic tubing onto the bleed screw. **Caution:** *The tube must be a tight fit on the screw or it will allow air to get into the system.*

3 Place the other end of the tube in a jar containing clean brake fluid, sufficient to keep the end of the tube submerged.

4 Have an assistant depress the clutch pedal. Tighten the bleed screw when the pedal is fully depressed. Now release the pedal, loosen the bleed screw and depress the pedal again. Continue this process until bubble-free fluid emerges from the end of the tube. Keep the reservoir level topped up, or air will enter the system through the reservoir and the bleeding procedure will have to be repeated.

5 When you're done, tighten the bleed screw and remove the plastic tubing.

7 Clutch components - removal, inspection and installation

Warning: *Dust produced by clutch wear and deposited on clutch components may contain asbestos, which is hazardous to your health. DO NOT blow it out with compressed air and DO NOT inhale it. DO NOT use gasoline or petroleum-based solvents to remove the dust. Brake system cleaner should be used to flush the dust into a drain pan. After the clutch components are wiped clean with a rag, dispose of the contaminated rags and cleaner in a covered, marked container.*

Removal

Refer to illustration 7.3

1 Remove the transmission (see Chapter 7, Part A).

2 Make alignment marks across the flywheel and pressure plate.

3 Remove the pressure plate-to-flywheel bolts **(see illustration)**, progressively and in a diagonal pattern, to reduce the possibility of warping the pressure plate.

4 Once the spring pressure is released, remove the bolts completely and lift off the pressure and driven plates. Note which side of the driven plate faces forward and which side faces to the rear.

Inspection

Refer to illustrations 7.6 and 7.8

5 Clean the clutch driven plate, cover and flywheel in a solvent to remove all grease and oil. Take precautions against dust inhalation, which may contain asbestos.

6 Inspect the friction surfaces of the flywheel and pressure plate for

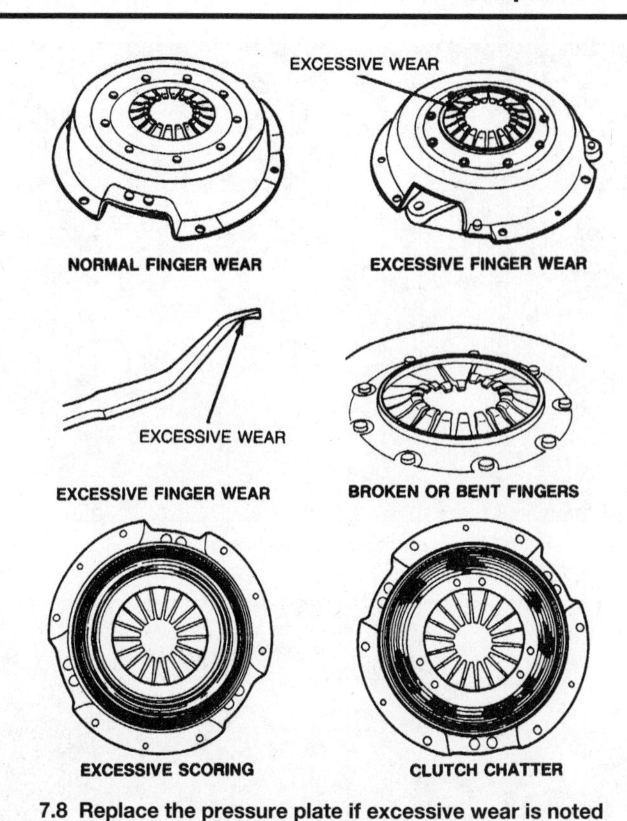

7.8 Replace the pressure plate if excessive wear is noted

scoring or cracks **(see illustration)**. Score marks of less than 0.040 inch are acceptable; anything deeper means it's time to replace the flywheel.

7 To check the pressure plate for warping, lay a straight edge across it and use a feeler gauge to determine if it's distorted. The maximum warpage is listed in this Chapter's Specifications.

8 Inspect the 'fingers' of the diaphragm spring for wear **(see illustration)**, especially where they contact the release bearing. Signs of bluing indicate overheating.

9 Inspect the driven plate linings for wear. If the lining is worn down to, or almost down to, the rivet heads, the plate must be replaced. If the linings are oil contaminated or show a hard black glaze, the plate should be replaced and the cause of the oil contamination rectified.

10 Normally, it's a good idea to replace the driven plate and pressure plate together, and also to install a new release bearing, if it is showing any signs of wear (see Section 8).

Installation

Refer to illustrations 7.14 and 7.15

Note: *On later models, the flywheel dowel pins have been repositioned. Because the pressure plate fits over these dowel pins, early clutch assemblies cannot be installed on later engines (approximately 1984 and later). Later clutch assemblies are drilled to fit both pin positions.*

11 Before installing the clutch components, clean any oil or grease off the friction surfaces of the pressure plate or flywheel. Make sure your hands are clean, to avoid contamination of the driven plate.

12 Inspect the pilot bearing in the crankshaft and replace it if necessary (see Section 9).

13 When installing the clutch assembly, make sure the driven plate is centered in the pressure plate so the transmission input shaft can be inserted. Use a universal alignment tool, or make your own from an old broom handle. It should be a snug fit in the pilot bearing and the splines of the driven plate.

14 Install the driven plate on the flywheel. Make sure the plate is facing the right direction. The plate is usually marked "Schwungrad," or "flywheel side" **(see illustration)**.

7.14 "Flywheel side" marking on driven plate (it might also say "Schwungrad")

7.15 Clutch centering tool in position

15 Place the pressure plate in position, observing the alignment marks if applicable, and insert the retaining bolts finger-tight, then insert the alignment tool **(see illustration)**.

16 Verify that the driven plate is concentric with the pressure plate, then tighten the pressure plate bolts, progressively and in a diagonal pattern, to the torque listed in this Chapter's Specifications.

17 Remove the alignment tool and visually check the alignment of the pilot bearing, the driven plate and the pressure plate. If they're not concentric, installing the transmission will be difficult.

18 Install the transmission (see Chapter 7, Part A).

8 Clutch release bearing - removal, inspection and installation

Refer to illustrations 8.5a, 8.5b and 8.6

1 Whenever the engine or transmission is removed, take the opportunity to inspect the release arm and bearing. Check the bearing for smoothness of operation, cracking or bluing caused by overheating. Replace as necessary.

2 If you haven't already done so, remove the transmission (see Chapter 7, Part A).

3 Remove the bellhousing dust boot from the release arm.

4 Disconnect the release arm from the ball stud pivot. On some models this may be held by a spring clip.

8.5a Release arm and bearing on a cable-operated clutch . . .

8.5b . . . and on a hydraulic type clutch

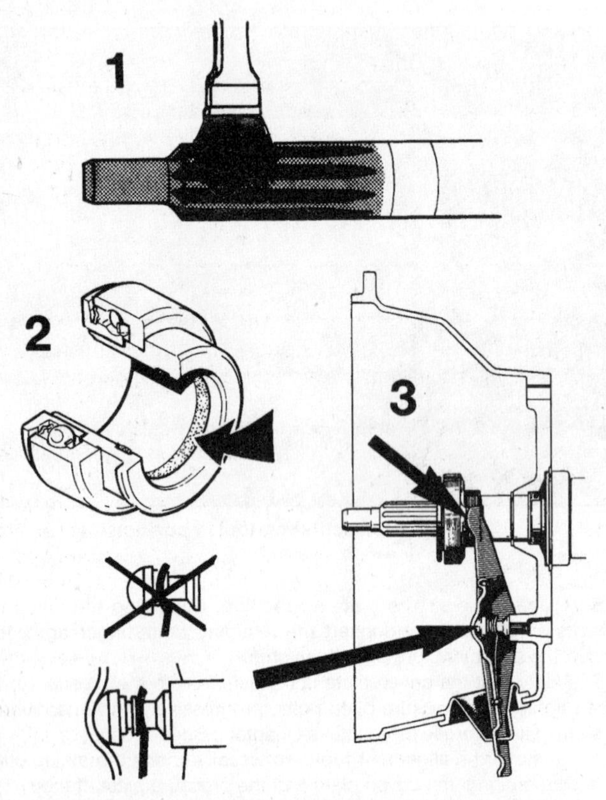

8.6 Clutch release component lubrication details (inset shows correct fitting of spring clip on ball stud pivot)

1 *Input shaft splines*
2 *Groove in release bearing*
3 *Release arm pivot points*

9.2 Pry out the pilot bearing retainer with a small screwdriver

9.3 Remove the pilot bearing with a slide hammer or similar tool

5 Slide the release bearing and arm off the guide sleeve and separate them **(see illustrations)**.

6 Clean the guide sleeve and splines of the transmission shaft, and grease them with a heat-resistant grease or anti-seize compound **(see illustration)**.

7 Fill the groove in the release bearing with the same grease, then engage the bearing with the release arm; also grease the bearing-to-arm pivot point **(see illustration 8.6)**.

8 Installation is the reverse of removal. Make sure the spring on the ball stud pivot (if applicable) is located correctly. Also be careful not to get grease onto the clutch friction surfaces.

9 Install the transmission (see Chapter 7, Part A).

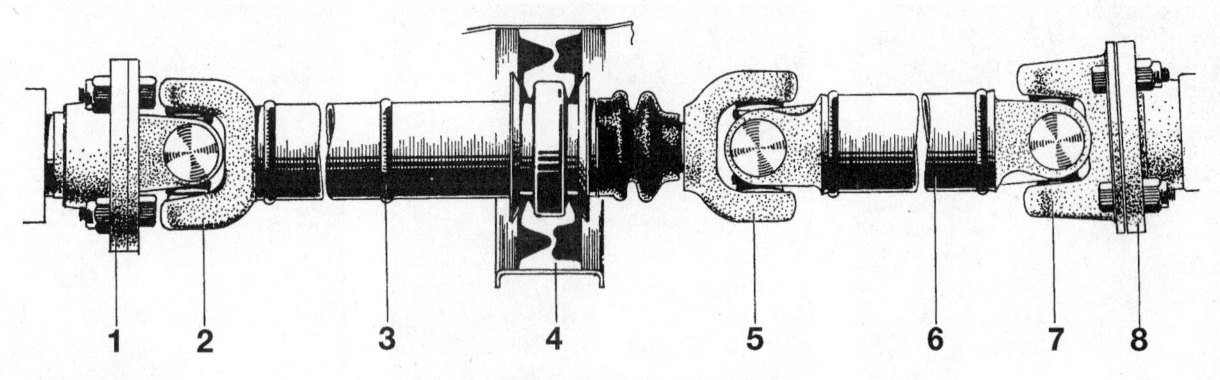

10.1 Typical driveshaft assembly

1 Driveshaft-to-transmission flange
2 Front universal joint
3 Front section of shaft
4 Center support bearing and splined sliding joint
5 Splined section universal joint
6 Rear section of shaft flange
7 Rear universal joint
8 Driveshaft-to-rear axle flange

12.2 Be sure to mark all U-joint flanges before removing and disassembling the driveshaft - these marks will preserve the driveshaft's dynamic balance after it's re-installed

9 Pilot bearing - inspection and replacement

Refer to illustrations 9.2 and 9.3

1 Remove the clutch (see Section 7).
2 Pry out the bearing retainer **(see illustration)**.
3 Remove the pilot bearing with a slide hammer or similar tool **(see illustration)**.
4 Clean and lightly grease the pilot bearing, then verify that it turns easily. If it doesn't, install a new bearing.
5 Install the new bearing with a large socket, or a piece of pipe, with an outside diameter slightly smaller than the outside diameter of the bearing.
6 Install a new bearing retainer.
7 Install the clutch (see Section 7).

10 Driveshaft and universal joints - general information

Refer to illustration 10.1

1 The driveshaft **(see illustration)** is a two-piece tube joined at the center by a sliding splined joint, which allows for fore-and-aft movement between the engine and the rear axle.
2 The front section is joined to the transmission by a flanged universal joint, and supported at its rear end by the center support bearing.
3 The rear section is splined into the rear end of the front section, and joined to the rear axle by a flanged universal joint.
4 A third universal joint is located just behind the splined center joint.
5 In 1985, a larger diameter driveshaft was installed, and in 1987, the front universal joint on manual transmission models was replaced by a special rubber joint.

11 Driveline inspection

1 Raise the rear of the vehicle and support it securely on jackstands. Block the front wheels to keep the vehicle from rolling off the stands.
2 Crawl under the vehicle and visually inspect the driveshaft. Look for any dents or cracks in the tubing. If any are found, the driveshaft must be replaced.
3 Check for oil leakage at the front and rear of the driveshaft. Leakage where the driveshaft enters the transmission extension housing indicates a defective extension housing seal (see Chapter 7). Leakage where the driveshaft enters the differential indicates a defective pinion seal (see Section 19).
4 While under the vehicle, have an assistant rotate a rear wheel so the driveshaft will rotate. As it does, make sure the universal joints are operating properly without binding, noise or looseness.
5 The universal joints can also be checked with the driveshaft motionless, by gripping your hands on either side of a joint and attempting to twist the joint. Any movement at all in the joint is a sign of considerable wear. Lifting up on the shaft will also indicate movement in the universal joints.
6 Check the driveshaft mounting bolts at the ends to make sure they're tight.
7 Check for play in the splined center joint by grasping the front and rear sections of the driveshaft and turning them against each other. Any excessive movement should be investigated (see Section 13).

12 Driveshaft - removal and installation

Refer to illustrations 12.2, 12.3a 12.3b, 12.4 and 12.6

1 Raise the vehicle and place it securely on jackstands.
2 Before removing either the front or rear section of the driveshaft, mark the relationship of the U-joints to the front and rear flanges **(see illustration)**. Also mark the two sections of the shaft relative to each other. The driveshaft is balanced as a unit, and incorrect reassembly can cause vibration.

12.3a Use two wrenches - one on the bolt, one on the nut - to remove the universal joint flange bolts

12.3b Pulling the rear shaft from the splined joint

12.4 Front rubber coupling on later models

3 Remove the four bolts from the pinion flange **(see illustration)**, lower the end of the driveshaft, then pull the shaft rearwards to disengage the rear section from the splined center joint **(see illustration)**.

4 Remove the bolts from the front flange or rubber coupling **(see illustration)**, then from the center support bearing frame, and lower the front section of the shaft. Support the shaft while removing the bolts for the support bearing frame.

5 Installation is the reverse of removal. Make sure the alignment marks you made in Step 2 are lined up. Coat the splines with moly-based grease.

6 When correctly installed, the yokes of both sections of the driveshaft should be in the same plane **(see illustration)**. Don't fully tighten the center support bearing bolts until the flange nuts and bolts have been tightened. The support bearing frame bolt holes are slotted to allow the bearing to assume an unstressed position.

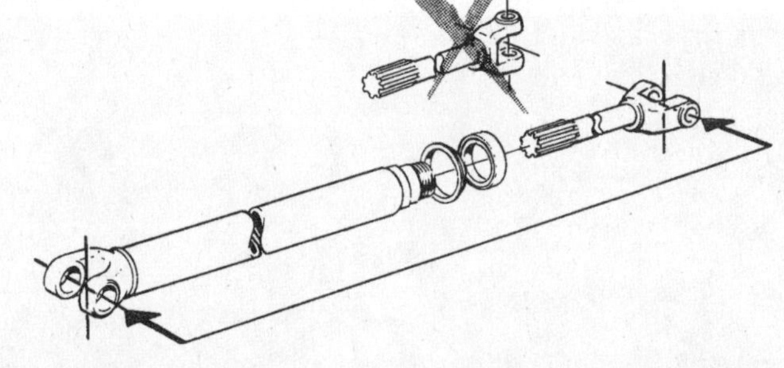
12.6 When reassembling the driveshaft, make sure the yokes are in the same plane

13 Center support bearing - removal, inspection and installation

Refer to illustrations 13.4a, 13.4b, 13.9 and 13.12

1 Remove the rear portion of the driveshaft (see Section 12).

2 Inspect the splines of the sliding joint for wear or cracking.

3 If only the male splines are worn or damaged, remove the splined section universal joint (see Section 14).

4 Install a new splined section, grease the splines and install a new rubber cover **(see illustrations)**.

5 If the female splines are also damaged, the driveshaft must be replaced.

6 Inspect the support bearing for wear and the rubber housing for deterioration and oil contamination (it will feel spongy).

7 The bearing should rotate freely without grating or harshness.

8 If either the bearing or rubber housing are in poor condition, replace them as follows.

9 Pull the rubber mounting from the driveshaft **(see illustration)**.

10 Place the driveshaft in a padded vise, and gently tap off the bearing, being careful not to damage the dust boot. Note which way the dust boot is installed.

11 Clean the dust boot and the bearing seat on the driveshaft.

12 Install the dust boot with its chamfered edge facing toward the bearing **(see illustration)**.

13 Tap the new bearing into place, using a large socket or a piece of

13.4a Internal splines of the front section

13.4b Install the rubber cover

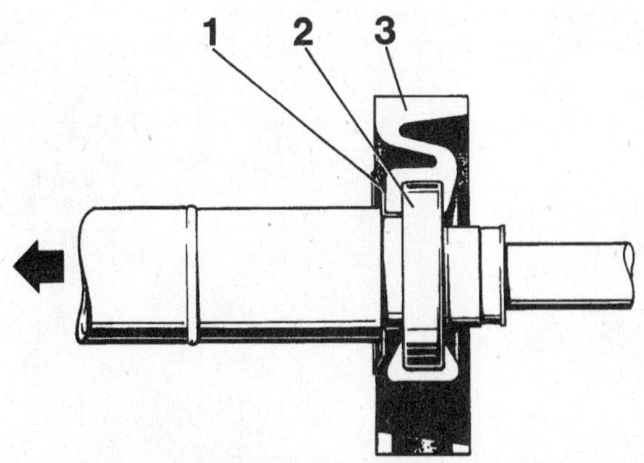

13.9 Center bearing and rubber mounting assembly (arrow points toward front of vehicle)

1 *Dust boot*
2 *Center bearing*
3 *Rubber mounting*

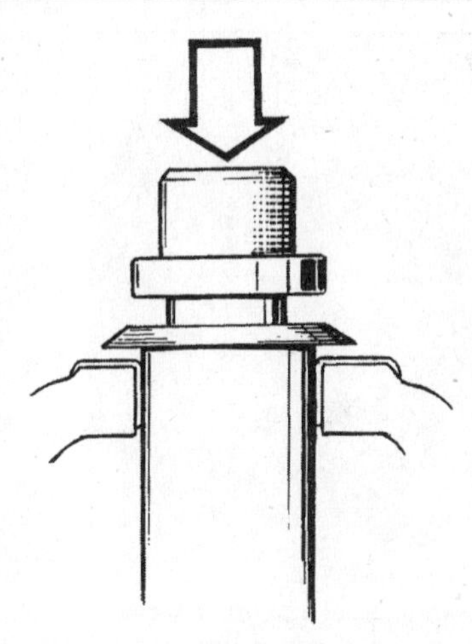

13.12 Make sure the dust boot's chamfered edge faces toward the center support bearing

14.4a Tap the flange yoke down . . .

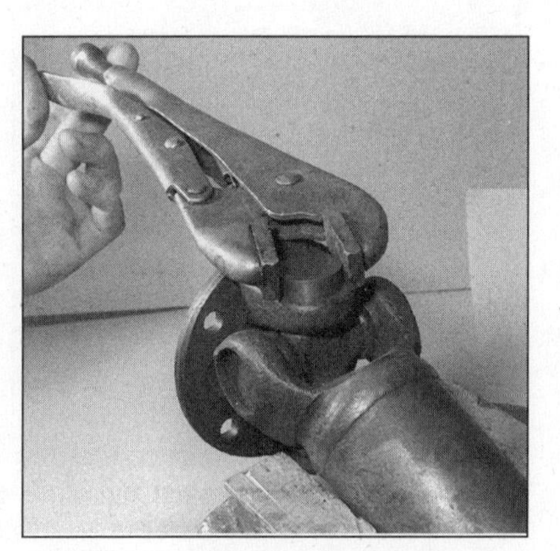

14.4b . . . and remove the bearing cup

14.8 Spider, bearings and snap-rings

tubing with an outside diameter matching the diameter of the inner race.
14 Install the rubber mount over the bearing; the spring and washer should be in the lower segment.
15 Check the condition of the rubber dust boot which fits over the splined center joint. Replace it if it's cracked or torn.
16 Install the driveshaft (see Section 12).

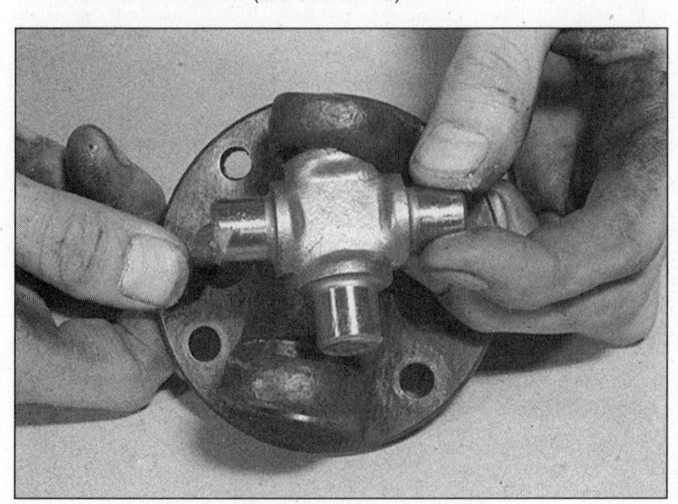

14.9a Place the spider in the yoke . . .

14 Universal joints and rubber coupling - replacement

U-joints

Refer to illustrations 14.4a, 14.4b, 14.8, 14.9a, 14.9b, 14.10a, 14.10b, 14.11a, 14.11b and 14.11c

1 Remove the driveshaft (see Section 12).
2 Remove the snap-rings from the bearing cups. Use a punch to free them if they're stuck.
3 Place the driveshaft in a vise with the yokes of the driveshaft (not the flange) resting on the open vise jaws (remember that the driveshaft is hollow and easily damaged by too much vise pressure).
4 Using a hammer and a socket which just fits over the bearing cups, gently tap the flange yoke down so that the uppermost bearing cup protrudes about 1/8 to 11/64-inch **(see illustration)**. Remove the cup **(see illustration)**.
5 Rotate the driveshaft 180-degrees and repeat this operation on the opposite bearing.
6 The flange can now be removed from the driveshaft and the same procedure used to remove the spider from the shaft.
7 Clean the bearing seats in the flange and driveshaft yokes.
8 Remove the bearings from the new spider **(see illustration)** and make sure that the needle bearings are well greased. If they aren't, thoroughly lubricate them with moly-based grease.
9 Place the spider in the flange yoke **(see illustration)** and install one bearing cup onto the yoke **(see illustration)**.

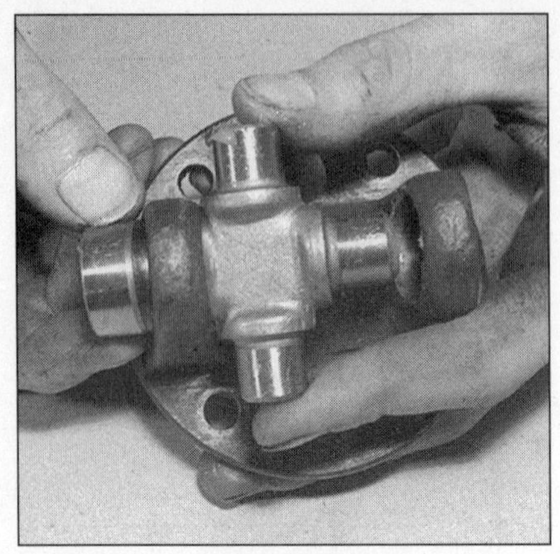
14.9b . . . and install a bearing cup

14.10a Pressing the bearing cup into the yoke

14.10b Installing the snap-ring

14.11a Tap the first cup into the shaft yoke . . .

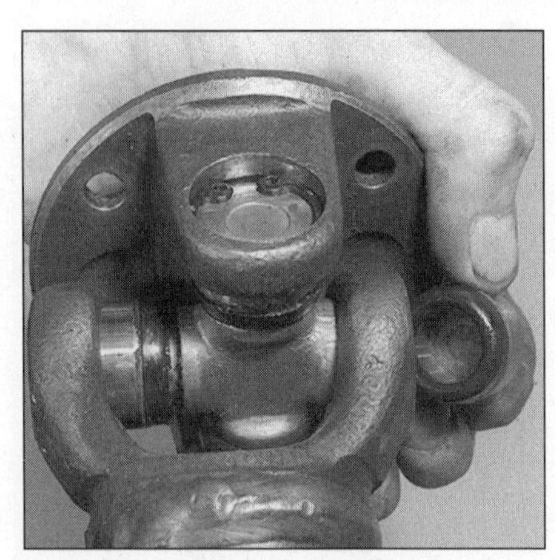
14.11b . . . install the other cup . . .

14.11c . . . and press it into place with the vise

10 Using the vise and a socket of suitable diameter, press the bearing cup into the yoke until it protrudes by about 1/8 to 11/64-inch **(see illustration)**. Install the snap-ring **(see illustration)**.

11 Repeat this procedure on the opposite side. Then install the flange yoke and spider on the driveshaft yoke (remember to match up your alignment marks) and install the remaining bearing cups **(see illustrations)**.

12 Install the rest of the bearing retaining snap-rings.

14.18 Rubber coupling locating plate

13 Check the joint for full and free movement. If it feels too tight, rest it on the vise and gently tap it with a plastic mallet. This should center the bearings and free the joint.

14 Install the driveshaft (see Section 12).

Rubber coupling

Refer to illustrations 14.18 and 14.19

15 Raise the vehicle and place it securely on jackstands.

16 Make alignment marks between the driveshaft and the transmission output flange.

17 Remove the six nuts and bolts which hold the flanges to the coupling (the forward-facing bolts stay with the flange - they can't actually be removed).

18 Pull the shaft rearwards and lower the front section. Remove the rubber coupling, the center sleeve and the locating plate **(see illustration)**.

19 Installation is the reverse of removal **(see illustration)**. Be sure to match up the alignment marks. Apply anti-seize compound to the locating plate pin.

15 Rear axle - description and check

Description

Refer to illustration 15.1

1 The rear axle **(see illustration)** is a semi-floating design, i.e. the weight of the vehicle is carried by the axleshafts. Each axleshaft rides on a tapered roller bearing pressed onto its outer end. The brake disc assemblies and the wheels are attached to the axleshaft flanges.

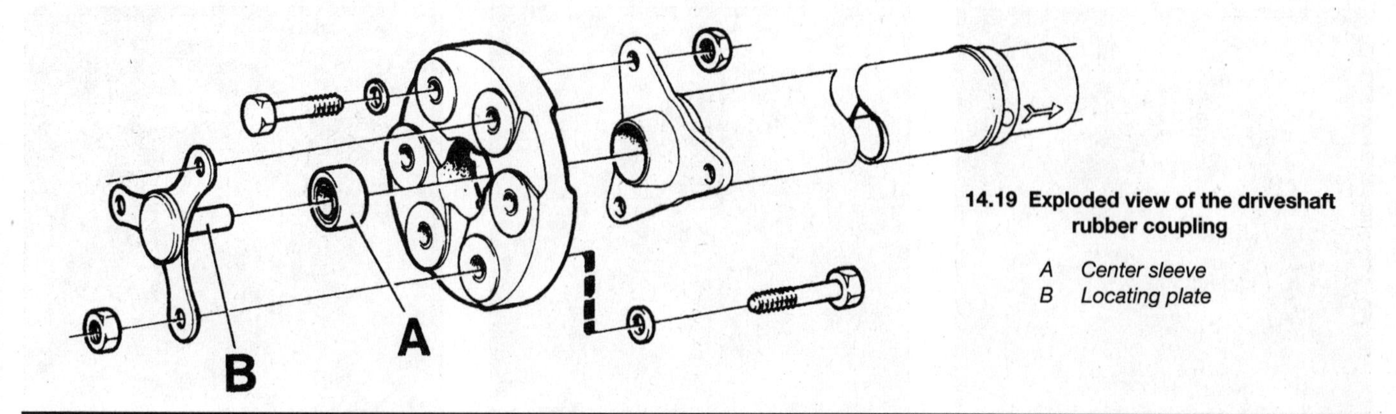

14.19 Exploded view of the driveshaft rubber coupling

A Center sleeve
B Locating plate

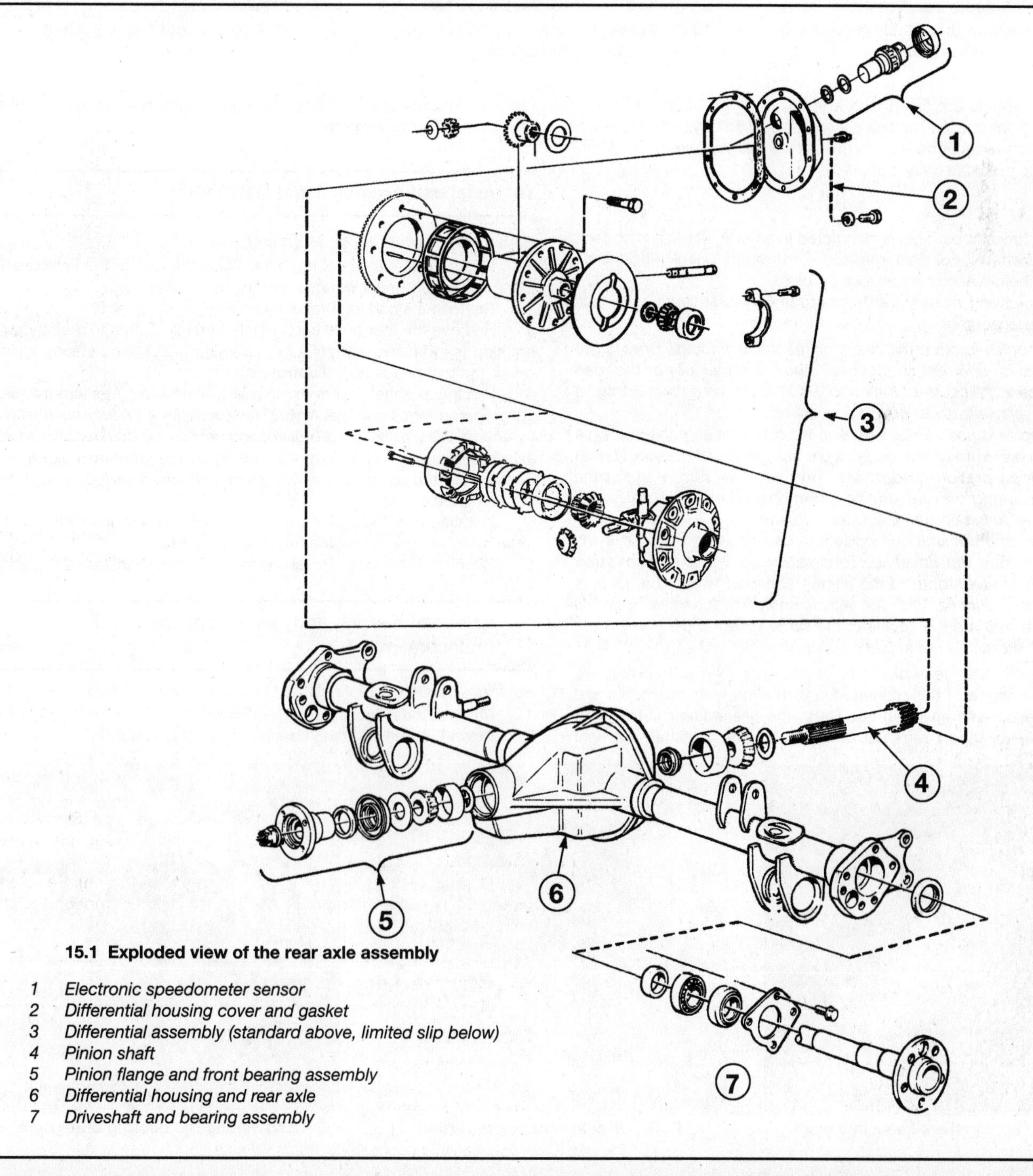

15.1 Exploded view of the rear axle assembly

1 *Electronic speedometer sensor*
2 *Differential housing cover and gasket*
3 *Differential assembly (standard above, limited slip below)*
4 *Pinion shaft*
5 *Pinion flange and front bearing assembly*
6 *Differential housing and rear axle*
7 *Driveshaft and bearing assembly*

16.3a Removing the axleshaft retaining plate bolts

16.3b Brake shoe steady spring under the bolts (arrow)

16.4 Pulling out the driveshaft

2 Most models are fitted with a standard, type 1030, rear axle. Some high performance models are equipped with a type 1031, which uses different reduction ratios in the differential. A limited slip differential is also available as an option.

Check

3 Many times, a problem is suspected in an axle area when, in fact, it lies elsewhere. For this reason, a thorough check should be performed before assuming an axle problem.

4 The following noises are those commonly associated with axle diagnosis procedures:

a) *Road noise is often mistaken for mechanical faults. Driving the vehicle on different surfaces will show whether or not the road surface is the cause of the noise. Road noise will remain the same if the vehicle is under power or coasting.*

b) *Tire noise is sometimes mistaken for mechanical problems. Tires which are worn or low on pressure are particularly susceptible to emitting vibrations and noises. Tire noise will remain about the same during varying driving situations, where axle noise will change during coasting, acceleration, etc.*

c) *Engine and transmission noise can be deceiving because it will travel along the driveline. To isolate engine and transmission noises, make a note of the engine speed at which the noise is most pronounced. Stop the vehicle and place the transmission in Neutral and run the engine to the same speed. If the noise is the same, the axle is not at fault.*

5 Overhaul and general repair of the rear axle differential are beyond the scope of this manual due to the many special tools and critical measurements required. Thus, the procedures listed here will involve axleshaft removal and installation, axleshaft oil seal replacement, axleshaft bearing replacement and removal of the entire unit for repair or replacement.

16 Axleshaft - removal and installation

Refer to illustrations 16.3a, 16.3b and 16.4

1 Loosen the rear wheel lug nuts. Raise the rear of the vehicle and place it securely on jackstands. Remove the wheels.

2 Remove the parking brake assembly (see Chapter 9).

3 Remove the four bolts securing the retaining plate to the rear axle housing **(see illustration)**. Note the parking brake shoe steady spring under the bolt heads **(see illustration)**.

4 Pull the axleshaft out from the rear axle housing **(see illustration)**. Be prepared for oil spillage. **Note:** *There is nothing to disconnect at the inner end. If the axleshaft is broken, and part of the shaft remains in the axle housing, you can remove it by taking off the differential cover and pushing out the broken piece with some stiff wire inserted through the differential.*

5 Installation is the reverse of removal. Be sure to install the brake shoe steady spring under the retaining bolt heads.

6 Check the rear axle oil level and top up if necessary (see Chapter 1).

17 Axleshaft bearing and seal - inspection and replacement

Refer to illustrations 17.7a, 17.7b, 17.7c and 17.8

1 Remove the axleshaft (see Section 16).

2 Clean the axleshaft and bearing thoroughly in solvent.

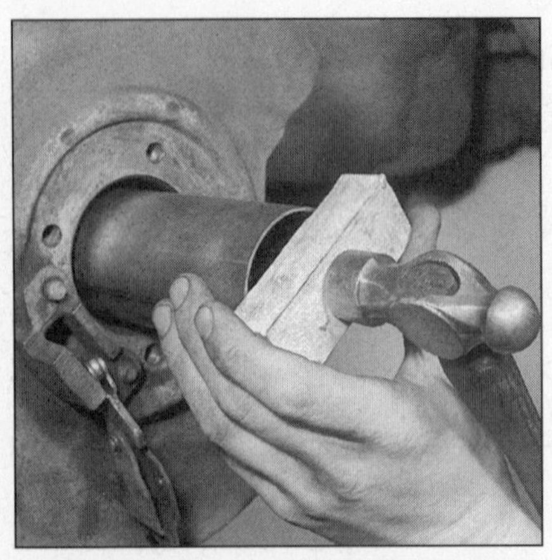
17.7a Tapping the oil seal into place

17.7b Oil seal in position (arrow)

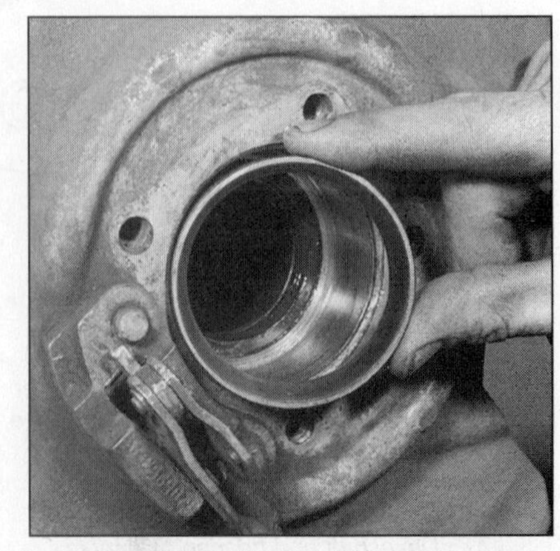
17.7c Fitting the bearing outer race

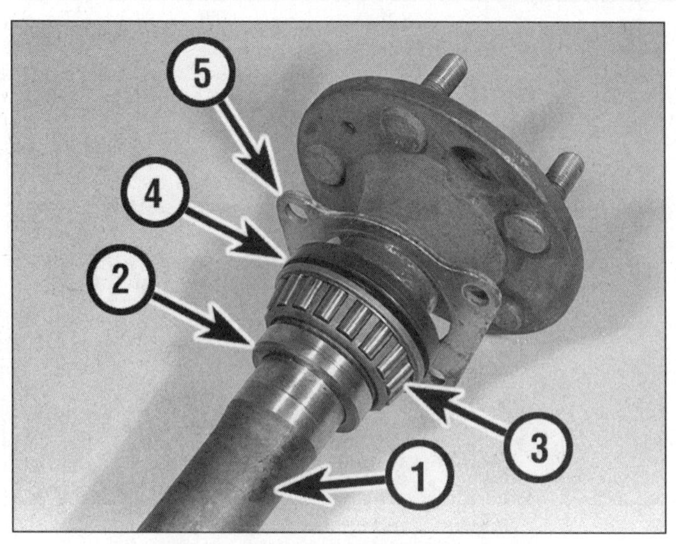

17.8 Axleshaft and bearing assembly

1 *Axleshaft*
2 *Locking ring*
3 *Bearing*
4 *Seal*
5 *Retaining plate*

3 Inspect the axleshaft splines for cracks or wear. Make sure the axleshaft is neither twisted nor bent. If any of these signs are evident, the axleshaft should be replaced.
4 Inspect the bearing for signs of wear or bluing. If it's damaged it must be replaced. If the outer seal is damaged, you'll still have to remove the bearing to replace the seal.
5 You can cut off the locking ring with a hammer and chisel or a grinder, but you run the risk of damaging the axleshaft; removing the bearing is also difficult unless you have a hydraulic press. And if you try to install the new bearing and locking ring without a press, you might damage the bearing. If you don't have access to a hydraulic press, the easiest way to replace the locking ring, bearing and seal is to take the axleshaft assembly and the new parts to an automotive machine shop, where the old ring can be cut off, the old bearing pressed off and the new seal, bearing and locking ring pressed on.
6 Remove the bearing outer race and inner oil seal from the axle housing.
7 Clean out the bearing seat area, then grease the new seal and install it, lip facing in, and the bearing outer race **(see illustrations)**.
8 Pack the new bearing and seal with grease, make sure that the retaining plate is in position on the axleshaft **(see illustration)** and install the axleshaft (see Section 16).
9 Check and, if necessary, top-off the rear axle housing with the specified lubricant (see Chapter 1).

18.5 Using a puller to remove the flange

18 Pinion oil seal - replacement

Refer to illustrations 18.5, 18.9a, 18.9b, 18.10, 18.11a and 18.11b

1 Raise the rear of the vehicle and place it securely on jackstands.
2 Drain the oil from the rear axle (see Chapter 1).
3 Disconnect the rear end of the driveshaft from the pinion flange (see Section 12).
4 To remove the pinion flange nut, insert two long bolts through the bolt holes in the flange and place a crowbar between them, then break loose the pinion nut.
5 Use a puller to remove the flange **(see illustration)**.
6 Pry out the oil seal with a screwdriver. Make sure you don't damage the inner face of the pinion housing. Any scoring or scratches in this area will cause an oil leak.
7 Clean off the inner surface of the pinion housing, and verify that there are no score marks or scratches (light marks can be removed with fine emery cloth). Also check the seal rubbing surface of the pinion flange.
8 Grease the lips and the spring recess of the new seal.
9 Install seal with the 'outside' mark facing out **(see illustrations)**.
10 Using a block of wood and a hammer, install the seal flush with the front of the housing **(see illustration)**.
11 Install the flange and pinion nut **(see illustrations)**.

Setting pinion bearing preload

Refer to illustration 18.14

Note: *The following procedure is intended only for tightening the pinion nut after seal replacement. It's not intended - nor is it appropriate for - the procedure necessary after the pinion bearing has*

8

18.9a 'Outside' mark on the pinion oil seal

18.9b Install the seal . . .

18.10 . . . and tap it into place

18.11a Install the flange . . .

18.11b . . . and tighten the pinion nut

been removed or replaced.

12 Whenever the pinion drive flange is removed, the pinion bearing preload must be checked as follows.

13 On differentials without a compression sleeve (axles without an "S" in their serial number), simply tighten the pinion nut to the torque listed in this Chapter's Specifications.

14 On differentials with a compression sleeve (axles with a serial number beginning with the letter "S"), you'll need an inch-pound torque wrench to measure the force required to turn the rear axle through the pinion shaft (pinion bearing preload).

15 Tighten the pinion nut to the *first* pinion nut torque specification listed in this Chapter's Specifications.

16 Rotate the pinion nut with the inch-pound torque wrench and verify that the pinion bearing preload is within the range listed in this Chapter's Specifications (make sure the brakes aren't binding).

17 If it's lower, tighten the pinion nut a little more, and re-check the preload again. Continue until the specified value is reached, but make sure you don't exceed the maximum torque listed in the Specifications.

18 If the pinion preload (the force required to turn the rear axle) is exceeded, a new compression sleeve will have to be installed by your Volvo dealer or other repair shop.

19 Install the driveshaft (see Section 12) and fill the rear axle with oil (see Chapter 1).

19 Rear axle assembly - removal and installation

1 Raise the rear of the vehicle and place it securely on jackstands positioned just in front of the rear jacking points. Remove the rear wheels.

2 Drain the oil from the rear axle (see Chapter 1).

3 If you're planning to install a new axle housing, remove the brake calipers, discs and parking brake assemblies (see Chapter 9).

4 If you're planning to install a new axle housing, remove the axleshafts (see Section 16).

5 Position a floor jack under the differential housing of the rear axle.

6 If the exhaust pipe runs underneath the rear axle, remove the rear section of the exhaust system (see Chapter 4).

7 Disconnect the driveshaft from the pinion flange (see Section 12) and position it out of the way.

8 Remove or disconnect the stabilizer bar, trailing arms, reaction rods, track rod, lower ends of the shock absorbers and spring mounts (see Chapter 10).

9 Remove the parking brake cable bracket from the top of the rear axle (see Chapter 10).

10 Disconnect all brake line brackets that are attached to the axle.

11 Disconnect the bracket for the rear axle breather.

12 On models equipped with an electronic speedometer, disconnect the electrical leads from the differential rear cover (see Chapter 12).

13 Lower the rear axle assembly slightly and make sure that all brake, electrical and suspension components have been disconnected, and that nothing is in the way. Lower the axle assembly all the way.

14 Maneuver the rear axle from under the vehicle.

15 Installation is the reverse of removal. Refer to the relevant Sections of this Chapter, and Chapters 9 and 10 for the brake and suspension components, respectively. Do not fully tighten any suspension fasteners until the vehicle has been lowered to the ground.

16 Fill the rear axle with the recommended lubricant (see Chapter 1).

17 Install the wheels, remove the stands, lower the vehicle and tighten all suspension fasteners to the torque listed in the Chapter 10 Specifications.

Chapter 9 Brakes

Contents

Specifications

General

Brake fluid type .... See Chapter 1

Front disc brakes

Disc thickness:	
Solid, new	0.56 inch
Solid, wear limit*	0.50 inch
Ventilated, new:	
ATE	0.94 inch
Girling	0.87 inch
Ventilated, wear limit:*	
ATE	0.90 inch
Girling	0.80 inch
Disc run-out (all types)	0.0039 inch maximum
Disc thickness variation	0.0008 inch maximum
Brake pad friction material	1/8 inch

* *Refer to marks stamped in the disc (they supercede information printed here).*

Rear disc brakes

Disc thickness	
New	0.38 inch
Wear limit	0.33 inch
Disc run-out	0.0039 inch maximum
Disc thickness variation	0.0008 inch maximum
Brake pad friction material - minimum thickness	1/8 inch

* *Refer to marks stamped in the disc (they supercede information printed here).*

Parking brake

Drum diameter	6.32 inches maximum
Runout	0.006 inch maximum
Out-of-round	0.008 inch maximum

Torque specifications

	Ft-lbs (unless otherwise indicated)
Front brake caliper mounting bolts	74
Master cylinder retaining nuts	22
Rear brake backing plate bolts	28
Rear brake caliper mounting bolts	43

1 General information

Refer to illustration 1.1

The brake system **(see illustration)** is hydraulically-operated, with disc brakes at all four wheels. Ventilated discs are used on the front brakes of all 1980 and later models.

The front brake caliper is either an ATE or a Girling, depending on year of manufacture (see note at beginning of Section 3). Both types are similar in operation, but differ slightly in construction. The caliper is a two-piece casting, bolted together and attached to the steering knuckle. The upper and lower cylinders and pistons are completely separate from each other, but both pairs of cylinders are joined together by internal borings in the caliper. The brake disc, which is bolted to the hub, rotates between the two halves of the caliper. When the brake pedal is applied, the pistons, under the action of the hydraulic fluid, force the brake pads against the brake disc.

The piston has a rubber seal around its circumference to seal the hydraulic fluid in the cylinder, and this seal also acts as a return spring

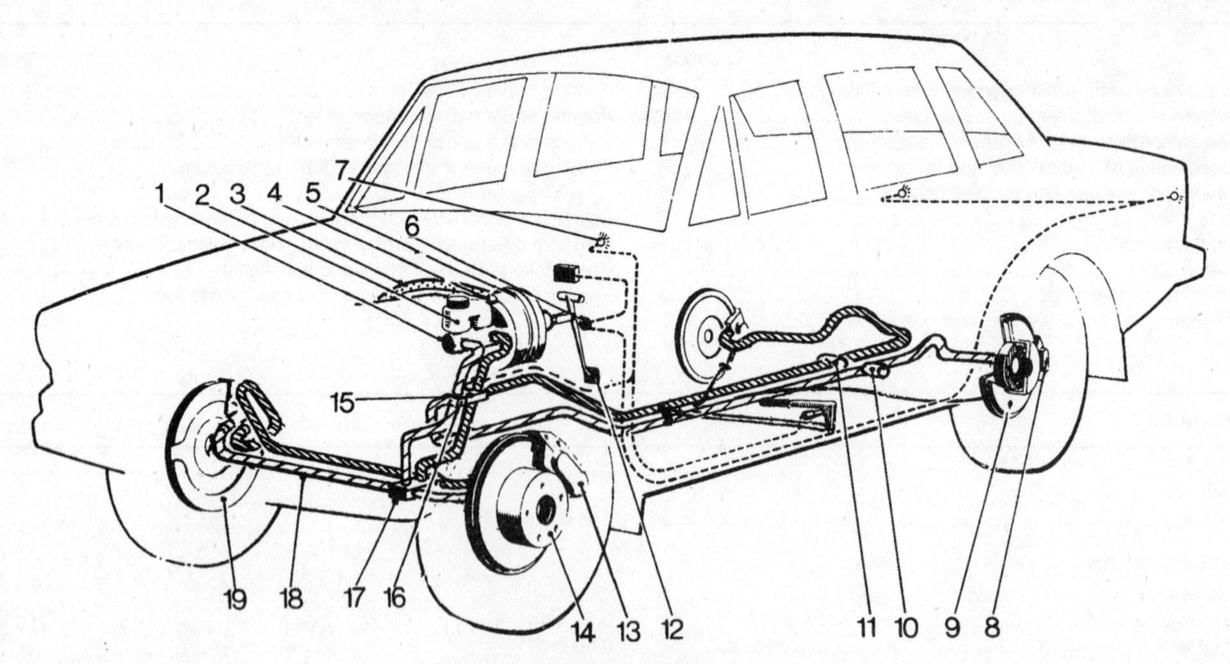

1.1 Brake system layout (non-ABS models)

1. *Master cylinder*
2. *Brake fluid reservoir*
3. *Vacuum line to booster*
4. *Check valve*
5. *Power brake booster*
6. *Brake light switch*
7. *Warning light*
8. *Rear brake caliper*
9. *Rear brake disc*
10. *Brake valve, secondary circuit*
11. *Brake valve, primary circuit*
12. *Brake pedal*
13. *Front brake caliper*
14. *Front brake disc*
15. *Pressure differential warning switch*
16. *Pressure differential warning valve*
17. *6-branch union (double 3-branch union)*
18. *Brake line*
19. *Cover plate*

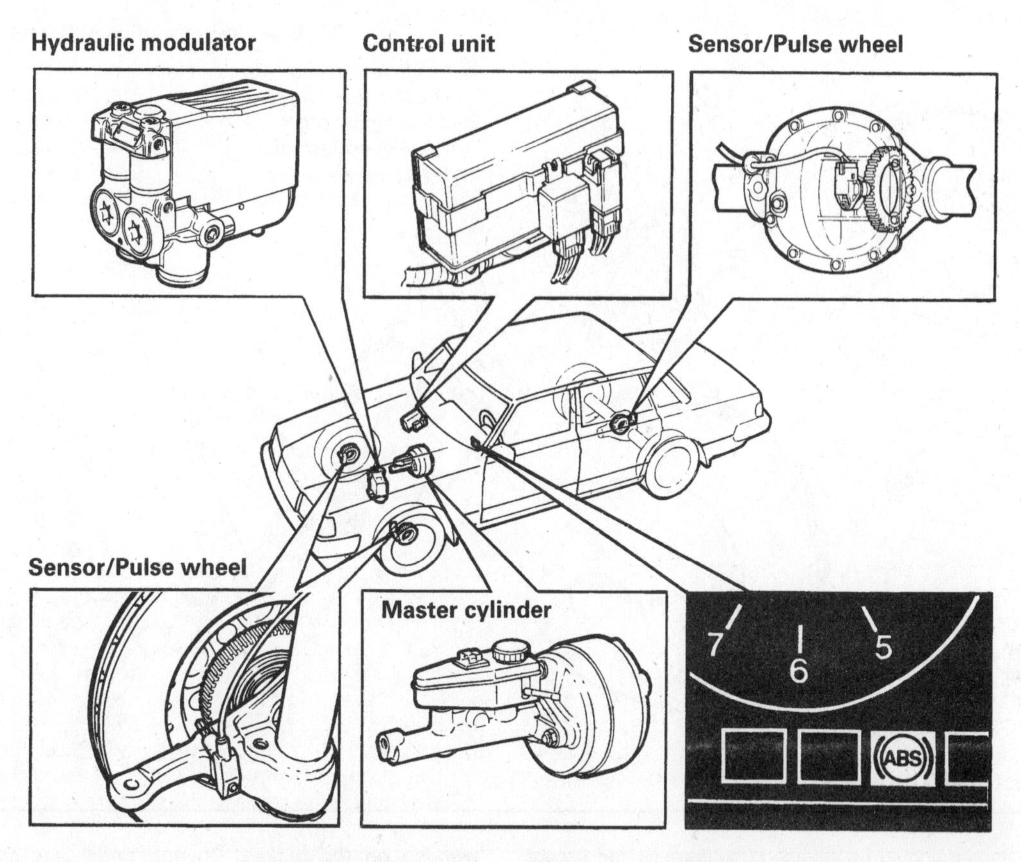

2.1 Anti-lock Brake System (ABS) main components

when the brakes are released. The seal also provides for automatic adjustment for brake pad wear. A further seal, fitted to the piston where it contacts the brake backplate, prevents dust entering the cylinder.

The rear brake caliper is also an ATE or a Girling, depending on model (see note at beginning of Section 3). Both types of caliper are similar to their front counterparts, except that they have single cylinders and pistons.

On non-ABS models the hydraulic system is a dual-circuit layout with each circuit operating two of the front pistons (there are four pistons in each front caliper) and a rear brake caliper. This arrangement ensures that half the braking system will still be effective in the event of a hydraulic pressure failure in one circuit. Should this situation arise, the driver is warned by an indicator light which is activated by a pressure differential valve in the brake system.

ABS models use a more conventional front-to-rear split, with the primary (rear) piston in the master cylinder serving the front brakes and the secondary (front) piston serving the rear brakes.

The power brake booster is actuated by the brake pedal. Vacuum assistance from the intake manifold on carbureted models, and from a vacuum pump on fuel-injected models, reduces the pedal pressure required for braking.

Pressure regulating valves are incorporated in each of the rear wheel brake lines to prevent rear wheel lock-up during heavy braking situations.

The parking brake consists of cable-operated brake shoes enclosed in small brake drums integral with the rear discs.

When you work on the brake system, be careful, methodical and scrupulously clean. Replacement parts should meet original equipment specifications and quality levels. **Warning:** *Dust created by the brake system may contain asbestos, which is harmful to your health. Never blow it out with compressed air and don't inhale any of it. An approved filtering mask should be worn when working on the brakes. Do not, under any circumstances, use petroleum-based solvents to clean brake parts. Use brake system cleaner only. Also, brake fluid is toxic and will attack paint. Never siphon it by mouth and be sure to wash thoroughly after skin contact.*

2 Anti-lock Brake System (ABS) - general information

Refer to illustration 2.1

Description

The Anti-lock Brake System (ABS) **(see illustration)** is designed to maintain vehicle steerability, directional stability and optimum deceleration under severe braking conditions on most road surfaces. It does so by monitoring the rotational speed of each wheel and controlling the brake line pressure to each wheel during heavy braking. This prevents the wheels from locking up.

The ABS system has three main components - the wheel speed sensors and pulse wheels, the electronic control unit and the hydraulic modulator. The sensors - one at each front wheel and one at the differential - send a variable voltage signal to the control unit, which monitors these signals, compares them to its program and determines whether or not a wheel is about to lock up. When a wheel is about to lock up, the control unit signals the hydraulic modulator to reduce hydraulic pressure (or not increase it further) at that wheel's brake caliper. Pressure modulation is handled by electrically-operated solenoid valves.

If a problem develops within the system, an "ABS" warning light will glow on the dashboard. Sometimes, a visual inspection of the ABS system can help you locate the problem. Carefully inspect the ABS wiring harness. Pay particularly close attention to the harness and connections near each wheel. Look for signs of chafing and other damage caused by incorrectly routed wires. If a wheel sensor harness

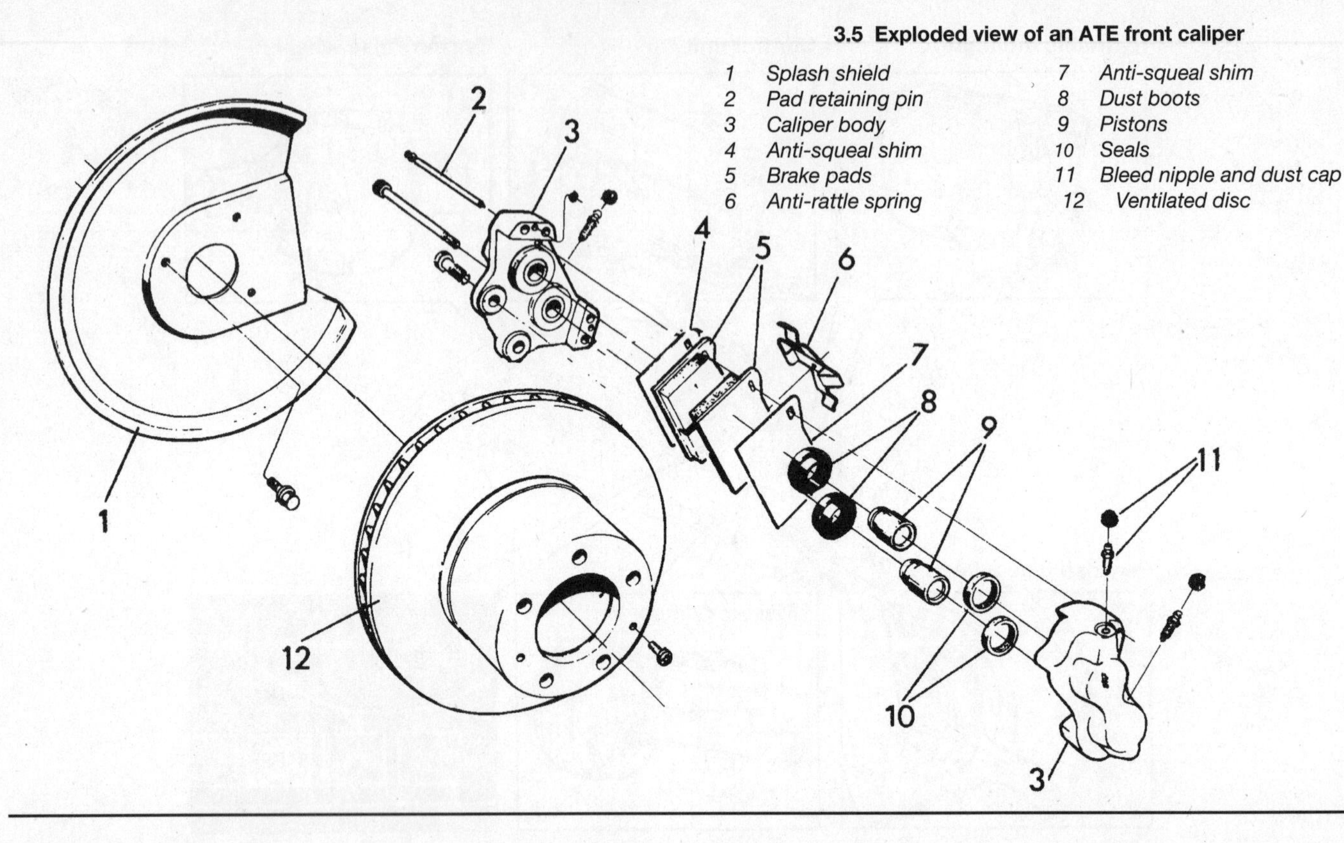

3.5 Exploded view of an ATE front caliper

1. *Splash shield*
2. *Pad retaining pin*
3. *Caliper body*
4. *Anti-squeal shim*
5. *Brake pads*
6. *Anti-rattle spring*
7. *Anti-squeal shim*
8. *Dust boots*
9. *Pistons*
10. *Seals*
11. *Bleed nipple and dust cap*
12. *Ventilated disc*

is damaged, the sensor should be replaced (the harness and sensor are integral). **Warning:** *Do NOT try to repair an ABS wiring harness. The ABS system is sensitive to even the smallest changes in resistance. Repairing the harness could alter resistance values and cause the system to malfunction. If any ABS wiring harness is damaged in any way, it must be replaced.* **Caution:** *Make sure the ignition is turned off before unplugging or reattaching any electrical connections.*

Diagnosis and repair

If a dashboard warning light comes on and stays on while the vehicle is in operation, the ABS system requires attention. Although special electronic ABS diagnostic testing tools are necessary to properly diagnose the system, you can perform a few preliminary checks before taking the vehicle to a dealer service department.

a) *Check the brake fluid level in the reservoir.*
b) *Check the electrical connectors at the control unit.*
c) *Check the electrical connectors at the hydraulic modulator.*
d) *Check the fuses.*
e) *Follow the wiring harness to each front wheel and to the differential sensor and verify that all connections are secure and that the wiring is undamaged.*

If the above preliminary checks do not rectify the problem, the vehicle should be diagnosed by a dealer service department or other qualified repair shop. Due to the complex nature of this system, all actual repair work must be done by a dealer service department or other qualified repair shop.

3 Brake pads - replacement

Warning: *Disc brake pads must be replaced on both front wheels at the same time - never replace the pads on only one wheel. Also, the dust created by the brake system may contain asbestos, which is harmful to your health. Never blow it out with compressed air and don't inhale any of it. An approved filtering mask should be worn when working on the brakes. Do not, under any circumstances, use petroleum-based solvents to clean brake parts. Use brake system cleaner only!*

Note: *Either Girling or ATE type brake calipers are installed; the type installed can be determined by the code numbers 1 and 2, which refer to Girling and ATE, respectively. These code numbers are located on a plate. On early models, the plate is on the B-pillar behind the right front door; from March 1978 to July 1979, it's on the right front door; on 1980 models, it's in front of the radiator; and on 1981 on models it's in the engine compartment (sedans) or luggage compartment (station wagons). On 1976 and 1977 models, the code applies only to the rear calipers - all front calipers are manufactured by Girling. On 1983 and later models, code number 2 indicates that the front brakes are Girling, and the rear brakes ATE.*

1 The brake pads should be replaced when they have reached the limit listed in this Chapter's Specifications. Always replace the pads on both front or both rear brakes at the same time. Replacing the pads on only one side can create braking imbalance.

2 High-friction rear pads should not be used in conjunction with early type front brake pads (Volvo part number DB818 or DB828), as this can cause braking imbalance between front and rear.

3 Loosen the wheel lug nuts, raise the front or rear of the vehicle, and place it securely on jackstands. Remove the wheels.

4 Remove about two-thirds of the fluid from the master cylinder reservoir and discard it. Position a drain pan under the brake assembly and clean the caliper and surrounding area with brake system cleaner.

Removal

ATE type (front)

Refer to illustration 3.5

5 Using a thin punch, knock out the pad retaining pins **(see illustration)**.

6 Lift out the anti-rattle spring.

7 Remove the pads, taking note of any anti-squeal shims to make sure they are installed in the same position. Also identify the pad positions if they are to be re-used.

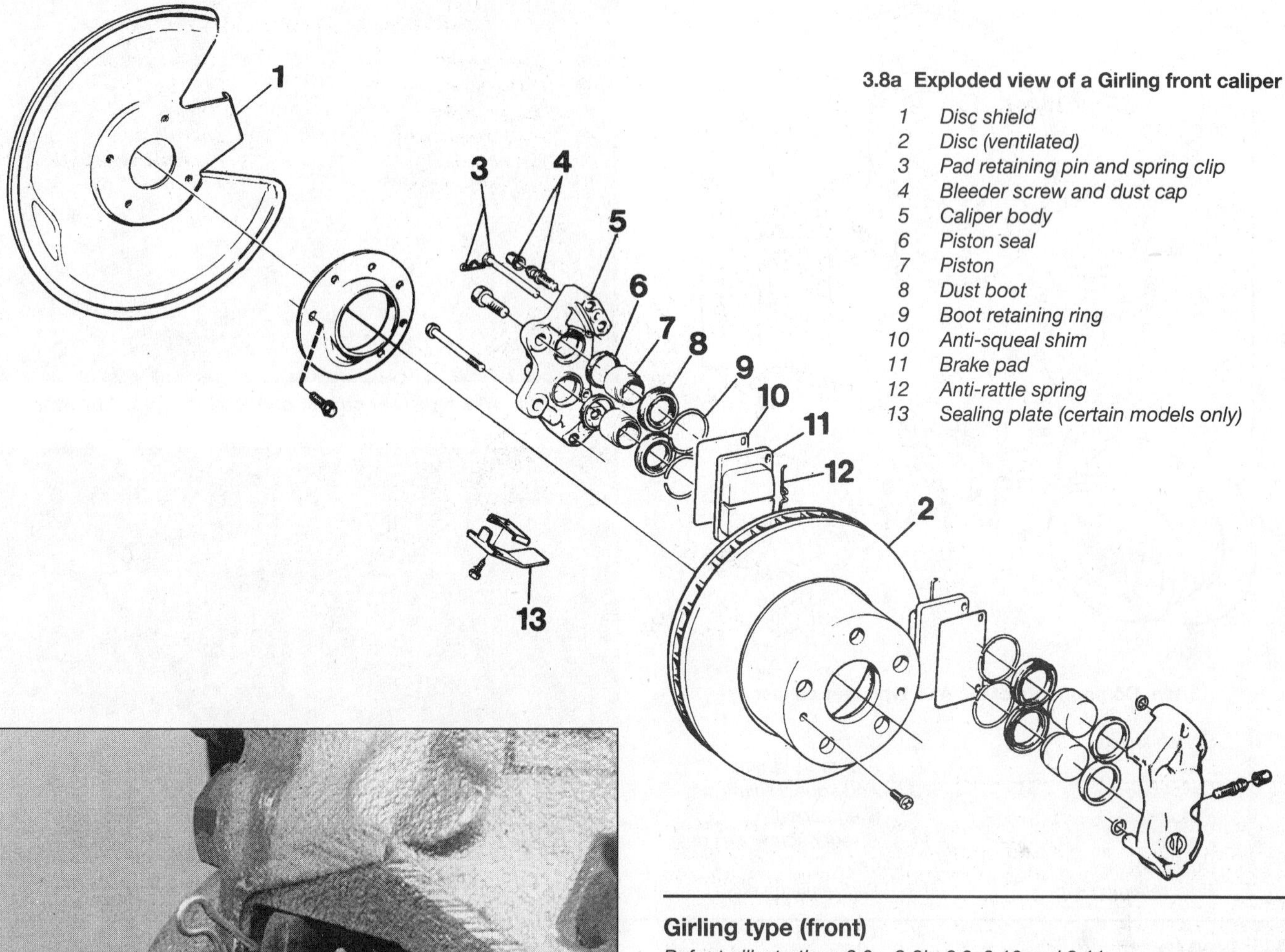

3.8a Exploded view of a Girling front caliper

1 Disc shield
2 Disc (ventilated)
3 Pad retaining pin and spring clip
4 Bleeder screw and dust cap
5 Caliper body
6 Piston seal
7 Piston
8 Dust boot
9 Boot retaining ring
10 Anti-squeal shim
11 Brake pad
12 Anti-rattle spring
13 Sealing plate (certain models only)

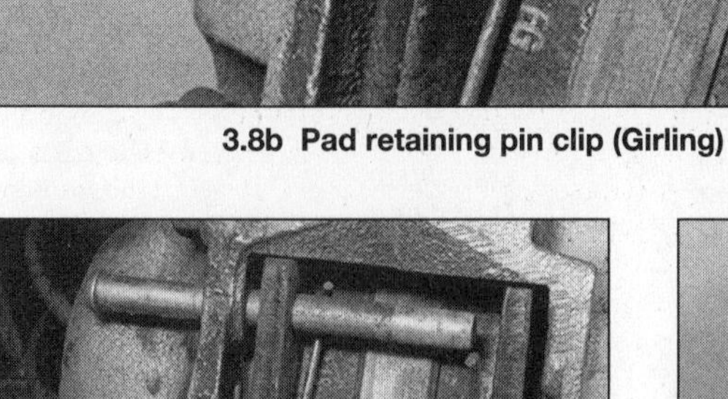

3.8b Pad retaining pin clip (Girling)

Girling type (front)

Refer to illustrations 3.8a, 3.8b, 3,9, 3.10 and 3.11

8 Remove the spring clips from the pad retaining pins **(see illustration)**.

9 Remove the upper pad retaining pin **(see illustration).**

10 Pry out the pad anti-rattle spring **(see illustration)**, then remove the lower pin.

11 Remove the brake pads, taking note of any anti-squeal shims which may be installed between the pad the piston, so that they can be installed in the same position **(see illustration)**. Also identify the pad positions if they are to be re-used.

3.9 Removing the upper pad retaining pin (Girling)

3.10 Anti-rattle springs and lower pin (Girling)

3.11 Removing the pads (Girling)

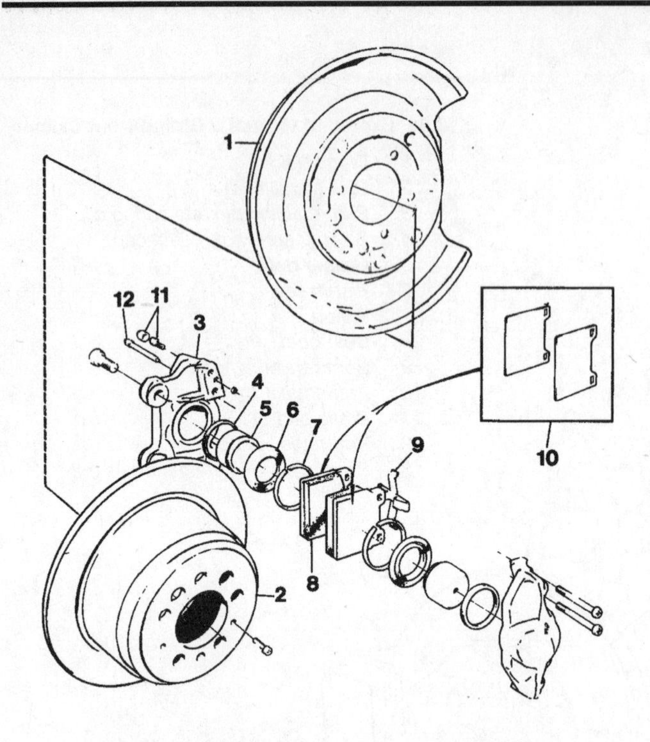

3.12a Components of the ATE type rear caliper

1 Splash shield
2 Disc
3 Caliper body
4 Piston seal
5 Piston
6 Dust boot
7 Boot retaining clip
8 Pad
9 Anti-rattle spring
10 Anti-squeal shims (if equipped)
11 Bleeder screw and dust cap
12 Pad retaining pin

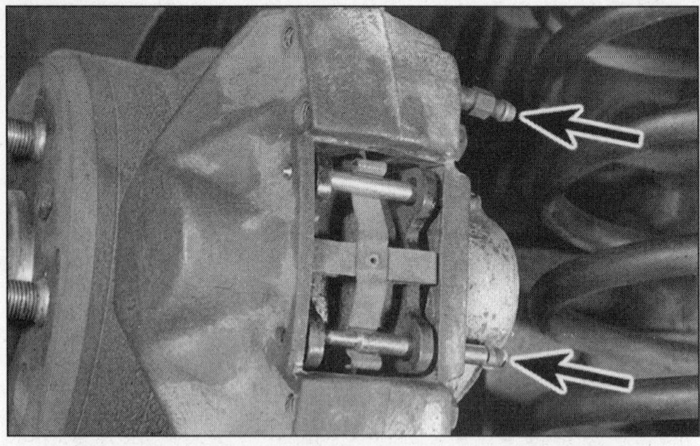

3.12b ATE type rear caliper pad retaining pins (arrows)

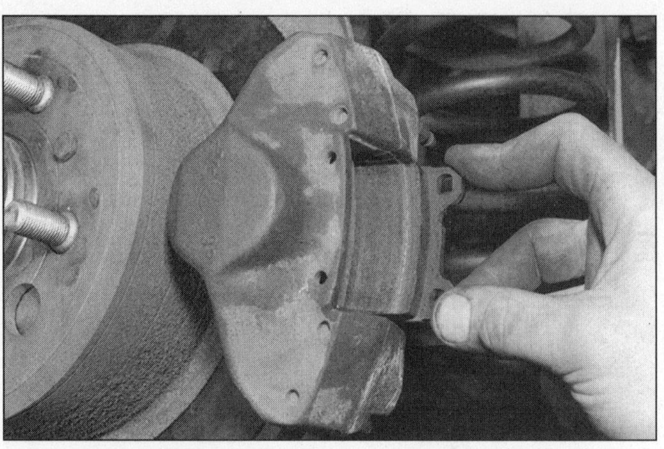

3.12c ATE type rear caliper pads being withdrawn

ATE type (rear)

Refer to illustrations 3.12a, 3.12b and 3.12c

12 The procedure is similar to that described for the front pads **(see illustrations)**.

Girling type (rear)

Refer to illustration 3.13

13 The procedure is similar to that described for the front pads, except that the spring-loaded cover plate must be removed first.

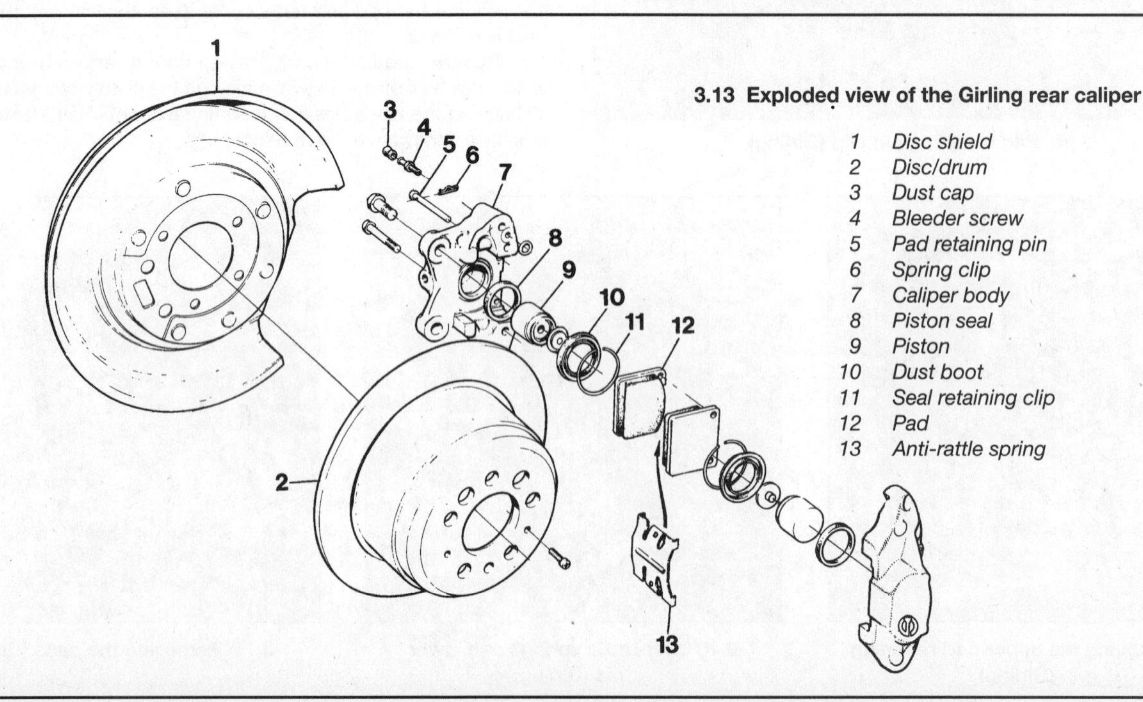

3.13 Exploded view of the Girling rear caliper

1 Disc shield
2 Disc/drum
3 Dust cap
4 Bleeder screw
5 Pad retaining pin
6 Spring clip
7 Caliper body
8 Piston seal
9 Piston
10 Dust boot
11 Seal retaining clip
12 Pad
13 Anti-rattle spring

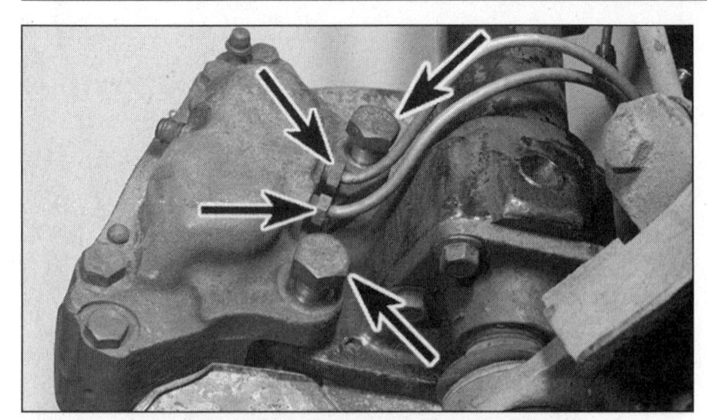

4.4 Disconnect both brake line fittings (left arrows), then remove the front caliper retaining bolts (right arrows)

Inspection (both types)

14 Inspect the piston seals for leakage and the dust boots for deterioration. If the seals are leaking or the dust boots are torn, refer to Section 4.

15 Use a block of wood as a lever to compress the pistons in the caliper (the new brake pads will otherwise be too thick). During this operation brake fluid may overflow from the reservoir, so be prepared for spillage. If the pistons cannot be compressed by this method, you can loosen the appropriate bleeder screw while compressing the piston, but you must bleed the brakes when you're done (see Section 10).

16 On ATE rear brakes, check the position of the stepped portion of the piston (see Section 4).

Installation (both types)

17 Apply anti-squeal compound to the brake pad backing plates, but make sure that it doesn't get on the pads or discs.

18 Installation of the pads is the reverse of removal. Replace items such as anti-rattle springs and spring clips as necessary.

19 When you're done, operate the brakes several times to bring the pads up to the disc.

20 If any bleeder screws were loosened, bleed the hydraulic system (see Section 10).

21 Install the wheels, remove the jackstands and lower the vehicle. Tighten the wheel lug nuts to the torque listed in the Chapter 1 Specifications.

22 Check the brake fluid level and top-off, if necessary (see Chapter 1).

23 Avoid harsh braking as far as possible for the next few hundred miles to allow the new pads to seat in.

4 Brake caliper - removal, overhaul and installation

Front caliper

Refer to illustrations 4.4, 4.8, 4.9 and 4.10

Warning: *Dust created by the brake system may contain asbestos, which is harmful to your health. Never blow it out with compressed air and don't inhale any of it. An approved filtering mask should be worn when working on the brakes. Do not, under any circumstances, use petroleum-based solvents to clean brake parts. Use brake system cleaner only. Also, brake fluid is toxic and will attack paint. Never siphon it by mouth and be sure to wash thoroughly after skin contact.*

Note: *The procedure given here covers both types of caliper, ATE and Girling. Before starting work, obtain a set of new seals and new caliper mounting bolts.*

1 Loosen the wheel lug nuts, raise the front of the vehicle and place it securely on jackstands. Remove the front wheel.

2 Place a drain pan under the caliper and thoroughly wash the caliper and pads with brake system cleaner.

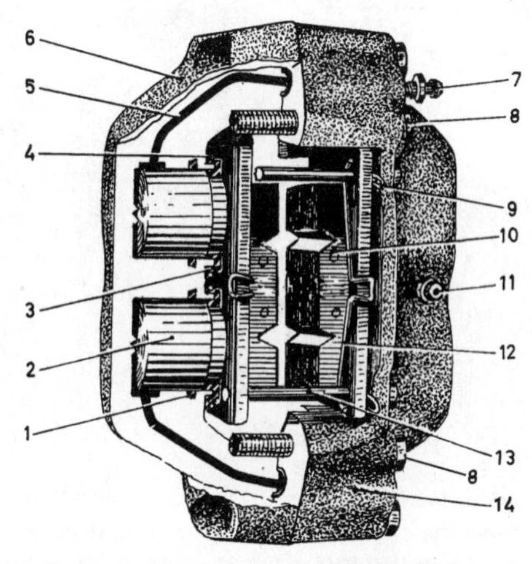

4.8 Cutaway view of a front caliper (Girling shown)

1 *Seal*
2 *Piston*
3 *Dust boot*
4 *Retaining ring*
5 *Brake fluid passage connecting left and right caliper halves*
6 *Outer caliper half*
7 *Upper bleeder screw*
8 *Bolts holding caliper halves together (do not remove)*
9 *Retaining clip*
10 *Brake pad*
11 *Lower bleeder screw*
12 *Anti-rattle spring*
13 *Retaining pin*
14 *Inner caliper half*

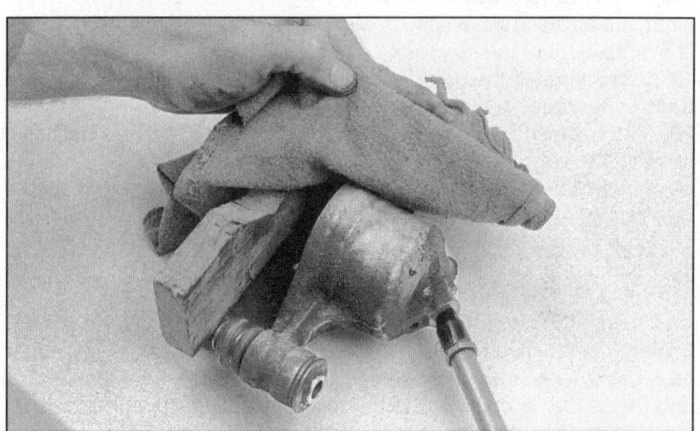

4.9 With the caliper padded to catch each piston, use compressed air to force the piston out of its bore - make sure your fingers are not between the piston and the caliper

3 Remove the cap from the brake fluid reservoir, place a square of cellophane or polyethylene over the filler neck and put the cap back on. This will reduce brake fluid loss when the brake lines are disconnected.

4 Disconnect the brake line fittings on the caliper **(see illustration)**. **Note:** *Use a flare-nut wrench, if available, which will prevent the corners of the fittings from being rounded-off.* Cap the ends of the lines, or clamp the flexible hoses, to reduce fluid loss.

5 Remove the brake pads (see Section 3).

6 Remove the caliper-to-steering knuckle bolts **(see illustration 4.4)** and lift the caliper away. Discard the caliper bolts.

7 Drain the fluid from the caliper. Using brake system cleaner, carefully clean all traces of dirt and dust from the caliper before placing it on a clean bench. **Warning:** *After removing the caliper, do not attempt to split the two halves of the caliper. Special tools are required, and the caliper halves cannot be reassembled outside of factory conditions.*

8 Remove the retaining clips and dust boots from the pistons **(see illustration)**.

9 Place a piece of wood between the pistons, to act as a cushion,

4.10 To remove the seal from the caliper bore, use a plastic or wooden tool, such as a pencil

then remove the pistons from the cylinders by applying low air pressure to the brake line fittings **(see illustration)**. **Warning:** *The pistons may be ejected with some force. Keep your fingers clear. Use only enough air pressure to ease the pistons from the bores.*

10 Remove the seals, being careful not to scratch the cylinder bores **(see illustration)**.

11 Clean the pistons in brake system cleaner and wash out the bores in the same manner, before inspecting them for scratches or scoring. Light imperfections can be removed with steel wool or very fine emery cloth, but anything more serious will require replacement of the pistons (if available), or of the complete caliper.

12 Blow through the cylinder bores with air, and verify that the fluid transfer passages in the caliper body are not clogged.

13 Apply clean brake fluid to the cylinder bores, pistons and new seals.

14 Install the seals in the grooves in the cylinder bores. Make sure they're fully seated in the groove all the way around.

15 Lubricate the pistons and install them into the cylinder bores.

16 Install the dust boot over the piston and the caliper housing. Install the dust boot retaining clips, if applicable.

17 Install the caliper on the steering knuckle with new bolts, and tighten them to the torque listed in this Chapter's Specifications.

18 Use a feeler gauge to verify that the gap between the brake disc and the caliper is equal on both sides, to within 0.010 inch. If it isn't, insert shims between the caliper and the steering knuckle to even up the gap.

19 Connect the brake lines, install the brake pads (see Section 3) and bleed the hydraulic system (see Section 10).

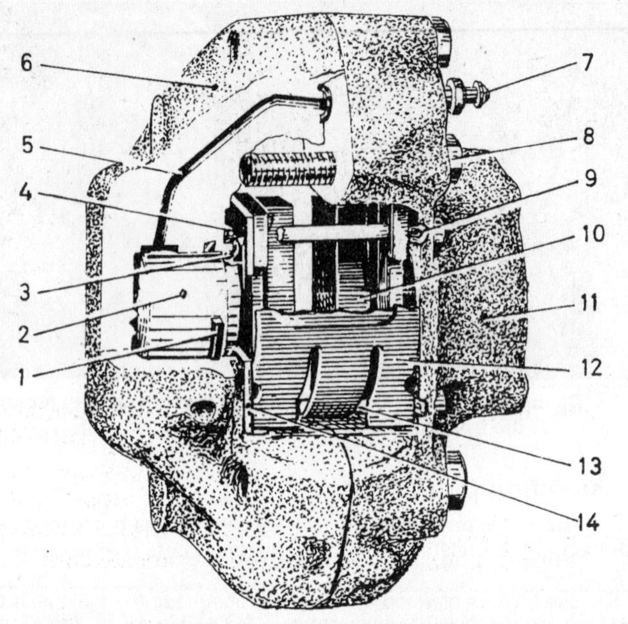

4.21a Cutaway view of a rear caliper (Girling shown)

1 Seal
2 Piston
3 Dust boot
4 Boot retaining ring
5 Brake fluid passage between caliper halves
6 Outer caliper half
7 Bleeder screw
8 Bolt
9 Retaining clip
10 Brake pad
11 Inner caliper half
12 Anti-rattle spring
13 Retaining pin
14 Washer

20 Install the wheels, remove the jackstands, lower the vehicle and tighten the wheel lug nuts to the torque listed in the Chapter 1 Specifications.

Rear brake caliper

Refer to illustrations 4.21a, 4.21b, 4.21c, 4.21d, 4.21e and 4.22

21 The procedure is similar to that described for the front calipers in Section 4 **(see illustrations)**. Again, be sure to use new bolts when installing the overhauled caliper.

22 **Note:** *ATE pistons require a special tool* **(see illustration)** *to set the angle of the stepped shoulder to the caliper.* The angle is 20 ± 2 degrees. If you don't have this tool, make one from stiff cardboard or sheetmetal.

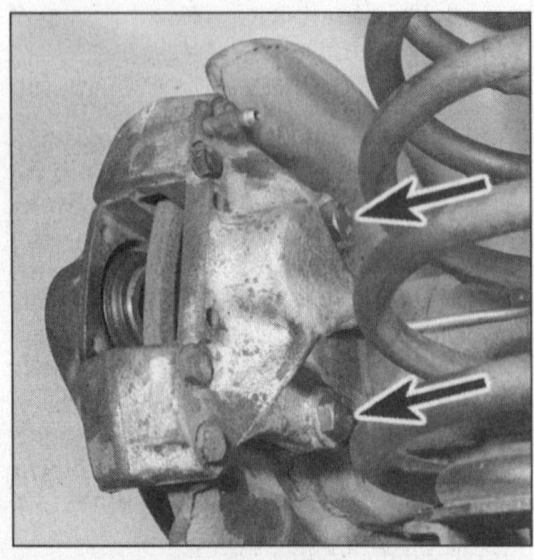

4.21b ATE rear caliper retaining bolts (arrows)

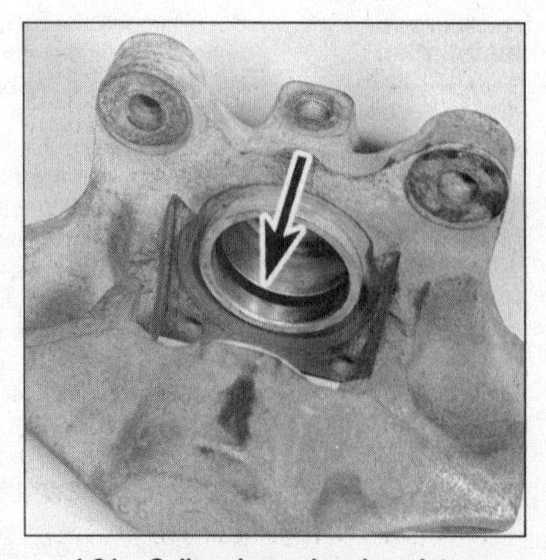

4.21c Caliper bore showing piston seal (arrow)

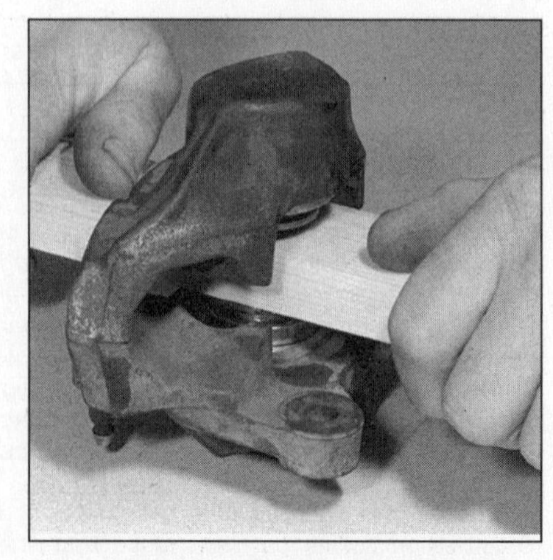

4.21d Pressing in a new piston

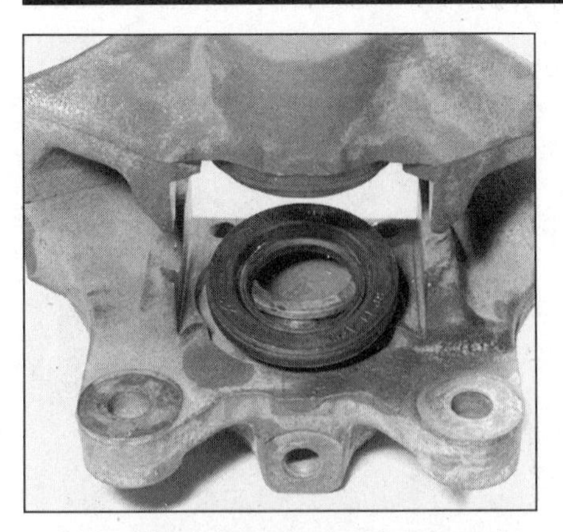

4.21e This is how the dust boot must be installed

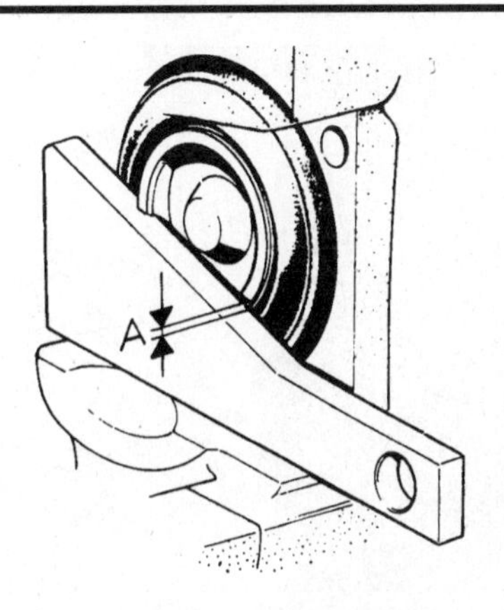

4.22 Using the Volvo special tool to align an ATE type caliper piston

A (clearance between tool and stepped portion) = 0. 04 inch (1. 0 mm)

5.3 The brake pads on this vehicle were obviously neglected, as they wore down to the rivets and cut deep grooves into the disc - wear this severe means the disc must be replaced

5 Brake disc - inspection, removal and installation

Note: *This procedure applies to both the front and rear brake discs.*

Inspection

Refer to illustrations 5.3, 5.4a, 5.4b and 5.5

1 Loosen the wheel lug bolts, raise the vehicle and support it securely on jackstands. Remove the wheel. If the rear brake disc is being worked on, release the parking brake.

2 If the disc is going to be removed, remove the brake caliper as outlined in Section 4. It is not necessary to disconnect the brake hose. After removing the caliper, suspend it out of the way with a piece of wire.

3 Visually inspect the disc surface for scoring or damage. Light scratches and shallow grooves are normal after use and may not always be detrimental to brake operation, but deep scoring - over 0.015 inch - requires disc removal and refinishing by an automotive machine shop. Be sure to check both sides of the disc **(see illustration)**. If pulsating has been noticed during application of the brakes, suspect disc runout.

4 To check disc runout, place a dial indicator at a point about 1/2-inch from the outer edge of the disc **(see illustration)**. Set the indicator to zero and turn the disc. The indicator reading should not exceed the specified allowable runout limit. If it does, the disc should be refinished by an automotive machine shop. **Note:** *It is recommended that the discs be resurfaced regardless of the dial indicator reading, as this will impart a smooth finish and ensure a perfectly flat surface, eliminating any brake pedal pulsation or other undesirable symptoms related to questionable discs. At the very least, if you elect not to have the discs resurfaced, remove the glazing from the surface with emery cloth or sandpaper using a swirling motion* **(see illustration).**

5.4a To check disc runout, mount a dial indicator as shown and rotate the disc

5 It is absolutely critical that the disc not be machined to a thickness under the specified minimum allowable thickness. The minimum wear (or discard) thickness is stamped into the hub of the disc. The disc thickness can be checked with a micrometer **(see illustration).**

Removal

Refer to illustrations 5.6a, 5.6b, 5.6c, 5.6d and 5.6e

6 Remove the disc retaining bolts **(see illustrations)** and remove

5.4b Using a swirling motion, remove the glaze from the disc with sandpaper or emery cloth

5.5 The disc thickness can be checked with a micrometer

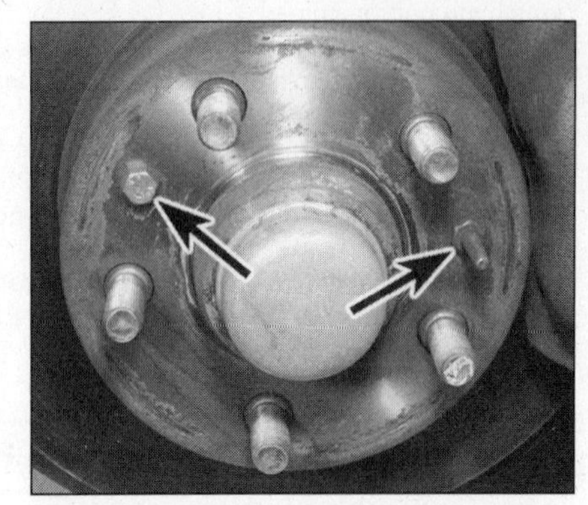

5.6a Front brake disc retaining bolts (arrows); note location of wheel locating pin (if equipped)

5.6b Remove the disc retaining bolt . . .

5.6c . . . and remove the brake disc

5.6d If the disc is stuck, spray plenty of penetrant onto the area between the hub and the disc

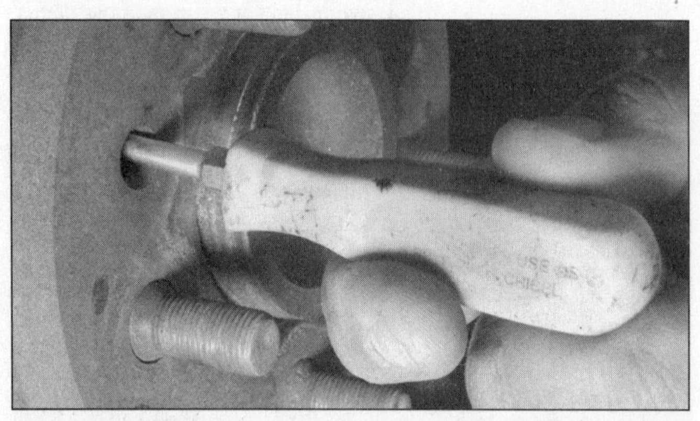
5.6e If penetrant fails to loosen the rust between the hub and the disc on a rear disc, insert a thin flat-bladed screwdriver through the hub flange, rotate the star wheel on the parking brake adjusting screw and contract the parking brake shoes

6.1 Brake master cylinder reservoir (1), master cylinder (2) and power brake booster (3)

the disc from the hub **(see illustration)**. If the disc is stuck to the hub, spray a generous amount of penetrant onto the area between the hub and the disc **(see illustration)**. Allow the penetrant a few minutes to loosen the rust between the two components and tap the disc off with a soft-faced hammer. If a rear disc still sticks, insert a thin, flat-bladed screwdriver through the hub flange, rotate the star wheel on the parking brake adjusting screw and contract the parking brake shoes **(see illustration)**.

Installation

7 Place the disc on the hub and install the disc retaining bolts. Tighten the bolts securely.

8 Install the brake pads and caliper (see Sections 3 and 4). Tighten the caliper mounting bolts to the torque listed in this Chapter's Specifications.

9 Install the wheel, then lower the vehicle to the ground and tighten the wheel lug nuts to the torque listed in the Chapter 1 Specifications. Depress the brake pedal a few times to bring the brake pads into contact with the disc.

10 Adjust the parking brake shoes, if necessary.

11 Check the operation of the brakes carefully before placing the vehicle into normal service.

6 Master cylinder - removal, overhaul and installation

Refer to illustrations 6.1, 6.7, 6.8a, 6.8b, 6.12a, 6.12b and 6.12c

1 The master cylinder and reservoir are bolted to the power brake booster unit **(see illustration)**. After a long service life, the seals in the cylinder can start to leak internally. One indication of this condition is a brake pedal that continues to sink under pressure, even though there are no exterior signs of leakage from other parts of the system.

2 Siphon as much brake fluid as possible from the master cylinder reservoir. Place some rags beneath the cylinder to catch any spilled fluid.

3 Disconnect the fluid level warning electrical lead (if equipped).

4 Remove the nuts securing the cylinder to the power brake booster.

5 On 1984 and 1985 models, disconnect the fluid supply line to the clutch hydraulic system.

6 Disconnect all brake fluid line fittings and remove the cylinder. Make sure no brake fluid drips on the paint. If it does, rinse it off with water immediately. **Caution:** *Once the master cylinder is removed, do not depress the brake pedal or damage to the brake booster unit may occur.*

7 Separate the reservoir from the cylinder - it pulls out of the seals **(see illustration)**. It may be necessary to pry it up.

8 Depress the pistons, then remove the locking ring from the end of the cylinder. Pull out the piston and spring **(see illustrations)**.

9 Clean all parts with brake system cleaner and blow through the equalizing and overflow holes in the cylinder with filtered, unlubricated compressed air.

10 Examine the cylinder bores. If there are any score marks or scratches, replace the master cylinder.

11 The pistons, connector sleeve and seals come as one unit in the repair kit, and should be lubricated with clean brake fluid before installation.

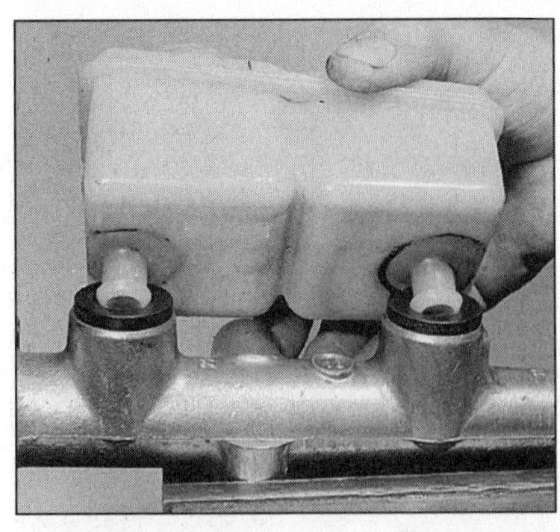

6.7 Removing the reservoir from the master cylinder

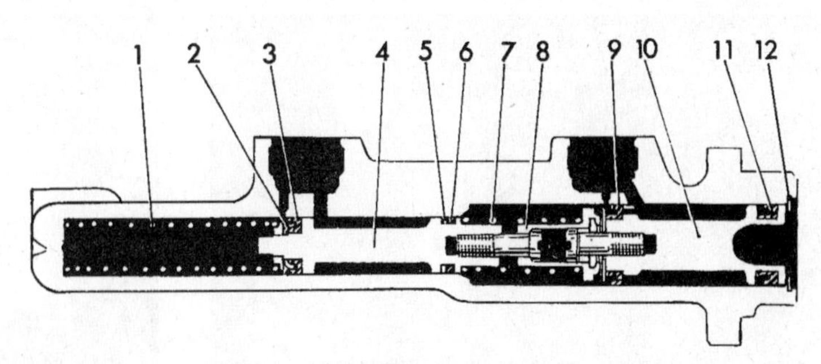

6.8a Sectional view of the master cylinder

1 *Spring*
2 *Spring seat*
3 *Seal*
4 *Secondary piston*
5 *Seal*
6 *Seal*
7 *Spring*
8 *Connector sleeve*
9 *Seal*
10 *Primary piston*
11 *Seal*
12 *Locking ring*

6.8b Remove the locking ring to release the pistons - you'll have to push the pistons in to accomplish this

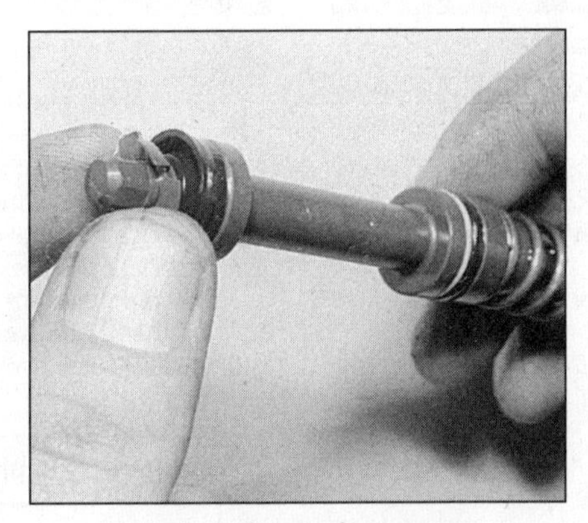

6.12a Install the spring seat . . .

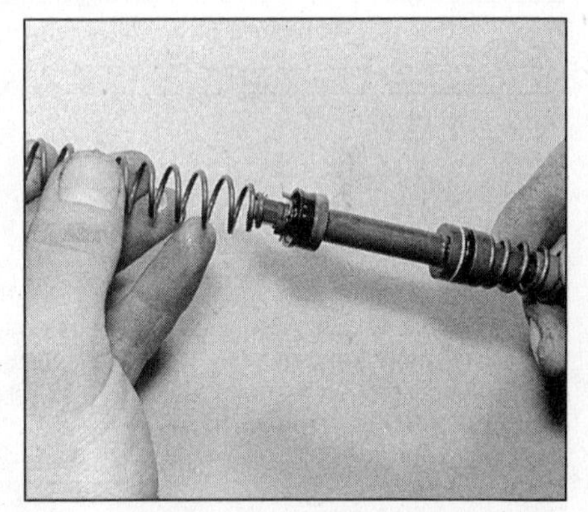

6.12b . . . and the spring

12 Pour clean brake fluid into the cylinder to lubricate the bore, then insert the piston and seal assembly, making sure the spring and spring seat are in position **(see illustrations)**.
13 Install the locking ring.
14 Install the reservoir - use new seals.
15 Install the master cylinder on the power brake booster and tighten the master cylinder mounting nuts to the torque listed in this Chapter's Specifications.
16 Connect all hydraulic line fittings.
17 Bleed the system (see Section 10).

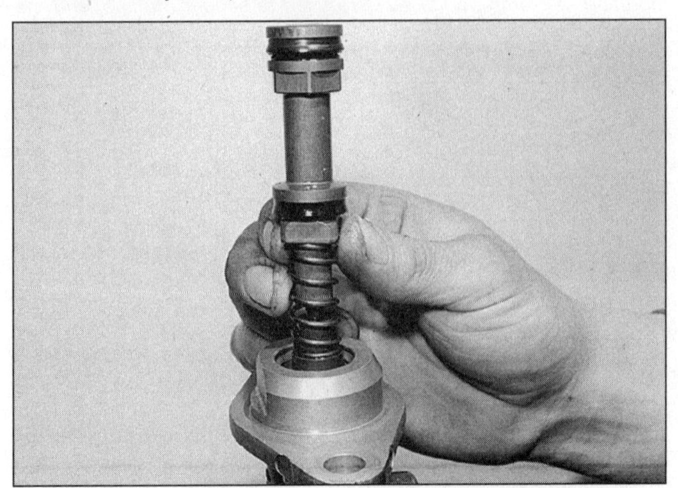

6.12c Installing the pistons into the master cylinder

7 Rear brake valves - removal and installation

Refer to illustration 7.3

1 Any fault in either of the two brake valves, which are integral with the rear brake lines, can only be rectified by replacing the valve assembly.
2 Raise the rear of the vehicle and place it securely on jackstands.
3 Disconnect the rigid brake line **(see illustration)** from the valve. Cap or plug the line to prevent loss of fluid.

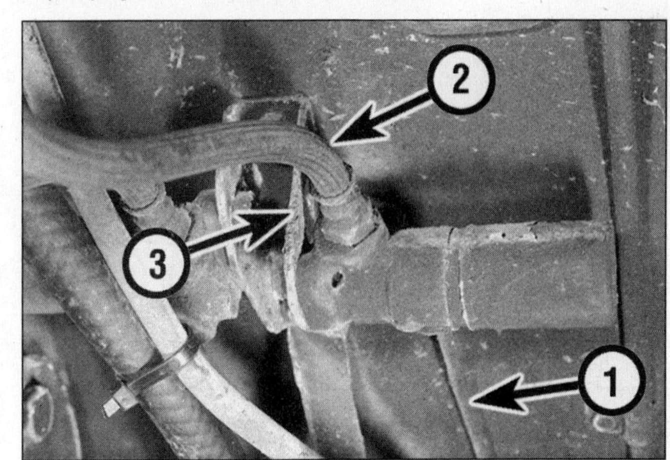

7.3 Rear brake valve: rigid line (1), flexible hose (2) and retaining bolt (3)

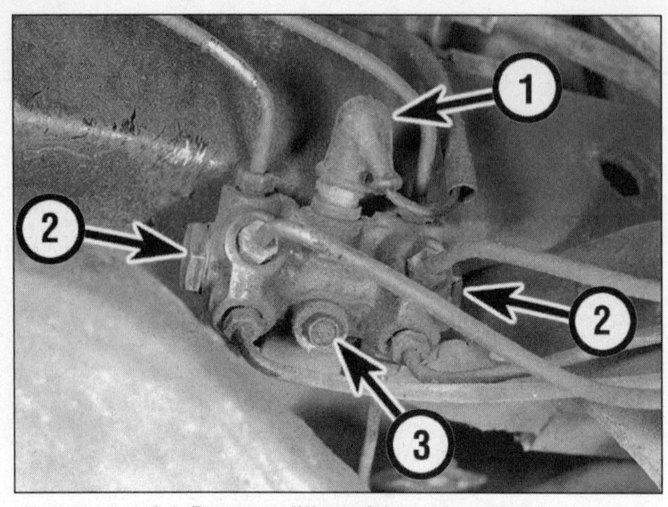

8.1 Pressure differential warning valve

1 *Electrical connection*
2 *End-plugs*
3 *Retaining bolt*

9.3 Typical brake line-to-brake hose line fitting

1 *Flexible brake hose*
2 *Rigid metal line*
3 *Spring clip*

4 Loosen, but don't try to disconnect, the flexible brake hose from the valve. Don't unscrew the hose fitting more than a quarter of a turn - you'll kink the hose.

5 Remove the brake valve retaining bolt, then unscrew the valve body from the flexible hose. Don't twist the hose.

6 Installation is the reverse of removal. **Note:** *Different models have different operating pressures. Make sure to replace the valve with the correct part number for your vehicle.*

7 When you're done, bleed the hydraulic circuit (see Section 10).

8 Pressure differential warning valve - removal and installation

Refer to illustration 8.1

1 The pressure differential warning valve, on most models, is located in the engine compartment, on the left chassis member **(see illustration)**. On early models, it may be bolted to the right side.

2 An overhaul kit is available on some models, but we don't recommend disassembly of the pressure differential warning valve. If it's malfunctioning, replace it.

3 Block the breather hole on the brake fluid reservoir filler cap with a piece of cellophane to prevent excess fluid loss.

4 Clean all dirt from around the valve and surrounding area.

5 Unplug the electrical connector from the valve.

6 Remove the retaining bolt and remove the valve.

7 Installation is the reverse of removal.

8 Bleed the system (see Section 10) when you're done.

9 To check your work, depress the brake pedal hard for 60 seconds and check the valve for leaks.

9 Brake hoses and lines - inspection and replacement

Refer to illustration 9.3, 9.6 and 9.9

1 Whenever leakage is suspected, inspect the hydraulic hoses and lines, hose connections and fittings.

2 Inspect the flexible hoses for chafing, kinking, cracking or swelling, and replace them as necessary.

3 Always unscrew rigid metal lines from the flexible hose part first, using two wrenches to avoid twisting, then release the flexible hose end from its bracket and clip **(see illustration)**.

4 When installing a new hose, make sure the hose will not chafe against adjacent components, and that it's not twisted or stretched.

5 Rigid metal lines should be cleaned off and inspected for corrosion, chafing and other damage.

6 Check the security of all clips and clamps. **Note:** *On post 1977 models, the routing of the brake line on the rear axle has been modified. A clamp and sleeve are used to secure the line, and the*

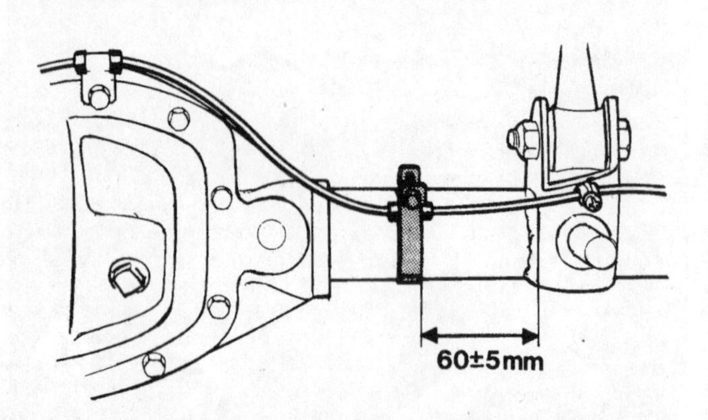

9.6 Position of rear brake line clamp on post-1977 vehicles

9.9 Using a flare-nut wrench prevents rounding off brake line fitting nuts

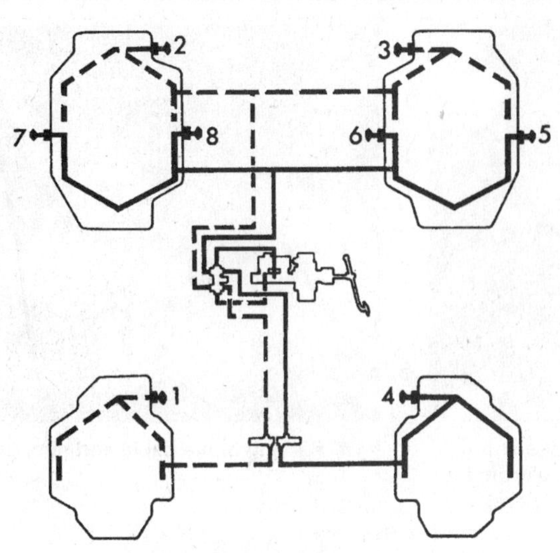

10.4 Brake bleeding sequence

1 *Left rear caliper bleed screw*
2 *Left front caliper, upper inner bleed screw*
3 *Right front caliper, upper inner bleed screw*
4 *Right rear caliper bleed screw*
5 *Right front caliper, outer bleed screw**
6 *Right front caliper, lower inner bleed screw**
7 *Left front caliper, outer bleed screw**
8 *Left front caliper, lower inner bleed screw**
**Bleed these in pairs (both right, then both left)*

clamp should be positioned about 2-3/8 inches from the reaction rod bracket **(see illustration)**. *If a new line is installed on an earlier model, the new line must be clamped to the axle.*

7 Any line which shows signs of corrosion should be replaced. Brake lines are generally available in two forms: They're either pre-cut to the correct length and bent to the right shape, or come in standard lengths of straight tubing with the ends flared and flare-nuts installed. Don't purchase straight tubing unless you have the right tools (a tubing bender and a tubing cutter), and the skill, to cut and bend them correctly. Have the job done by a professional.

8 If you decide to do it yourself, use the old line as a pattern to bend the new line into shape. Be careful not to bend a line any more than necessary, or it may kink or collapse.

9 Special flare-nut wrenches **(see illustration)** should be used instead of regular wrenches to avoid rounding off line fitting nuts.

10 When you're done, bleed the hydraulic system (see Section 10).

10 Hydraulic system - bleeding

Refer to illustration 10.4

Non-ABS models

1 Whenever the hydraulic system has been overhauled, a part has been replaced, or the level in the reservoir has become too low, air will enter the system. This causes some or all of the pedal travel to be used up in compressing air rather than pushing fluid against brake pistons. If only a little air is present, the pedal will have a 'spongy' feel, but if an appreciable amount has entered, the pedal will not offer any substantial resistance to the foot and the brakes will hardly work at all.

2 To overcome this, brake fluid must be pumped through the hydraulic system until all the air has been passed out in the form of bubbles in the fluid.

3 If only one hydraulic circuit has been disconnected, only that circuit needs to be bled. If both circuits have been disconnected, or the brake fluid is being replaced, the whole system must be bled. The circuits in a non-ABS system are arranged as follows:

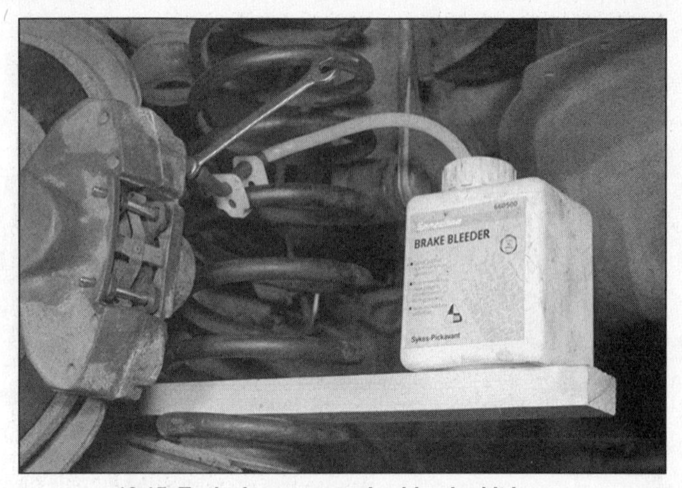

10.15 Typical one-way-valve bleeder kit in use

Primary*: Upper cylinders in both front calipers, and left-hand rear.

Secondary*: Lower cylinders in both front calipers, and right-hand rear.

***Note:** *On pre-1977 models, the primary and secondary circuits are reversed.*

4 Bleed the system as shown **(see illustration)**.

5 There are three bleed screws on each front caliper, and one on each rear caliper. When bleeding the front caliper lower cylinders, both the lower bleed screws should be opened and closed together.

Bleeding with an assistant

6 Obtain a glass jar, a supply of fresh brake fluid and two clear plastic tubes to fit over the bleed screws.

7 Top-off the master cylinder reservoir. Keep it topped-off throughout the operation.

8 Attach the tube(s) to the bleed screw(s) of the first caliper to be bled (see Step 4). Pour a little brake fluid into the jar and place the open ends of the tube(s) in the jar, dipping into the fluid.

9 Loosen the bleed screw(s). Have the assistant depress and release the brake pedal five times, stopping on the fifth downstroke. Tighten the bleed screw(s) and have the assistant release the pedal.

10 Top-off the master cylinder reservoir.

11 Repeat Steps 9 and 10 until clean fluid, free from air bubbles, emerges from the bleed screw(s).

12 Repeat the process on the remaining calipers in the order given.

13 When you're done, verify that the brake pedal feels firm. Top-off the master cylinder reservoir and install the cap.

14 Discard the fluid bled from the system. Dispose of it in a sealed container.

Bleeding with a one-way-valve brake bleeder

Refer to illustration 10.15

15 A one-way-valve brake bleeder kit **(see illustration)** simplifies the bleeding process and reduces the risk of expelled air or fluid being drawn back into the system.

16 Connect the outlet tube to the bleed screw and then open the screw half a turn. If possible, position the tube so that it can be viewed from inside the car. Depress the brake pedal as far as possible and slowly release it. The one-way valve in the bleed kit will prevent expelled air or fluid from returning to the system at the end of each pedal return stroke. Repeat this operation until clean hydraulic fluid, free from air bubbles, can be seen coming through the bleed tube. Tighten the bleed screw and remove the tube.

17 Repeat the operations on the remaining bleed screws in the correct sequence. Make sure that throughout the process the fluid reservoir level never falls so low that air can be drawn into the master cylinder, otherwise the work up to this point will have been wasted.

11.3 Parking brake cable adjuster bolt (arrow)

11.4 Turning a parking brake shoe adjuster through the hole in the drum (not all models have adjusters here)

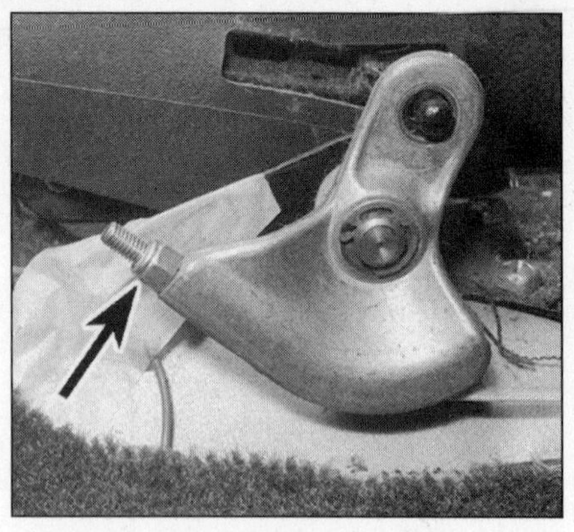
11.7 Parking brake cable end nuts (arrow)

Bleeding with a pressure bleeding kit

18 This device is usually operated by air pressure.
19 By connecting a pressurized container to the master cylinder fluid reservoir, bleeding is then carried out by simply opening each bleed screw in turn and allowing the fluid to run out, rather like turning on a tap, until air bubbles are no longer visible in the fluid being expelled.
20 Using this system, the large reserve of hydraulic fluid provides a safeguard against air being drawn into the master cylinder during the bleeding process.

Models with ABS

21 Bleeding the hydraulic system on models equipped with an Anti-lock Brake System (ABS) is carried out in the same manner as for non-ABS models, except the order in which the calipers are bled is different. Begin with the left rear caliper, followed by the right rear, left front and, finally, the right front caliper.

11 Parking brake - adjustment

Note: *Parking brake adjustment is normally necessary only to compensate for cable stretch, or after installing a new cable or parking brake shoes. Some models are equipped with shoe adjusters; some aren't. The shoe adjuster mechanism's main use is for initial set-up after installing new shoes. On models without shoe adjusters, adjustment is handled at the parking brake lever.*
1 Loosen the rear wheel lug nuts. Raise the rear of the vehicle and place it securely on jackstands. Remove the rear wheels. Release the parking brake.

Vehicles with shoe adjusters

Refer to illustrations 11.3, 11.4 and 11.6
2 Remove the ashtray from the center console (see Chapter 11).
3 Loosen the adjuster bolt **(see illustration)** and release all tension from the cables.
4 Back off the adjuster mechanism in the rear drum by inserting a thin screwdriver through the hole in the drum and rotating the notched wheel **(see illustration)**.
5 Expand the adjuster until the shoes just begin to bind on the drum, then back it off four or five notches.
6 Tighten the adjuster bolt on the parking brake lever until it takes two-to-eight clicks of the lever to obtain full braking efficiency. Make sure that the wheels are still free to turn with the parking brake released.
7 Adjust the tension between the cables, if necessary, by using the cable end nuts, until the yoke at the rear of the parking brake is square with the parking brake lever when the brake is applied **(see illustration)**. **Note:** *It will be necessary to remove the center console for this operation.*

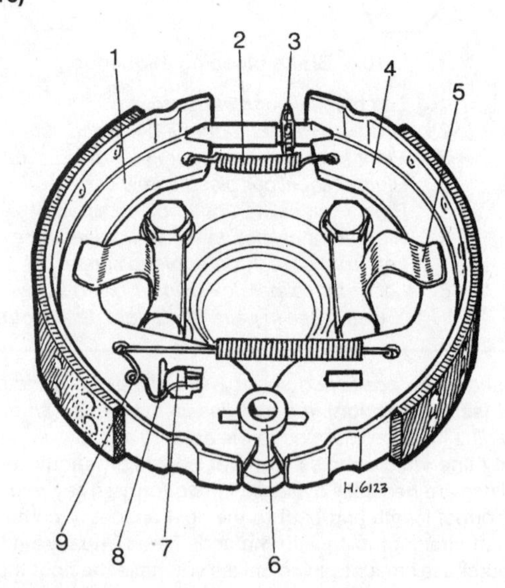

12.4a Parking brake shoe arrangement

1	*Parking brake shoe*	6	*Anchor bolt*
2	*Return spring*	7	*Lever*
3	*Adjuster (some models)*	8	*Washer*
4	*Parking brake shoe*	9	*Spring*
5	*Shoe steady clip*		

8 Install the ashtray.
9 Install the wheels, remove the jackstands, lower the vehicle and tighten the wheel lug nuts to the torque listed in the Chapter 1 Specifications.

Models without shoe adjusters

10 Remove the ashtray from the center console.
11 Release the parking brake.
12 Refer to Step 6 and proceed as described above.

12 Parking brake shoes - replacement

Refer to illustrations 12.4a, 12.4b, 12.4c and 12.9
Warning: *Dust created by the brake system may contain asbestos, which is harmful to your health. Never blow it out with compressed air and don't inhale any of it. An approved filtering mask should be worn when working on the brakes.*
1 The parking brake shoes must be replaced when they have worn down to the rivet heads, or in the case of bonded linings, when only 1/16-inch of lining remains (in practice, the shoe linings are unlikely to

12.4b Brake shoe return spring (arrow)

12.4c Brake shoes and spreader bar being removed

12.9 Parking brake lever in position

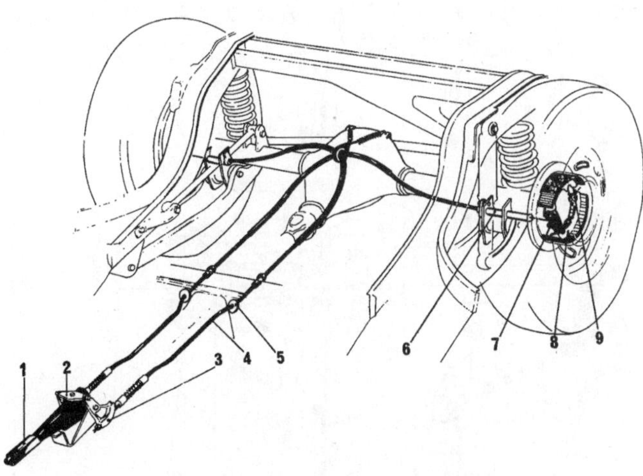

13.3 Parking brake cable layout

1 *Lever*
2 *Yoke*
3 *Lever*
4 *Cable*
5 *Rubber grommet*
6 *Plastic tube*
7 *Levers*
8 *Brake shoes*
9 *Adjuster mechanism (some models)*

13.8 Pressing out the parking brake cable-to-operating lever pin

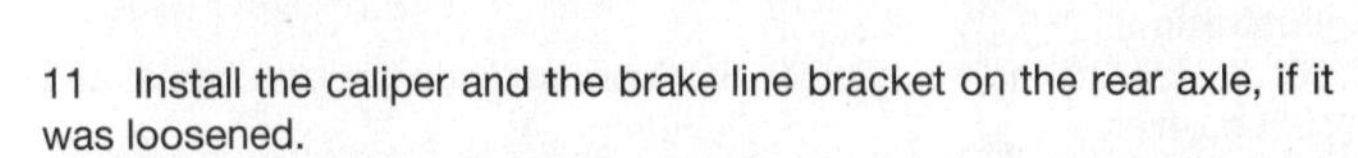

undergo significant wear unless they're habitually applied while the vehicle is in motion). The shoes must also be replaced if they become contaminated with oil as a result of an axle oil seal failure.

2 Loosen the parking brake adjuster at the rear of the parking brake lever (see Section 11).

3 Remove the rear disc (see Section 5).

4 Note which holes the return springs are hooked into **(see illustrations)**, then unhook the springs and remove the brake shoes and adjuster (if equipped) or the spreader bar **(see illustration)**.

5 Clean off all dust from the backing plate with brake system cleaner. **Warning:** *Do NOT blow off brake dust with compressed air.*

6 Inspect the adjuster mechanism (if equipped), the cable end and the lever for wear.

7 Apply brake anti-seize compound to the adjuster or the spreader bar and the cable pivot points, and to the parts of the backing plate in contact with the parking brake shoes. Do NOT allow anti-seize compound to contaminate the linings or the drum friction surface.

8 Install the lower return spring between the shoes, install the shoes back over the hub, and install the upper return spring, locating the adjuster mechanism or spreader bar as you do so.

9 Make sure that the parking brake lever is slotted into the hole in the upper brake shoe **(see illustration)**, and that the steady springs are located over the brake shoes.

10 Back off the adjuster and install the brake disc (see Section 5).

11 Install the caliper and the brake line bracket on the rear axle, if it was loosened.

12 Verify that the disc turns freely - it may bind until the shoes are centered. Apply the parking brake a few times to center the shoes.

13 Adjust the parking brake (see Section 11).

14 Install the wheels, remove the jackstands, lower the vehicle and tighten the wheel lug nuts to the torque listed in the Chapter 1 Specifications.

13 Parking brake cables - replacement

Refer to illustrations 13.3, 13.8, 13.9, 13.12 and 13.15

Note: *The following procedure describes the replacement of only one cable, but it applies to both.*

1 Remove the center console (see Chapter 11).

2 Loosen the adjuster bolt at the rear end of the parking brake lever (see Section 11).

3 Loosen the nut at the end of the cable to be replaced. **Note:** *The cables cross over underneath the vehicle* **(see illustration)**, *so the cable on the left of the parking brake lever operates the right brake, and vice versa.*

4 Lift up the front end of the rear seat cushion and fold back the carpet. This will reveal some cable clips which must be released.

5 Free the grommet where the cable runs through the floorpan.

6 Remove the rear brake disc (see Section 5).

7 Remove the parking brake shoes (see Section 12).

8 Press out the pin securing the cable to the operating lever **(see illustration)**.

9 Remove the bolt from the cable bracket and guide tube on the

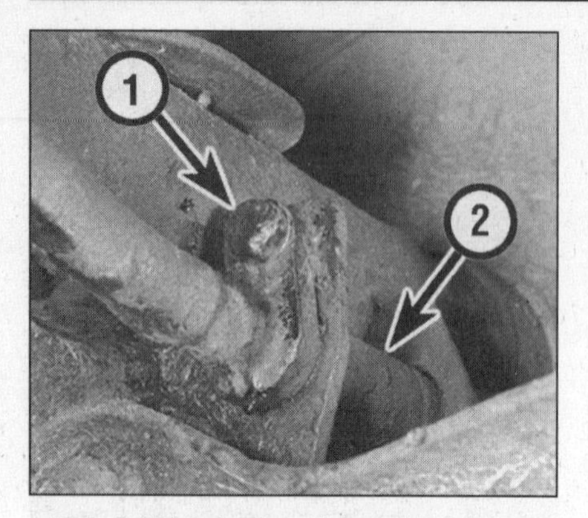

13.9 Parking brake cable guide tube on the trailing arm

1 Retaining bolt *2 Guide tube*

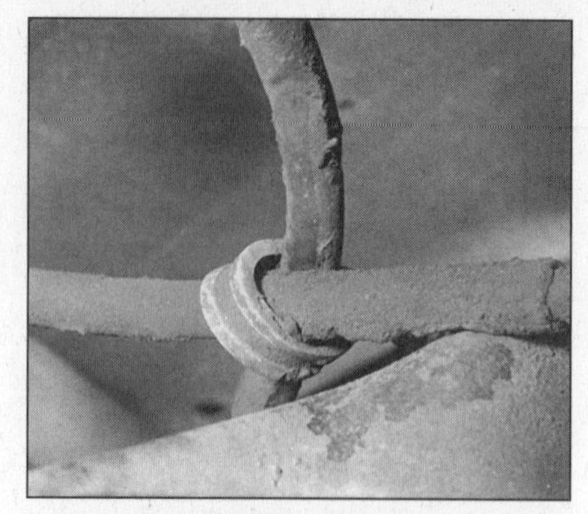

13.12 Route the cables correctly

13.15 Plastic tubes (arrows) through which the cables pass

trailing arm **(see illustration)**, and pull out the cable, plastic collar and rubber seal.

10 Pull the other end of the cable from the center support and interior of the vehicle.

11 Install the rubber seal and plastic collar in the guide tube on the trailing arm.

12 Route the cable through the center support and grommet, and through the hole in the floorpan. Install the grommet after both ends of the cable are connected. **Note:** *The cable from the left wheel should pass through the hole to the right of the driveshaft, and vice-versa. The cable from the left wheel should pass under the right cable where they pass through the center support* **(see illustration).**

13 Push the cable through the plastic guide tube on the trailing arm and reconnect it to the operating lever on the brake backplate.

14 Position the operating lever behind the backing plate, then install the brake shoes, disc/drum and caliper.

15 Pull the other end of the cable into the interior of the vehicle, through the plastic tube, and connect it to the parking brake lever **(see illustration).**

16 Install the cable clamps on the floorpan, and install the carpet and seat cushion.

17 Adjust the parking brake cable (see Section 11).

18 Install the console.

19 Install the rear wheels, remove the jackstands, lower the vehicle and tighten the wheel lug nuts to the torque listed in this Chapter's Specifications.

14 Power brake booster - description, check and replacement

Description

Refer to illustration 14.2

1 The power brake booster is bolted to the engine firewall, with the brake master cylinder bolted to it.

2 The master cylinder is connected by a pushrod to the power brake booster, which in turn is connected to the brake pedal, again by a pushrod **(see illustration).**

3 With the engine running, a constant vacuum is maintained in the power brake booster, derived from the intake manifold on carbureted models, and from a vacuum pump on some fuel-injected models.

4 This vacuum pulls a large diaphragm, which is connected to the pushrods, when the brake pedal is depressed, providing assistance to the driver. (On some vehicles, a twin-diaphragm unit is used.)

5 If the power brake booster should fail while the vehicle is in operation, the brakes will still operate, but will be much harder to apply.

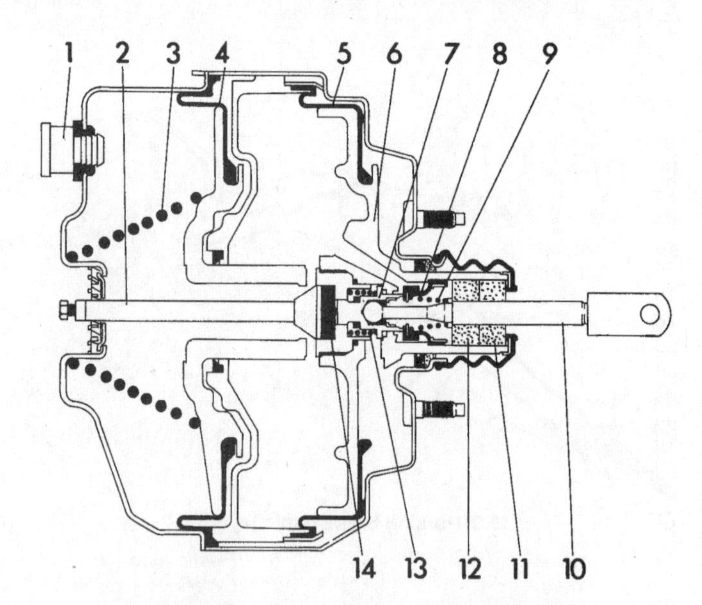

14.2 Cutaway view of the power brake booster

1	*Check valve*	8	*Seal assembly*
2	*Front pushrod*	9	*Spring*
3	*Return spring*	10	*Rear pushrod*
4	*Diaphragm, front*	11	*Filter*
5	*Diaphragm, rear*	12	*Filter*
6	*Guide housing*	13	*Spring*
7	*Valve piston seat*	14	*Reaction disc*

Check

6 Operate the brake pedal several times to release residual pressure in the system.

7 Keeping the pedal depressed, start the engine.

8 The pedal should move down slightly if the booster is correctly functioning. Keep the pedal depressed for another 15 seconds - it should not move any further.

9 If the booster fails this test, check the booster and all vacuum hoses for leaks and replace the non-return valve(s) in the system (see below). When applicable, overhaul the vacuum pump (see Section 15). If the booster still fails to function correctly, replace it.

Replacement

Refer to illustrations 14.11 and 14.13

10 Remove the brake master cylinder (see Section 6). If you're careful, you can leave the brake line fittings connected, but be careful

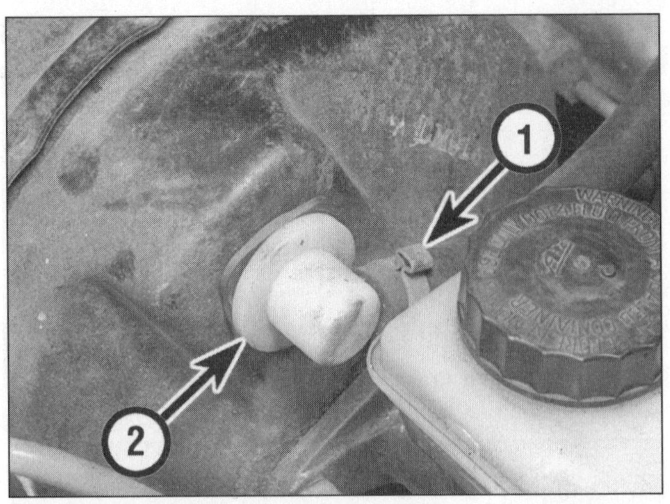

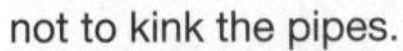
14.11 Vacuum hose connection (1) and check valve (2)

14.13 Pushrod link retaining pin (arrow)

not to kink the pipes.
11 Disconnect the vacuum hose from the check valve on the booster **(see illustration)**.
12 Remove the under-dash sound insulation, and the floor covering from behind the brake pedal.
13 Disconnect the brake pedal from the pushrod **(see illustration)**.
14 Remove the four nuts securing the booster unit to the firewall.
15 Lift the power brake booster from the engine compartment.
16 Installation is the reverse of removal with one exception: If there is no sealing ring fitted to the power brake booster-to-firewall contact area, apply a bead of RTV sealant.
17 When you're done, bleed the brake system if the master cylinder was disconnected (see Section 10).
18 Check the operation of the brakes before putting the vehicle back into service.

Check valve and seal replacement

19 Disconnect the vacuum hose **(see illustration 14.11)**.
20 Use two flat-bladed screwdrivers to pry out the check valve from the booster unit.
21 Remove and discard the check valve seal.
22 Installation is the reverse of removal. Be sure to use a new seal and grease the seal before installing it. Make sure it's not dislodged when pushing in the valve. Also, replace the band-type hose clamp with a screw-type clamp.

15 Vacuum pump - removal, overhaul and installation

Refer to illustration 15.6

Note: *This procedure applies only to some fuel-injected vehicles.*

1 Disconnect both hoses from the vacuum pump.
2 Unscrew the mounting bolts and withdraw the pump.
3 Clean off all external dirt and secure the pump in a vise.
4 Remove the valve housing cover.
5 Scribe a line across the edges of the upper and lower body flanges to facilitate reassembly
6 Unscrew the flange screws and separate the upper and lower bodies **(see illustration)**.
7 Unscrew the center bolt and remove the diaphragm, washers and spring from the lower housing.
8 Invert the pump and remove the lower cover.
9 Extract the lever shaft, lever and pump rod.
10 Replace worn components as necessary.
11 Reassembly is a reversal of disassembly. Make sure that the diaphragm center bolt is installed with its washer and O-ring. Clean the center bolt threads and coat them with thread-locking compound.

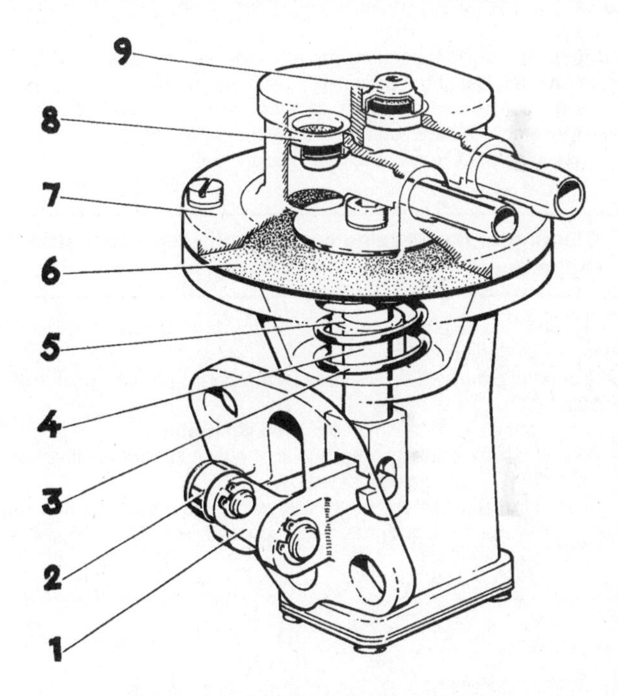

15.6 Cutaway view of the vacuum pump

1 Lever
2 Roller
3 Pump spring
4 Pump rod
5 Nylon bushing
6 Pump diaphragm
7 Valve housing
8 Suction valve
9 Discharge valve

12 Make sure that the raised side of the diaphragm is facing up and the dished sides of the washers are against the diaphragm.
13 Tighten the center bolt. Make sure that the hole in the diaphragm is opposite the one in the housing.
14 Install the pump on the engine and connect the hoses.

16 Brake light switch - check and adjustment

Refer to illustrations 16.1a and 16.1b

1 Remove the panel under the dashboard. Make sure that the brake pedal is fully released, then measure the distance between the switch and the pedal arm. The distance should be as shown **(see illus-**

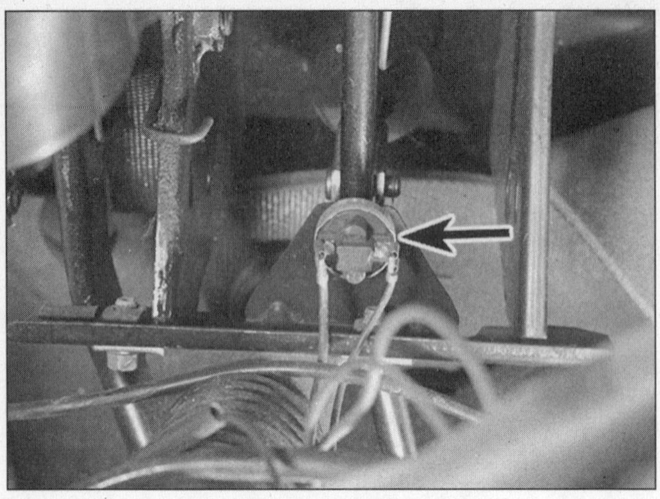

16.1a Brake light switch (arrow)

trations). Adjust by loosening the retaining screws and moving the switch bracket. Retighten the retaining screws.

2 When adjustment is correct, the stop-lights should come on when pedal movement is between about 5/16 and 9/16-inch.

3 Install the panel under the dashboard.

17 Parking brake warning light switch - replacement and adjustment

1 The parking brake warning light switch is mounted on a bracket under the parking brake.

2 To remove the switch, first remove the parking brake console (see Chapter 11).

3 Unplug the electrical connector **(see illustration)**.

4 Remove the retaining screw and lift the switch from the bracket.

5 Installation is the reverse of removal.

6 To adjust the switch, verify that, with the parking brake fully released, the warning light goes out, and comes on when the parking brake lever is pulled up.

7 Bend the switch mounting bracket as necessary to achieve this condition.

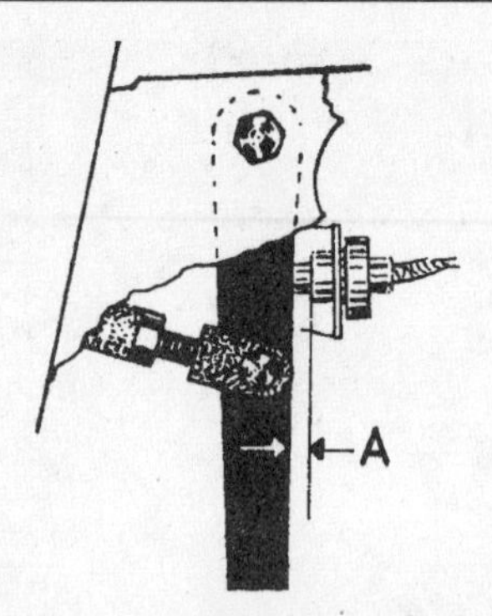

16.1b Brake light switch adjustment

A = 5/32-inch (4.0 mm)

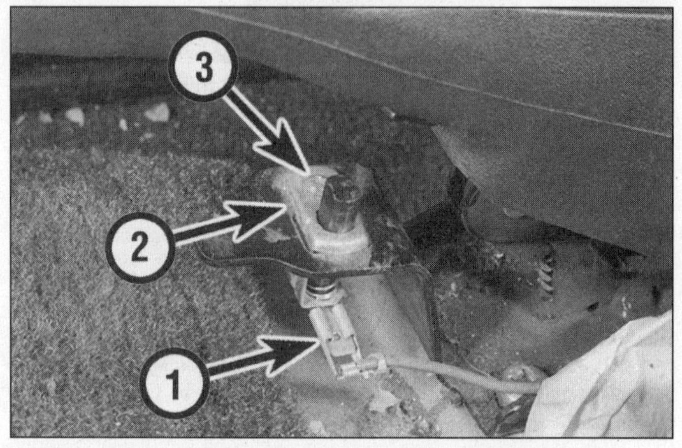

17.3 Parking brake "On" warning light switch

1 Electrical connector
2 Switch body
3 Retaining screw

Chapter 10 Steering and suspension

Contents

Specifications

General

Power steering fluid type ... See Chapter 1

Torque specifications

	Ft-lbs
Balljoint housing-to-strut	13 to 21
Control arm balljoint nut	37 to 52
Control arm front bushing nut/bolt	55
Control arm rear bracket bolts	30
Control arm rear bushing nut	41
Control arm-to-balljoint nuts	74 to 96
Coupling pinch-bolts	15
Pinion cover bolts	14
Pinion nut (ZF type)	17
Pre-tension housing cover bolts	14
Reaction rod mounting nuts/bolts	63
Shock absorber nuts/bolts	63
Spring lower mounting nut	14
Spring upper mounting to body	33
Stabilizer front nuts	33
Stabilizer rear nuts	63
Steering column lower bracket bolts	15
Steering gear U-bolt nuts	15
Steering lug nut	44
Steering shaft coupling flange bolts	17
Steering tie-rod locknuts	52
Suspension strut upper housing nuts	15
Track rod to axle	44
Track rod to body	63
Track rod-to-spindle arm balljoint nut	44
Trailing arm mounting nut/bolt	83

1.1a Details of the front suspension

1 *Stabilizer bar*
2 *Stabilizer bar bushing/clamp*
3 *Coil spring/strut assembly*
4 *Balljoint*
5 *Control arm*
6 *Rear control arm bracket*
7 *Tie-rod end*
8 *Tie-rod*
9 *Dust boot*
10 *Steering gear*

1 General information

Front suspension

Refer to illustrations 1.1a and 1.1b

The front suspension **(see illustrations)** consists of MacPherson type struts located by the vehicle body and by a pair of control arms. Each strut assembly consists of a shock, a coil spring and a strut tube. The upper mount (the big rubber bushing at the upper end of the strut assembly) is attached to the strut tower (the body) and the lower end is attached to the control arm by a balljoint. The coil spring collar (seat) and the spindle (front wheel axle) are welded to the strut tube. There is no steering knuckle. A stabilizer bar, supported in rubber bushings, connects the two control arms. The stabilizer is attached to the control arms by a pair of links.

Rear suspension

Refer to illustration 1.2

The rear suspension **(see illustration)** consists of a solid axle located by two upper arms, known as "reaction rods," two lower trailing arms and a track bar (or Panhard rod). The reaction rods provide longitudinal stability, and the track bar provides lateral stability. The trailing arms, track bar and reaction rods are attached to the body by rubber type mountings. On some vehicles, a stabilizer connects the trailing arms.

The rear axle is suspended by coil springs. Damping is handled by hydraulic or gas-filled shock absorbers, mounted independently of the springs. Self-leveling gas-filled shock absorbers are installed on some vehicles.

Steering

Refer to illustration 1.3

The steering gear **(see illustration)** is a rack-and-pinion design. Vehicles are equipped with either manual or power steering. There are two types of manual steering gears, Cam Gear and ZF. Cam Gear units

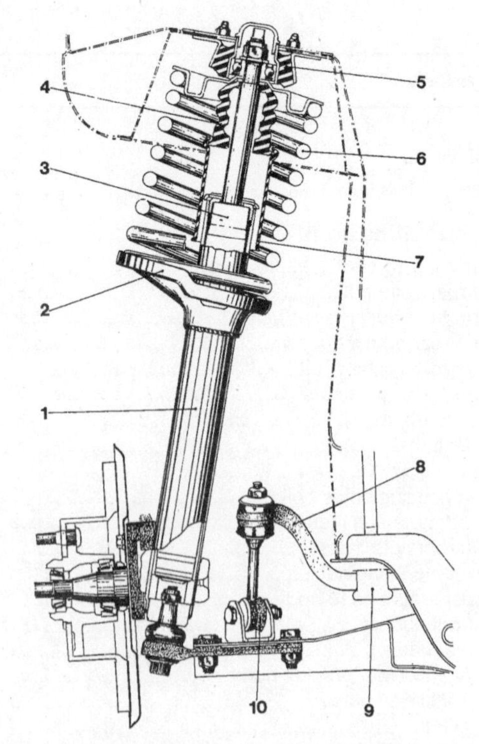

1.1b Components of the strut assembly

1 *Strut tube*
2 *Lower spring seat*
3 *Shock absorber*
4 *Rubber bumper*
5 *Upper housing*
6 *Coil spring*
7 *Shock absorber dust cover*
8 *Stabilizer bar*
9 *Stabilizer bushing clamp*
10 *Stabilizer link*

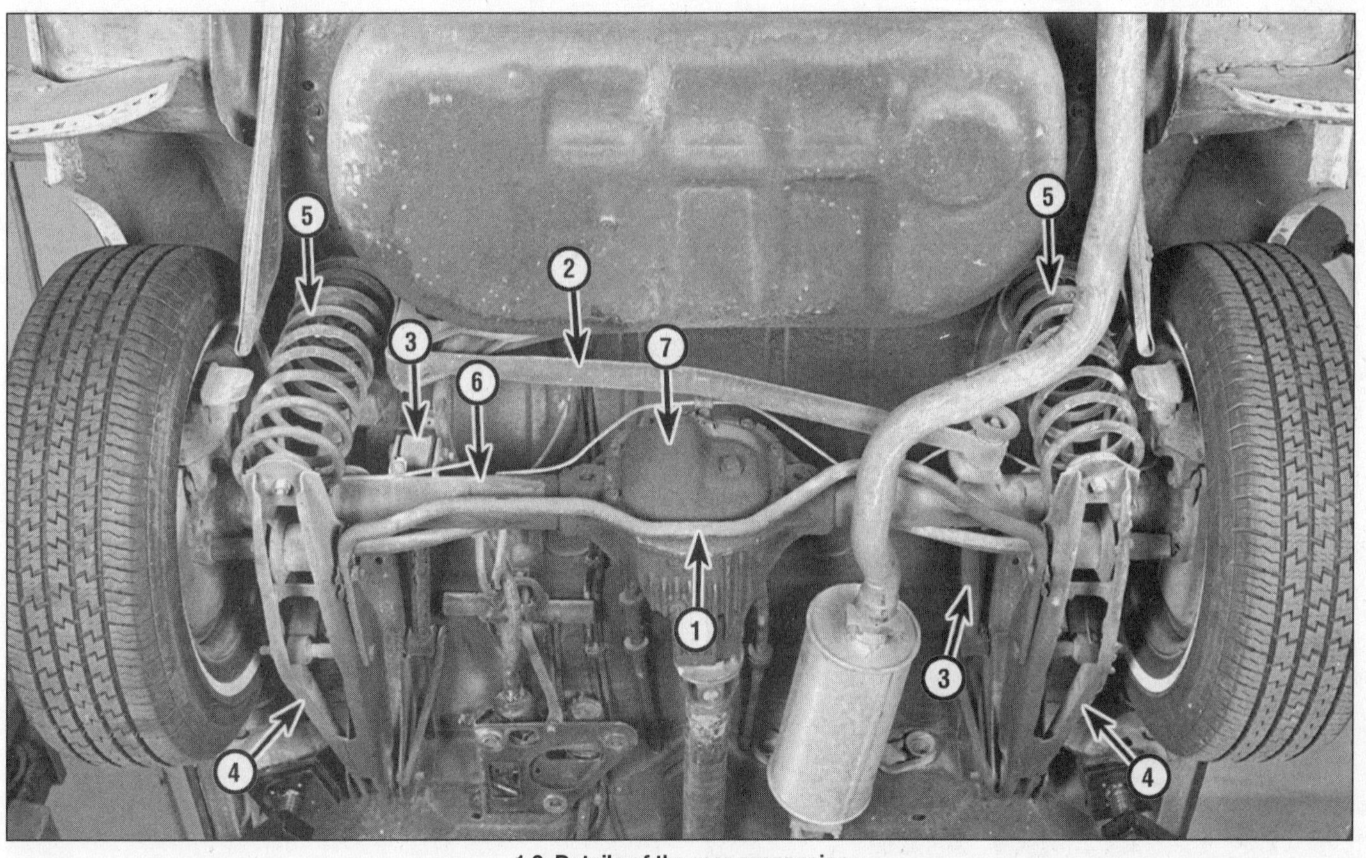

1.2 Details of the rear suspension

1 *Stabilizer bar*
2 *Track bar (Panhard rod)*
3 *Reaction rod*
4 *Trailing arm*
5 *Coil spring*
6 *Axle tube*
7 *Differential*

1.3 Typical steering gear assembly (manual shown, power similar)

1 *Steering column shaft*
2 *Steering gear*
3 *Tie-rod*
4 *Tie-rod end*
5 *Tie-rod end balljoint*
6 *Steering arm*

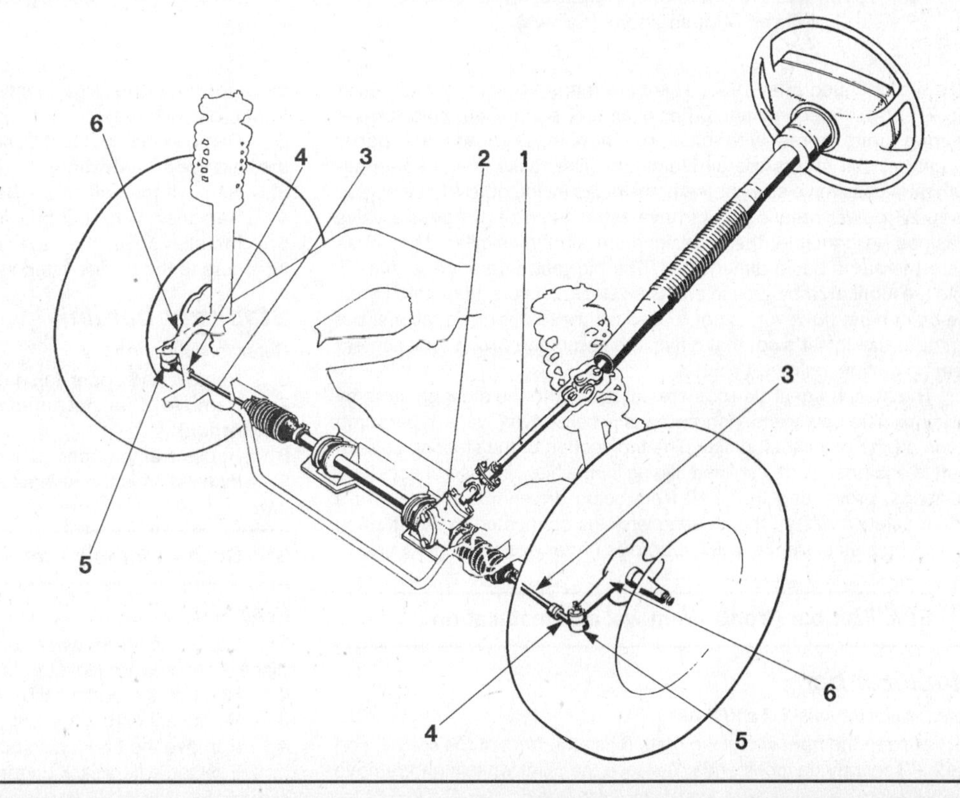

2.2 Stabilizer bar-to-link nut (arrow)

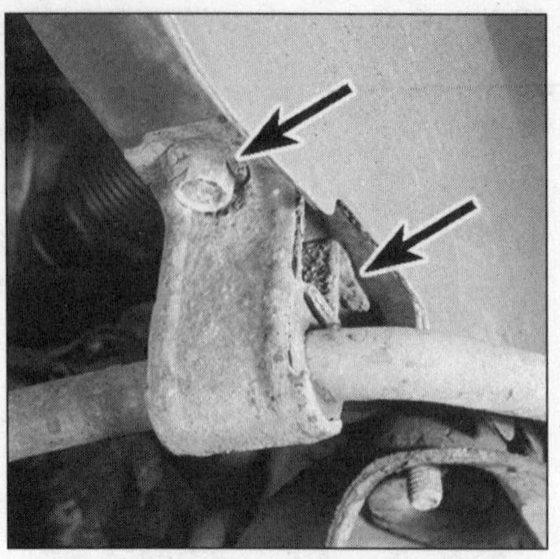
2.3 Stabilizer bar bushing clamp bolts (arrows)

2.7 Lower link bracket nut and bolt

3.4 To remove the brake backing plate, remove these three retaining bolts (arrows)

3.9 Pry out the grommet . . .

were discontinued after 1989. There are many versions of Cam Gear and ZF power steering units. For example, some Cam Gear power steering units come with cast iron housings; others are partly aluminum; still others are all-aluminum. The racks inside some all-aluminum units have straight teeth; the racks inside others have angled teeth. ZF power steering units have either fixed or removable valve housings. In principle, these designs are all fairly similar. They even share the same basic dimensions. The biggest difference is that ZF units are lubricated by grease and Cam Gear units are lubricated by oil. The point here, however, is not to discuss these design variations, but to emphasize that it's not that difficult to obtain the wrong replacement steering gear if you're not careful.

The steering gear tie-rods are connected to the steering arms by balljoints. The power steering pump is a belt-driven, vane-type pump driven off the crankshaft pulley. The top section of the steering column shaft is equipped with a splined sliding joint which, in case of accident, collapses, preventing the shaft from being driven into the car. As a further safety measure the steering wheel is connected to the steering column by a steel sleeve which crumples under excessive pressure.

2 Stabilizer bar (front) - removal and installation

Stabilizer bar

Refer to illustrations 2.2 and 2.3

1 Loosen the front wheel lug nuts. Raise the front of the vehicle and place it securely on jackstands. Remove the front wheels and remove the undertray (the large plastic cover bolted to the frame and lower front crossmember).

2 Remove the nuts that attach the outer ends of the stabilizer bar to the links **(see illustration)**.

3 Remove the bolts from the bushing clamps **(see illustration)**.

4 Replace any rubber bushings which are worn or damaged.

5 Installation is the reverse of removal. Be sure to tighten all fasteners to the torque listed in this Chapter's Specifications.

Stabilizer bar link

Refer to illustration 2.7

6 Remove the upper link nut **(see illustration 2.2)**.

7 Remove the nut and bolt from the lower bracket **(see illustration)**.

8 Replace any rubber bushings which are damaged or worn.

9 Installation is the reverse of removal.

3 Strut assembly - removal and installation

Refer to illustrations 3.4, 3.9, 3.10, 3.11, 3.13, 3.15a and 3.15b

1 Loosen the front wheel lug nuts. Raise the front of the vehicle and place it securely on jackstands. Remove the front wheels.

2 Remove the brake caliper and disc (see Chapter 9).

3 Remove the front hub and wheel bearing assembly (see Chapter 1).

4 Remove the brake disc backing plate bolts **(see illustration)**.

5 Position a floor jack under the control arm and raise the jack head

3.10 . . . and loosen the center nut

3.11 Make an alignment mark (arrow) at the dimple (arrow) in the top of the strut tower before removing the mounting nuts

3.13 You'll have to remove these balljoint-to-control arm retaining nuts (arrows) to remove the control arm, to remove the strut assembly or to replace a balljoint

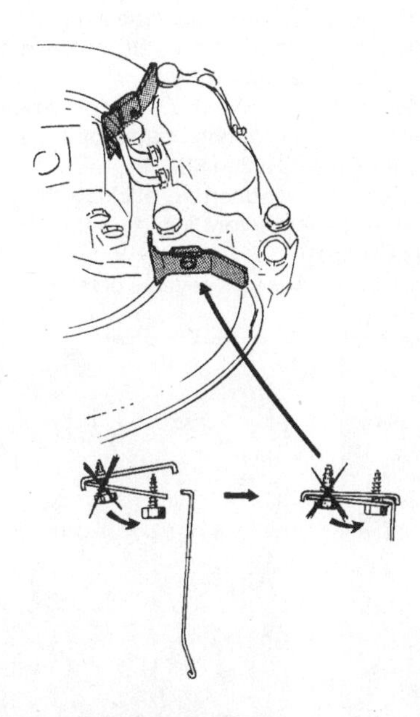

3.15a Pre-1979 vehicles use these small sealing plates between the backing plates and the calipers

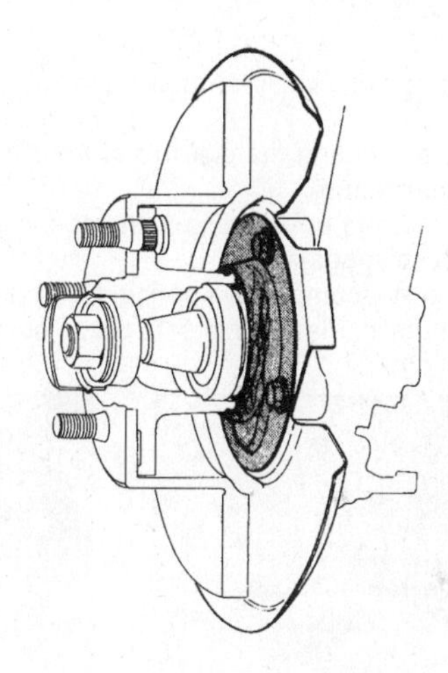

3.15b Some vehicles also have a dished collar between the backing plate and the spindle

just enough to take the weight of the strut.

6 Disconnect the tie-rod end from the steering arm (see Section 13).

7 Disconnect the stabilizer link upper nut (see Section 2).

8 Detach the brake line bracket.

9 Pry out the grommet from the upper end of the strut upper mount **(see illustration)**.

10 If you plan to replace the strut's shock absorber, coil spring or the strut body, loosen (but don't remove) the center nut while holding the piston rod stationary **(see illustration)**.

11 Make an alignment mark on the plate opposite the dimple on the strut tower **(see illustration)**, then remove the upper mounting nuts.

12 Lower the jack under the control arm and guide the strut assembly out from under the fender. It may be necessary to remove the jack completely and actually pull down on the strut to get it out.

13 Remove the three balljoint retaining nuts **(see illustration)** and separate the strut assembly from the lower control arm (see Section 5).

14 If you're removing the strut assembly in order to replace the upper mount, spring, strut cartridge or strut tube, refer to Section 4.

15 Installation is the reverse of removal. Be sure to tighten all fasteners to the torque listed in this Chapter's Specifications. **Note:** *On models built up to 1979, separate sealing plates were bolted to the backing plate* **(see illustration)**. *These sealing plates may still be available, but newer backing plates, which don't use the sealing plates, will work on older vehicles. Some vehicles also use a separate dished collar bolted to the backing plate* **(see illustration)**. *If no dished collar is installed, it's a good idea to install one.*

4 Strut shock absorber - replacement

Refer to illustrations 4.2, 4.3, 4.4a, 4.4b, 4.4c, 4.4d, 4.6, 4.7, 4.9a, 4.9b and 4.10

Warning: *Disassembling a strut or coil-over shock absorber assembly is a potentially dangerous undertaking and utmost attention must be directed to the job, or serious injury may result. Use only a high quality spring compressor and carefully follow the manufacturer's instructions furnished with the tool.*

Note: *Later models may be equipped with gas-filled shock absorbers. The procedure for replacing them is similar to that described here.*

1 Remove the strut assembly (see Section 3) and mount it in a vise. Line the vise jaws with wood or some other soft material to prevent damage to the unit. Don't tighten the vise excessively on the strut tube.

2 Install a spring compressor on the coil spring **(see illustration)**

4.2 Following the tool manufacturer's instructions, install the spring compressor on the spring and compress it sufficiently to relieve all pressure from the upper mount

4.3 Remove the center nut . . .

and compress the spring sufficiently to relieve tension on the upper mount.

3 Remove the nut, preventing the piston rod from turning with a backup wrench **(see illustration)**.

4 Remove the washer, the upper mount, the spring seat and the rubber bumper **(see illustrations)**.

5 Carefully lift off the still-compressed spring. Don't drop it, bang it into anything or slam it down. Place it in a safe place, out of the way.

6 Loosen the shock absorber nut **(see illustration)**.

7 Lift out the shock absorber and retrieve the rubber grommet from inside the strut tube **(see illustration)**.

8 If you wish to replace the strut tube on an earlier vehicle, you'll have to remove the nut inside the bottom of the tube that secures the balljoint to the tube (see Section 6).

9 Install a new shock absorber, sticking the rubber grommet in place with a little grease before inserting the shock absorber in the tube **(see illustrations)**.

10 Installation is otherwise the reverse of removal. However, note the

4.4a . . . the washer, . . .

4.4b . . . the upper mount, . . .

4.4c . . . the spring seat . . .

4.4d . . . and the rubber bumper

4.6 Remove the shock absorber nut

4.7 Remove the rubber grommet

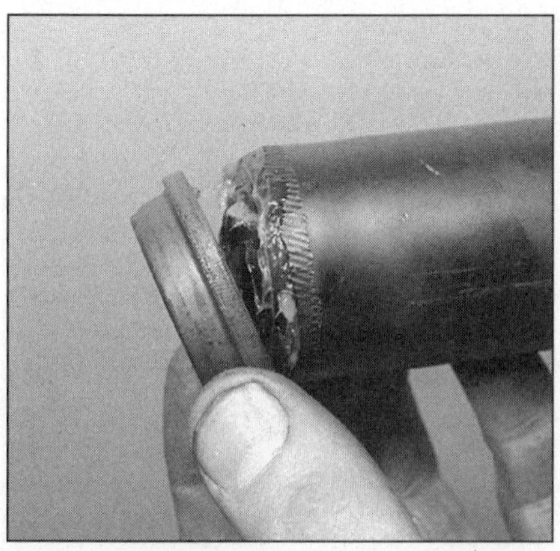
4.9a Put some grease on the end of the shock absorber so the grommet will stay in place, . . .

4.9b . . . then insert the new shock absorber into the strut tube

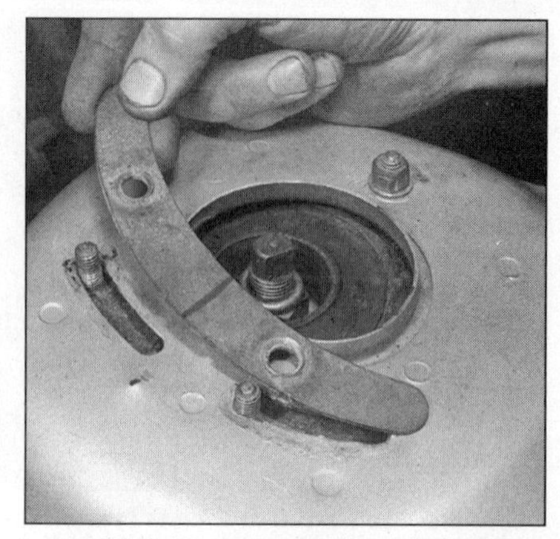
4.10 Match up the mark you made with the dimple on the strut tower

5.4 Control arm front retaining nut (arrow)

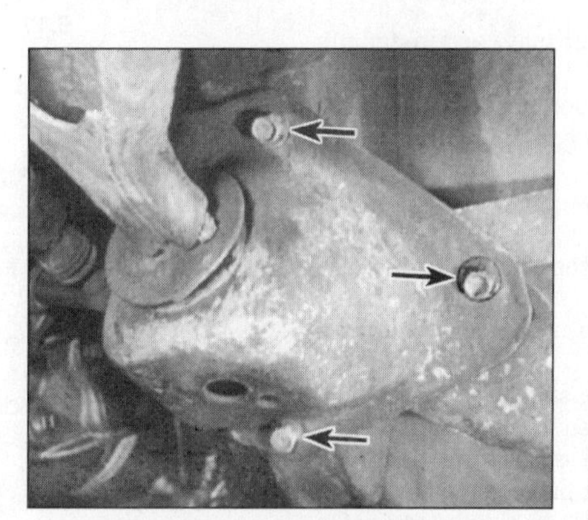
5.5a Rear control arm bracket bolts (arrows)

5.5b Bracket-to-control arm nut (arrow)

following points:

(a) *Tighten all nuts and bolts to the torque listed in this Chapter's Specifications.*

(b) *Make sure the spring is correctly positioned in the spring seat.*

(c) *Don't loosen the coil spring compressor until the center nut has been tightened.*

(d) *When installing the top plate, align your mark with the dimple on the strut tower before tightening the nuts* **(see illustration)**.

5 Control arm - removal and installation

Refer to illustrations 5.4, 5.5a, 5.5b and 5.6

Note: *You can remove the rear control arm bracket for bushing replacement without removing the control arm itself, but note the following points: The stabilizer links on* **both** *sides must be disconnected to allow the control arms to move freely; and final tightening of the control arm-to-chassis fasteners must be done with the control arm in a loaded condition (see Step 15).*

1 Loosen the front wheel lug nuts. Raise the vehicle and place it securely on jackstands. Remove the wheel.

2 Disconnect the stabilizer bar from the link (see Section 2).

3 Remove the three retaining nuts and disconnect the balljoint from the control arm **(see illustration 3.13)**.

4 Remove the control arm front retaining nut and bolt **(see illustration)**.

5 Remove the bolts securing the rear control arm bracket **(see illustration)** and lower the control arm. Remove the bracket from the control arm **(see illustration)**.

6 Press out the bushings from the control arm and the rear control arm bracket. If you don't have a press, have this done by an automotive machine shop. When installing the new bushings in the control arm and rear control arm bracket, make sure the flanged ends of the bushings face toward the front **(see illustration)**.

7 After the bushings are replaced, attach the bracket to the control

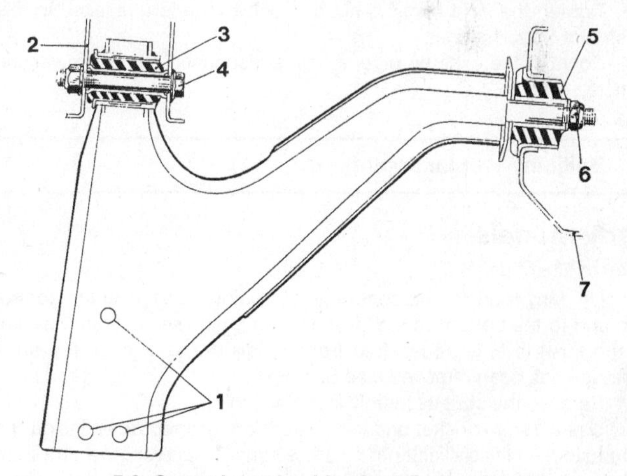

5.6 Control arm bushing installation details

1 *Holes for balljoint*
2 *Front crossmember*
3 *Front rubber bushing (flange faces to the front)*
4 *Bolt*
5 *Rear rubber bushing (flange faces to the front)*
6 *Nut*
7 *Rear control arm bracket*

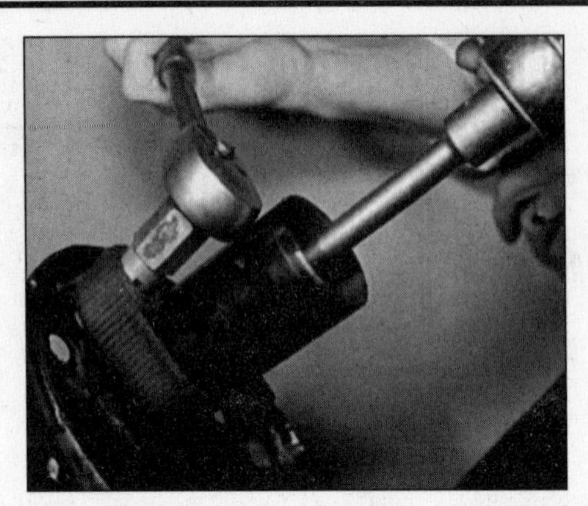

6.3 Use a strap wrench to prevent the strut tube from turning while removing the nut

arm, but don't tighten the nut yet.
8 Place the rear end of the control arm in position and install the three bracket bolts, but don't tighten them yet.
9 Install the front retaining nut and bolt, but don't tighten it yet.
10 Attach the balljoint to the control arm and tighten the nuts to the torque listed in this Chapter's Specifications.
11 Position a jack under the outer end of the control arm, raise the jack and compress the spring.
12 Connect the stabilizer bar to the link rod (see Section 2). Remove the jack.
13 Tighten the rear bracket bolts to the torque listed in this Chapter's Specifications.
14 Install the wheels, lower the vehicle and push it a few yards back and forth, at the same time bouncing the front end.
15 Tighten the rear bushing nut to the torque listed in this Chapter's Specifications. The weight of the vehicle must be on its wheels, or a jack raised under the control arm (see Step 13) to simulate this. On vehicles with a B21 engine, disconnect the exhaust downpipe for access on the right-hand side.
16 Tighten the front bushing nut and bolt to the torque listed in this Chapter's Specifications.
17 Connect the exhaust pipe, if it was disconnected, and lower the vehicle.

6 Balljoint - replacement

Early models

Refer to illustration 6.3

1 On early models, the control arm balljoint is bolted to the control arm and to the bottom of the strut; the nut which secures the balljoint to the strut is only accessible from inside the strut after the strut cartridge has been removed (see Section 4).
2 Remove the strut assembly (see Section 3).
3 Use a 19mm socket and long extension inserted down inside the strut to loosen the balljoint nut by a few turns. To prevent the strut from turning while loosening the nut, use a strap wrench around the strut, installed close to the weld mark at the top **(see illustration)**.
4 With the nut loosened, use a long brass drift to tap the end of the balljoint stud to free the socket from the ball.
5 Apply some grease to the inside of the socket and remove the nut. (The grease will hold the nut in the socket when it comes off the threads.)
6 Remove the balljoint-to-control arm nuts and remove the balljoint.
7 Installation is the reverse of removal. Note the following points:

(a) *Before installing the balljoint, clean all grease from the balljoint seat, or the indicated torque value will be inaccurate.*
(b) *Tighten all nuts and bolts to the torque listed in this Chapter's Specifications.*

6.11 Balljoint-to-strut bolts (arrows)

Later models

Refer to illustration 6.11

Note: *On 1979 and later models with power steering, the balljoints for left and right sides are different and not interchangeable.*

8 On later models, the balljoint is connected to the strut by a housing bolted to the bottom of the strut. It's accessible from the outside, so there's no need to remove the shock absorber.
9 Loosen the front wheel lug nuts. Raise the vehicle and support it securely on jackstands. Remove the front wheel.
10 On pre-1978 vehicles, loosen the strut retaining nut a couple of turns.
11 Remove the four balljoint-to-strut bolts **(see illustration)**. On some models, these bolts may have lock-washers which must be bent down.
12 Remove the balljoint-to-control arm nuts **(see illustration 3.13)**.
13 Ease the balljoint housing from the bottom of the strut.
14 Place the housing in a vise and remove the balljoint nut, then use a drift to drive the balljoint from the housing.
15 Clean all grease from the housing and new balljoint, then attach the new balljoint to the housing and tighten the nut to the torque listed in this Chapter's Specifications.
16 The remainder of installation is the reverse of removal. Be sure to use new balljoint housing bolts (and lock-washers, if applicable). Tighten all nuts and bolts to the torque listed in this Chapter's Specifications.

7 Stabilizer bar (rear) - removal and installation

Refer to illustration 7.3

1 A rear stabilizer bar is bolted between the two trailing arms on some models.
2 To remove the stabilizer, raise the rear of the vehicle and support it securely on jackstands. Support the rear axle with a floor jack to unload the shock absorber mountings.
3 Remove the shock absorber lower mounting bolts and the other nuts and bolts connecting the stabilizer bar to the trailing arm on both sides of the vehicle **(see illustration)**.
4 Remove the stabilizer bar.
5 Installation is the reverse of removal. After lowering the vehicle back onto its wheels and bouncing the rear end a few times, tighten all fasteners to the torque listed in this Chapter's Specifications

8 Track bar - removal and installation

Refer to illustrations 8.2a, 8.2b and 8.4

1 Raise the rear of the vehicle and support it securely on jackstands.

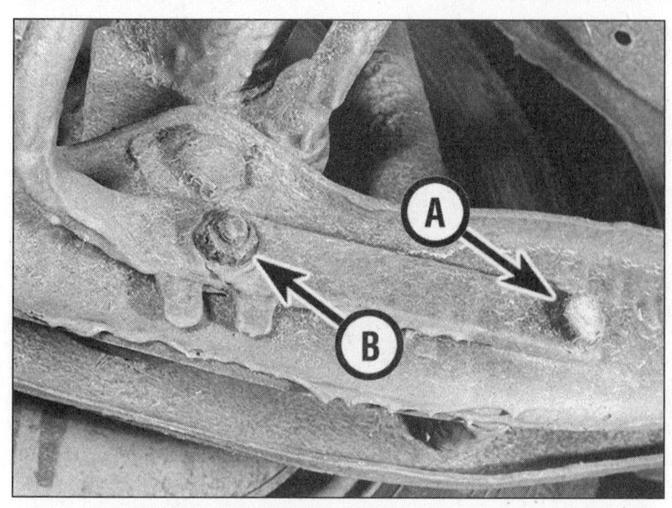

7.3 Stabilizer bar installation details - (A) shock absorber lower bolt and (B) stabilizer bar nut

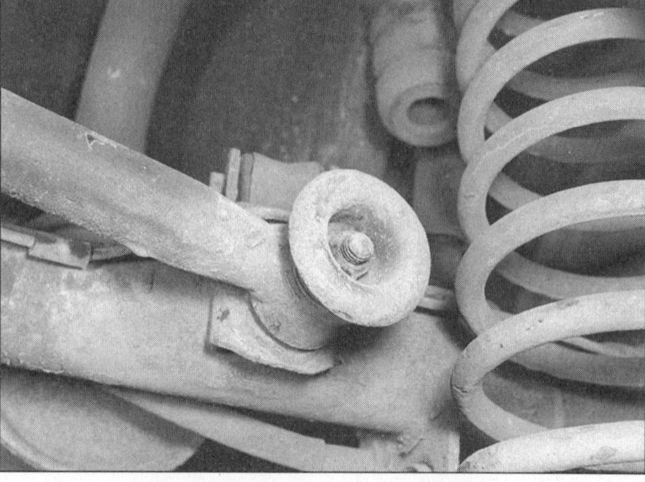

8.2a Right track bar mounting nut

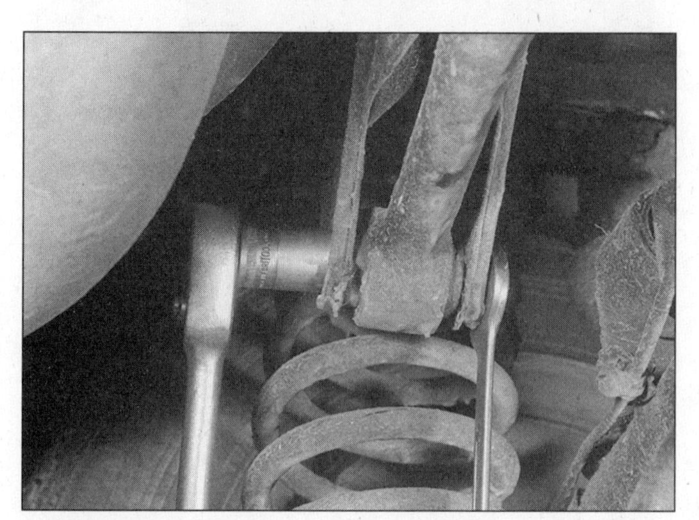

8.2b Removing the bolt from the left track bar mounting nut

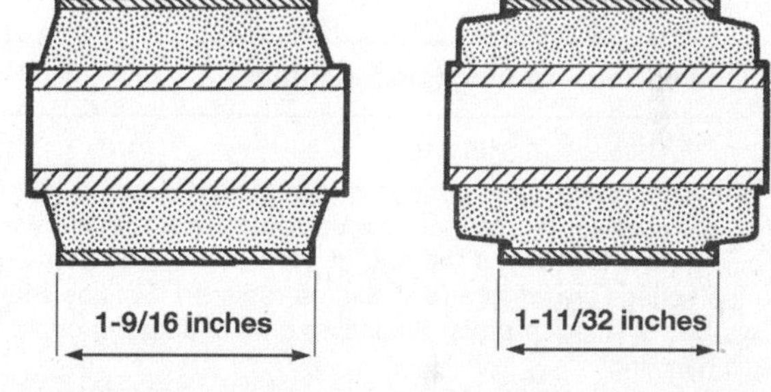

8.4 Track bar bushings come in two different diameters, but they're functionally the same

2 Remove the track bar mounting nuts **(see illustrations)**, then tap out the bolts with a soft drift.

3 Press out the rubber bushings with a mandrel of a diameter just smaller than the bushings. Or, if you don't have the proper tools, have this done by an automotive machine shop.

4 Press in new bushings with a little liquid soap as a lubricant. Bushing dimensions vary **(see illustration)**, but this has no effect on their performance.

5 Install the track bar, but do not tighten any bolts yet.

6 Lower the vehicle to the ground, then tighten both fasteners to the torque listed in this Chapter's Specifications.

9 Shock absorber - removal and installation

Refer to illustrations 9.3a and 9.3b

Note: *The procedure for replacing self-leveling gas-filled shock absorbers is similar to the procedure described below.*

1 The shock absorber can be functionally checked using the 'bounce' method. Push down hard on the corner of the vehicle nearest to the shock absorber to be tested and release it immediately. Generally speaking, the vehicle should bounce once and return to rest. If it keeps bouncing, the shock absorbers should be replaced. Shocks should always be replaced in pairs. Replacing only one shock could affect vehicle handling.

2 Loosen the rear wheel lug nuts, raise the rear of the vehicle and place it securely on jackstands. Remove the rear wheels. Position a floor jack under the differential and raise the jack just enough to take the weight off the axle.

3 Remove the shock absorber top and bottom mounting nuts/bolts **(see illustrations)**, and remove the shock absorber.

9.3a Shock absorber lower mounting details

1 *Trailing arm*
2 *Shock absorber*
3 *Bolt*
4 *Spacer*

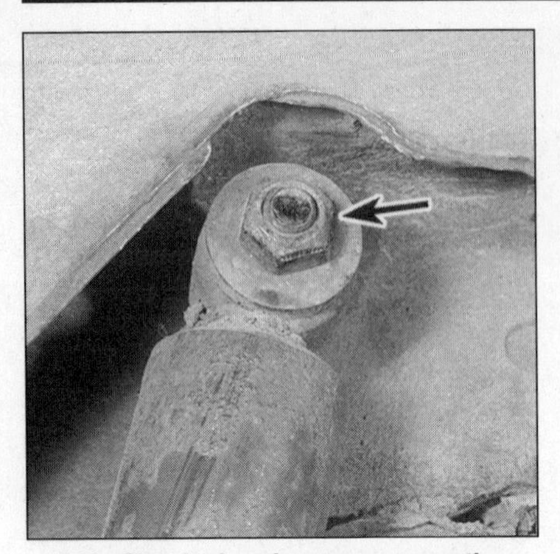
9.3b Shock absorber upper mounting nut (arrow)

10.3 Coil spring lower mounting nut (arrow)

10.5 Make sure the coil spring is fully seated in the upper mount

4 Installation is the reverse of removal. Note that the spacer is installed on the inner side of the shock absorber. Tighten the mounting nuts and bolts to the torque listed in this Chapter's Specifications.

10 Coil spring - removal and installation

Refer to illustrations 10.3 and 10.5

1 The best way to test springs is to remove them and have them tested in a spring rate tester at your Volvo dealer or a Volvo specialist shop. As a general rule, if the rear of the vehicle sags permanently, then the springs are weak and should be replaced. Springs should be always be replaced in pairs. Replacing only one spring could affect vehicle handling.

2 Loosen the rear wheel lug nuts, raise the rear of the vehicle and support it securely on jackstands placed under the frame. Remove the rear wheels. Position a floor jack under the differential and raise the jack just enough to take the weight off the axle.

3 Remove the nut from the spring lower mount **(see illustration)**. Also disconnect the shock absorber lower mount (see Section 9) and disconnect the stabilizer bar, if equipped (see Section 7).

4 Lower the floor jack until all spring tension is released, then lift the spring off its lower mount and guide it out from under the vehicle.

5 Guide the spring into location in the upper mount **(see illustration)** while raising the floor jack. Installation is otherwise the reverse of removal. Tighten all nuts and bolts to the torque listed in this Chapter's Specifications.

11 Suspension arms - removal and installation

1 Loosen the wheel lug nuts, raise the rear of the vehicle and support it securely on jackstands. Remove the rear wheels.

Reaction rods

Refer to illustration 11.2

Note: *Reaction rods come in three different lengths. If the reaction rod is being replaced, make sure you obtain a rod of the same length.*

2 Remove the nuts from the reaction rod attachment points **(see illustration)**, then use a soft drift to tap out the bolts, and remove the rod.

3 The rubber bushings can be removed from the rod with a mandrel with a diameter slightly smaller than the bushing, and installed in the same manner.

4 Use a little liquid soap as a lubricant, and make sure the flat sides of the rubber bushing are parallel to the rod before pressing the bushing home.

5 Installation is the reverse of removal. Place the weight of the vehicle back on the ground, then tighten the fasteners to the torque listed in this Chapter's Specifications.

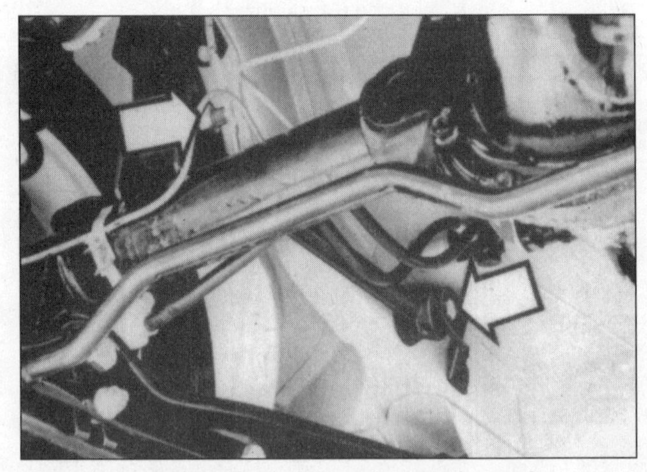
11.2 Reaction rod mounting nuts (arrows)

Trailing arms

Refer to illustrations 11.11, 11.12 and 11.13

6 Loosen the rear wheel lug nuts, raise the rear of the vehicle and place it securely on jackstands. Remove the wheels. Position a floor jack under the differential and raise the jack just high enough to take the weight off the axle.

7 Remove the bolts from the stabilizer, if equipped (see Section 7).

8 Remove the shock absorber lower bolt (see Section 9).

9 Remove the nut from the lower spring seat (see Section 10).

10 Lower the rear axle on the floor jack and remove the spring (see Section 10).

11 Remove the nut and bolt from the trailing arm-to-rear axle bracket **(see illustration)** and lower the rear end of the trailing arm.

12 Remove the nut and bolt from the trailing arm front mounting bracket **(see illustration)** and lift the arm away.

13 If the front bushings are cracked or damaged, replace them. If you have a hydraulic press and suitable mandrels, press out the bushings and install new ones. If you don't have a press, take the arm to an automotive machine shop and have the bushing replaced there. **Note:** *An improved bushing* **(see illustration)** *is installed on later models; if an early type bushing has to be replaced, the bushings on both sides must be replaced with the later bushing (The early type is no longer available).* Use a little liquid soap as a lubricant when installing the new bushings.

14 The rear bushings remain in the brackets on the rear axle. Again, if they're cracked or damaged, they must be replaced. However, they may be difficult to remove and install without special tools, so we recommend having this job done by an automotive machine shop.

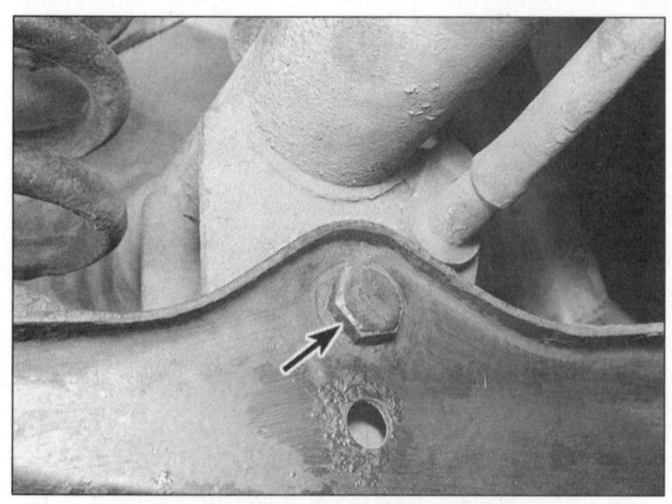

11.11 Trailing arm-to-rear axle bolt (arrow)

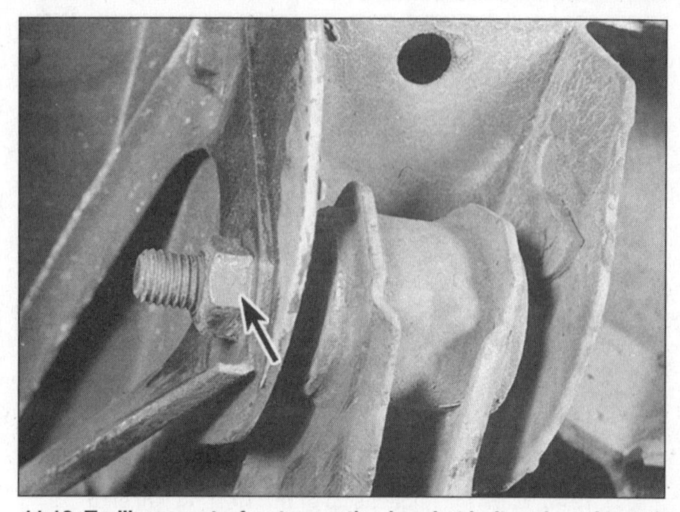

11.12 Trailing arm-to-front mounting bracket bolt and nut (arrow)

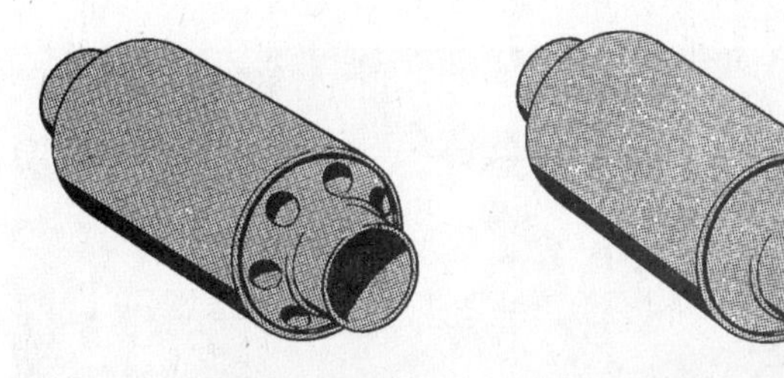

11.13 The old type trailing arm bushings (left) are no longer available; if you have to replace either bushing, and it's the old type, replace it with two of the new ones (right)

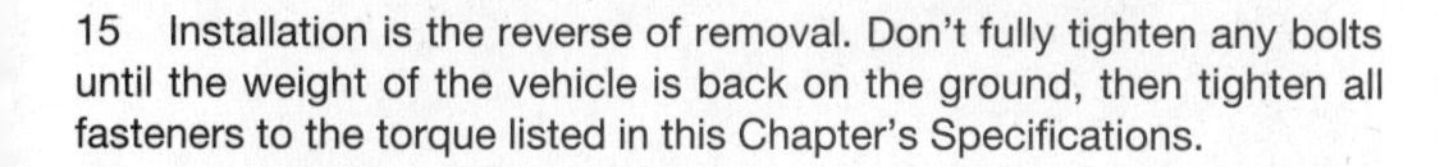

15 Installation is the reverse of removal. Don't fully tighten any bolts until the weight of the vehicle is back on the ground, then tighten all fasteners to the torque listed in this Chapter's Specifications.

12 Steering wheel - removal and installation

Refer to illustrations 12.2. 12.3, 12.4a, 12.4b and 12.5

Warning: *This procedure is for vehicles without SRS (airbag). If your vehicle is equipped with an airbag we recommend any work involving steering wheel removal be performed by a dealer service department.*

1 Disconnect the cable from the negative battery terminal. **Caution:** *Make sure the radio is turned off before disconnecting the cable to avoid damaging the microprocessor in the radio.*

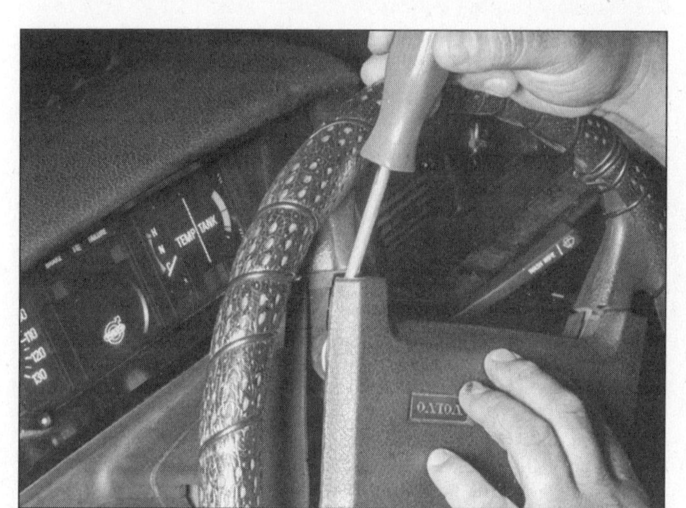

12.2 Pry off the horn pad

2 Pry out the center impact pad (early models) **(see illustration)** or center plate (later models without SRS).

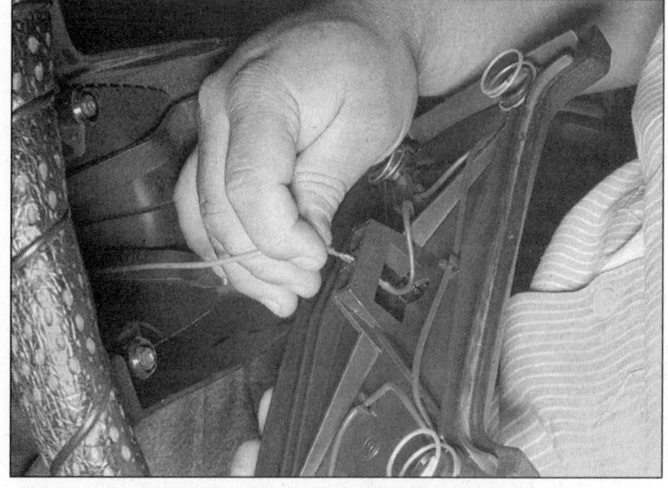

12.3 Disconnect the horn lead

12.4a Remove the steering wheel nut

12.4b Mark the relationship of the steering wheel to the steering shaft

12.5 Remove the steering wheel on pre-1979 models with a steering wheel puller (a puller isn't necessary on 1980 and later models)

13.3 Loosen the tie-rod end locknut - hold the tie-rod with another wrench on the "flats" of the tie-rod

13.4 Mark the relationship of the tie-rod end to the tie-rod threads

3 Unplug the horn wire **(see illustration)**.

4 Remove the steering wheel nut **(see illustration)**. Mark the relationship of the steering wheel to the steering wheel shaft **(see illustration)**. On post-1979 models, you'll need a 27 mm deep socket.

5 On pre-1979 models, use a puller to ease the steering wheel from its taper seat **(see illustration)**. Don't try to use a hammer to knock the wheel off - you might damage the upper crumple shaft. On post-1979 models, you can pull off the steering wheel with both hands under the wheel - but be careful, or you could pull the wheel off into your face!

6 On some models there's a spring and spring seat under the steering wheel nut - don't lose them.

7 Installation is the reverse of removal. Be sure to align the match marks and tighten the steering wheel nut to the torque listed in this Chapter's Specifications.

13 Tie-rod end - removal and installation

Refer to illustrations 13.3, 13.4 and 13.5

1 The tie-rod end balljoints require no routine lubrication. When wear occurs, the balljoints on the tie-rod ends must be replaced.

2 Loosen the front wheel lug nuts, raise the vehicle and place it securely on jackstands. Remove the wheels.

3 Using a wrench on the flats provided to prevent the tie-rod from rotating, break loose the locknut at the tie-rod end **(see illustration)**.

4 Mark the position of the tie-rod end on the tie-rod **(see illustration)**.

5 Loosen the balljoint stud nut and separate the ball stud from the steering arm with a balljoint separator or two-jaw puller **(see illustration)**.

6 Unscrew the tie-rod end from the tie-rod.

7 Screw the new tie-rod end onto the tie-rod up to the mark you made, then insert the balljoint stud into the steering arm. Tighten the balljoint nut to the torque listed in this Chapter's Specifications. Tighten the locknut securely.

8 Install the wheel, remove the jackstands, lower the vehicle and tighten the wheel lug nuts to the torque listed in the Chapter 1 Specifications.

9 Have the wheel alignment checked and, if necessary, adjusted.

14 Steering gear boots - replacement

Refer to illustrations 14.3a, 14.3b and 14.7

1 If the steering rack bellows are torn or damaged, they should be replaced as soon as possible.

2 Remove the tie-rod end (see Section 13).

3 Loosen the clamps securing the boot to the steering gear and tie-rod **(see illustrations)**. On early models, be prepared for oil spillage. On later models, the clamps are plastic and must be cut off.

13.5 Use a balljoint separator to separate the tie-rod end ball stud from the steering arm

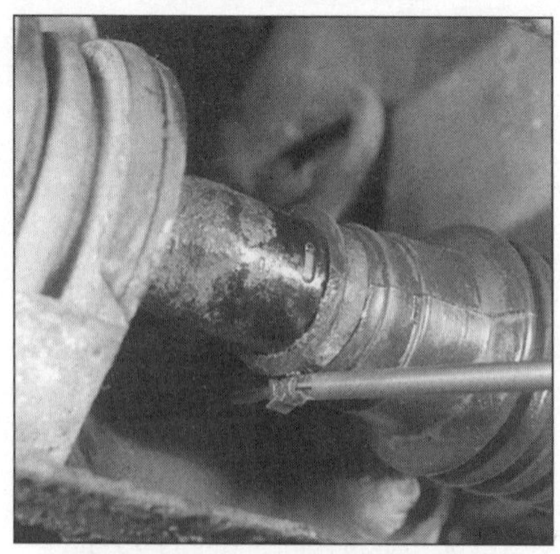

14.3a Loosen and remove the clamp which attaches the dust boot to the steering gear . . .

14.3b . . . then loosen and remove the clamp which attaches the dust boot to the tie-rod

14.7 Before you slide a new boot onto the steering gear, be sure to coat the inner and outer lips with grease to protect them

15.1 Mark the relationship of the U-joint flange to the pinion shaft, then remove the pinch bolt (arrow)

4 Clean off all oil or grease from the steering gear, and inspect the rack and inner balljoint for damage or wear.

5 If the steering rack uses grease as a lubricant (ZF units), smear the rack and inner balljoint with the grease recommended in this Chapter's Specifications. **Caution:** *The amount of grease given in the Specifications is for a new, 'dry' rack, so modify the amount of grease to prevent overfilling, which could split the dust boot under pressure, when the steering is on full lock.*

15.6a Left steering gear U-bolt clamp (arrow) . . .

6 If the steering rack is lubricated with oil (Cam Gear units), coat the rack and inner balljoint with oil, and squirt a little more into the dust boot after installing it, before tightening the outer clamp. Bear in mind the Caution in Step 5.

7 Lubricate the necks of the dust boot before sliding it over the tie-rod and onto the steering gear **(see illustration)**.

8 Install the inner clamp on the steering gear housing, then center the steering wheel and allow the boot to assume its normal, relaxed position.

9 Refer to Steps 5 or 6, then install and tighten the outer clamp.

10 Reconnect the tie-rod end to the steering arm (see Section 13).

11 Have the wheel alignment checked and, if necessary, adjusted.

15 Steering gear - removal and installation

10

Refer to illustrations 15.1, 15.6a, 15.6b and 15.6c

1 On later models, the joint is protected by a plastic sleeve. Push the sleeve up for access to the U-joint flange. Mark the relationship of the U-joint flange to the pinion shaft, then remove the clamp bolt **(see illustration)**. Pry the flange open slightly with a screwdriver. **Warning:** *On models equipped with SRS (airbag), disconnect the cable from the negative battery terminal, lock the steering wheel in the straight ahead position and DO NOT rotate the steering wheel with the steering column-to-steering gear coupling disconnected.*

2 Loosen the front wheel lug nuts, raise the front end of the vehicle, place it securely on jackstands and remove the wheels.

3 Disconnect the tie-rod ends from the steering arms (see Sec-

15.6b . . . and right U-bolt clamp (arrow)

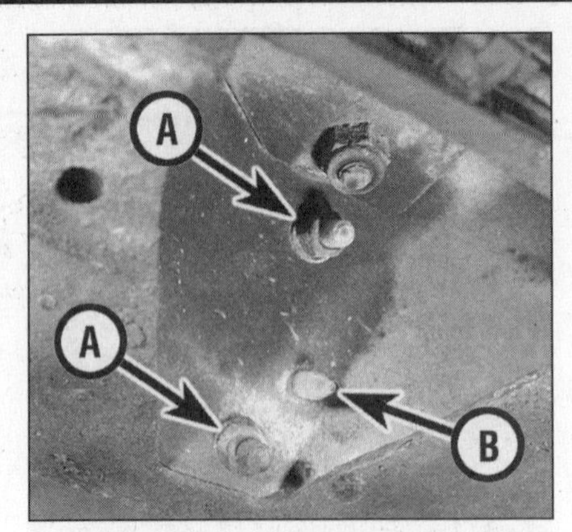

15.6c U-bolt retaining nuts (A) and locating pin (B) (not all units have a locating pin)

16.9 A typical pre-1985 Saginaw power steering pump assembly

tion 13).

4 Remove the splash guard from under the car.

5 On models equipped with power steering, disconnect and plug the hydraulic hoses.

6 Remove the two 'U' bolts securing the steering gear to the front axle member **(see illustrations)**.

7 Disconnect the steering gear from the steering shaft U-joint flange and remove the steering gear, complete with tie-rods.

8 Installation is the reverse of removal. **Note:** *The mounting pads on some steering gears have locating pins, such as (B) in illustration 15.6c. Don't try to install a steering gear with a locating pin on the mounting pads onto a vehicle without a locating hole - the pin will be cut off.* Be sure to tighten all fasteners to the torque listed in this Chapter's Specifications. Check the power steering fluid level and add some, if necessary (see Chapter 1).

9 Have the wheel alignment checked when you're done.

16 Power steering pump - removal and installation

ZF-type (some early models)

1 This pump has a separate fluid reservoir.

2 Either clamp the hoses from the reservoir to prevent fluid loss before disconnecting the hoses, or place a container beneath the pump, then disconnect the hoses and allow the fluid to drain.

3 Remove the nuts from the two long bolts through the mounting bracket.

4 Remove the tensioner locking screws on both sides of the pump, push the pump in and lift off the drivebelt.

5 Swing the pump up and remove the three bolts securing the mounting bracket to the engine block, then remove the pump and bracket.

6 If a new pump is being installed, transfer the mounting bracket to the new pump.

7 Installation is the reverse of removal.

8 When you're done, adjust the drivebelt tension (see Chapter 1), and fill and bleed the system (see Section 17).

Saginaw type (pre-1985 models)

Refer to illustrations 16.9 and 16.12

9 The pre-1985 Saginaw power steering pump **(see illustration)** has an integral reservoir.

10 Place a container beneath the pump, then disconnect the hoses and allow the fluid to drain.

11 Loosen the adjustment bolt (it's below the pulley) and, if necessary, loosen the pivot bolt, then push the pump back and remove the drivebelt from the pulley.

12 Remove the pump bracket bolts **(see illustration)** and remove the pump and bracket as a single assembly.

13 Take the pump/bracket assembly to a work bench and remove the pulley. You may need a pulley removal tool to get the pulley off the hub.

14 Unbolt the pump from the bracket.

15 Installation is the reverse of removal.

Saginaw type (1985 and later models)

Refer to illustrations 16.16 and 16.23

16 The 1985 and later Saginaw **(see illustration)** is a lightweight pump, with a separate reservoir mounted on the left inner fender, in front of the suspension strut tower.

17 Removal and installation procedures are again similar to earlier types, but the belt tensioner is the turnbuckle type.

18 If a replacement pump is to be installed, the pulley and the mounting bracket have to be transferred to the new pump as follows.

19 Remove the center bolt from the pulley.

20 Screw a 3/8-inch x 2-3/4 inch UNC bolt into the empty bolt hole, and use a puller against the bolt head, with the puller arms hooked under the pulley center boss, to pull the pulley from the pump shaft.

21 Transfer the mounting bracket to the new pump.

22 Apply a few drops of oil to the pump shaft and install the pulley to the shaft.

23 The pulley now has to be pressed onto the shaft, which can be done by using the same bolt as in removal, with a suitable socket or piece of tubing arranged as shown **(see illustration)**.

24 Press the pulley on until its outer face is flush with the shaft end.

25 Make sure the bolt does not contact the inner face of the shaft during this operation by adding more washers as necessary.

26 Remove the press arrangement and install the pulley center bolt.

27 Installation is the reverse of removal. Adjust the drivebelt tension (see Chapter 1). Fill and bleed the system (see Section 17).

17 Power steering system - bleeding

1 Following any operation in which the power steering fluid lines have been disconnected, the power steering system must be bled to remove all air and obtain proper steering performance.

2 With the front wheels in the straight ahead position, check the power steering fluid level and, if low, add fluid until it reaches the Cold mark on the dipstick.

3 Start the engine and allow it to run at fast idle. Recheck the fluid level and add more if necessary to reach the Cold mark on the dipstick.

4 Bleed the system by turning the wheels from side-to-side, without hitting the stops. This will work the air out of the system. Keep the reservoir full of fluid as this is done.

5 When the air is worked out of the system, return the wheels to the straight ahead position and leave the vehicle running for several more

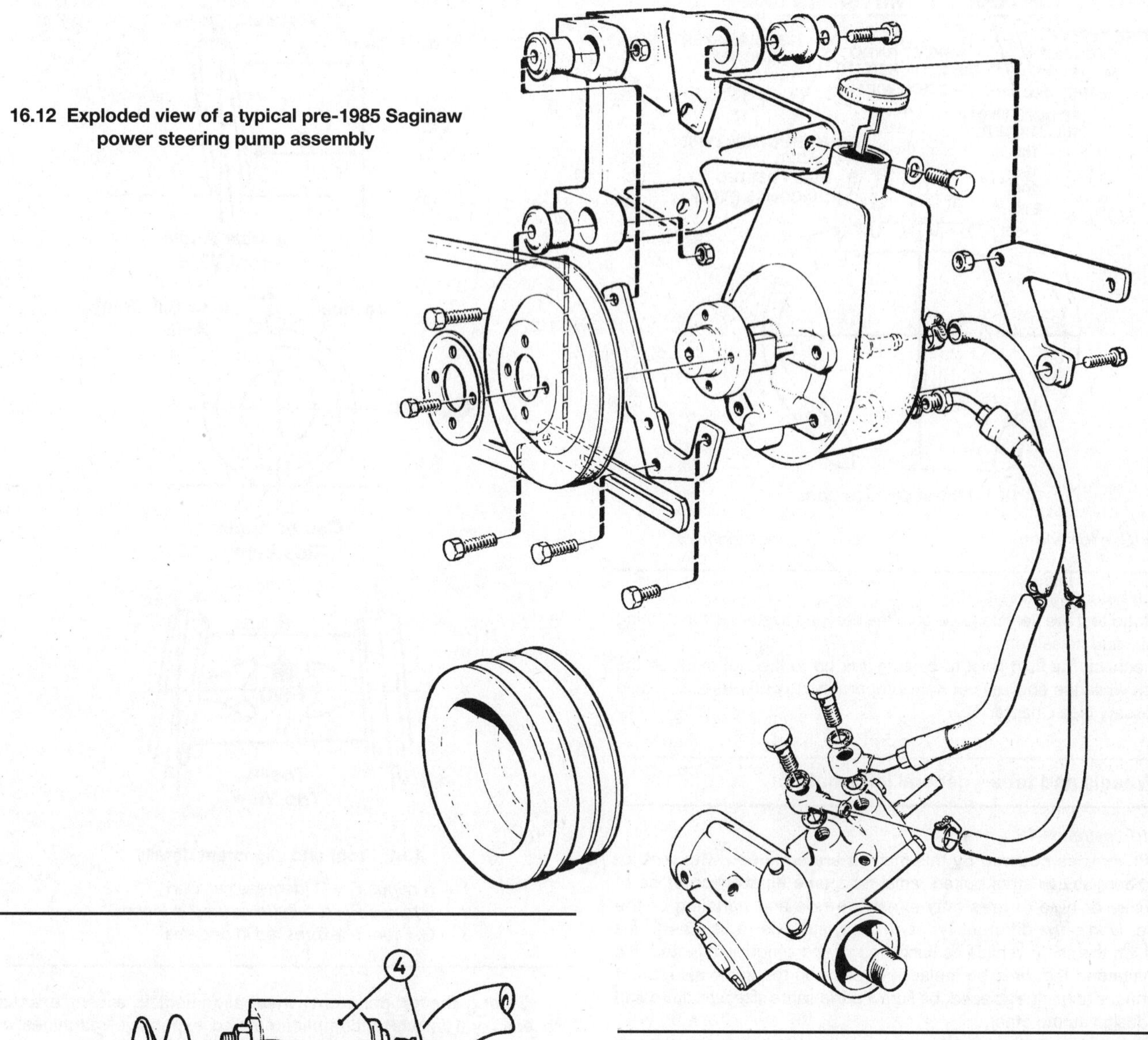

16.12 Exploded view of a typical pre-1985 Saginaw power steering pump assembly

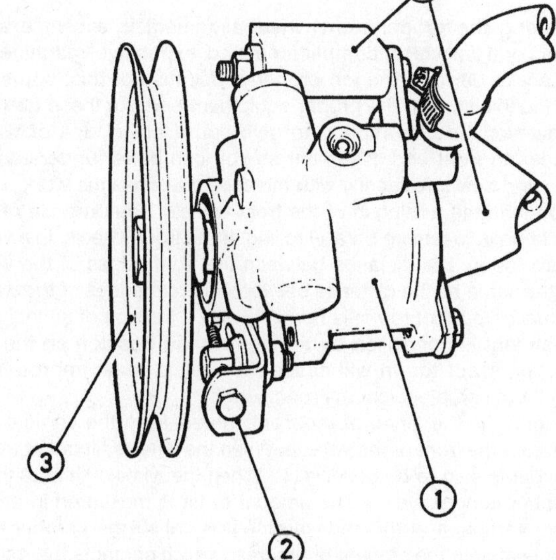

16.16 Saginaw power steering pump (1985 and later models)

1 Pump
2 Belt adjuster
3 Pulley
4 Mounting bracket

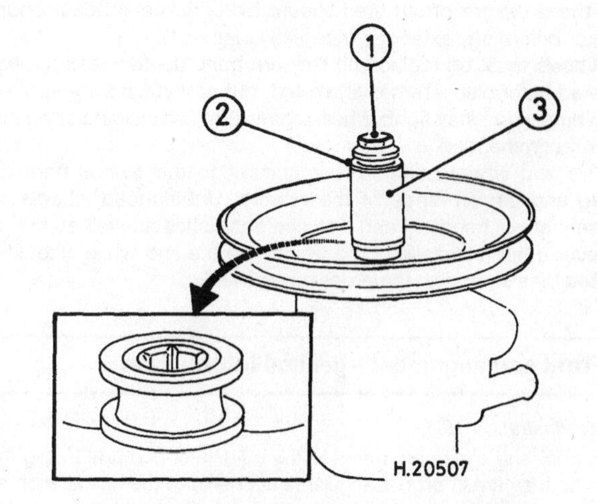

16.23 Pressing a new pulley onto the shaft

1 Bolt
2 Washers
3 Socket or tube

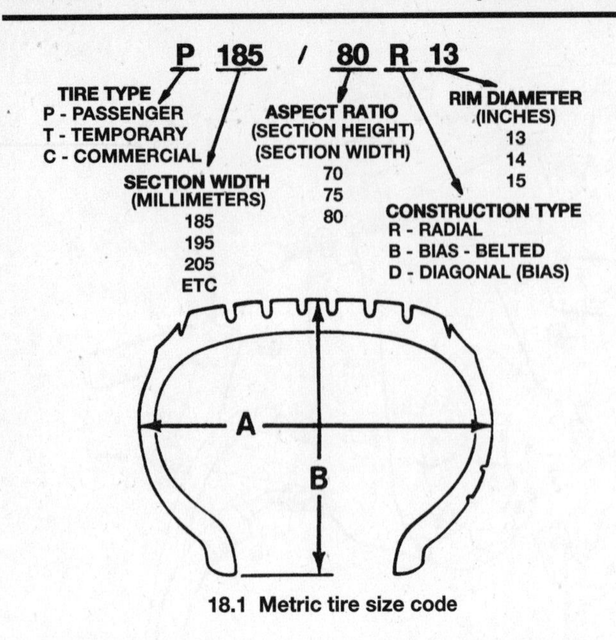

18.1 Metric tire size code

A = Section width *B = Section height*

minutes before shutting it off.

6 Road test the vehicle to be sure the steering system is functioning normally and noise free.

7 Recheck the fluid level to be sure it is up to the Hot mark on the dipstick while the engine is at normal operating temperature. Add fluid if necessary (see Chapter 1).

18 Wheels and tires - general information

Refer to illustration 18.1

All vehicles covered by this manual are equipped with metric-sized fiberglass or steel belted radial tires **(see illustration)**. Use of other size or type of tires may affect the ride and handling of the vehicle. Don't mix different types of tires, such as radials and bias belted, on the same vehicle as handling may be seriously affected. It's recommended that tires be replaced in pairs on the same axle, but if only one tire is being replaced, be sure it's the same size, structure and tread design as the other.

Because tire pressure has a substantial effect on handling and wear, the pressure on all tires should be checked at least once a month or before any extended trips (see Chapter 1).

Wheels must be replaced if they are bent, dented, leak air, have elongated bolt holes, are heavily rusted, out of vertical symmetry or if the lug nuts won't stay tight. Wheel repairs that use welding or peening are not recommended.

Tire and wheel balance is important to the overall handling, braking and performance of the vehicle. Unbalanced wheels can adversely affect handling and ride characteristics as well as tire life. Whenever a tire is installed on a wheel, the tire and wheel should be balanced by a shop with the proper equipment.

19 Front end alignment - general information

Refer to illustration 19.1

A front end alignment refers to the adjustments made to the front wheels so they are in proper angular relationship to the suspension and the ground. Front wheels that are out of proper alignment not only affect steering control, but also increase tire wear. The front end adjustments normally required are camber and toe-in **(see illustration)**. Caster should be checked to determine if any of the suspension components are bent.

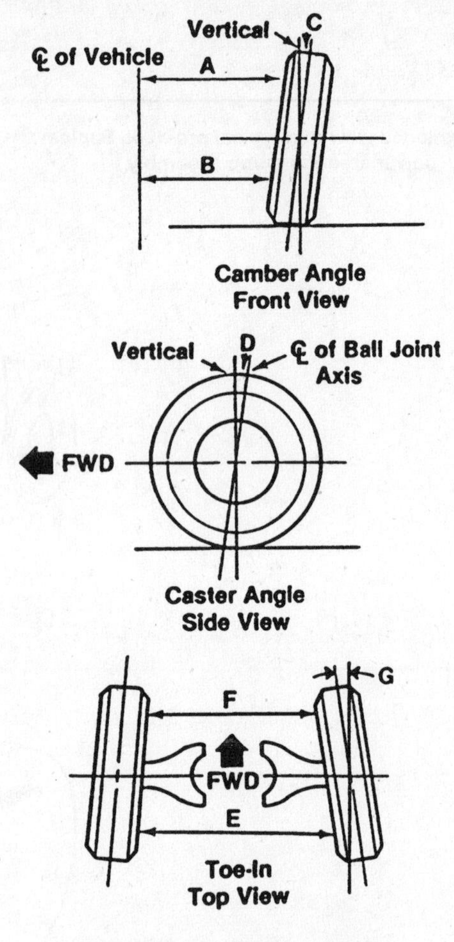

19.1 Front end alignment details

1 *A minus B = C (degrees camber)*
2 *E minus F = toe-in (measured in inches)*
3 *G = toe-in (expressed in degrees)*

Getting the proper front wheel alignment is a very exacting process, one in which complicated and expensive machines are necessary to perform the job properly. Because of this, you should have a technician with the proper equipment perform these tasks. We will, however, use this space to give you a basic idea of what is involved with front end alignment so you can better understand the process and deal intelligently with the shop that does the work.

Toe-in is the turning in of the front wheels. The purpose of a toe specification is to ensure parallel rolling of the front wheels. In a vehicle with zero toe-in, the distance between the front edges of the wheels will be the same as the distance between the rear edges of the wheels. The actual amount of toe-in is normally only a fraction of an inch. Toe-in adjustment is controlled by the tie-rod end position on the inner tie-rod. Incorrect toe-in will cause the tires to wear improperly by making them scrub against the road surface.

Camber is the tilting of the front wheels from the vertical when viewed from the front of the vehicle. When the wheels tilt out at the top, the camber is said to be positive (+). When the wheels tilt in at the top the camber is negative (-). The amount of tilt is measured in degrees from the vertical and this measurement is called the camber angle. This angle affects the amount of tire tread which contacts the road and compensates for changes in the suspension geometry when the vehicle is cornering or traveling over an undulating surface.

Caster is the tilting of the top of the front steering axis from the vertical. A tilt toward the rear is positive caster and a tilt toward the front is negative caster.

Chapter 11 Body

Contents

1 General information

The vehicles covered by this manual are of unitized construction. The body is designed to provide rigidity so that a separate frame is not necessary. Front and rear frame side rails integral with the body support the front end sheet metal, front and rear suspension systems and other mechanical components. Due to this type of construction, it is very important that, in the event of collision damage, the underbody be thoroughly checked by a facility with the proper equipment.

Certain components are particularly vulnerable to accident damage and can be unbolted and repaired or replaced. Among these parts are the body moldings, bumpers, the hood, doors and all glass.

Only general body maintenance practices and body panel repair procedures within the scope of the home mechanic are included in this Chapter.

2 Body - maintenance

1 The condition of your vehicle's body is very important, because the resale value depends a great deal on it. It's much more difficult to repair a neglected or damaged body than it is to repair mechanical components. The hidden areas of the body, such as the wheel wells, the frame and the engine compartment, are equally important, although they don't require as frequent attention as the rest of the body.

2 Once a year, or every 12,000 miles, it's a good idea to have the underside of the body steam cleaned. All traces of dirt and oil will be removed and the area can then be inspected carefully for rust, damaged brake lines, frayed electrical wires, damaged cables and other problems. The front suspension components should be greased after completion of this job.

3 At the same time, clean the engine and the engine compartment

These photos illustrate a method of repairing simple dents. They are intended to supplement *Body repair - minor damage* in this Chapter and should not be used as the sole instructions for body repair on these vehicles.

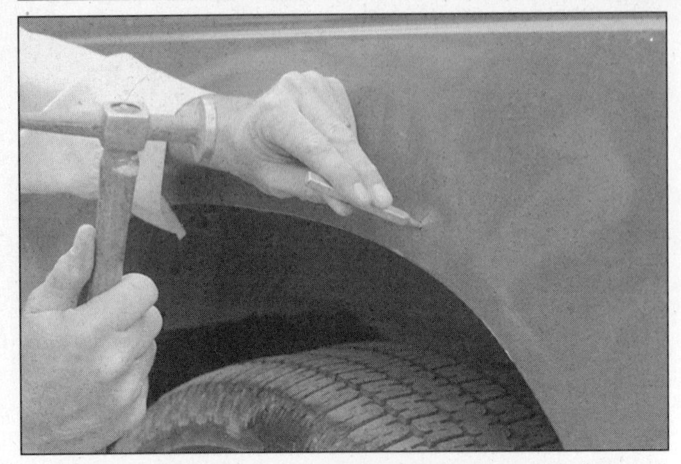

1 If you can't access the backside of the body panel to hammer out the dent, pull it out with a slide-hammer-type dent puller. In the deepest portion of the dent or along the crease line, drill or punch hole(s) at least one inch apart . . .

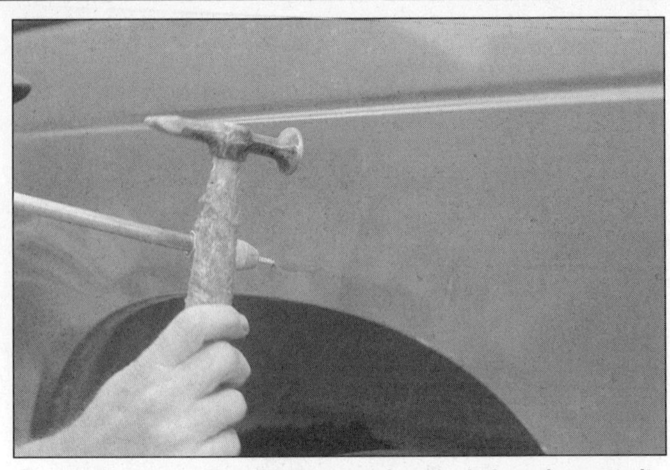

2 . . . then screw the slide-hammer into the hole and operate it. Tap with a hammer near the edge of the dent to help 'pop' the metal back to its original shape. When you're finished, the dent area should be close to its original contour and about 1/8-inch below the surface of the surrounding metal

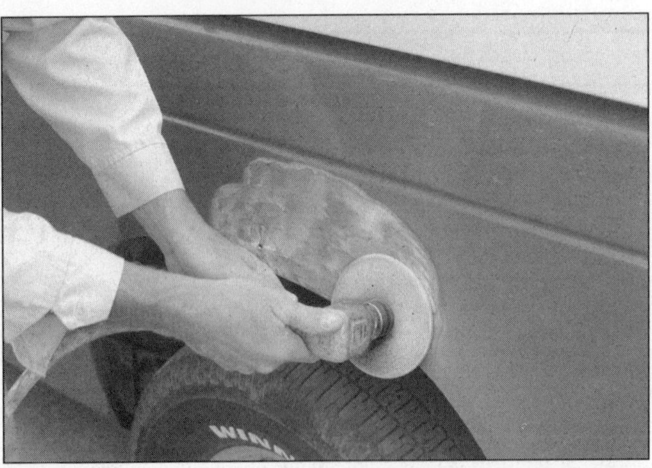

3 Using coarse-grit sandpaper, remove the paint down to the bare metal. Hand sanding works fine, but the disc sander shown here makes the job faster. Use finer (about 320-grit) sandpaper to feather-edge the paint at least one inch around the dent area

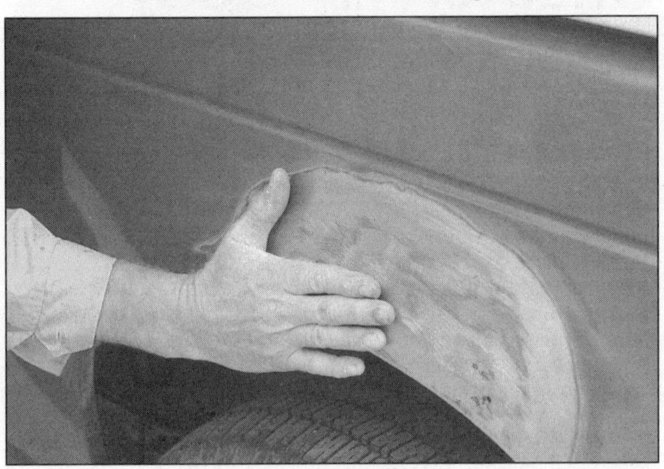

4 When the paint is removed, touch will probably be more helpful than sight for telling if the metal is straight. Hammer down the high spots or raise the low spots as necessary. Clean the repair area with wax/silicone remover

5 Following label instructions, mix up a batch of plastic filler and hardener. The ratio of filler to hardener is critical, and, if you mix it incorrectly, it will either not cure properly or cure too quickly (you won't have time to file and sand it into shape)

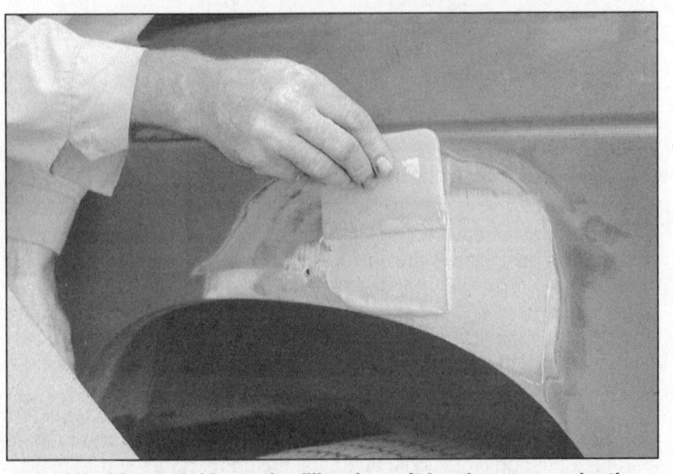

6 Working quickly so the filler doesn't harden, use a plastic applicator to press the body filler firmly into the metal, assuring it bonds completely. Work the filler until it matches the original contour and is slightly above the surrounding metal

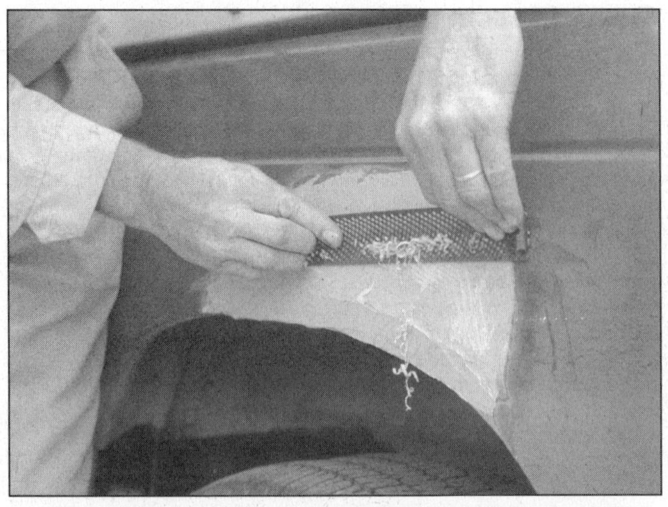

7 Let the filler harden until you can just dent it with your fingernail. Use a body file or Surform tool (shown here) to rough-shape the filler

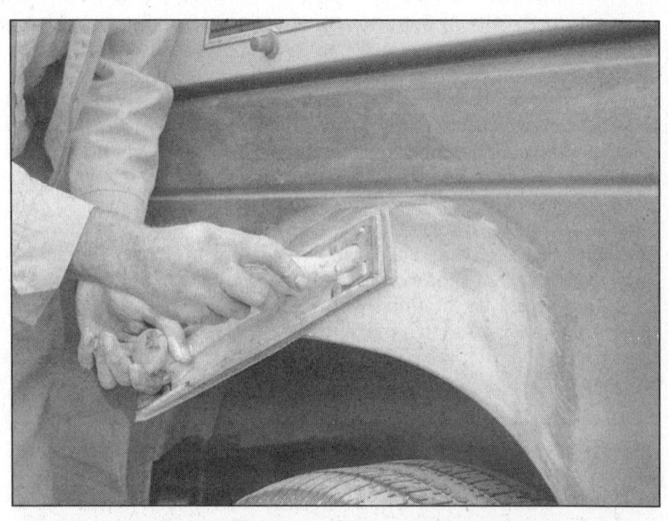

8 Use coarse-grit sandpaper and a sanding board or block to work the filler down until it's smooth and even. Work down to finer grits of sandpaper - always using a board or block - ending up with 360 or 400 grit

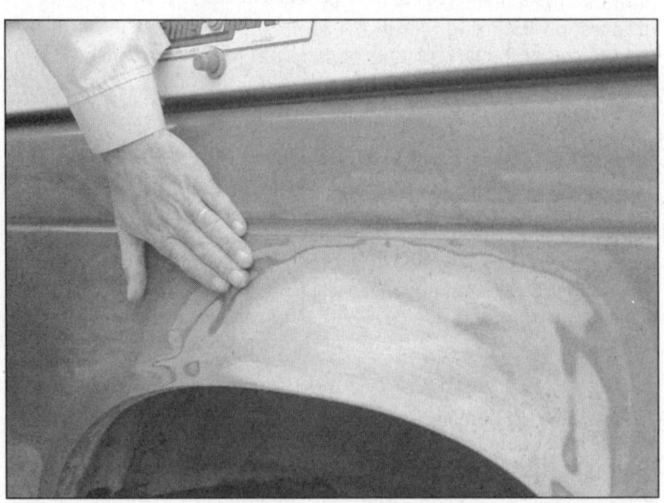

9 You shouldn't be able to feel any ridge at the transition from the filler to the bare metal or from the bare metal to the old paint. As soon as the repair is flat and uniform, remove the dust and mask off the adjacent panels or trim pieces

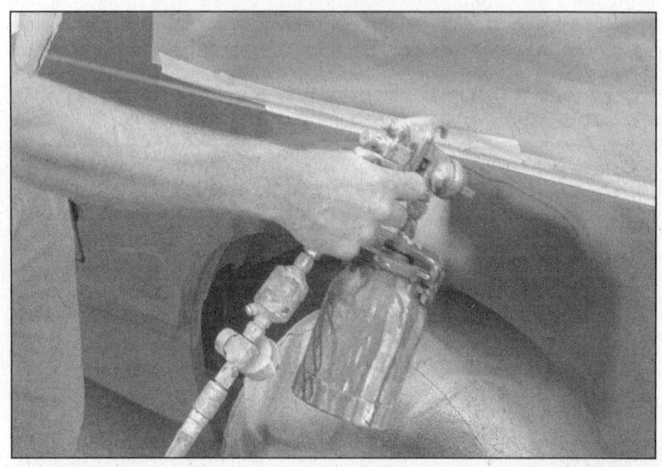

10 Apply several layers of primer to the area. Don't spray the primer on too heavy, so it sags or runs, and make sure each coat is dry before you spray on the next one. A professional-type spray gun is being used here, but aerosol spray primer is available inexpensively from auto parts stores

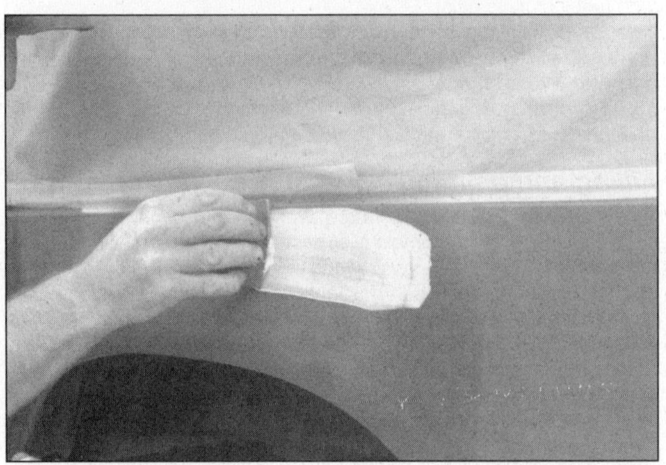

11 The primer will help reveal imperfections or scratches. Fill these with glazing compound. Follow the label instructions and sand it with 360 or 400-grit sandpaper until it's smooth. Repeat the glazing, sanding and respraying until the primer reveals a perfectly smooth surface

12 Finish sand the primer with very fine sandpaper (400 or 600-grit) to remove the primer overspray. Clean the area with water and allow it to dry. Use a tack rag to remove any dust, then apply the finish coat. Don't attempt to rub out or wax the repair area until the paint has dried completely (at least two weeks)

with a steam cleaner or water soluble degreaser.

4 The wheel wells should be given close attention, since undercoating can peel away and stones and dirt thrown up by the tires can cause the paint to chip and flake, allowing rust to set in. If rust is found, clean down to the bare metal and apply an anti-rust paint.

5 The body should be washed about once a week. Wet the vehicle thoroughly to soften the dirt, then wash it down with a soft sponge and plenty of clean soapy water. If the surplus dirt is not washed off very carefully, it can wear down the paint.

6 Spots of tar or asphalt thrown up from the road should be removed with a cloth soaked in solvent.

7 Once every six months, wax the body and chrome trim. If a chrome cleaner is used to remove rust from any of the vehicle's plated parts, remember that the cleaner also removes part of the chrome, so use it sparingly.

3 Vinyl trim - maintenance

1 Don't clean vinyl trim with detergents, caustic soap or petroleum-based cleaners. Plain soap and water works just fine, with a soft brush to clean dirt that may be ingrained. Wash the vinyl as frequently as the rest of the vehicle.

2 After cleaning, application of a high quality rubber and vinyl protectant will help prevent oxidation and cracks. The protectant can also be applied to weather-stripping, vacuum lines and rubber hoses, which often fail as a result of chemical degradation, and to the tires.

4 Upholstery and carpets - maintenance

1 Every three months remove the carpets or mats and clean the interior of the vehicle (more frequently if necessary). Vacuum the upholstery and carpets to remove loose dirt and dust.

2 Leather upholstery requires special care. Stains should be removed with warm water and a very mild soap solution. Use a clean, damp cloth to remove the soap, then wipe again with a dry cloth. Never use alcohol, gasoline, nail polish remover or thinner to clean leather upholstery.

3 After cleaning, regularly treat leather upholstery with a leather wax. Never use car wax on leather upholstery.

4 In areas where the interior of the vehicle is subject to bright sunlight, cover leather seats with a sheet if the vehicle is to be left out for any length of time.

5 Body repair - minor damage

See photo sequence

Repair of minor scratches

1 If the scratch is superficial and does not penetrate to the metal of the body, repair is very simple. Lightly rub the scratched area with a fine rubbing compound to remove loose paint and built up wax. Rinse the area with clean water.

2 Apply touch-up paint to the scratch, using a small brush. Continue to apply thin layers of paint until the surface of the paint in the scratch is level with the surrounding paint. Allow the new paint at least two weeks to harden, then blend it into the surrounding paint by rubbing with a very fine rubbing compound. Finally, apply a coat of wax to the scratch area.

3 If the scratch has penetrated the paint and exposed the metal of the body, causing the metal to rust, a different repair technique is required. Remove all loose rust from the bottom of the scratch with a pocket knife, then apply rust inhibiting paint to prevent the formation of rust in the future. Using a rubber or nylon applicator, coat the scratched area with glaze-type filler. If required, the filler can be mixed with thinner to provide a very thin paste, which is ideal for filling narrow scratches. Before the glaze filler in the scratch hardens, wrap a piece of smooth cotton cloth around the tip of a finger. Dip the cloth in thinner and then quickly wipe it along the surface of the scratch. This will ensure that the surface of the filler is slightly hollow. The scratch can now be painted over as described earlier in this section.

Repair of dents

4 When repairing dents, the first job is to pull the dent out until the affected area is as close as possible to its original shape. There is no point in trying to restore the original shape completely as the metal in the damaged area will have stretched on impact and cannot be restored to its original contours. It is better to bring the level of the dent up to a point which is about 1/8-inch below the level of the surrounding metal. In cases where the dent is very shallow, it is not worth trying to pull it out at all.

5 If the back side of the dent is accessible, it can be hammered out gently from behind using a soft-face hammer. While doing this, hold a block of wood firmly against the opposite side of the metal to absorb the hammer blows and prevent the metal from being stretched.

6 If the dent is in a section of the body which has double layers, or some other factor makes it inaccessible from behind, a different technique is required. Drill several small holes through the metal inside the damaged area, particularly in the deeper sections. Screw long, self tapping screws into the holes just enough for them to get a good grip in the metal. Now the dent can be pulled out by pulling on the protruding heads of the screws with locking pliers.

7 The next stage of repair is the removal of paint from the damaged area and from an inch or so of the surrounding metal. This is easily done with a wire brush or sanding disk in a drill motor, although it can be done just as effectively by hand with sandpaper. To complete the preparation for filling, score the surface of the bare metal with a screwdriver or the tang of a file or drill small holes in the affected area. This will provide a good grip for the filler material. To complete the repair, see the Section on filling and painting.

Repair of rust holes or gashes

8 Remove all paint from the affected area and from an inch or so of the surrounding metal using a sanding disk or wire brush mounted in a drill motor. If these are not available, a few sheets of sandpaper will do the job just as effectively.

9 With the paint removed, you will be able to determine the severity of the corrosion and decide whether to replace the whole panel, if possible, or repair the affected area. New body panels are not as expensive as most people think and it is often quicker to install a new panel than to repair large areas of rust.

10 Remove all trim pieces from the affected area except those which will act as a guide to the original shape of the damaged body, such as headlight shells, etc. Using metal snips or a hacksaw blade, remove all loose metal and any other metal that is badly affected by rust. Hammer the edges of the hole inward to create a slight depression for the filler material.

11 Wire brush the affected area to remove the powdery rust from the surface of the metal. If the back of the rusted area is accessible, treat it with rust inhibiting paint.

12 Before filling is done, block the hole in some way. This can be done with sheet metal riveted or screwed into place, or by stuffing the hole with wire mesh.

13 Once the hole is blocked off, the affected area can be filled and painted. See the following subsection on filling and painting.

Filling and painting

14 Many types of body fillers are available, but generally speaking, body repair kits which contain filler paste and a tube of resin hardener are best for this type of repair work. A wide, flexible plastic or nylon applicator will be necessary for imparting a smooth and contoured finish to the surface of the filler material. Mix up a small amount of filler on a clean piece of wood or cardboard (use the hardener sparingly). Follow the manufacturer's instructions on the package, otherwise the filler will set incorrectly.

15 Using the applicator, apply the filler paste to the prepared area. Draw the applicator across the surface of the filler to achieve the desired contour and to level the filler surface. As soon as a contour that approximates the original one is achieved, stop working the paste. If you continue, the paste will begin to stick to the applicator. Continue to add thin layers of paste at 20-minute intervals until the level of the

9.3 Tie a piece of string to the hood light wire so you can pull it back into place when the hood is reinstalled

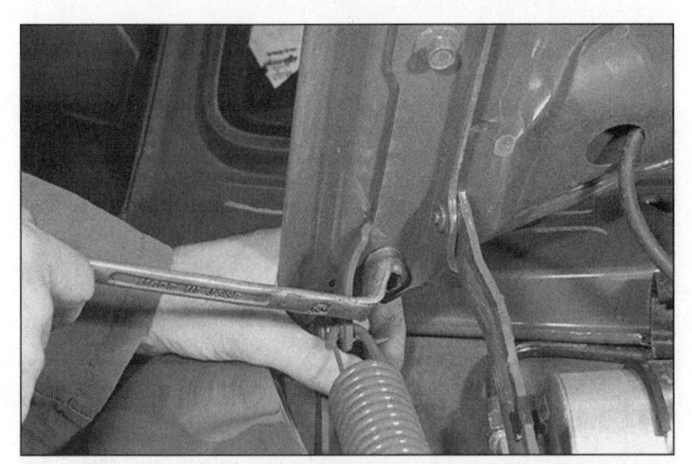

9.4 Mark the hood bolt locations before removing them

filler is just above the surrounding metal.

16 Once the filler has hardened, the excess can be removed with a body file. From then on, progressively finer grades of sandpaper should be used, starting with a 180-grit paper and finishing with 600-grit wet-or-dry paper. Always wrap the sandpaper around a flat rubber or wooden block, otherwise the surface of the filler will not be completely flat. During the sanding of the filler surface, the wet-or-dry paper should be periodically rinsed in water. This will ensure that a very smooth finish is produced in the final stage.

17 At this point, the repair area should be surrounded by a ring of bare metal, which in turn should be encircled by the finely feathered edge of good paint. Rinse the repair area with clean water until all of the dust produced by the sanding operation is gone.

18 Spray the entire area with a light coat of primer. This will reveal any imperfections in the surface of the filler. Repair the imperfections with fresh filler paste or glaze filler and once more smooth the surface with sandpaper. Repeat this spray-and-repair procedure until you are satisfied that the surface of the filler and the feathered edge of the paint are perfect. Rinse the area with clean water and allow it to dry completely.

19 The repair area is now ready for painting. Spray painting must be carried out in a warm, dry, windless and dust free atmosphere. These conditions can be created if you have access to a large indoor work area, but if you are forced to work in the open, you will have to pick the day very carefully. If you are working indoors, dousing the floor in the work area with water will help settle the dust which would otherwise be in the air. If the repair area is confined to one body panel, mask off the surrounding panels. This will help minimize the effects of a slight mismatch in paint color. Trim pieces such as chrome strips, door handles, etc., will also need to be masked off or removed. Use masking tape and several thickness of newspaper for the masking operations.

20 Before spraying, shake the paint can thoroughly, then spray a test area until the spray painting technique is mastered. Cover the repair area with a thick coat of primer. The thickness should be built up using several thin layers of primer rather than one thick one. Using 600-grit wet-or-dry sandpaper, rub down the surface of the primer until it is very smooth. While doing this, the work area should be thoroughly rinsed with water and the wet-or-dry sandpaper periodically rinsed as well. Allow the primer to dry before spraying additional coats.

21 Spray on the top coat, again building up the thickness by using several thin layers of paint. Begin spraying in the center of the repair area and then, using a circular motion, work out until the whole repair area and about two inches of the surrounding original paint is covered. Remove all masking material 10 to 15 minutes after spraying on the final coat of paint. Allow the new paint at least two weeks to harden, then use a very fine rubbing compound to blend the edges of the new paint into the existing paint. Finally, apply a coat of wax.

6 Body repair - major damage

1 Major damage must be repaired by an auto body shop specifically equipped to perform body and frame repairs. These shops have the specialized equipment required to do the job properly.

2 If the damage is extensive, the body and frame must be checked for proper alignment or the vehicle's handling characteristics may be adversely affected and other components may wear at an accelerated rate.

3 Due to the fact that all of the major body components (hood, fenders, etc.) are separate and replaceable units, any seriously damaged components should be replaced rather than repaired. Sometimes the components can be found in a wrecking yard that specializes in used vehicle components, often at considerable savings over the cost of new parts.

7 Hinges and locks - maintenance

Once every 3000 miles, or every three months, the hinges and latch assemblies on the doors, hood and trunk should be given a few drops of light oil or lock lubricant. The door latch strikers should also be lubricated with a thin coat of grease to reduce wear and ensure free movement. Lubricate the door locks with spray-on graphite lubricant.

8 Windshield and fixed glass - replacement

Replacement of the windshield and fixed glass requires the use of special fast-setting adhesive/caulk materials and some specialized tools and techniques. These operations should be left to a dealer service department or a shop specializing in glass work.

9 Hood - removal, installation and adjustment

Refer to illustrations 9.3, 9.4 and 9.10

Note: *The hood is heavy and somewhat awkward to remove and install - at least two people should perform this procedure.*

Removal and installation

1 Use blankets or pads to cover the cowl area of the body and the fenders. This will protect the body and paint as the hood is lifted off.

2 Scribe or draw alignment marks around the bolt heads to insure proper alignment during installation.

3 Disconnect any cables or wire harnesses which will interfere with removal **(see illustration)**.

4 Have an assistant support the weight of the hood. Remove the hinge-to-hood bolts **(see illustration)**.

5 Lift off the hood.

6 Installation is the reverse of removal.

Adjustment

7 Fore-and-aft and side-to-side adjustment of the hood is done by moving the hood in relation to the hinge plate after loosening the bolts or nuts.

8 Scribe a line around the entire hinge plate so you can judge the amount of movement.

9 Loosen the bolts or nuts and move the hood into correct alignment. Move it only a little at a time. Tighten the hinge bolts or nuts and carefully lower the hood to check the alignment.

10 If necessary after installation, the entire hood latch assembly can be adjusted from side-to-side on the hood and the lock pin screwed in-or-out so it closes securely and is flush with the fenders. To do this, scribe a line around the hood latch mounting bolts to provide a reference point. Then loosen the bolts and reposition the latch assembly **(see illustration)**. Following adjustment, retighten the mounting bolt. The hood closing height can be adjusted by using a screwdriver to screw the lock pin in-or-out as necessary.

11 Finally, adjust the hood bumpers on the radiator support so the hood, when closed, is flush with the fenders.

12 The hood latch assembly, as well as the hinges, should be periodically lubricated with white lithium-base grease to prevent sticking and wear.

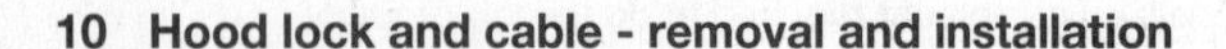

10 Hood lock and cable - removal and installation

Refer to illustrations 10.2 and 10.3

Lock

1 Remove the radiator grille (see Section 11)

2 Working through the grille opening, detach the cable from the hood lock lever **(see illustration)**.

3 Remove the retaining bolts, lower the hood lock and remove it from the vehicle **(see illustration)**.

Cable

4 Detach the cable from the lock and attach a string or wire to the end.

5 Disconnect the cable clips, separate it from the handle and pull the cable through the firewall and into the vehicle interior.

6 Connect the string or wire to the new cable, pull it back through the firewall into the engine compartment, then secure it with the cable clips and connect it to the lock and handle.

11 Radiator grille - removal and installation

Refer to illustration 11.1

1 Open the hood and rotate the plastic quick-release fasteners 90-degrees to release the top of the grille **(see illustration)**.

2 Rotate the top of the grille out, lift it from the mounts and remove it from the vehicle.

3 Installation is the reverse of removal.

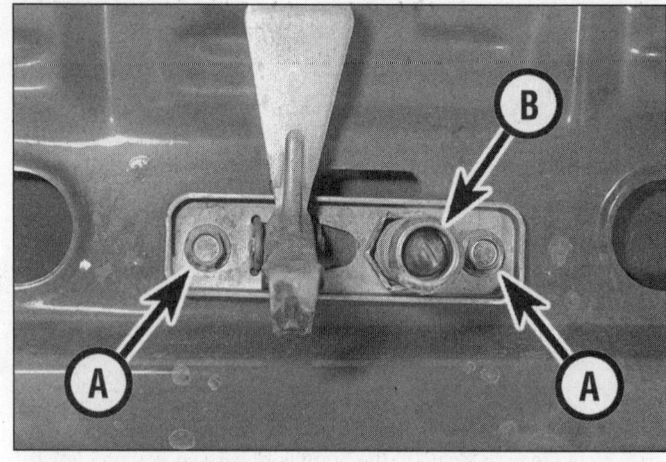

9.10 Loosen the bolts (A) to adjust the hood from side-to-side and screw the lock pin (B) in-or-out to adjust the closed height

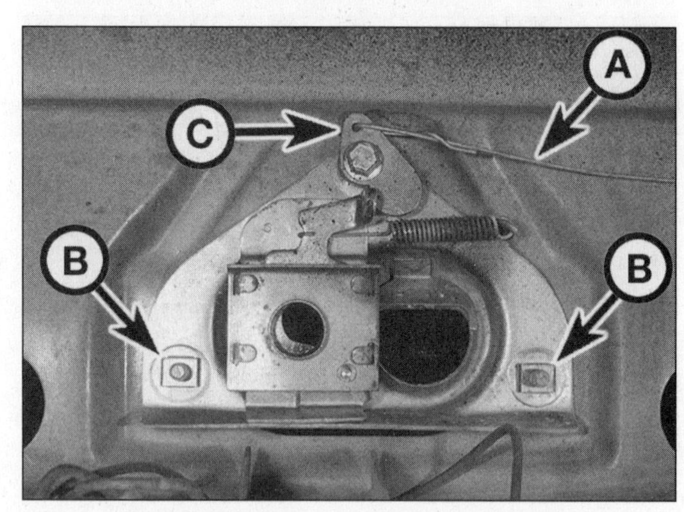

10.2 Bottom view of the hood lock assembly

A) Release cable
B) Retaining nuts
C) Release cable

10.3 Hood lock retaining screw locations (arrows)

11.1 Rotate the plastic retainers 90-degrees to release the grille

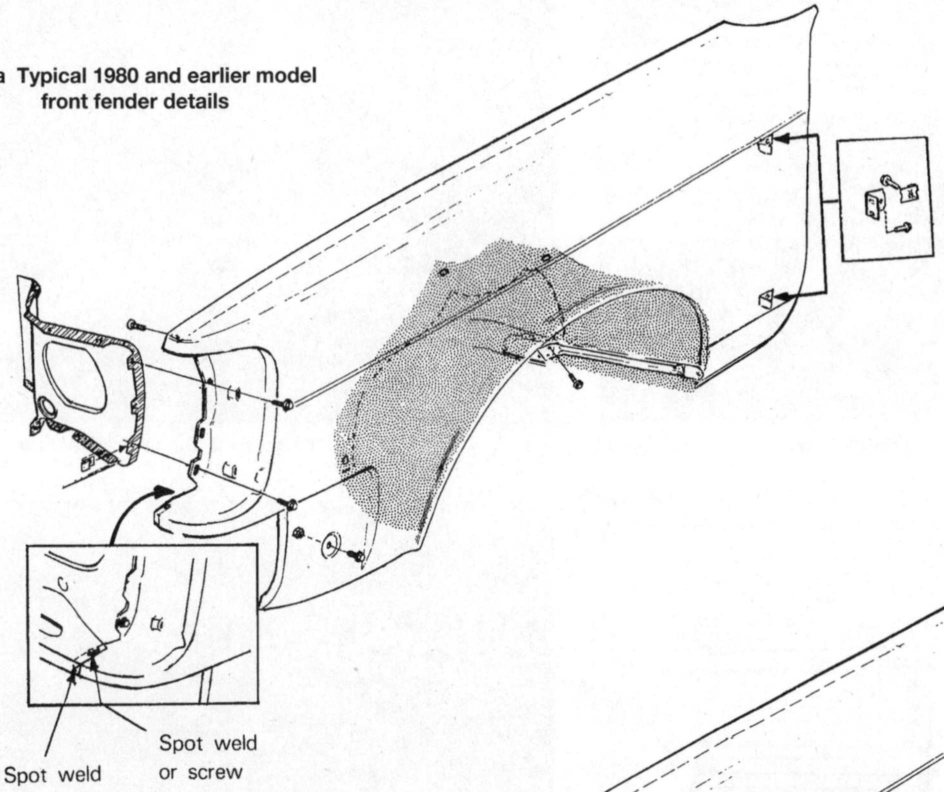

12.4a Typical 1980 and earlier model front fender details

Spot weld

Spot weld or screw

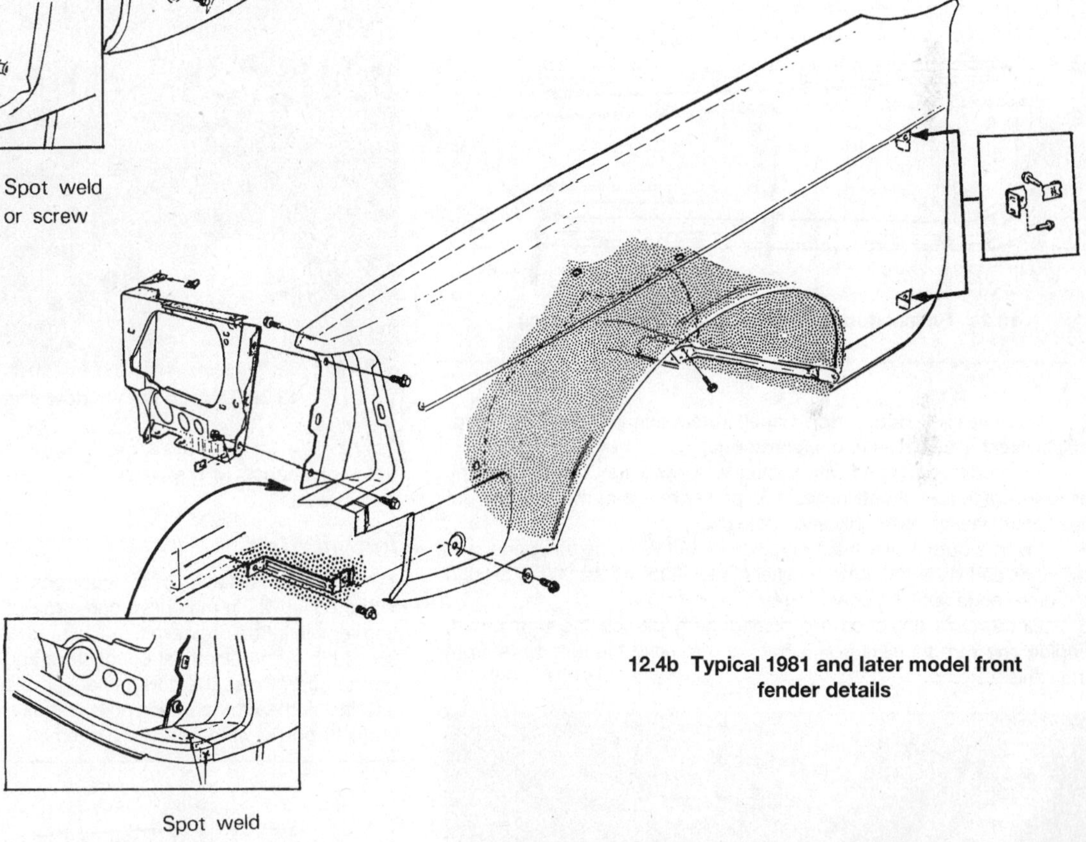

12.4b Typical 1981 and later model front fender details

Spot weld

12 Front fender - removal and installation

Refer to illustrations 12.4a and 12.4b

1 Raise the vehicle, support it securely on jackstands and remove the front wheel.

2 Disconnect the antenna and all light bulb wiring harness connectors and other components that would interfere with fender removal.

3 Remove the front bumper (see Section 9).

4 Remove the splash shield and fender mounting bolts **(see illustrations)**.

5 Detach the fender. It may be necessary to use a hammer and chisel to separate the fender from the spot weld at the front lower edge **(see illustrations 12.4a and 12.4b)**. It is a good idea to have an assistant support the fender while it's being moved away from the vehicle to prevent damage to the surrounding body panels.

6 Installation is the reverse of removal. Tighten all nuts, bolts and screws securely.

13 Door trim panel - removal and installation

Refer to illustrations 13.2a, 13.2b, 13.2c, 13.3a, 13.3b and 13.4

Removal

1 Disconnect the negative cable from the battery.

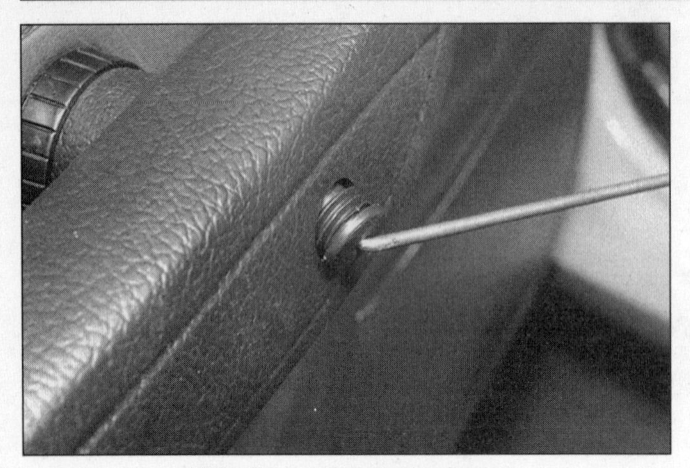

13.2a Pry out the plastic plug . . .

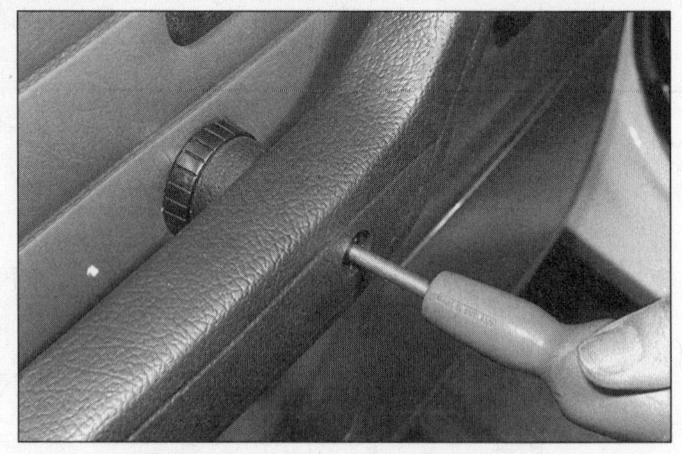

13.2b . . . for access to the armrest screw

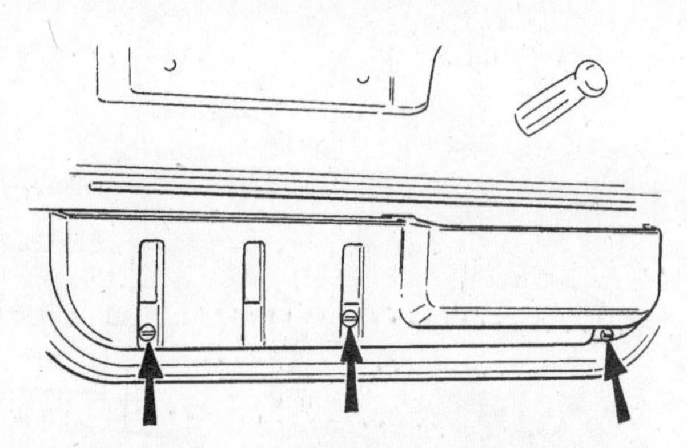

13.2c Typical door pocket screw locations (arrows)

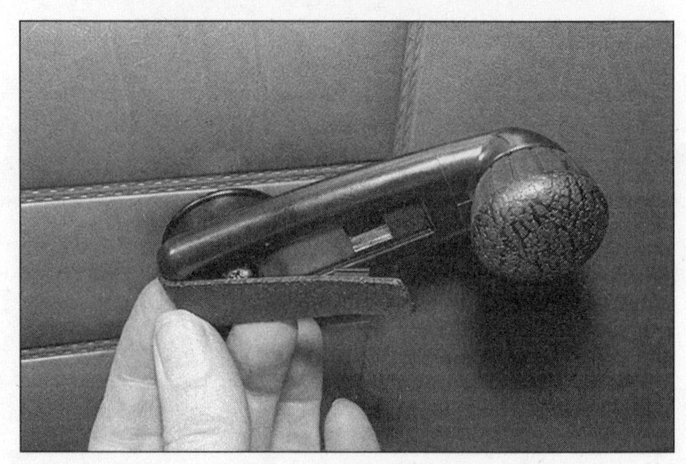

13.3a Detach the window crank trim cover . . .

2 Remove all door trim panel retaining screws and door pull/armrest assemblies **(see illustrations)**.

3 On models equipped with manual window regulators, remove the window crank **(see illustrations)**. On power regulator models, pry out the control switch assembly and unplug it.

4 Insert a putty knife or flat screwdriver between the trim panel and the door and disengage the retainers **(see illustration)**. Work around the outer edge until the panel is free.

5 Once all of the clips are disengaged, detach the trim panel, unplug any wire harness connectors and remove the trim panel from the vehicle.

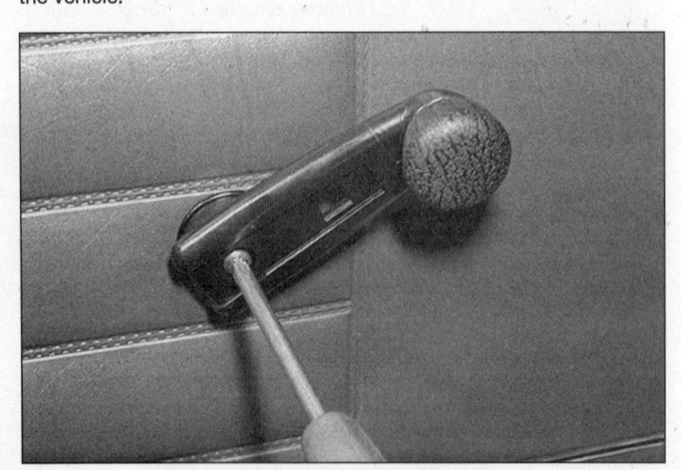

13.3b . . . for access to the retaining screw

6 For access to the inner door, carefully peel back the plastic watershield.

Installation

7 Prior to installation of the door panel, be sure to reinstall any clips in the panel which may have come out during the removal procedure and remain in the door itself.

8 Plug in the electrical connectors and place the panel in position on the door. Press the door panel into place until the clips are seated and install the armrest/door pulls. Install the manual regulator window crank or power window switch assembly.

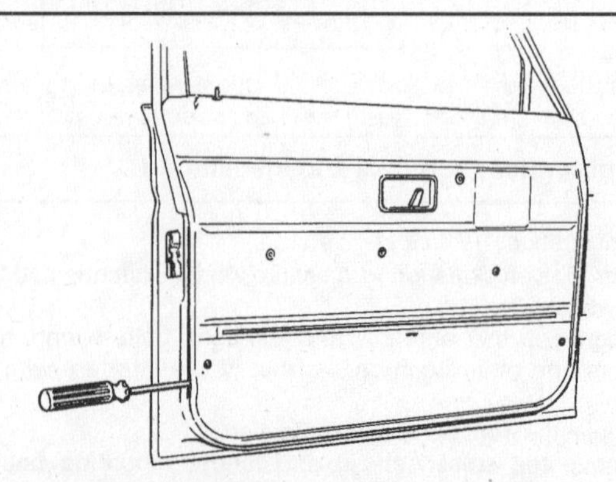

13.4 Pry carefully, working around the door trim panel until all of the clips are disengaged

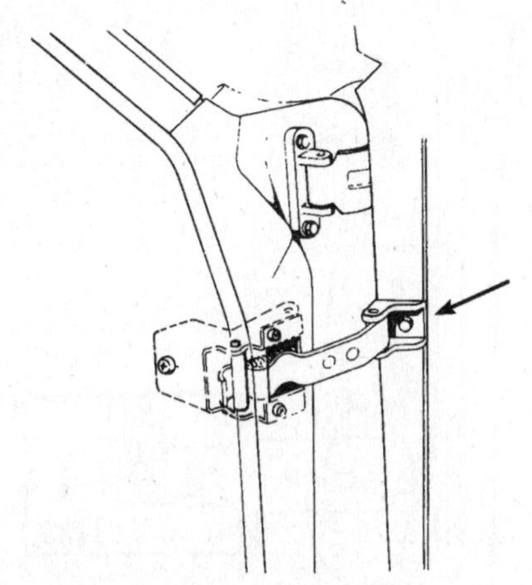

14.3 The early model door check strap can be detached after removing the bolt (arrow)

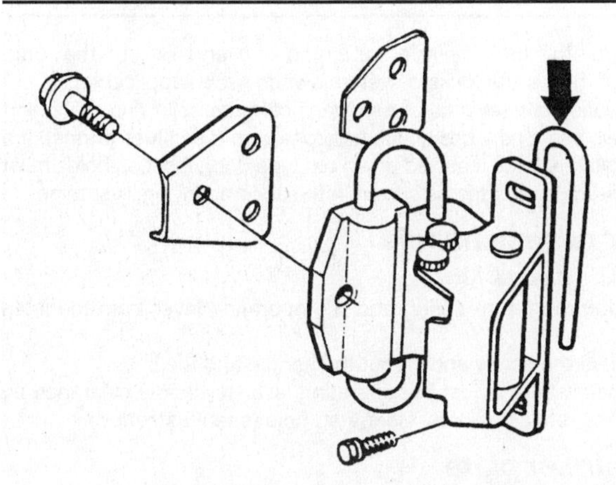

14.6 On later models, the door check strap and hinge position can be adjusted using a shim (arrow)

14 Door - removal, installation and adjustment

Refer to illustrations 14.3 and 14.6

Removal and installation

1 Remove the door trim panel. Disconnect any wire harness connectors and push them through the door opening so they won't interfere with door removal.

2 Place a jack or jackstand under the door or have an assistant on hand to support it when the hinge bolts are removed. **Note:** *If a jack or jackstand is used, place a rag between it and the door to protect the door's painted surfaces.*

3 Scribe alignment marks around the door hinges. On early models, remove the bolt and detach the door check strap **(see illustration)**.

4 Remove the hinge-to-door bolts and carefully lift off the door.

5 Installation is the reverse of removal.

Adjustment

6 Following installation of the door, check the alignment and adjust it if necessary as follows:

a) *Up-and-down and forward-and-backward adjustments are made by loosening the hinge-to-body bolts and moving the door as necessary. On later models, shims can be used to adjust the position of the hinge* **(see illustration)**.

b) *The door lock striker can also be adjusted both up-and-down and sideways to provide positive engagement with the lock mechanism. This is done by loosening the striker, moving it as necessary, then retightening it.*

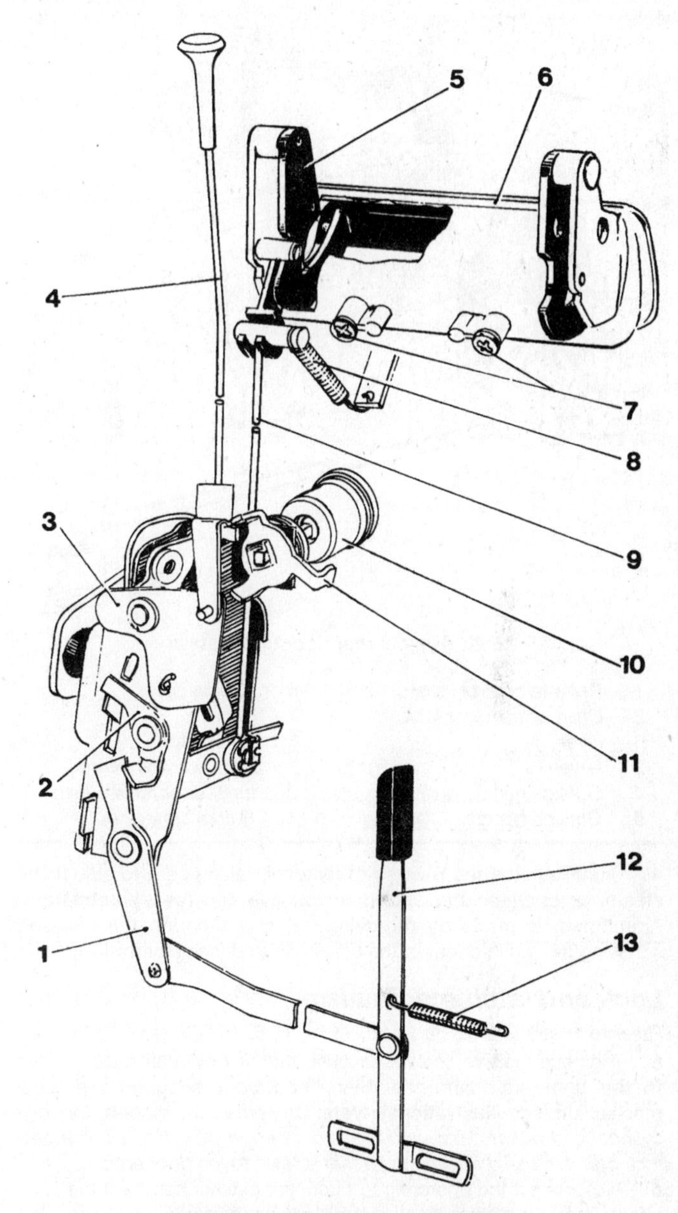

15.2a Typical front door lock details

1	*Lever*	8	*Return spring*
2	*Lever*	9	*Handle pullrod*
3	*Lever*	10	*Lock cylinder*
4	*Lock button pullrod*	11	*Lock cylinder lever*
5	*Outside handle*	12	*Inner door handle lever*
6	*Handle cover*	13	*Return spring*
7	*Screws*		

15 Door lock, handles and latches - removal and installation

Exterior handle

Refer to illustrations 15.2a 15.2b, 15.4a and 15.4b

1 Raise the glass fully and remove the door trim panel and plastic watershield.

2 Detach the handle return spring and disconnect the handle-to-latch operating rod **(see illustrations)**.

3 Remove the two screws securing the handle to the door and lift off the handle.

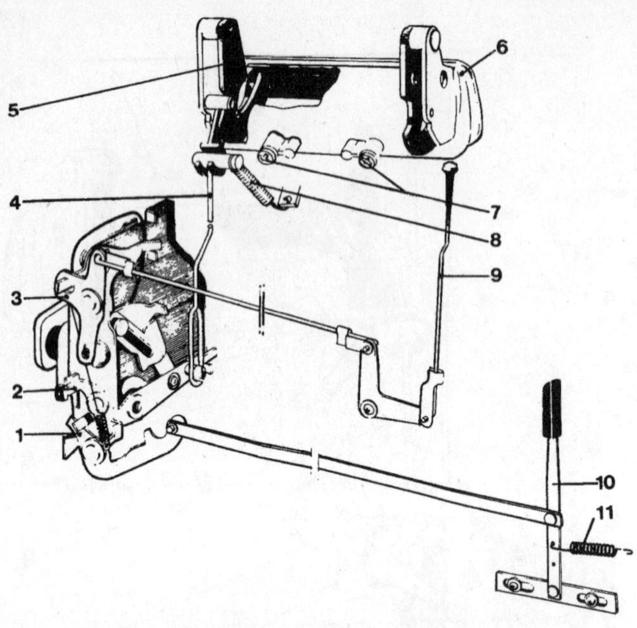

15.2b Typical rear door lock details

1	*Remote control lever*	6	*Handle cover*
2	*Child safety door lock lever*	7	*Screws*
3	*Lever*	8	*Return spring*
4	*Outside handle pullrod*	9	*Lock button pullrod*
5	*Outside handle*	10	*Inner door handle lever*
		11	*Return spring*

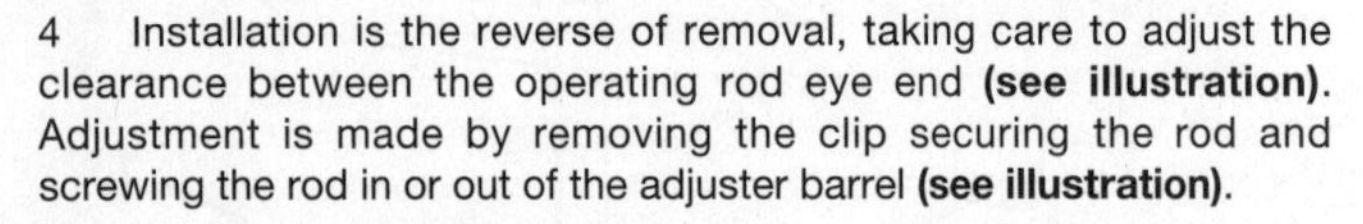

4 Installation is the reverse of removal, taking care to adjust the clearance between the operating rod eye end **(see illustration)**. Adjustment is made by removing the clip securing the rod and screwing the rod in or out of the adjuster barrel **(see illustration)**.

Lock and latch mechanism

Refer to illustrations 15.5a 15.5b, 15.5c, 15.6, 15.10a and 15.10b

5 Remove the two screws securing the lock cylinder retaining clip to the door edge and the Allen head bolts securing the latch mechanism **(see illustrations)**. **Note:** *On some early models, the lock cylinder clip was not screwed to the door edge. When installing a new lock cylinder, use a retaining clip that screws to the door edge.*

6 Disconnect the operating rod from the exterior handle at the latch lever, and if installed, the central locking motor control rod (see Chapter 12) **(see illustration)**.

7 On rear doors, disconnect the remote lock button operating rod.

8 Lift out the latch assembly.

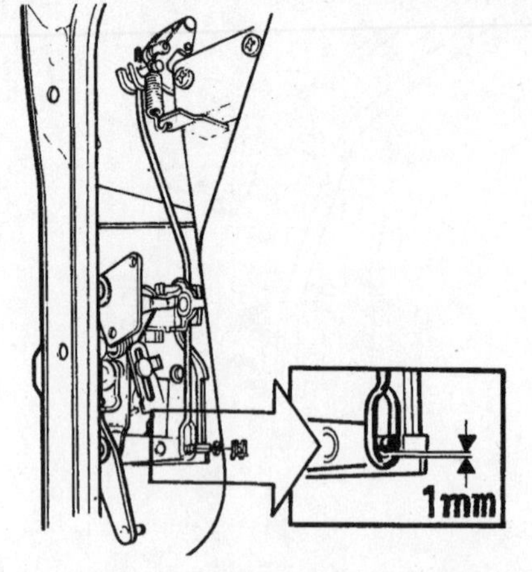

15.4a Adjust the clearance between the operating rod eye and latch lever as shown

9 Pry out the lock cylinder retaining clip and lift out the lock. Disconnect the central locking system switch (when applicable).

10 Installation is a reversal of removal, noting the following: 1978 and later models use a new design latch mechanism **(see illustrations)**. If a new type latch is to be installed on an old type door, location holes must be drilled in the door edge as shown in the accompanying illustration.

Interior release handle

Refer to illustration 15.13

11 Unhook the return spring and the operating lever from the latch assembly.

12 Remove the screw and lift out the handle and the lever.

13 On installation, adjust the operating lever-to-handle clearance by means of the retaining screw elongated hole **(see illustration)**.

Latch striker plate

14 Remove the striker plate securing screws and lift off the plate.

15 Installation is the reverse of removal. Adjust the plate as follows.

16 Loosen the securing screws so that the striker can just be moved by hand.

17 Close the door, holding the exterior handle in so that it cannot operate: the striker plate will take up the correct position as the door is closed.

18 Open the door again. Tighten the striker plate screws, outboard screws first, followed by the inner screws.

19 Check that the door closes smoothly without undue force and does not lift as the latch engages with the striker.

20 If the latch striker plate is too far in, loosen the screws slightly and pry it outwards, keeping it at the same horizontal plane.

15.4b Adjust the outside handle operating rod by turning the adjustment barrel (arrow) in-or-out

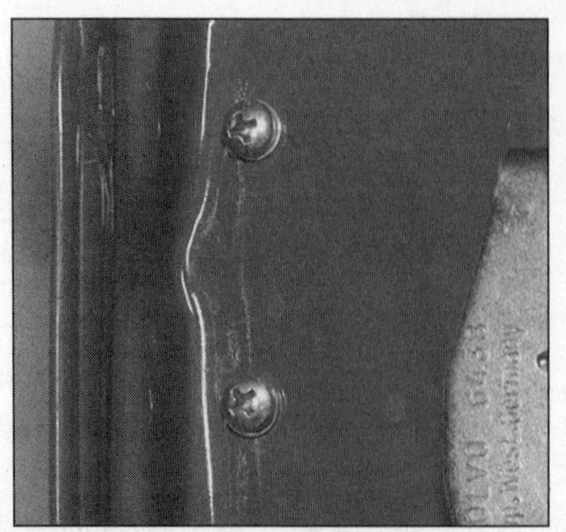

15.5a The lock cylinder screws are accessible on the end of the door

15.5b The latch mechanism is held in place by two Allen-head screws

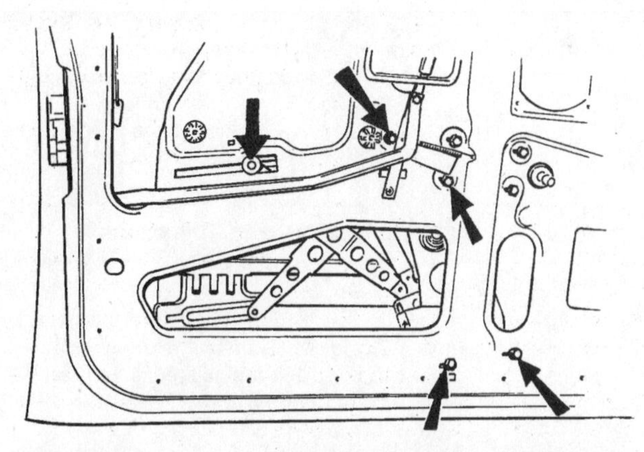

16.12 Power window regulator retaining bolt locations (arrows)

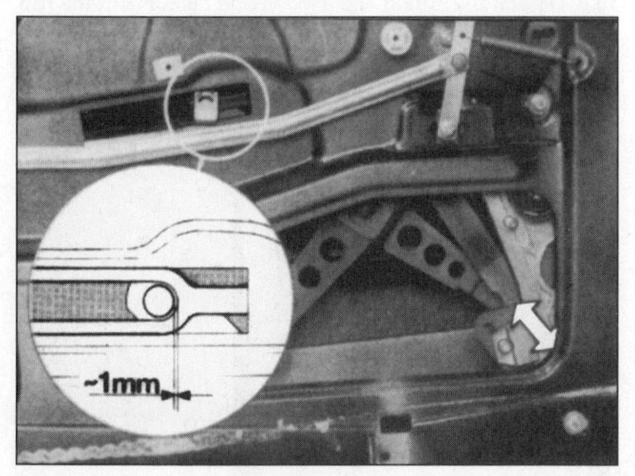

16.17 With the power window full lowered, the gap in the fork should be as shown

Installation

5 Installation is the reverse of removal.

Adjustment

6 The regulator position should be adjusted so that it requires minimum effort to raise and lower the window. Loosen the regulator mounting bolts slightly **(see illustration 16.4)**, so that the regulator position can be moved with some effort. Adjust the regulator position while raising and lowering the glass until a smooth action is achieved, then tighten the bolts.

Power regulator

Removal

7 Fully lower the window to its stop.

8 Release the regulator arms from the glass rail. Do this by pushing the safety brackets to loosen them, extracting the washers and prying the arms towards you.

9 Raise the glass to the top of the door opening by hand and tape, prop or wedge it securely in place.

10 Disconnect the battery negative cable.

11 Disconnect the electrical leads from the regulator motor. On 1979 and earlier models, this will mean detaching the fusebox for access to the leads. On later models, the motor lead is disconnected at the switch itself, located on the arm rest.

12 Release the lifting arm from the side of the rail in the door, remove the regulator mounting bolts and withdraw the mechanism through the access hole in the door **(see illustration)**.

13 If it is necessary to remove the motor from the regulator, clamp the regulator sector gear and backplate securely together in a vise so that they can't move because they are spring-loaded. **Warning:** *The regulator is under considerable spring pressure and can cause personal injury if the spring suddenly releases when the mechanism is disassembled.*

16.16 Loosen the upper stop locknut (arrow), raise the window all the way up, then tighten the locknut

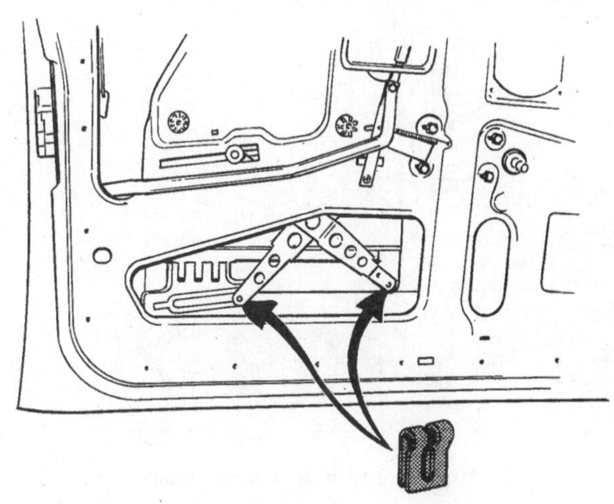

17.3 Detach the glass retaining clips

Installation

14 Installation is the reverse of removal.

Adjustment

15 The regulator operation should be adjusted so that it requires minimum effort to raise or lower the window. It should take no more than five seconds to lower the window. Loosen the mounting bolts slightly so the regulator can be moved with some effort **(see illustration 16.12)**. Adjust the regulator position while raising and lowering the glass until a smooth action is achieved, then tighten the bolts.

16 Raise the window all the way up, then loosen the upper stop locknut **(see illustration)**. Try to raise the window further, then adjust the stop lug against the toothed quadrant and tighten the lug locknut.

17 Lower the window fully, then check to make sure that the lifting arm does not bottom in the slide fork. Adjust the stop lug nut if necessary to provide a clearance in the fork of approximately 1/32 in (1.0 mm) **(see illustration)**.

17 Door window glass - removal and installation

Refer to illustrations 17.3 and 17.4

1 Remove the door trim panel and plastic watershield (see Section 13).

2 Lower the window.

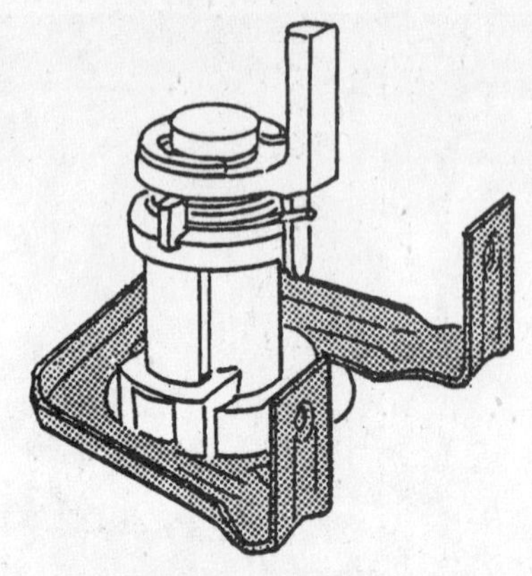

15.5c Use this type of bracket when replacing an earlier type lock cylinder

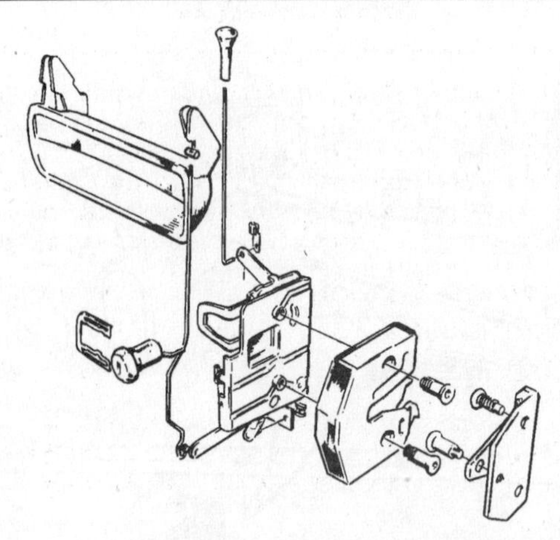

15.10a Later type door latch details

21 Fore-and-aft adjustment can be made by placing shims under the plate.

16 Door window regulator - removal, installation and adjustment

Refer to illustrations 16.4. 16.12, 16.16 and 16.17

1 Remove the door panel and watershield (see Section 13).

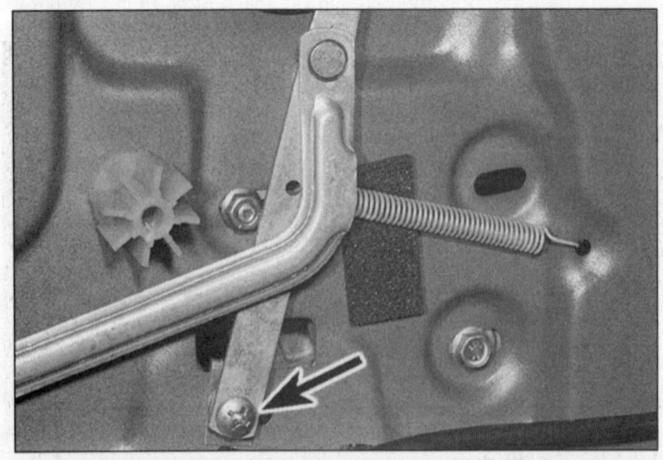

15.13 Adjust the operating lever-to-handle clearance (it should be around 1/4-inch) by loosening this screw (arrow)

15.6 Detach the spring clip (arrow) and disconnect the rod

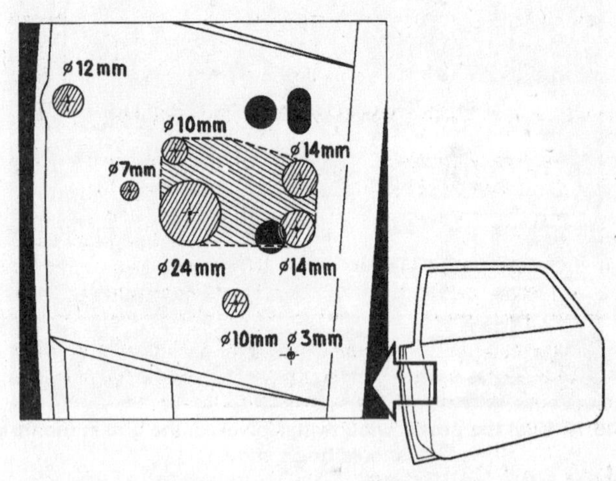

15.10b When installing a later type latch assembly on an older model, the door will have to be modified as shown

Manual regulator

Removal

2 Raise the window glass fully and wedge or tape it securely so that it doesn't fall.

3 Detach the glass from the regulator (see Step 3, Section 17).

4 Remove the attaching bolts, then lower the regulator and remove it through the door opening **(see illustration)**.

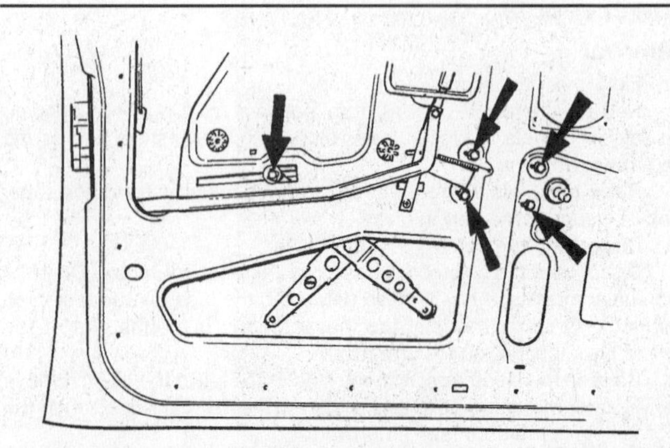

16.4 Remove the bolts (arrows) and lower the regulator out through the door opening

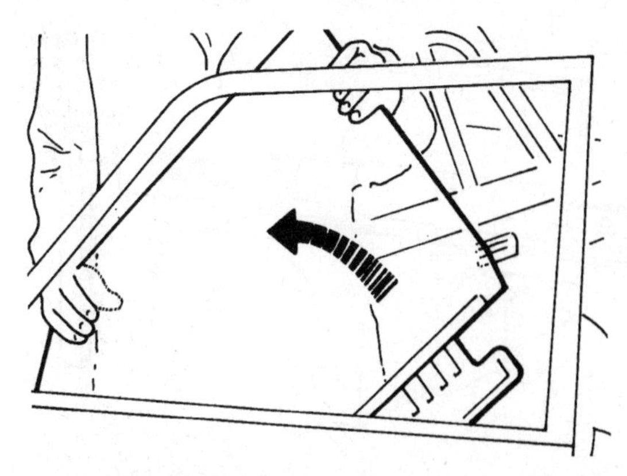
17.4 Lift the glass up and rotate it out of the door

3 Remove the locking clips from the window lift arms, then detach the lift arms from the lift channel **(see illustration)**.

4 Lift the window and lift channel up and out of the door, swinging it to the inside of the door **(see illustration)**.

5 Installation is the reverse of removal, adjusting the window fore and aft in the lift channel so that it closes smoothly without binding.

18 Trunk lid - removal, installation and adjustment

Refer to illustrations 18.1a, 18.1b and 18.5

1 The trunk lid is mounted on two hinges attached to the trunk lid by two bolts and to the rear pillars by three bolts and the assembly is counter-balanced by shock absorber type springs **(see illustrations)**.

2 To remove the trunk lid, make alignment marks across the lid and hinge before undoing the lid-to-hinge bolts and lifting off the lid.

3 Installation is a reversal of removal.

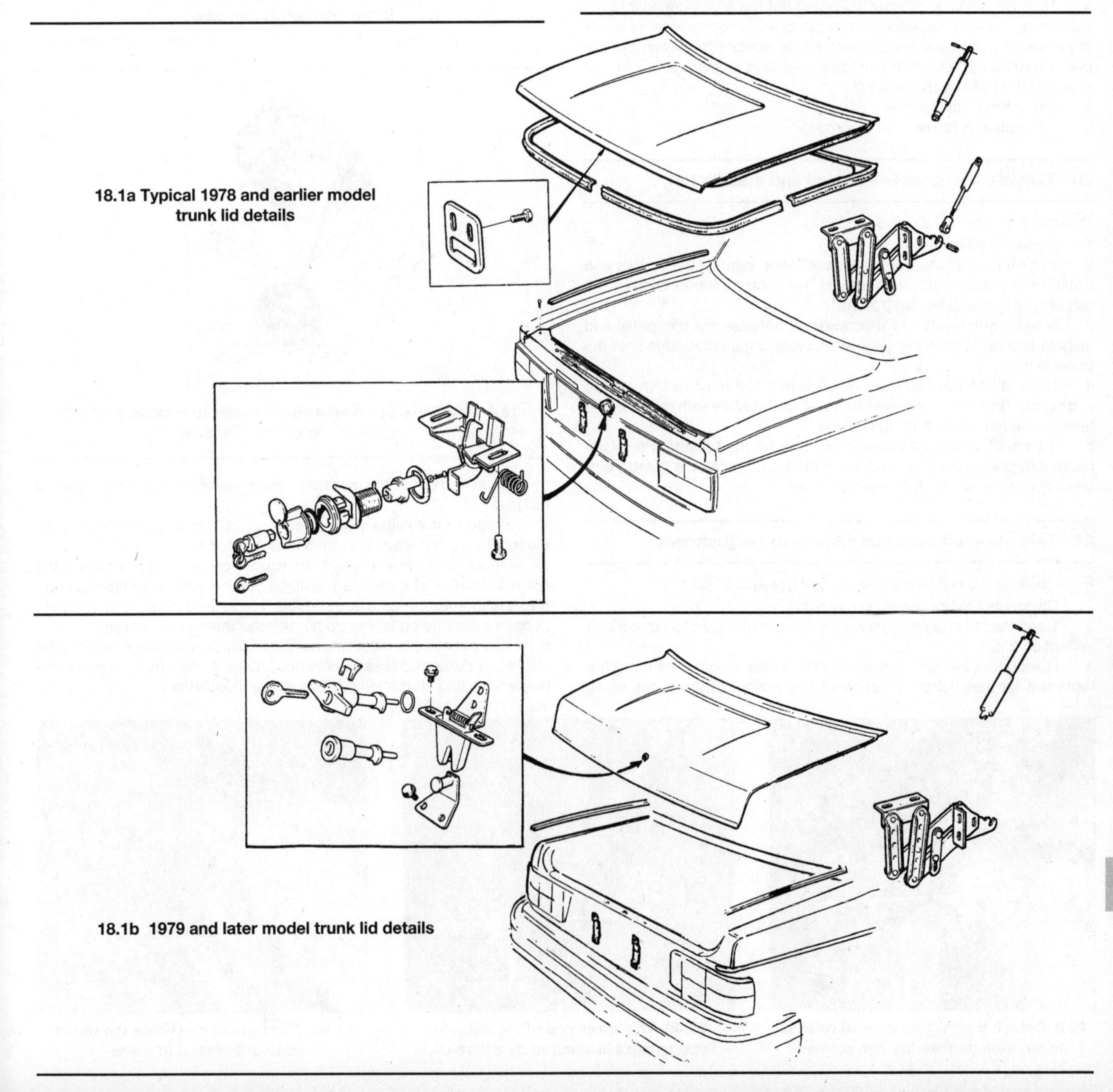
18.1a Typical 1978 and earlier model trunk lid details

18.1b 1979 and later model trunk lid details

4 The position of the lid can be adjusted using the elongated holes in the hinge attachment.
5 Removal of the hinge and balance spring assembly will require a special a tool (available from a dealer) to keep the spring compressed before removing the mounting bolts **(see illustration)**.

19 Trunk latch and lock - removal and installation

Refer to illustration 19.4

1 On early models, the lock and latch are in the rear body valance and the striker plate is in the trunk lid, while on later models this is reversed **(see illustrations 18.1a and 18.1b).**
2 Remove the bolts securing the latch or striker to the body or trunk lid and detach the plate or striker.
3 Install in reverse order, then adjust the latch or striker in the elongated holes to obtain correct closure of the trunk lid.
4 Remove the lock cylinder, by prying out the clip securing it to the valance or trunk lid. On later models the lock cylinder is located inside the trunk lid panel so a special tool will be needed to detach the clip **(see illustration)**. Also, some models may have an extra screw or bolt securing the cylinder to the latch.
5 Lift out the lock cylinder.
6 Installation is the reverse of removal.

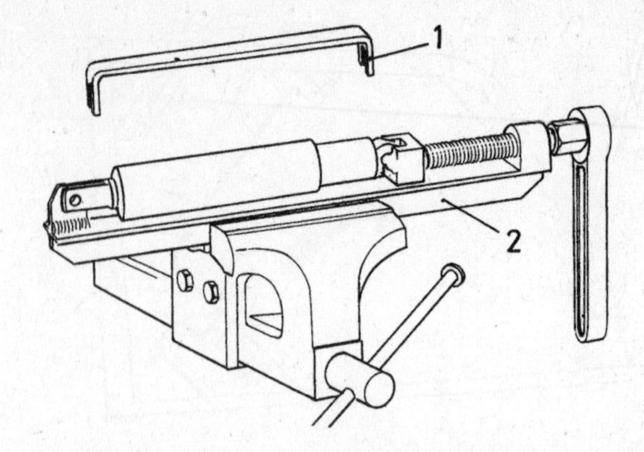

18.5 This type of tool is necessary for holding the spring in the compressed position
1) Holding tool 2) Spring compressor

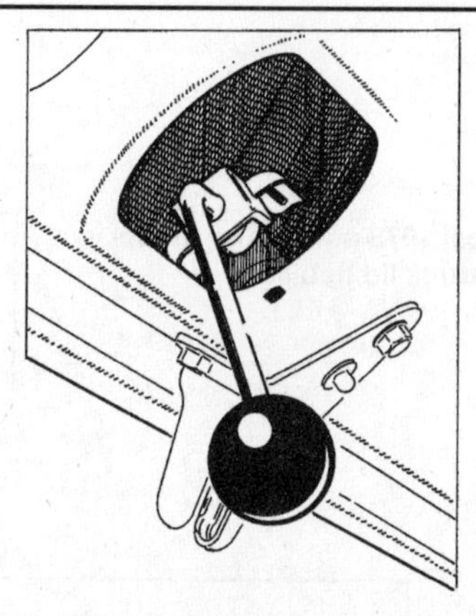

19.4 A tool like this one makes it easier to remove the lock cylinder clip on later models

20 Tailgate trim panel – removal and installation

Refer to illustration 20.2

1 Open the tailgate.
2 Pull off the tailgate internal release cover, remove the screws and detach the plastic trim piece around the outer circumference of the tailgate trim panel **(see illustration)**.
3 Insert a putty knife or flat screwdriver between the trim panel and tailgate and disengage the clips. Work around the outer edge until the panel is free.
4 Once all of the clips are disengaged, detach the trim panel, unplug any wire harness connectors that will interfere with removal and remove the trim panel from the vehicle.
5 To install, place the panel in position on the tailgate. Press the panel into place until the clips are seated and install the plastic trim piece and screws. Install the release cover.

21 Tailgate - removal, installation and adjustment

Refer to illustrations 21.4a, 21.4b, 21.5, 21.6a and 21.6b

1 Disconnect the negative battery cable.
2 Open the tailgate and remove the interior trim panel as described in Section 20.
3 Disconnect all electrical equipment in the tailgate, marking the terminals for installation. Disconnect the washer tube. All the wires must be pulled out of the tailgate at the point where they enter near the hinges.
4 Remove the clips from each end of the tailgate struts **(see illustrations)**, but leave the struts attached for now.
5 On early models, pry out the rubber covers which conceal the hinge bolts. On later models the hinge bolt is visible **(see illustration)**. Make alignment marks around the hinge bolt mounting flanges. Loosen the hinge bolts, but do not remove them at this stage.
6 You will need at least two people to support the tailgate while the struts are detached **(see illustration)**. Detach the struts, remove the hinge bolts and lift the tailgate away **(see illustration)**.

20.2 Detach the tailgate internal release cover, then remove the trim screws

21.4a The lower end of the tailgate support strut is secured by a wire clip

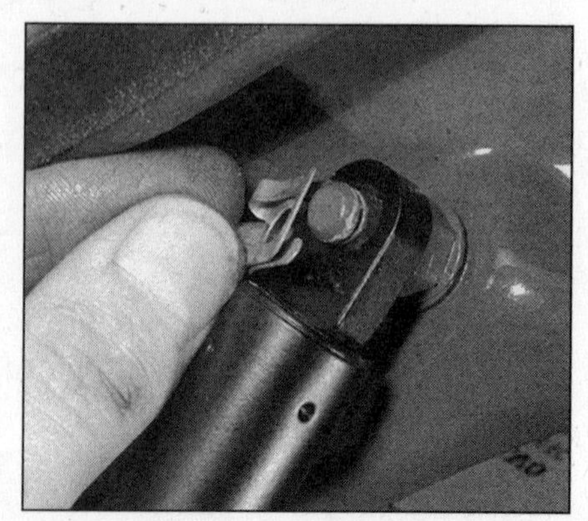

21.4b This type of clip holds the upper end of the strut in place

21.5 The tailgate hinge bolt is hidden behind the weather-stripping

21.6a Pull out sharply to detach the lower end of the strut

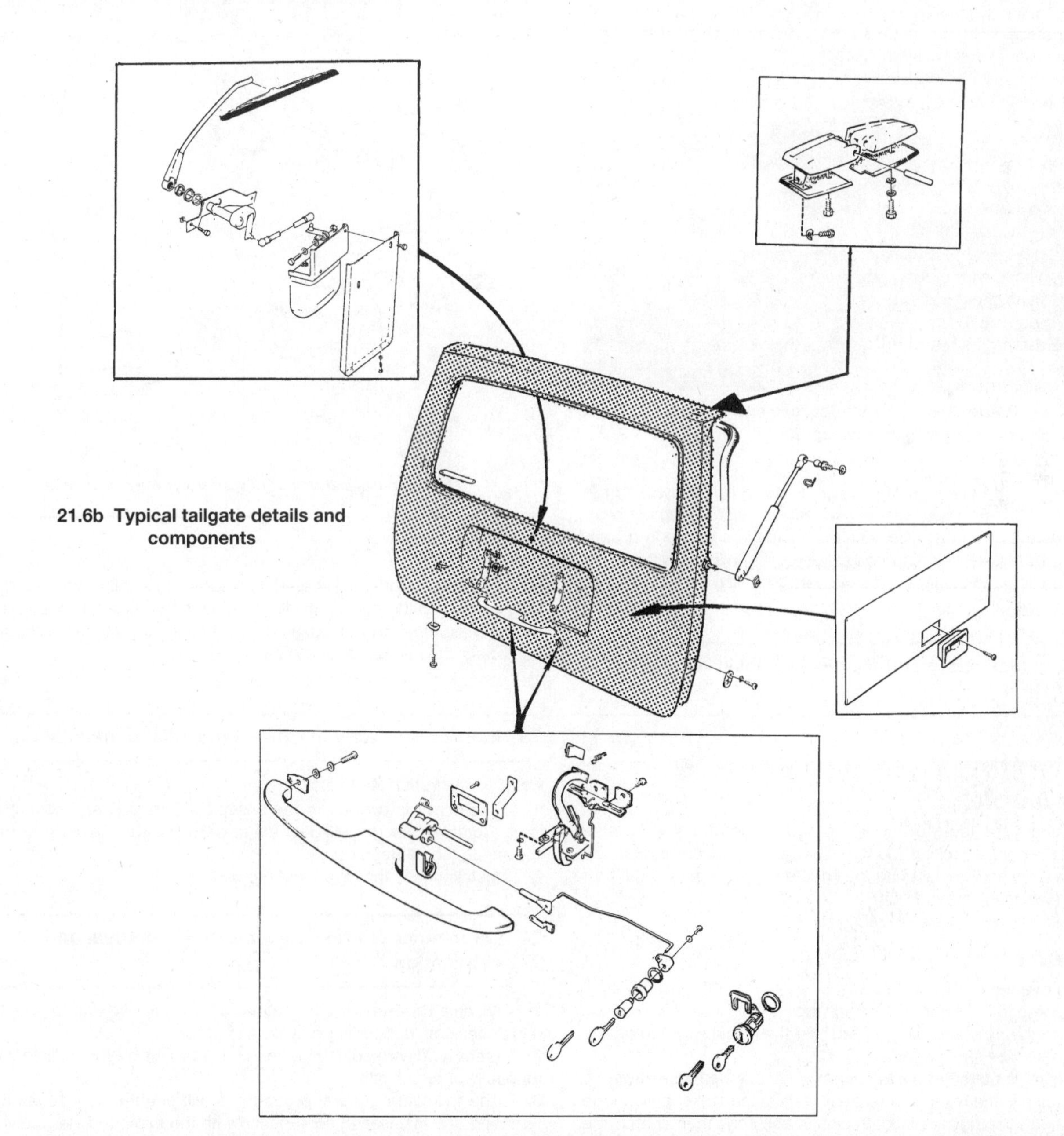
21.6b Typical tailgate details and components

22.6 Tailgate latch bolt locations

22.7 The tailgate striker is held in place by two bolts

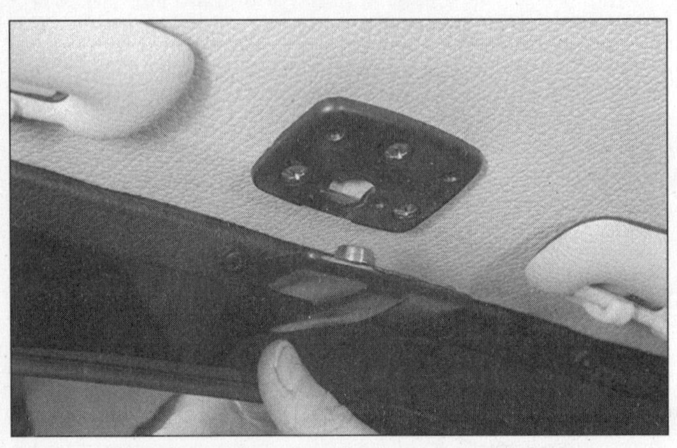
23.1 Rotate the mirror 90-degrees and detach it from the baseplate

7 Installation is the reverse of removal. **Note:** *When reinstalling the tailgate, align the lid-to-hinge bolts with the marks made during removal.*
8 After installation, close the tailgate and make sure it's in proper alignment with the surrounding body. Adjust the tailgate position by loosening the hinge to tailgate bolts and gently moving the tailgate into correct alignment.

22 Tailgate lock, latch and handle – removal and installation

Refer to illustrations 22.6 and 22.7
1 Remove the interior trim panel as described in Section 20.

Exterior handle

2 Disconnect the lock-to-handle operating rod. If installed, also disconnect the central locking rod from the motor (refer to Chapter 12).
3 Remove the screws securing the handle to the tailgate and lift off the handle **(see illustration 21.6b)**
4 Installation is the reverse of removal.

Latch and striker

5 To remove the latch, disconnect the operating rod from the lock, and if equipped, the rod from the central locking motor.
6 Remove the bolts securing the latch to the tailgate and lift off the latch assembly **(see illustration)**.
7 The striker is bolted to the luggage area floor **(see illustration)**.
8 Installation is the reverse of removal. Adjust the striker plate in the elongated holes so that the tailgate closes securely, then tighten the screws.

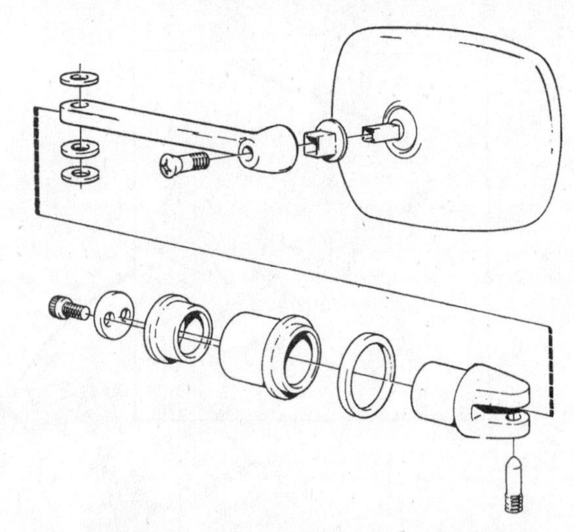
24.4 Typical early model rear view mirror details

Lock cylinder

9 The lock cylinder is secured by a screw accessible from inside, or by a locking spring plate in similar fashion to the door latches, and can be removed after disconnecting the operating rods to the latch, and if equipped, the central locking motor.
10 Installation is the reverse of removal.

23 Interior rear view mirror – removal and installation

Refer to illustration 23.1
1 The mirror is clipped to the baseplate by a ball-and-socket type joint. Rotate the mirror 90-degrees to detach it then lower it from the baseplate **(see illustration)**.
2 Installation is the reverse of removal.

24 Door-mounted rear view mirrors - removal and installation

1 Models covered by this manual were equipped with a variety of mirror designs, depending on year and model.
2 Later types are adjustable remotely from inside the vehicle, either manually or electrically.
3 The following general procedures will enable most types to be removed and installed in conjunction with the exploded views given in the various illustrations.

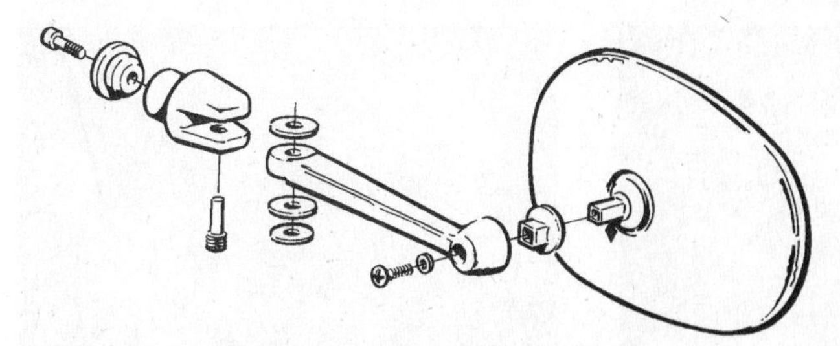

24.5 Typical mirror used on 1978 and 1979 models

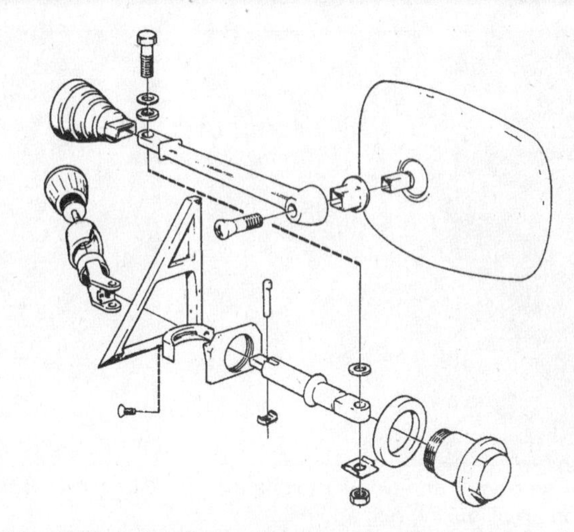

24.6a 1980 and earlier model manually adjusted mirror details

24.6b Pry off the cover and use an Allen wrench to remove the mount

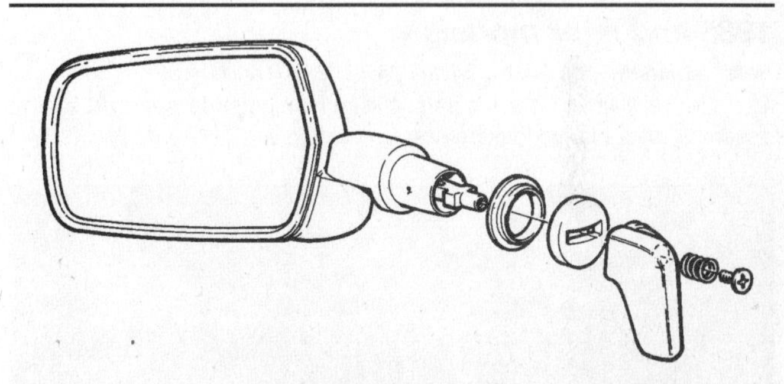

24.7a Typical 1980 and later model manually adjusted mirror details

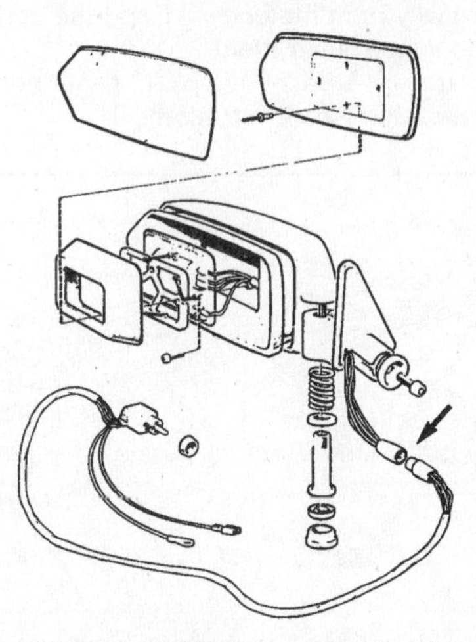

24.7b Typical power mirror details

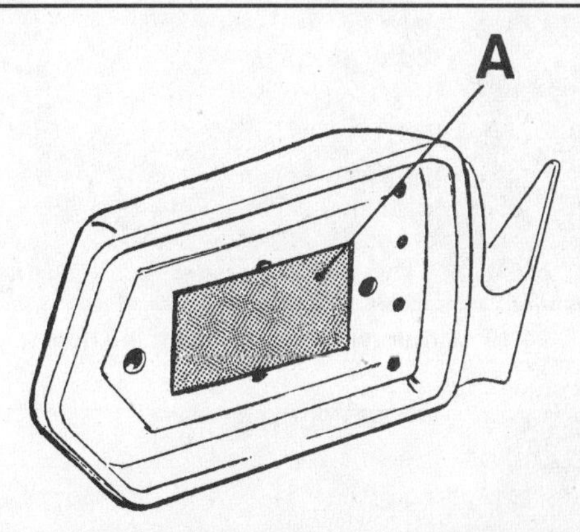

24.11 Apply the adhesive pad (A) to the center of the mirror assembly

1980 and earlier models

Refer to illustrations 24.4, 24.5, 24.6a and 24.6b

4 On 1977 and earlier models, the mirror is held in place on the arm by a securing screw **(see illustration)**.

5 On 1978 and 1977 models, the arm is secured to the door mount by an Allen head screw **(see illustration)**.

6 To remove the manually adjusted mirror, pry off the trim panel from the inside front corner of the window and remove the Allen head screw securing the bolt to the window **(see illustrations)**. Installation is the reverse of removal.

1980 through 1985 models

Refer to illustrations 24.7a, 24.7b and 24.11

7 Two types of remote control mirrors are used on 1980 to 1985 models **(see illustrations)**.

8 The mirror glass is attached to the backplate by an adhesive pad.

9 The mirror glass can be replaced without removing the housing from the door. Use a wide-bladed tool to pry off the mirror glass. This job can be made easier by using a heat gun or hair drier to heat the mirror glass and soften the adhesive. Take precautions against being cut by wearing heavy gloves and eye protection.

10 Clean off the old adhesive using rubbing alcohol or a similar solvent.

11 A new adhesive pad should be supplied with the new mirror glass. Peel off the backing paper and apply the pad to the backplate **(see illustration)**. If any air bubbles are trapped beneath the pad, puncture them with a pin.

12 Press the new glass into position, pressing only on the center of the glass to avoid breaking it.

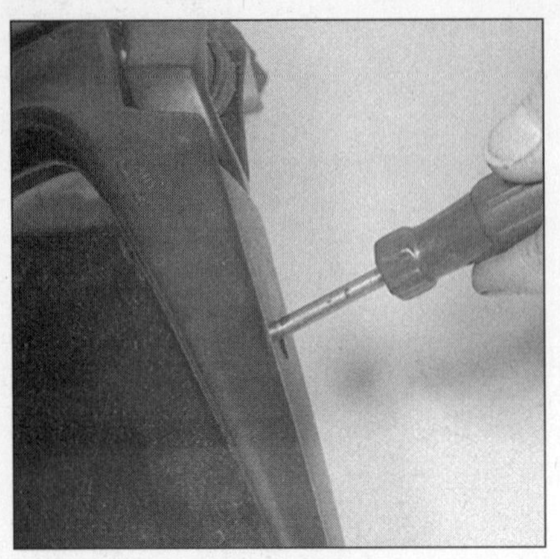
24.14 Use a small screwdriver to release the internal locking ring and detach the mirror glass

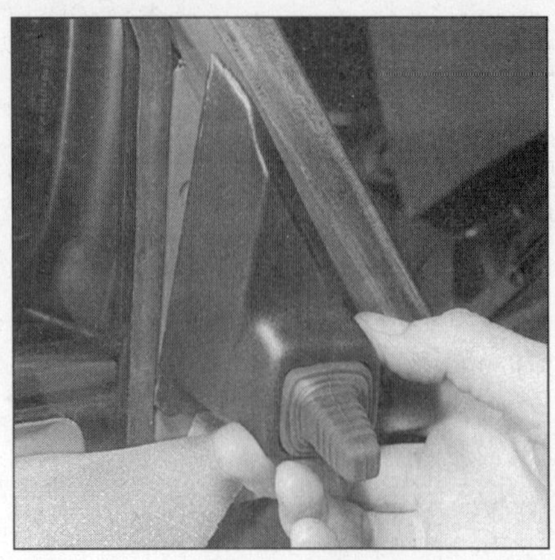
24.17 Pry off the mirror control trim piece

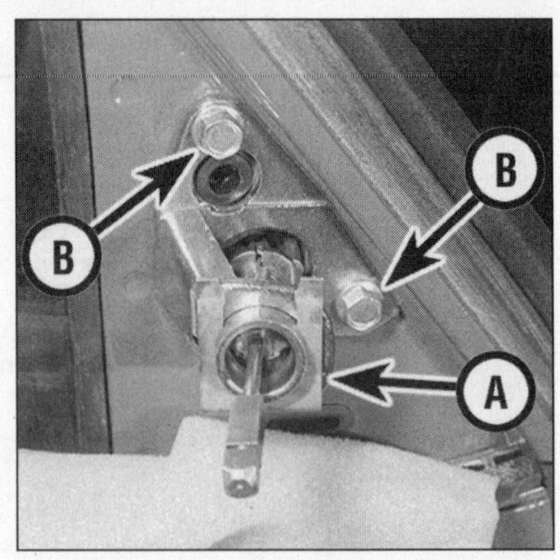

24.18 Remove clip (A) and the bolts (B)

1985 and later models

Refer to illustrations 24.14, 24.17, 24.18, 24.19 and 24.20

13 On 1985 and later models, the mirror glass is secured to the assembly by a plastic locking ring.

24.19 Withdraw the mirror from the door

24.20 Make sure the rubber grommet (arrow) is in place before inserting the mirror

14 To remove the mirror, press the lower edge of the glass inwards to line up the locking ring with the hole in the lower edge of the mirror assembly, then insert a narrow screwdriver into the hole and pry the locking ring to the right to unlock it and release the mirror glass **(see illustration)**.

15 Fitting the new glass is the reverse of this procedure.

16 To remove the complete assembly from the door, first remove the door interior trim panel (see Section 13).

17 Pull or pry off the trim piece from the mirror control **(see illustration)**.

18 Remove the spring clip and securing bolts **(see illustration)**.

19 Pull the mirror away from the door, guiding the control through the opening in the door **(see illustration)**.

20 Installation is the reverse of removal, ensuring the rubber grommet is correctly located **(see illustration)**.

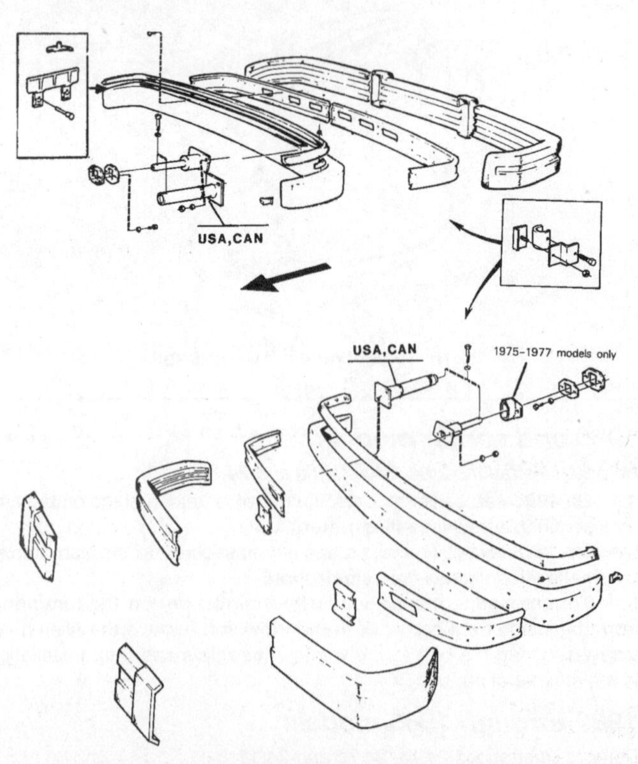

25.1a Typical 1980 and earlier model front and rear bumper details

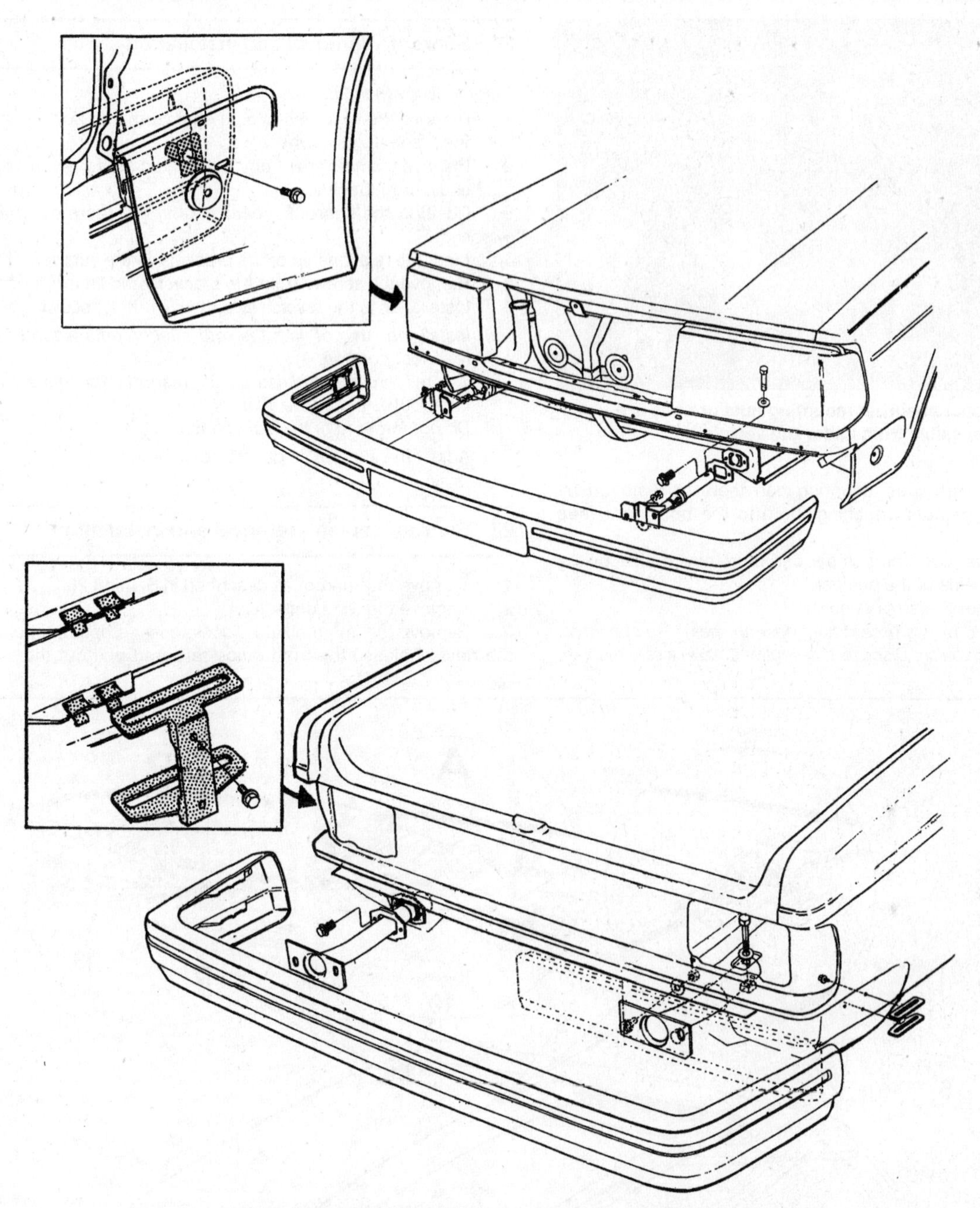

25.1b Typical 1981 and later model front and rear bumper details

All power-remote models

21 The motor is located in the mirror assembly, behind the glass.
22 On 1979 and earlier models, the connector for the electrical leads is located in the door, on later types it is located behind the fuse block.
23 Removal and installation procedures are similar to that described for the manually-adjusted mirrors, with the addition of disconnecting the wiring and feeding the wires through as the mirror assembly is removed. Note that the mirrors may be fixed with adhesive, or on later types, with a locking ring.
24 The motor can be removed without removing the complete assembly from the door, by removing the glass and undoing the screws securing the motor to the backing plate, after removing the rubber cover over the motor.
25 Installation is the reverse of removal.

25 Bumpers - removal and installation

Refer to illustrations 25.1a, 25.1b and 25.5

1 The bumpers may be attached to the body by rubber type impact absorbers (early models) or by gas-filled absorbers (later models) **(see illustrations)**.
2 To remove the bumper, first pull off the rubbing strip, being careful not to damage the rubber studs by which it is attached. Some types of rubbing strip use clips and a retaining strip, and may also have studs, secured by a nut accessible from inside the trunk/luggage area after the flooring is removed.
3 Remove the nuts/bolts from the side wrap-around mounting.
4 Remove the nuts securing the bumper to the impact absorber or mounting bracket, and remove the bumper.

25.5 The impact absorber mounting nuts or bolts are accessible from under the vehicle

5 The impact absorber or mounting can then be removed by removing the bolts or nuts attaching them to the bodywork **(see illustration)**.

6 The rubber cover over the bumper can be removed after prying off the clips on the inside of the bumper.

7 Installation is the reverse of removal.

Warning: *Do not weld or use excess heat near any gas-filled absorbers as there is a risk of explosion. Dispose of defective units in a safe manner.*

26 Sunroof - removal and installation

Refer to illustration 26.1

1 Remove the wind deflector (if equipped) and position the sunroof half-open **(see illustration)**.

2 Release the headliner from the front edge of the sunroof and slide the headliner to the rear.

3 Position the sunroof forwards until the rearmost brackets are exposed.

4 Mark the relationship of the brackets to the sunroof.

5 Remove the screws from the brackets and lift off the sunroof.

6 Installation is the reverse of removal, noting following points:

a) *Install the sunroof with the roof attachments wound back to the rearmost position.*

b) *The reinforcement plates on the rear attachments are inserted in the grooves below the sunroof.*

c) *Do not forget to fit the leaf springs.*

7 Adjust the sunroof as described in Section 27.

27 Sunroof cables - removal and installation

1 Remove the sunroof as described in Section 26.

2 Remove the wind deflector.

3 Remove the intermediate pieces, cover strips and holders above the drive. Release the front guide rails and pull out the cables **(see**

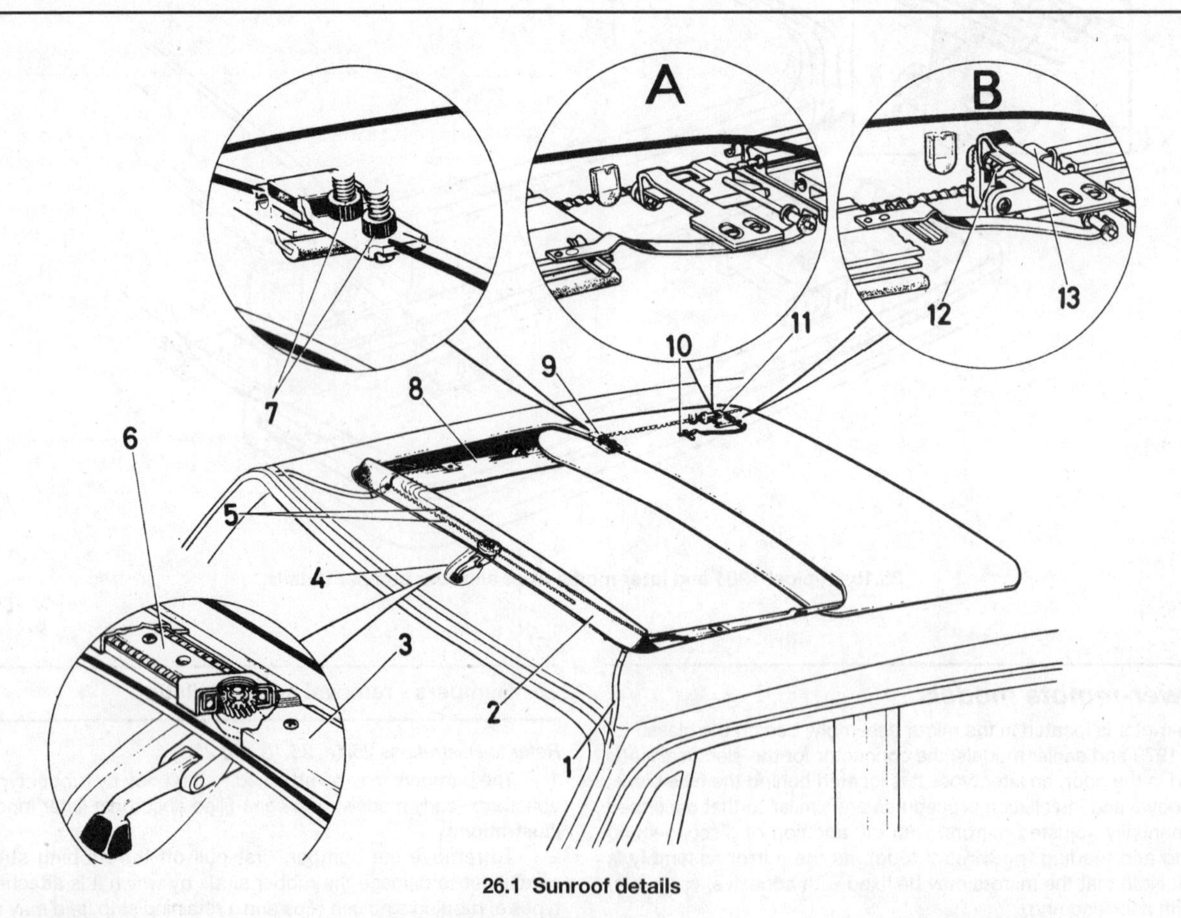

26.1 Sunroof details

1 *Drain hose*
2 *Wind deflector*
3 *Cover strip*
4 *Crank and housing*
5 *Cables*
6 *Front guide rail*
7 *Front adjustment*
8 *Intermediate piece*
9 *Front attachment*
10 *Leaf spring*
11 *Rear attachment*
12 *Rear adjustment*
13 *Reinforcing plate*
A *Rear attachment - roof open*
B *Rear attachment - roof closed*

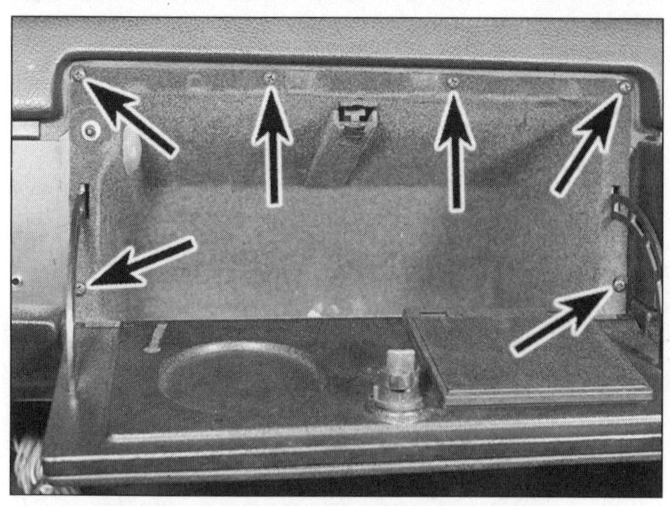

29.2 Typical glovebox retaining screw locations

illustration 26.1).

4 Install the replacement cables with the attachments for the sunroof opposite each other at the rear end of the roof opening. Screw the front guide rails on securely.

5 Install the intermediate pieces, holders, cover plates and the wind deflector.

6 Install the sunroof securely and reinstall the leaf springs.

7 Crank the sunroof forwards until it's completely closed and check that it's level with the main roof. If necessary, adjust at the front and rear attachments (**see illustration 26.1**). Make sure that the lifts on the rear attachments are pushed up when the roof is closed.

8 When the roof is closed the crank should point straight forwards in the vehicle. If it does not, unscrew the crank and gear housing, turn the crank to the stop position and replace.

28 Under-dash panels - removal and installation

Warning: *Some models are airbag-equipped. Always disconnect the negative battery cable and unplug the yellow electrical connector under the steering column when working in the vicinity of the crash sensor or steering column to avoid the possibility of accidental deployment of the airbag, which could cause personal injury* (see Chapter 12).

1 The soundproofing panels under the dash and to each side of the

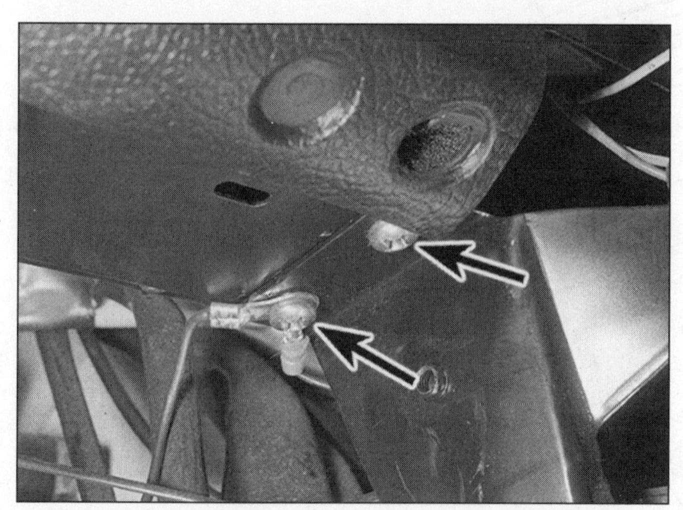

30.8a Typical console support frame upper mounting screw locations (arrows)

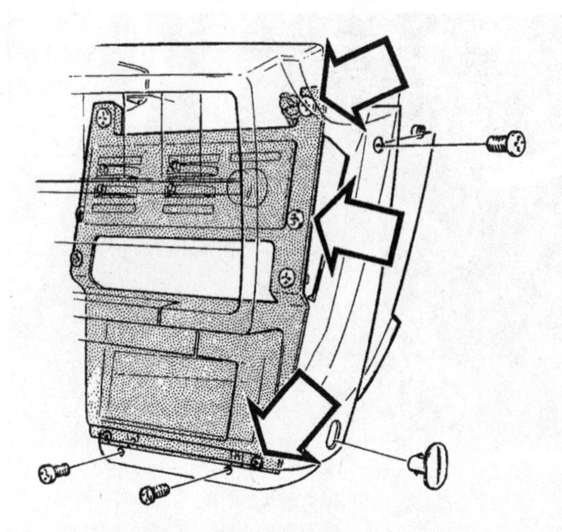

30.4 Center console trim plate screw locations (arrows)

center console/heater assembly are secured either by large plastic screws, self-tapping screws or clips. In some cases all three methods may be utilized.

2 Remove whichever securing method is encountered and lift out the panels.

3 Installation is the reverse of removal.

29 Glovebox - removal and installation

Refer to illustration 29.2

Warning: *Some models are airbag-equipped. Always disconnect the negative battery cable and unplug the yellow electrical connector under the steering column when working in the vicinity of the crash sensor or steering column to avoid the possibility of accidental deployment of the airbag, which could cause personal injury* (see Chapter 12).

1 Open the glovebox door.

2 Remove the retaining screws **(see illustration)**.

3 Pull the glovebox forward sufficiently to enable the wires to the light and microswitch (if equipped) to be disconnected.

4 Remove the glovebox.

5 Installation is the reverse of removal.

30 Center console - removal and installation

Refer to illustrations 30.4, 30.8a and 30.8b

Warning: *Some models are airbag-equipped. Always disconnect the negative battery cable and unplug the yellow electrical connector under the steering column when working in the vicinity of the crash sensor or steering column to avoid the possibility of accidental deployment of the airbag, which could cause personal injury* (see Chapter 12).

1 Disconnect the negative battery cable.

2 Remove the ashtray.

3 Remove the radio.

4 Remove the screws securing the plastic outer panel to the inner metal support frame **(see illustration)**.

5 Pull the plastic outer panel off the support frame and disconnect all switches noting their connections (one way to do this is to remove each switch in turn and reconnect it to the leads after removal.

6 Remove the plastic outer panel.

7 If further dismantling of the support frame is required, the heater control panel should be removed as described in Chapter 3.

8 Remove the support frame mounting screws, noting that some of them may have ground wires under them **(see illustrations)** and pull out the frame.

9 Installation is the reverse of removal.

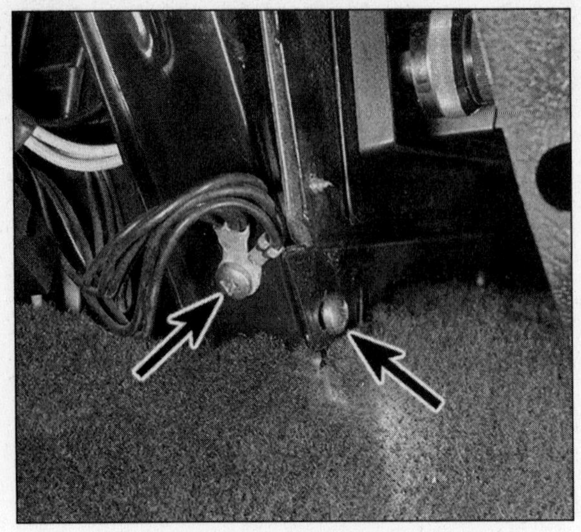
30.8b Lower console bracket screws (arrows)

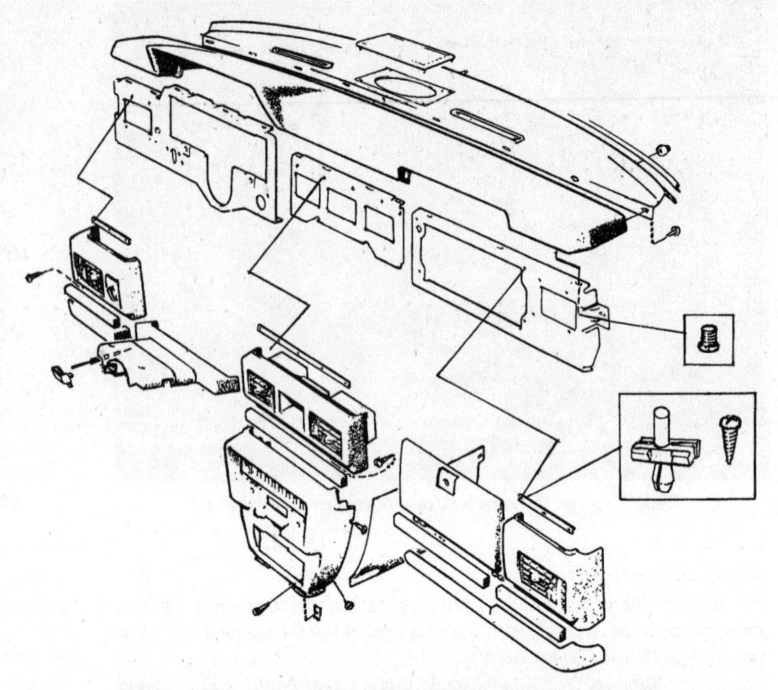
31.4a 1980 and earlier models instrument panel details

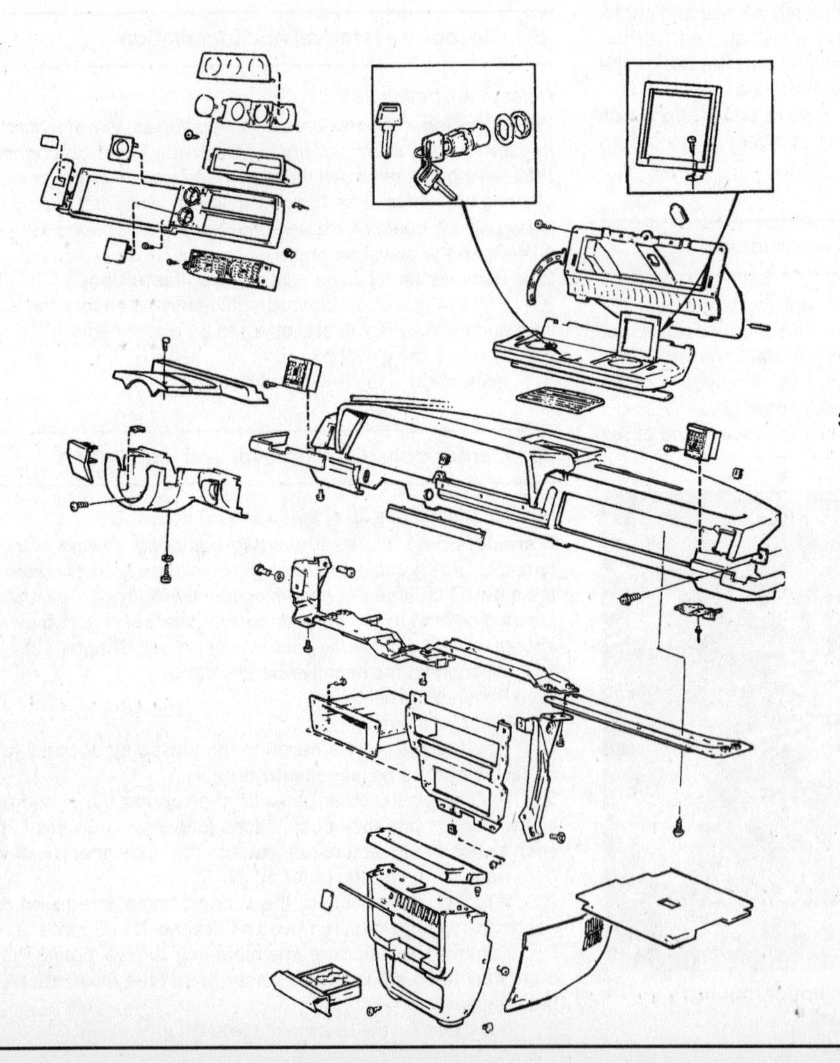
31.4b 1981 and later model instrument panel details

31.8a Left side vent screw location

31.8b Right side vent screw location

31.8c After removing the screws, detach the vent

31 Instrument panel - removal and installation

Refer to illustrations 31.4a, 31.4b, 31.8a, 31.8b, 31.8c and 31.9

Warning: *Some models are airbag-equipped. Always disconnect the negative battery cable and unplug the yellow electrical connector under the steering column when working in the vicinity of the crash sensor or steering column to avoid the possibility of accidental deployment of the airbag, which could cause personal injury* (see Chapter 12).

1 Disconnect the negative battery cable.

2 Remove the steering column cover, and if required for greater access, the steering wheel (see Chapter 10).

3 Remove the center console (see Section 30).

4 Remove the instrument cluster **(see illustrations)**.

5 Remove the auxiliary instrument panel, if equipped.

6 Remove the glovebox (see Section 29).

7 Pull off the padding strip from along the front edge of the dashboard panel.

8 This will reveal the screws which secure the air vent outlets to the dashboard. Remove these screws and pull out the vents **(see illustrations)**.

9 The remainder of the procedure consists of working around the instrument panel, removing the screws and clips which secure the crash padding to the under-frame **(see illustration)**. The under-frame can then be removed if desired by removing the securing screws/bolts.

10 Take careful note of all the screws and bolts you remove, as they can be of different sizes for different locations, and some will have ground wires or relay clips secured under them.

11 Installation is the reverse of removal.

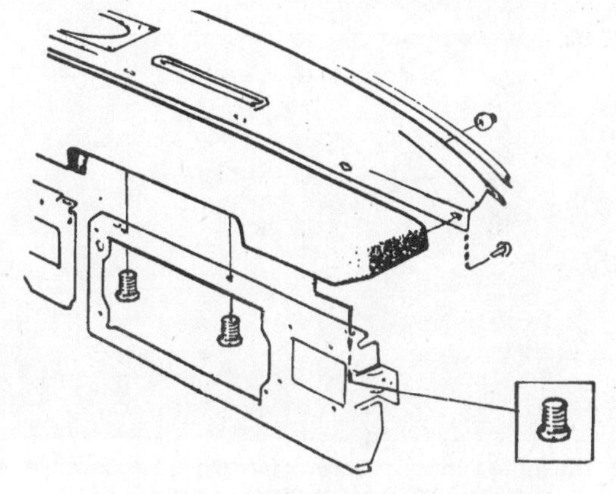

31.9 Crash padding attachment details

32 Parking brake console - removal and installation

Refer to illustrations 32.1a, 32.1b, 32.2, 32.5 and 32.6

Warning: *Some models are airbag-equipped. Always disconnect the negative battery cable and unplug the yellow electrical connector under the steering column when working in the vicinity of the crash sensor or steering column to avoid the possibility of accidental deployment of the airbag, which could cause personal injury* (see Chapter 12).

1 Remove the rear ashtray and pry out the insert **(see illustrations)**.

2 Push out the seat belt warning light and rear courtesy light and disconnect the wiring **(see illustration)**.

32.1a Remove the ashtray . . .

32.1b . . . pry out the insert

32.2 Disconnect the electrical connectors

32.5 Detach the console from the locating notch at the front edge

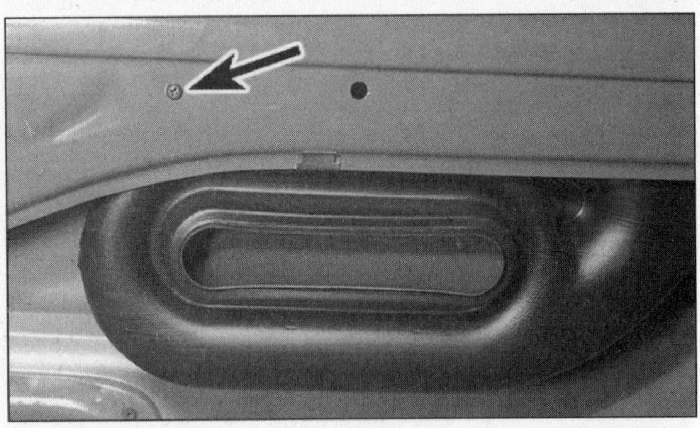

33.2 The inner vent duct screws (arrow) are accessible after removing the trim panel

3 Pry out and disconnect the seat heating pad switches (if equipped).
4 Slide both front seats forward as far as they will go and remove the screws from the rear sides of the console.
5 Lift the console rear edge up, slide it forward off the locating notch at the front and lift the console up over the parking brake **(see illustration)**.
6 On some models, the console must be detached from the seat belt mounts **(see illustration)**.
7 Installation is the reverse or removal.

33 Rear air extractor vent (station wagon models) - removal and installation

Refer to illustration 33.2

1 The outer grille is secured by four screws. Remove the screws and remove the vent grille.
2 The inner vent duct can be reached after removing the luggage compartment trim panel. It is secured by screws **(see illustration)**. Remove the screws and remove the vent duct.
3 Installation is the reverse of removal.

34 Seat headrests - removal and installation

Refer to illustration 34.1

1 On later models, the head restraints can be removed by pushing against the spring retainers while applying upward pressure to the head restraint to release them **(see illustration)**.
2 Installation is the reverse of removal.

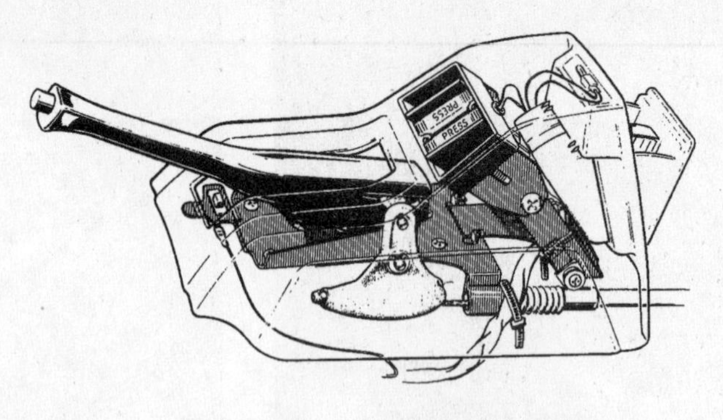

32.6 Parking brake console details

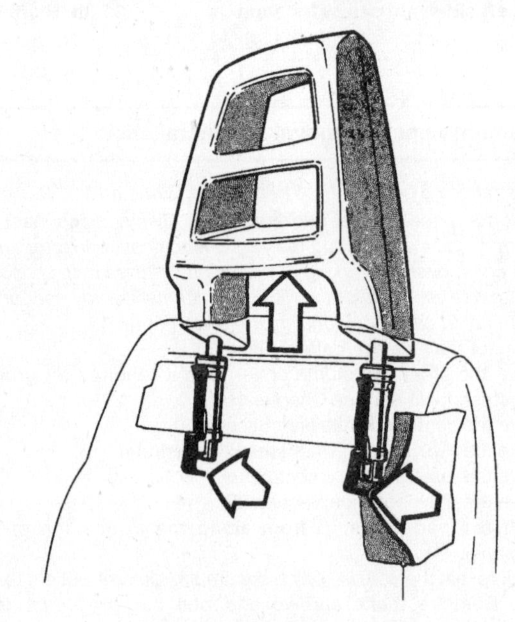

34.1 Press through the covering at the areas shown (arrows) to release the retaining clips on later models

35 Seats - removal and installation

Warning: *Some models are airbag-equipped. Always disconnect the negative battery cable and unplug the yellow electrical connector under the steering column when working in the vicinity of the crash sensor or steering column to avoid the possibility of accidental deployment of the airbag, which could cause personal injury* (see Chapter 12).

Front

1 If equipped, disconnect the seat heating pad connector.
2 Disconnect the warning light connector under the front passenger seat.
3 Remove the bolts securing the seat runners to the floorpan, there are four of them, and lift out the seat.

Rear

Sedan

4 Disengage the clips at the lower front edge of the seat cushion, pull the cushion forwards, turn it sideways and remove it from the vehicle.
5 The seat back should be pushed up to disengage the locating clips at its rear, after which it can be removed.

37.2 Detach the cover and remove the seat belt bolt

37.9 Remove the bolts and detach the inertia reel mounting bracket

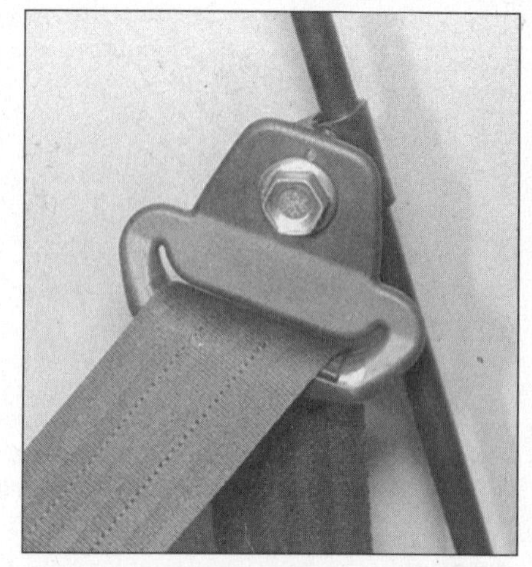
37.10 Seat belt upper mounting bolt location (station wagon)

Station wagon

6 Push down the release lever below the front edge of the seat and raise the cushion to the vertical position.

7 Push the seat cushion rearwards to unhook the hinge, then remove cushion from the vehicle.

8 To remove the seat back, operate the release handle and fold the seat forwards. Detach the fold-down panel.

9 Ease the spring-loaded swivel bars rearwards to release the seat back so that it can be lifted out.

Installation

10 Installation of all seats is the reverse of removal.

36 Seat belt check

1 Check the seat belts, buckles, latch plates and guide loops for any obvious damage or signs of wear.

2 Make sure the seat belt reminder light comes on when the key is turned on.

3 The seat belts are designed to lock up during a sudden stop or impact, yet allow free movement during normal driving. The retractors should hold the belt against your chest while driving and rewind the belt when the buckle is unlatched.

4 If any of the above checks reveal problems with the seat belt system, replace parts as necessary.

37 Seat belts - removal and installation

Refer to illustrations 37.2, 37.9, 37.10, 37.11a and 37.11b

Warning: *Some models are airbag-equipped. Always disconnect the negative battery cable and unplug the yellow electrical connector under the steering column when working in the vicinity of the crash sensor or steering column to avoid the possibility of accidental deployment of the airbag, which could cause personal injury* (see Chapter 12).

Front belts

1 The lower mountings are bolted to the inner floorpan at the bottom of the B-pillar. Remove the bolt to release the belt.

2 The upper mounting is bolted to the B-pillar at head height. Pry off the plastic cover and remove the bolt **(see illustration)**.

3 The inertia reel is concealed behind the trim panel on the B-pillar. Remove the trim and unscrew the inertial reel securing bolt.

4 The inboard buckles are bolted to the transmission tunnel at each side of the parking brake console.

37.11a The seat belt buckle mounting bolt is hidden by the seat back

Note: *On some models, the buckles may be inside the parking brake console, which must be removed for access to the bolts* (see Section 32).

Rear belts

Sedan

5 The lower mountings are bolted to the floorpan in much the same way as the front belts. The bolts are accessible after removal of the seat cushion.

6 The upper mountings for the two outer belts are bolted to the C-pillars, again using a similar method as the front belts.

7 The inertia reels are located under plastic covers at each end of the package shelf, and are similar to the units fitted to the rear of the station wagon.

Station wagon

8 Remove the trim panels from the side of the luggage area to gain access to the inertia reels.

9 The reel is mounted on a bracket, which is secured by three bolts **(see illustration)**.

10 The upper mounting is clamped to a support bar **(see illustration)**.

11 The lower mountings are bolted to the floorpan and wheel arches, and are accessible after folding back the seat cushion **(see illustrations)**.

37.11b Center lap belt mounting bolt

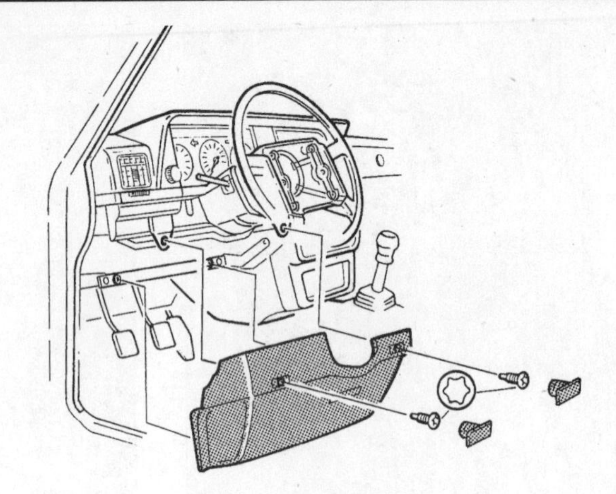

39.1 Knee bolster details

All belts

12 Installation is the reverse of removal, observing the originally fitted sequence of any washers, spacers etc. Tighten all bolts securely.

38 Headliner - removal and installation

Sedan

1 The roof lining is a molded fiber panel.
2 Remove the interior fittings secured to the roof-grab handles, rear view mirror, sun visors and courtesy light.
3 On 1979 and earlier models, the rear window has to be removed. See Section 8.
4 On 1980 and later models, remove the complete rear seat to gain access to the C-pillar trim panels.
5 Disconnect the heated rear window electrical connector (if equipped).
6 Remove the trim panel from above the windshield, and the trim panels above the B-pillar.
7 Remove the retaining screws, detach the headliner and remove it through the front passenger's door. Place the gear selector in reverse (manual) or Park (automatic) for clearance. Handle the headliner with care as it is easily damaged.

Station wagon

8 The headliner on station wagon models is made of a cloth material, stretched over metal bows tensioned across the roof panel.
9 Carry out the preliminary operations as for the Sedan version, except that there is no need to remove the rear seats.
10 Pull the edge of the headliner from the lip of the roof panel, working around the roof until it's free.
11 Starting from the rear, disengage the roof bows by bending them gently in the middle, working forwards to the front. Lift out the headliner and bows.
12 When installing a new headliner or replacing the bows, make sure the bows have protective plastic end caps installed to prevent damage to the headliner material.

All models

13 Installation is the reverse of removal. Start at the front, attaching the bows to the roof, and work to the rear.

39 Steering column cover - removal and installation

Refer to illustrations 39.1, 39.2a and 39.2b

Warning: *Some models are airbag-equipped. Always disconnect the negative battery cable and unplug the yellow electrical connector under the steering column when working in the vicinity of the crash sensor or steering column to avoid the possibility of accidental deployment of the airbag, which could cause personal injury* (see Chapter 12).

1 On airbag-equipped models, remove the knee bolster **(see illustration)**.
2 Remove the cover retaining screws, separate the two halves and remove the covers from the column **(see illustrations)**.
3 Installation is the reverse of removal, making sure that the two halves are clipped together correctly before tightening the screws.

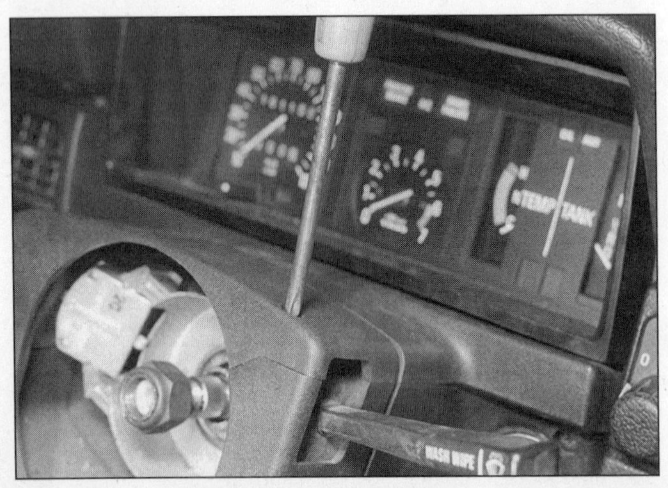

39.2a Remove the upper . . .

39.2b . . . and lower steering column cover screws

Chapter 12 Chassis electrical system

Contents

Specifications

Electronic speedometer sensor tip-to-pickup clearance	0.03 inch

Torque Specifications

	Ft-lbs (unless otherwise indicated)
Steering wheel nut	44
Steering column lower bracket bolts	15
Steering shaft flange bolts	17
Coupling pinch bolts	15

1 General information

The electrical system is a 12-volt, negative ground type. Power for the lights and all electrical accessories is supplied by a lead/acid-type battery which is charged by the alternator.

This Chapter covers repair and service procedures for the various electrical components not associated with the engine. Information on the battery, alternator, distributor and starter motor can be found in Chapter 5.

It should be noted that when portions of the electrical system are serviced, the negative battery cable should be disconnected from the battery to prevent electrical shorts and/or fires.

2 Electrical troubleshooting - general information

A typical electrical circuit consists of an electrical component, any switches, relays, motors, fuses, fusible links or circuit breakers related to that component and the wiring and connectors that link the component to both the battery and the chassis. To help you pinpoint an electrical circuit problem, wiring diagrams are included at the end of this book.

Before tackling any troublesome electrical circuit, first study the appropriate wiring diagrams to get a complete understanding of what makes up that individual circuit. Trouble spots, for instance, can often be narrowed down by noting if other components related to the circuit are operating properly. If several components or circuits fail at one time, chances are the problem is in a fuse or ground connection, because several circuits are often routed through the same fuse and ground connections.

Electrical problems usually stem from simple causes, such as loose or corroded connections, a blown fuse, a melted fusible link or a bad relay. Visually inspect the condition of all fuses, wires and connections in a problem circuit before troubleshooting it.

If testing instruments are going to be utilized, use the diagrams to plan ahead of time where you will make the necessary connections in order to accurately pinpoint the trouble spot.

The basic tools needed for electrical troubleshooting include a circuit tester or voltmeter (a 12-volt bulb with a set of test leads can also be used), a continuity tester, which includes a bulb, battery and set of test leads, and a jumper wire, preferably with a circuit breaker incorporated, which can be used to bypass electrical components. Before attempting to locate a problem with test instruments, use the wiring diagram(s) to decide where to make the connections.

Voltage checks

Voltage checks should be performed if a circuit is not functioning properly. Connect one lead of a circuit tester to either the negative battery terminal or a known good ground. Connect the other lead to a connector in the circuit being tested, preferably nearest to the battery or fuse. If the bulb of the tester lights, voltage is present, which means that the part of the circuit between the connector and the battery is problem free. Continue checking the rest of the circuit in the same fashion. When you reach a point at which no voltage is present, the problem lies between that point and the last test point with voltage. Most of the time the problem can be traced to a loose connection. **Note:** *Keep in mind that some circuits receive voltage only when the ignition key is in the Accessory or Run position.*

Finding a short

One method of finding shorts in a circuit is to remove the fuse and connect a test light or voltmeter in its place to the fuse terminals. There should be no voltage present in the circuit. Move the wiring harness from side-to-side while watching the test light. If the bulb goes on, there is a short to ground somewhere in that area, probably where the insulation has rubbed through. The same test can be performed on each component in the circuit, even a switch.

Ground check

Perform a ground test to check whether a component is properly grounded. Disconnect the battery and connect one lead of a self-powered test light, known as a continuity tester, to a known good ground. Connect the other lead to the wire or ground connection being tested. If the bulb goes on, the ground is good. If the bulb does not go on, the ground is not good.

Continuity check

A continuity check is done to determine if there are any breaks in a circuit - if it is passing electricity properly. With the circuit off (no power in the circuit), a self-powered continuity tester can be used to check the circuit. Connect the test leads to both ends of the circuit (or to the -power" end and a good ground), and if the test light comes on the circuit is passing current properly. If the light doesn't come on, there is a break somewhere in the circuit. The same procedure can be used to test a switch, by connecting the continuity tester to the switch terminals. With the switch turned On, the test light should come on.

Finding an open circuit

When diagnosing for possible open circuits, it is often difficult to locate them by sight because oxidation or terminal misalignment are hidden by the connectors. Merely wiggling a connector on a sensor or in the wiring harness may correct the open circuit condition. Remember this when an open circuit is indicated when troubleshooting a circuit. Intermittent problems may also be caused by oxidized or loose connections.

Electrical troubleshooting is simple if you keep in mind that all electrical circuits are basically electricity running from the battery, through the wires, switches, relays, fuses and fusible links to each electrical component (light bulb, motor, etc.) and to ground, from which it is passed back to the battery. Any electrical problem is an interruption in the flow of electricity to and from the battery.

3 Fuses - general information

Refer to illustration 3.1

The electrical circuits of the vehicle are protected by a combination of fuses, circuit breakers and fusible links. The fuse block is located at the rear corner of the drivers side kick panel, under a cover **(see illustration)**.

Each of the fuses is designed to protect a specific circuit, and the various circuits are identified on the fuse panel itself.

If an electrical component fails, always check the fuse first. A blown fuse is easily identified through the glass tube or clear plastic body. Visually inspect the element for evidence of damage. If the fuse looks good, but you still suspect it's faulty, perform a continuity check

3.1 The main fuse block is located under the left side kick panel

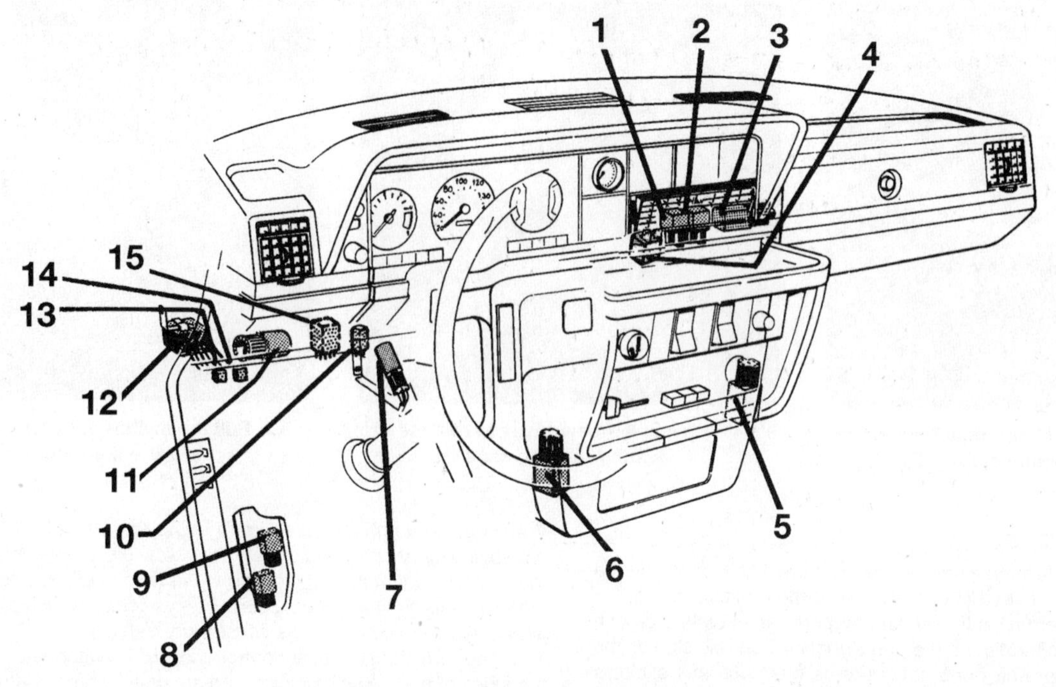

5.2 Typical relay locations

1 *Central door lock opening relay*
2 *Central door lock closing relay*
3 *Overdrive relay*
4 *Power window relay (1980 and later)*
5 *Air conditioning relay*
6 *Turn signal relay*
7 *Interior lighting time delay relay*
8 *Rear wiper intermittent relay*
9 *Windshield wiper intermittent relay*
10 *Rear window defogger relay*
11 *Bulb failure warning sensor*
12 *Fuel pump relay*
13 *Rear foglamp relay (1986 and later)*
14 *Main lighting relay (1985 and later)*
15 *Seatbelt reminder relay*

at the metal ends on glass tube fuses or the blade terminal tips exposed in the fuse body.

Be sure to replace blown fuses with the correct type. Fuses of different ratings are physically interchangeable, but only fuses of the proper rating should be used. Replacing a fuse with one of a higher or lower value than specified is not recommended. Each electrical circuit needs a specific amount of protection. The amperage value of each fuse is molded into the fuse body.

If the replacement fuse immediately fails, don't replace it again until the cause of the problem is isolated and corrected. In most cases, the cause will be a short circuit in the wiring caused by a broken or deteriorated wire.

4 Fusible links - general information

Some circuits are protected by fusible links. The links are used in circuits which are not ordinarily fused, such as the ignition circuit.

Although the fusible links appear to be a heavier gauge than the wire they are protecting, the appearance is due to the thick insulation. All fusible links are several wire gauges smaller than the wire they are designed to protect. The fusible links are color coded and a link should always be replaced with one of the same color obtained from a dealer.

Fusible links cannot be repaired, but a new link of the same size wire can be put in its place. The procedure is as follows:

a) *Disconnect the negative cable from the battery.*
b) *Disconnect the fusible link from the wiring harness.*
c) *Cut the damaged fusible link out of the wiring just behind the connector.*
d) *Strip the insulation back approximately 1-inch.*
e) *Position the connector on the new fusible link and twist or crimp it into place.*
f) *Use rosin core solder at each end of the new link to obtain a good solder joint.*
g) *Use plenty of electrical tape around the soldered joint. No wires should be exposed.*
h) *Connect the battery ground cable. Test the circuit for proper operation.*

5 Relays - general information

Refer to illustration 5.2

Several electrical accessories in the vehicle use relays to transmit the electrical signal to the component. If the relay is defective, that component will not operate properly.

The various relays are grouped together under the dash, as well as in other locations **(see illustration)**.

If a faulty relay is suspected, it can be removed and tested by a dealer service department or a repair shop. Defective relays must be replaced as a unit.

6 Turn signal/hazard flasher - check and replacement

Warning: *Some models are airbag-equipped. Always disconnect the negative battery cable and unplug the yellow electrical connector under the steering column when working in the vicinity of the crash sensor or steering column to avoid the possibility of accidental deployment of the airbag, which could cause personal injury (see Section 28).*

7.4 Column switch electrical connector (A) and retaining screws (B)

8.1 Use a small screwdriver to release the switch retaining clips

8.2 Pull the switch out and unplug the electrical connector

Turn signal flasher

1 The turn signal flasher, a small canister-shaped unit located in the left side of the center instrument panel, flashes the turn signals.

2 When the flasher unit is functioning properly, an audible click can be heard during its operation. If the turn signals fail on one side or the other and the flasher unit does not make its characteristic clicking sound, a faulty turn signal bulb is indicated.

3 If both turn signals fail to blink, the problem may be due to a blown fuse, a faulty flasher unit, a broken switch or a loose or open connection. If a quick check of the fuse box indicates that the turn signal fuse has blown, check the wiring for a short before installing a new fuse.

4 To replace the flasher, simply pull it out of the wiring harness.

5 Make sure that the replacement unit is identical to the original. Compare the old one to the new one before installing it.

6 Installation is the reverse of removal.

Hazard flasher

7 The hazard flasher flashes all four turn signals simultaneously when activated.

8 The hazard flasher is checked in a fashion similar to the turn signal flasher (see Steps 2 and 3).

9 The hazard flasher is actuated by the turn signal flasher unit (see Steps 4 and 5).

10 Installation is the reverse of removal.

7 Steering column switches - removal and installation

Refer to illustration 7.4

Warning: *Some models are airbag-equipped. Always disconnect the negative battery cable and unplug the yellow electrical connector under the steering column when working in the vicinity of the crash sensor or steering column to avoid the possibility of accidental deployment of the airbag, which could cause personal injury (see Section 28).*

1 The turn signal/ cruise control and wiper switches are located on the sides of the steering column and are operated by control stalks.

2 Disconnect the negative cable from the battery.

3 Remove the screws and detach the steering column cover (see Chapter 11).

4 Remove the retaining screws, unplug the switch electrical connector and lift the switch off **(see illustration)**.

5 Installation is the reverse of removal.

8 Instrument panel switches - removal and installation

Refer to illustrations 8.1, 8.2, 8.3a and 8.3b

Warning: *Some models are airbag-equipped. Always disconnect the negative battery cable and unplug the yellow electrical connector under the steering column when working in the vicinity of the crash sensor or steering column to avoid the possibility of accidental deployment of the airbag, which could cause personal injury (see Section 28).*

1 The toggle type panel switches can easily be detached by prying the retaining clips loose with a small screwdriver **(see illustration)**.

2 Pull the switch out, unplug the electrical connector and remove the switch from the dash **(see illustration)**.

3 Illuminated switches have a removable connection/socket, and the bulb is a press fit Installation **(see illustrations)**.

4 Installation is the reverse of removal.

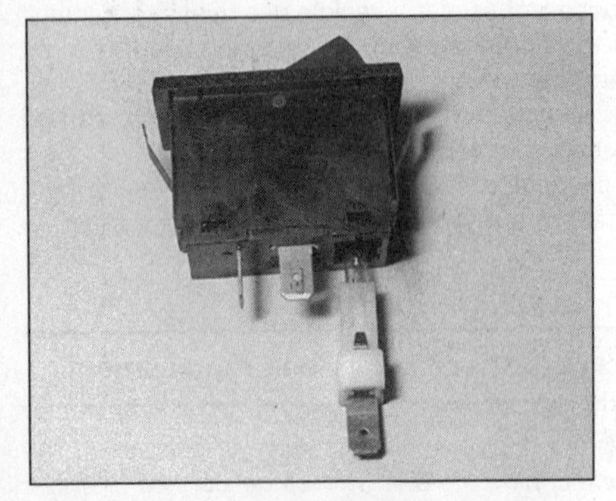

8.3a On illuminated switches, the bulb and holder are inserted from the back

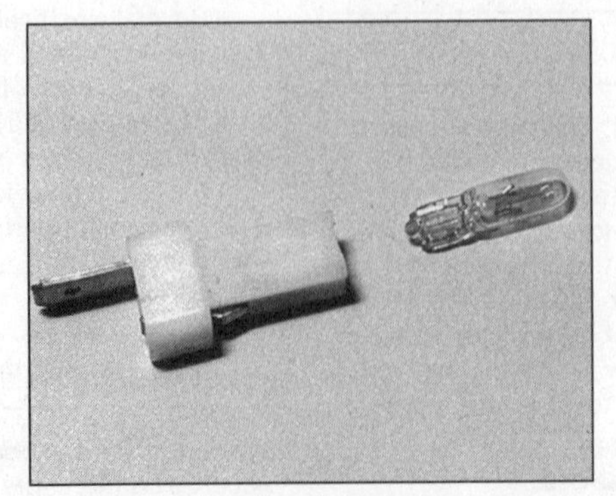

8.3b The bulb pulls straight out of the holder

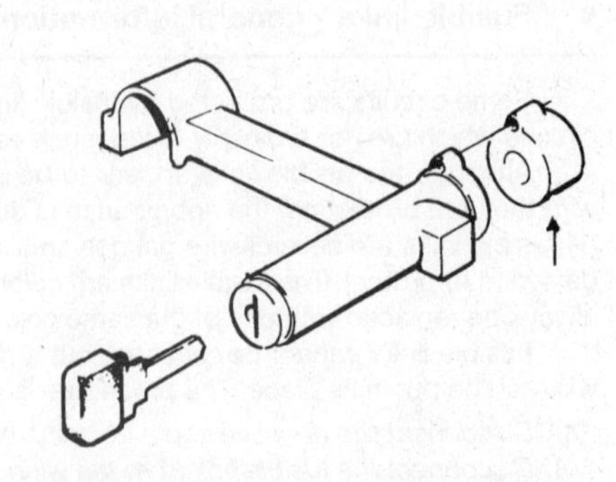

9.3 Unplug the electrical connector, remove the screw and pull the switch (arrow) out of the housing

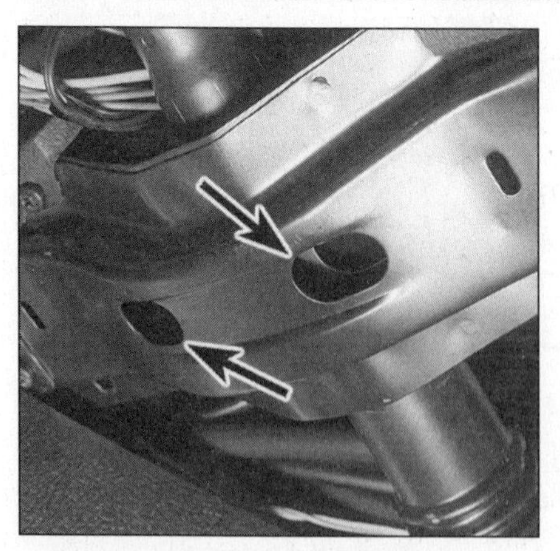

10.5 Working through the access holes (arrows) in the column support bracket, remove the shear-head bolts

10.6 The steering column lower support bracket is held in place by bolts on either side

10.7 Remove the lock pin (arrow) and unscrew the pinch-bolt

9 Ignition switch - removal and installation

Refer to illustration 9.3

Warning: *Some models are airbag-equipped. Always disconnect the negative battery cable and unplug the yellow electrical connector under the steering column when working in the vicinity of the crash sensor or steering column to avoid the possibility of accidental deployment of the airbag, which could cause personal injury (see Section 28).*

1 Disconnect the negative cable from the battery.

2 Remove the steering column covers and under dash covers (see Chapter 11).

3 Unplug the electrical connector and remove the retaining screw, then withdraw the switch from the housing **(see illustration).**

4 Installation is the reverse of removal.

10 Steering column and column lock - removal and installation

Refer to illustrations 10.5, 10.6, 10.7, 10.8, 10.10, 10.11, 10.12, 10.16, 10.23 and 10.26

Warning: *This procedure is for vehicles without SRS (airbag). We recommend any work involving the SRS be performed by a dealer service department.*

1 Disconnect the negative cable from the battery.

2 Place the key in the ignition and turn it to the unlock position (the key will remain in the cylinder for the entire procedure). Remove the steering wheel (see Chapter 10). After the steering wheel has been removed place the steering wheel retaining nut on the shaft to ensure the shaft does not fall out of the upper mount when the column is removed.

3 Remove the steering column cover and column switches. Unplug the ignition switch electrical connector.

4 Remove the underdash panels.

5 Working through the access holes in the steering column support bracket, drill out or extract the steering lock shear-head bolts **(see illustration).**

6 Disconnect the heater duct for access, and remove the steering column lower bracket **(see illustration).**

7 Pull out the lock pin and remove the pinch-bolt from the upper coupling joint **(see illustration).**

8 Remove the bolts from the flange, pry the flange apart and then remove the upper coupling assembly **(see illustration).** If no flange is installed, separate the lower coupling.

9 Push the steering column forward through the bulkhead to free the rubber seal, then guide it down, and out from underneath the support bracket. Remove the steering column from the vehicle and place it on a workbench.

10 Remove the two retaining screws and separate the lock from the steering column **(see illustration).** If equipped with a torque limiting column, break the plastic washers from the shear-bolts and remove the bolts with vise-grips or other suitable tool. Press the steering lock from the steering column with a hydraulic press and suitable fixtures.

11 Check the steering column to make sure the collapsible unit has not been compressed or damaged and there should be no axial

10.8 Remove the nuts and separate the steering shaft flange

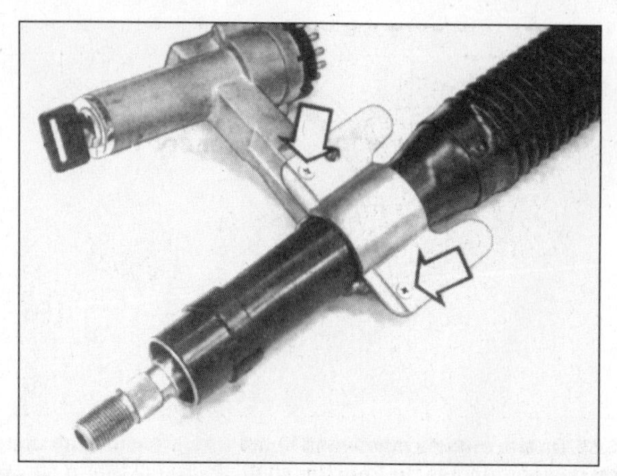

10.10 Remove the two screws (arrows) securing the steering lock to the column

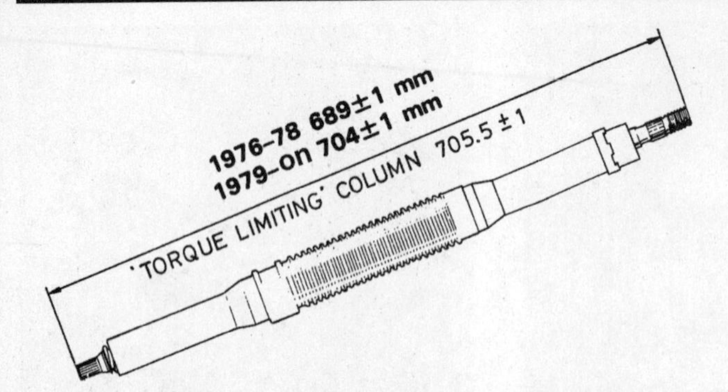

10.11 Measure the steering column to make sure it isn't collapsed

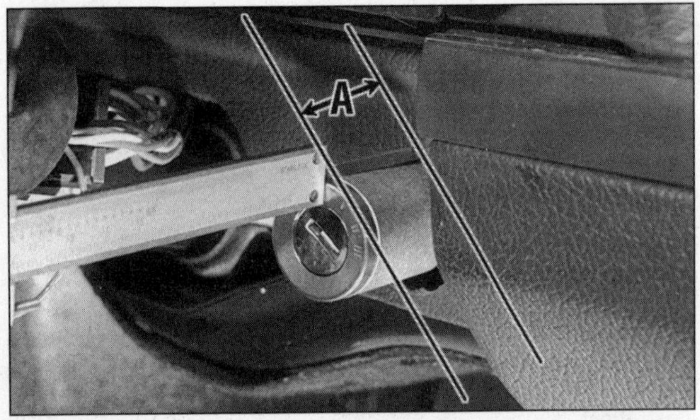

10.16 Measure from the 'III' mark on the lock to the dash to determine how far the lock protrudes - A=5/8-inch

movement between the upper and lower ends of the shaft **(see illustration)**. If there is any damage, replace the column assembly. Later models use a torque limiting column design which, while somewhat different in appearance, is checked in the same manner.

12 Install the lock assembly aligning the lock barrel with the center of the steering column opening. After the lock assembly has been installed in the specified location **(see illustration)**, remove the key and check for proper lock operation.

13 Make sure the plastic guides are in position in the column support frame.

14 Install the column, pushing its lower, end through the engine compartment bulkhead, and its upper end back over the support.

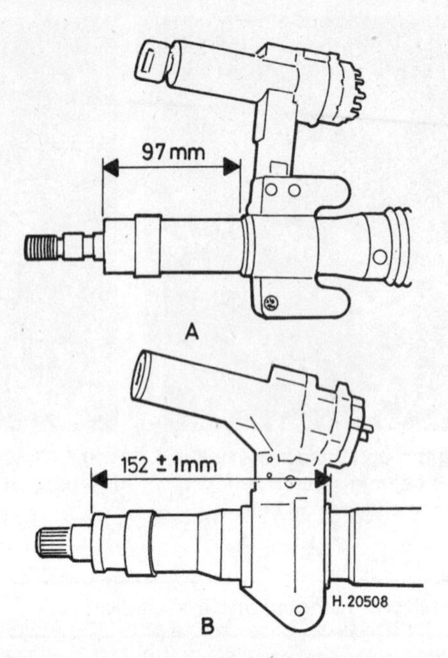

10.12 Make sure the lock assembly is installed with the specified clearance

A Standard column (earlier models)
B Torque limiting column (later models)

Install the shear-head bolts finger-tight only.

15 Install the lower bracket finger-tight.

16 Adjust the position of the column up or down, so that the steering lock protrudes from the dashboard by the specified amount **(see illustration)**.

17 Make sure the steering column is not in contact with the plastic guides. If it is, loosen the bolts and move the support bracket, tightening the bolts on completion.

18 Tighten the shear-head bolts in the upper attachment, without shearing them.

19 Tighten the lower bracket bolts to the torque listed in this Chapter's Specifications.

20 Install the heater duct.

21 Make sure the rubber grommet is installed in the bulkhead, then loosely install the upper coupling joint.

22 If equipped, tighten the coupling flange bolts to the torque listed in this Chapter's Specifications.

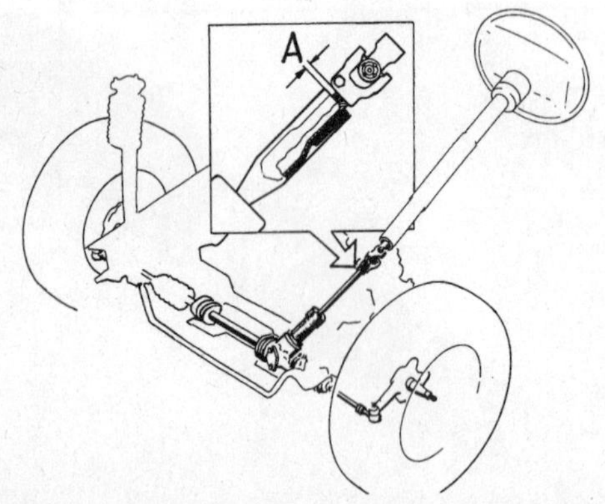

10.23 On the models mentioned in the text, measure the upper steering shaft coupling-to-lower shaft clearance (A) - it should around 1/2-inch

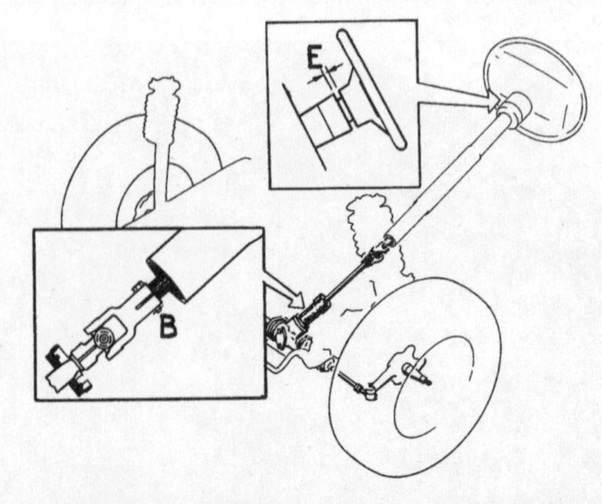

10.26 The steering wheel-to-housing clearance (E) should be about 7/64-inch - adjust by loosening the pinch bolt (B) and moving the steering shaft up or down

11.8 Pull the radio knobs off

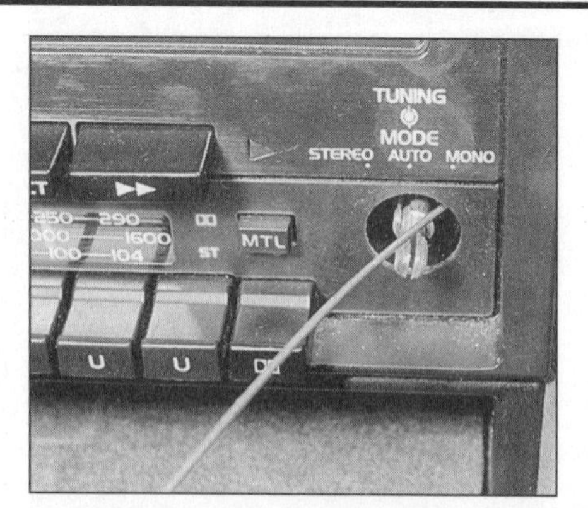

11.9a Insert a wire hook into the opening . . .

11.9b . . . and pull back the release clips (radio removed for clarity)

23 On 1979 and later models, or 1978 and later models with ZF type power steering, measure the distance between the upper steering shaft coupling and the stop on the lower shaft **(see illustration)**.

24 Adjust by loosening the pinch-bolt in the lower coupling and moving the lower shaft up or down.

25 Tighten the pinch-bolts to the torque listed in this Chapter's Specifications.

26 Temporarily install the steering wheel, and check the distance between the steering wheel and the case and adjust by loosening the lower coupling pinch-bolt and moving the steering shaft up or down **(see illustration)**. Tighten the pinch-bolt securely after adjustment.

27 The remaining procedure is the reverse of removal. Check the operation of the steering lock before finally tightening the shear-head bolts until they shear, then install the coupling joint pinch-bolt locking pins.

11 Radio - removal and installation

Refer to illustrations 11.8, 11.9a, 11.9b and 11.9c

Warning: *Some models are airbag-equipped. Always disconnect the negative battery cable and unplug the yellow electrical connector under the steering column when working in the vicinity of the crash sensor or steering column to avoid the possibility of accidental deployment of the airbag, which could cause personal injury (see Section 28).*

1 The radio may be housed in either the center console or in the dashboard panel above the center air vents, depending on model year.

2 The type of radio installed depends on the model year and personal choice. The procedure given here is for original equipment only.

Early models

3 Remove the center console trim panel retaining screws.

4 Pull off the radio control knobs.

5 Unscrew the locknuts which will be found under the knobs.

6 Pull the center console forward slightly. Pull the radio mounting frame forward and lift out the radio, disconnecting the electrical connectors and antenna lead as you do so.

7 Installation is the reverse of removal.

Later models

8 Pull off the radio control knobs **(see illustration)**.

9 Bend a stiff piece of wire into a hook and use it to pull back the side clips as shown in the photographs, working through the space left by the knob **(see illustrations)**. Release each side clip and withdraw the radio **(see illustration)**.

10 Install by reconnecting the electrical connectors and antenna, then pushing the radio back into its housing until the clips snap into place.

11 Reinstall the control knobs.

12 Headlamp wiper motor - check and replacement

Refer to illustrations 12.2 and 12.5

1 On earlier models the motors are mounted on the sides of the headlamp units. On later models they're mounted on the undersides of the headlamp units, but the procedure for removal on both types is similar.

2 Remove the wiper arm **(see illustration)**.

3 Remove the radiator grille, and on early models the headlamp bezel. On later models, remove the headlamp unit (Section 13).

4 Disconnect the electrical connector to the motor.

5 Release the spindle locknut and lift the motor from the bracket, pulling the spindle through the framework **(see illustration)**.

6 Installation is the reverse of removal.

11.9c Pull the radio out and disconnect the electrical connectors and antenna lead

12.2 Flip up the headlamp wiper arm cover and remove the nut

12.5 Release the locknut (arrow) and detach the wiper motor

13.2 Detach the headlamp bezel and lift it out

13.3 Rotate the headlamp retainer counterclockwise until released

13.4 Unplug the connector from the back of the headlamp

13.5a Release the clip . . .

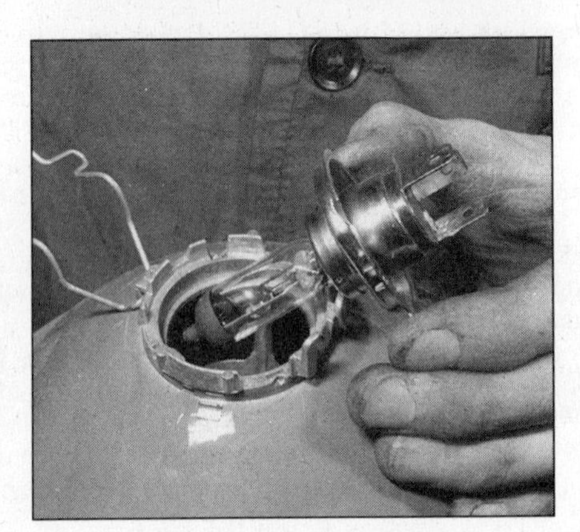

13.5b . . . and lift the bulb out

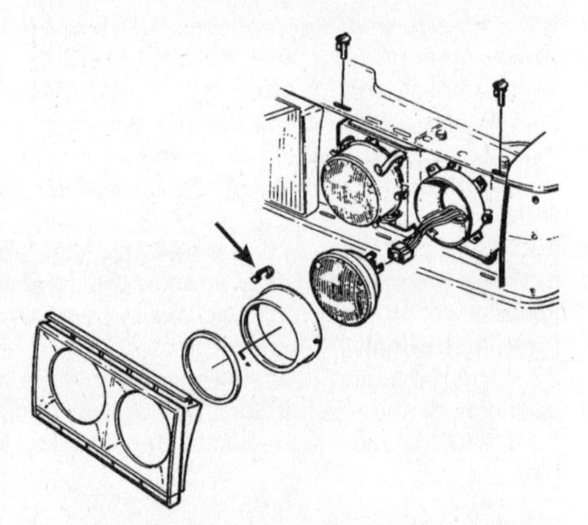

13.8 Remove the four retaining clips (arrow) and remove the retaining ring and sealed beam

13 Headlamps - removal and installation

Refer to illustrations 13.2, 13.3, 13.4, 13.5a, 13.5b, 13.8, 13.12 and 13.15

1 Two types of headlamp bulbs are used on Volvo 240 models; the composite type, which consists of a lens/housing and replaceable bulb and a sealed beam type which is replaced as a unit. **Caution:** *Handle the composite type bulb with extreme care. Do not touch the bulb with your bare fingers. If touched, wipe the bulb clean with rubbing alcohol. Failure to do so will cause premature failure of the bulb.*

Round headlamp models

Composite type

2 Remove the two headlamp bezel quick-release fasteners and lift off the trim bezel **(see illustration)**.

3 Turn the chromed ring counterclockwise to release it and lift it and the headlamp out. **(see illustration)**

4 Disconnect the electrical connector and remove the rubber cover **(see illustration)**.

5 Release the spring clip securing the bulb and lift out the bulb **(see illustrations)**.

6 Install a new bulb in the reverse order of removal.

Sealed beam type

7 Remove the two headlamp bezel quick-release fasteners and lift off the trim bezel.

8 Remove the four retaining clips and the retaining ring **(see illustration)**.

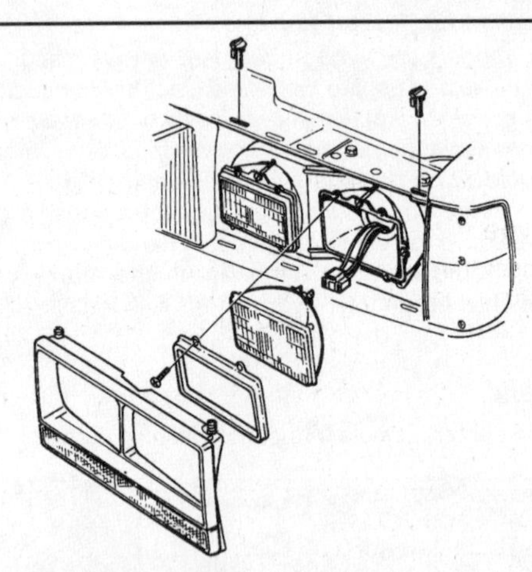

13.12 Rectangular sealed beam headlamp details

9 Remove the sealed beam and disconnect the electrical connector.

10 Install a new sealed beam in the reverse order of removal.

Rectangular headlamp models

Sealed beam type

11 Remove the two headlamp bezel quick-release fasteners and lift off the trim bezel.

12 Remove the four screws and the sealed beam retaining ring **(see illustration)**.

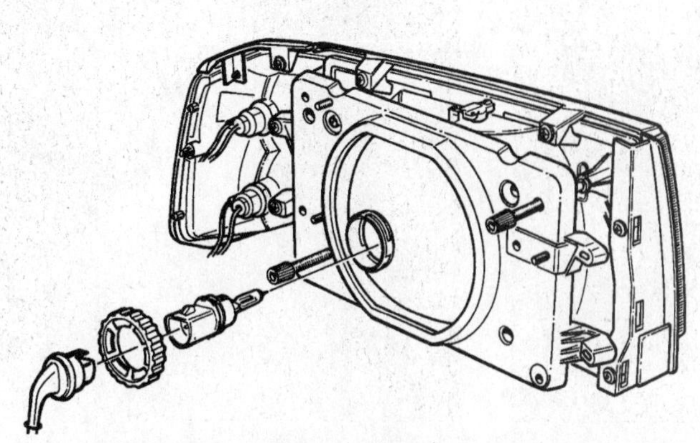

13.15 The composite headlamp bulb assembly used on later models

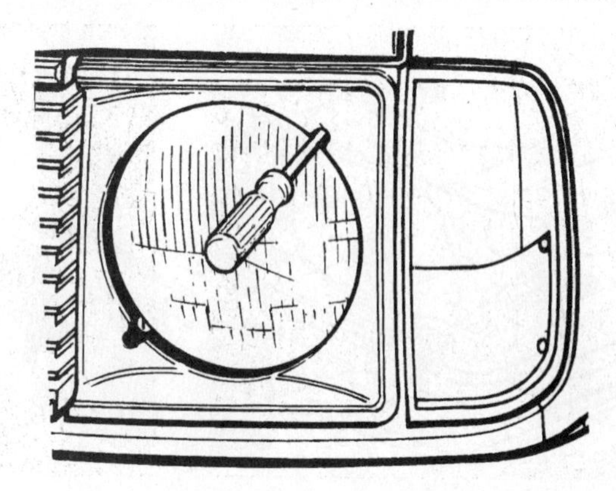

14.1a Use a screwdriver to turn the adjustment screws on round headlamp models

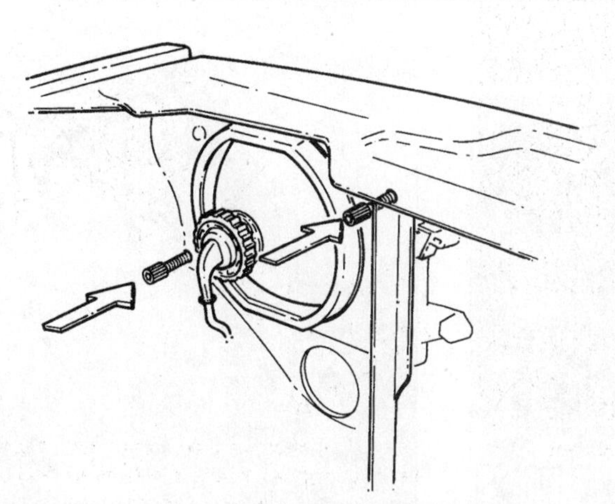

14.1b On later models, the adjuster knobs (arrows) are accessible from behind the headlamp

13 Remove the sealed beam and disconnect the electrical connector.
14 Install a new sealed beam in the reverse order of removal.

Composite type

15 Working under the hood, locate the rear of the headlamp housing and disconnect the electrical connector to the bulb **(see illustration)**.
16 Rotate the retaining ring counter-clockwise and remove the bulb from the housing.
17 Install a new composite bulb in the reverse order of removal (see Step 1, Caution).

14 Headlamps - adjustment

Refer to illustrations 14.1a, 14.1b and 14.1c

Note: *The headlamps must be aimed correctly. If adjusted incorrectly they could blind the driver of an oncoming vehicle and cause a serious accident or seriously reduce your ability to see the road. The headlamps should be checked for proper aim every 12 months and any time a new headlamp is installed or front end body work is performed. It should be emphasized that the following procedure is only an interim step which will provide temporary adjustment until the headlamps can be adjusted by a properly equipped shop.*

1 Headlamps on these models have two spring loaded adjusting screws, one on the top controlling up-and-down movement and one on the side controlling left-and-right movement **(see illustrations)**.
2 There are several methods of adjusting the headlamps. The

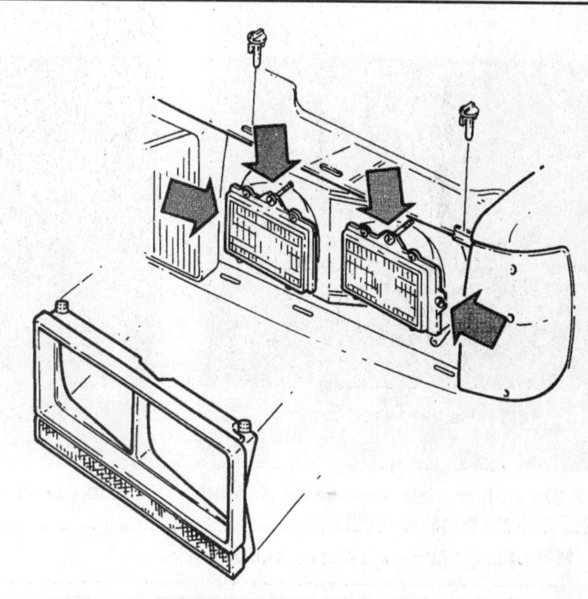

14.1c Remove the bezel on four-headlamp models to access the adjuster screws (arrows)

simplest method requires a blank wall 25 feet in front of the vehicle and a level floor.
3 Position masking tape vertically on the wall in reference to the vehicle centerline and the centerlines of both headlamps.
4 Position a horizontal tape line in reference to the centerline of all the headlamps. **Note:** *It may be easier to position the tape on the wall with the vehicle parked only a few inches away.*
5 Adjustment should be made with the vehicle sitting level, the gas tank half-full and no unusually heavy load in the vehicle.
6 Starting with the low beam adjustment, position the high intensity zone so it is two inches below the horizontal line and two inches to the right of the headlamp vertical line. Adjustment is made by turning the top adjusting screw clockwise to raise the beam and counterclockwise to lower the beam. The adjusting screw on the side should be used in the same manner to move the beam left or right.
7 With the high beams on, the high intensity zone should be vertically centered with the exact center just below the horizontal line. **Note:** *It may not be possible to position the headlamp aim exactly for both high and low beams. If a compromise must be made, keep in mind that the low beams are the most used and have the greatest effect on driver safety.*
8 Have the headlamps adjusted by a dealer service department or service station at the earliest opportunity.

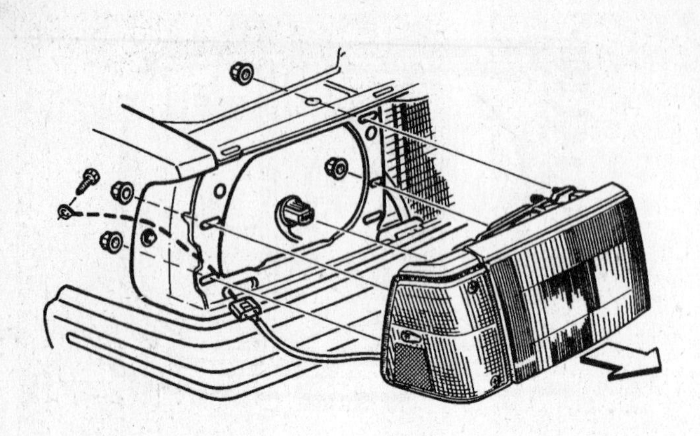

15.6a Later model composite headlamp housing mounting details

15.6b The parking light housing inserts into locating holes in the body

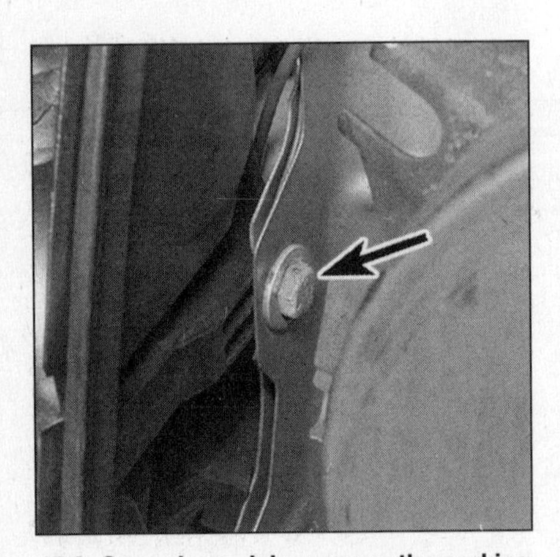

15.8 On early models, remove the parking light housing bolt (arrow)

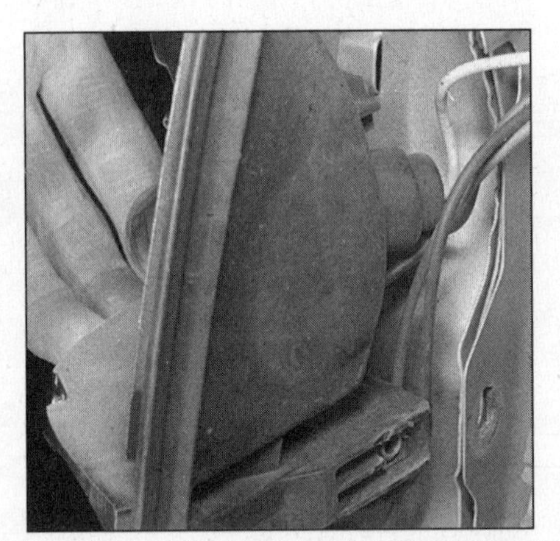

15.9 Detach the light housing and lift it off

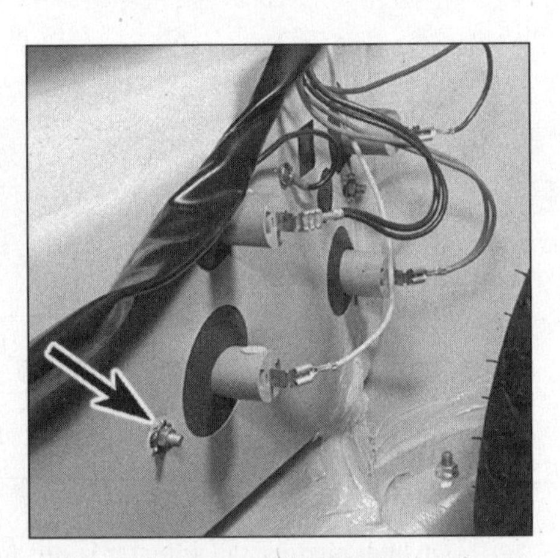

15.12 Sedan model tail light housing retaining nut (arrow)

15 Light housings- removal and installation

Refer to illustrations 15.6a, 15.6b, 15.8, 15.9, and 15.12

Later model composite headlamp housing

1 Disconnect the negative cable from the battery.
2 If necessary, remove the battery for clearance.
3 Remove the ground terminal screw and remove the ground connector to the headlamp assembly.
4 Disconnect the headlamp, turn signal and parking light bulb electrical connectors.
5 If equipped, flip up or remove the headlamp wiper arms.
6 From inside the engine compartment, remove the nuts securing the headlamp unit to the framework, and pull the complete headlamp and parking light assembly from the vehicle **(see illustration)**. The parking light has locating dowels which fit into the fender, so ease the unit out inboard side first **(see illustration)**.

Parking light housing

7 On early models, remove the headlamp bezel.
8 Remove the bolt securing the parking light housing to the sheetmetal. This bolt is located between the headlamp and the parking light **(see illustration)**.
9 Ease the unit forwards, inboard edge first, to release it **(see illustration)**.
10 On later models with composite headlamps, remove the headlamp housing and remove the two screws retaining the parking light housing.

High level brake light housing

11 Should the lens become cracked and need replacement, consult your Volvo dealer. The lens is bonded to the back glass. It may be possible to release it by applying gentle heat with a heat gun to soften the adhesive, and your Volvo dealer will also advise on the adhesive for installation.

Tail light housing

12 The tail light housing can be removed after disconnecting the leads to the unit and removing the nuts which secure the housing to the rear panel **(see illustration)**.
13 These are accessible from inside the vehicle as described for later model bulb replacement.

16 Bulb replacement

Warning: *Some models are airbag-equipped. Always disconnect the negative battery cable and unplug the yellow electrical connector under the steering column when working in the vicinity of the crash sensor or steering column to avoid the possibility of accidental deployment of the airbag, which could cause personal injury (see Section 28).*
Note: *The bulb failure warning system used on these models is very sensitive. Sometimes the warning light will remain on after a bulb on*

16.1 On early models, the turn signal bulb is accessible after removing the two screws (arrows) and detaching the lens

16.2 Later models have three lens retaining screws (arrows)

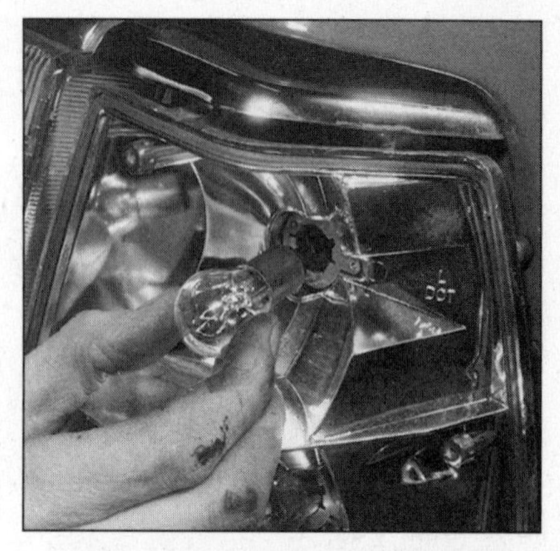

16.3 Rotate the parking light bulb counterclockwise and withdraw it from the housing

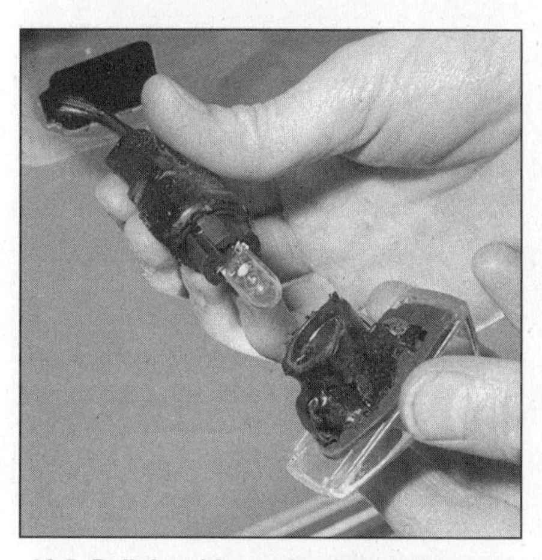

16.5 Pull the side marker bulb holder out of the housing

16.7 Pull the outer cover off

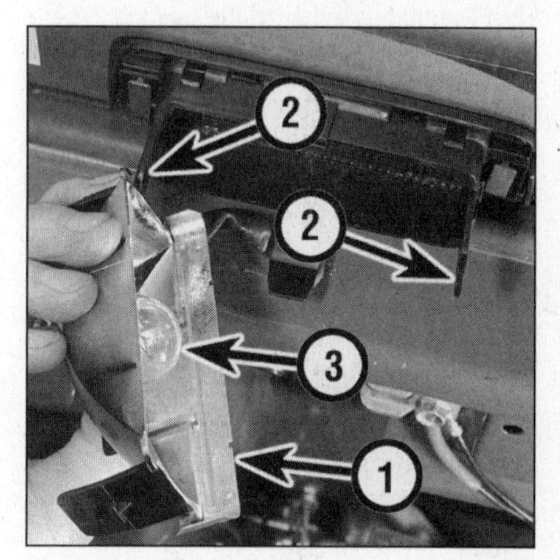

16.8 The high level brake light housing (1), clips (2) and bulb (3)

only one side of the vehicle is replaced. If this happens, replace the corresponding bulb on the other side because the original bulb will use less current than the new one, causing the system to sense an imbalance.

Parking and turn signal lights

Refer to illustrations 16.1, 16.2 and 16.3

1 On early models, remove the two screws from the lens and remove the lens **(see illustration)**.

2 On later models there are three screws **(see illustration)**.

3 Rotate the bulb and withdraw it from the holder **(see illustration)**.

Side marker lights

Refer to illustration 16.5

4 Push the sidemarker lens unit forward, at the same time prying up the rear edge.

5 Pull the socket from the lens unit **(see illustration)**.

6 The bulb is pressed in the holder.

High level brake light

Refer to illustrations 16.7 and 16.8

7 Pull off the outer cover **(see illustration)**.

8 Pry the housing from the plastic clips on the lens **(see illustration)**.

9 Push in on the bulb and rotate it counterclockwise to remove it.

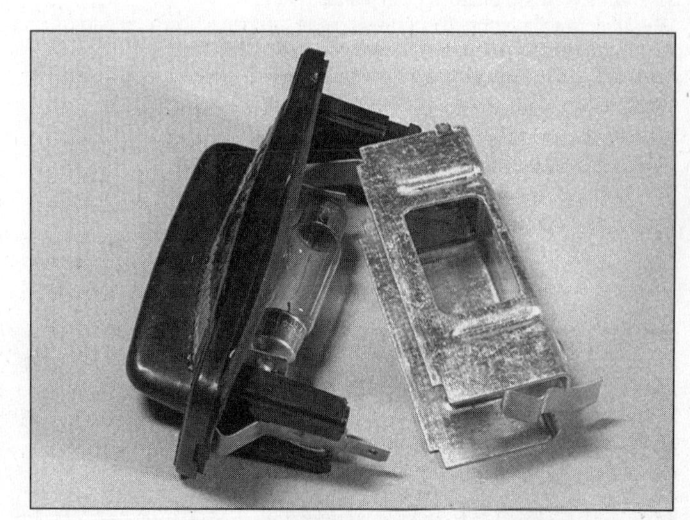

16.11 Remove the license light inner cover for access to the bulb

License plate light

Refer to illustrations 16.11 and 16.12

10 Pry the light unit from the housing (station wagons and early sedans), or slide it rearwards to free it (later sedans).

11 If equipped, remove the inner cover **(see illustration)**.

16.12 Replace this type of bulb by spreading the contacts and detaching the bulb

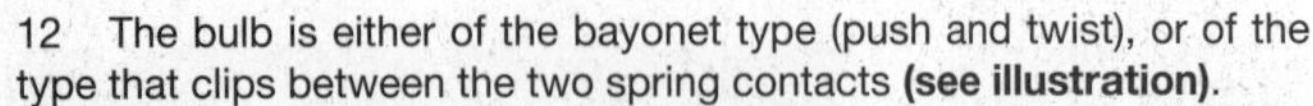

12 The bulb is either of the bayonet type (push and twist), or of the type that clips between the two spring contacts **(see illustration)**.

13 To remove the light unit, disconnect the wires, noting which terminal they serve.

Tail light bulbs

Early models

Sedan

Refer to illustrations 16.14 and 16.15

14 Remove the four screws securing the lens to the baseplate **(see illustration)**.

15 The bulbs are bayonet-type (push in and rotate counterclockwise to remove) **(see illustration)**.

Station wagon

16 The procedure is similar to that for the sedan, the light cluster being a different shape.

Later models

Sedan

17 Access to the bulbs is from inside the trunk.

18 Unscrew the knurled knob and hinge the panel upwards and off.

19 Turn the bulbholders counterclockwise to remove them.

20 The bulbs are removed by pushing in and rotating counterclockwise.

Station wagon

Refer to illustration 16.23

21 For access to the left-hand light cluster, unclip the side cover in the luggage compartment and remove the spare tire.

22 For the right-hand light cluster, lift up the floor panel, unclip and remove the side panel.

23 Turn the bulbholders counterclockwise to release them **(see illustration)**.

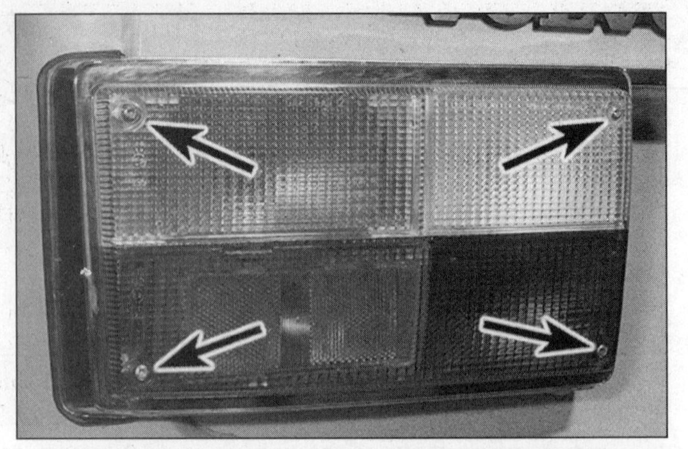

16.14 The tail light lens is held in place by screws (arrows)

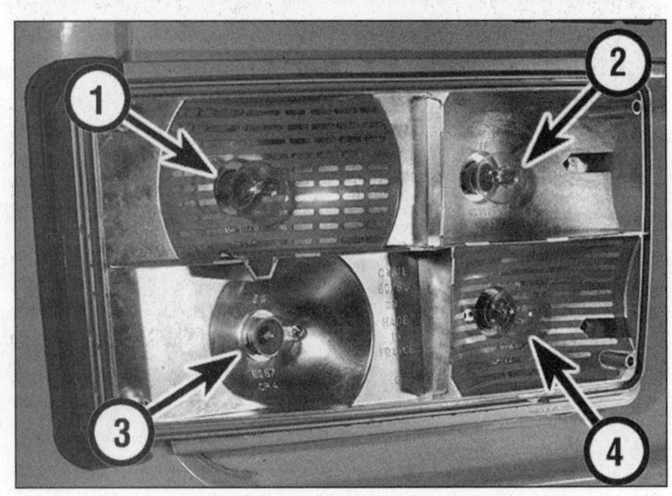

16.15 With the lens removed, remove the bulbs

1 *Turn signal*
2 *Backup light*
3 *Tail light*
4 *Brake light*

24 The bulbs are bayonet-type (push in and rotate counterclockwise to remove).

Hood light

Refer to illustrations 16.25, 16.26 and 16.28

25 Remove the securing screw **(see illustration)**.

26 Pull out the holder and disconnect the electrical connector **(see illustration)**.

27 The bulb is held between two spring contacts.

16.23 Typical station wagon bulb holder removal

16.25 Use a Phillips head screwdriver to remove the hood light housing

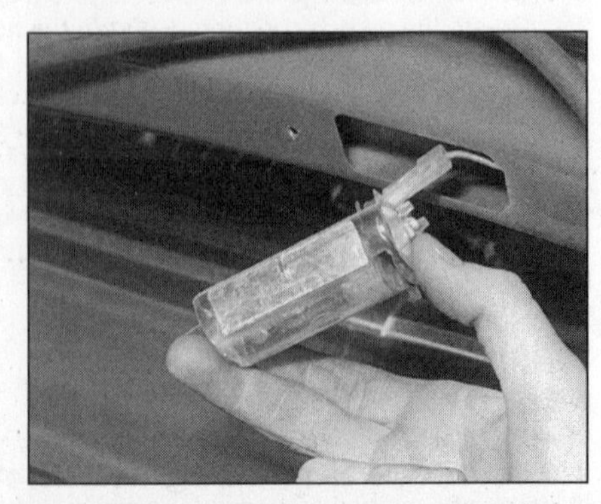

16.26 Rotate the housing out of the opening

16.28 Hood light bulb (1) and mercury switch (2)

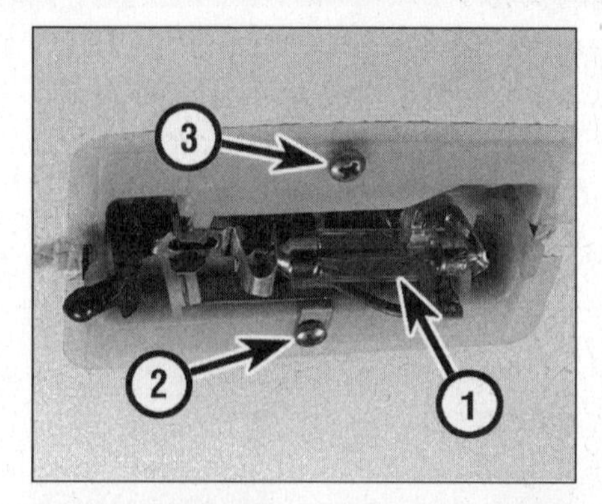

16.31 Courtesy light details - bulb (1) and retaining screws (2) and (3)

16.34 The instrument cluster bulbs are accessible after removing the cluster

17.1 The horns are bolted to the brace in front of the radiator

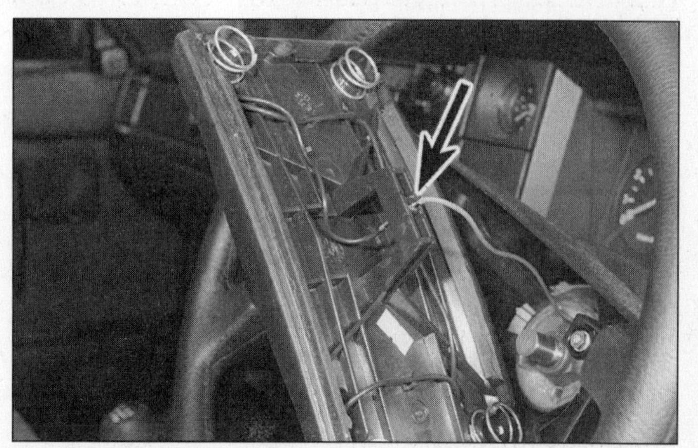
17.3 Check the horn wiring connector (arrow) and the four spring contacts

28 The light is operated by a mercury switch which switches the light on when the hood is opened, and off when it is closed **(see illustration)**.

29 Installation is the reverse of removal.

Courtesy light

Refer to illustration 16.31

30 Pry off the plastic lens.

31 The bulb is held between the spring contacts **(see illustration)**.

32 To remove the complete unit, remove the two screws and lift the unit from the roof panel, disconnecting the electrical connector.

33 Installation is the reverse of removal.

Instrument cluster bulbs

Refer to illustration 16.34

34 The instrument cluster bulbs are accessible after removing the cluster **(see illustration)**.

Sedan trunk light

35 The procedure is similar to that described for the hood light.

Station wagon luggage area light

36 The procedure is similar to that described for the courtesy light.

17 Horns - check and replacement

Refer to illustrations 17.1 and 17.3

Warning: *Some models are airbag-equipped. Always disconnect the negative battery cable and unplug the yellow electrical connector under the steering column when working in the vicinity of the crash sensor or steering column to avoid the possibility of accidental deployment of the airbag, which could cause personal injury (see Section 28).*

1 Twin horns are bolted to the paneling behind the radiator grille **(see illustration)**.

2 If the horn fails to operate, first check the circuit fuse and wiring at the horn units.

3 If the fuse and wiring are in order, pry out the center crash pad from the steering wheel (non-airbag models only), disconnect the wire, and check the contacts **(see illustration)**.

4 Replace the horns by unplugging the electrical connectors, removing the bolts and lifting them off.

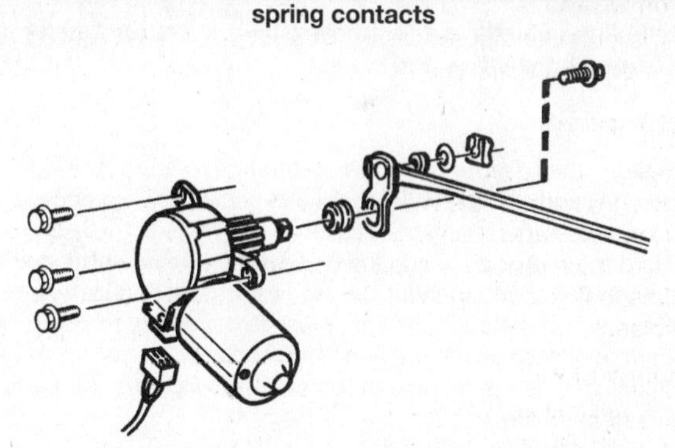
18.2 Windshield wiper motor details

18 Windshield wiper motor - removal and installation

Refer to illustration 18.2

1 Disconnect the negative cable from the battery.

2 In the engine compartment, unplug the wiper motor electrical connector and remove the mounting bolts **(see illustration)**.

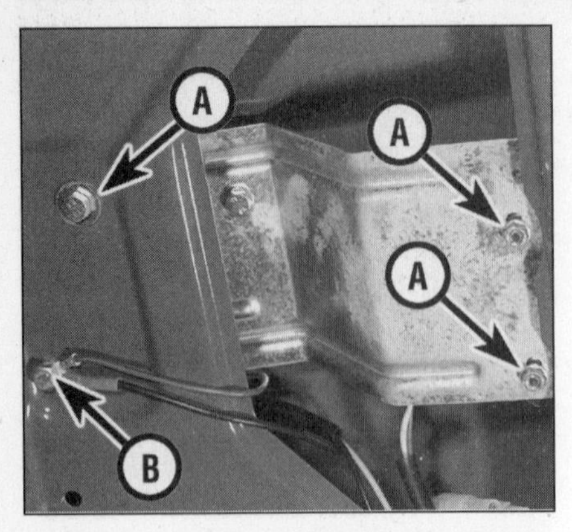

19.6 Wiper motor retaining bolts (A) and ground terminals (B)

20.3 Instrument cluster retaining screws (arrows)

20.4a The top edge of the cluster is retained by clips

20.4b Pull the cluster out for access, then detach the wires and cables

21.3 Unplug the electrical connector from the back of the clock

21.4 The clock is held in place in the housing by nuts

3 Inside the passenger compartment, remove the glovebox (Chapter 11) and sound insulator panel if necessary, for access to the motor drive-to-wiper crank connection.

4 Hold the motor drive crank with a wrench, remove the crank bolt and detach the crank, then lift the motor out **(see illustration 18.2)**. If the motor isn't in the park position, it will be necessary to pry the motor crank arm clip and detach the arm at that point.

5 Installation is the reverse of removal, making sure the motor is in the parked position.

19 Tailgate wiper motor - removal and installation

Refer to illustration 19.6

1 Lift up the cap, remove the nut and detach the wiper arm.

2 Remove the wiper spindle nut, washer and sealing ring, then open the tailgate.

3 Remove the interior trim panel from the tailgate (see Chapter 11).

4 Reaching inside the double skin, pry off the wiper arm crank balljoint.

5 Disconnect the electrical and ground leads from the wiper motor.

6 Remove the remaining bolts from the motor support plate **(see illustration)**.

7 Withdraw the motor and support plate from inside the tailgate.

8 Installation is the reverse of removal.

9 When installing the wiper arm, make sure to position it on the splines of the drive spindle so that in the park position, the arm comes to rest approximately 1 in (25 mm) above the lower edge of the glass.

20 Instrument cluster - removal and installation

Refer to illustrations 20.3, 20.4a and 20.4b

Warning: *Some models are airbag-equipped. Always disconnect the negative battery cable and unplug the yellow electrical connector under the steering column when working in the vicinity of the crash sensor or steering column to avoid the possibility of accidental deployment of the airbag, which could cause personal injury (see Section 28).*

1 Disconnect the negative battery cable.

2 Remove the steering column cover (see Chapter 11).

3 Remove the two screws securing the lower edge of the panel **(see illustration)**. On some models these screws may be on the sides of the panel.

4 Ease the lower edge forwards, while detaching the clips at the top, then unplug the electrical connectors **(see illustrations)**.

5 Reach behind the cluster and disconnect the speedometer cable. There may be a plastic locking collar or a lead seal on the coupling to prevent speedometer tampering, and this will have to be broken.

6 Withdraw the instrument cluster from the panel.

7 Installation is the reverse of removal.

21 Clock - removal and installation

Refer to illustrations 21.3 and 21.4

Warning: *Some models are airbag-equipped. Always disconnect the negative battery cable and unplug the yellow electrical connector under the steering column when working in the vicinity of the crash sensor or*

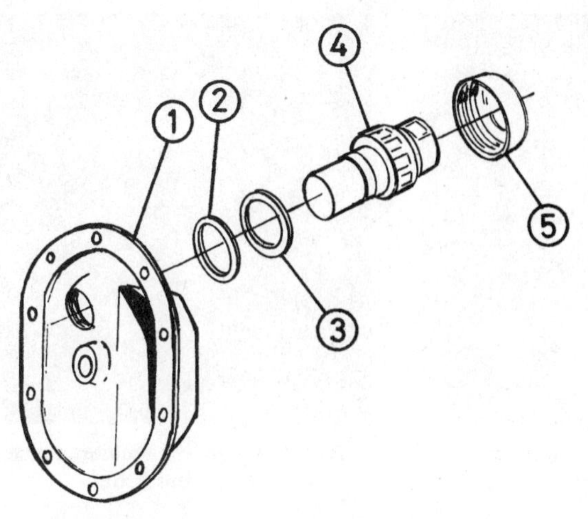

24.3 Electronic speedometer sensor details

1 *Rear axle cover*
2 *Adjusting shim*
3 *O-ring*
4 *Sensor*
5 *Retaining collar*

steering column to avoid the possibility of accidental deployment of the airbag, which could cause personal injury (see Section 28).

1 The clock may be located in various positions in the dashboard or console depending on model and year.

2 To remove the clock, refer to Chapter 11 for details of the dashboard or console section removal in which the clock is Installed.

3 Pull the section forward sufficiently to disconnect the wiring to the clock, then remove the clock and panel section **(see illustration)**.

4 The clock is generally held in place by nuts **(see illustration)**, and can be removed after these are removed.

5 Installation is the reverse of removal.

22 Shift indicator - general information

1 Some later models are equipped with a shift indicator system as an aid to fuel economy.

2 A lamp indicates to the driver when a shift to a higher gear is required.

3 The main components of the system are a control unit and a clutch pedal switch.

4 The control unit receives engine speed signals from terminal one on the ignition coil, and speed information from a speed transmitter on the rear axle and, on vehicles so equipped, from the overdrive relay.

5 At engine start, the warning lamp comes on but goes out as soon as the vehicle is driven.

6 By taking into account engine rpm and speed, the control unit calculates the most desirable gear ratio, and if shifting to another gear is required, the signal lamp illuminates. Operation of the clutch pedal switch signals the control unit that a change of gear has occurred.

Memory reprogramming

7 If the battery is disconnected or the power supply to the control unit is interrupted, then the memory will be erased.

8 To reprogram, drive the vehicle in 2nd gear and each higher gear for an eight second period (each gear). The signal lamp will flicker once as each gear programming is completed, Make sure that the foot is lifted completely from the clutch pedal after each shift.

23 Speedometer cable - removal and installation

Warning: *Some models are airbag-equipped. Always disconnect the negative battery cable and unplug the yellow electrical connector under the steering column when working in the vicinity of the crash sensor or steering column to avoid the possibility of accidental deployment of the airbag, which could cause personal injury (see Section 28).*

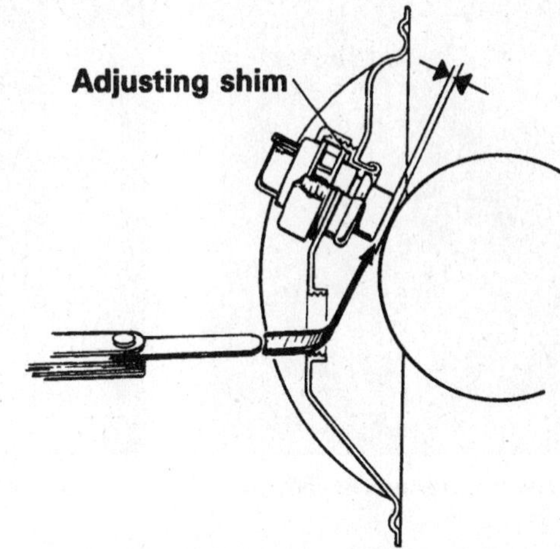

24.7 Use a feeler gauge to measure between the sensor tip and the pickup

1 Remove the instrument cluster (see Section 20).

2 Tie a piece of string to instrument cluster end of the speedometer cable.

3 Working underneath the vehicle, detach the speedometer from the rear (exhaust side) of the transmission/automatic transmission/overdrive unit.

4 Detach the cable from the retaining clips, and withdraw it from beneath the car.

5 Detach the cord and tie it to the new cable.

6 Installation is the reverse of removal, but it will help if an assistant pulls the cable through from the instrument end while the cable is fed through from underneath.

24 Electronic speedometer sensor - removal, installation and adjustment

Refer to illustrations 24.3. and 24.7.

1 Raise the rear of the vehicle and support it securely on jackstands. Draining the rear axle lubricant shouldn't be necessary for this procedure because with the rear raised, the level will be below the fill plug level.

Removal

2 Unplug the electrical connector from the sensor. Some models are equipped with a tamper-proof seal, which must removed first.

3 Unscrew the retaining collar, then detach the sensor and lift it out of the housing **(see illustration)**.

4 Clean and inspect the sensor, O-ring and any adjusting shims. It is a good idea to replace the O-ring with a new one if it shows any wear.

Installation

5 Apply a light coat of grease to the O-ring, make sure to install any shims, then insert the sensor into position, tighten the collars securely and plug in the electrical connector.

Adjustment

6 Adjustment is accomplished using adjustment shims (available at your dealer) under the sensor. Adjustment shouldn't be necessary if the original sensor and shims are reinstalled. Check the adjustment as follows.

7 Remove the fill plug and insert a long feeler gauge between the sensor tip and the pickup ring on the differential **(see illustration)**.

25.8 Detach the spring clip (arrow)

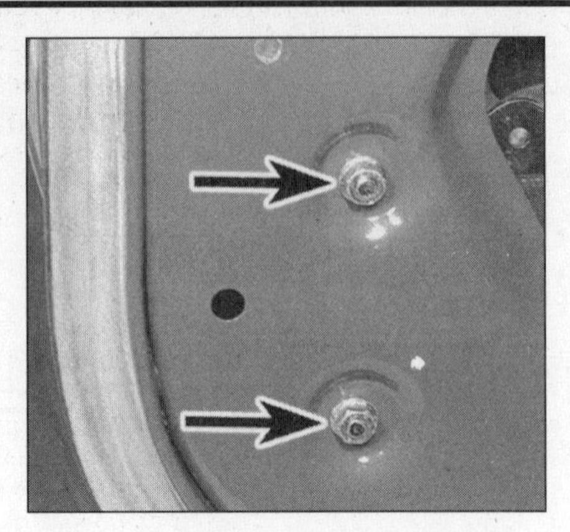

25.10a Remove the two securing nuts (arrows)

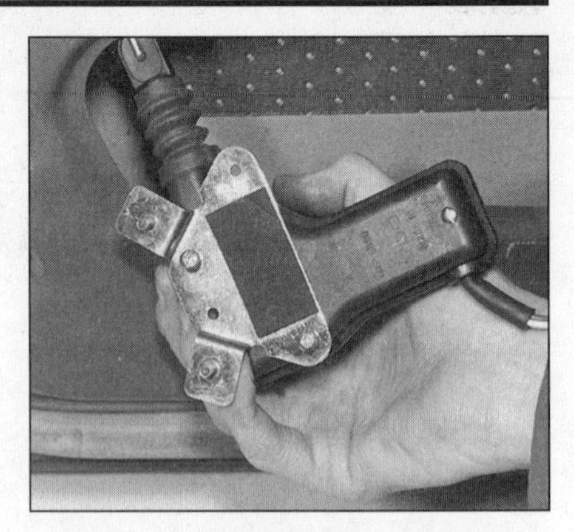

25.10b Lower the locking motor from the door

Check the measurements with those listed in this Chapter's Specifications.

25 Central door locking components - removal and installation

General information

1 On early models, only the four passenger doors operated on the system, but later models also include the tailgate/trunk lock.
2 The system is operated by turning the key in the driver's door lock or pushing down the interior lock button.
3 All doors are automatically unlocked by unlocking the driver's door. The driver's door is actually locked mechanically, but this operation causes the switching gear in the door to operate the lock motors in the other doors.
4 The system motors actuate the door mechanical lock mechanisms.
5 The system control relays are located in the instrument panel behind the above the center console.

Door lock motors - front and rear doors

Refer to illustrations 25.8, 25.10a and 25.10b

6 Remove the door trim panel (see Chapter 11).
7 If equipped, remove the plastic cover from the door lock link rod. This is an anti-theft device installed on later models.
8 Disconnect the spring clip from the eye end, and disconnect the link rod **(see illustration)**.
9 Unplug the electrical connector from the motor.
10 Remove the two securing nuts and lift out the motor **(see illustrations)**.
11 Installation is the reverse of removal.

Tailgate lock motor

12 The procedure is similar to that described above, except that on installation the motor should be adjusted as follows.
13 Attach the motor to the door loosely.
14 Move the lock cylinder lever to the locked position (towards the motor).
15 Move the motor towards the lock cylinder, compressing the rubber bellows.
16 Tighten the motor securing nuts securely.

Driver's door switches

Refer to illustrations 25.20 and 25.21

17 One door switch is clipped to the door lock cylinder, the other is in the lock link rod.
18 Remove the door trim panel and anti-theft device.
19 Disconnect the wiring at the connector.
20 Unclip the switch from the lock cylinder **(see illustration)**.
21 Remove the link rod switch by detaching the spring clips from the link rods, then lifting out the rods and switch **(see illustration)**.
22 Installation is the reverse of removal, making sure the lock

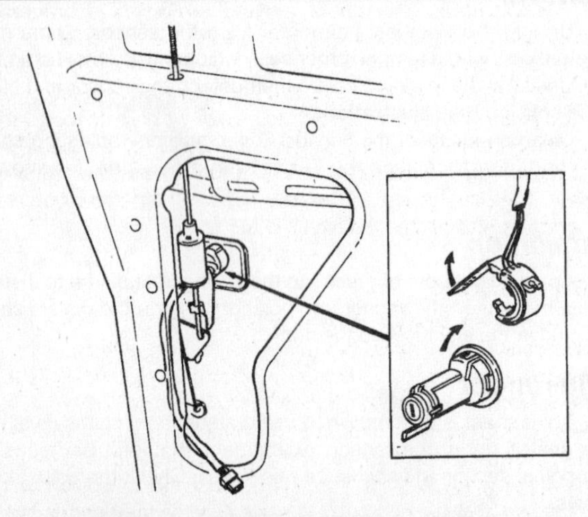

25.20 Door lock switch removal details

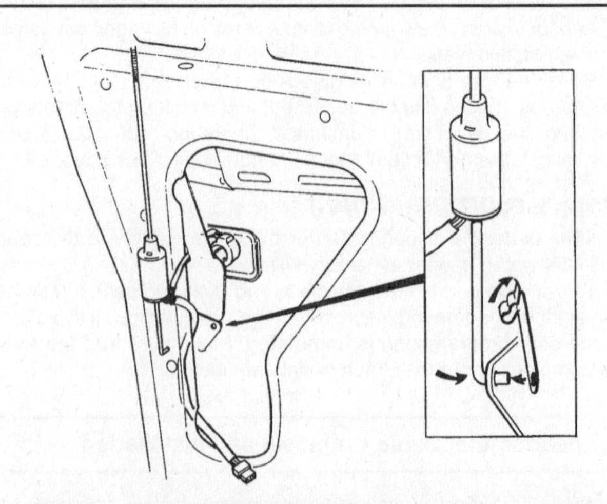

25.21 Lock rod and switch details

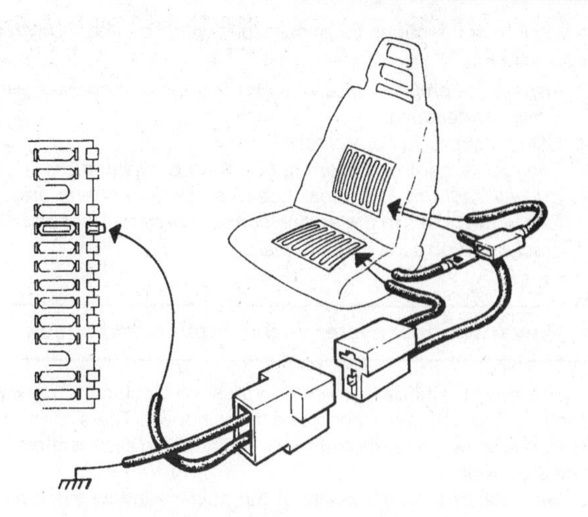

26.2 Heated seat electrical connector details

cylinder is positioned correctly before installing the switch and linkrod **(see illustration 25.20)**.

Control relays

23 The control relays are located behind the clock panel above the center console.
24 To remove them, remove the center console and clock panel and pull the relays from their clips, disconnecting the wiring.
25 Installation is the reverse of removal.

26 Heated seat elements - removal and installation

Refer to illustration 26.2

Warning: *Some models are airbag-equipped. Always disconnect the negative battery cable and unplug the yellow electrical connector under the steering column when working in the vicinity of the crash sensor or steering column to avoid the possibility of accidental deployment of the airbag, which could cause personal injury (see Section 28).*

1 On some models the driver's seat and front passenger seat are heated. The elements are controlled by switches in the parking brake console.
2 Disconnect the leads to the seat heating pads at the fuse block **(see illustration)**.
3 Unbolt and remove the seat, complete with runners, from the vehicle floor.
4 Disconnect the leads between the seat back panel and the seat cushion pad.
5 Remove the screws which secure the seat cushion and remove the cushion.
6 Place the seat upside down on the bench, and cut and remove the upholstery front retaining clamps.
7 Release the plastic hooks and pull out the heater pad from the seat back.
8 When installing the new heater pad to the seatback, make sure that the barbed side of the elements faces the seat padding and the electrical leads run on the inner side of the seat back.
9 To replace the pad in the seat cushion, place the cushion upside-down and remove the cover plate.
10 Cut and remove the clamp at the rear end of the cushion. Withdraw the heater pad. Note that a thermostat is installed in the cushion heater pad.
11 When installing the new pad, make sure that the barbed side of the heater element is towards the padding, and the electrical leads run on the inner side of the cushion.
12 Connect the upholstery using new clamps.

27 Rear window defogger - Check and repair

1 The rear window defogger consists of a number of horizontal elements baked onto the glass surface.
2 Small breaks in the element can be repaired without removing the rear window.

Check

3 Turn the ignition switch and defogger system switches On.
4 Place the positive lead of a voltmeter on the heater element closest to the incoming power source.
5 Wrap a piece of aluminum foil around the negative lead on the positive side of the suspected broken element and slide it slowly toward the negative side. Watch the voltmeter needle and when it moves from zero, you have located the break.

Repair

6 Repair the break in the element using a repair kit specifically recommended for this purpose, such as Dupont paste No. 4817 (or equivalent). Included in this kit is plastic conductive epoxy.
7 Prior to repairing a break, turn off the system and allow it to de-energize for a few minutes.
8 Lightly buff the element area with fine steel wool, then clean it thoroughly with rubbing alcohol.
9 Use masking tape to mask off the area being repaired.
10 Thoroughly mix the epoxy thoroughly, following the instructions provided with the repair kit.
11 Apply the epoxy material to the slit in the masking tape, overlapping the undamaged area about 3/4-inch on either end.
12 Allow the repair to cure for 24 hours before removing the tape and using the system.

28 Airbag - general information

Later models are equipped with a Supplemental Restraint System (SRS), more commonly known as an airbag. This system is designed to protect the driver from serious injury in the event of a head-on or frontal collision. It consists of an airbag module in the center of the steering wheel, a crash sensor and a standby power supply mounted under the driver's seat. A padded knee bolster mounted under the steering column designed to protect the driver's legs in the event of an accident is also part of the airbag system.

Airbag module

The airbag inflator module contains a housing incorporating the cushion (airbag) and inflator unit, mounted in the center of the steering wheel. The inflator assembly is mounted on the back of the housing over a hole through which non-toxic nitrogen gas is expelled, inflating the bag in approximately 0.2 seconds when an electrical signal is sent from the system. The airbag contact reel assembly is mounted on the steering column under the module and carries this signal to the module. The contact reel assembly is designed to transmit an electrical signal regardless of steering wheel position.

Crash sensor

The crash sensor incorporates a mercury switch and a microprocessor and measures deceleration. If the deceleration is high enough, the mercury switch closes, sending an electrical signal to the gas generator in the airbag module, inflating the airbag. The crash sensor microprocessor checks this system every time the vehicle is started, causing the "SRS" light to go on then off, if the system is operating properly. If there is a fault in the system, the light will go on and stay on and the microprocessor will store fault codes indicating the nature of the fault. If the SRS light goes on and stays on, the vehicle should be taken to your dealer immediately for service.

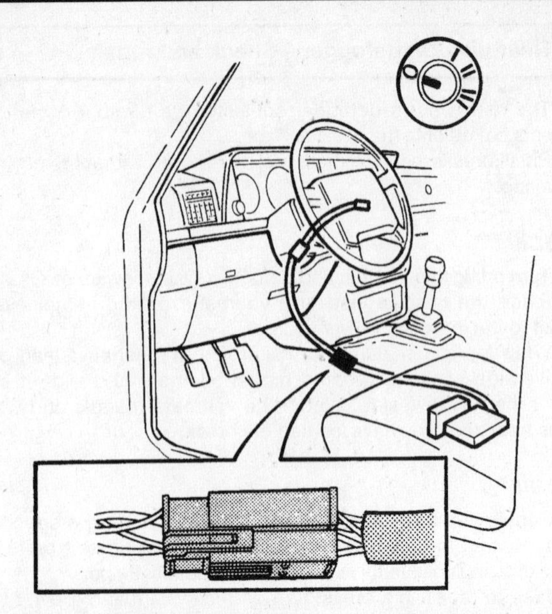

28.6 With the ignition Off, disconnect the battery negative cable and unplug the yellow airbag connector (shown)

Standby power unit

The standby power unit consists of a voltage converter and a capacitor that stores sufficient energy to trigger the airbag even if the battery power is cut off. Because the standby power unit powers the system with the battery disconnected, the system must be disabled when working in the vicinity of the airbag module.

Disabling the system

Refer to illustration 28.6

a) Turn the ignition switch to Off.
b) Detach the cable from the negative battery cable.
c) Unplug the yellow connector at the base of the steering column **(see illustration).**

Enabling the system

a) Turn the ignition switch to the Off position.
b) Plug in the yellow connector at the base of the steering column.
c) Connect the negative battery cable.

29 Cruise control system - description and check

The cruise control system maintains vehicle speed with a vacuum actuated servo motor located in the engine compartment, which is connected to the throttle linkage by a cable. The system consists of the servo motor, clutch switch, brake switch, control switches, a relay and associated vacuum hoses.

Because of the complexity of the cruise control system and the special tools and techniques required for diagnosis, repair should be left to a dealer service department or a repair shop. However, it is possible for the home mechanic to make simple checks of the wiring and vacuum connections for minor faults which can be easily repaired. These include:

a) Inspect the cruise control actuating switches for broken wires and loose connections.
b) Check the cruise control fuse.
c) The cruise control system is operated by vacuum so it's critical that all vacuum switches, hoses and connections are secure. Check the hoses in the engine compartment for tight connections, cracks and obvious vacuum leaks.

30 Power window system - description and check

The power window system operates the electric motors mounted in the doors which lower and raise the windows. The system consists of the control switches, the motors (regulators), glass mechanisms and associated wiring.

Because of the complexity of the power window system and the special tools and techniques required for diagnosis, repair should be left to a dealer service department or a repair shop. However, it is possible for the home mechanic to make simple checks of the wiring connections and motors for minor faults which can be easily repaired. These include:

a) Inspect the power window actuating switches for broken wires and loose connections.
b) Check the power window fuse/and or circuit breaker.
c) Remove the door panel(s) and check the power window motor wires to see if they're loose or damaged. Inspect the glass mechanisms for damage which could cause binding.

31 Wiring diagrams - general information

Refer to illustration 31.4

Since it isn't possible to include all wiring diagrams for every year covered by this manual, the following diagrams are those that are typical and most commonly needed.

Prior to troubleshooting any circuits, check the fuse and circuit breakers (if equipped) to make sure they're in good condition. Make sure the battery is properly charged and check the cable connections (see Chapter 1).

When checking a circuit, make sure that all connectors are clean, with no broken or loose terminals. When unplugging a connector, do not pull on the wires. Pull only on the connector housings themselves.

Refer to the accompanying chart for the wire color codes applicable to your vehicle **(see illustration).**

Color code

SB = black	BN = brown	GN = green
GR = gray	Y = yellow	OR = orange
W = white	P = pink	VO = violet
R = red	B = blue	

31.4 Wiring diagram color codes

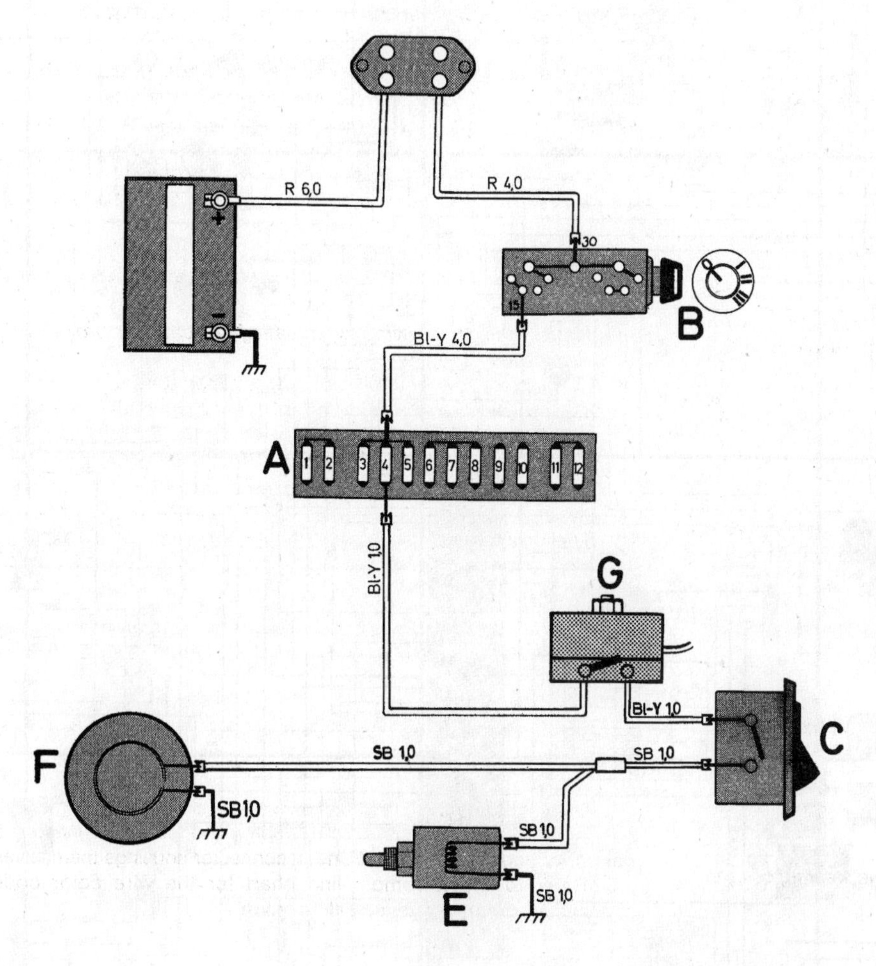

A Fusebox
B Ignition switch
C AC switch
E Solenoid valve
F Compressor solenoid valve
G Thermostat

Fuse no. 4

Heated driver seat
Reversing (Back-up) light
Air conditioning
Electric windows (relay) 1977 only
Glove compartment light

1976 and 1977 air conditioning system circuit diagram

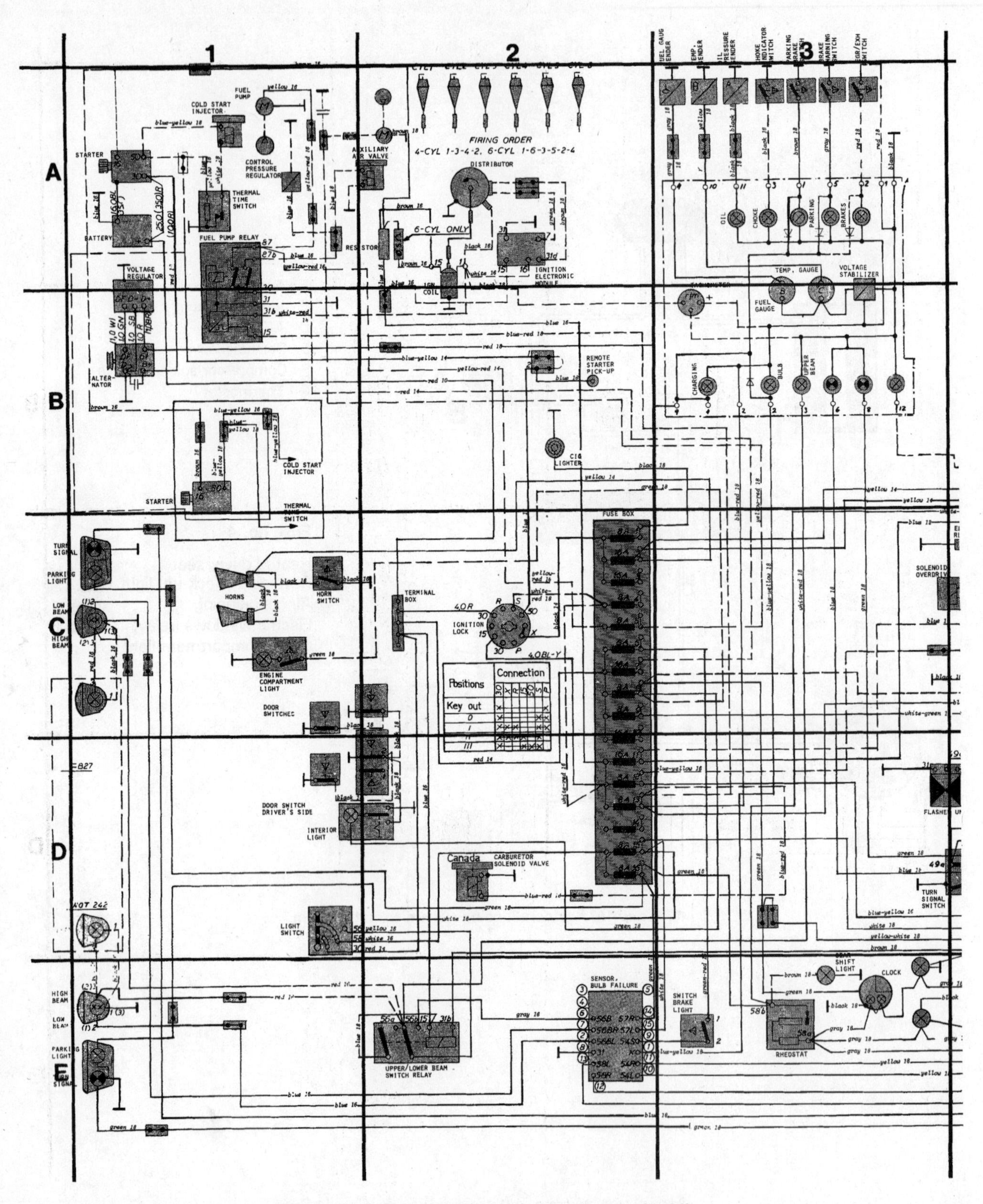

Typical 1976 through 1979 model wiring diagram (1 of 6)

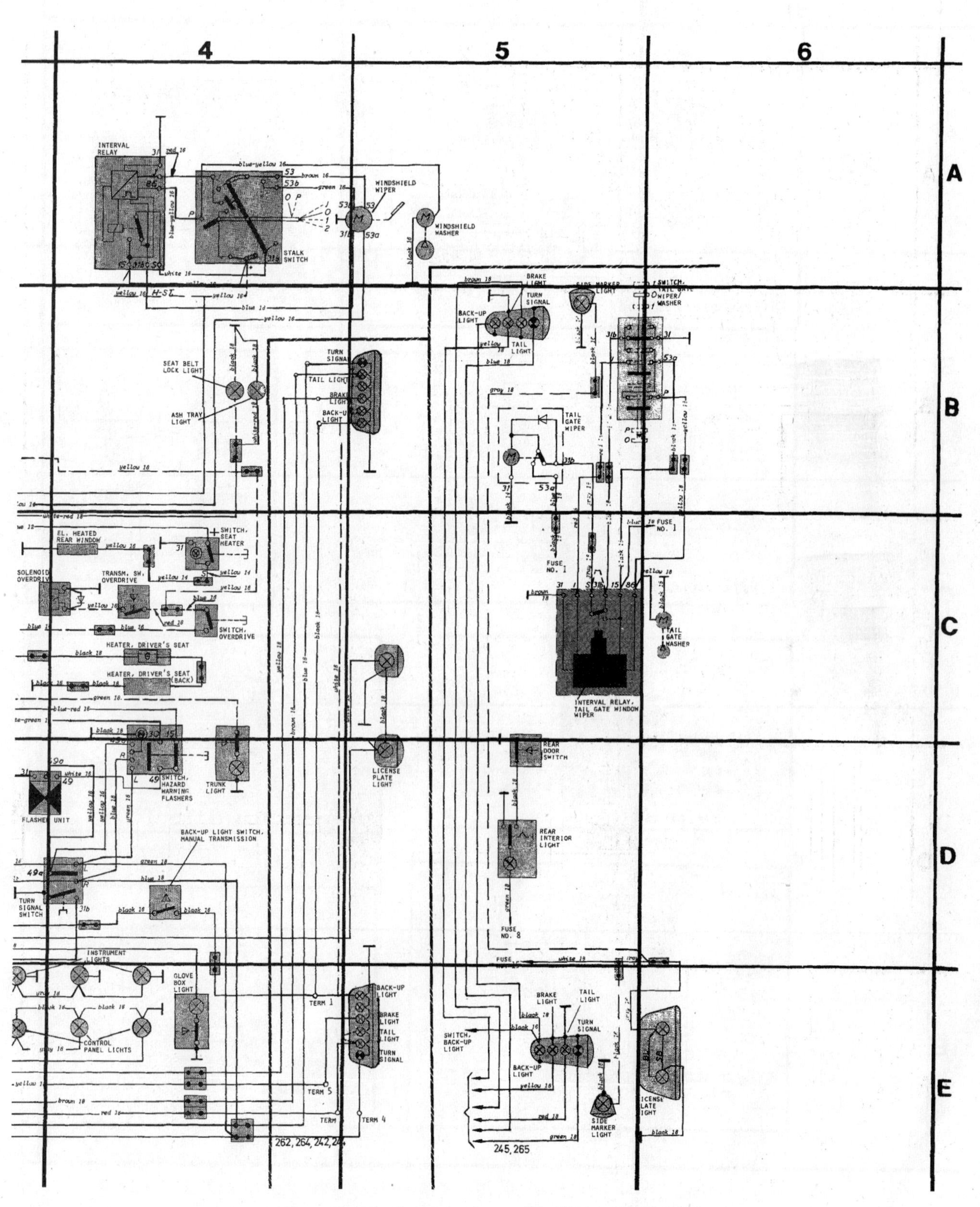

Typical 1976 through 1979 model wiring diagram (2 of 6)

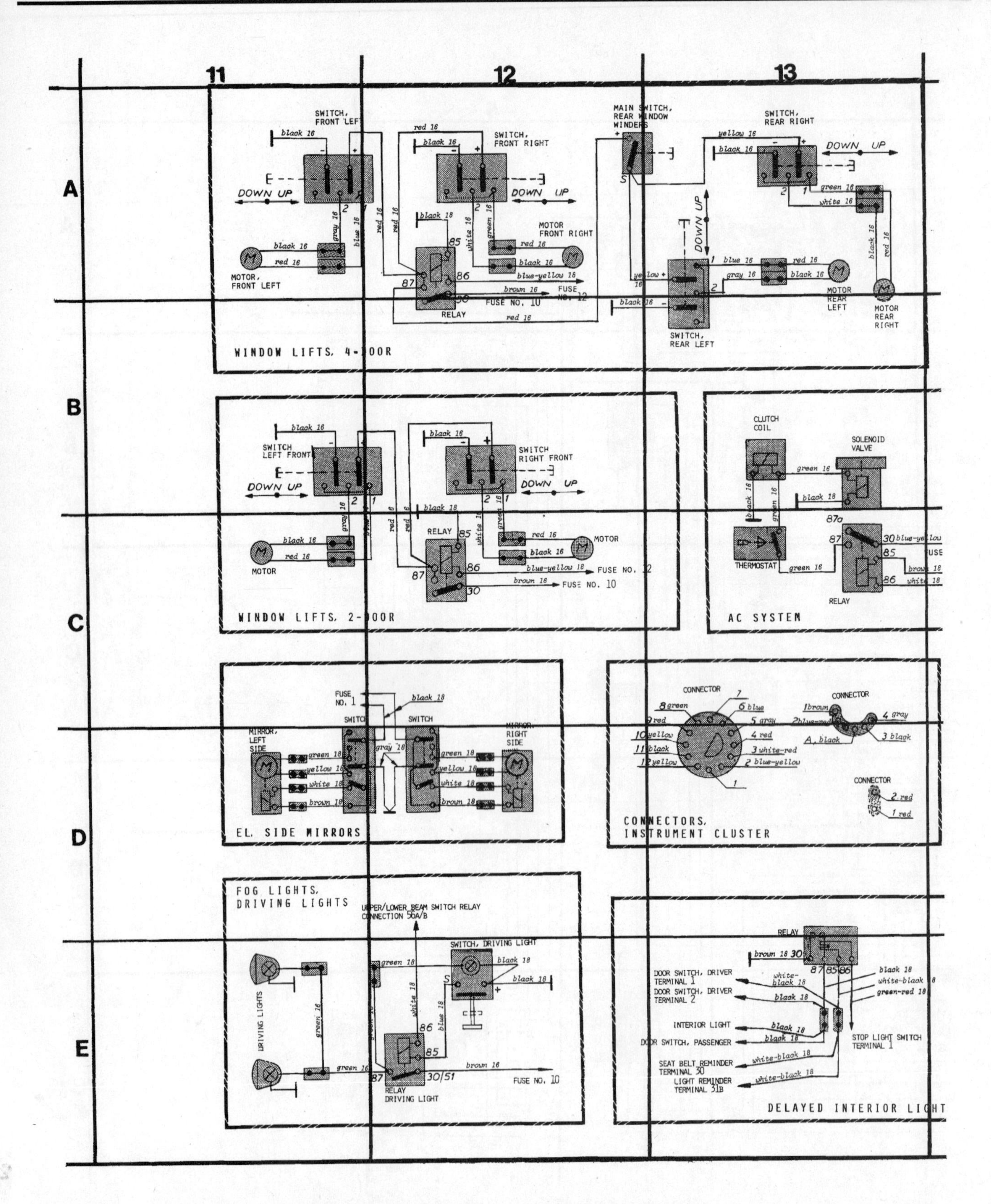

Typical 1976 through 1979 model wiring diagram (3 of 6)

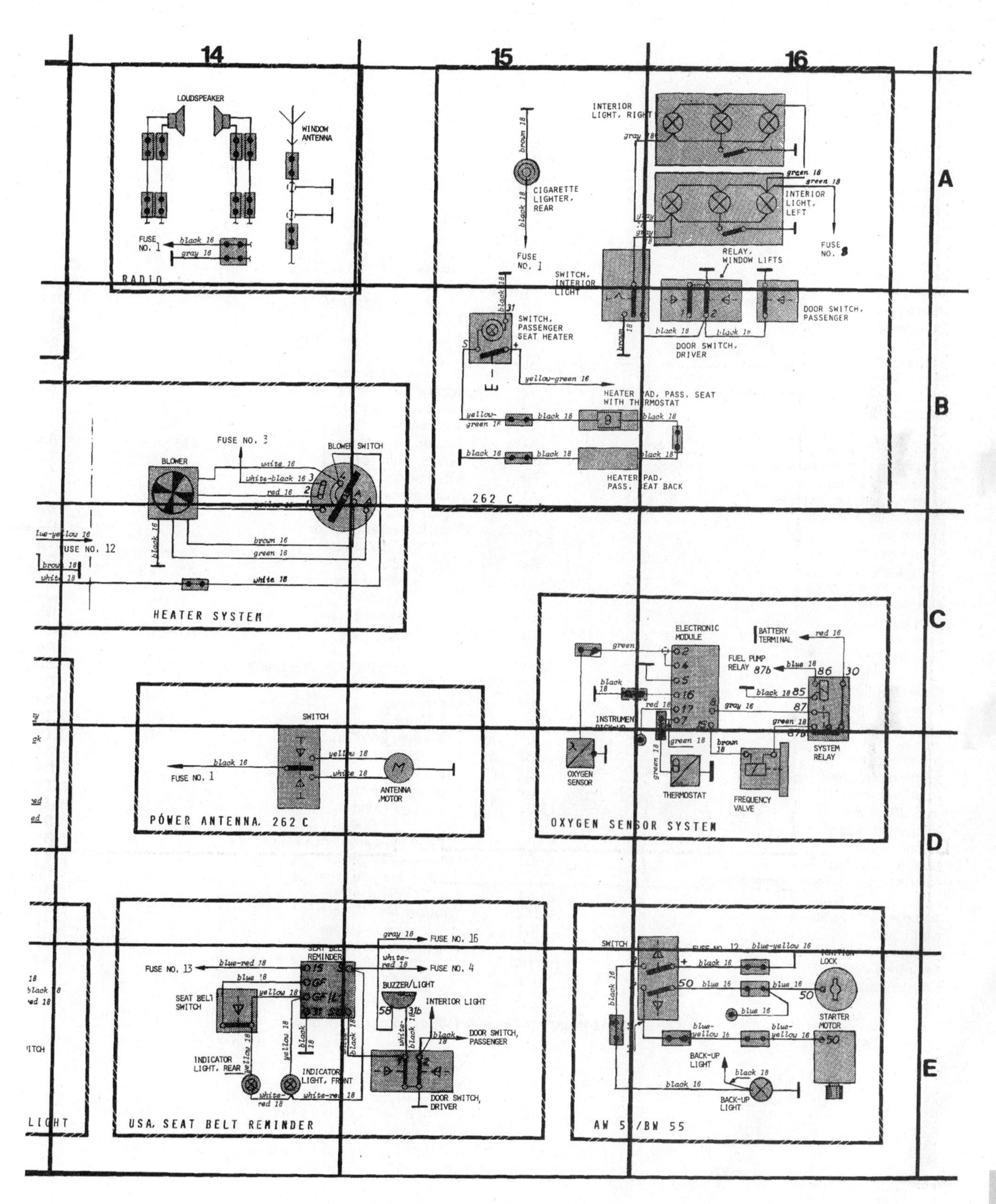

Typical 1976 through 1979 model wiring diagram (4 of 6)

Component	Location	Component	Location
AC system	C13	Parking lights: front	C1,E1
Alternator	B1	rear	B5,E5
Ashtray light	B4	Power antenna, 262C	D14
Automatic transmission	E15	Radio	A14
Auxiliary air valve	A2	Relay, upper/lower beam	E2
Battery	A1	Remote starter pick-up	B2
Brake lights	B5,E5	Rheostat, instrument lights	E3
Carburettor solenoid relay (Canada)	D2	Seat belt lock light	B4
Cigarette lighter	B2	Seat belt reminder	E14
Clock	E3	Seat heater	C4
Cold start injector	A1	Sender, for gauges	A3
Connectors, instrument cluster	D12	Sensor, bulb failure	E2
Control panel lights	E4	Side mirrors	D11
Control pressure regulator	A1	Spark plugs	A2
Delayed interior light	E13	Starter	B1
Distributor	A2	Switch: back-up lights	D4
Door switches	D2	brake lights	E3
Electrically heated rear window	C4	door	D2
Engine compartment light	C1	hazard warning flashers	D4
Flasher unit	D4	overdrive	C4
Fuel pump relay	B1	rear door	D5
Fuse box	C1	seat heater	C4
Gauges: fuel	B3	stalk	A4
temperature	B3	tailgate wiper/washer	B6
Gear shift light	E3	turn signal	D4
Glovebox light	E4	Tachometer	B3
Heater system	C14	Tail lights	B5,E5
Horns	C1	Tail lights, 245/265	B6,E6
Ignition coil	B2	Thermal time switch	A1
Ignition electronic module	B2	Trunk light	D4
Ignition lock	C2	Turn signal: front	C1,E1
Indicator lights	A3,B3	rear	B5,E5
Instrument lights	E6	switch	D4
Interior light	D2	Voltage regulator	B1
Interior light, rear 245	D5	Voltage stabilizer	B3
Interval relay: tailgate window wiper	C5	Washer: tailgate window	C5
wiper	A4	windshield	A4
License plate light	C5,E5	Window lifts: 2-door	C11
Light switch	D1	4-door	B11
Overdrive solenoid	C4	Wiper: tailgate window	B5
Overdrive switch	C4	windshield	A4
Oxygen sensor system	D15		

Key to 1976 through 1979 model wiring diagram (5 of 6)

Component	Location	Component	Location
AC system	C13	Parking lights, front	C1,E1
Alternator	B1	rear	B5,E5
Ash tray light	B4	Power antenna, 262C	D14
Automatic transmission	E15	Radio	A14
Auxiliary air valve	A2	Relay, upper/lower beam	E2
Battery	A1	Remote starter pick-up	B2
Brake lights	B5,E5	Rheostat, instrument lights	E3
Carburetor solenoid relay (Canada)	D2	Seat belt lock light	B4
Cigarette lighter	B2	Seat belt reminder	E14
Clock	E3	Seat heater	C4
Cold start injector	A1	Sender, for gauges	A3
Connectors, instrument cluster	D12	Sensor, bulb failure	E2
Control panel lights	E4	Side mirrors	D11
Control pressure regulator	A1	Spark plugs	A2
Delayed interior light	E13	Starter	B1
Distributor	A2	Switch, back-up lights	D4
Door switches	D2	brake lights	E3
Electrically heated rear window	C4	door	D2
Engine compartment light	C1	hazard warning flashers	D4
Flasher unit	D4	overdrive	C4
Fuel pump relay	B1	rear door	D5
Fuse box	C1	seat heater	C4
Gauges, fuel	B3	stalk	A4
temperature	B3	tail gate wiper/washer	B6
Gear shift light	E3	turn signal	D4
Glove box light	E4	Tachometer	B3
Heater system	C14	Tail lights	B5,E5
Horns	C1	Tail lights, 245/265	B6,E6
Ignition coil	B2	Thermal time switch	A1
Ignition electronic module	B2	Trunk light	D4
Ignition lock	C2	Turn signal, front	C1,E1
Indicator lights	A3,B3	rear	B5,E5
Instrument lights	E6	switch	D4
Interior light	D2	Voltage regulator	B1
Interior light, rear 245	D5	Voltage stabilizer	B3
Interval relay, tail gate window wiper	C5	Washer, tail gate window	C5
wiper	A4	windshield	A4
License plate light	C5,E5	Window lifts, 2-door	C11
Light switch	D1	4-door	B11
Overdrive solenoid	C4	Wiper, tail gate window	B5
Overdrive switch	C4	windshield	A4
Oxygen sensor system	D15		

Key to 1976 through 1979 model wiring diagram (6 of 6)

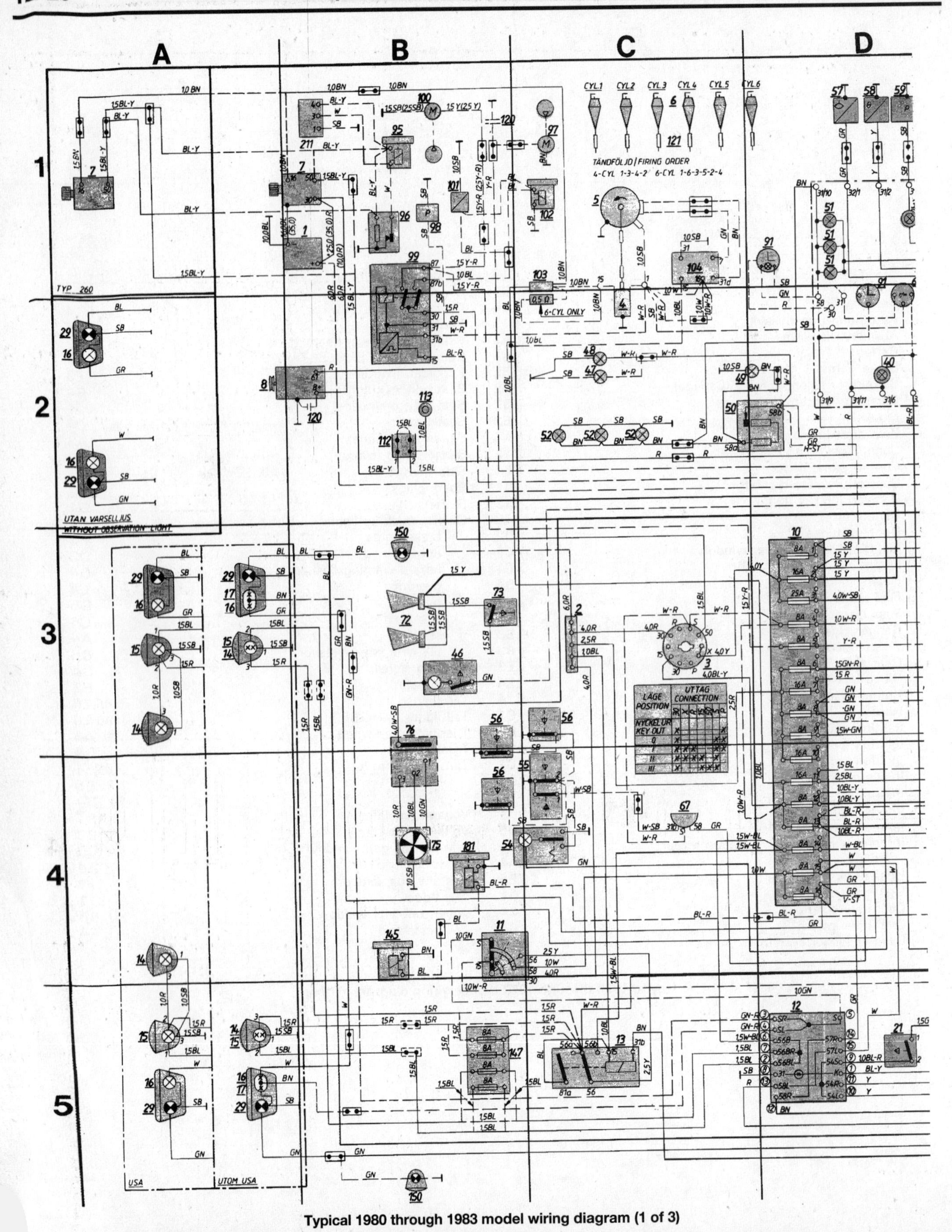

Typical 1980 through 1983 model wiring diagram (1 of 3)

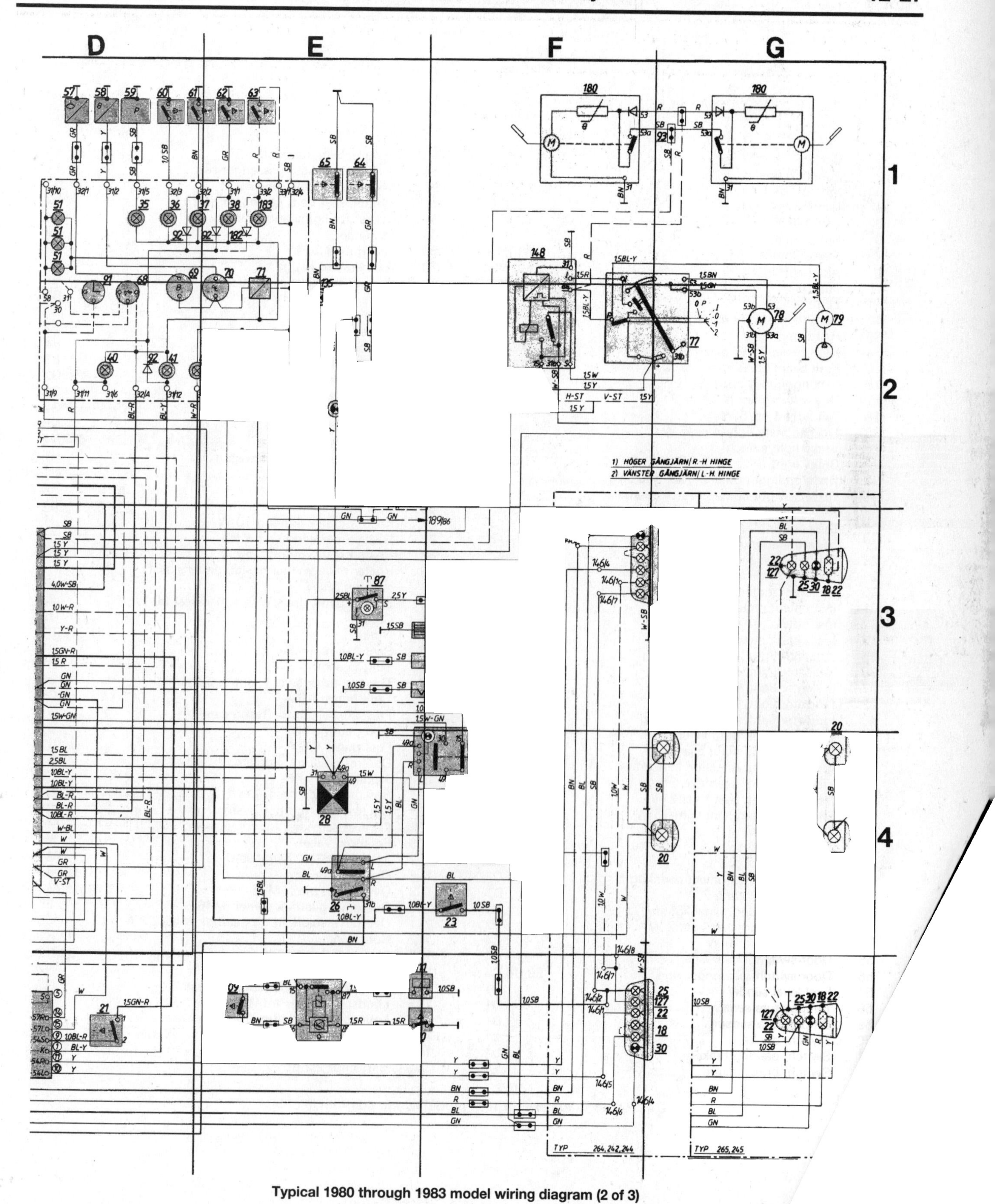

Typical 1980 through 1983 model wiring diagram (2 of 3)

No.	Component	Location
1	Battery	B1
2	Junction box	C3
3	Ignition switch	C3,N2
4	Ignition coil 12A	C2
5	Distributor	C1,N3
6	Spark plug	C1
7	Starter motor	A1,B1,
8	Alternator	A2 N3
10	Fusebox	D3,K5
11	Light switch	B4
12	Bulb failure warning sensor	D5
13	Headlamp relay	C5
14	High beam 75 W max	A3,A4,A5
15	Low beam 55 W max	A3,A5
16	Parking light 4 cp/5 W	A2,A3,A5
17	Day running lights 32 cp/21 W	A3,A5
18	Tail light 4 cp/5 W	G3,G5
20	License plate light 4 cp/5 W	G3,G4
21	Brake light switch	D5,H4
22	Brake light	G3,G5
23	Back-up light contact, man. gearbox	F4
24	Back-up light contact, auto. gearbox	N2
25	Rear spotlight 32 cp/21 W	G3,G5
26	Direction indicator stalk	E4
27	Hazard warning light switch	F4
28	Flasher device	A4
29	Direction indicator, front	A2,A3
30	Direction indicator, rear	G3,G5
31	Instrument panel connection	M4
32	Instrument panel connection	M4
33	Instrument panel connection	M5
35	Indicator light, oil pressure	D1
36	Indicator light, choke	D1
37	Indicator light, parking brake	E1
38	Indicator light, brake failure	E1
40	Indicator light, battery charging	D2
41	Indicator light, bulb failure	D2
42	Indicator light, main beams	D2
43	Indicator light, direction indicator	E2
44	Indicator light, overdrive	E2
45	Indicator light, seat belts 2 W	E2, N1
46	Engine compartment light 15 W	B3
47	Seat belt lock	C2
48	Ashtray light	C2
49	Gear selector light	C2
50	Rheostat for instrument panel light	C2
[illegible]	Instrument lighting 2 W	D1
[illegible]	Instruments and panel lighting	C2
[illegible]	[illegible]love compartment light 2 W	E2
[illegible]	[illegible]tesy light 10 W	B4
[illegible]	[illegible]itch, driver side	C4
[illegible]	[illegible]h, passenger side	B3,B4,C3
[illegible]	[illegible]ler	D1
[illegible]	[illegible]sor	D1
[illegible]	[illegible]r	D1
[illegible]	[illegible]	D1
[illegible]	[illegible]e	D1
[illegible]	[illegible]	E1

No.	Component	Location
63	Switch, Lambda-sond	E1
64	Switch, seat belt, passenger seat	E1
65	Switch, seat belt, driver seat	E1,N1
66	Switch, seat belt, passenger seat	E2
67	Buzzer for lights and key	C4
68	Tachometer	D2
69	Fuel gauge	D2
70	Temperature gauge	E2
71	Voltage stabilizer 10 ± 0.2 V	E1
72	Horn, 7.5 A	B3
73	Horn switch	B3
74	Cigar lighter 7 A	F2
75	Heater fan (standard) 115 W	B4
76	Fan switch	B3
77	Windshield wash/wipe switch	G2
78	Windshield wiper 3.5A	G2
79	Windshield washer 2.6 A	G2
80	Trunk lock motor 0.6 A	N5
81	Trunk lock switch	B4,N4
82	Tailgate washer/wipe switch	H1
83	Tailgate wiper 1A	H1
84	Tailgate washer 3.4 A	H2
85	Rear door switch	J1
86	Rear courtesy lighting 10 W	J1
87	Heated rear window switch	E3
88	Heated rear window 150 W	F3
89	Heater pads + thermostat, driver's seat cushion 30 W	F3
90	Heated pads, driver's seat, backrest 30 W	F3
91	Clock	D1, D2
92	Diode	D1,D2,E1
94	Seat belt reminder	N1
95	Start injector	B1
96	Thermal timer switch, start injector	B1
97	Tank pump 1.6 A	C1
98	Pressure sensor for Turbo cars	B1
99	Fuel pump relay	B1
100	Fuel pump 6.5 A	B1
101	Control pressure regulator	B1
102	Auxiliary air valve	C1
103	Resistance 0.9 ohm/4 cyl.	C1
104	Control unit, ignition system	C1
105	Control solenoid, for compressor, AC 3.9 A	M4
106	Solenoid valve	N4
107	AC switch (+ thermostat)	M4
108	Overdrive relay	E5,M3
109	Overdrive switch M46	E5,M3
110	Overdrive gearbox switch M46	E5
111	Overdrive solenoid on gearbox M46 2.2 A	E5
112	Coupling	B2,H1
113	Power socket for running starter motor	B2,N2
114	Thermostat, Lambda system	N2
115	Pressure switch, Lambda system	N2
116	Loudspeaker, LH front door 4	M2
117	Loudspeaker, RH front door 4	M2
118	Antenna, windshield	M3
119	TDC sensor for monotester	N4
120	Capacitor 2.2 μF	B1,B2

Key to 1980 through 1983 model wiring diagram (3 of 3)

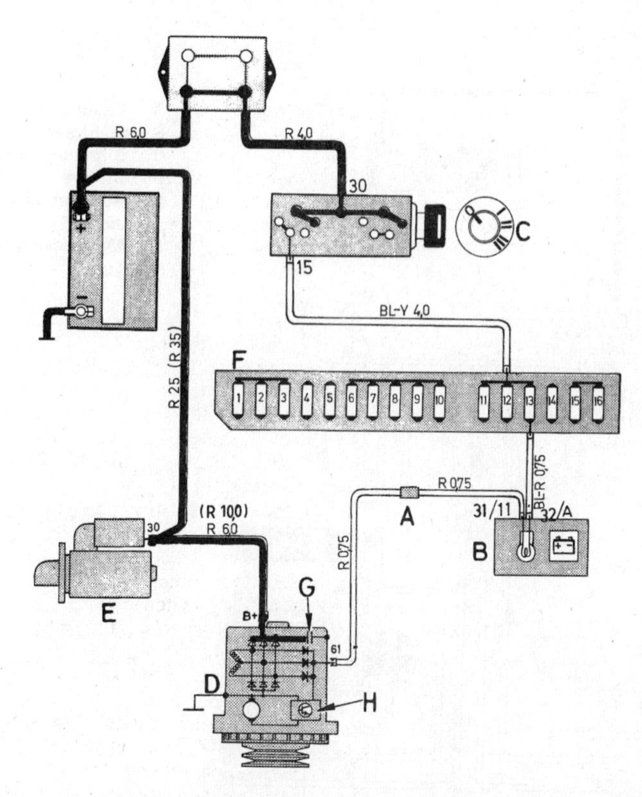

A Connector
B Battery charge failure warning light
C Ignition switch
D Alternator
E Starter motor
F Fusebox
G Capacitor
H Integral regulator

Fuse No. 13
Direction indicators

Typical 1984 and later model charging circuit diagram

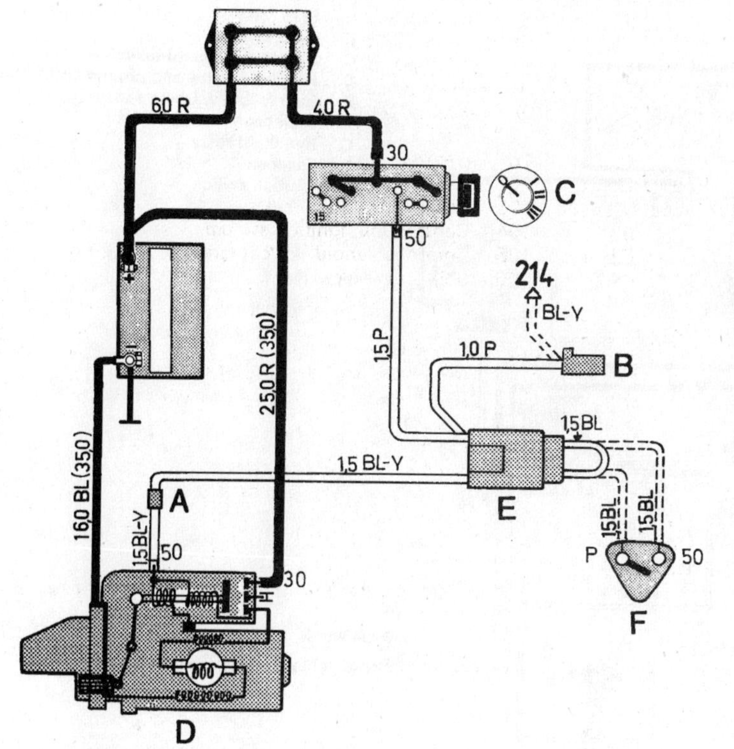

A Connector
B Service socket (for starter motor)
C Ignition switch
D Starter motor
E 1 junction (manuals)
2 junction (automatics)
F Start inhibitor switch (automatics)
214 Control unit, LH-Jetronic 2.4

On automatics the start inhibitor switch F must be closed if the starter motor is to be energized.

Typical 1984 and later model starter circuit diagram

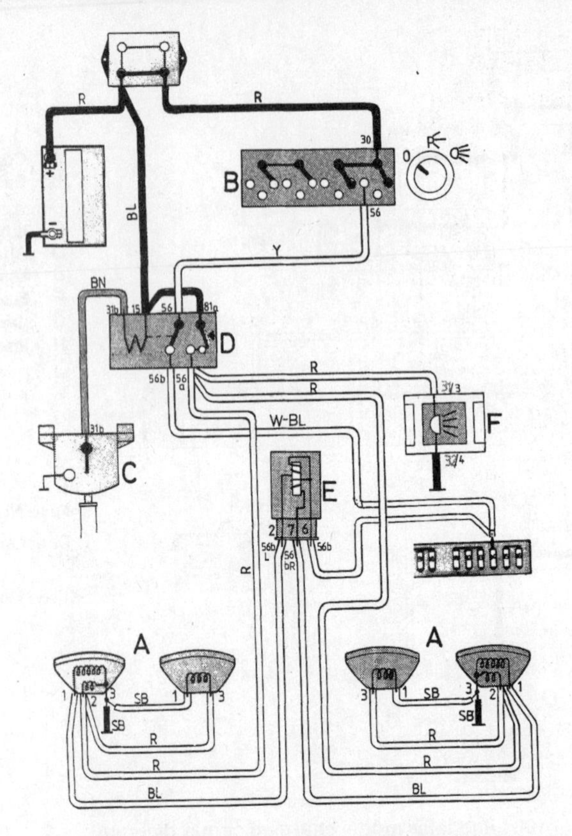

A Headlamps + ground connections
B Light switch
C Direction indicator switch
D Step relay
E Bulb failure warning sensor
F Main beam indicator light

Typical 1984 through 1986 model headlight circuit diagram

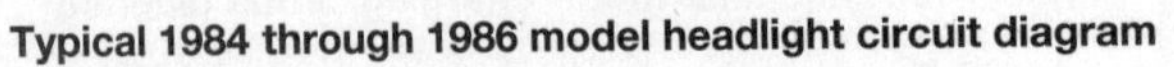

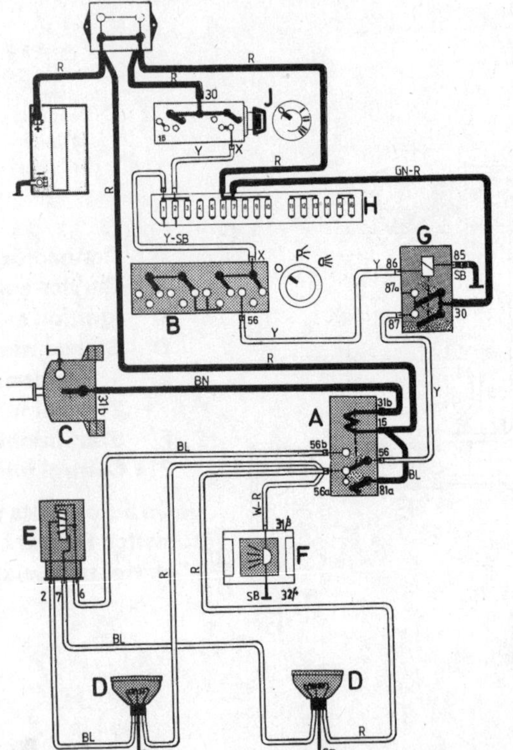

A Step relay
B Light switch
C Direction indicator switch
D Headlight bulbs and ground connections
E Bulb failure warning sensor
F High beam indicator light
G Headlight relay
H Fusebox
J Ignition switch

Fuse No. 1

Cigar lighter

Fuse No. 7

Brake lights

Typical 1987 and later model model headlight circuit diagram

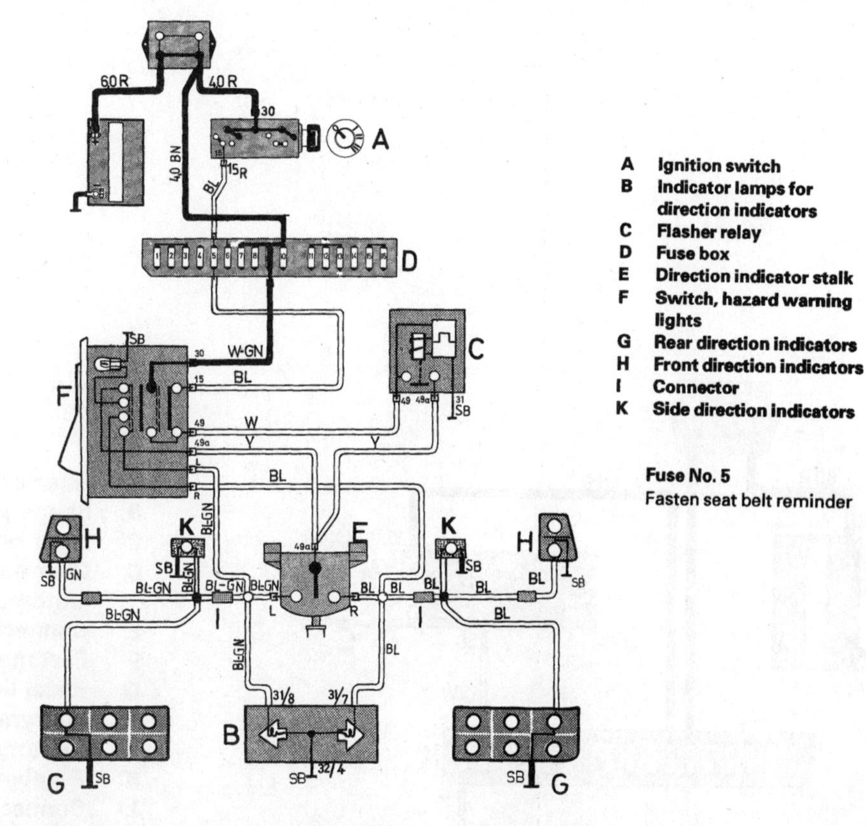

A Ignition switch
B Indicator lamps for direction indicators
C Flasher relay
D Fuse box
E Direction indicator stalk
F Switch, hazard warning lights
G Rear direction indicators
H Front direction indicators
I Connector
K Side direction indicators

Fuse No. 5
Fasten seat belt reminder

Typical 1984 and later model parking, tail and license plate lights circuit diagram

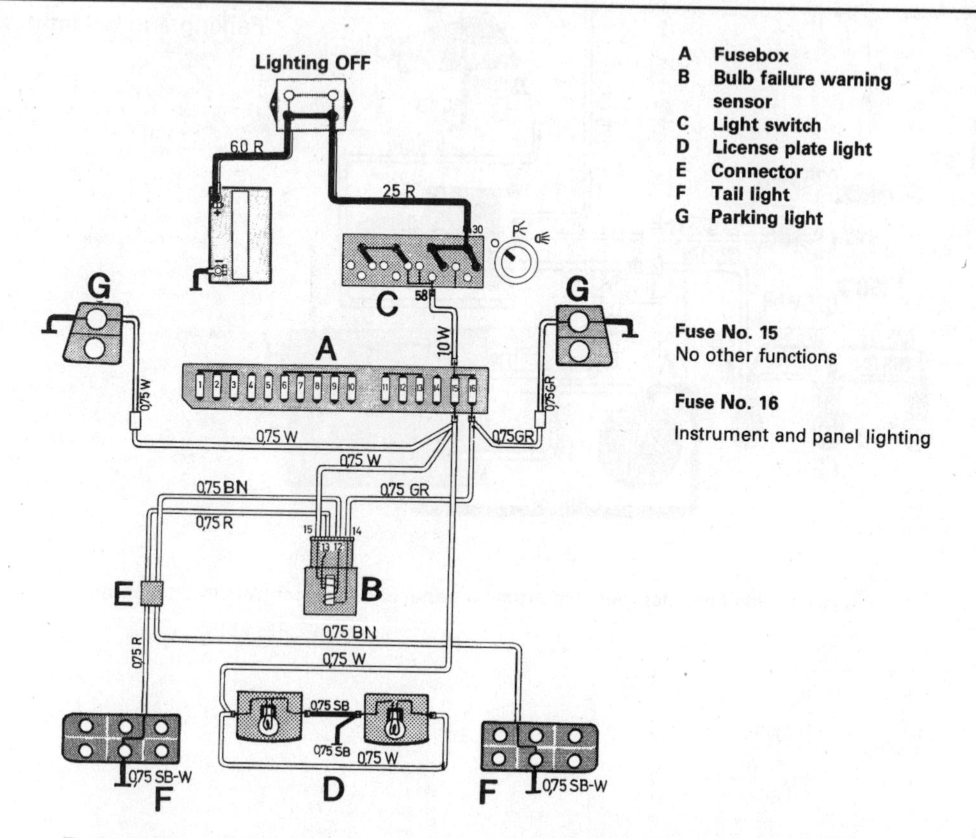

A Fusebox
B Bulb failure warning sensor
C Light switch
D License plate light
E Connector
F Tail light
G Parking light

Fuse No. 15
No other functions

Fuse No. 16
Instrument and panel lighting

Typical 1984 and later model turn signal and hazard warning lights circuit diagram

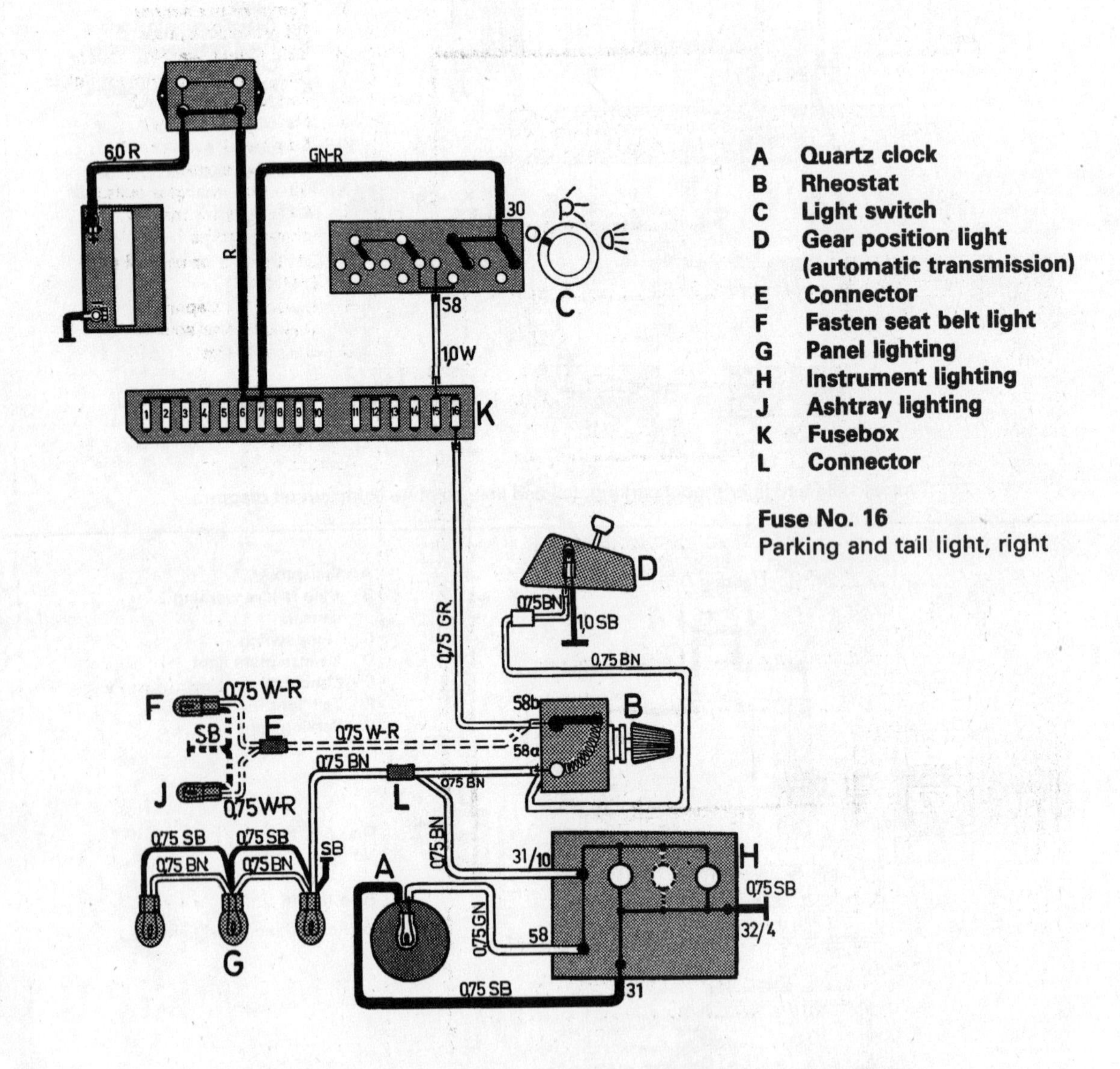

Typical 1984 and later model instrument panel and light control circuit diagram

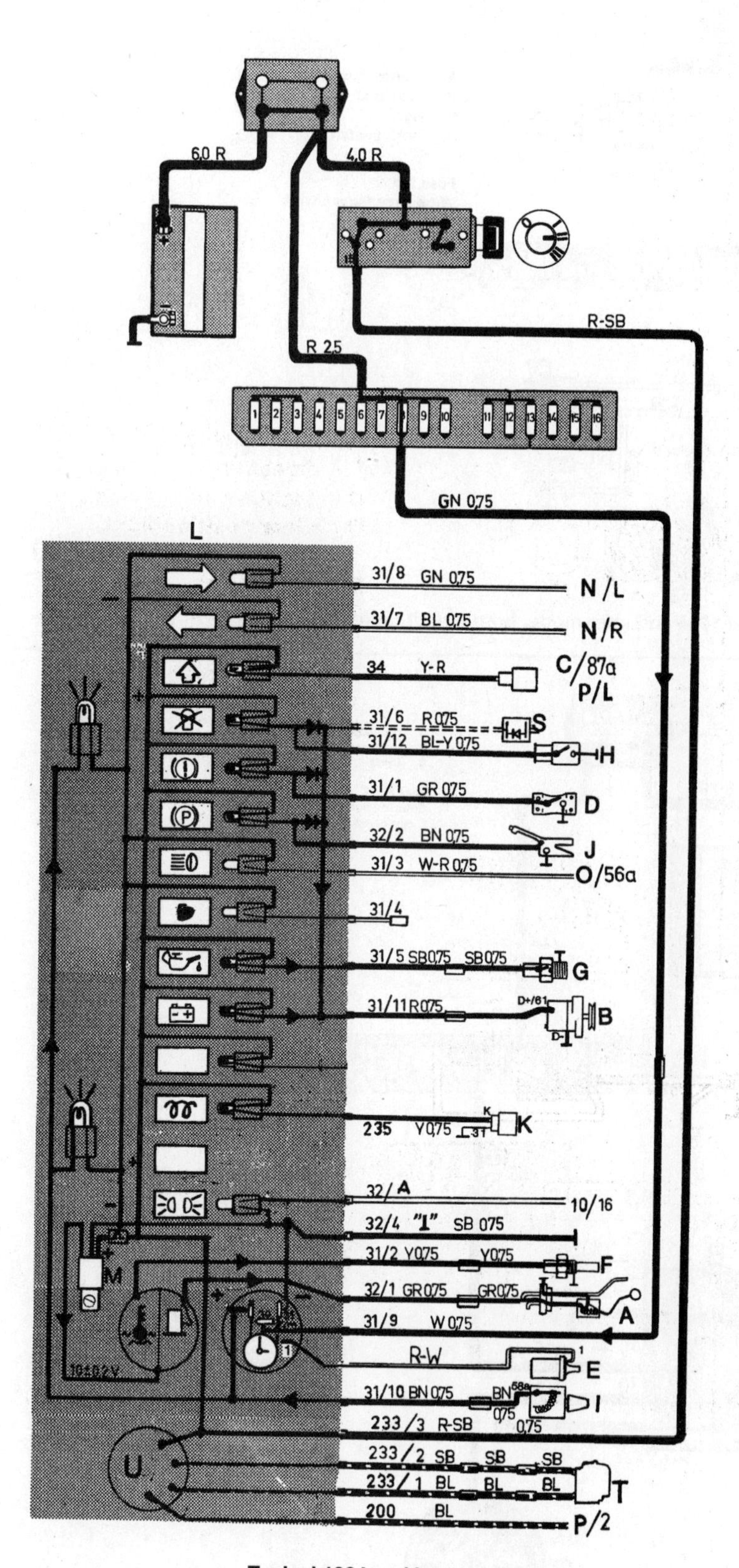

Typical 1984 and later model instrument cluster circuit diagram

- **A** Fuel level sender
- **B** Alternator
- **C** Overdrive relay AW 71
- **D** Brake failure light contact
- **E** Ignition coil
- **F** Temperature sensor
- **G** Oil pressure sensor
- **H** Bulb failure warning sensor
- **I** Rheostat for instrument lights
- **J** Parking brake switch
- **K** Control unit, diesel
- **L** Combined instrument
- **M** Voltage regulator
- **N** Direction indicator switch
- **O** Step relay for main/dipped beams
- **P** Shift indicator control unit (214/26)
- **S** Diode box (Japan)
- **T** Speedometer sensor
- **U** Speedometer

Fuse No. 8
Courtesy lighting

Ignition ON, engine not running

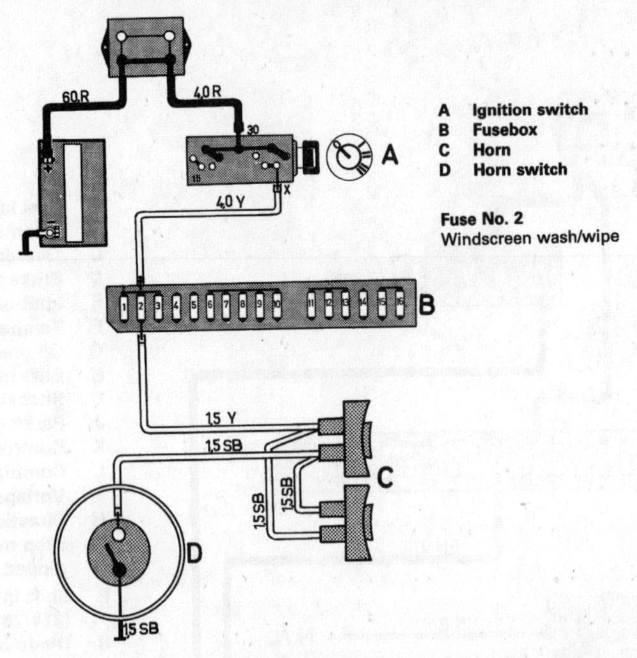

Typical 1984 and later model horn circuit diagram

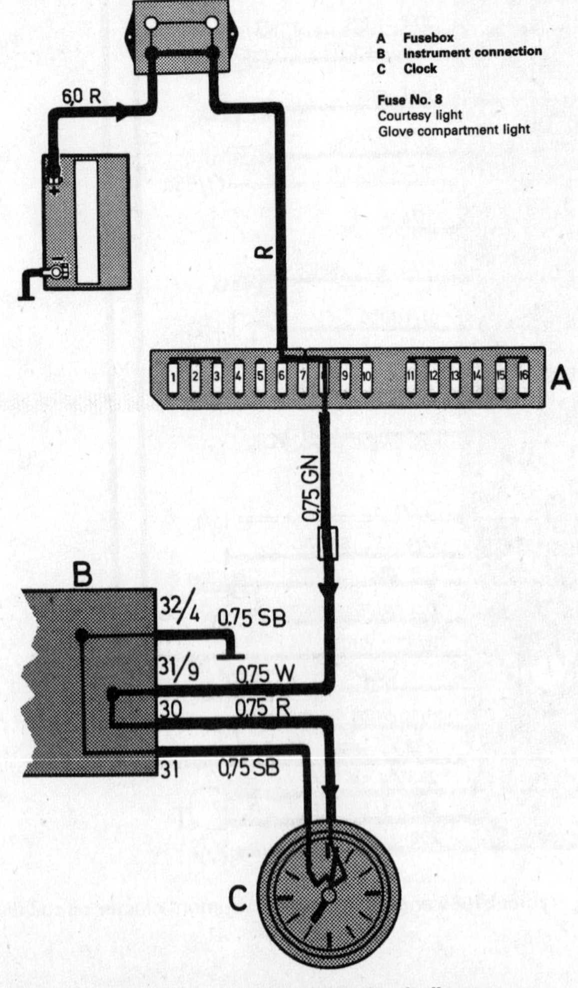

Typical 1984 and later model clock circuit diagram

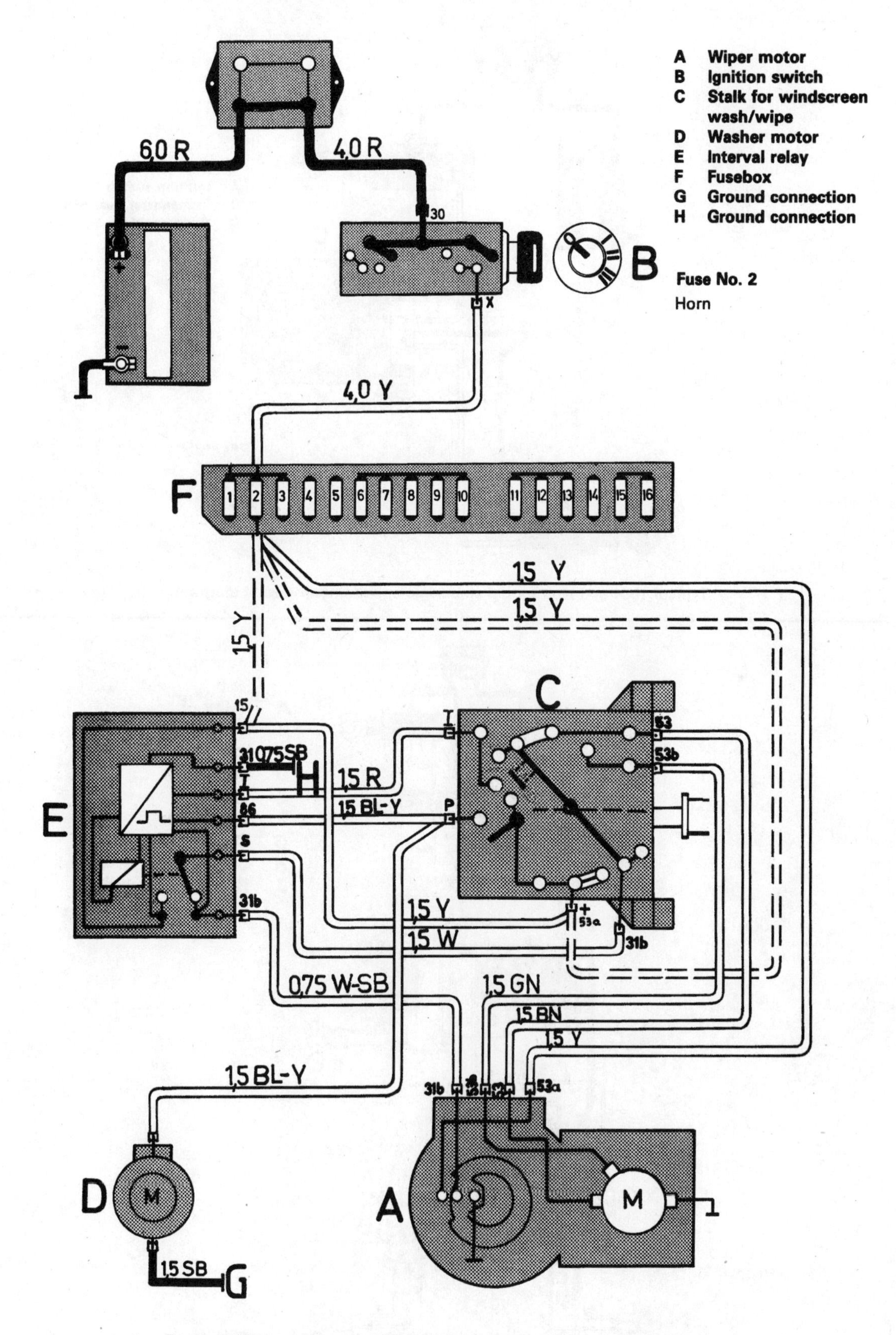

Typical 1984 and later model windshield wiper/washer circuit diagram

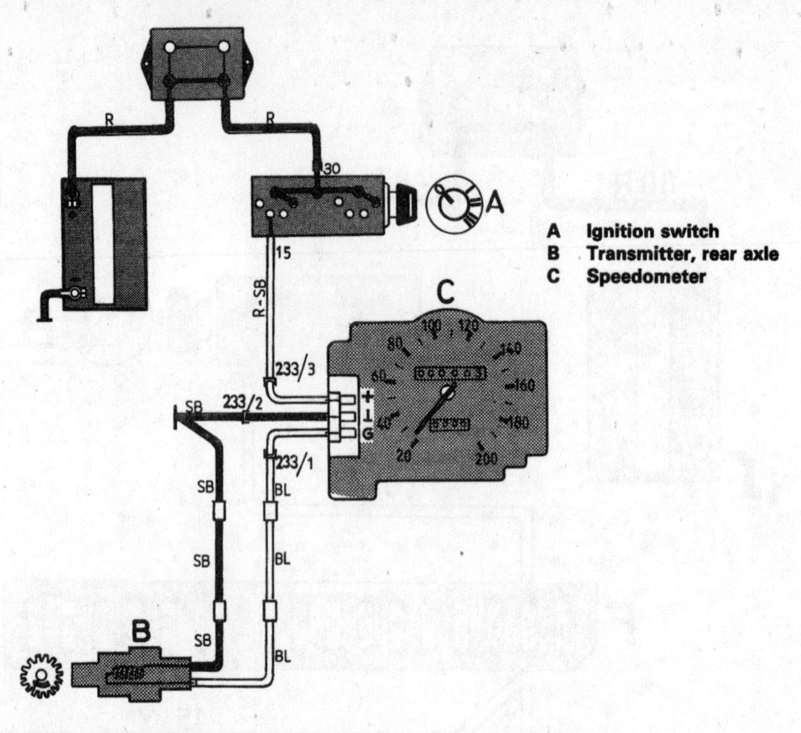

Typical 1984 and later model electronic speedometer circuit diagram

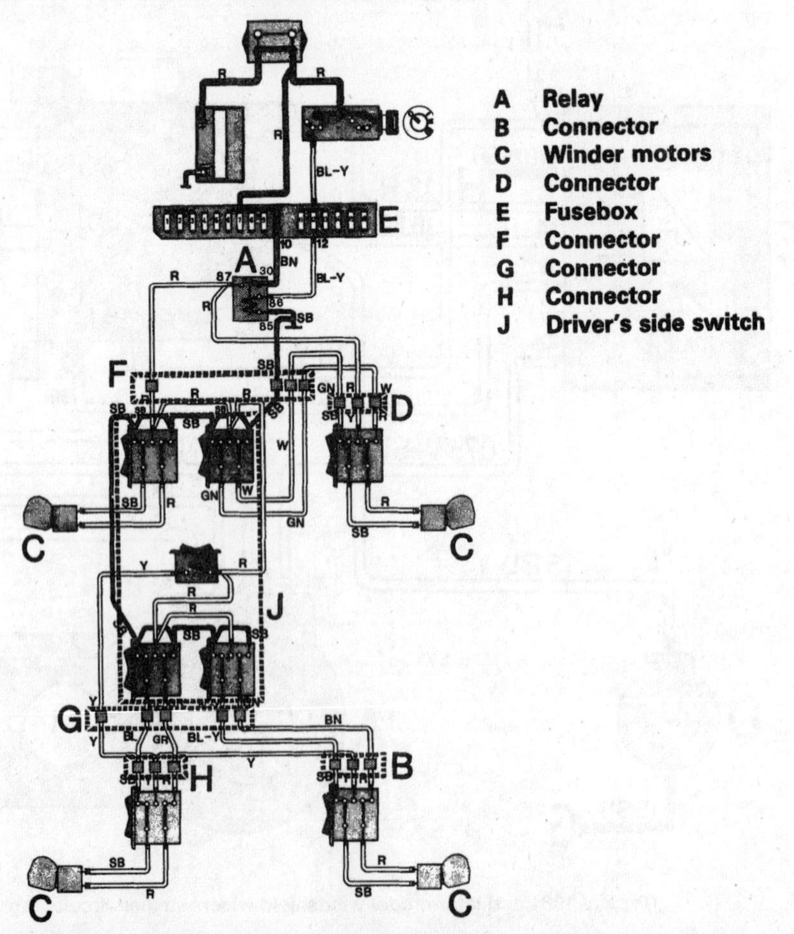

Typical 1984 and later model power window circuit diagram

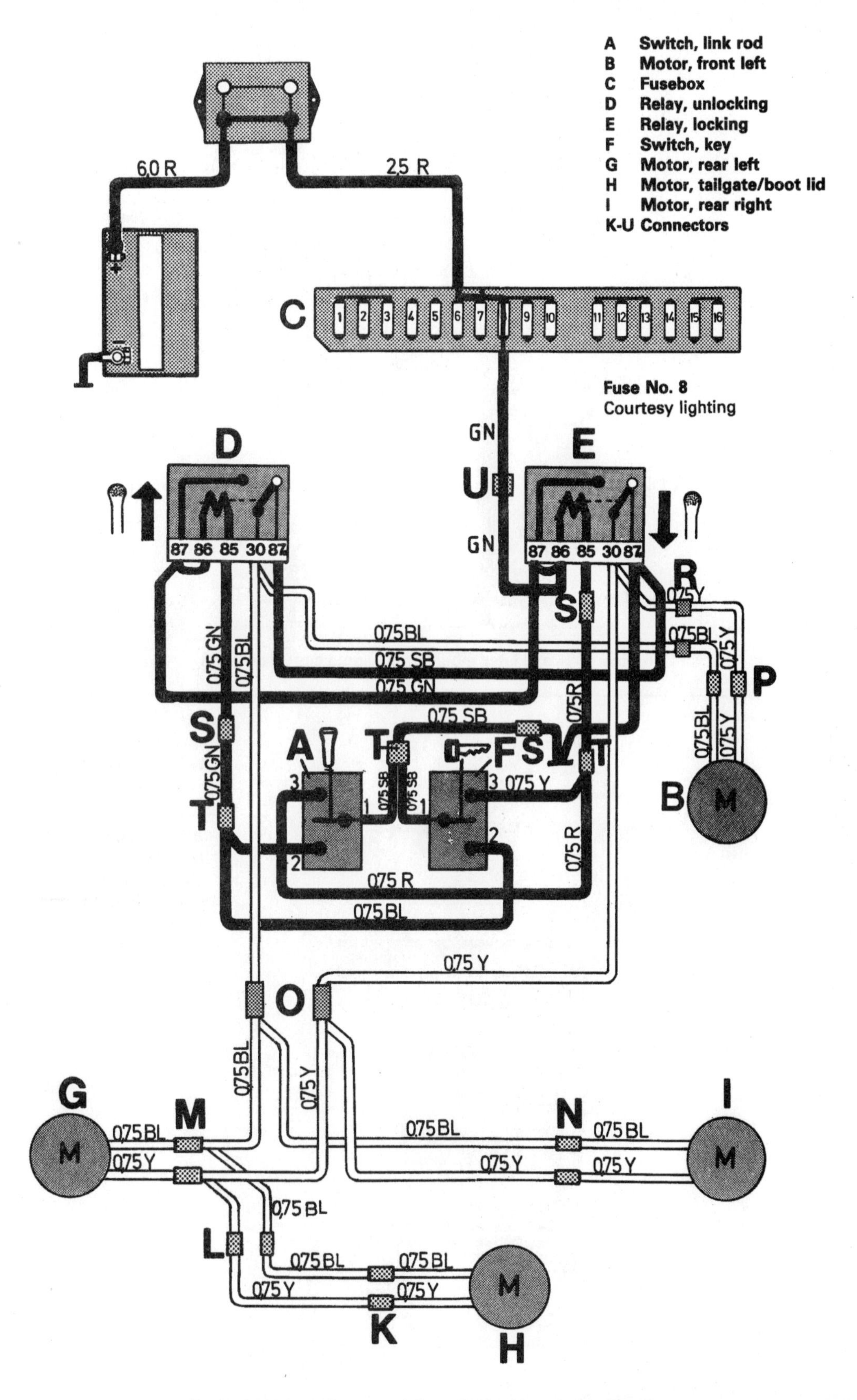

Typical 1984 and later model central locking circuit diagram

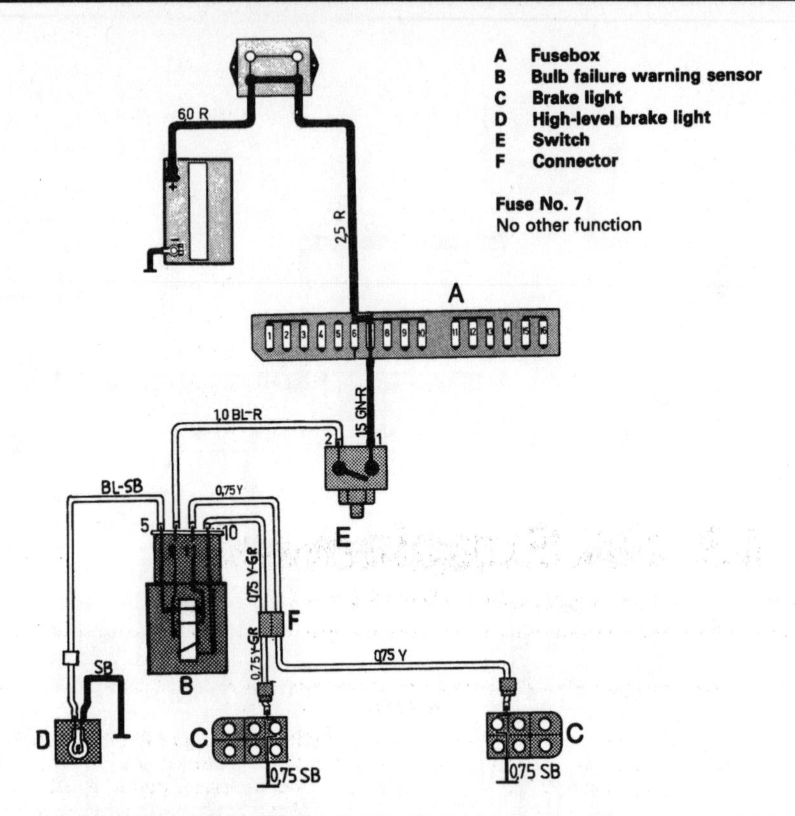

Typical 1984 and later model brake lights circuit diagram

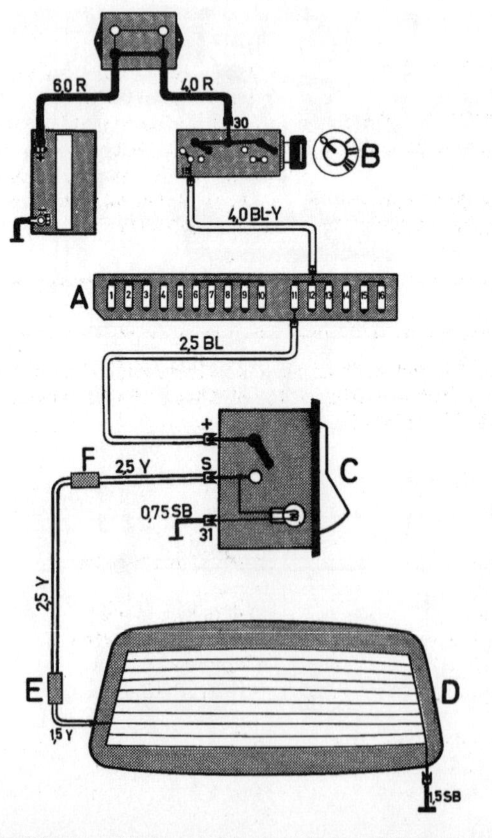

A Fusebox
B Ignition switch
C Switch for heated rear window
D Heated rear window
E Connector
F Connector

Typical 1984 and later model rear window defogger circuit diagram

Chapter 13 UK Supplement

***Note:** The following Sections contain information applicable to UK models only.*

Contents

1 Introduction

This Supplement contains information applicable to models imported into the UK only.

The Sections in the Supplement follow the same order as the Chapters to which they relate in the main part of the manual. The Specifications are all grouped together for convenience, but they too follow Chapter order.

It is recommended that before any particular operation is undertaken, reference be made to the appropriate Section(s) of this Supplement before reference is made to the main Chapters of the manual. In this way, any procedures which apply to UK models only can be noted first.

2 Specifications

Note: *The following Specifications are applicable to UK models, and supersede the relevant information appearing in the main Chapter Specifications. In cases where no information is given below, the main Chapter Specifications can be regarded as being applicable to UK models. Refer also to the "Notes for UK readers" Section at the start of this manual.*

B21 engine

General

Type	4-cylinder, in-line, ohc
Application:	
B21A	Carburettor models
B21E	Fuel injection models
Bore	3.622 in (92.0 mm)
Stroke	3.150 in (80.0 mm)
Compression ratio:	
B21A up to 1978	8.5:1
B21A 1979 to 1983	9.3:1
B21A 1984	10.0:1
B21E	9.3:1
Oil capacity:	
With oil filter	6.8 Imp pints (3.85 litres)
Without oil filter	5.9 Imp pints (3.35 litres)

Cylinder bores

Standard sizes:	
C	3.6220 to 3.6224 in (92.00 to 92.01 mm)
D	3.6224 to 3.6228 in (92.01 to 92.02 mm)
E	3.6228 to 3.6232 in (92.02 to 92.03 mm)
G	3.6236 to 3.6240 in (92.04 to 92.05 mm)
1st oversize	3.6417 in (92.5 mm)
2nd oversize	3.6614 in (93.0 mm)
Wear limit	0.004 in (0.10 mm)

Cylinder head

Maximum warp (for use without machining):	
Across diagonals	0.020 in (0.50 mm)
Across width	0.010 in (0.250 mm)

Renew cylinder head if warp is greater than 0.040 in (1.0 mm) longitudinally or 0.020 in (0.5 mm) crosswise

Camshaft

Journal diameter	1.1791 to 1.1799 in (29.950 to 29.970 mm)
Inlet valve opening timing (cold) at 0.028 in (0.7 mm) clearance:	
B21A 1975	5° BTDC
B21A 1976 to 1983	13° BTDC
B21A 1984	10° BTDC
B21E	15° BTDC
Diameter of camshaft bearings	1.1811 to 1.819 in (30.000 to 30.021 mm)

Intermediate shaft

Diameter bearing journal:	
Front	1.8494 to 1.8504 in (46.975 to 47.000 mm)
Intermediate	1.6939 to 1.6949 in (43.025 to 43.050 mm)
Rear	1.6900 to 1.6909 in (42.925 to 42.950 mm)
Diameter of bearing in block:	
Front	1.8512 to 1.8524 in (47.020 to 47.050 mm)
Intermediate	1.6957 to 1.6968 in (43.070 to 43.100 mm)
Rear	1.6917 to 1.6929 in (42.970 to 43.000 mm)

B23 engine (where different to B21)

General

Bore	3.780 in (96.0 mm)
Compression ratio:	
B23A 1981 to 1984	10.3:1
B23E 1979 to 1984	10.0:1

Cylinder block

Bore grades	As for B230F (Chapter 2B)

Camshaft

Inlet valve opening timing (cold) at 0.028 in (0.7 mm) clearance:	
B23A 1975	5° BTDC
B23A 1976 on	13° BTDC
B23E 1979 to 1980	21° BTDC
B23E 1981 to 1983	15° BTDC
B23E 1984	13° BTDC

B200 engine (where different to B21)

General

Bore	3.5 in (88.9 mm)
Capacity	121 cu in (1986 cc)
Compression ratio	10:1

Cylinder block

Bore grade:	
C	3.5000 to 3.5004 in (88.90 to 88.91 mm)
D	3.5004 to 3.5008 in (88.91 to 88.92 mm)
E	3.5008 to 3.5012 in (88.92 to 88.93 mm)
G	3.5016 to 3.5020 in (88.94 to 88.95 mm)
Wear limit	0.004 in (0.1 mm)

Pistons

Clearance in bore:	
1985 models	0.0001 to 0.0010 in (0.003 to 0.027 mm)
1986 on models	0.0004 to 0.0011 in (0.010 to 0.030 mm)

Piston rings

Height:	
Top compression	0.0681 to 0.0685 in (1.728 to 1.740 mm)
Second compression	0.0681 to 0.0685 in (1.728 to 1.740 mm)
Oil scraper	0.1368 to 0.1374 in (3.475 to 3.490 mm)
Piston ring-to-groove clearance:	
Top compression	0.0024 to 0.0036 in (0.060 to 0.092 mm)
Second compression	0.0012 to 0.0025 in (0.030 to 0.062 mm)
Oil scraper	0.0008 to 0.0022 in (0.020 to 0.055 mm)
Piston ring end gap:	
Top compression	0.012 to 0.020 in (0.30 to 0.50 mm)
Second compression	0.012 to 0.022 in (0.30 to 0.55 mm)
Oil scraper	0.010 to 0.020 in (0.25 to 0.50 mm)

Camshaft

B200K:	
Marking	L
Maximum lift height	0.386 in (9.8 mm)
Inlet valve opens	5° BTDC
B200E:	
Marking	A
Maximum lift height	0.41 in (10.5 mm)
Inlet valve opens	13° BTDC
Endfloat	0.008 to 0.020 in (0.2 to 0.5 mm)

Crankshaft

Endfloat	0.0032 to 0.0106 in (0.080 to 0.270 mm)
Main bearing running clearance	0.0009 to 0.0028 in (0.023 to 0.072 mm)
Big-end bearing:	
Running clearance	0.0009 to 0.0026 in (0.023 to 0.067 mm)
Diameter:	
Standard	2.1654 in (55.0 mm) nominal
First undersize	2.1555 in (54.75 mm) nominal
Big-end crankpins:	
Diameter:	
Standard	1.9291 in (49.0 mm) nominal
First undersize	1 9193 in (48.75 mm) nominal
Second undersize	1.9094 in (48.50 mm) nominal
Maximum ovality	0.0001 in (0.003 mm)
Maximum taper	0.0002 in (0.005 mm)

Connecting rods

Endfloat on crankshaft	0.010 to 0.018 in (0.25 to 0.45 mm)

B230 engine (where different to B200)

General

Bore	3.780 in (96.0 mm)
Capacity	141.2 cu in (2316 cc)
Compression ratio:	
B230A and B230E	10.3:1
B230K	10.5:1

Cylinder block

Bore grades	As for B230F (Chapter 2B)

Piston rings

Piston ring-to-groove clearance	As for B230F (Chapter 2B)
Piston ring end gap	As for B230F (Chapter 2B)

Camshaft

B230A:	
Marking	A
Maximum lift height	0.413 in (10.5 mm)
Inlet valve opens	13° BTDC
B230K:	
Marking	T
Maximum lift height	0.390 in (9.9 mm)
Inlet valve opens	7° BTDC
B230E:	
Marking	V
Maximum lift height	0.448 in (11.37 mm)
Inlet valve opens	11° BTDC

Torque wrench setting	lbf ft	Nm
Crankshaft pulley/sprocket centre bolt (B21/B23 engines **only**)	122	165

Fuel system - general

Fuel tank capacity:

Total	13.2 Imp gal (60.0 litres)
Expansion chamber	1.1 Imp gal (5.0 litres)
Reserve (red section of gauge)	1.76 Imp gal (8.0 litres)

Idle and CO content adjustment

Note: *Pulsair system disconnected and plugged where applicable.*

	Idle speed (rpm)	CO content (%)	
		Setting	Checking
B21A:			
1975 to 1977	850	2.5	1.5 to 4.0
1978	900	2.5	2.0 to 3.5
1979 to 1983	900	2.0	1.5 to 3.0
1984	900	1.5	1.0 to 2.5
B23A 1981 to 1984	900	2.0	1.5 to 3.0
B21E 1975 to 1980	900	2.0	1.0 to 3.0*
B21E 1981 to 1984	900	1.0	0.5 to 2.0
B23E 1979 to 1980	950	2.0	1.5 to 2.5
B23E 1981 to 1984	900	1.0	0.5 to 2.0
B200K	900	1.5	1.0 to 2.5
B200E	900	1.0	0.5 to 2.0
B230A	900	2.0	1.5 to 3.0
B230K (manual)	800	1.0	0.5 to 1.5
B230K (automatic)	900	1.0	0.5 to 1.5
B230E	900	1.0	0.5 to 2.0

** 4.0, 1975 to 1977*

Fuel system - fuel injection engines

General

Type	Continuous injection (CI)

Injectors

B21E up to 1978	007 (or use 015)
B21E, B23E, B200E and B230E 1979 on	015
Opening pressure:	
007	43.5 to 52 lbf/in^2 (300 to 360 kPa)
015 (up to August 1982)	46.4 to 55 lbf/in^2 (320 to 380 kPa)
015 (September 1982 on)	50.7 to 59.4 lbf/in^2 (350 to 410 kPa)

Auxiliary air valve

Resistance	40 to 60 ohms
Fully open at	-22°F (30°C)
Fully closed at	158°F (70°C)

Control pressure regulator

Resistance	20 to 30 ohms
Airflow sensor plate rest position (maximum control pressure, engine warm and fuel pump running)	0 to 0.012 in (0 to 0.3 mm) below venturi waist

Main fuel pump

Capacity (at 72 lbf/in^2) (500 kPa), 12 volt supply and at 68°F (20°C):	
Up to 1979	1.4 pints (0.8 litre) per 30 seconds
1980 on	1.76 pints (1.0 litre) per 30 seconds

Tank pump

Consumption (12 volt supply):	
B21B, B23B and early B200B/B230B	1 to 2 A
Later models	3 to 4 A

Pressures

Line pressure	62 to 77 lbf/in^2 (450 to 530 kPa)
Rest pressure (after 20 minutes)	22 lbf/in^2 (150 kPa) minimum
Control pressure (warm engine)	50 to 54 lbf/in^2 (345 to 375 kPa)
Control pressure (cold engine)	See graph (in text)

Fuel system - carburettor engines

Fuel pump

Type	Mechanical diaphragm, operated from intermediate shaft
Pump pressure	2.2 to 4.0 lbf/in^2 (15.0 to 27.0 kPa) at 1000 rpm

SU HIF 6

Application	B21A engine
Metering needle designation	BDJ
Damping piston endfloat	0.043 to 0.067 in (1.1 to 1.7 mm)
Float needle valve diameter	1.75 mm

Solex (Zenith) 175 CD

Application	B21A engine
Metering needle designation:	
1975	B2BB (early) B1ED (late)
1976 on	B1EE
Damping piston endfloat	0.039 to 0.071 in (1.0 to 1.8 mm)
Float needle valve diameter	2.0 mm
Temperature compensator designation	60L
Temperature compensator opening commences	68°F (20°C)

Pierburg (DVG) 175 CDUS

Application	B21A, B23A and B230A engines
Metering needle designation:	
B21A	PN
B23A and B230A	DC
Damping piston endfloat	0.020 to 0.060 in (0.5 to 1.5 mm)
Float needle valve diameter	2.5 mm
Float height	0.28 to 0.36 in (7.0 to 9.0 mm)

Solex Cisac (B200K)

Main jet (primary)	145
Main jet (secondary)	140
Air correction jet (primary)	160
Air correction jet (secondary)	135
Idle fuel jet	35
Part load enrichment jet	60
Idling solenoid jet	43
Float level	1.33 in (33.8 mm)
Choke flap gap	0.122 in (3.1 mm)
Fast idle cam gap	0.075 in (1.90 mm)

Solex Cisac (B230K) - where different to B200K

Main jet (primary)	142
Main jet (secondary)	125
Air correction jet (primary)	130
Air correction jet (secondary)	160
Idling solenoid jet	46

3 Routine maintenance

The following maintenance schedule is that recommended by the manufacturer, and is the minimum required. Where the vehicle is operated under adverse conditions that include frequent stop-start driving, use as a taxi, frequent and prolonged motorway driving and driving on unmade roads, the intervals should be reduced - consult your dealer.

Every 250 miles (400 km), weekly, or before a long journey

Check the engine oil level (Chapter 1, Section 4)
Check the coolant level (Chapter 1, Section 4)
Check the brake fluid (and where applicable, clutch fluid) level (Chapter 1, Section 4)
Top up the washer fluid reservoirs (Chapter 1, Section 4)
Check the tyre pressures and condition of the tyre treads (Chapter 1, Section 5)
Check the function of main-line electrical services - lights, horn, wipers, etc. (Chapter 12)
Inspect the engine and the underside of the vehicle for leaks

Every 6000 miles (10 000 km) or six months, whichever comes first

All items listed above, plus:
Tyres - inspect thoroughly (Chapter 1, Section 5)
Renew the engine oil and oil filter (Chapter 1, Section 6)
Check the power steering system oil level and top up as necessary (Chapter 1, Section 7)
Check the battery electrolyte level and top up as necessary (Chapter 1, Section 12)
Check and adjust the ignition timing (mechanical) (Chapter 1, Section 16)
Check and renew (as necessary) the contact breaker points (mechanical ignition) (Chapter 1, Section 18)
Check the braking system for leaks and rectify as necessary (Chapter 1, Section 24)
Check the cooling system for leaks and rectify as necessary (Chapter 3)
Check and adjust the idle speed and CO content (Chapter 4, Part A or B, or Chapter 13, Section 6)
Check the carburettor dashpot fluid and top up as necessary (Chapter 4, Part B, Section 9)
Check the operation of the warm start valve (B200K and B230K) (Chapter 4, Part B, Section 11)
Check and adjust the choke (Chapter 4, Part B or Chapter 13, Section 6)
Selector linkage (automatics) check and adjust (Chapter 7, Part B, Section 3)

Every 12 000 miles (20 000 km) or twelve months, whichever comes first

All items listed above, plus:
Check and adjust accessory drivebelts - alternator, power steering pump, etc. (Chapter 1, Section 11)
Check battery charge (Chapter 1, Section 12)
Check the gearbox/automatic transmission oil/fluid level and top-up as necessary (Chapter 1, Section 14 or 8)
Check the oil level in the rear axle and top up as necessary (Chapter 1, Section 15)
Lubricate the (mechanical) distributor felt pad (Chapter 1, Section 18)
Check the fuel lines for damage and leaks (Chapter 1, Section 20)
Check the exhaust system (Chapter 1, Section 22)
Check the front and rear suspension (Chapter 1, Section 23)
Check the brake pads for wear (Chapter 1, Section 24)
Check the brake hoses and pipes for leaks and damage (Chapter 1, Section 24)
Check and adjust the handbrake (Chapter 1, Section 24)
Renew the spark plugs (Chapter 1, Section 27)
Check and adjust the front wheel bearings (Chapter 1, Section 37)
Check the concentration of coolant anti-freeze (Chapter 3, Section 2)
Check and adjust the kickdown cable (automatics) (Chapter 7, Part B, Section 5 or Chapter 13, Section 7)
Check and adjust the clutch (mechanical) (Chapter 8, Section 3)
Check the propeller shaft and bearings for damage (Chapter 8, Section 11)
Check the hydraulic clutch (if applicable) for leaks and rectify as necessary (Chapter 8)
Check the brake servo (Chapter 9, Section 14)
Have the steering alignment checked and adjusted as necessary (Chapter 10, Section 19)
Check the underseal/paintwork (Chapter 11, Section 2)
Lubricate the doors, bonnet, boot and tailgate hinges (Chapter 11, Section 7)

Every 24 000 miles (40 000 km) or two years, whichever comes first

All items listed above, plus:
Renew the air filter (Chapter 1, Section 26)
Check and adjust the valve clearances (Chapter 1, Section 28)
Renew the automatic transmission fluid (Chapter 1, Section 29)
Renew the coolant (Chapter 1, Section 30)
Renew the gearbox oil (Chapter 1, Section 31)
Renew the in-line fuel filter (Chapter 1, Section 35)
Check and clean the crankcase ventilation system (Chapter 1, Section 36)
Check the engine compression (Chapter 2, Part B, Section 3)
Renew the brake/clutch fluid by bleeding (Chapter 9, Section 10 and Chapter 8, Section 6)
Check and adjust the rear brake band (BW 35 automatics only) (Chapter 13, Section 7)

Every 48 000 miles (80 000 km) or four years, whichever comes first

All items listed above, plus:
Renew the camshaft drivebelt (Chapter 2, Part A)

4 Engine

General information

The main difference between the UK and US markets is that the US did not have the 1986 cc (B200) engine. However, as this engine, and indeed all engines fitted to the 240 are derived from the B21 unit, the basic servicing and overhaul procedures described in the main part of this manual are applicable for all types. When following any procedure, always refer to the Specifications section at the beginning of this Chapter for any differences between the UK and US versions.

5 Fuel system (fuel injection engines)

General information

Two types of fuel injection have been fitted to the 240 series with early models having a mechanical system, known as CIS or K-Jetronic, and later models having an electronic system (LH-Jetronic). Both these systems were used on US models and, as such, the information given in Chapters 4A and 6 is applicable to UK models. Where necessary, additional specifications for the CIS system are shown at the beginning of this Chapter.

6 Fuel system (carburettor engines)

General information

With the exception of the B200 and B230 K series engines, the carburettors fitted to UK versions of the Volvo 240 are the same as those fitted to Canadian models. The UK reader can, therefore, assume that where the Canadian market is mentioned specifically, this also applies to UK models. These carburettors are covered in Chapter 4B with further information on emission control equipment in Chapter 6 and additional specifications shown at the beginning of this Chapter. The carburettor fitted to the B200 and B230 K series engines is covered later in this Section.

Solex Cisac carburettor (B200K)

Refer to illustrations 6.1, 6.23, 6.26

General description

1 Fitted to B200K engines, the Solex Cisac carburettor is a fixed-jet, twin-barrel, downdraught type **(see illustration).**

2 An idling solenoid valve is fitted to the carburettor to shut off fuel supply when the ignition is turned off, preventing engine run-on.

3 The primary and secondary throttle valves are preset in the factory, and should not be adjusted in service.

4 The carburettor also incorporates a part-load enrichment valve, an accelerator pump, and a vacuum-operated choke control valve (although the choke operation is still manual).

5 The idling channels in the carburettor are heated by an electric thermistor plate in the base of the carburettor.

Idle speed and CO adjustment

6 Remove the air inlet duct from the top of the carburettor.

7 Check that the throttle valves operate smoothly, and that the secondary valve does not start to move until the primary valve is 2/3 open.

8 Check and adjust the accelerator cable so that the throttle quadrant contacts both the closed and open stops, with the cable just taut when the pedal is released.

9 Check and adjust the operation of the choke as described in paragraphs 18 to 25 of this Section.

10 Check and adjust the fast idle cam as described in paragraphs 26 and 27.

11 Refit the air cleaner duct, and ensure that all crankcase ventilation hoses are connected correctly. Disconnect and plug the Pulsair hoses, when applicable.

12 Connect a tachometer and CO meter to the engine, in accordance with the meter manufacturer's instructions.

13 Start the engine and allow it to reach normal operating temperature.

14 Adjust the idle speed to the specified value with the idling (volume) control screw.

15 Check the CO content. If adjustment is required, prise out the plug from the CO adjustment screw.

16 Adjust the CO content to that specified by screwing in to decrease CO, and out to increase. Re-adjust the idle speed if necessary, and repeat until both idle speed and CO content are as specified.

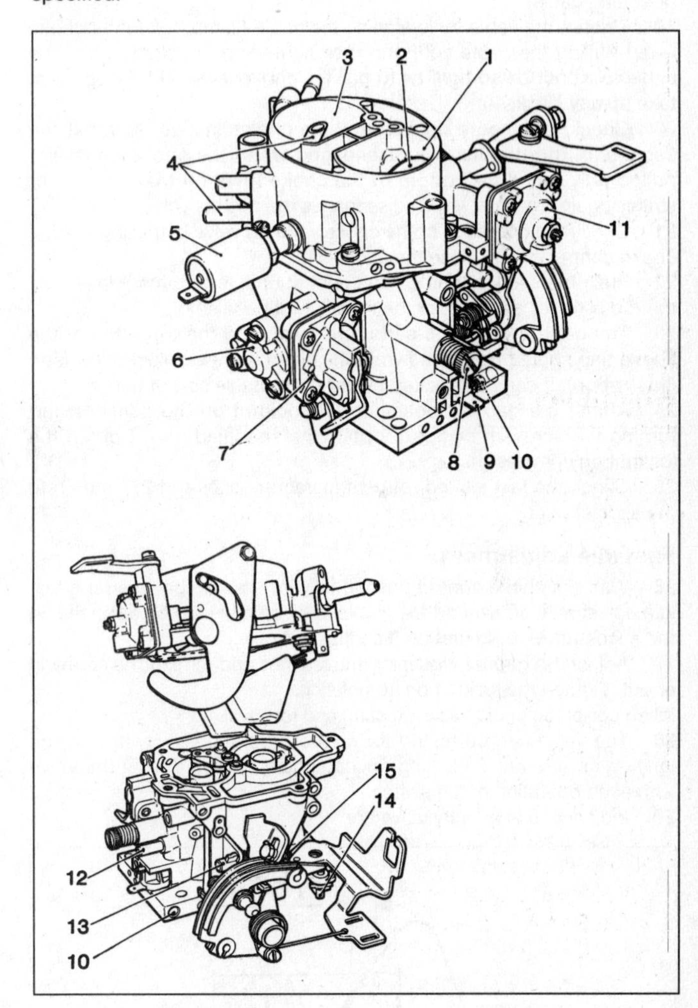

6.1 Exploded view of Solex Cisac carburettor

1 *Primary barrel*
2 *Choke (cold start) flap*
3 *Secondary choke*
4 *Float chamber vents*
5 *Idling (fuel) cut-off solenoid*
6 *Part load enrichment device*
7 *Accelerator pump*
8 *Idle speed screw*
9 *Thermistor heater for idling channels*
10 *CO (mixture) screw*
11 *Choke control vacuum unit*
12 *Vacuum take-off (distributor advance)*
13 *Primary throttle valve adjustment screw*
14 *Secondary throttle valve adjustment screw*
15 *Fast idle adjustment screw*

Note: *13 and 14 should not be adjusted in service*

17 On completion, stop the engine, fit a new tamperproof plug to the screw and remove the test equipment. Unplug and re-connect the Pulsair hoses, when applicable.

Note: *If no CO meter is available, a reasonable setting can be achieved by turning the adjuster screw in until the idle speed becomes erratic, then screwing the adjuster out until the idle becomes smooth. It may now be necessary to adjust the idle speed, and this setting should be regarded as a temporary measure until the CO content can be adjusted by your dealer. This should be done at the earliest opportunity.*

Choke adjustment

18 Remove the air inlet duct from the top of the carburettor. Check that, with the choke pushed fully in, there is no slackness in the operating cable.

19 Tighten the cable by loosening the cable clamp bolt in the choke lever, pulling the cable tight and then tightening the clamp bolt. The cable must not be so tight as to pull the choke lever off its stop - just take up any slack.

20 Check for smooth operation of the choke linkage, and that the secondary throttle valve is disengaged when the choke control is pulled fully out. (The bottom of the choke lever on the carburettor strikes against a latch, which disengages the throttle valve.)

21 Check the operation of the choke vacuum valve by pulling out the choke control to fully close the choke flap.

22 Push the vacuum unit pullrod right in until it bottoms. Make sure the rod is not at an angle, or the reading will be false.

23 The gap between the carburettor wall and the top edge of the choke flap should be as shown in the accompanying illustration **(see illustration).** It can be checked by using a suitable size of drill.

24 Adjust the gap by loosening the locknut on the adjuster and turning the screw in or out to obtain the specified gap. Tighten the locknut on completion.

25 Check the fast idle adjustment (paragraphs 26 and 27), then refit the air inlet duct.

Fast idle adjustment

26 With the choke control pushed fully in, the gap between the fast idle adjustment screw and the choke lever cam should be as shown in the accompanying illustration **(see illustration).**

27 Adjust the gap by loosening the locknut and turning the screw in or out. Tighten the locknut on completion.

Idling cut-off solenoid valve - testing and removal

28 The valve can be tested for correct operation by switching the ignition on and off. A clicking sound should be heard from the valve with each operation of the switch.

29 Indications of a faulty valve are:

a) *If the valve does not open, the engine will stall at idle.*
b) *If the valve fails to close, the engine may run-on after the ignition is switched off*

30 The valve is removed by simply unscrewing it from the side of the carburettor. When refitting, it should be tightened by hand only.

Carburettor removal and refitting

31 Disconnect the battery negative terminal.

32 Remove the air inlet duct from the carburettor.

33 Disconnect the fuel and ventilation hoses.

34 Disconnect the throttle and choke control cables.

35 Disconnect the idling solenoid and idling channel heater leads.

36 Remove the four securing nuts, and lift the carburettor from the manifold.

37 Refit in reverse order, using a new gasket on either side of the insulation flange, and ensure the crankcase ventilation and warm start hoses are connected correctly, or performance will be affected.

38 Check and adjust the accelerator and choke control cables as described in paragraphs 8 and 18 to 25 respectively.

Dismantling and reassembly

A carburettor repair kit, containing the gaskets and O-rings necessary for overhaul, should be obtained prior to dismantling.

Note: *Both the primary and secondary throttle valves are preset in the factory, and should not be adjusted in service. Do not remove the throttle valves or spindles. If wear is suspected in this area (which should only be evident after very high mileage), consult your Volvo dealer.*

39 Remove the carburettor as described in paragraphs 31 to 36, and wash down the exterior surfaces in solvent.

40 Remove the screws from the top cover and lift off the cover and float assembly.

41 Extract the float hinge pin and remove the float.

42 Shake the float to see if there is any fuel inside, indicating a leak, and renew as necessary.

43 Remove the idling solenoid valve from the side of the carburettor. It should only be screwed in finger-tight.

44 Check that the fuel inlet valve (float valve) is free to move and not sticking. Renew the valve, or remove it and clean it in clean petrol. Use a new washer under the valve when refitting.

45 Note that there is a filter screen in the inlet pipe union. Remove the union and take out the filter. Clean the filter in petrol. Partial blockage of this filter will cause poor performance at full throttle.

46 Remove the cover from the choke pull-down vacuum unit (do not lose the spring), and inspect the diaphragm for splits. Renew as necessary before refitting the unit.

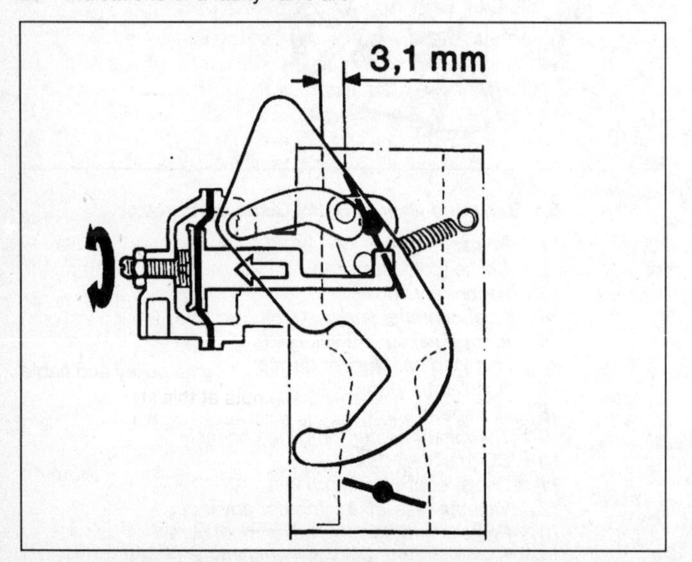

6.23 Choke vacuum control adjustment diagram - Solex Cisac carburettor

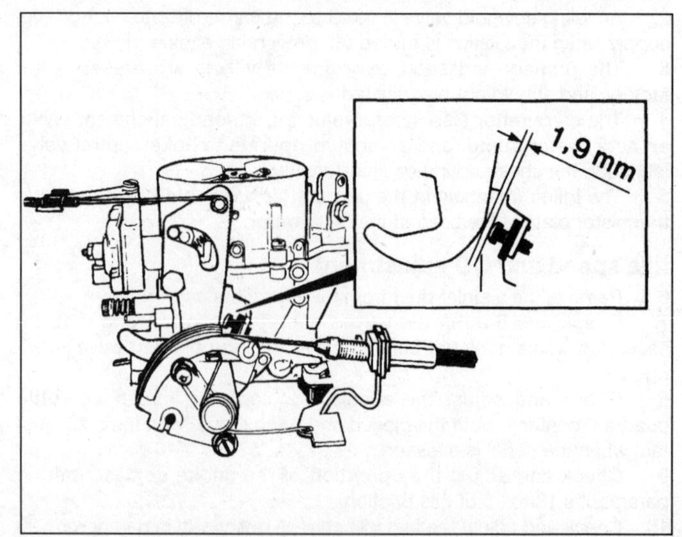

6.26 Fast idle adjustment - Solex Cisac carburettor

47 If the choke lever is removed for any reason, do not lose the ball and spring fitted beneath it, and refit them before fitting the lever.
48 If the choke flap is removed, the screws should be refitted using thread-locking fluid.
49 Tip out any fuel remaining in the float chamber and dispose of it safely. Remove the jets (one at a time to avoid mixing), and blow through them and their channels using compressed air. Do not probe the jets with wire. Use a bristle from a nylon brush to clear stubborn blockages, or soak the jets in a cleaning solvent.
50 If the idle (volume) screw and CO content adjusting screws are removed, on refitting them they should be adjusted as follows. Screw both screws in until they bottom, then screw the idle screw out 5 full turns, and screw the CO content screw out 8 full turns. This is a basic setting, both screws being re-adjusted when the carburettor is refitted.
51 Remove the covers from the accelerator pump and part-load enrichment valve, retrieving their springs, and inspect the diaphragms for splitting. Renew as necessary.
52 Clean the mating faces of the cover and carburettor body, making sure no traces of gasket are left sticking to them.
53 Check that all linkages are working smoothly, and apply a little engine oil to the pivot points.
54 Assemble the float to the top cover, fit a new cover gasket, and check the float level by inverting the cover and measuring the distance between the face of the gasket and the high point of the float, which should be as shown in the Specifications.
55 Adjust the height by bending the lug which contacts the float valve needle.
56 Fit the top cover and float assembly to the carburettor body and tighten the securing screws.
57 Fit the idle solenoid valve, screwing it in finger-tight only.
58 Refit the carburettor as described in paragraph 37, then check and adjust the choke and accelerator cables, idle speed, CO content, and the fast idle speed, as described earlier in this Section.

Idling channel heater - testing

59 This is a thermistor type heating element which keeps the idling channels warm to prevent icing.
60 To check the operation of the device, an ammeter will be required. Switch on the ignition, connect the ammeter between the connector plug and the thermistor.
61 At 68°F (20°C) the ammeter should read 1 amp. As the thermistor warms up, the reading should drop. If there is no initial reading, withdraw the thermistor retaining roll pin and cap, and thoroughly clean all contact surfaces. If there is still no reading, use a test lamp or voltmeter to verify the presence of supply voltage at the connector.

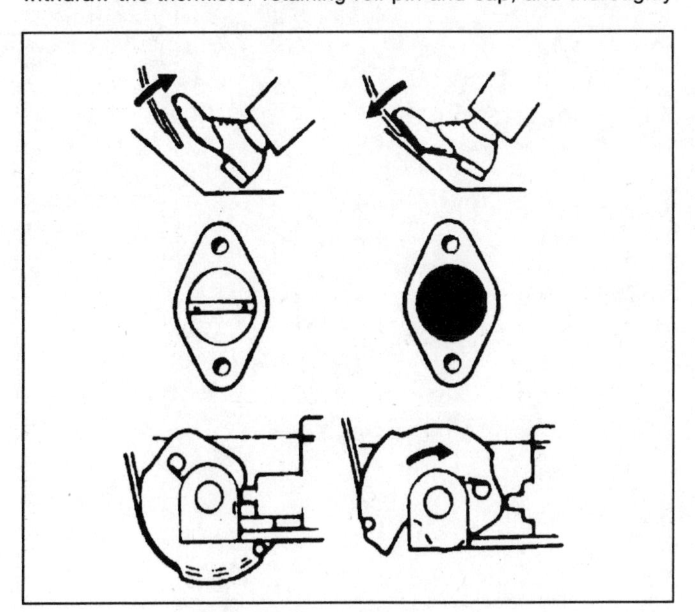

7.8 Kickdown cam-to-plunger relationship

A Accelerator pedal released
B Accelerator pedal fully depressed

Solex Cisac carburettor (B230K)

General description

62 The Solex Cisac carburettor fitted to B230K engines is a slightly modified version of that fitted to B200K engines.
63 The choke control valve is vacuum-operated, opening in two stages in response to the vacuum in the inlet manifold.
64 The throttle is modified to reduce spring force. The idle speed adjustment screw acts directly on the throttle linkage.
65 An electro-magnetic hot start valve, wired through the ignition switch, operates as follows:

a) *When the ignition is switched off, the valve opens the float chamber vent to atmosphere.*
b) *When the ignition is switched on, the valve closes the float chamber vent.*
c) *When the starter motor is energised, the valve opens the float chamber vent.*

66 A fuel cut-off valve reduces fuel consumption by cutting the idle circuit fuel supply when decelerating from engine speeds above 1350 rpm. A micro-switch, mounted on the throttle control pulley, operates the cut-off solenoid valve control unit, interrupting fuel supply during these conditions. The solenoid re-opens when engine speed drops below 1350 rpm, or the accelerator pedal is depressed. The solenoid valve also acts as a fuel cut-off valve when the ignition is switched off, preventing run-on.
67 The accelerator jet is only fitted to stage one, since stage two is isolated during moderate acceleration.
68 The second stage throttle valve is operated by vacuum, engine speed and load determining its opening.

7 Automatic transmission

BW 35 transmission

Refer to illustration 7.8 and illustrations 5.10 and 5.11 (Chapter 7, Part B).

General description

The three-speed BW 35 automatic transmission was only fitted to early models with the B21A and B21E engines. With the exception of that shown below, the servicing procedures are the same as those described for the BW 55 in Chapters 1 and 7, Part B.

Kickdown cable - renewal and adjustment

1 Drain the transmission fluid then remove the sump.
2 At the engine end of the cable, loosen the cable adjuster locknuts and release the cable from the bracket.
3 At the transmission end, rotate the cam of the kickdown valve until the cable end can be removed from the cam.
4 Unscrew the cable sheath bracket from the transmission casing and remove the cable.
5 The new cable is supplied with the stop on the cable at the engine end loose, which is crimped onto the cable once the adjustments are carried out. Do not lubricate the cable as it is already lubricated with silicone.
6 Screw the cable sheath bracket into the transmission casing, and attach the cable end to the cam on the kickdown valve.
7 Attach the engine end of the cable to the throttle pulley and cable adjuster bracket, but do not tighten the locknuts at this stage.
8 Check that the accelerator cable is fully released, then adjust the kickdown cable using the adjuster, so that the kickdown valve cam-to-plunger relationship is as (A) shown in the accompanying illustration **(see illustration).**
9 In this position, the stop at the engine end should be gently crimped onto the cable at the specified distance from the end of the sheath as shown in illustration 5.10 (Chapter 7, Part B). Do not crimp it too tightly, as it may have to be moved.

10 Have an assistant fully depress the accelerator pedal, and check that the cam-to-plunger relationship is now as (B) shown in illustration 7.8.
11 The cable stop adjustment (A in illustration 5.11, Chapter 7, Part B) should now be between 1.694 to 1.852 in. (43 to 47 mm).
12 Adjust as necessary on the adjuster, then tighten the locknuts and finally crimp the stop tight on to the cable.
13 Refit the sump, using a new gasket, and refill the transmission with the specified fluid (Chapter 1).
14 In normal service, it should only be necessary to check and adjust the position of the crimped stop using the cable adjuster as described earlier. Further adjustment may be performed by a Volvo dealer or transmission specialist, using a transmission fluid pressure gauge.

Starter inhibitor switch - adjustment

15 The switch is located on the left-hand side of the transmission unit, and is operated by internal mechanism.
16 To remove the switch, first select 'P'. Raise and support the vehicle, then disconnect the electrical leads to the switch.
17 Remove the switch by unscrewing it from the transmission casing. A cranked spanner will be needed to clear the lip of the casing.
18 Check that the operating pin projects from the threaded portion by 0.55 in (14.0 mm). This check should also be carried out on new switches.
19 If the protrusion is less than that specified, press the switch pin in and out and measure the protrusion again. Each time the switch is pressed in and released, the pin will protrude a little more. Be careful not to let the pin come out too far.
20 If the pin protrusion is greater than that specified, fit a new switch.
21 Refitting is a reversal of removal, but use a new sealing washer under the switch.
22 On completion, check that the engine can only be started in 'P' or 'N', and that the reversing light comes on when 'R' is selected.

Brake band adjustment

23 The brake band adjuster is located on the right-hand side of the transmission casing, towards the rear end.
24 Access is difficult from underneath, and it may be better to remove the carpet from the front footwell and remove the transmission tunnel panel.
25 Loosen the locknut on the adjuster, then loosen the adjuster screw.
26 Tighten the adjuster screw to 10 lbf ft (14 Nm), then back off the screw one turn.
27 Tighten the locknut, being careful not to disturb the adjuster screw.
28 Refit the transmission tunnel panel and the carpet.

Index

D

E

F

G

H

I

J

T

U

V

W